Kirk-Othmer

ENCYCLOPEDIA OF CHEMICAL TECHNOLOGY

Second Edition

VOLUME 21

Uranium and
Uranium Compounds

to

Water (Analysis)

Interscience Publishers
a division of John Wiley & Sons, Inc.
New York · London · Sydney · Toronto

Kirk-Othmer

ENCYCLOPEDIA

OF CHEMICAL

TECHNOLOGY

Second completely revised edition

VOLUME 21

Uranium and
Uranium Compounds
to
Water (Analysis)

CONTENTS

EDITORIAL STAFF FOR VOLUME 21

CONTRIBUTORS TO VOLUME 21

A. L. Barney, *E. I. du Pont de Nemours & Co., Inc.,* Vinylidene polymers (Fluoride)

Manfred J. R. Cantow, *Air Reduction Co.,* Vinyl polymers (Poly(vinyl chloride))

O. N. Carlson, *Ames Laboratory of the U.S. Atomic Energy Commission,* Vanadium and vanadium alloys

B. B. Dayton, *The Bendix Corporation,* Vacuum technology

Carlos Del Rio, *Universidad Nacional de Mexico,* Vitamins (Nicotinic acid)

Donald G. Diddams, *Sterling Drug Inc.,* Vanillin

Harry W. Dougherty, *Merck, Sharp & Dohme,* Vitamins (Nicotinic acid; Biological aspects of nicotinic acid)

R. G. Dressler, *Trinity University,* Water (Sources and utilization)

J. P. Dux, *American Viscose Division,* FMC Corporation, Vinyon and related fibers

F. G. Edwards, *The Dow Chemical Company,* Vinylidene polymers (Chloride)

G. W. A. Fowles, *University of Reading, Reading, England,* Vanadium compounds

Peter W. Fryth, *Hoffmann-La Roche Inc.,* Vitamins (Survey)

Roger Gilmont, *Roger Gilmont Instruments, Inc.* and *Polytechnic Institute of Brooklyn,* Vapor-liquid equilibria

David A. Gordon, *Geigy Industrial Chemicals Division of Geigy Chemical Corporation,* UV absorbers

Sheldon B. Greenbaum, *Diamond Shamrock Chemical Co.,* Vitamins (Vitamin B_{12}; Vitamin D)

David C. Herting, *Tennessee Eastman Co.,* Vitamins (Vitamin E)

Conrad E. Hoffman, *E. I. du Pont de Nemours & Co., Inc.,* Viral infections, chemotherapy

Ralph A. Horne, *Woods Hole Oceanographic Institute,* Water (Properties)

Maynard A. Joslin, *University of California,* Vinegar

Jack K. Krum, *The R. T. French Co.,* Vanillin

Morton Leeds, *Air Reduction Co.,* Vinyl polymers (Poly(vinyl alcohol))

Richard J. Magee, *American Cyanamid Co.,* Veterinary drugs

V. L. Mattson, *Kerr-McGee Corporation,* Uranium and uranium compounds

Ivo Mavrovic, *Consulting Engineer,* Urea and urea derivatives

S. B. Mecca, *Schuylkill Chemical Company,* Uric acid

George O. Morrison, *Technical Consultant,* Vinyl polymers (Poly(vinyl acetals))

B. W. Nies, *United States Gypsum Co.,* Wallboard (Gypsum board)

William Oroshnik, *Chemo Dynamics,* Vitamins (Vitamin A)

K. A. Pigott, *Mobay Chemical Company,* Urethan polymers

J. R. Roberts, *United States Gypsum Co.,* Wallboard

David Rhum, *Air Reduction Co.,* Vinyl polymers (Poly(vinyl acetate))

R. D. Ross, *Thermal Research and Engineering Corp.,* Wastes, industrial

C. E. Schildknecht, *Gettysburg College,* Vinyl polymers (Vinyl ethers, monomers and polymers)

A. Louis Shor, *American Cyanamid Co.,* Veterinary drugs

Marvin W. Skougstad, *U.S. Geological Survey,* Water (Analysis)

Richard S. Stearns, *Sun Oil Company,* Viscometry

E. R. Stevens, *Ames Laboratory of the U.S. Atomic Energy Commission,* Vanadium and vanadium alloys

D. C. Thompson, *E. I. du Pont de Nemours & Co., Inc.,* Vinylidene polymers (Fluoride)

Arthur F. Wagner, *Merck, Sharp & Dohme Research Laboratories, Merck & Co., Inc.,* Vitamins (Vitamin K)

R. Wessling, *The Dow Chemical Company,* Vinylidene polymers (Chloride)

A. S. Wood, *GAF Corporation,* Vinyl polymers (Poly(pyrrolidone))

ABBREVIATIONS AND SYMBOLS

A	ampere(s)
A	anion (eg, HA)
Å	Angstrom unit(s)
AATCC	American Association of Textile Chemists and Colorists
abs	absolute
ac	alternating current
ac-	alicyclic (eg, ac-derivatives of tetrahydronaphthalene)
accel(d)	accelerated(d)
acceln	acceleration
ACS	American Chemical Society
addn	addition
AEC	Atomic Energy Commission
AGA	American Gas Association
Ah	ampere-hour(s)
AIChE	American Institute of Chemical Engineers
AIME	American Institute of Mining and Metallurgical Engineers
AIP	American Institute of Physics
AISI	American Iron and Steel Institute
alc	alcohol(ic)
alk	alkaline (not alkali)
Alk	alkyl
AMA	American Medical Association
A-min	ampere-minute(s)
amt	amount (noun)
anhyd	anhydrous
AOAC	Association of Official Analytical (formerly Agricultural) Chemists
AOCS	American Oil Chemists' Society
APHA	American Public Health Association
API	American Petroleum Institute
app	apparatus
approx	approximate(ly)
aq	aqueous
Ar	aryl
as-	asymmetric(al) (eg, as-trichlorobenzene)
ASA	American Standards Association. Later (1966) called USASI
ASHRAE	American Society of Heating, Refrigerating and Air-Conditioning Engineers
ASM	American Society for Metals
ASME	American Society of Mechanical Engineers
ASTM	American Society for Testing and Materials
atm	atmosphere(s), atmospheric
at. no.	atomic number
at. wt	atomic weight
av	average
b	barn(s)
b (as in b_{11})	boiling (at 11 mm Hg)
bbl	barrel(s)
bcc	body-centered cubic
Bé	Baumé
Bhn	Brinell hardness number
bp	boiling point
BP	*British Pharmacopoeia* (General Medical Council in London)
Btu	British thermal unit(s)

bu	bushel(s)
C	Celsius (centigrade); coulomb(s)
C-	denoting attachment to carbon (eg, *C*-acetyl-indoline)
ca	circa, approximately
CA	Chemical Abstracts
cal	calorie(s)
calcd	calculated
cfm, ft³/min	cubic foot (feet) per minute
cg	centigram(s)
cgs	centimeter-gram-second
Ci	curie(s)
CI	Colour Index (number); the CI numbers given in *ECT*, 2nd ed., are from the new *Colour Index* (1956) and Suppl. (1963), *Soc. Dyers Colourists*, Bradford, England, and *AATCC*, U.S.A.
CIE	Commission Internationale de l'Eclairage (see also ICI)
cif	cost, insurance, freight
cl	carload lots
cm	centimeter(s)
coeff	coefficient
compd, cpd	compound (noun)
compn	composition
concd	concentrated
concn	concentration
cond	conductivity
const	constant
cont	continued
cor	corrected
cp	chemically pure
cP	centipoise(s)
cpd, compd	compound (noun)
cps	cycles per second
crit	critical
cryst	crystalline

crystd	crystallized
crystn	crystallization
cSt	centistokes
cu	cubic
d	density (conveniently, specific gravity)
d	differential operator
d-	*dextro*-, dextrorotatory
D	Debye unit(s)
D-	denoting configurational relationship (as to *dextro*-glyceraldehyde)
db	dry-bulb
dB	decibel(s)
dc	direct current
dec, decomp	decompose(s)
decompd	decomposed
decompn	decomposition
den	denier(s)
den/fil	denier(s) per filament
deriv	derivative
detd	determined
detn	determination
diam	diameter
dielec	dielectric (adj.)
dil	dilute
DIN	Deutsche Industrienormen
distd	distilled
distn	distillation
dl	deciliter(s)
dl-, DL	racemic
dm	decimeter(s)
DOT	Department of Transportation
dp	dewpoint
dyn	dyne(s)
e	electron; base of natural logarithms
ed.	edited, edition, editor
elec	electric(al)
emf	electromotive force
emu	electromagnetic unit(s)
eng	engineering
equil	equilibrium(s)
equiv	equivalent
esp	especially

esr, ESR	electron spin resonance
est(d)	estimate(d)
estn	estimation
esu	electrostatic unit(s)
eu	entropy unit(s)
eV	electron volt(s)
expt(l)	experiment(al)
ext(d)	extract(ed)
extn	extraction
F	Fahrenheit; farad(s)
F	faraday constant
FAO	Food and Agriculture Organization of the United Nations
fcc	face-centered cubic
Fed, fedl	federal (eg, Fed Spec)
fl oz	fluid ounce(s)
fob	free on board
fp	freezing point
frz	freezing
ft	foot (feet)
ft-lb	foot-pound(s)
ft³/min, cfm	cubic foot (feet) per minute
g	gram(s)
g	gravitational acceleration
G	gauss(es)
G	Gibbs free energy
gal	gallon(s)
gal/min, gpm	gallon(s) per minute
g/den	gram(s) per denier
gem-	geminal (attached to the same atom)
g-mol	gram-molecular (as in g-mol wt)
g-mole	gram-mole(s)
G-Oe	gauss-oersted(s)
gpm, gal/min	gallon(s) per minute
gr	grain(s)
h, hr	hour(s)
hl	hectoliter(s)
hmw	high-molecular-weight(adj.)
hp	horsepower(s)
hr, h	hour(s)
hyd	hydrated, hydrous
hyg	hygroscopic
Hz	hertz(es)
i, insol	insoluble
^{i} (eg, Pr^i)	iso (eg, isopropyl)
i-	inactive (eg, *i*-methionine)
IACS	International Annealed Copper Standard
ibp	initial boiling point
ICC	Interstate Commerce Commission
ICI	International Commission on Illumination (see also CIE); Imperial Chemical Industries, Ltd.
ICT	International Critical Tables
ID	inner diameter
IEEE	Institute of Electrical and Electronics Engineers
in.	inch(es)
insol, i	insoluble
IPT	Institute of Petroleum Technologists
ir	infrared
ISO	International Organization for Standardization
IU	International Unit(s)
IUPAC	International Union of Pure and Applied Chemistry
J	joule(s)
K	Kelvin
K	dissociation constant
kbar	kilobar(s)
kc	kilocycle(s)
kcal	kilogram-calorie(s)
keV	kilo electron volt(s)
kg	kilogram(s)
kG	kilogauss(es)
kgf	kilogram force(s)
kJ	kilojoule(s)
kp	kilopond(s) (equals kilogram force(s)
kV	kilovolt(s)
kVa	kilovolt-ampere(s)
kW	kilowatt(s)
kWh	kilowatt-hour(s)

l	liter(s)		mm	millimeter(s)
l-	*levo*-, levorotatory		mM	millimole(s)
L-	denoting configurational relationship (as to *levo*-glyceraldehyde)		m*M*	millimolar
			mo(s)	month(s)
lb	pound(s)		mol	molecule, molecular
LC$_{50}$	concentration lethal to 50% of the animals tested		mol wt	molecular weight
			mp	melting point
			mph	miles per hour
lcl	less than carload lots		MR	molar refraction
LD$_{50}$	dose lethal to 50% of the animals tested		mV	millivolt(s)
			mμ	millimicron(s) (10^{-9} m)
liq	liquid		^{n} (eg, Bu^{n}),	
lm	lumen		*n*-	normal (eg, normal butyl)
lmw	low-molecular-weight (adj.)		n (as, n_{D}^{20})	index of refraction (for 20°C and sodium light)
ln	logarithm (natural)			
log	logarithm (common)		*n*-, ^{n}	normal (eg, *n*-butyl, Bu^{n})
m	meter(s)		*N*	normal (as applied to concentration)
m	molal			
m-	meta (eg, *m*-xylene)		*N*-	denoting attachment to nitrogen (eg, *N*-methylaniline)
M	metal			
M	molar (as applied to concentration; not molal)			
			NASA	National Aeronautics and Space Administration
mA	milliampere(s)			
mAh	milliampere-hour(s)		ND	*New Drugs* (NND changed to ND in 1965)
manuf	manufacture			
manufd, mfd	manufactured		NF	*National Formulary* (American Pharmaceutical Association)
manufg, mfg	manufacturing			
max	maximum		nm	nuclear magneton; nanometer(s) (10^{-9} m)
Mc	megacycle(s)			
MCA	Manufacturing Chemists' Association		nmr, NMR	nuclear magnetic resonance
			NND	*New and Nonofficial Drugs* (AMA) (1958–1965). Later called ND
mcal	millicalorie(s)			
mech	mechanical			
meq	milliequivalent(s)		NNR	*New and Nonofficial Remedies* (1907–1958). Later called NND
MeV	million electron volt(s)			
mfd, manufd	manufactured			
			no.	number
mfg, manufg	manufacturing		NOIBN	not otherwise indexed by name (DOT specification for shipping containers)
mg	milligram(s)			
min	minimum; minute(s)			
misc	miscellaneous		*o*-	ortho (eg, *o*-xylene)
mixt	mixture		*O*-	denoting attachment to oxygen (eg, *O*-acetylhydroxylamine)
ml	milliliter(s)			
MLD	minimum lethal dose		Ω	ohm(s)

Ω-cm	ohm-centimeter(s)	rad	radian
OD	outer diameter	Rep	roentgen(s) equivalent physical
Oe	oersted(s)		
o/w	oil-in-water (eg, o/w emulsion)	resp	respectively
		rh	relative humidity
owf	on weight of fiber	Rhe	unit of fluidity (1/P)
oz	ounce(s)	RI	Ring Index (number); from *The Ring Index*, Reinhold Publishing Corp., N.Y., 1940. See also RRI
p-	para (eg, *p*-xylene)		
P	poise(s)		
pdr	powder		
PhI	*Pharmacopoeia Internationalis*, 2 vols. and Suppl., World Health Organization, Geneva, 1951, 1955, and 1959		
		rms	root mean square
		rpm	revolutions per minute
		rps	revolutions per second
		RRI	Revised Ring Index (number); from *The Ring Index*, 2nd ed., American Chemical Society, Washington, D.C., 1960
phr	parts per hundred of rubber or resin		
pos	positive (adj.)		
powd	powdered	RT	room temperature
ppb	parts per billion (parts per 10⁹)	s, sol	soluble
		ˢ (eg, Buˢ), *sec-*	secondary (eg, *sec*-butyl)
ppm	parts per million	*s-*, *sym-*	symmetrical (eg, *s*-dichloroethylene)
ppt(d)	precipitate(d)		
pptn	precipitation		
Pr. (no.)	Foreign prototype (number); dyestuff designation used in *AATCC Year Books* for dyes not listed in the old *Colour Index* (1924 ed.; 1928 Suppl.); obsolete since new *Colour Index* was published (1956 ed.; 1963 Suppl.)	*S-*	denoting attachment to sulfur (eg, *S*-methylcysteine)
		SAE	Society of Automotive Engineers
		satd	saturated
		satn	saturation
		scf, SCF	standard cubic foot (feet) (760 mm Hg, 63°F)
		scfm	standard cubic feet per minute
prepd	prepared		
prepn	preparation	Sch	Schultz number (designation for dyes from *Farbstofftabellen*, 4 vols., Akademie Verlag, Leipzig, 1931–1939)
psi	pound(s) per square inch		
psia (psig)	pound(s) per square inch absolute (gage)		
pt	point		
pts	parts	sec	second(s)
qual	qualitative	*sec-*, ˢ	secondary (eg, *sec*-butyl; Buˢ)
quant	quantitative		
qv	which see (quod vide)	SFs	Saybolt Furol second(s)
R	Rankine; roentgen; univalent hydrocarbon radical (or hydrogen)	sl s, sl sol	slightly soluble
		sol, s	soluble
		soln	solution

soly	solubility		(ASA changed to USASI in 1966)
sp	specific		
sp, spp	species (sing. and pl.)	USP	(*The*) *United States Pharmacopeia* (Mack Publishing Co., Easton, Pa.)
Spec	specification		
sp gr	specific gravity		
SPI	Society of the Plastics Industry	uv	ultraviolet
sq	square	V	volt(s)
St	stokes	*v-*, *vic-*	vicinal (attached to adjacent atoms)
STP	standard temperature and pressure (760 mm Hg, 0°C)	var	variety
		vic-, *v-*	vicinal (attached to adjacent atoms)
subl	sublime(s), subliming		
SUs	Saybolt Universal second(s)	vol	volume(s) (not volatile)
		v s, v sol	very soluble
sym, *s-*	symmetrical (eg, *sym*-dichloroethylene)	vs	versus
		v/v	volume per volume
t (eg, Bu^t), *t-*, *tert-*	tertiary (eg, tertiary butyl)	W	watt(s)
		Wh	watt-hour(s)
t-, *tert-*, ^t	tertiary (eg, *t*-butyl)	w/o	water-in-oil (eg, w/o emulsion)
TAPPI	Technical Association of the Pulp and Paper Industry		
		wt	weight
tech	technical	w/v	weight per volume
temp	temperature	w/w	weight per weight
tert-, *t-*, ^t	tertiary (eg, *tert*-butyl)	xu (ca 10^{-11} cm)	x unit(s)
theoret	theoretical		
Twad	Twaddell	yd	yard(s)
USASI	United States of America Standards Institute	yr	year(s)

Quantities

Some standard abbreviations (prefixes) for very small and very large quantities are as follows:

deci (10^{-1})	d	deka (10^1)	dk
centi (10^{-2})	c	hecto (10^2)	h
milli (10^{-3})	m	kilo (10^3)	k
micro (10^{-6})	μ	mega (10^6)	M
nano (10^{-9})	n	giga (10^9)	G (or B)
pico (10^{-12})	p	tera (10^{12})	T
femto (10^{-15})	f		
atto (10^{-18})	a		

U continued

URANIUM AND URANIUM COMPOUNDS

Uranium, symbol U, at. no. 92, at. wt. 238.03, is a member of the actinide series of transition elements (see Actinides).

Historical Background

Prior to the Discovery of Fission. Three names stand out in the early history of the element uranium. First, Martin Heinrich Klaproth, the Berlin apothecary, who refused to believe that the heavy black mineral in the Joachimsthal ores was a compound of zinc. Klaproth was convinced that the "pechblende" from Bohemia contained a previously undiscovered element. He named the element, which he believed he had isolated, "uranit" in honor of the recently discovered planet, Uranus. This was in 1789 (1). Klaproth had treated pitchblende first with nitric acid and then with potassium chloride to produce uranium tetrachloride. He obtained a yellow precipitate which he attempted to reduce with carbon and borax. The black metallike powder of uranium oxide he obtained was mistakenly believed to be elemental.

Nearly a half century later the French chemist, Peligot, questioned the conclusions of Klaproth regarding the elemental composition of "uranit" (2,3). Peligot succeeded first in producing a uranium chloride from "uranit," and he was then able to produce true elemental uranium by reacting the chloride with elemental potassium. After leaching away the potassium chloride, he correctly identified the remaining heavy black powder as elemental uranium. Peligot also developed the classical purification process which is based on the solubility of uranium nitrate in diethyl ether. Thus, the element whose existence Klaproth suspected as early as 1780 was not isolated until 1841. Credit for the proper classification of the element uranium goes to Mendeleev who corrected Peligot's inaccurate estimate of its atomic weight.

In the early history of uranium Becquerel was also important, who, in 1896, discovered that uranium was radioactive (4). This discovery led to the rapid development of the element that, perhaps more than any other, will mold the political, sociological, and economical shape of the future. Becquerel was apparently more interested in radium than in uranium, but the discovery of radioactivity in the more abundant of these closely related elements was unquestionably one of the most important events in the early history of the element uranium. A short time after Becquerel's discovery the Curies isolated the elements radium and polonium from uranium ore.

Prior to the discovery of fission in 1939, the uses known for uranium or uranium compounds were relatively unimportant. The first plant to treat uranium ore was built in Czechoslovakia in 1906. Here the product sought was radium, not uranium. A plant was built in Denver, Colo., in 1913 to process handpicked, high-grade carnotite ore from the Colorado Plateau. Here again, the radium was recovered and most of the uranium was wasted.

In the early 1920s, a plant was built in Africa at the Shinkolobwe mine in the Republic of the Congo to process the high-grade pitchblende ore for its radium content. The Canadian deposits at Great Bear Lake were first worked by Eldorado Mining & Refining, Ltd., in 1933.

In all of the above early operations, uranium ores were mined primarily for their radium content. It has been reliably estimated that between 1906 and 1939 the combined production of radium from all of these operations was not more than 1 kg.

The recovery of uranium in these early plants was very low. The amount of uranium that was concentrated as a by-product of radium was probably somewhere between 4,000 and 5,000 tons of contained uranium. Great efforts were made to find profitable markets for uranium compounds or alloys. Limited markets were found in the ceramic industry where uranium was used to impart yellow and green color to fired clay products and to glassware. Efforts to utilize uranium compounds as a catalyst for the production of ammonia showed promise at one time, but gave way to cheaper and more efficient materials. During and after World War I, there was serious effort to replace wolfram (tungsten) with uranium in certain work-hardening tool steels. Clarke B. Carpenter of the Colorado School of Mines made notable contributions to the knowledge of the behavior of uranium when alloyed with iron in tool steels, as well as in other types of steel.

The combined research effort of many workers failed to develop markets based on the chemical or purely physical properties of uranium. Many tons of uranium by-product from radium extraction operations were wasted during this period. Some uranium products were placed in storage ponds or piles with the hope that an ultimate market would someday develop. These waste piles did indeed prove to be a valuable source of uranium in the early days of the fission program in World War II.

Since the Discovery of Fission. The work of Becquerel led rapidly to a basic understanding that the phenomenon of radioactivity is the emission of both particles and energy from the nucleus of the atom. In 1934, Enrico Fermi (5) produced trace amounts of new radioactive elements by bombarding uranium nuclei with neutrons. In 1939 L. Meitner and O. R. Frisch (6) recognized the true significance of Fermi's earlier work and during this same year the nature of the fission reaction was first scientifically described. After the uranium nucleus absorbs a neutron, the nucleus splits into two nearly equal fragments with the emission of more neutrons which are then available to react with additional uranium. Thus the elements of a chain reaction are set in motion. Efforts to achieve an actual chain reaction began at Columbia University, but were moved to the University of Chicago in 1942. In December 1942 this work, under the general direction of Dr. Fermi, resulted in operation of the first self-sustaining nuclear reactor. In the following year, the government-directed operation at Hanford, Wash., started the neutron irradiation of uranium-238 to produce neptunium-239, which decays by β emission to produce plutonium-239. Many initial operational difficulties were encountered in this new field, but by 1945, there was a large dependable production capability.

The extremely high degree of purification required in nuclear feed material presented challenges of a type then unknown to the mining and mineral-processing industries. The dedicated efforts of many now-famous scientists succeeded in solving these problems in an incredibly short time. Recourse was taken to the uranyl nitrate–diethyl ether process of Peligot for the purification of uranium oxide (7) and plants based on this process were built for the government and operated by four prominent U.S. chemical companies.

While development of processing was proceeding at a rate previously unknown to the world, a fantastic program of discovery and development of new supplies of ore was under way. This program was carried out in every part of the world not under the control of the enemies of the United States and its allies. The United States Atomic Energy Commission (USAEC) and its forerunner, The Manhattan Project, apparently felt during the early days that only the high-grade pitchblende ores of the Shinkolobwe type would be suitable for large-scale development (8). It was only when it became apparent that the supply of this type of ore was very limited that real progress was made toward the finding of substantial reserves of low-grade ore. A number of new uranium-bearing minerals were then identified, and new tools were rapidly developed for finding and tracing the radioactive deposits. New and better understanding of the principles of ore deposition and methods of concentration were developed in record time under the impetus and pressure of the AEC program. Additional references to these programs can be found in the section on ore deposits (see p. 7).

By the late 1950s, it became evident that the U.S. Government requirements for uranium could be adequately met by the then existing mining and processing facilities. Many domestic contracts for the purchase of uranium concentrates by the AEC were scheduled to terminate in 1966 and it appeared that the slack to be created by termination of AEC uranium-purchase contracts could not be promptly absorbed by private contracts for uranium to be used for nuclear-power generation. There was a danger in shutting down a substantial portion of the nation's uranium mines because flooding might cause irreparable damage and it was felt that it would not be in the best interest of our national defense to permit these mines to become idle for an indefinite period. Accordingly, a "stretch-out" provision was offered to all producers of uranium concentrates who held AEC contracts. Those producers who accepted the "stretch-out" amendment were permitted to make smaller deliveries to the AEC during the period remaining before December 1966. Then they were permitted to increase the amount of their original contract by stretching deliveries over the period January 1, 1967 through December 31, 1970. This action provided relief to the uranium-mining industry during the period of transition from a completely government-dominated market to a market provided by the privately owned nuclear-power industry.

Many mining companies, in the late 1950s and early 1960s, apparently anticipated a period of oversupply of uranium. There was a resultant letdown in exploration and development activity during this period. However, the rate of increased demand for uranium products for the generation of nuclear power has greatly exceeded the estimates of both government and private experts. Those companies that correctly assessed the future market situation and continued their exploration and development programs are today in a very strong raw-material position. The search for new deposits of uranium in 1968 resembled the early boom days. Exploration drilling activity exceeded that of the most active years of the AEC program. At the beginning of 1969, it was estimated that American mining companies hold firm orders for delivery of uranium products well in excess of $1 billion. It is difficult to conceive of an industry based on a single element, uranium, developing in so short a span of years. For all practical purposes, the uranium industry did not exist in 1940. In 1967, the AEC estimated that new orders received for "selected atomic energy products" amounted to $1,687,728,000. The AEC also estimated the cost of nuclear-electric utilities, either operating, under construction, or in the planning stage, as of October 1968, at over $12 billion (9).

Properties

Physical Properties. Uranium is a dense lustrous metal which, in appearance, resembles iron. It is ductile and malleable. It tarnishes rapidly in air, and in a very short time even a polished surface tarnishes.

High-purity uranium metal, free of porosity, has a density of 19.05 ± 0.02 g/ml at 25°C (10,11). Uranium fuel rods have a slightly lower density.

The melting point of uranium metal has been accepted as 1130°C (2071°F) (12). More recent measurements indicate that the true melting point may be 3°C higher. The boiling point of uranium has not been experimentally determined with accuracy, but by extrapolation of vapor pressure data, is estimated to be 3818°C (11). The heat of fusion of uranium has been calculated to be 4.7 kcal/g-at.

The specific heat of uranium has been observed to be about 0.028 cal/(g)(°C) at 100°C and about 0.044 cal/(g)(°C) at 600°C (13,14).

Uranium metal exists in three distinct allotropic modifications or crystallographic forms. The α form is stable below 660°C. Between 660 and 770°C the β form persists, and above 770°C, uranium exists only in the γ form. The physical and crystallographic properties of each phase of uranium have been studied in great detail. The literature of the USAEC and the United Kingdom Atomic Energy Authority (UKAEA) contains numerous well-documented references to these properties (15). The surface tension of uranium at close to its melting point has been calculated at 1070 ergs/cm² (UKAEA).

The mechanical properties of elemental uranium have been well documented and reported in the literature of the United States, Great Britain, and Russia (7,11,16).

The elastic properties of single and polycrystals have been measured in various orientations. The elastic properties of cast, rolled, and swaged bars have been determined over wide temperature ranges. Elastic properties have also been determined under compression conditions. These properties have been reported by both the USAEC and the UKAEA (16).

The tensile properties of the element, in both crystalline and fabricated form, have been reported over wide temperature ranges (16–19).

The hardness of uranium metal is increased substantially on transforming from α to β. Brinell hardness of "as cast" uranium at room temperature is about 187. Slight carbon impurities cause an increase in hardness (19).

Impact (17), creep (16,18), and fatigue (20,21) properties have been studied widely and the above cited references to the American, British, and Russian literature contain extensive and well-documented data on these properties.

Electrical and Magnetic Properties. The electrical conductivity of uranium is about one-half that of iron and, as is true with most metals, improves with purity. Actual measurements are far below those calculated from a purely theoretical basis (22). The conductivity in single crystals varies with axis orientation (23). At slightly above 1°K, uranium of high purity becomes superconductive (24).

Uranium is weakly paramagnetic (7,19). Its magnetic susceptibility has been measured by the UKAEA to be 1.720×10^6 emu/g at 15°C, and increasing to 1.891×10^6 emu/g at 654°C (25).

The thermoelectric potential of uranium is in about the same position as molybdenum in the thermoelectric series. Copper–uranium thermocouples lie in the range of 4.5 μV/°C (26).

Only slight discontinuities are reported in the thermoelectric curves at the points of phase transformation (27).

Chemical Properties. There are few elements whose chemical and physiochemical properties have been more intensively studied than those of uranium. Literature references in this area are abundant (11,19). Of the four oxidation states, +3, +4, +5, and +6, only two, the +4 and +6, are stable enough to be of practical importance. Aqueous solutions of +3 uranium are readily oxidized to the +4 state with evolution of hydrogen. The +5 state readily disproportionates into the +4 and +6 states. The alternation possible between the +4 and the +6 states has great economic significance. The highly stable and disseminated grains of uraninite in igneous rock formations are in the +4 oxidation state, but when altered to the +6 state they are soluble enough to dissolve in circulating ground water. The relative solubility of uranium in this state accounts for its wide distribution in seawater, in fresh-water rivers, lakes, and ground water. It is in this very stable hexavalent stage that uranium may be introduced into such suitable host rocks as porous sandstones. In these sedimentary host rocks, reducing conditions are, or were, in the geologic past quite common, as hydrogen and carbon monoxide may have escaped from deeper petroliferous horizons. Carbonaceous residues from entrained organic trash may have provided a local reducing environment that altered the +6 state back to the insoluble +4 state of oxidation. It was in this manner that many commercially important deposits of uranium were formed.

The alternation of uranium back and forth between the +4 and +6 oxidation states is of importance in the extraction of uranium from its ores and also in the purification in the refineries. The relative ease of changing oxidation state, coupled with the markedly amphoteric properties of uranium, are most important factors in the metallurgy of uranium. Uranium is one of the few elements which can form an anionic complex in sulfate solution, $UO_2(SO_4)_2{}^{2-}$. This important property is useful in the recovery of uranium by the solvent-extraction process from sulfuric acid leach liquor.

Unlike many metals, uranium does not occur as a sulfide mineral in nature, since uranium has a very weak affinity for sulfur.

Isotopes.   There are eleven known isotopes of uranium.   Three of these, ^{234}U, ^{235}U, and ^{238}U, exist in nature.   All isotopes of uranium are unstable and, as they decay, emit α or β particles. ^{238}U is the most stable isotope and, as would be expected, it is the most naturally abundant.   The composition of natural uranium is approx 99.28% ^{238}U; 0.71% ^{235}U; and 0.005% ^{234}U.   Slight variations in the ratio of the three natural isotopes in equilibrium rock samples have been the subject of much discussion. The formation of additional ^{235}U by exposure of ^{238}U to naturally occurring neutrons from cosmic-ray absorption or spontaneous fission has been postulated (19).

The eight artificial uranium isotopes all have a very short half-life as compared to the three natural uranium isotopes, as shown in Table 1.   These half-lives were determined by Glenn T. Seaborg and Perlman in 1948.   Seaborg has also positively identified trace amounts of plutonium in certain rich uranium ores from Africa.

Table 1.   Half-Lives of Natural and Artificial Uranium Isotopes

Isotope	Half-life	Isotope	Half-life
natural		artificial	
^{234}U	235,000 yr	^{230}U	20.8 days
^{235}U	700 million yr	^{231}U	4.2 days
^{238}U	4.5 billion yr	^{232}U	70 yr
artificial		^{233}U	160,000 yr
^{228}U	9.3 min	^{237}U	6.8 days
^{229}U	58 min	^{239}U	23.5 min

The transuranium elements are produced by bombarding uranium-238 with neutrons to produce an element with a higher atomic number.   This new element, in turn, may be further irradiated to produce a transuranium element with a still higher atomic number, and so on.   The transuranium elements, with their atomic numbers, are listed in Table 2.   It is possible to convert the lower numbers of the series to higher atomic numbers through prolonged neutron bombardment in an environment of intense neutron flux.   Such conversions are being made at the USAEC facility at Oak Ridge, Tenn. (28).

Table 2.   Transuranium Elements and Their Atomic Numbers

Element	At. no.	Element	At. no.
neptunium	93	einsteinium	99
plutonium	94	fermium	100
americium	95	mende–	
curium	96	levium	101
berkelium	97	nobelium	102
californium	98	lawrencium	103

Battelle Memorial Institute (Pacific Northwest Laboratories) estimates that by 1980 civilian power reactors in the United States will be producing approx 501 kg annually of plutonium-238, 220 kg of americium-241, 373 kg of americium-243, and 73 kg of curium-244.   Isotopes of the transuranium elements have found many useful roles in industry and medicine.   They have even been useful in the analysis of samples from the moon surface collected in the Surveyor V and Surveyor VI projects.   Indus-trial uses of the transuranium isotopes include radiography for nondestructive inspection of metals, thickness measurements, density gages, hydrogen:carbon ratios, soil-moisture and compaction measurements, and for many other applications. See also Radioisotopes.

Occurrence and Minerals

Uranium compounds are widely distributed in most of the rock types, in ground water as well as ocean water, in living matter, and even in meteorites. Uranium is not an excessively rare element in nature, but deposits of sufficient concentration to be of commercial value are not common. The earth's crust contains about forty times as much uranium as silver, and about one-tenth as much uranium as copper.

The igneous rocks of the earth's crust are estimated to contain about 1 ppm of uranium in the low-silica igneous rocks, and about 4 ppm in the high-silica igneous rocks. The sedimentary rocks are estimated to contain an average of about 2 ppm of uranium; ocean water contains about 0.002 ppm and groundwater averages about 0.0002 ppm. Uranium appears to be present in the earth's crust in about the same amount as molybdenum and lead.

Uranium, as encountered in ore, is nearly always present in the form of the tetravalent ion. When in solution in groundwater, it is in the hexavalent state. This explains the importance of carbonaceous matter as an effective agent in precipitating uranium from circulating groundwater. The list of known uranium minerals is extensive. Over 250 of these are listed in the appendix to Vol. I of *Uranium Metallurgy* by Wilkinson (28a). Relatively few of the known minerals which contain uranium are of economic importance.

Anhydrous Oxides. Pitchblende, $UO_2 + UO_3$, is a black to dark-gray mineral with a hardness of 4.7–6.3, and a specific gravity of 4.5–7. The other anhydrous oxide minerals of uranium are uraninite and the amorphous powder frequently called "uranium black." It appears that the so-called uranium black is primarily a chalky powder formed on the weathered surface of pitchblende. The uranium content of these oxides varies from about 66% uranium to as high as 92%, depending on the ratio of UO_2 to UO_3, and the presence of impurities or alteration products. Closely related anhydrous oxides are brannerite and cleveite which contain, in addition to uranium oxide, variable amounts of thorium and rare-earth oxides. The anhydrous oxides of uranium may be classed as either endogenetic (primary) or exogenetic (secondary). Pitchblende was the mineral originally mined for its radium content in the Republic of the Congo and in Canada. It is present in small amounts in many granitic pegmatites which have crystallized directly from magmas or in hydrothermal deposits derived from magmatic sources.

Pitchblende or uraninite deposits with commercial production records are located in the Rum Jungle district in the Northern Territory of Australia, in the Republic of the Congo at the famous Shinkolobwe mine, at Joachimsthal in Czechoslovakia, in the Central Massif and other locations in France, and in Urgeirica in Portugal. Small quantities of pitchblende occur in Cornwall. The Eldorado mine at Port Radium in northwest Canada was one of the first important producers of pitchblende. Uraninite is also the principal uranium mineral in the Bancroft, Beaverlodge, and Blind River Districts of Ontario, Canada. The principal uranium mineral encountered in the gold mines of the Witwatersrand of the Republic of South Africa is also uraninite. There are a number of small pitchblende mines in the Front Range of the Rockies in Colorado. There are numerous pitchblende occurrences in Argentina, but none yet has been commercially important. Uraninite is a very common, though not necessarily the dominant, uranium mineral in many of the mines in the Colorado Plateau area. It is also a common mineral in the ores of the Gas Hills and Shirley Basin areas of Wyoming. Pitchblende, along with other uranium minerals, occurs near Marysvale, Utah, and in northeastern Arizona. Regardless of the identity of the predominant uranium mineral in any mine, pitchblende or uraninite seems always to be present.

Hydrous Oxides. The hydrated oxide minerals of uranium constitute a very important group of minerals.

Gummite, $UO_3.nH_2O$, and becquerelite, $2UO_3.3H_2O$, are typical hydrated oxides. They are fairly common minerals, but not commercially important.

Carnotite, $K_2O.2UO_3.V_2O_5.3H_2O$, and tyuyamunite, $CaO.2UO_3.V_2O_5.8H_2O$, are commercially important hydrated oxides of uranium and vanadium associated with either calcium or potassium. These minerals are quite soft and have a Mohs' hardness of 2–2.5. The specific gravity ranges from 3.5 to about 4.5. They are usually found in the weathering zone of sedimentary rocks which have been enriched with organic residues. Other uranium minerals frequently associated with the hydrated oxides are the hydrated calcium–uranium silicates, uranophane, $CaO.2UO_3.2SiO_2.6H_2O$, and coffinite, a hydrated silicate, $U(SiO_4)_{1-x}(OH)_{4x}$.

The hydrated oxide type of uranium minerals are important constituents of the ores produced in the Colorado Plateau, Wyoming, Utah, and Arizona, and in Katanga in the Congo. At Radium Hill in South Australia, the predominant mineral is davidite which is a complex hydrated oxide of uranium with minor amounts of iron, titanium, vanadium, and chromium.

Uranium Phosphates. The uranium phosphates are not at present major sources of uranium, but they are widespread and are locally important.

The two most important uranium phosphates are autunite, $CaO.2UO_3.P_2O_5.$-$8H_2O$ and torbernite, $CuO.2UO_3.P_2O_5.12H_2O$. These minerals are soft with a Mohs' hardness varying from 2–2.5. The specific gravity ranges from 3 to about 3.6 and is usually associated with the oxidation zones of pegmatites and certain hydrothermal deposits. They are of some commercial significance in Cornwall, in Madagascar, and in France in the Haute-Loire region south of the Massif Central. Parsonite, $2PbO.UO_3.P_2O_5.H_2O$, is another lead–uranium phosphate mineral that occurs in significant quantity in Katanga. A very complex barium–uranium phosphate mineral, uranocircite, $BaU_2(PO_4)_2O_4.8H_2O$, is found at Rosmaneira, Spain.

Organic Uranium Compounds. There are a number of urano–organic complexes which are quite common in many of the sedimentary deposits. Thucholite (carburan), typical of this class of compounds, is found as small, irregularly rounded asphaltic nodules embedded in feldspar, quartz, or mica, often associated with uraninite and cyrtolite in pegmatitic veins. These compounds are predominantly carbon with variable amounts of water, uranium, thorium and rare-earth oxides, and silica.

Miscellaneous Uranium-Bearing Minerals. There is a long list of minerals in which uranium occurs as an impurity or as a substituted metal. Few, if any, of these minerals are of economic importance.

Uranium Mining Districts, Production, and Reserves

The United States. In 1968, the USAEC (29) reported that 98% of all uranium reserves of the United States were in sedimentary rock formations. Over 96% of the reserves are in Mesozoic or early Tertiary formations. Only 2% are reported to be in rocks of Precambrian age. Nearly 30% of the known reserves are in ore running between 0.20 and 0.25% (4–5 lb of U_3O_8 per ton). About 25% of the reserves are in the range of 0.25 to 0.35% and about 11% higher than 0.35%. There were an estimated 1300 properties producing uranium ore in the United States during 1968.

Sedimentary sandstones and conglomerates account for 94% of the known reserves. About 2.5% are in finely crystalline meta-sedimentary rocks. About 1% are associated with coal deposits and slightly over 0.5% are in igneous rocks. Of all known reserves 90% are contained in less than 10% of the known deposits, and 80% of the production in 1968 came from less than 10% of the operating mines.

The Colorado Plateau Area. This province covers the southeast third of Utah, the western one-sixth of Colorado, the northeast third of Arizona, and the northwest quarter of New Mexico. Nearly 60% of all known uranium reserves of the United States are in the Colorado Plateau area.

It is estimated that production of uranium from the Colorado Plateau, up to 1967, was about 122,600 tons of U_3O_8. Known reserves of this province, as of the beginning of 1967, have been estimated by the AEC (29) at 88,408 tons of U_3O_8. These estimates are based on ore that can be profitably mined and milled at a selling price of $8/lb of U_3O_8 in yellow-cake mill concentrate. A selling price of $10/lb of U_3O_8 would probably double these reserves.

The original mining of uranium ore on the Colorado Plateau dates back to 1913 when small quantities of hand-picked, high-grade ore were mined in the Uravan District and shipped to Denver for recovery of radium values.

The discovery of higher-grade radium ores in Africa nearly put an end to the Colorado Plateau uranium industry. The carnotite ores of the Uravan District contained over 1% vanadium pentoxide values, and these were needed for the production of tool steel during World War I. U.S. Vanadium Corporation (later acquired by Union Carbide) and Vanadium Corporation of America (later merged with Foote Mineral Co.) each operated mines and concentrating plants in the Uravan District for many years during which the principal value was vanadium, and the secondary value was uranium. The great search for uranium ores during the latter months of World War II, and during the decade following the end of the war, expanded the limits of the original Uravan District. Uranium values were found in sandstones and mudstones as far north as Moab, Utah, and south and west into the Lukachukai Mountains of northeast Arizona.

In the early 1950s, the area had extended as far west as the Grand Canyon and to the southeast almost to Albuquerque. The greatest reserves were found in the Ambrosia Lake, Jackpile, and Gallup regions of northwest New Mexico.

During the development of the huge ore bodies in the Ambrosia Lake area the extreme importance of structural features in the control of uranium deposits was first recognized. At the end of 1968, there were seven uranium mills operating in the Colorado Plateau. Three of those were in New Mexico, three in Colorado, and one in Utah. Another mill is projected for completion in 1970 near Gallup, N. Mex. Four mills operating in this area during the 1950s were closed at the end of 1968.

The ores in the Uravan District of Colorado, Utah, and in northern Arizona are relatively small and there are very few mines in this part of the Plateau with substantial proven reserves. In extreme southern Utah, and in Arizona, the deposits are mostly in the Moenkopi, Shinarump, and Chinle sandstones; they are small and spotty, but numerous. The mines in the Grants area of New Mexico are larger, and a number of these mines have very substantial reserves (30) in the bleached areas of coarse red Westwater sandstone. Most of the ore in the Ambrosia Lake area occurs in anticlinal traps and along faults around a structural dome. Ore thickness in some of the faulted zones is over 100 ft. Most of the sandstone ore in the ore zones is weakly cemented with clay and calcite. The following processing plants were operating in the Colorado Plateau area in 1968: Anaconda Company at Bluewater, N. Mex.; Atlas Company at Moab, Utah; Climax Uranium Company at Grand Junction, Colo.; Homestake-Sapin Partners at Grants, N. Mex.; Kerr-McGee Corporation at Grants, N. Mex.; and Union Carbide Nuclear Corporation at Uravan and Rifle, Colo.

The Wyoming Basins Area. This area embraces nearly all of southcentral Wyoming and a small area of northwest Colorado. Production from this area up to the beginning of 1967 was about 27,000 tons of U_3O_8. Reserves that can be profitably mined at \$8/lb of U_3O_8 were estimated by the AEC at approx 53,000 tons at the beginning of 1968. This tonnage was increased during 1968 by a very active exploration-drilling program. It is likely that a price increase for yellow cake to \$10/lb would more than double the Wyoming reserves.

At the beginning of 1968 (31), Wyoming's proven uranium reserves were 36% of the United States total. In 1967, Wyoming produced 25% of the nation's uranium. Over 50% of all exploration drilling in the United States in 1967 was in the Wyoming Basins province.

Production from the Wyoming Basins is presently centered in the Gas Hills and Shirley Basin areas. Extensive exploratory drilling and other development work in the Powder River Basin and elsewhere is indicative of expansion of uranium activity in Wyoming.

At the end of 1968, there were five active uranium mills in Wyoming. The Wind River and White River formations of Tertiary age are important host rocks in this area. Most of the known uranium reserves of the Wyoming Basins area are relatively shallow, and most of the mining is by open-pit methods. There are a few underground mines.

Other U.S. Areas. About 96% of all known U.S. uranium reserves are located in either the Colorado Plateau or in the Wyoming Basins areas. The remaining 4% are widely scattered. Reserves in this category were estimated by the USAEC (29) at the beginning of 1968 to a total of 6,399 tons of U_3O_8. These estimates are based on \$8/lb for U_3O_8 and would be much larger if the price should escalate. The areas covered in this "other" category include, the Northern Plains, the Colorado and the Southern Rocky Mountains, Gulf Coastal Plains, the Northern Rockies, the Sierra Nevada Range, Alaska, Northern and Central Basin and Range, the Columbia River Plateau, the Southwestern Basin and Range area, and the Southern Plains area.

The area covered by this category of deposits is huge, and includes many hundreds of known occurrences of uranium. There are two relatively small mills operating on ores from this area. One mill at Cañon City, Colo., operates on ores produced from the Colorado Front Range. Another mill near Falls City, Tex., operates on ore produced in that area. Two mills, one in the State of Washington, and one in Oregon, which formerly operated on ore produced in the Far West, are now idle. Lack of sufficient ore is said to be responsible for the closing of these mills.

Canada. Proven uranium ore reserves for the Dominion of Canada, at a price of less than \$10/lb, are estimated at 200,000 tons of U_3O_8.

Blind River. The uranium ore reserves of the Blind River district in western

Ontario are among the largest known in the Free World. Reserves of ore ranging between 0.08 and 0.11% U_3O_8 have been estimated in excess of 300 million tons (11). Blind River ore is in Precambrian conglomerates that are cemented with silica and sericite (a variety of mica). The principal uranium mineral is brannerite, with smaller amounts of uraninite, thucholite, and monazite. There are also sizeable reserves of thorium associated with the uranium. The ratio of uranium to thorium is about 2 to 1. The conglomerate veins dip steeply and all mining is from underground workings.

Bancroft District. These deposits occur in pegmatites in the southern part of the Canadian Shield. The ore bodies are relatively small and average about 0.1% U_3O_8. The principal uranium mineral is uraninite and fairly large amounts of calcite and fluorite occur in the ore. The ratio of uranium to thorium is about the same as in the Blind River ores, 2 to 1. There was no active processing plant in the Bancroft area at the end of 1968.

Northwest Canada. On the east shore of Great Bear Lake in the Northwest Territory, pitchblende and uraninite occur in Precambrian rocks. These ores were worked as early as 1930 for their radium content. The processing plant at Port Radium was not treating uranium ores in 1968.

There has been extensive production of uranium from the Beaverlodge district along the north shore of Lake Athabasca. Pitchblende in veins and as a dissemination in the Archean rocks of the area account for most of the ore in this area. Some production has come from pegmatite dikes. Most of the known higher-grade ore has been mined in this district and known reserves are now estimated to be about 40,000 tons of U_3O_8.

The Republic of South Africa. The Republic of South Africa ranks third among the Western nations in magnitude of proven uranium reserves. The European Nuclear Energy Agency, at the beginning of 1968, estimated the reserves of this country at about 200,000 tons of U_3O_8. This is ore that could be produced when uranium is sold at not less than $10/lb of U_3O_8.

The principal uranium mineral is uraninite; thucholite is locally important. These minerals occur together with gold in the Precambrian conglomerates of the Witwatersrand Series. Mineralization occurs for more than 200 miles, stretching from the Far-East Range to the Orange Free State (32). The ore varies between 0.1 and 0.008% U_3O_8, and most of the ore would not be of commercial grade if it were not for the gold content. The value of uranium production has dropped sharply in recent years. As contracts with the AEC and the UKAEA have expired, they have not been renewed. The economic pressure on the gold mines, because of the fixed price of gold and higher operating costs, has caused a number of the mines to cease operations.

The reserves of uranium in South Africa are large, but production is determined by the profitability of the gold mines. There are large areas in the southern part of the Republic that are geologically favorable for the occurrence of uranium deposits. Since 1967 there has been an increase in the amount of exploratory drilling, as well as other prospecting activity. No important discoveries have been announced. There is also at present considerable prospecting activity in the potentially productive areas of Southwest Africa. There were neither proven reserves nor production from Southwest Africa in 1968.

Uranium Reserves of the Rest of the World. About 83% of the Free World reserves of proven uranium ore are located in the United States, Canada, and the

Republic of South Africa. Another 12% are located in France, Australia, Spain, Portugal, the Republic of the Congo, and Argentina. The remaining 5% are widely scattered in many locations.

The rapid increase in the number of nuclear-power plants scheduled for construction within the next decade has accelerated uranium prospecting all over the world. The estimate of about 700,000 tons of proven uranium reserves in the noncommunist countries in 1968 (33) may well be doubled by 1975. This will be almost certain if the price of U_3O_8 reaches or exceeds $10/lb. Numerous areas in the Communist nations are known to be geologically favorable for the occurrence of uranium deposits. Reliable data on proven ore reserves in these countries are not available.

Extraction of Uranium from its Ores

The methods used to extract uranium values from its ores vary widely, and composition of the uranium minerals is only one of several factors affecting the choice of milling methods.

The ore itself may vary from hard igneous rock to soft, weakly-cemented sedimentary rock. The principal gang mineral may be quartz, which is relatively inactive chemically, or there may be fairly large amounts of acid-consuming minerals, such as calcite, present. Some of the ores are highly refractory and require intensive treatment, while others literally fall apart while being transported from the mine to the mill.

Preconcentration. Conventional ore-dressing techniques have not been generally successful in the preconcentration of uranium minerals. Studies continue in this field and there is a tremendous economic incentive for a satisfactory preconcentration method. In most uranium ores, it is quite possible to produce concentration ratios that are acceptable. It is seldom possible to produce a tailing that can be discarded without sacrificing acceptable recoveries.

In areas where the uranium values occur as masses in pegmatitic rock with large areas of unmineralized pegmatite separating the ore minerals, there is a field for preconcentration by electronic sorting devices. These are effective only with ore coarser than 2 in. with the fines removed. Such devices have been used with some success in Canada at the Bicroft mine in eastern Ontario and at Beaverlodge on Lake Athabasca. A similar device was used effectively in Australia at the Mary Kathleen mine. In these devices a coarse ore from which the fine particles have been removed is fed onto a relatively small, slow-moving conveyor or "picking belt." A Geiger counter mounted above the belt can be programmed to actuate a cylinder-operated "pusher" to remove either barren or radioactive pieces of ore from the conveyor.

Heavy-media concentration systems (see Vol. 10, p. 701) are also effective for preconcentration, if a clean separation of waste of throw-away grade and uranium mineral concentrations is possible at fairly coarse sizes. Heavy-media systems are useful down to about 0.25 in. in standard static cones or drum separators. Heavy-media cyclone separation is feasible down to about 16 mesh. Heavy-media installations in Sweden are used to remove barren limestone from the ore.

Methods based on flotation (qv) have been thoroughly investigated on a wide variety of ores. A reasonably satisfactory ratio of concentration is usually attainable, but it has been most difficult to produce tailings low enough in uranium to discard.

In some of the more important uranium mining districts, if even a very low ratio of concentration could be achieved, this would greatly expand the reserves which can be recovered economically. Preconcentration presents a real challenge, and research continues on a world-wide basis. At the beginning of 1969, there were no plants operating in the United States with preconcentration units. There were a few in operation ten years earlier, but these are now idle.

Crushing and Grinding. The sand grains in the sandstone ores are essentially barren. Since they are the hardest and, in most cases, the coarsest component of the rock, there is no point in grinding the ore finer than the average grain size of the sand.

Coarse crushing is almost universally done in jaw crushers. Secondary crushing may be done with smaller jaw crushers, gyratories, or hammer mills, depending on the clay content of the ore. Grinding is done in rod mills, ball mills, and hammer mills. In at least one plant, autogenous grinding is accomplished in large-diameter mills of the air-swept type. The scrubbing action in the grinding mill can be of great importance. The uranium values are concentrated in the cementing material and in the coating of the sand grains which are separated from the barren sand during the grinding action. In many cases, the ore is so poorly consolidated that there is no need to close the grinding circuit with screens or classifiers.

Leaching. After grinding and preconcentration (if preconcentration is practiced) the uranium values are leached from the ore with a suitable solvent. Leaching is normally done in mechanically agitated tanks or in air-agitated columns. Leaching is followed by conventional liquid–solid separation using classifiers and thickeners. It is customary to clarify the final solution by filtration. Dissolution of the uranium in the ground ore may be accomplished in either acid or alkaline leaching circuits. There is no fixed rule governing the choice between acid and carbonate leaching. Ore containing limestone or sandstone ores with a high percentage of calcite as grain-cementing material is generally leached in alkaline circuits. In ores cemented with clay, silica, or organic material, the acid leach is preferable.

The Acid Process. If acid consumption exceeds 150 lb/ton of ore treated, the alkaline system would normally be used. The delivered costs of acid and sodium carbonate and sodium hydroxide vary widely in different areas and these costs may be the determining factor.

Capital costs for acid circuits are somewhat lower than those for alkaline circuits. The carbonate leach is usually performed under pressure and the cost of autoclaves is an important factor. Recoveries are in general slightly higher in acid circuits which are easier to control than the carbonate circuits. As a general rule, the acid-leach circuit is preferred unless the lime content of the ore is so high that acid leaching is economically precluded.

In the Colorado Plateau area, most of the ores are amenable to acid leaching. Sulfuric acid is universally used and it is satisfactory with ores containing oxidizable components, such as organic matter and ferrous oxide. Very high leaching efficiencies with sulfuric acid are common. Most plants operate at above 95% dissolution of uranium values and a number of plants exceed 98% extraction. To obtain these high efficiencies, it is necessary to oxidize any tetravalent uranium present in the ore to the hexavalent state. The lowest cost and most generally used oxidizers are MnO_2 and $NaClO_3$. Typical leach reactions are the following:

$$6\,H_2SO_4 + 3\,MnO_2 + 3\,UO_2 \rightarrow 3\,UO_2SO_4 + 3\,MnSO_4 + 6\,H_2O$$

$$3\,H_2SO_4 + NaClO_3 + 3\,UO_2 \rightarrow 3\,UO_2SO_4 + NaCl + 3\,H_2O$$

In practice, the oxidation potential of the solution is determined by measuring the ferric to ferrous ratio. The role of ferric iron in the oxidation of tetravalent uranium is important. The internal reaction that makes possible the two oxidation reactions illustrated above involves the conversion of ferrous iron to ferric iron. The ferric iron then oxidizes the UO_2. The reaction (7) is probably taking the following course:

$$2\,Fe^{2+} + MnO_2 + 4\,H^+ \rightarrow 2\,Fe^{3+} + Mn^{2+} + 2\,H_2O$$
$$UO_2 + 2\,Fe^{3+} \rightarrow UO_2{}^{2+} + 2\,Fe^{2+}$$

In most ores, sufficient iron is present for this reaction. It is necessary to add metallic iron to some ores to ensure an adequate supply of iron. Most of the ores of the Colorado Plateau can be leached in 8 hr or less with adequate agitation slightly above room temperature.

The Carbonate Process. Alkaline or "carbonate" leaching is generally used in the treatment of uranium ores when the lime content results in excessive acid consumption. It is difficult to dissolve tetravalent uranium with the carbonate leach without the use of oxidants. The reaction between uranium and aqueous carbonate solutions results in readily soluble complex uranium compounds, such as $Na_4UO_2(CO_3)_3$. With proper oxidizing conditions, extractions with the carbonate leach process run between 90 and 95%.

Since the carbonate-leach process involves virtually no reaction of the gang minerals, it is imperative that all uranium-bearing minerals be directly exposed. This usually means that ores must be ground somewhat finer if the carbonate leach is to be employed. The acid leach is much more effective in removing oxidation film of metallic salts that may effectively protect the uranium mineral surface from the leach solution.

Carbonate leaching at atmospheric pressure is slow and recoveries are relatively low. Typically, the ore is leached in autoclaves or in Pachuca tanks with air providing most of the needed oxygen. Potassium permanganate is frequently used when an additional oxidant is required. Separation of the leach liquor from the undissolved solid is usually effected in a countercurrent decantation system using a series of thickeners. The liquor is clarified in pressure filters, and the solids are washed on drum filters before being discarded as tailings. The uranium is precipitated from the clarified sodium carbonate solution with sodium hydroxide.

In the precipitation step, there is partial regeneration of the sodium carbonate as shown by the following reaction:

$$2\,Na_4(UO_2(CO_3)_3) + 6\,NaOH \rightarrow Na_2U_2O_7 + 6\,Na_2CO_3 + 3\,H_2O$$

The excess sodium hydroxide normally present is converted to the carbonate in packed towers through which carbon dioxide is circulated.

If vanadium is present in the ore being leached by the carbonate process, a variation in the precipitation process is necessary. The clarified leach liquor is neutralized by acidification to a pH of about 6. This causes sodium uranylvanadate to precipitate. The precipitate is fused with soda ash and a convenient form of carbon. The uranium is reduced to the insoluble UO_2 state and the water-soluble sodium vanadate is leached away.

There have been many modifications of both the acid-leach and the carbonate-leach process. Local conditions and unusual mineral combinations have resulted in numerous ingenious mechanical and chemical innovations.

Extraction of Uranium from Leach Liquors. Initially all uranium produced had to meet the purchasing specifications of the USAEC. The more recent specifications developed by the privately owned producers of nuclear fuel are not greatly different from those of the USAEC. All of the extraction processes have been designed with the primary objectives of high recovery and products that meet the USAEC specifications.

Direct Precipitation. Methods involving direct or selective precipitation have not been commercially successful in acid-leach systems. The precipitation from carbonate-leach liquors is discussed below (see p. 18). The highly complex composition of the acid-leach liquor makes it virtually impossible to obtain specification products by this route. Precipitation with phosphates and with caustic soda are useful in treating side or internal streams for further purification, but have not been satisfactory for making a finished product.

Ion Exchange. The ion-exchange (IX) process has been extensively used for recovery of uranium from acid-leach liquors. Although ion-exchange plants have been replaced in some localities with solvent-extraction (SX) plants, the ion-exchange process is still important to the uranium industry.

The resins used as adsorbents are usually strongly basic anion-exchange resins with quaternary ammonium bases. Such resins may be used with either acid- or carbonate-leach liquors, but their commercial application has been generally limited to acid liquors. These resins extract uranium in the form of anionic complexes.

The physical structure of the resin is most important. The grains must be approximately spherical and hard enough to resist the abrasion of handling, as well as the action encountered within the column. The volume change in loading and stripping tends to crack the resin and hasten its degradation.

The ion-exchange process is necessarily a batch process involving sorption, washing, elution, and regeneration. The development of highly automated valving equipment has so reduced manual operation that a modern IX plant can almost be regarded as a continuous process.

In the treatment of liquors from sulfuric acid leaching, the IX process hinges on the rather unusual capacity of uranium to form anions in sulfate solutions. These anions exchange readily with the anions present in the prepared resin. Elution of the uranium from the resin is usually accomplished with sodium chloride or ammonium nitrate.

Gittus (7) illustrates the equilibrium in a pregnant acid medium in the following manner:

$$UO_2^{2+} + n\,SO_4^{2-} \rightleftharpoons (UO_2(SO_4)_n)^{2-2n}$$

where $n = 1$, 2 or 3. He further illustrates the adsorption of uranium from the acid medium by the following two reactions:

$$4\,RCl + (UO_2(SO_4)_3)^{4-} \rightleftharpoons R_4UO_2(SO_4)_3 + 4\,Cl^-$$
$$2\,RCl + (UO_2(SO_4)_2)^{2-} \rightleftharpoons R_2UO_2(SO_4)_2 + 2\,Cl^-$$

The nitrate reaction would be similar. These reactions indicate that from an acid medium uranium can be adsorbed as either the disulfate complex or the trisulfate complex. The high specificity of the anionic process for uranium makes it the most desirable process. A satisfactory resin is produced by the amination of a copolymer of chloromethylstyrene and divinylbenzene (34).

The rate of adsorption on such a resin is rapid and about 7 min is adequate for conditions near equilibrium. The resin must have a higher specific gravity than the pregnant liquor or the liquor could not permeate the resin and it must be highly resistant to both solution and attrition. The precipitate obtained from the eluate from such a resin contains normally about 80% U_3O_8. Sorption capacities and other characteristics of various anion-exchange resins manufactured in the United States, Russia, Germany, and Great Britain have been published (11).

Resin-in-Pulp Ion Exchange. The so-called resin in pulp, or RIP, process has found limited application on the Colorado Plateau. The advantage of this process is that it eliminates the need for separating the undissolved solids from the leach liquor. The slurry, after treatment with acid, flows through a series of baskets made of perforated stainless steel and filled with strongly basic anion-exchange resin beads of carefully selected size. The baskets are alternately raised and lowered in the slurry so as to cause intimate contact of the slurry with the resin. Equilibrium conditions are usually reached in about 40 min. Elution procedure is similar to that used in fixed-bed IX systems.

The mechanical features of the RIP process have resulted in fairly high maintenance costs. The many ingenious devices intended to minimize these problems have been only partially successful. The cost of resin regeneration to remove "poisons" has been a serious deterrent to the ion-exchange process.

Solvent Extraction. The first commercial application of solvent extraction to the recovery of uranium from ore was at the Kerr-McGee mill at Shiprock, N. Mex. in 1956. Since that time, the SX process has been adapted to the treatment of a wide variety of ores. A number of IX installations have been replaced by the SX process and, in some cases, this has resulted in lower operating costs and improved recovery. The features that have contributed to the rapid acceptance of the SX process include simplicity and ease of control and operation, low installation cost (as compared to IX), high recovery of uranium values, and high purity of product. Other features of importance are the ease of regeneration of the solvent, and the ability to recover by-product values from the uranium-barren solution.

The solvents in general use are alkylamines and organophosphorus compounds. The process is based on the property of these solvents, which are immiscible with water, to form complexes with uranium salts. These complexes are soluble in an excess of the solvent. When clarified pregnant leach solution is brought into contact with the organic solvent, the uranium is distributed between the aqueous and organic phases. Under proper process conditions, the uranium can be extracted almost quantitatively into the organic phase while most of the other constituents of the leach liquor remain in the aqueous phase. After mixing to ensure contact, the mixture of solutions is removed to a quiet settling tank where the lighter, but uranium-loaded organic layer rises to the surface and is decanted for stripping. After the solvent is stripped of its uranium content, it is returned to the mixers to meet incoming pregnant leach liquor. The uranium is stripped in an aqueous solution from which it is precipitated, usually as a very pure "yellow cake." These operations are normally conducted in batteries of cells in a countercurrent system. The flowsheet of a typical large Colorado Plateau uranium-ore processing plant, using a sulfuric acid leach, followed by sand-slime separation, clarification, and subsequent extraction, is shown in Figure 1 (35).

Two organophosphorus compounds are in general use as solvents for extracting

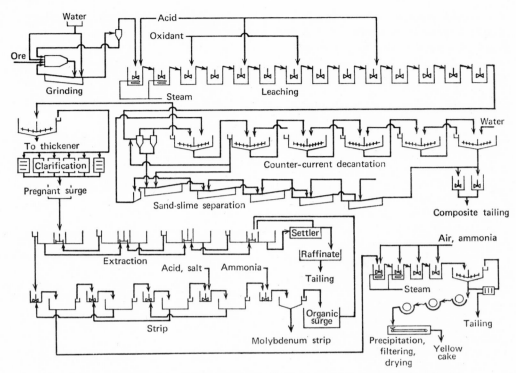

Fig. 1. Flowsheet of a typical Colorado Plateau uranium-ore-processing plant.

uranium from acid-leach liquors. The original, and most generally used, is di(2-ethylhexyl)phosphoric acid, D2EHPA. This is a dialkylphosphoric acid. The second solvent is dodecylphosphoric acid, DDPA, which is a monoalkylphosphoric acid. These two solvents have about equal selectivity for uranium. With both, it is necessary to reduce all iron present in the feed liquor quantitatively to the Fe^{2+} state. Solubility losses with DDPA are substantially higher than with D2EHPA and strong acid solutions are necessary to strip DDPA. Carbonate stripping of DDPA is not feasible because of the high solubility of the sodium salt of the acid in the aqueous phase (36). Kerosene, with a high flash point, is usually employed as a carrier for the solvent. To improve the miscibility, certain additives, such as tributyl phosphate and long-chain alcohols, are generally used. The organophosphorus compounds are preferred to the alcohols because of a synergistic effect in increasing the distribution coefficient. This is particularly true when used with D2EHPA. Solubility and entrainment losses of the organophosphorus extractants are very low in well-designed and properly operated plants (11).

The cation-exchange reaction (7) is illustrated in the following general equation:

$$2\ HOP(OR)_2(org) + UO_2{}^{2+}(aq) \rightleftarrows UO_2(OP(OR)_2)_2(org) + 2\ H^+$$

Stripping of loaded D2EHPA is usually carried out with sodium carbonate. DDPA is stripped with either hydrochloric or hydrofluoric acids. The D2EHPA solvent-extraction process is generally referred to as the Dapex process.

The alkylamines were first used as uranium extractants in Oak Ridge and the process is generally known as the Amex process. Tertiary aliphatic amines with high molecular weight and a normal chain are generally satisfactory uranium extractants. The extraction of uranium from an acid-leach liquor with a weak-base alkylamine is illustrated by the following reaction (7):

$$(R_3NH)_2SO_4(org) + UO_2(SO_4)_2{}^{2-}(aq) \rightleftharpoons (R_3NH)_2UO_2(SO_4)_2(org) + SO_4{}^{2-}(aq)$$

It is customary to use a high-flash-point kerosene as the carrier or diluent for the alkylamine solvent. Low-cost, long-chain alcohols are used to improve the compatibility of the solvent and the carrier.

Stripping of the loaded alkylamine extractants is a relatively simple operation and a variety of stripping agents may be used. Nitrates and sulfates are satisfactory with close pH control. Soda ash and caustic soda have also been used commercially.

Attempts have been made to apply solvent-extraction technique to unclarified slurries, but there have been no commercial applications of this interesting concept. In this case, judging from pilot-plant results, the dialkylphosphoric acids are preferred over the alkylamines. An efficient system for solvent recovery from the aqueous raffinate must be developed before this process becomes commercially attractive (37).

Precipitation of Uranium Concentrates. The end product of all of the extraction processes, whether acid or carbonate leach, is a purified uranium solution that may or may not have been upgraded by ion exchange or solvent extraction. The uranium in this solution must be precipitated and dewatered prior to shipment.

Solutions resulting from carbonate leaching are usually precipitated directly from the clarified leach liquor without concentration. The only remaining active plants using the carbonate-leach process on the Colorado Plateau are the Homestake-Sapin Partners plants near Grants, N. Mex., and the Atlas Corporation plant at Moab, Utah. The plant of Eldorado Mining & Refining, Ltd., at Beaverlodge near Lake Athabasca in Saskatchewan also precipitates directly from carbonate leach liquors. At these plants yellow cake is precipitated with caustic soda. At the Homestake-Sapin plant the thickened slurry is filtered and then roasted at 1600°F. The roasting converts the contained vanadium to a soluble sodium vanadate which is leached away from the uranium compound. The insoluble uranium compound is filtered and dried for shipment. The precipitation reaction (neglecting the presence of vanadium) is illustrated by the following equation:

$$2\ (UO_2(CO_3)_3)^{4-} + 2\ Na^+ + 6\ OH^- \rightarrow Na_2U_2O_7 + 6\ CO_3{}^{2-} + 3\ H_2O$$

The precipitation is not quantitative, but losses are held to a minimum by carbonation of the mother liquor with CO_2 and return of the carbonated product to the leach system.

Precipitation of uranium from acid solutions obtained from either ion-exchange or solvent-extraction strip liquors is usually accomplished by neutralization with either ammonia or magnesia. Lime may be used to precipitate uranium, but the resulting product has too low a uranium content to meet AEC specifications.

Precipitation with ammonia produces a very acceptable product in the form of a complex salt of the probable composition:

$$(NH_4)_2((UO_2)_2SO_4(OH)_4(H_2O)_n)$$

The ammonium salt is preferred if the product is to be used for the manufacture of fuel element.

Yellow cake with a higher uranium content can be obtained by precipitation with magnesia. The magnesium sulfate formed in the reaction is water soluble and the uranium compound can be separated by filtration. The reaction with magnesium hydroxide probably proceeds in the following way:

$$2\ UO_2SO_4 + 3\ Mg(OH)_2 \rightarrow MgU_2O_7 + 2\ MgSO_4 + 3\ H_2O$$

Ore Processing Plants of the World

In the United States. *The Colorado Plateau Area.* At the end of 1968, there were seven uranium mills operating in the Colorado Plateau area. Virtually all ore treated in the Colorado Plateau area is sandstone or mudstone containing varying amounts of calcite. The ore minerals are principally uraninite, carnotite, and coffinite. The cementing material in the sandstone is usually clay or calcite, resulting in a structurally weak or incompetent rock.

Of the seven uranium mills in operation in the Colorado Plateau area at the end of 1968, five were using sulfuric acid leach systems and two plants were using a carbonate leach. Table 3 gives processing data of mills in the Colorado Plateau area.

Table 3. Processing Data of Mills in the Colorado Plateau Area

Company	Location	Capacity,[a] tons/day	Type leach	Type concn
American Metal Climax	Grand Junction, Colo.	500	acid	SX
Anaconda Co.	Grants, N. Mex.	3000	acid	RIP
Atlas Corp.	Moab, Utah	1500	carbonate	RIP
Kerr-McGee Corp.	Grants, N. Mex.	6000	acid	SX
United Nuclear Homestake Partners	Grants, N. Mex.	3500	carbonate	direct pptn
Union Carbide Corp.	Uravan and Rifle, Colo.	1000	acid	SX

[a] Approximate.

Wyoming Basins Processing Mills. The only mill operating in the Shirley Basin of Wyoming at the end of 1968 was that of the Petrotomics Company. This plant has a capacity of 1000 tons per day. A construction program beginning in 1969 will increase the capacity of this plant to 1500 tons per day. The plant utilizes an acid leach followed by solvent extraction.

Utah Construction Company plans a 1000-ton mill in the Shirley Basin which is expected to be in production by the end of 1970. This will probably be an acid-leach solvent-extraction mill.

In the Gas Hills district of Wyoming, four mills were in operation at the end of 1968.

The Federal-American Partners' mill in Fremont County has a daily capacity of 900 tons of ore. This plant operates an acid-leach system followed by a resin-in-pulp IX concentration.

Union Carbide Corporation in Natrona County has a mill with a rated daily capacity of 800 tons of ore. This mill is equipped with an acid-leach circuit and resin-in-pulp IX concentration system.

The Utah Construction Company mill in the Gas Hills area has a rated capacity of 1000 tons of ore per day. It is equipped with an acid-leach circuit, followed by an IX circuit utilizing the Eluex process for treatment of the IX strip liquor.

The Western Nuclear mill at Jeffrey City treats ores from the Gas Hill district. It has a daily capacity of 1200 tons of ore. This mill employs an acid-leach circuit followed by a resin-in-pulp and solvent-extraction (Eluex) system. A 10% sulfuric acid solution is used as the eluate in the RIP circuit. The solvent in the SX plant is a tertiary amine in a kerosene carrier. "Isodecanol" (see Vol. 1, p. 567) is used to control formation of emulsion; ammonium sulfate is used for stripping.

Other Areas. There are three uranium processing plants located outside of the Colorado Plateau and the Wyoming Basins areas. One of them is owned by the Cotter Corporation and is at Cañon City, Colo. This plant treats ores mined from the Front Range of the Rocky Mountains. The mill has a capacity of 400 tons of ore per day. It has a carbonate-leach system followed by precipitation. This plant also operates an acid-leach circuit for treatment of residues from African ores which were processed in the early days of the AEC program. This circuit includes solvent extraction.

At Edgemont, S. Dak., Susquehanna-Western, Inc., operates a processing plant with a rated capacity of 650 tons of ore per day. This plant operates an acid-leach circuit followed by solvent extraction.

Susquehanna-Western also operates a uranium processing plant at Falls City, Tex. This plant has a rated capacity of 1000 tons of ore per day. It utilizes an acid-leach circuit followed by a solvent-extraction circuit.

Canada. The ores of the Blind River or Elliot Lake districts of western Ontario are conglomerates consisting of well-rounded quartz pebbles. The ore minerals are brannerite, uraninite, and thucholite. Fine grinding is necessary to expose the uranium minerals for acid leaching.

Brannerite is difficult to dissolve and excess acidity and long leaching time are required to attain high recovery. Leaching is carried out in air-agitated Pachuca tanks. Both resin-in-pulp and static-bed ion-exchange systems are used to upgrade the uranium-bearing leach solutions.

In the Bancroft area of eastern Ontario, the ore occurs in pegmatitic rock. Very fine grinding is required to liberate the uranium minerals. Leaching was accomplished in air-agitated Pachucas with sulfuric acid and sodium chlorate. Static ion-exchange columns were utilized for upgrading the leach liquor. No mills were operating in the Bancroft area at the end of 1968.

In the Beaverlodge district on the north shore of Lake Athabasca, the Eldorado Mining & Refining, Ltd., operates Canada's only carbonate-leach plant. Autoclave carbonate leaching is followed by direct uranium precipitation.

In the latest expansion of the Beaverlodge mill, provision has been made to handle a pyrite concentrate from the Verna mine. This pyrite is acid leached, filtered, and the uranium precipitated with magnesia. The precipitate is then combined with the flotation tailing and fed to the carbonate-leach system (38).

Eldorado Mining and Refining, Ltd., has also operated a uranium processing plant at Port Radium on Great Bear Lake in the Northwest Territory. Acid leaching was followed with a solvent-extraction plant. This plant was not in operation at the end of 1968.

Republic of South Africa. The extraction of uranium values from the gold ores of South Africa begins after the ore has been ground and the gold values removed. In a typical plant, the residue from the gold extraction plant is filtered, washed, and repulped with sulfuric acid and manganese dioxide and leached in air-agitated pachuca tanks. The filtered leach liquor then passes through an ion-exchange circuit. It is eluted from the columns and the uranium is precipitated as ammonium diuranate. There are minor variations in the flowsheets of the various South African plants (38).

France. There are four uranium mining districts in France with ore processing plants. In the Verdée District there is a mill at L'Ecarpière with a capacity for treating 300,000 tons of ore per year. Magnesium uranate is produced with an acid-leach circuit followed with an ion-exchange concentration circuit.

The processing plant near Bessines in the Crouzille District has a reported capacity of 600,000 tons of ore per year. The acid-leach circuit is followed by three separate processing lines. Two of these produce magnesium uranate by more or less standard ion-exchange circuits and the third uses a solvent-extraction circuit for concentration of uranium values.

Two processing plants are in the Forez District. The mill near Gueugnon treats ore from the nearby Grury mining area, as well as concentrates from Mounana (in the Republic of Gabon, Africa). This plant has a reported capacity to produce 400 tons of uranium per year as a high-grade uranyl nitrate solution. Another processing plant near Thiers also produces a high-grade uranyl nitrate solution. This plant has a capacity to treat 180,000 tons of ore per year. The Forez District plants use acid-leach circuits followed by a lime precipitation step. The precipitate is redissolved with acid and further concentrated by solvent extraction (39).

All of the processing plants in France are operated under the direct control of the French Atomic Energy Commission. Some of the plants were designed, partially owned by and operated by Société Industrielle des Mineraux de L'Ouest (S.I.M.O.). SIMO is owned jointly by the Commissariat à l'Énergie Atomique (CEA), the French government, and the private firm of Kuhlman.

Australia. The Mary-Kathleen processing plant has been closed and has been on a "stand-by" basis since 1963. Preconcentration at the mine upgraded ore containing about 4 lb of U_3O_8 to about 20 lb of U_3O_8 per ton of concentrate. This was accomplished by a combination of heavy-media cones and by froth flotation. These concentrates were shipped to a chemical treatment plant in Port Pirie in South Australia.

At Port Pirie, the concentrates were digested at elevated temperatures with sulfuric acid. The unclarified slurry from the digestors was discharged into tanks and neutralized with lime. The solids were subsequently settled in a countercurrent decantation (CCD) thickener circuit and the pregnant liquor clarified by filtration. This treatment was followed by a four-column ion-exchange circuit. Uranium was precipitated as magnesium diuranate and shipped to the UKAEA. The Mary-Kathleen mine produced approximately $100 million in uranium values before it was closed.

A small mill was built to treat ores from the Rum Jungle district in the Northwest Territory, equipped with acid-leach and solvent-extraction circuits. This plant and another small mill at Moline on the South Alligator River are presently not in operation.

The Processing of Uranium Feed Material

The uranium in yellow cake, which is the concentrate from the ore-processing plant, must pass through a complex series of processes before the uranium is in a form suitable for reactor fuel.

The fuel cycle for enriched uranium is illustrated in Figure 2.

The cycle for nonenriched fuel is very similar to the above except that it does not include the enrichment step.

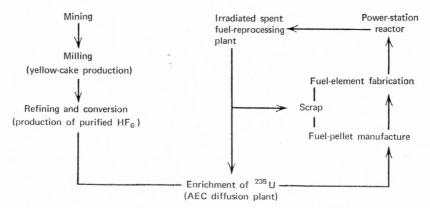

Fig. 2. Fuel cycle for enriched uranium.

The availability of mineral and other raw-material, power-load centers, and economic factors dictate the location of plants employing the fuel cycle.

Most of the uranium ores in the United States are in the west or southwest and most of these contain less than 5 lb of U_3O_8 per ton. The ratio of concentration from ore to yellow cake is about 400 to 1, which means the ore-processing plant must be located as close to the mine as possible.

The original refineries and conversion plants were built by the AEC in the Midwest near the Mississippi River or its tributaries. The gaseous-diffusion enrichment plants and the fuel-fabrication plants were also located in this same general area. The power reactors are located all over the country. Thus, even though the tonnages of nuclear materials are not large by comparison with other fuels, the transportation of these products has become an important factor. Criticality considerations are important in the movement of enriched products and elaborate precautions are taken to minimize the hazard that might result from transportation accidents.

Refining and Conversion of Yellow Cake. Concentrates from the ore-processing plants must be purified before they are suitable feed material for either isotopic enrichment plants or for direct production of unenriched metal or oxide of reactor quality.

Yellow-cake concentrates from most of the ore-processing plants in the United States are either ammonium diuranate or magnesium diuranate. Sodium uranate may be the end product from carbonate-leach plants. The concentrates vary from about 70–85% in their U_3O_8 content. Moisture in the products from the more modern plants is usually less than 1%.

AEC specifies tolerable limits of impurities in concentrates that have been, in general, acceptable for industrial use. Some of the maximum tolerable limits are listed in Table 4. By comparison with the refinery end product, yellow cake is indeed a very crude product.

The end products of the refinery may be uranium oxide or metal if the product is to be used as nonenriched fuel in power reactors. Much of the product from Canadian, British, and French refineries falls into this category.

As of the end of 1968, virtually all of the production from U.S. ore-processing plants was being shipped to the refinery operated by Allied Chemical Corporation at Metropolis, Ill., for conversion to a highly purified uranium hexafluoride suitable for

Table 4. Tolerable Impurities in Yellow-Cake Concentrates

Impurity	Percent	Impurity	Percent
arsenic	0.80	phosphates[b]	6.00
boron	0.03	rare earths	0.20
carbonate	1.00	sulfates	15.00
copper	1.70	uranium	
fluorides	0.10	insol in HNO_3	0.05
halides[a]	0.10	water	10.00
molybdenum	0.60	V_2O_5	2.00
organic matter	0.10	thorium	2.00

[a] Other than fluorides.

[b] Must be less than $1.31 \times \%$ Fe plus $0.002 \times \%$ U_3O_8.

isotopic enrichment at one of the USAEC gaseous-diffusion enrichment plants. The Allied Chemical refinery was reopened in early 1968 after a shutdown of $3\frac{1}{2}$ yr. Its capacity will be expanded to 10,000 tons per year by the end of 1969.

The USAEC refinery at Weldon Springs, Mo., was closed in 1966 and will probably not be reopened. The Weldon Springs plant and the AEC plant at Fernald, O., were capable of producing uranium metal, UO_3, and UF_4. The AEC plants at Oak Ridge, Tenn., and Paducah, Ky., have the capability to convert UO_3 to UF_6 and the plant at Portsmouth, O., can convert UF_4 to UF_6. None of these plants refine yellow cake for industrial use.

Kerr-McGee Corporation will have a refinery and conversion plant with a capacity of 5000 tons of uranium per year in operation at Sallisaw, Okla., during 1970. This and the Allied Chemical plant at Metropolis are the only privately owned refineries in the United States.

The Canadian refinery at Port Hope, Ont., is being remodeled to produce high-purity uranium hexafluoride. Similar operations are under way at the UKAEA in England. Concentrates from the various ore-processing plants in France are shipped to the refineries at Bouchet or at Malvési where products of nuclear purity are manufactured.

The flowsheets of the USAEC refineries, as well as those of Allied Chemical and Kerr-McGee, have numerous minor variations, but the basic steps are quite similar. The yellow cake must first be purified, then fluorinated to the green salt or tetra-fluoride. At this point, the process may be directed to production of nonenriched oxide or metal suitable for fuel fabrication, or it may be directed toward the production of high-purity hexafluoride gas to be used as feed material to one of the government-owned enrichment plants. The process in use at the Allied Chemical Company plant has been described in detail in the technical literature (40).

All the plants receive yellow-cake concentrates in steel drums. Receiving lots are carefully sampled and thoroughly blended. In the AEC plant, the yellow cake is then taken into solution with nitric acid and purified in a solvent-extraction process using tributyl phosphate in a hexane carrier as the uranium solvent. The uranium, which is present in the yellow cake in the UO_3 or U_3O_8 form, emerges from the denitration step as a purified UO_3. In the Allied Chemical process, the uranium in the yellow cake is first reduced with hydrogen obtained by cracking anhydrous ammonia. This is accomplished in a fluid-bed reactor in a single-stage reaction. If yellow cake with a high sodium content is to be processed, the sodium is removed by washing with ammonium sulfate and filtering before the reduction reaction. Reduction of the UO_3 in the AEC plant is accomplished by cracking ammonia and feeding the hydrogen and nitrogen into fluid-bed reactors.

Hydrofluorination of UO_2 to the green salt UF_4 is common to all three refinery flowsheets. In each process, anhydrous hydrofluoric acid is vaporized and contacted with the UO_2 in hydrofluorinators in a two-step process. The fluid-bed hydrofluorinators in the AEC and Kerr-McGee plants are equipped with slow-sweep agitators. There is a tendency for the fine UO_2 particles to sinter in the fluid-bed reactors and stick to the reactor walls. The use of the slow-speed stirring devices prevents the build-up of this sintered material. Virtually all equipment in process contact in this area is of Inconel construction. Inconel (International Nickel Co.) 600 and Inconel X-750 are most generally used in the hydrofluorination unit.

In the hydrofluorination operation anhydrous hydrofluoric acid is fed from storage into a vaporizer, then into a superheater and preheated. The hot HF gas enters the bottom of the second-stage reactor at about 550°C. The overflow from the second-stage reactor flows by gravity into a hopper in which a gas–solids separation is made. The settled solids are removed through suitable gas locks and are conveyed to UF_4 storage bins. The gas phase from the UF_4 hopper enters the bottom of the first-stage reactor. UO_2 enters the side of the first-stage reactor. The overflow from this reactor flows by gravity to an interstage hopper which acts as a gas–solids separator. The gaseous phase contains unreacted HF, water vapor, and certain volatile products resulting from the reaction of HF with such impurities as silicon and boron that may have escaped the solvent-extraction purification step that precedes hydrofluorination. These gases first pass through a filter in the top of the interstage hopper and then through a condenser that collects aqueous HF. The remaining gas is scrubbed and then escapes through a vent. The solids collected in the interstage hopper are fed through suitable locks and screw conveyors to the second-stage hydrofluorinator fluid-bed reactor.

The uranium tetrafluoride normally contains not less than 96% UF_4. Small amounts of unreacted UO_2 and some UO_2F_2 are the principal impurities. Only traces of the other metals are tolerable in the UF_4 product.

The reaction involved in the hydrofluorination of UO_2 is reversible. For this reason it is necessary to maintain a high concentration of HF as compared to that of water vapor. If the water-vapor ratio becomes high it is necessary to drop the reactor temperature to prevent a reversal of the hydrofluorination reaction. Several factors affect the rate of reaction.

The effects of temperature and concentration of HF are well established. The physical properties of the UO_2 also have an effect on reaction rate. A porous UO_2 particle with a large surface area reacts more quickly than a dense particle. These factors not only affect the rate of reaction, but they also have a bearing on the physical characteristics of the UF_4 that is produced in the reactors. Under normal conditions coarse UF_4 particles are formed as the reactor temperature is increased. For further information on anhydrous HF, see Vol. 9, p. 610.

The final processing step is the conversion of UF_4 to UF_6. Fluorine for this step of the process is generated as a very pure gas in electrolytic cells located at the refinery. For information on elemental fluorine production and properties, see Vol. 9, p. 507. Anhydrous HF is decomposed in the cells; the hydrogen offgas is scrubbed and vented. The fluorine is filtered, preheated, and fed to the primary fluorination reactors. The product from the second reactor is condensed, filtered, and placed in product cylinders. Through recycling and distillation the UF_6 gas is quantitatively separated from traces of other gases that may be present in the system. Solids or ash that may settle to the hoppers under the reactors are recycled to the primary reactor. Virtually all materials used in the fluorination of UF_4 are of Monel (International Nickel Co.) 400.

Uranium hexafluoride may be prepared by the fluorination of a large number of uranium compounds. The conversion of the tetrafluoride, UF_4, requires the least amount of fluorine and is therefore the preferred commercial process. Fluorination of UF_4 to UF_6 takes place in the temperature range of 250–400°C. The kinetics involved in the heterogeneous reactions in the fluorination of UF_4 are discussed in detail by Galkin, reference 11, p. 322.

In a typical operation, the powdered UF_4 is fed through suitable locks into the top of the primary reactor. Filtered and preheated fluorine is introduced into the side of the reactor. Unreacted UF_4 is collected in a hopper at the bottom of the reactor. This material is periodically removed and recycled back to the process feed. The gaseous product of the reaction, UF_6, leaves the reactor through a cooler near the bottom of the primary reactor. The gas then passes through a cyclone for removal of fine solids that were not collected in the hopper at the bottom of the primary reactor and, after filtration, is passed through a cold trap or condenser for final removal of solids. The gas then flows to the cylinder loading station for shipping. Utilization of fluorine and conversion to UF_6 is nearly quantitative in this process.

The refinery of the Eldorado Mining and Refining Ltd., at Port Hope, Ont., where refined metal and purified UO_2 are the end products, has a quite different process. The British and French refineries use somewhat similar flowsheets. All of these refineries are presently being modified to provide for the production of UF_6.

At Port Hope, the yellow cake is first sampled and then dissolved in nitric acid. A highly purified uranyl nitrate is the product of the solvent-extracting system that follows the digestion step. At this point the process splits. One direction leads to the production of ceramic-grade UO_2 or to UO_2 for conversion to uranium carbide. The other route leads to the production of UO_3 or to the metal.

If the end product is to be reactor-grade UO_2, the uranyl nitrate is treated with aqueous ammonia and ammonium diuranate is precipitated. This product is washed, filtered, and dried, then fed to a reduction furnace and reduced to UO_2 which is of suitable purity for the production of unenriched reactor-fuel pellets. If uranium carbide is to be produced, the UO_2 is ground, thoroughly mixed with graphite, and placed in an induction furnace which produces uranium carbide "skulls" or high-density slugs which may be ground or otherwise machined for use in fuel rods.

If the refinery product is to be uranium trioxide or metal, the uranyl nitrate is first denitrated by heating. This produces UO_3 which may be packaged as "orange oxide" or it may be cast into pellets that are fed into a vertical moving-bed reactor. Hydrogen is introduced into the bottom of the reactor and the pellets are reduced to UO_2 (brown oxide). The UO_2 pellets are fed into a second reactor where a stream of HF converts the UO_2 to UF_4 or "green salt." The latter is filtered, mixed with magnesium, and reduced to uranium metal in a thermite-type bomb. The reduction reaction requires about 3 min and reaches a temperature of about 1900°C. The slag and metal separate and are allowed to cool over a two-day period. The cylindrical metal ingots are about 20 in. in diam and about 20 in. long. They weigh about 2 tons each. The metal may be remelted under vacuum conditions and cast into slugs for fuel-rod fabrication, or the ingots may be forged or rolled into convenient shapes for machining.

Uranium-235 Enrichment. Most of the nuclear reactors built for the generation of electric power in the United States and in many other countries are based on the use of uranium fuel enriched in isotope ^{235}U. The gaseous diffusion process is the only method used for large-scale separation of ^{235}U from ^{238}U.

The natural uranium power reactors which have been quite popular in England require no isotopic enrichment. The purified natural uranium may be fabricated in the form of the metal, oxide, or carbide for direct use in these reactors.

The Gaseous-Diffusion Process. In this process, highly purified gaseous uranium hexafluoride is pumped through a series of porous membranes that are arranged in cells in a cascade pattern. The ^{238}U and ^{235}U diffuse through the membranes at slightly different rates, so that a great many stages are required to obtain high concentrations of ^{235}U. The membranes have a pore diameter of about 100 Å (41). Membranes produced by deposition of an emulsion of polytetrafluoroethylene (PTFE) on a metallic grid are generally satisfactory, but have also been constructed of various materials, such as fritted alumina, arc-sputtered aluminum, sintered nickel, and other metallic and ceramic powders. Much of the operating data for gaseous-diffusion plants was highly classified for defense purposes until the late 1950s. Since then most of this information has been released for publication (41).

Fantastically large amounts of electric power are required to operate gaseous-diffusion plants (42). It has been stated that the USAEC gaseous-diffusion plant at Oak Ridge, Tenn., consumes more electric power than the entire Republic of France. For further information and details of the gaseous diffusion process for isotopic concentration, see Vol. 7, p. 79.

Centrifugal Isotopic Separation. The high capital cost and huge power requirements of gaseous-diffusion plants have led to extensive investigation into the possibility of centrifugal separation of ^{238}U and ^{235}U.

Evaporative concurrent and countercurrent centrifuges with long rotors and with ultrahigh-speed capability have been extensively studied (43). In these centrifuges, the hollow rotor is partially filled with the unseparated compound in liquid form. High velocity is attained under temperature and pressure conditions that favor evaporation. The lighter isotope may be drawn axially from the rotor and the enriched, heavier isotope remains in the rotor.

A concurrent type of centrifuge which does not employ the evaporative principle has also been used experimentally. It is claimed that this device is more efficient than the evaporative centrifuge. Increases in the abundance ratio of ^{235}U to ^{238}U of as much as 5.6% have been reported with a single centrifugation. Claims of spectacular technical breakthroughs in centrifuge design and operation have been frequently reported during the past decade. Evidence of their commercial success is not yet apparent in large-scale installations.

Electromagnetic Separation. This process is based on the concept of accelerating a stream of an ionized gaseous uranium compound through a powerful magnetic field which deflects the path of the isotopes to a degree somewhat proportional to their respective masses. A plant, which has since been abandoned, was built at Oak Ridge to demonstrate this process. This principle of isotopic separation has been applied to many other elements and has been extensively used in Russia for the preparation of various metallic isotopes (44).

Fuel-Pellet Manufacture. Uranium hexafluoride, that has been enriched in the USAEC diffusion plants, is the starting material for the manufacture of fuel pellets. Procedures vary in detail, but the basic steps are standard.

Cylinders containing UF_6, as received from the AEC, are placed under hoods and heated with steam to vaporize the UF_6. The gaseous UF_6 is then reacted with ammonium hydroxide solution to form ammonium diuranate, ADU, which precipitates as bright yellow crystals. These are separated from the solution in a continuous centrifuge. The filtrate is usually treated in an ion-exchange system to recover traces of uranium that are not precipitated. The ADU solids are dried with steam to remove water and excess ammonia and then enter a closed indirect rotary calciner, operating at 760°C under reducing conditions. The ADU is converted to a UO_2 powder which is ground to break up agglomerates. A uniform size of the UO_2 particles at this point in the process is of extreme importance as it affects the density of the finished product. The UO_2 powder is then thoroughly blended, sampled, and stored as feed for the pellet-production line. The upper limit of enrichment for this process is about 5% ^{235}U.

The UO_2 produced in the ceramic line is the feed material to the pelletizing line. The powder is precisely measured into hydraulic presses which produce cylinders ranging in diameter from 0.3 in. to 0.5 in. Lengths vary from 0.5 to 1.0 in., depending on customer specifications. The pellets are thoroughly inspected for soundness, size, and density.

The pellets are loaded on trays made of molybdenum metal and fed into a sintering furnace. The furnace cycle requires 10 hr and involves holding the pellets for 4 hr at 1600°C. There is no loss of weight in the sintering operation, but the volume is reduced by about one-half. After a second rigid inspection the pellets pass through a centerless grinder which produces the final pellet—a cylinder with a slight "hour glass" configuration. The slight decrease in diameter of the pellet near the center of the long axis permits it to pass through extensive temperature changes while in reactor service with minimal danger of rupturing the metal case in which the pellets are contained. After grinding, the pellets are again inspected and packed in paper board tubes for shipment to the fuel-element fabricators. Rejected pellets and the grinding dust are sent to a scrap purification system. A single pellet measuring about 0.5 in. in diameter and 1 in. in length weighs about 30 g and has a value of about $7.

Highly enriched uranium is processed in a somewhat different manner. Highly enriched fuel usually contains 93% ^{235}U and is used in special test reactors and for research and development purposes. This will probably be the grade used in the "breeder" reactors now under development.

Highly enriched UF_6 from cylinders is vaporized with steam under hoods and then reacted with an organic liquid to produce UF_4. The solid UF_4 is recovered from filters and dried. It may then be converted to UO_2 or U_3O_8 by high-temperature reaction with steam. Metal can be produced from the tetrafluoride by the conventional thermite-reduction process using calcium metal granules. The highly enriched metal or oxides must be handled with great care to avoid criticality hazards. They are usually packed in 5-in.-diam "coffee cans" that are securely spaced, one to a 55-gal steel drum, for shipment. The value of enriched processed uranium may run as high as $5500/lb.

The scrap-recovery system involves dissolving the scrap in nitric acid and purifying it in a solvent-extraction circuit using tributyl phosphate in a hexane carrier.

The uranium is stripped with distilled water and precipitated with ammonia as ammonium diuranate, ADU. This material can then be fed back into the pellet-manufacturing system, or if necessary, it can be returned to the AEC for enrichment.

The principal producers of fuel pellets are the following: General Electric Co., San Jose, Calif.; Gulf General Atomic, Inc., San Diego, Calif.; Kerr-McGee Corp., Oklahoma City, Okla.; National Lead Co., Albany, N.Y.; Nuclear Materials & Equipment Corp., Apollo, Pa.; and United Nuclear Corp., Hematite, Mo.

Fuel-Element Fabrication. Fuel slugs or pellets must be placed in containers of suitable cladding material before they can be placed in power reactors. The cladding metal protects the pellets from the reactor coolant and retains the fission products within the fuel element. The cladding also permits the fuel rod or the accumulation of pellets to be maneuvered as a unit for control purposes.

The cladding material must be capable of withstanding high temperature and pressure over a long period of time. It must also have high strength and excellent corrosion resistance. A most essential property is a low neutron cross section. In fact, it must be nearly transparent to neutrons. Certain stainless steels and zirconium alloys have proved to be the most desirable metals for cladding or jacketing, but metals that are more refractory will be needed for some of the high-temperature, fast reactors now being developed.

Zirconium is usually associated with hafnium in its ores. Hafnium has an exceptionally high neutron cross section and therefore must be quantitatively removed from zirconium that is to be used for cladding.

Typical power-reactor cores contain between 50 and 150 tons of fuel elements. These must be very precisely sized and end-sealed. The cladding must be thick enough to provide structural reliability but thin enough to minimize neutron resistance. After the pellets are in place in the tubes, the air is evacuated and replaced with helium.

At the end of 1968 there were thirteen active manufacturers of fuel elements in the United States. There were eight manufacturers of oxide-fuel and eight manufacturers of metal-clad fuel elements. Three manufacturers produce both oxide and metal elements.

Power Reactors. Nuclear power is generated by conventional use of heat to activate turbogenerators. The heat is provided by nuclear fission. (See Nuclear reactors.)

Light-water reactors utilize enriched fuel (2–5% ^{235}U). In the pressurized type (PWR), distilled water serves as moderator and coolant. The operating temperature is usually about 260°C and the pressure about 2500 psi. A secondary coolant system is necessary and this is used through heat exchangers to produce low-pressure steam to operate the turbogenerators. The other version of the light-water reactor is designated the boiling-water reactor (BWR). The BWR uses a direct cooling cycle without heat exchangers.

Gas-cooled reactors use natural unenriched uranium as fuel. They are cooled by CO_2 gas and moderated by graphite. The hot gases boil water through heat exchangers and the steam drives the turbogenerators.

Heavy-water reactors (HWR) are largely a Canadian development and can utilize either natural or enriched fuel. The HWR is both moderated and cooled with heavy water (deuterium oxide). This type of reactor requires no pressure vessel. Steam is generated in thick-walled heat exchangers.

Breeder reactors, now under development, will probably follow closely the so-called "advanced converters" expected to be in operation by 1975. The advanced converters will utilize high temperatures and operate at higher pressures than those in use today. Helium will be used as the primary coolant in one advanced reactor now being designed.

Four types of breeder reactors are under active consideration:

1. Plutonium-239 is the initial fissile material and uranium-238 is the fertile material. The fissile material produced will be plutonium-239. The fission neutrons will have high energy and are termed "fast" neutrons.

2. Uranium-235 is the initial fissile material and, as in 1., the fertile material is Uranium-238. Plutonium-239 is the fissile material produced and the fission neutrons are "fast."

3. Plutonium-239 is the initial fissile material, but thorium-232 is the fertile material, and uranium-233 is produced. The fission neutrons are "fast."

4. Uranium-233 is the initial fissile material and thorium-232 is the fertile material. Uranium-233 is produced and the neutrons are "slow" or thermal.

Moderators are not used in fast breeder reactors. Power densities are thereby greatly increased and reactor-core sizes are decreased.

Numerous metallurgical problems and the development of high-temperature coolants have to be resolved before the breeder reactor is commercially feasible.

Table 5. Nuclear Power Reactors and Their Output

Country	No. of stations	Output, kWe[a]
Belgium	2	1,500
Canada	4	5,228
France	10	4,181
West Germany	8	3,719
India	6	1,180
Israel	1	200
Italy	4	1,197
Japan	32	21,580
Netherlands	1	350
New Zealand	1	100
Pakistan	2	325
Spain	7	2,303
Sweden	4	2,144
Switzerland	6	2,556
United Kingdom	35	14,934
United States	90[b]	74,748
total	213	136,245

[a] Kilowatts, electrical.   [b] Approx.

An idea of the magnitude of the power-reactor business can be gained from a consideration of the status of reactor construction at the beginning of 1969. Table 5 lists reactors actually in service, under construction, and planned for service before 1980.

It is estimated that the cost of building nuclear-power plants in the United States in 1968 is about $136,000 per kWe. This indicates that the cost of all nuclear power stations in the world by 1980 will be in excess of $21 billion based on 1968 dollars.

Reprocessing of Irradiated Fuel. Unless fuel elements are removed from nuclear reactors before the fuel has been completely consumed, trouble may develop from two sources. Fission products will build up to the point where they absorb the neutrons necessary to maintain the critical reaction. If the fuel elements are permitted to remain in the reactor for their entire life the effect of the prolonged exposure to radiation and heat would cause them to warp or rupture. It is therefore necessary periodically to remove the partially spent fuel and reprocess it to recover the valuable fissionable materials, uranium-233, uranium-235, and plutonium-239. These must then be separated from the fission products neptunium, americium, strontium, etc.

Since the types of fuels used in the government production reactors are quite different from those used in power reactors, different types of reprocessing plants are needed for treating power reactor fuel.

The fuel from the defense reactors is processed at the AEC Hanford works near Richland, Wash., and at Savannah River, Ga. The Redox process plant at Hanford was closed in 1966. The Purex reprocessing plant at the Hanford works handles all of the fuel load from the Richland production reactor.

Reprocessing of power-station reactor fuel was originally done on a toll basis under AEC contracts. There is now one active, privately owned, reprocessing plant in the United States, owned by the Nuclear Fuels Services Company (subsidiary of W. R. Grace, Inc.), and located at West Valley, N.Y. This plant went into operation in 1966.

Allied Chemical Corporation plans a reprocessing plant in South Carolina. Reprocessing plants are also under consideration by the General Electric Company and other firms. The United Kingdom Atomic Energy Authority offers reprocessing service for Free-World countries, but the demand for these services is still very small. It is estimated that by 1980 the Free-World requirement for fuel reprocessing from power reactors will amount to about 15 tons/day. In 1968 the cost for reprocessing spent light-water reactor fuels was about $32/kg of contained uranium.

When irradiated cores are removed from reactors they are allowed to "cool" under water for about four months. This "cooling" period permits the highly radioactive fission products to decay to lower levels of activity. During this period the neptunium-239, which was produced from uranium-238 in the reactor, is transformed to uranium-239. The fuel elements are then placed in heavily shielded containers and are shipped to the reprocessing plant. Strict regulations govern these shipments to minimize the hazard of critical transportation accidents. The lead-lined containers are normally equipped with crash frames. Cooling systems must be provided while the spent fuel is being transported. Upon arrival at the reprocessing plant the fuel is immediately unloaded and placed in underwater storage.

The first step in treatment involves removal of the cladding metal. This is done in heavily shielded cubicles with remote-operated tools. The elements are then sheared into short lengths and dissolved in nitric acid. Pieces of cladding that remain attached to the fuel are not dissolved, but are removed from the nitrate solution by settling and filtering. The clarified nitrate solution is fed to a solvent-extraction circuit where the uranium and plutonium are separated from the fission products. A second solvent-extraction circuit then separates the uranium and plutonium. The uranium can be recovered as UO_2 and blended directly with enriched material, or it can be converted to UF_6 and returned to an AEC plant for reenrichment.

The recovered plutonium cannot be placed back in the fuel cycle in the same manner as the recovered uranium. It is very toxic and hard to handle and process. It is usually reused by mixing it with natural or depleted uranium to bring the total fissile content to between 2.5 and 5%. It can then be processed and returned to the fuel cycle.

Compounds of Uranium

The Oxides. Both economic and military pressures have contributed to exhaustive studies of the uranium–oxygen system (19). The oxidation-reduction potential of uranium permits an almost continuous oxygen variation in a given uranium oxide crystal. The presence of nonstoichiometric oxygen within the crystal lattice of certain of the oxides has been observed.

It has been pointed out above (see p. 5) that alternations in the oxidation state of uranium have significance in winning the values from the ores and also in refining the ore concentrates to standards of nuclear-grade purity. Varying solubility and ion-complexing properties depend largely on the oxidation state of the uranium.

Uranium dioxide, UO_2, melts at approx 3000°C. Its crystalline form is normally face-centered cubic. Its density is about 10.97 g/cm³.

Uranium dioxide is very stable and occurs in nature as uraninite and pitchblende. It is of particular importance in the uranium refining operation and is the intermediate oxide produced in the manufacture of the metal, the tetrafluoride, or the hexafluoride (see p. 22). Uranium dioxide in both natural and isotopically enriched forms is widely used in the manufacture of fuel pellets for use in power reactors. Its value in this application is tied to its desirable physical properties. Average linear coefficient of expansion in the temperature range of 20–946°C is 10.8×10^{-6} and is relatively unaffected by crystallographic orientation. Volume expansion coefficients are also quite satisfactory in reactor-temperature ranges. The low heat conductivity of uranium dioxide makes it difficult to remove heat generated within the fuel elements of reactors. Pressed and sintered, uranium dioxide is very stable when exposed to air or water below 300°C. It is also quite inert to superheated steam.

The reaction of uranium dioxide with fluorides to produce uranium tetrafluoride is of great importance as an intermediate step in the production of the hexafluoride. Reaction of the dioxide with fluorine at elevated temperature also produces the hexafluoride. Nonoxidizing chlorinating agents may be used to convert the dioxide to uranium tetrachloride. Uranyl nitrate is formed when the dioxide is heated with concentrated nitric acid. Thermal reduction of the dioxide with reactive metals, such as magnesium, is the common procedure for making uranium metal. Peroxyuranate, $UO_4 . 2H_2O$, may be formed by reacting uranium dioxide with an alkaline solution of hydrogen peroxide.

There is not complete agreement as to the maximum oxygen content possible in a UO_2 crystal. Katz (19) indicates the upper limit may be $UO_{2.15}$ or possibly higher.

Triuranium octoxide, U_3O_8, may be obtained by roasting the higher or lower oxides. It crystallizes in the orthorhombic system and has a density of about 8.4 g/cm³. The color may be anywhere between an olive green and black, depending on the roasting temperature. This oxide is thermally stable in the temperature range of 650–900°C. Evolution of oxygen is fairly rapid at about 1000°C, but complete conversion to UO_2 does not take place below 2000°C.

Triuranium octoxide, U_3O_8, reacts with HF or HCl to form mixtures of tetravalent and hexavalent halides. In the presence of reducing agents, only the tetravalent states are formed.

Work at the Ames Materials Project (45) indicates the potential for gradual transition from U_3O_8 to UO_3. Zachariasen has offered a plausible explanation of the transition from the hexagonal uranium lattice in a UO_2 crystal to the orthorhombic lattice in UO_3 by a gradual change in the parameters of uranium atoms (46).

Triuranium octoxide has been reduced to UO_2 with methane (47) and also with sulfur and hydrogen sulfide. U_3O_8 may be reduced to UO_2 with oxalic acid, ethanol vapor, or carbon at elevated temperature; magnesium reduces it to the metal. U_3O_8 may be formed by thermal decomposition of UO_3 at temperatures above 650°C.

Uranium trioxide, UO$_3$, is also an intermediate frequently produced in the purification of uranium compounds. It exists as an amorphous powder and in at least four crystalline forms. It is readily obtained from the thermal decomposition of uranyl nitrate or ammonium diuranate in ore extraction and purification operations. Uranium trioxide crystallizes in the hexagonal system and has a density of 5.15 g/cm³. It may be reduced to the dioxide with hydrogen or carbon. The reactions with fluorinating and chlorinating compounds are almost identical to those described for U_3O_8.

Uranyl salts may be converted to UO_3 without formation of U_3O_8 in an oxygen stream under carefully controlled temperature conditions.

Uranium monoxide, UO, has no practical importance at present. It is known to exist only as a thin film on the surface of very pure uranium metal and its existence has been verified only by x ray. The color is brown to black. Uranium monoxide forms very slowly above 2000°C in mixtures of UO_2 and uranium.

Uranium peroxide, UO$_4$, is not known to exist in the anhydrous state, but the hydrated oxide is important in purification technology. It may be produced by reacting hydrogen peroxide with an aqueous solution of uranyl nitrate. Upon dehydrating, the peroxide is converted to the trioxide, UO_3.

Other oxides. Several hydrated uranium oxides have received extensive study. Among these are $UO_3 \cdot 2H_2O$, $UO_3 \cdot H_2O$, and $UO_3 \cdot 0.5H_2O$. Most of the work on these hydrates has been reported in USAEC studies conducted at the University of California and at Columbia University (48,49).

The Fluorides. *Uranium trifluoride, UF$_3$,* is not a compound of economic importance. Its melting point has been observed to be 1495°C. The trifluoride is insoluble in water and is rather inert to weak acids. It dissolves in strong nitric acid with the evolution of nitrogen oxides. It is also soluble in hot sulfuric and perchloric acids.

Uranium tetrafluoride, UF$_4$, is a very important intermediate in the production of uranium metal and uranium hexafluoride. The melting point has been determined to be 969°C. It is a green solid at low temperature and has a specific gravity of 6.70. UF_4 is insoluble in water.

Uranium tetrafluoride is always prepared from uranium compounds in the tetravalent state. UF_4 is generally prepared by reacting uranium dioxide with HF at temperatures on the order of 550°C. The reaction rate is relatively slow and is not particularly dependent on temperature. A large excess of HF is usually present in the reaction vessel. This is the only important chemical process for the production of UF_4, since the simplicity of the reaction at moderately high temperature and the purity of the product obtained have made this the preferred process.

Other routes have also been investigated, such as (a) dissolving sodium uranate in HF and reducing with stannous chloride; (b) the electrolytic reduction of uranyl sulfate solutions; (c) dehydration of hydrated UF_4; and (d) reaction of uranium oxides with fluorocarbons at high temperature.

Uranium tetrafluoride is reduced by hydrogen to UF_3 at a temperature on the order of 1000°C. Water vapor reacts with UF_4 to produce UO_2 and HF above 200°C.

Uranium metal is usually made by the reduction of the tetrafluoride with magnesium. The tetrafluoride is also the starting compound for the production of non-enriched UO_2 of reactor purity.

Uranium tetrafluoride finds its most important use in the commercial uranium industry as an intermediary in the production of uranium hexafluoride for feed to isotopic enrichment plants.

A number of crystalline hydrates are obtained by precipitation from aqueous solution. At room temperature the normal hydrate is $UF_4 \cdot 2\frac{1}{2}H_2O$; while at about 50°C, the hydrate $UF_4 \cdot 1\frac{1}{2}H_2O$ precipitates. At 92°C the hemihydrate $UF_4 \cdot \frac{1}{2}H_2O$ is formed. Uranium ammonium pentafluoride, NH_4UF_5, may be formed in the preparation of UF_4. Two intermediate uranium fluorides are known to exist between the tetrafluoride and the hexafluoride. These have been identified as U_4F_{17} and U_2F_9. All are unstable and decompose on heating to form either UF_4 or UF_6.

A long list of double salts of UF_4 and metal fluorides are known to exist and have been reported in USAEC and UKAEA literature. These include such compounds as $NaUF_5$, $NaUF_6$, K_2UF_6, and many others (19).

Uranium hexafluoride, UF_6, as the only suitable feed material to the gaseous-diffusion enrichment plants, has become one of the world's important chemical compounds.

Uranium hexafluoride is a clear, colorless, crystalline substance which vaporizes without passing through the liquid stage at atmospheric pressure. The physical and chemical properties of the crystalline solid, of the liquid, and of the gas have been extensively studied and reported in the literature by various governmental agencies and private investigators (7,11,15,19,50,51). Some of the properties are listed in Table 6.

The boiling point, or more correctly, the temperature at which the sublimation pressure of UF_6 reaches 760 mm Hg, has been reported by numerous investigators from slightly over 55°C to nearly 57°C. A value of 56.4 is considered to be correct within about one-tenth of one degree. The vapor pressure of solid UF_6 is difficult to determine with precision because of the difficulty in removing final traces of HF.

Table 6. Physical and Thermodynamic Properties of Uranium Hexafluoride (15)

Property	Value
triple point, at 1134 mm Hg, °C	64.052
sublimation point, °C	56.4
density	
solid, g/cm³	5.09
liquid, g/ml	6.63
heat of formation, solid, at 25°C, kcal/mole	−516
heat of vaporization, at 64.01°C, kcal/mole	6.907
heat of fusion, at 64.01°C, kcal/mole	4.588
heat of sublimation, at 64.01°C, kcal/mole	11.495

Values measured at 0°C vary between 17 and 17.9 mm Hg. At 15°C, the vapor pressure is about 55 mm Hg, at 80°C, measurements indicate a vapor pressure of approx 1800 mm Hg (11). Vapor pressures for liquid uranium hexafluoride have been observed with greater accuracy since thermal equilibrium is easier to achieve. Measurements indicate a vapor pressure of about 1170 mm Hg at 65°C; this increases to about 1600 at 75°C, and to approx 2455 at 90°C.

The critical temperature of UF_6 has been reported to be as high as 245°C; other observers have reported it as low as 232°C. Calculations indicate a theoretical critical temperature at 249°C.

Uranium hexafluoride is highly reactive. It is instantly hydrolyzed by water, forming UO_2F_2 with the release of considerable heat (about 50 kcal/mole at 25°C). UF_6 forms an azeotropic mixture with HF. Pressure determines the ratio of HF to UF_6 in the mixture.

Uranium hexafluoride reacts with numerous inorganic solids at high temperature. Reduction reactions are typical. Amorphous carbon reduces the hexafluoride to the tetrafluoride. Glass and quartz are not attacked by UF_6 if there are no traces of moisture present.

Gold and platinum are resistant to attack by UF_6 at room temperature, but tarnish rapidly when heated. Lead, zinc, tin, and iron react rapidly, but copper, aluminum, and nickel are more resistant. UF_6, under ordinary conditions, does not react with dry air, oxygen, or nitrogen.

For methods of preparation of UF_6, see above under the processing of uranium feed material (p. 21). The commercial process employs direct fluorination of uranium tetrafluoride, $UF_4 + F_2 \rightarrow UF_6$, a reaction which takes place rapidly at 250°C.

Elemental fluorine converts most uranium compounds to the hexafluoride at higher temperature. UF_6 has been produced without the use of elemental fluorine by the following reaction:

$$2\ UF_4 + O_2 \xrightarrow{800°C} UF_6 + UO_2F_2$$

OTHER URANIUM COMPOUNDS

Nitrates. In addition to the oxides and fluorides of uranium, another compound of importance is uranyl nitrate, $UO_2(NO_3)_2$. The hexahydrate is orthorhombic and may be obtained by crystallization from dilute nitric acid solutions. Its heat of formation is equal to -764.3 kcal/mole. The nitrate is soluble in organic solvents, such as diethyl ether. It was this property that led to the first known purification of uranium by Peligot in 1841. The ability of water-immiscible organic solvents to remove uranium from aqueous nitrate solutions is the basis for most refining processes in use today.

Uranyl nitrate is also of importance in the recovery of uranium from process wastes and in the reprocessing of irradiated uranium.

Sulfates. Uranyl sulfate is of great importance to the ore-processing plant using sulfuric acid for dissolution of uranium values (52). The sulfate is also suitable as a fuel in homogeneous reactors. The monohydrate and the trihydrate are stable. In ores which are difficult to digest, the high heat stability of uranyl sulfate is used in a rather unique process (11). The ore is sulfatized at high temperature converting the iron and aluminum, as well as the uranium, to sulfates. Upon heating to the 600°C range, the iron and aluminum sulfates are converted to oxides which are insoluble in water. The soluble uranyl sulfate may be leached away.

Carbonates. Uranyl carbonates are of importance in high-lime ores which may be leached by the carbonate process. The compound, $Na_4UO_2(CO_3)_3$, is obtained when ores are pressure leached at elevated temperatures with soda ash solutions. The high solubility of uranium tricarbonate in aqueous solutions makes the removal of impurities possible, including iron, aluminum, nickel, and chromium, which may be precipitated as oxycarbonates and hydroxides. Hexavalent uranium can be extracted from sodium carbonate solutions with basic anion-exchange resins. Sodium uranyl tricarbonate decomposes at 400°C to sodium uranate and sodium carbonate.

It is possible to reduce hexavalent uranium in carbonate solutions to the tetravalent state with zinc, hydrosulfate, and ammonia.

Chlorides. Various uranium chlorides are known such as UCl_3, UCl_4, UCl_5, and UCl_6. For the properties of the various uranium chlorides, see reference 19 which also contains a very extensive bibliography citing both British and American government agencies on the subject of uranium chlorides.

Others. The bromides, oxyhalides, selenides, and tellurides of uranium have all received extensive study during the period of uranium process development (19).

Bibliography

"Uranium and Uranium Compounds," in *ECT* 1st ed., Vol. 14, pp. 432–458, by J. J. Katz, Argonne National Laboratory.

1. M. H. Klaproth, "The Chemical Investigation of Uraninite, a Newly Discovered Metallic Substance," *Chem. Ann.* **2,** 387–403 (1789).
2. E. Peligot, "Research on Uranium," *Compt. Rend.* **13,** 417–426 (1841).
3. E. Peligot, "Research on Uranium," *Ann. Chim. Phys.* **5,** 5–47 (1842).
4. H. Becquerel, "On the Radiations Emitted in Phosphorescence," *Compt. Rend.* 122, 420–421 (1896).
5. E. Fermi, et al., "Artificial Radioactivity Produced by Neutron Bombardment," *Proc. Roy. Soc. London,* **A 146,** 483–500 (1934).
6. L. Meitner and O. R. Frisch, "Disintegration of Uranium by Neutrons, a New Type of Nuclear Reaction," *Nature* 143, 239–240 (1939).
7. J. H. Gittus, *Uranium, Metallurgy of the Rarer Metals,* Butterworth & Co., Inc., Washington, D.C., 1963.
8. C. J. Roden, "Analytical Chemistry of the Manhattan Project," *National Nuclear Energy Series, Div VIII,* Vol. 2, McGraw-Hill Book Co., New York, 1950.
9. *The Nuclear Industry, 1968,* USAEC, U.S. Govt. Printing Office, Washington, D.C., 1968, pp. 2–6.
10. B. Blumenthal, *USAEC ANL-5019,* U.S. Govt. Printing Office, Washington, D.C., 1953, p. 54.
11. N. P. Galkin, B. N. Sudarikov, et al., *Technology of Uranium,* Moscow, 1964, p. 15. English translation for the USAEC by Israel Program for Scientific Translations, Jerusalem, 1966.
12. A. L. Dahl and H. E. Cleaves, "The Freezing Point of Uranium," *J. Res. Natl. Bur. Std.* **43,** 513 (1949).
13. J. M. North, *UKAEA Document AERE M/R 1016,* London, 1952.
14. D. C. Ginnings and R. J. Corruccini, *USAEC, A 3947,* Natl. Bur. Std., Washington, D.C., 1946, p. 8.
15. C. D. Harrington and A. E. Ruehle, *Uranium Production Technology,* D. Van Nostrand Co., Inc., New York, 1959.
16. E. L. Francis, *UKAEA Document IGR/R-287,* London, May 20, 1958.
17. S. J. Paprocki and H. A. Saller, *USAEC Report BMI-753,* U.S. Govt. Printing Office, Washington, D.C., 1943.

18. *Proc. Intern. Conf. on the Peaceful Uses of Atomic Energy, Geneva, Aug. 1955*, United Nations, N.Y., 1956.

18a. A. B. McIntosh and T. J. Heal, in reference 18, Vol. 6, 1955, p. 48.

19. J. J. Katz and E. Rabinowitch, "The Chemistry of Uranium," *National Nuclear Energy Series, Div VIII*, Vol. 5, McGraw-Hill Book Co., New York, 1951.

20. J. R. Bonn and G. Murphy, *Fourth Nuclear Engineering and Scientific Conference, Chicago, Ill., 1958*.

21. L. R. Gardner, "The Creep of Alpha Uranium," *Uranium and Graphic Symposium*, Inst. of Metals, Commonwealth Council of Mining & Metallurgical Inst., London, 1962.

22. J. G. Thompson, *Natl. Bur. Std. Rept. CT-890*, U.S. Govt. Printing Office, Washington, D.C., 1943.

23. S. T. Konobeevsky, in reference 18, Vol. 7, 1955, p. 443.

24. T. G. Berlincourt, *Phys. Rev.* **114** (1959).

25. L. F. Bates and D. Hughes, *UKAEA Document NSA 8-1769*, London, January 1954.

26. H. Ebert and A. Schultz, *Z. Metalk.* **38**, 46 (1947).

27. A. N. Holden, *Physical Metallurgy of Uranium*, Addison-Wesley Publishing Co., Reading, Mass., 1958.

28. *Chem. Eng. News* **45**, 15–16 (July 31, 1967).

28a. W. D. Wilkinson, *Uranium Metallurgy*, Vol. I., Interscience Publishers, a div. of John Wiley & Sons, Inc., New York, 1962.

29. "Statistical Data of the Uranium Industry," *USAEC*, Grand Junction, Colo., 1968.

30. R. T. Zitting et al., *U.S. Bur. Mines Bull.* **57**, 54–57 (1957).

31. D. A. McGee, "Atomic Energy in America," *Governor's Uranium Conference, Casper, Wyo. Nov. 1968*.

32. Reference 18, Vol. 2, Physics; Research Reactors, p. 54.

32a. L. T. Nel, in reference 32, Vol. 2, 1958, p. 54.

33. *Joint Rept. European Nuclear Energy Agency, Paris, and International Atomic Energy Agency, Vienna*, December 1967.

34. U.S. Pat. 2,591,573 (April 1, 1952), C. H. McBurney (to Rohm & Haas).

35. *Uranium Mining & Processing at Grants, New Mexico, Operation*, Kerr-McGee Corp., Oklahoma City, Okla., 1964.

36. E. T. Pinkey, W. Lurie, and P. C. N. Van Zyl, *Chemical Processing of Uranium Ores*, International Atomic Energy Agency, Vienna, 1962, p. 42.

37. D. A. Ellis, in reference 32, Vol. 3, 1958, p. 499.

38. J. W. Clegg and D. D. Foley, *Uranium Ore Processing*, Addison-Wesley Publishing Co., Reading, Mass., 1958.

39. *The Uranium Industry of France*, Commissariat à l'énergie atomique (CEA), Paris, 1968.

40. *Electron. World* **169** (19), (1968).

41. J. Charpin et al., in reference 32, Vol. 4, 1958, p. 380.

42. *The Guardian*, Manchester, England (Sept. 1960).

43. J. W. Beams, in reference 32, Vol. 4, 1958, p. 428.

44. V. S. Zolatarev, A. I. Iljin, and E. G. Komar, in reference 43, p. 471.

45. R. E. Rundel, N. C. Baenziger, et al., *Materials Project Ames, Report CC-2397, USAEC*, Washington, D.C., 1945.

46. W. H. Zachariasen, *Materials Project Chicago, Reports CK-2667 and CC-2768, USAEC*, Washington, D.C., 1945.

47. G. E. MacWood, *Union California UCRL Rept. 4.7.602, USAEC*, Washington, D.C., 1945.

48. M. J. Polissar and S. B. Kilner, *Union California UCRL Rept. 4.6.42, USAEC*, Washington, D.C., 1942.

49. D. T. Vier, *SAM Columbia Rept. A-1277, USAEC*, Washington, D.C., Dec. 8, 1955.

50. M. Brodsky and P. Pagny, in reference 43, p. 69.

51. R. D. Ackley et al., *Rept. AECD-3475, USAEC*, Washington, D.C., 1951.

52. C. H. Secoy, *J. Am. Chem. Soc.* **72**, 3343 (1950).

V. L. Mattson
Kerr-McGee Corporation

UREA AND UREA DERIVATIVES

Rouelle first discovered urea in urine in 1773. His discovery was followed by the synthesis of urea from ammonia and cyanic acid by Woehler in 1828. This is considered to be the first synthesis of an organic compound from an inorganic compound. In 1870, Bassarow (1) produced urea by heating ammonium carbamate in a sealed tube in what was the first synthesis of urea by dehydration.

The chemical formula, NH_2CONH_2, indicates that urea can be considered to be the amide of carbamic acid NH_2COOH, or the diamide of carbonic acid $CO(OH)_2$. At room temperature urea is colorless, odorless, and tasteless. When it is dissolved in water it hydrolyzes very slowly to ammonium carbamate and eventually decomposes to ammonia and carbon dioxide. This is the basis for the use of urea as a fertilizer (qv).

Urea is commercially produced by the direct dehydration of ammonium carbamate NH_2COONH_4 at elevated temperature and pressure. Ammonium carbamate is ob-

Table 1. General Properties of Urea

Property	Assigned value
melting point	132.7, °C
index of refraction, n_D^{20}	1.484, 1.602
specific gravity, d_4^{20}	1.335
crystalline form and habit	tetragonal, needles or prisms
free energy of formation, at 25°C	−47,120 cal/g mole
heat of fusion	60 cal/g, endothermic
heat of solution, in water	58 cal/g, endothermic
heat of crystallization,	
70% aqueous urea-solution	110 cal/g, exothermic
bulk density	0.74 g/cm³

Table 2. Specific Heat of Urea

Temperature, °C	Specific heat, cal/(g)(°C)
0	0.344
50	0.397
100	0.451
150	0.504

Table 3. Properties of Saturated Aqueous Solutions of Urea

Temperature, °C	Solubility in water, g urea/100 g solution	Density, g/cm³	Viscosity, cP	H_2O vapor pressure, mmHg
0	41.	1.120	2.63	4
20	51.6	1.147	1.96	13
40	62.2	1.167	1.72	40
60	72.2	1.184	1.72	90
80	80.6	1.198	1.93	160
100	88.3	1.210	2.35	220
120	95.5	1.221	2.93	135
130	99.2	1.226	3.25	7

tained by direct reaction of ammonia, NH_3, and carbon dioxide, CO_2. The two reactions are usually carried out simultaneously in a high-pressure reactor. Recently urea has been used commercially as a cattle feed supplement. Other important applications for urea are the manufacture of resins (see Amino resins and plastics), glues, solvents, and some medicines.

Tables 1, 2, 3, 4, and 5 refer to the physical and chemical properties of urea.

Table 4. Properties of Saturated Solutions of Urea in Ammonia (2)

Temperature, °C	Urea in solution, wt %	Vapor pressure of solution, atm
0	36	4
20	49	7
40	68	9.4
60	79	10.8
80	84	13.3
100	90	12.5
120	96	5

Table 5. Properties of Saturated Solutions of Urea in Methanol and Ethanol (3)

Temperature, °C	Methanol		Ethanol	
	Urea, wt %	Density, g/cm³	Urea, wt %	Density, g/cm³
20	22	0.869	5.4	0.804
40	35	0.890	9.3	0.804
60	63	0.930	15.0	0.805

Chemical and Physical Properties

CHEMICAL PROPERTIES

Werner, in his studies on urea (4), advocates the following cyclic formula and its transient tautomeric enol form:

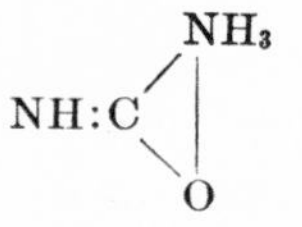

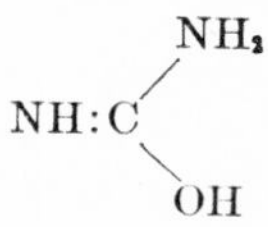

However, the work of Walker and Wood (5) indicates that urea is a weak base and that it lacks any acidic property. At atmospheric pressure and at its melting point urea decomposes to ammonia, biuret, $HN(CONH_2)$, cyanuric acid, $C_3N_3(OH)_3$, ammelide, $NH_2C_3(OH)_2$, and triuret, $NH_2(CONH)_2COHN_2$. Biuret is in practice the main and the least desirable by-product present in the commercially synthesized urea. An excessive amount (more than 2 wt %) of biuret in fertilizer-grade urea is detrimental to plant growth.

Urea acts as a monobasic substance and forms salts with acids (10). With nitric acid it forms urea nitrate, $CO(NH_2)_2 \cdot HNO_3$, which decomposes explosively when heated. Solid urea is rather stable at room temperature and atmospheric pressure. Heated under vacuum and at its melting point, urea sublimes without change. At

180–190°C urea will sublime, under vacuum, and be converted to ammonium cyanate, NH_4OCN (6). When solid urea is rapidly heated in a stream of gaseous ammonia at elevated temperature and at a pressure of several atmospheres it sublimes completely and decomposes partially to cyanic acid, HNCO, and to ammonium cyanate. Also solid urea will dissolve in liquid ammonia and will form the very unstable compound urea-ammonia, $CO(NH_2)_2NH_3$, which decomposes above 45°C (2). Urea–ammonia reacts with alkali metals to form salts such as NH_2CONHM or $CO(NHM)_2$.

An aqueous urea solution slowly hydrolyses to ammonium carbamate at room temperature or at its boiling point (7). Traces of cyanate are found in solution. Prolonged heating of aqueous urea solutions will cause the formation of biuret;

$$2\ NH_2CONH_2 \rightleftarrows NH_2CONHCONH_2 + NH_3$$
$$\text{urea} \qquad\qquad\qquad \text{biuret}$$

This reaction is promoted by low pressure, high temperature, and prolonged heating time. At a pressure of 100–200 atm, biuret will revert to urea when heated in the presence of ammonia (8,9).

Urea reacts with silver nitrate, $AgNO_3$, in the presence of sodium hydroxide, NaOH, and forms a diargentic derivative, $CON_2H_2Ag_2$, of a pale-yellow color. Sodium hydroxide promotes the change of urea into the form $HNC(NH_2)\ OH$, which then reacts with silver nitrate. Oxidizing agents in the presence of sodium hydroxide convert urea to nitrogen and carbon dioxide. The latter reacts with sodium hydroxide to form sodium carbonate (11):

$$NH_2CONH_2 + 3\ NaBr + 2\ NaOH \rightarrow N_2 + 3\ NaBr + Na_2CO_3 + 3\ H_2O$$

The reaction of urea with alcohols yields carbamic acid esters, commonly called urethans:

$$NH_2CONH_2 + ROH \rightarrow NH_2COOR + NH_3$$

Urea reacts with formaldehyde and forms compounds such as monomethylolurea, $NH_2CONHCH_2OH$, dimethylolurea, and others, depending upon the molar ratio of formaldehyde to urea and upon the pH of the solutions. See Amino resins and plastics. In the reaction with hydrogen peroxide and urea, a white crystalline powder, urea peroxide, $CO(NH_2)_2H_2O_2$, is formed. It is known under the trade name of Hyperol, and is generally employed as an oxidizing agent.

Urea derivatives are employed in medicine as sedatives and hypnotics. The basic compound from which these derivatives are synthesized is barbituric acid, obtained from urea and malonic acid (more exactly, from its ester):

$$
\begin{array}{ccccc}
NH_2 & HOCO & & NH{-}CO & \\
| & | & & & \diagdown \\
CO\ + & CH_2 \rightarrow CO & & & CH_2 + 2\ H_2O \\
| & | & & & \diagup \\
NH_2 & HOCO & & NH{-}CO &
\end{array}
$$
$$\text{malonyl urea or}$$
$$\text{barbituric acid}$$

See Barbituric acid and barbiturates.

PHYSICAL PROPERTIES

Urea has the remarkable property of forming crystalline complexes or adducts with straight chain organic compounds. These crystalline complexes consist of a hollow channel, formed by the crystallized urea molecules, in which the straight-chain hydrocarbon is completely occluded. Such compounds are known as clathrates. The type of hydrocarbon occluded, on the basis of the length of its carbon chain, will be determined by the temperature at which the clathrate is formed. This property of urea clathrates is widely used in the petroleum refining industry for the production of jet aviation fuels and for dewaxing of lubricant oils. The clathrates are broken down by simply dissolving urea in water or in alcohol and by decanting the hydrocarbons from the urea solution.

Manufacture

Urea Synthesis. All modern urea synthesis processes are based on the principle of reacting ammonia and carbon dioxide in a high-pressure and high-temperature reactor to form ammonium carbamate, and simultaneously dehydrating the ammonium carbamate to urea;

$$2\,NH_3 + CO_2 \leftrightarrows NH_2COONH_4 \qquad\qquad NH_2COONH_4 \leftrightarrows NH_2CONH_2 + H_2O$$

$$\text{ammonium carbamate} \qquad\qquad\qquad\qquad \text{urea}$$

$$\text{reaction 1} \qquad\qquad\qquad\qquad \text{reaction 2}$$

Reaction 1 is a strongly exothermic one. The heat of reaction, starting from gaseous ammonia and carbon dioxide at 25°C and atmospheric pressure, is in the range of 38,000 cal/g mole of solid carbamate formed at 25°C (12). The rate and the equilibrium of reaction 1 depend greatly upon pressure and temperature, because it involves large changes in volume. This reaction may only occur at a temperature which is below the dissociation pressure of ammonium carbamate, or conversely, the reactor operating pressure must be maintained above the vapor pressure of carbamate at a given operating temperature. Reaction 2 is endothermic by about 7500 cal/g mole of urea formed.

Table 6. Dissociation Pressure of Pure Carbamate

Temperature, °C	Ammonium carbamate vapor pressure, atm absolute
40	0.31
60	1.05
80	3.1
100	8.5
120	21
140	46
160	98
180	150, 190[a]
200	200, 360[a]

[a] Extrapolated by the author. Above 170°C the rate of reaction 2 rapidly increases and exact carbamate vapor pressure determinations are very difficult to perform due to the formation of water and urea and the consequent lowering of the carbamate partial pressure.

It is more or less constant because it can only take place in the liquid phase, or at a much slower rate in the solid phase, with minor variations in volume. The overall heat of reaction of urea from NH_2 and CO_2 is strongly exothermic and the excess heat must be removed from the reactor.

The dissociation pressure of pure carbamate has been investigated quite extensively (13–15) and the average value of all the investigations is shown in Table 6.

Properties of Ammonium Carbamate. Ammonium carbamate is a white crystalline solid, which is soluble in water (2). It can be formed at room temperature by passing ammonia gas over dry ice. In an aqueous solution it will slowly revert, at room temperature, to ammonium carbonate, $(NH_4)_2CO_3$, by the addition of one molecule of water. Above 60°C the aqueous solution of ammonium carbonate will be converted back to carbamate solution more or less rapidly. At 100°C no carbonate will be present in solution, only carbamate. A further increase in temperature above 150°C will cause ammonium carbamate to lose an additional molecule of water and thus to form urea. The specific heat of solid ammonium carbamate is given in Table 7. Ammonium carbamate has a melting point of about 150°C, and has a heat of fusion in the range of 4000 cal/g mole. The conversion of carbamate to urea begins at a relatively low temperature, at 100°C or lower. The heating time required to obtain any appreciable amount of urea at 100°C is on the order of magnitude of 20–30 hr. The rate of conversion increases as the temperature is raised (16–18). And at 185°C about 50% of the original ammonium carbamate is converted to urea in about 30 min.

Table 7. Specific Heat of Solid Ammonium Carbamate

Temperature, °C	Specific heat, cal/(g)(°C)
20	0.40
60	0.46
100	0.52
140	0.58
180	0.62

Equilibrium Conversion. The maximum equilibrium urea conversion attainable at 185°C is in the range of 53%, at infinite heating time. The equilibrium conversion can be increased either by increasing the reactor temperature or by dehydrating ammonium carbamate in the presence of excess ammonia. Excess ammonia has the effect of shifting the reaction to the right side of the overall equation:

$$2\,NH_3 + CO_2 \rightleftarrows NH_2CONH_2 + H_2O$$

Water, however, has the opposite effect; in the presence of excess water the reaction is shifted to the left.

Frèjacques (19) made a very accurate investigation of the equilibrium conversion of ammonium carbamate to urea with excess ammonia and with excess water, in the temperature range from 130 to 210°C. He derived the following formula for the equilibrium constant from his investigation:

$$K = \frac{x\,(b + x)(1 + a + b - x)}{(1 - x)(a - 2x)}$$

where K = equilibrium constant

x = mol fraction of total initial CO_2 converted to urea

a = mol ratio of NH_3 to CO_2 in total mixture

b = mol ratio of H_2O to CO_2 in the total initial mixture before the urea reaction takes place

The overall formula given below will serve to clarify the terms of the above derivation:

$$aNH_3 + bH_2O + CO_2 \rightleftarrows xNH_2CONH_2 + (1 - x)NH_4CO_2NH_2 + (a - 2)NH_3 + (b + x)H_2O$$

However, the values for the equilibrium constant K obtained by Frèjacques are somewhat lower than the values calculated from empirical data obtained from tests performed in commercial large-scale urea plants. This slight discrepancy might be explained on the basis of the fact that Frèjacques' investigations were performed on a batch autoclave, whereas in practice, a continuous reactor is used in commercial urea plants. It seems that a continuous reactor performs better than a batch autoclave.

Mavrovic (20) published a monographic presentation of the reaction equilibrium. He finds the actual equilibrium constant, at various temperatures, given in Table 8. Also a detailed study of the effect of pressure on the conversion of urea was performed by Kawasumi (21).

Table 8. Reaction Equilibrium Constant

Temperature, °C	Reaction equilibrium constant, K
140	0.695
150	0.850
160	1.075
170	1.375
180	1.800
190	2.380
200	3.180

COMMERCIAL UREA SYNTHESIS

The urea synthesis reactor always contains unreacted carbamate and more or less excess ammonia, depending upon the composition of the feeds. This poses the practical problem of separating the unreacted material from the urea solution and of reutilizing this unreacted material. Depending upon the method of reutilization of the unreacted material, the commercial urea synthesis processes are divided into the following main categories:

1. Once-through urea process. The unconverted carbamate is decomposed to NH_3 and CO_2 gas by heating the urea synthesis reactor effluent mixture at low pressure. The NH_3 and CO_2 gas is separated from the urea solution and utilized to produce ammonium salts by absorbing NH_3, either in sulfuric or nitric acid.

2. Solution recycle urea process. The NH_3 and CO_2 gas recovered from the reactor effluent mixture either in one or in several pressure staged decomposition sections is absorbed in water and recycled back to the reactor in the form of an ammoniacal aqueous solution of ammonium carbamate.

3. Internal carbamate recycle urea process. The unreacted carbamate and the excess ammonia are stripped from the urea synthesis reactor effluent by means of gaseous hot CO_2 or NH_3 at the reactor pressure, instead of letting the reactor effluent down to a much lower pressure. The NH_3 and CO_2 gas, thus recovered at reactor pressure, is condensed and returned to the reactor by gravity flow for recovery.

There are various alternative processes not yet commercially established or of lesser economic value within each group, with some variations as to the extent and the mode of recycling the unreacted NH_3 and CO_2 to the reactor. The most important features of these variants are as follows:

1a. Once-through urea process with partial excess NH_3 recycle. Excess NH_3 is recovered from the reactor effluent mixture at an intermediate pressure level prior to the low-pressure carbamate decomposition and recovery stage. The recovered excess

NH_3 gas is condensed with cooling water and pumped back to the reactor for recovery (22–24).

2a. Slurry recycle urea process. The NH_3 and CO_2 gas recovered from the reactor effluent mixture is condensed to form crystalline ammonium carbamate and the latter is pumped back into the reactor as a suspension in oil, or in liquid NH_3 (25,26).

2b. Urea process with NH_3 and CO_2 separation. The NH_3 and the CO_2 in the recovered gas are separated by means of a selective absorbent (monoethanolamine, MEA, or ammonium nitrate) and individually recycled back to the reactor for recovery (27).

2c. Hot gas recycle urea process. The gaseous mixture of NH_3 and CO_2 recovered from the reactor effluent is compressed in a multistage centrifugal compressor to the reactor pressure and introduced into the reactor for recovery (28). At each compression stage the NH_3 and CO_2 gas mixture is maintained above the dissociation temperature of carbamate to avoid solids formation within the compressor.

2d. Integrated ammonia-urea process. In this combined process the carbon dioxide formed in the ammonia plant by the shift reaction is scrubbed from the ammonia synthesis loop gas and absorbed by the urea synthesis mixture to produce urea (29–31).

DESCRIPTION OF PROCESSES

Once-Through Urea Process. This is the simplest urea process of all. Liquid NH_3 is pumped through a high pressure plunger pump and gaseous CO_2 is compressed through a compressor up to the urea synthesis reactor pressure at an NH_3 to CO_2 feed mol ratio of 2.0–3.0 to 1. The reactor usually operates in a temperature range from

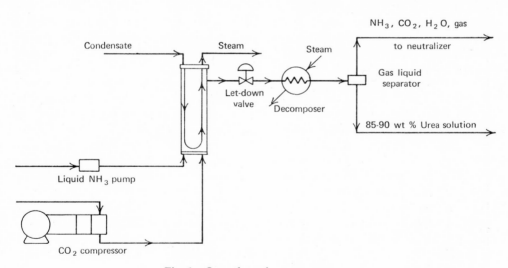

Fig. 1. Once-through urea process.

175 to 190°C. The reactor effluent is let down in pressure to about 2 atm and the carbamate decomposed and stripped from the urea-product solution in a steam heated shell-and-tube heat exchanger. The moist gas, separated from the 85–90% urea-product solution, and containing about 0.6 tons of gaseous NH_3 per ton of urea produced is usually sent to an adjacent ammonium nitrate or ammonium sulfate producing plant for recovery. An average conversion of carbamate to urea of about 60% is at-

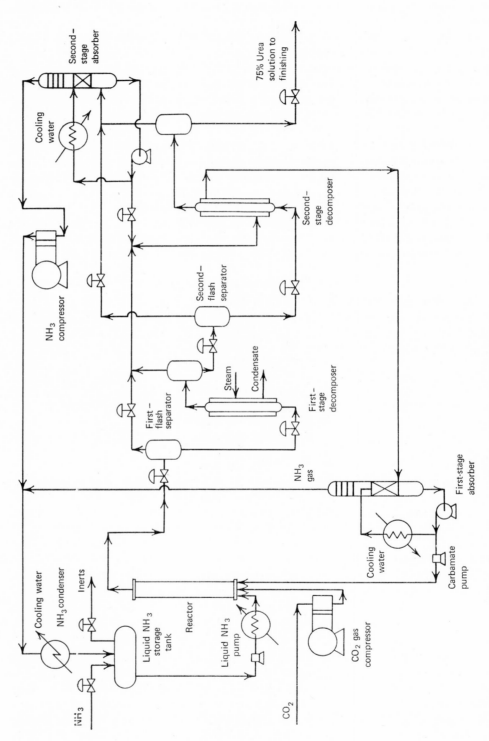

Fig. 2. Chemico total-recycle urea process.

tained. Excess heat is removed from the reactor by means of a low-pressure steam-producing coil in an amount of about 280,000 cal/kg urea produced.

Figure 1 is the flow diagram for the once-through urea process. The larger licensing companies who employ this process are as follows: Chemical Construction Corporation (U.S.A.); Mitsui Toatsu (previously Toyo Koatsu) (Japan); Montecatini (Italy); Stamicarbon (Holland); Vulcan Copper and Supply Co. (U.S.A.); and Lonza and Inventa (Switzerland). One of the drawbacks of this type of urea process is the fact that relatively large amounts of ammonium salts, not always readily saleable, must accompany the production of urea.

Solution Recycle Urea Process. The aqueous carbamate solution recycle urea process is the most commonly used technique in the urea manufacturing industry. Several large companies developed and perfected their own urea processes and they are listed below.

Allied Chemical Corporation (U.S.A.). The gaseous mixture of NH_3 and CO_2 obtained from the decomposition of ammonium carbamate contained in the reactor effluent solution, is treated with a solution of MEA, which selectively absorbs CO_2. The pure NH_3 gas is condensed to liquid with cooling water and pumped back into the reactor. The MEA solution, rich in CO_2, is regenerated in a stripping column by indirect steam heating. The lean MEA solution is then reused for further CO_2 absorption (27). The recovered pure CO_2 is recompressed to high pressure and introduced into the reactor. Due to the fact that the recovered NH_3 and CO_2 are recycled back to the reactor as pure and dry substances, and thus the bad effect of water on the conversion of carbamate to urea eliminated, a relatively high conversion of carbamate is attained. For instance, 79–82% of the carbamate is converted to urea per pass through the reactor, compared to 63–67% for the carbamate solution recycle methods. One of the drawbacks of such a process is the relatively high cooling water and steam consumption of the MEA section, its investment cost, the cost of the MEA losses, and the operating and maintenance cost of the gas selective absorption unit.

Chemical Construction Corporation (Chemico) (U.S.A.). The urea synthesis reactor is operated at about 220 atm pressure, at about 190°C, and at an overall NH_3 to CO_2 molar ratio in the reactor of about 4 to 1. Under these operating conditions about 64–67% of the carbamate present in the reactor is converted to urea. The unconverted carbamate and the excess NH_3 are recovered in a two-pressure staged decomposition and recovery sections (32). See Figure 2. The first decomposition and absorption stage operates at about 20 atm pressure. The reactor effluent, let down in pressure from 220 atm is heated in the first decomposition stage at about 155°C to decompose and strip from the urea product solution most of the unconverted carbamate and excess NH_3. The recovered gas is cooled in the first absorption stage; all the CO_2 is reacted with the stoichiometric amount of NH_3 to carbamate and dissolved in water along with some of the excess NH_3. The remaining portion of the excess ammonia gas is separated from the aqueous carbamate solution, purified from the last traces of CO_2 in a bubble cap column, and condensed to liquid with cooling water. The liquid ammonia is recycled back to the reactor along with the stoichiometric amount of fresh makeup of NH_3. The recovered ammoniacal aqueous solution of ammonium carbamate, containing about 25 wt % of water, together with the stoichiometric amount of fresh gaseous CO_2 are introduced into the reactor.

The urea-product solution, leaving the first decomposition stage and still containing some unreacted carbamate and excess NH_3, is let down in pressure and steam heated

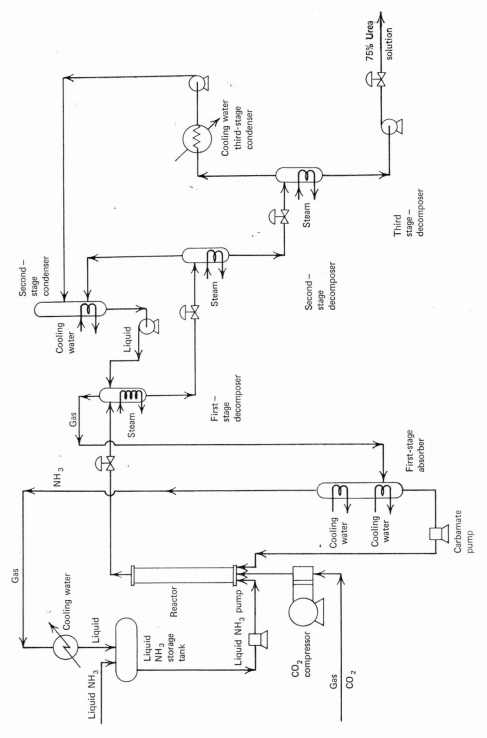

Fig. 3. Lonza total-recycle urea process.

in the second-staged decomposition section, which operates at about 2 atm and 120°C. Practically all of the residual carbamate is decomposed and stripped from the urea-product solution together with the residual excess ammonia. The 74–75 wt % urea solution thus obtained is further processed to solid urea in one of the urea finishing sections described below.

The NH_3 and the CO_2 gas, recovered from the second decomposition stage, is cooled in the second absorption stage where all the CO_2 is reacted to carbamate and dissolved in water along with some of the excess NH_3. The solution thus obtained is pumped back to the first absorption stage and thence to the reactor for total recovery. Only part of the excess ammonia is dissolved in the solution recovered from the second absorption stage. The remaining portion of the excess ammonia gas is separated from the aqueous carbamate solution, purified from the last traces of CO_2, in a bubble cap column (33), compressed to about 20 atm, condensed to liquid with cooling water, and pumped back to the reactor. A tight overall water-balance control system (34), within the urea-synthesis section, ensures a relatively high conversion in the reactor. Heat is internally recovered in the second-stage decomposer (35) by utilizing part of the heat of carbamate formation in heat exchange with the urea-product solution undergoing NH_3 and CO_2 degassing.

Inventa A. G. (Switzerland). The urea-synthesis reactor is maintained at a pressure of about 230 atm and at a temperature of about 185°C. The reactor mixture overall molar ratio of NH_3 to CO_2 is maintained between 3.5 and 4.0 to 1, and a conversion of about 62–65% is attained. The reactor effluent is let down to about 5 atm and heated to about 120°C (36). The recovered NH_3 and CO_2 gas is fed to an absorber, operating at about 5 atm and 50°C, where it is almost completely condensed to form an ammoniacal aqueous solution of ammonium carbamate. The heat of reaction is dissipated by circulating the solution through a heat exchanger. The condensed solution is pumped into a desorber, operating at about 140–150°C and 50 atm, in which the carbamate is decomposed once again to NH_3 and CO_2 gas and stripped from the solution together with excess NH_3. The gas, issuing from the desorber, contains water vapor in a much smaller amount than the gas obtained in the 5 atm reactor-effluent decomposer, described above. Thus the purpose of the desorber is to reduce the water content of the recovered NH_3 and CO_2 prior to recycling to the reactor. The desorber gas is fed to an absorber, operating at about 50 atm and 90°C, and condensed again to an ammoniacal aqueous solution of carbamate, which is recycled back to the reactor by means of a high-pressure plunger pump, together with the stoichiometric amount of fresh makeup NH_3 and CO_2. The fact that the unconverted carbamate originally contained in the reactor effluent is decomposed and condensed twice prior to recycling to the reactor, makes this process relatively uneconomical with respect to steam and cooling-water consumption.

Lonza A.G. (Switzerland). The urea-synthesis reactor is operated at about 300 atm, 200°C, and at the reactor mixture molar ratio of NH_3 to CO_2 of about 4.5 to 1 (37,38). About 70% of the total carbamate present in the reactor is converted to urea. Figure 3 represents a flow diagram of this process. The unconverted carbamate is decomposed to NH_3 and CO_2 in a steam-heated vessel operated at about 15 atm and 150°C. The recovered gas from the first decomposer prior to condensation in the first-stage absorber and recycling to the reactor, is contacted in countercurrent flow with an aqueous ammoniacal solution of carbamate issuing from the second-stage condenser for total recovery of NH_3 and CO_2.

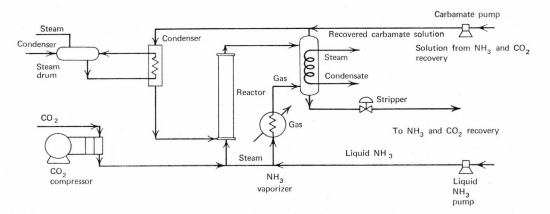

Fig. 4. Snam Progetti urea process.

The urea-product solution, separated from the gaseous phase in the first-stage decomposer and still containing in solution some residual unconverted carbamate and excess NH_3, is further degassed in two subsequent and pressure-staged decomposers. The NH_3 and CO_2 gas recovered in the second-stage decomposer is condensed in the second-stage condenser and the heat of reaction dissipated with cooling water. The solution thus formed is pumped into the first-stage decomposer for total recovery of NH_3 and CO_2.

The urea-product solution from the second-stage decomposer is further degassed from residual amounts of NH_3 and CO_2 in a steam-heated atmospheric decomposer. The degassed urea solution is sent to the evaporation and prilling sections for further processing to solid urea. The moist NH_3 and CO_2 gas mixture recovered from the third-stage decomposer is condensed to a liquid solution in the water-cooled third-stage condenser and pumped into the second-stage condenser for recovery. The first-stage decomposer of off gas is cooled in the first-stage absorber, all the CO_2 reacted to ammonium carbamate and dissolved in water along with part of the excess NH_3. The solution thus formed is pumped back to the reactor. The remaining part of the recovered gaseous excess NH_3 is stripped from the last traces of CO_2 in a column, condensed to liquid with cooling water, and recycled to the reactor.

Montecatini (Italy). The carbamate solution recycle urea process is basically similar to the urea processes described above. The unconverted carbamate is recovered in a series of three pressure staged decomposition absorption systems (39,40). In a most recent variation of their process developed in a pilot plant scale operation, the liquid NH_3 reactor feed is added to the first absorber, operating at about 15 atm, and crystallized ammonium carbamate is pumped in a suspension of liquid NH_3 to the reactor, instead of an aqueous solution. High savings on steam and cooling-water consumption are claimed (40).

Snam Progetti (Italy). This process is based on the principle of the internal carbamate recycle technique and is commonly called the Snam NH_3 stripping process, as opposed to the analogous Stamicarbon CO_2 stripping process. The basic difference between the SNAM process and the conventional carbamate solution recycle urea processes is the fact that the unconverted carbamate is stripped and recovered from the urea synthesis reactor effluent solution at reactor pressure, condensed to an aqueous

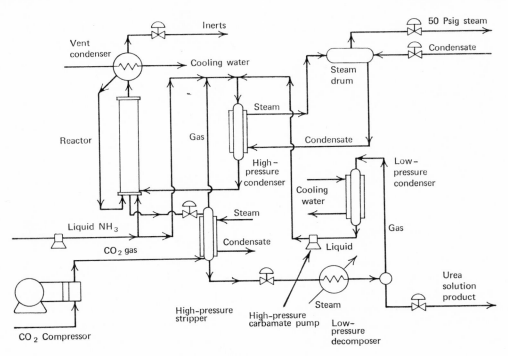

Fig. 5. CO₂ stripping process.

solution in a steam producing high-pressure condenser, and recycled back to the reactor by gravity (20). See Figure 4. Part of the liquid NH₃ reactor feed, vaporized in a steam heated exchanger, is used as inert gas to decompose and strip ammonium carbamate in the steam heated high-pressure stripper.

The reactor operates at about 130 atm and 180–190°C. The stripper operates at about 130 atm and 190°C. The stripper off-gas is condensed in a vertical shell-and-tube condenser, operating at about 130 atm and 148–160°C. Low-pressure steam is produced in the high-pressure carbamate condenser. The urea-product solution, leaving the stripper and still containing 2–3% of residual unreacted carbamate, is further degassed in a low-pressure decomposition-absorption system. The recovered ammoniacal solution of ammonium carbamate is pumped back to the reactor.

Stamicarbon (Holland). Their newest process also is based on the principle of the internal carbamate recycle technique and it is commonly called the Stamicarbon CO₂ stripping process. The novelty of the CO₂ stripping process consists of the fact that the reactor effluent is not let down to a lower pressure as in the conventional liquid recycle urea process, but is stripped at synthesis pressure by the gaseous CO₂ reactor feed stream in a steam heated vertical heat exchanger. (20,41,42). See Figure 5 for the diagram of this process.

The high pressure stripper operates at about 140 atm and 190°C. The stripped urea solution still contains about 15% of the unconverted carbamate, and it is let down to about 3 atm for further degassing in the steam heated low pressure decomposer at about 120°C. The off-gas recovered is condensed with cooling water in the low-pressure condenser, operating at about 65°C and 3 atm. The solution thus obtained is pumped to the high-pressure condenser by means of the high-pressure carbamate pump.

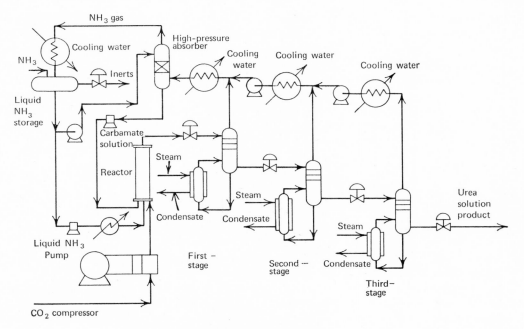

Fig. 6. Mitsui Toatsu total-recycle urea process.

The off-gas recovered from the high-pressure stripper is condensed in the high-pressure condenser, which operates at about 170°C and 140 atm. The heat of condensation is removed on the shell side of the condenser by vaporizing the equivalent amount of condensate. The 50 psig steam thus produced can be reused in another section of the plant.

The mixture of gas and liquid issuing from the high-pressure condenser is fed to the reactor for total CO_2 condensation to carbamate and subsequent conversion to urea. The inerts are vented from the reactor through a water-cooled vent condenser. The reactor effluent, at about 185°C and 140 atm, is fed to the high-pressure stripper as described above.

Mitsui Toatsu (previously Toyo Koatsu) (Japan). This total recycle urea process is a conventional carbamate solution recycle process with three pressure-staged carbamate decomposition and recovery systems. (43–46). The reactor is operated at about 195°C, 240 atm, and with a NH_3 to CO_2 molar ratio of about 4.3. About 67% of the total ammonium carbamate present in the reactor is converted to urea. See Figure 6.

The unconverted carbamate is decomposed and stripped from the urea solution together with excess NH_3 in a series of three pressure-stage decomposers, operating respectively at about 18 atm and 150°C, 45 psig and 130°C, and atmospheric pressure and 120°C. The main feature of the Mitsui Toatsu process is the fact that the gaseous phase in each decomposition stage is contacted in counter current flow with the urea-product solution issuing from the preceding decomposition stage. Either a packed section or a sieve tray section is used for this purpose. The effect is that the NH_3 and CO_2 gaseous mixture obtained from the decomposition of carbamate is considerably reduced in water vapor content. Thus the amount of water recycled to the reactor is maintained at a relatively low level and a relatively high conversion in the reactor is attained. The off-gas from each decomposition stage is condensed to solution in its respective water cooled condenser and the solution thus obtained is pumped to the next higher pressure-staged condenser.

Excess NH_3 is separated from the aqueous solution of carbamate and scrubbed from the last traces of CO_2 in countercurrent flow with reflux liquid, NH_3, fed to the top of the high-pressure absorber. The pure excess NH_3 thus obtained is condensed to liquid with cooling water and recycled to the reactor. The carbamate solution is recycled back to the reactor for total recovery.

Urea Finishing

Urea is normally obtained from the synthesis section as a 70–75 wt % aqueous solution, and it is usually further processed to solid urea, which is more economical for shipping and distribution. Two different routes are used to reduce urea to its solid form: evaporation and crystallization with crystal remelt. In both cases molten pure urea is the final product. The molten urea thus obtained is sprayed and solidified into small spherical particles called prills.

Evaporation. Water is evaporated from the steam-heated urea solution either under vacuum (47,48), or under reduced pressure with the addition of hot air as a drying agent (49) or by a process of atmospheric air-sweep evaporation (Whitlock Manufacturing Co.). Due to the fact that at low pressure and high temperature the formation of biuret from urea is accelerated, the urea evaporation route is utilized for the production of agricultural-grade urea only. The average biuret increase across the evaporation unit is about 0.4 percentage point. Typical urea product analyses at the inlet and at the outlet of the evaporation section are shown in Table 9.

Table 9. Urea Product Analyses

Product	Inlet,[a] wt %	Outlet,[b] wt %
urea	73.0	99.0
H_2O	26.7	0.3
biuret	0.3	0.7
total	100.0	100.0

[a] At 120°C.   [b] At 140°C.

Crystallization with Crystal Remelting. The presence of biuret in high-grade urea is detrimental, especially in the technical-grade urea used in the plastic manufacturing industry. Urea obtained by the crystallization method is relatively pure and quite well suited for industrial use. Such urea usually contains about 0.3 wt % biuret. See Figure 7. The urea-product solution containing about 73 wt % urea is fed to a vacuum crystallizer, operated at about 60 mm Hg absolute pressure and about 60°C. The water vapor, evaporated from the solution, is condensed in a water-cooled vacuum condenser provided with a vacuum jet.

The crystallizer slurry, containing about 30 wt % urea crystals in suspension, is sent to a continuous pusher type centrifuge. The urea crystals are separated from the mother liquor, washed with water, dried, elevated to the top of the prilling tower, and remelted in a steam-heated crystal melter. The urea melt thus obtained contains about 0.3 wt % biuret and about 0.2 wt % moisture. The mother

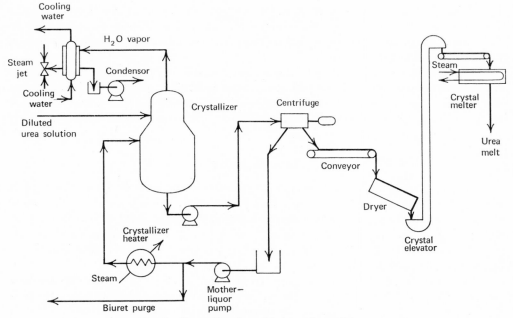

Fig. 7. Crystallization with crystal remelting.

liquor separated in the centrifuge is steam heated in the crystallizer heater in order to supply the equivalent amount of heat required to evaporate the water from the solution, and returned to the crystallizer to maintain the proper magma concentration. A small portion of the mother liquor, rich in biuret, is purged from the system in order to prevent biuret buildup in the mother liquor solution.

Prilling. Molten urea, obtained either by evaporation or by crystal melting, is sprayed as droplets down from the top of a tall cylindrical tower about 50 m high and allowed to fall countercurrently to a stream of ambient air. The urea droplets freeze into small spherical particles, about 1.5 mm in diameter, and are collected at the bottom of the prilling tower.

Pan Granulation. The technical staff of the Tennessee Valley Authority piloted a system in which the granulation is effected by spraying the urea melt onto a cascading bed of recycling fines in an inclined rotating pan granulator (50). Pan granulation has the advantage of being able to produce large granules, 2–3 mm in diameter, suitable for bulk blending, whereas large prills are more difficult to make.

MATERIALS OF CONSTRUCTION

Ammonium carbamate in aqueous solution is quite corrosive, especially at elevated temperature and pressure. For this reason, all the process equipment in contact with the urea-synthesis solution is made of stainless steel, types 304 and 316. Below 130°C steel of type 304 ss with low carbon content may be used. Above 130°C only 316 ss with low carbon content will withstand the corrosive action of the process solution. Stamicarbon developed a method of preventing corrosion of 316 ss walls of the urea-synthesis reactor by the addition of air to the reactor mixture (51). Allied Chemical Corporation uses a zirconium-lined reactor (52) in order to combat corrosion, and Mitsui Toatsu uses a titanium liner.

Economic Aspects

The average investment cost of a commercial total recycle urea plant producing prilled urea is given in Table 10. The average raw material and utility consumption per metric ton of prilled urea produced more or less approaches the same values for all commercially proven urea processes (Table 11). The total conversion cost of NH_3 to urea,

Table 10. Average Cost of a Urea Plant

Plant capacity, T/D	Investment cost, $
300	3,700,000
600	6,000,000
1200	10,000,000

Table 11. Raw Material and Utility Consumption[a]

Material	Consumption amount
NH_3, metric ton	0.58
CO_2, metric ton	0.75
steam, metric ton	1.3
cooling water, m^3	75
electric power, kWh	145

[a] Per metric ton of prilled urea produced.

including investment amortization, fixed and variable costs, etc but excluding the cost of NH_3, is in the range of $18.00 per metric ton of prilled urea. The average wholesale price for urea on the international market is in the range of $70/ton for agricultural-grade urea. The present world production of urea is shown in Table 12.

Table 12. World Production of Urea

Country	Production amount, million tons/year
Europe	3.2
U.S.A.	2.0
Japan	2.0
total	7.2

Specifications and Standards

Urea is shipped domestically in four-ply polyethylene-coated 45 lb paper bags, and in six-ply polyethylene-coated 80 lb bags for export. Urea prills obtained by the crystal melting route have better anticaking properties than the evaporation route prills, and their average crushing strength is about 500 g per prill, compared to 250 g per prill for the evaporation prills. The current product specifications for fertilizer-grade urea prills and for low-biuret technical-grade urea are listed in Tables 13 and 14.

Analytical and Test Methods

The total nitrogen is determined by the Kjeldahl method. The biuret is determined spectrophotometrically by adding NaOH and a 3% solution of $CuSO_45H_2O$,

Table 13. Specifications for Fertilizer-Grade Urea

Product	Amount
total nitrogen, wt %, min	46.3
H₂O, wt %, max	0.3
biuret, wt %, max	0.9
free NH₃, ppm, max	150
iron, ppm, max	2
prill size, mm	through 3.4 on 1.0 screen
granulometric analysis, wt %, min	95

Table 14. Specifications for Low-Biuret Technical-Grade Urea

Product	Amount
total nitrogen, wt %, min	46.3
H₂O, wt %, max	0.3
biuret, wt %, max	0.4
free NH₃, ppm, max	100
iron, ppm, max	1
ash, ppm, max	20
oil, ppm, max	20
color,[a] APHA units, max	15
turbidity,[b] APHA units, max	20
pH,[c] min	6.6

[a] 100 g in 100 ml methanol.   [b] 50 g in 100 ml H_2O.   [c] Borden Test.

Free NH_3 is determined by titration with 0.01 N hydrochloric acid using methyl purple as an indicator. Iron is determined spectrophotometrically by dissolving iron and measuring the intensity of the pink color upon the addition of thioglycolic acid. Standard analytical methods are employed in the determination of ash. Oil is extracted with carbon tetrachloride and determined by evaporation of the solvent. Spectrophotometrical methods are used to measure color and turbidity.

In the Borden pH Test, 15 g of urea prills are dissolved in 80 ml of freshly boiled and cooled distilled water. Then 2 ml of 20% formaldehyde solution containing less than 0.01% formic acid is added. The solution is mixed and allowed to stand for 10 min at 20°C. If the sample contains impurities, the pH falls, and a minimum pH of 6.6 is the empirical standard accepted by the resin manufacturers. It is believed that ammonium salts (sulfate, chloride, or nitrate) react with the formaldehyde, forming hexamethylenetetramine and free acid, which lowers the pH.

Uses

Urea is used mainly as a fertilizer (qv). In recent years the demand for urea as a supplement in cattle feed (see Feeds, animal) has been steadily growing. Another important use is in urea-formaldehyde resins (see Amino resins and plastics).

Bibliography

"Urea" in *ECT* 1st ed., Vol. 14, pp. 458–466, by F. A. Wolff and D. J. O'Flynn, E. I. du Pont de Nemours & Co., Inc., Polychemicals Department.

1. E. Blasiak, *Technology of Nitrogen Compounds*, Vol. II, State Technical Publisher, Warsaw, 1956, pp. 596–642.
2. E. Janecke, "Uber Das System: H_2O, CO_2, NH_3 . . ." *Z. Elektrochem* **36**, 645–654 (1930).

3. A. Seidell, *Solubility of Inorganic Compounds*, 2nd ed. D. Van Nostrand Co., New York.

4. E. A. Werner, *Chemistry of Urea*, Longmans, Green & Co., London, 1923.

5. J. Walker and J. K. Wood, *J. Chem. Soc.* **83**, 484–491 (1903).

6. R. Escales and H. Kopke, "Ammonium Cyanate and Carbamide," *Chem. Ztg.* **33**, 595 (1911).

7. R. C. Warner: "The Kinetics of the Hydrolysis of Urea and Arginine," *J. Biol. Chem.* **143**, 705–723 (1942).

8. U.S. Pat. 3,232,984 (Feb. 1, 1966), J. A. Finneran (to Pullman Inc.).

9. U.S. Pat. 3,255,246 (June 7, 1966), I. M. Singer, Jr. (to E. I. du Pont de Nemours & Co.).

10. D. F. Du Toit, "Combination of Carbamide with Acids," *Proc. Koninkl. Akad. Wetenschap, Amsterdam* **16**, 555–556 (1913).

11. Knop and Hufner in A. E. Werner, ed., *Chemistry of Urea*, Longmans, Green and Co., London, 1923, p. 161.

12. C. Matignon and M. Frèjacques, *Bull. Soc. Chim. France* **29**, 21 (1921); **31**, 394 (1922).

13. T. R. Briggs and V. Migrdichian, *J. Phys. Chem.* **28**, 1121 (1924).

14. E. Briner, *J. Chim. Phys.* **4**, 267 (1906).

15. N. F. I. Isambert, *Compt. Rend.* **92**, 919 (1881).

16. F. Fichter and B. Becker, "The Formation of Urea on Heating Ammonium Carbamate," *Ber.* **44**, 3470–3480 (1911).

17. C. Matignon and M. Frèjacques, "The Transformation of Ammonium Carbamate," *Bull. Soc. Chim.* **31**, 394–412 (1922).

18. K. K. Clark, V. L. Gaddy, and C. E. Rist, *Ind. Eng. Chem.* **25**, 1092 (1933).

19. M. Frèjacques, *Chim. Ind. (Paris)* **60**, 22 (1948).

20. Process Survey, Urea; European Chemical News (London), *Jan. 17, 1969*.

21. Schohachi Kawasumi, *Bull. Chem. Soc. Japan*, **24**, 148–151; **25**, 227–238; **26**, 218–227; **27**, 254–259 (1954).

22. U.S. Pat. 2,894,878 (July 14, 1959), L. H. Cook (to Chemical Construction Corporation).

23. U.S. Pat. 3,305,582 (Feb. 21, 1967), L. H. Cook and I. Mavrovic (to Chemical Construction Corporation).

24. U.S. Pat. 3,107,149 (May 12, 1961), T. O. Wentworth and L. W. Nisbet, Jr. (to Chemical Processes).

25. U.S. Pat. 3,072,721 (Jan. 8, 1963), S. W. Grossmann (to Allied Chemical Corp.).

26. U.S. Pat. 2,913,493 (Nov. 17, 1959), M. C. Sze and H. W. Wang (to Hydrocarbon Research).

27. U.S. Pat. 3,236,888 (Feb. 22, 1966), T. O. Wentworth (to Allied Chemical).

28. U.S. Pat. 3,200,148 (Aug. 10, 1965), L. H. Cook and I. Mavrovic (to Chemical Construction Corp.).

29. U.S. Pat. 3,303,215 (Feb. 7, 1967), E. Otsuka and T. Saida (to Toyo Koatsu Industries).

30. U.S. Pat. 3,310,376 (March 21, 1967), L. H. Cook and I. Mavrovic (to Chemical Construction Corp.).

31. U.S. Pat. 3,349,126 (Oct. 24, 1967), H. Hsu and W. C. Jaeschke (to Pullman Inc.).

32. U.S. Pat. 3,172,911 (March 9, 1965), I. Mavrovic (to Chemical Construction Corp.).

33. U.S. Pat. 3,155,722 (Nov. 3, 1964), I. Mavrovic (to Chemical Construction Corp.).

34. U.S. Pat. 3,270,050 (Aug. 30, 1966), I. Mavrovic (to Chemical Construction Corp.).

35. U.S. Pat. 3,137,725 (June 16, 1964), L. H. Cook and T. E. Gilbert (to Chemical Construction Corp.).

36. U.S. Pat. 3,248,425 (April 26, 1966), A. Ledergerber (to Inventa, A.G. für Forschung und Patent Ververtung, Zürich, Switzerland).

37. *Nitrogen* **33**, p. 31, Jan. 1965.

38. U.S. Pat. 3,193,353 (July 6, 1965), P. Matile, E. Peterhans, W. Zollinger (to Lonza Ltd., Basel, Switzerland).

39. U.S. Pat. 3,347,915 (Oct. 17, 1967), G. Fauser (to Montecatini Società Generale per l'Industria Mineraria e Chimica, Milan, Italy).

40. F. Bellia, *La Chimica e L' Industria;* **48** (11), 1180–1188 (1966).

41. *Nitrogen* **38**, pp. 29–34, Nov. 1965.

42. U.S. Pat. 3,356,723 (Dec. 5, 1967), P. J. C. Kaasenbrood (to Stamicarbon N.V., Heerlen, Netherlands).

43. U.S. Pat. 3,005,849 (Oct. 24, 1961), E. Otsuka (to Toyo Koatsu Industries, Inc.).

44. U.S. Pat. 3,090,811 (May 21, 1963), E. Otsuka, T. Takahashi, and H. Watanabe (to Toyo Koatsu Industries, Inc.).

45. U.S. Pat. 3,317,601 (May 2, 1967), E. Otsuka, S. Yoshimura, and K. Kanai (to Toyo Koatsu Industries).
46. U.S. Pat. 3,146,263 (Aug. 25, 1964), E. Otsuka (to Stamicarbon).
47. U.S. Pat. 3,171,770 (Nov. 22, 1960), H. J. B. Biekart, M. Bongard, and P. J. C. Kaasenbrood (to Stamicarbon).
48. U.S. Pat. 2,961,464 (Nov. 22, 1960), P. J. C. Kaasenbrood (to Stamicarbon).
49. U.S. Pat. 3,147,174 (Sept. 1, 1964), L. H. Cook (to Chemical Construction Corp.).
50. *Nitrogen* **44**, p. 31 (Nov. 1966).
51. U.S. Pat. 2,727,069 (Dec. 13, 1955), Joseph P. M. Van Waes (to Stamicarbon).
52. U.S. Pat. 3,236,888 (Feb. 22, 1966), T. O. Wentworth (to Allied Chemical Corp.).
53. U.S. Pat. 3,321,399 (May 23, 1967), Joseph Versteeg and Max W. Hill (to Esso Research and Eng. Co.)
54. Brit. Pat. 1,070,782 (June 1, 1967), John G. Simmons.
55. Neth. Appl. 6,513,730 (April 25, 1966), A. G. Knapsack.
56. Neth. Appl. 6,616,258 (May 21, 1967), W. A. Scholten (to Chemische Fabrieken N.V., Holland).
57. Fr. Pat. 1,468,882 (Feb. 10, 1967), Vladimir Wassilkow (to Establissements Liem, France).
58. Roscoe R. Braham, Jr. (Univ. of Chicago), NASA Accession N. 66-15689, p. 37 (1965).

Ivo Mavrovic
Consulting Engineer

UREA–FORMALDEHYDE RESINS.   See Amino resins and plastics.

URETHAN POLYMERS

Since the discovery in 1937, by Bayer (1,2) and co-workers, of the diisocyanate addition polymerization used for the preparation of urethan polymers (polyurethans) and polyureas, the group of compounds classified as polyurethans have grown to substantial commercial importance. The "polyurethans" do not actually contain primarily urethan groups, but are those polymers which contain significant numbers of urethan groups, regardless of what the rest of the molecule may be. Usually these polymers are obtained by the reaction of a polyisocyanate with polyhydroxy compounds, such as polyethers, polyesters, castor oil, or glycols. Compounds containing groups such as amino and carboxyl may also be used. Thus a typical polyurethan may contain, in addition to urethan groups, aliphatic and aromatic hydrocarbon residues, ester, ether, amide, and urea groups.

Urethans, which have the characteristic structure (**1**), can be considered either esters of the unstable carbamil acid, or amide esters of carbonic acid. The urethan

$$\diagup_{\diagup} N - \overset{\overset{\displaystyle O}{\|}}{C} - O -$$

(**1**)

polymers have a significant number of these groups, although not necessarily repeating in a regular order. The polyurethans are also sometimes called simply "urethans" or, less desirably, "isocyanate polymers."

The terms "urethane" and "polyurethane" are employed invariably in commercial, and quite widely in scientific, usage; they are not, however, in accordance with accepted chemical nomenclature, according to which it is preferable to reserve the termination "-ane" for saturated hydrocarbons and certain fully saturated heterocyclic compounds. Thus the terms "urethan" and "polyurethan" are preferable and will be used here.

The most common commercial method of forming polyurethans is by the reaction of di- or polyfunctional hydroxy compounds, such as hydroxyl-terminated polyesters or polyethers, with di- or polyfunctional isocyanates. The general structure of a polyurethan derived from a dihydroxy compound, HO—R—OH, and a diisocyanate, OCN—R′—NCO, is given by structure (2).

$$\text{mw-}\left[\text{R—O—}\overset{\overset{\text{O}}{\|}}{\text{C}}\text{—NH—R′—NH—}\overset{\overset{\text{O}}{\|}}{\text{C}}\text{—O—}\right]_n\text{-mw}$$

(2)

Although the technology of polyurethans is of recent origin, the chemistry of organic isocyanates dates back to 1849 when Wurtz (3) first synthesized aliphatic isocyanates by reacting organic sulfates with cyanates (eq. 1). It was not until approxi-

$$\text{R}_2\text{SO}_4 + 2\,\text{KCNO} \rightarrow 2\,\text{RNCO} + \text{K}_2\text{SO}_4 \tag{1}$$

mately 1937 that a commercial use of diisocyanates was explored. The diisocyanate addition polymerization was discovered during an investigation of means of preparing polyamides. Although the polyureas which were formed by reacting diamines with diisocyanates did not have any commercial use, this work led to the reaction of diisocyanates with glycols such as 1,4-butylene glycol (eq. 2). The polymers obtained by this

$$\text{HO(CH}_2)_4\text{OH} + \text{OCN(CH}_2)_6\text{NCO} \rightarrow \text{mw-}\left[\text{O(CH}_2)_4\text{O—}\overset{\overset{\text{O}}{\|}}{\text{C}}\text{—NH(CH}_2)_6\text{NH—}\overset{\overset{\text{O}}{\|}}{\text{C}}\right]_n\text{-mw} \tag{2}$$

reaction possessed interesting properties as plastics and fibers. During this same time investigators at E. I. du Pont de Nemours & Co., Inc., disclosed the use of hexamethylene diisocyanate as a more effective way to dry alkyd resins (4). Since that time the patent literature covering the use of polyurethans has grown at an extremely rapid rate. During World War II, the Bayer laboratories actively developed many fields of application; early promising results appeared in rigid foams, adhesives, and coatings. Following World War II and the publication of the Farbenfabriken Bayer technology, United States industry became extremely interested in urethan chemistry. During the late 1940s Du Pont and Monsanto Company began supplying 2,4-tolylene diisocyanate (methylphenylene diisocyanate; toluene diisocyanate) in pilot-plant quantities. Development of polyurethan elastomers (5) and of flexible foams (6) based on polyesters was disclosed by Farbenfabriken Bayer in 1950 and 1952, respectively, and shortly thereafter they came into commercial production. In 1953, Du Pont announced a poly(ether–urethan) flexible foam (7).

Full-scale commercial isocyanate manufacture began in the United States during 1954–1955. The primary use for tolylene diisocyanate was in flexible foam based on polyesters; the flexible foam obtained from these products, however, had two major drawbacks, high cost and poor hydrolysis resistance, which limited their commercial growth. In 1957, polyethers based on ethylene oxide and propylene oxide were introduced commercially into the polyurethan industry. These polyols lower the cost and improve the hydrolysis resistance of the products. Initially, flexible foam prepared from the poly(alkylene oxide) polyols was prepared through a "prepolymer" technique by which a prepolymer was formed from polyether and diisocyanate, and then catalyst, water, and stabilizers were added to produce the foam. In 1958, so-called "one-shot" foaming was developed in which polyether, diisocyanate, water, catalysts, and foam stabilizers were mixed in one step. In 1961 polymeric polyisocyanates with low vapor pressure were introduced primarily for use in rigid foams. During this time of growth of the foam industry, other areas of application were also growing, including surface coatings, wire enamels, elastomeric compounds, caulks and sealants, and elastomeric fibers. These products were based essentially on tolylene diisocyanate and, to a lesser extent, on diphenylmethane diisocyanate (methylenediphenylene diisocyanate). Many of these products, especially surface coatings, were not light stable, tending to yellow rather badly when exposed to ultraviolet light. In the early 1960s, means were found to use aliphatic or arylaliphatic isocyanates to prepare light-stable products. The initial work in this was done with hexamethylene diisocyanate. Recent years have seen a rapid growth in the number of available aliphatic isocyanates designed specifically for preparation of light-stable polymers.

From the conception of the new polymer in 1937, the area of polyurethan chemistry and applications has grown until in 1968 approx 500 million pounds of finished polymer were produced in the United States alone. The major portion of this market is still in flexible and rigid foams; comparatively speaking, all other areas are relatively small consisting of 5–10% of the total market. See also Polyurethan, in Encyclopedia of Polymer Science and Technology.

Raw Materials

Isocyanates. Although a number of methods of preparing isocyanates have been reported in the literature (9–12), the only method of commercial importance is the phosgenation of primary amines (eq. 3). The commercial preparation of diiso-

$$RNH_2 + COCl_2 \rightarrow \left[\begin{array}{cc} H & Cl \\ | & | \\ R—N: & \rightarrow C{=}O \\ | & | \\ H & Cl \end{array} \right] \rightarrow RNCO + 2\ HCl \qquad (3)$$

cyanates usually involves two steps. An amine solution is mixed with phosgene at a low or moderate temperature. The resulting slurry is then treated with more phosgene at a temperature around 120–150°C, and the product purified by distillation. Properties of some commercially available isocyanates are listed in Table 1. See also Isocyanates, organic.

It may be noted that in industry the use of the name "toluene diisocyanate" is widespread; the name is incorrect, however, because an ester is always named from a radical, not a hydrocarbon—"ethyl acetate," not "ethane acetate." The name "tolylene diisocyanate" is preferable, and will be used here, although the name approved by IUPAC and CA is "methylphenylene diisocyanate."

Table 1. Properties of Some Isocyanates (8)

Compound	Melting point, °C	Boiling point, °C	Sp gr	n_D^t	Mol wt	Flash pt, open cup, °C
2,4-tolylene diisocyanate[a,b,c,d,e,f]	19.5–21.5	120/10 mm	1.22(25/15.5°)	1.5654[25]	174	132
(65:35) tolylene diisocyanate[a,b,c,d,e,f]	3.5–5.5	120/10 mm	1.22(25/15.5°)	1.5666[25]	174	132
(80:20) tolylene diisocyanate[a,b,c,d,e,f]	11.5–13.5	120/10 mm	1.22(25/15.5°)	1.5663[25]	174	132
4,4'-diphenylmethane diisocyanate[a,b,c]	37–38	194–199/5 mm	1.19(50°)		250	202
dianisidine diisocyanate[g]	119–122	200–210/0.5 mm			296	
tolidine diisocyanate[g]	69–71.5				264	
hexamethylene diisocyanate[g]		140–142/21 mm	1.04(25/15.5)	1.4516[25]	168	140
m-xylylene diisocyanate[g]		150–152/3.5 mm	1.16(20°)		188	
phenyl isocyanate[a,g]	~-30	165	1.09(20/4°)	1.5367[19.6]	119	110
p-chlorophenyl isocyanate[a,h]	28	87.5/10 mm			154	
o-chlorophenyl isocyanate[h]		86/10 mm			154	
m-chlorophenyl isocyanate[h]	-4	72/10 mm			154	
3,4-dichlorophenyl isocyanate[a,h]	42	113/10 mm			188	
2,5-dichlorophenyl isocyanate[h]	27	108/10 mm			188	
methyl isocyanate[h]		38			57	
ethyl isocyanate[h]		59–60	0.90(20°)	1.3801[20]	71	<-7
n-butyl isocyanate[g]		115	0.88(20/4°)	1.4043[25]	99	<27
n-propyl isocyanate[g]		83–84	0.90(20/20°)		85	<27
octadecyl isocyanate[a,g]	10–20	170/2 mm	0.86(25/15.5°)	1.450[25]	~295	185

a Mobay Chemical Co.

b E. I. du Pont de Nemours & Co., Inc.

c National Aniline Division, Allied Chemical Corp.

d Union Carbide Corp.

e Olin-Mathieson Corp.

f Wyandotte Chemicals Corp.

g Carwin Organic Chemicals, Upjohn Co.

h Ott Chemical Co.

Table 2. Major Suppliers of Polyethers (8)

Chemical identification	Type of polyol	Trade name	Mol-wt range	Supplier
poly(oxypropylene) glycols	diols	Niax Diol	400–4000	Union Carbide Corp.
		Pluracol P	400–4000	Wyandotte Chemicals Corp.
		Voranol P	2000	Dow Chemical Co.
		Poly-G	400–2000	Olin-Mathieson Corp.
		PPG	400–2000	Jefferson Chemical Co.
		Actol 21–56	2000	Allied Chemical Corp.
		Fomrez ED	2000	Witco Chemical Co.
poly(oxypropylene-b-oxyethylene) glycols (block copolymers)	diols	Pluronic	1000–2800	Wyandotte Chemicals Corp.
poly(oxypropylene) adducts of glycerol	triols	Niax Triol LG	1000–3000	Union Carbide Corp.
		Pluracol GP	3000	Wyandotte Chemicals Corp.
		Voranol GP	2700–5000	Dow Chemical Co.
		Poly G	1000–4000	Olin-Mathieson Corp.
		Triol G	400–3000	Jefferson Chemical Co.
		Actol 31–56	3000	Allied Chemical Corp.
		Fomrez ET	1500–3000	Witco Chemical Co.
poly(oxypropylene) adducts of trimethylolpropane	triols	Pluracol TP	300–4000	Wyandotte Chemicals Corp.
poly(oxypropylene-b-oxyethylene) adduct of trimethylol-propane	triols	Pluracol TPE	4500	Wyandotte Chemicals Corp.
poly(oxypropylene) adducts of 1,2,6-hexanetriol	triols	Niax Triol		
		LHT	700–4400	Union Carbide Corp.
poly(oxypropylene) adducts of pentaerythritol	tetrols	Pluracol PeP	400–600	Wyandotte Chemicals Corp.
poly(oxypropylene-b-oxyethylene) adducts of ethylenediamine (block copolymers)	tetrols	Tetronic	1000–5000	Wyandotte Chemicals Corp.
poly(oxypropylene) adducts of sorbitol	hexols	Niax Herol LS	700	Union Carbide Corp.
		Pluracol SP	500–700	Wyandotte Chemicals Corp.
		Atlas G	500–5000	Atlas Chemical Co.

Polyethers. The polyethers are commercially the most important of the polyhydroxy compounds used to prepare polyurethans. The first polyether designed specifically for preparing polyurethans was a poly(oxytetramethylene) glycol derived from tetrahydrofuran (7). At present, most of the polyethers used for the manufacture of flexible polyurethan foam are derived from propylene oxide and/or ethylene oxide. Rigid foams and coatings are based on poly(oxyalkylene) derivatives of polyhydric alcohols such as glycerol, 1,1,1-trimethylolpropane (2-ethyl-2-hydroxymethyl-1,3-propanediol, see Vol. 1, p. 595), 1,2,6-hexanetriol, and sorbitol. Nitrogen-containing polyols such as ethylene oxide and propylene oxide block copolymers with ethylenediamine are also used commercially.

The block copolymers of ethylene oxide and propylene oxide (13,14) used in the manufacture of polyurethans can be represented by the general structure (3). In the

$$
\underset{\textbf{(3)}}{HO(CH_2CH_2O)_a(CH_2\overset{\overset{\displaystyle CH_3}{|}}{C}HO)_b(CH_2CH_2O)_cH}
$$

preparation, propylene oxide is reacted with propylene glycol or water in the presence of a basic catalyst to form a poly(oxypropylene) homopolymer, which is then further reacted with ethylene oxide to form the block copolymer. The resulting polyether has a higher percentage of primary hydroxyl groups than does a poly(oxypropylene) glycol of comparable molecular weight.

Poly(oxytetramethylene) glycols are prepared by the polymerization of tetrahydrofuran. See Vol. 10, p. 249.

Table 3. Typical Physical Properties of Commercially Available Polyethers (8)

Av mol wt	Hydroxyl no.	Sp gr (20°/20°)	Total unsaturation, meq/g	Viscosity, at 100°C, cP	Flash pt,[a] °C
		Poly(oxypropylene) glycols			
150	750	1.025		3	121
250	450	1.011		3.4	168
400	280	1.008		5	199
750	150	1.006	0.01	7.6	215
1000	112	1.005	0.01	11.5	227
1200	93	1.004	0.015	13.7	232
1850	61	1.003	0.03		232
2000	56	1.003	0.04	23.0	232
3000	37	1.002	0.09	48.0	232
4000	28	1.002	0.11	74.0	232
2000[b]	56	1.014	0.03	31.1	232
		Trimethylolpropane-based polyether triols			
300	561		0.005		
400	422		0.005	625	
700	240	1.027	0.005	325	
1500	112		0.02	290	
2500	67	1.005	0.04	440	
4000	42	1.004	0.07	670	
4500[b]	37		0.08	500	

[a] Cleveland open cup method.   [b] Modified with ethylene oxide.

Poly(oxypropylene) triols are at present the most important class of polyethers used in the manufacture of polyurethans. These triols are made by the same general reaction as are poly(oxypropylene) glycols, but with low-molecular-weight triols, such as trimethylolpropane, glycerol, and 1,2,6-hexanetriol, rather than propylene glycol. Polyethers having functionality greater than three have significant use in polyurethans, particularly in rigid foam. Polyols used in the preparation of these products include sorbitol, mannitol, pentaerythritol, and sucrose. Suppliers and properties of some commercially available polyethers are listed in Tables 2 and 3.

Polyesters. During the initial development of polyurethans, polyesters were the most commonly used type of polyol. Since unsaturated polyesters were found undesirable for use in polyurethans, completely saturated polyesters containing terminal hydroxyl groups rather than carboxyl groups are used. An excellent review of their preparation is given in Müller (15). See also Polyesters.

The most common monomers used in polyesters for the urethan polymers are adipic acid, phthalic anhydride, ethylene glycol, propylene glycol, 1,3-butylene glycol, 1,4-butylene glycol, and diethylene glycol. For preparation of branched polyesters, triols such as 1,2,6-hexanetriol, trimethylolpropane, and 1,1,1-trimethylolethane (2-hydroxymethyl-2-methyl-1,3-propanediol, see Vol. 1, p. 594) are commonly used. Recently, lactones such as caprolactone have been used to prepare polyesters for polyurethans (16,17).

Since it is desirable to use polyesters containing only hydroxyl groups as reactive sites, the polyesters of choice are those with very low acid numbers and very low water content. For use in elastomers, linear polyesters having molecular weights close to 2000 are preferred, whereas slightly branched ones of similar molecular weight are used for flexible foams and elastic coatings, and highly branched polyesters are used for rigid foam and chemically resistant coatings.

These polyesters are prepared by heating the glycol or triol in a reactor to approx 60–90°C. The carboxylic acid is added and heated at such a rate that water distils out rapidly. The ratio of acid to hydroxyl component is such that the hydroxyl is

Table 4. Typical Characteristics of a Range of Polyesters

| Components | | | | Degree of | |
Acid	Glycol	Triol	Hydroxyl no.	branching	Use
adipic	ethylene, propylene, or diethylene	none	50–60	none	elastomers
adipic or "dimer" acid	diethylene	glycerol or others	50–65	slight	flexible foams and coatings
adipic plus phthalic	ethylene or propylene	glycerol or others	150–200	moderate	semiflexible foams and coatings
adipic plus phthalic	ethylene or propylene	glycerol or others	250–300	high	rigid foam, chemically resistant coatings
adipic plus phthalic	ethylene or propylene	glycerol or others	400–450	very high	rigid foam, wire coatings

Table 5. Typical Properties of Castor Oil (8)

Oil grade	Hydroxyl value	% volatile (moisture)	Acid value	Color (Gardner)
DB oil (urethan grade)	163	0.02	1	1+
no. 1 oil	163	0.2	2	1+
no. 3 oil	154	0.3	12	6

in excess, so that nearly all the acid groups are consumed and the polyester is prepared in the desired molecular-weight range. It is usually necessary to force the reaction to completion by heating to about 200°C while flushing with nitrogen or carbon dioxide under reduced pressure. Typical characteristics of polyesters suitable for a range of applications are indicated in Table 4.

    Castor Oil. Castor oil is important as a raw material in polyurethans, especially in the field of coatings and potting encapsulation compounds. Three principal grades of castor oil are commercially available; some of their properties are listed in Table 5 (see also Castor oil). Castor oil is particularly important in the preparation of urethan-modified alkyd resins. Castor oil is transesterified with a polyhydroxy compound such as glycerol to form a combination of glycerides, as in equation 4. The resulting polyhydroxy compounds can then be reacted with polyisocyanates.

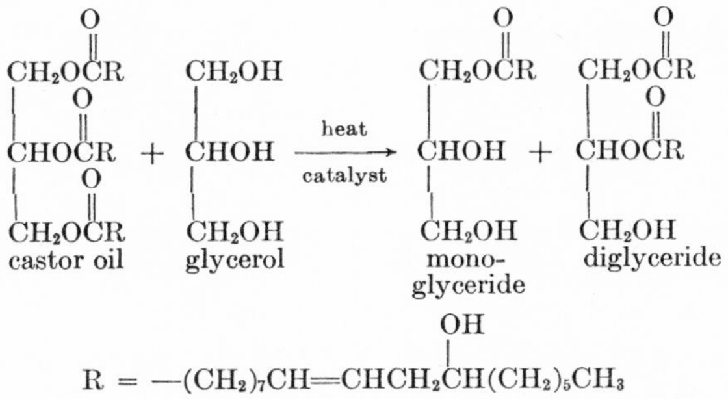

$$R = -(CH_2)_7CH\!=\!CHCH_2CH(CH_2)_5CH_3$$

with OH substituent, equation (4)

Reactions of Isocyanates

    Isocyanates contain the highly unsaturated —N=C=O group and are extremely reactive with a large number of compounds; they may also react with themselves. The reaction can proceed with almost any compound possessing "active hydrogen," ie, one that may be replaced by sodium, and with a few compounds having hydrogen atoms not readily replaced by sodium.

    The electronic structure of the isocyanate group indicates that it should have the resonance possibilities shown in structures (**4a**)–(**4c**). The normal reaction ultimately

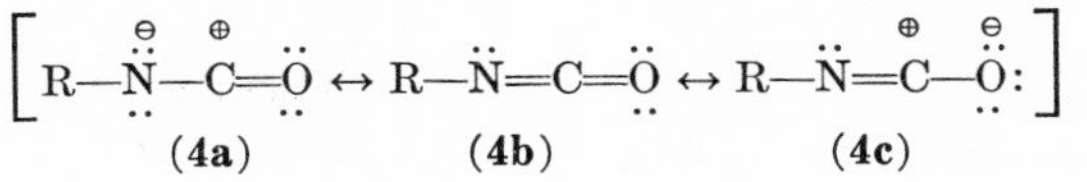

$$\left[R\!-\!\overset{\ominus}{\underset{..}{N}}\!-\!\overset{\oplus}{C}\!=\!\overset{..}{\underset{..}{O}} \leftrightarrow R\!-\!\overset{..}{N}\!=\!C\!=\!\overset{..}{\underset{..}{O}} \leftrightarrow R\!-\!\overset{..}{N}\!=\!\overset{\oplus}{C}\!-\!\overset{\ominus}{\underset{..}{O}}\!: \right]$$

      (**4a**)   (**4b**)   (**4c**)

produces addition to the carbon–nitrogen double bond. In the reaction with those compounds having an active hydrogen, the hydrogen atom becomes attached to the nitrogen of the isocyanate and the remainder of the active-hydrogen compound becomes attached to the carbonyl carbon (eq. 5). In most cases, this addition product is

$$R-N{=}C{=}O + H-A \rightarrow R-NH-\overset{\overset{\displaystyle O}{\|}}{C}-A \tag{5}$$

stable, but in some special cases the addition product is only moderately stable and may either dissociate to re-form the initial reactants or decompose to other products.

The more important groups that react with the isocyanates are the amino, hydroxyl, and carboxyl groups (eqs. 6–10).

$$RNCO + R'NH_2 \rightarrow \underset{\text{urea}}{RNH-\overset{\overset{\displaystyle O}{\|}}{C}-NHR'} \tag{6}$$

$$RNCO + HOR' \rightarrow \underset{\text{urethan}}{RNH-\overset{\overset{\displaystyle O}{\|}}{C}-OR'} \tag{7}$$

$$RNCO + HOH \rightarrow \left[RNH-\overset{\overset{\displaystyle O}{\|}}{C}-OH \right] \rightarrow \underset{\text{amine}}{RNH_2} + CO_2 \tag{8a}$$

$$RNH_2 + RNCO \rightarrow \underset{\text{urea}}{RNH-\overset{\overset{\displaystyle O}{\|}}{C}-NHR} \tag{8b}$$

$$RNCO + R'COOH \rightarrow \left[RNH-\overset{\overset{\displaystyle O}{\|}}{C}-O-\overset{\overset{\displaystyle O}{\|}}{C}-R' \right] \rightarrow \underset{\text{amide}}{RNHCOR'} + CO_2 \tag{9}$$

$$ArNCO + R'COOH \rightarrow \left[ArNH-\overset{\overset{\displaystyle O}{\|}}{C}-O-\overset{\overset{\displaystyle O}{\|}}{C}-R' \right]$$

$$\rightarrow \left[ArNH-\overset{\overset{\displaystyle O}{\|}}{C}-O-\overset{\overset{\displaystyle O}{\|}}{C}-NHAr \right] + R'-\overset{\overset{\displaystyle O}{\|}}{C}-O-\overset{\overset{\displaystyle O}{\|}}{C}-R' \tag{10a}$$

$$\left[ArNH-\overset{\overset{\displaystyle O}{\|}}{C}-O-\overset{\overset{\displaystyle O}{\|}}{C}-NHAr \right] \rightarrow \underset{\text{urea}}{ArNH-\overset{\overset{\displaystyle O}{\|}}{C}-NHAr} + CO_2 \tag{10b}$$

Secondary reactions that are important in the formation of urethan polymers are those with urea (eq. 11) and urethans (eq. 12).

$$RNCO + RNH-\overset{\overset{\displaystyle O}{\|}}{C}-NHR \rightarrow \underset{\underset{\text{biuret}}{\text{CONHR}}}{RN-CONHR} \tag{11}$$

$$RNCO + RNH-\overset{\overset{\displaystyle O}{\|}}{C}-OR' \rightarrow \underset{\underset{\text{allophanate}}{\text{CONHR}}}{RN-COOR'} \tag{12}$$

Primary aliphatic *amines* are extremely reactive at 0–25°C, the most basic generally being the most reactive. Secondary aliphatic amines, as well as primary aromatic amines, react similarly although not quite as readily. Secondary aromatic amines are still less reactive. These reactions are sufficiently rapid so that they are not strongly influenced by the many compounds that are strong catalysts for other, slower, isocyanate reactions.

Primary *alcohols* react readily at 25–50°C; secondary alcohols usually react only about 0.3 as fast. Tertiary alcohols react much more slowly, approx 0.005 as fast as the primary (18). The normal reaction of an alcohol and an isocyanate is readily catalyzed by mild and strong bases, by many metals, and by acids. Phenols react more slowly with isocyanates than do the alcohols. Moderately elevated temperatures (50–75°C) and a catalyst, such as a tertiary amine or aluminum chloride, are used to promote the reaction (19,20). The reactivity of *water* with isocyanates is similar to that of the secondary alcohol. The reaction is not as simple as the formation of urethan, however, since the initial product is not stable, splitting out carbon dioxide. The amine thus formed reacts with another isocyanate to give a good yield of a disubstituted urea.

Carboxylic acids react readily with isocyanates through the hydroxyl group, but are somewhat less reactive than the primary alcohols and water. This reaction is catalyzed by tertiary amines and many other bases, as well as numerous metal compounds. The initial addition product is not stable.

Aromatic isocyanates can react with themselves to form *dimers*, or "uretidine diones" (eq. 13). Dimerization is catalyzed vigorously by trialkylphosphines and

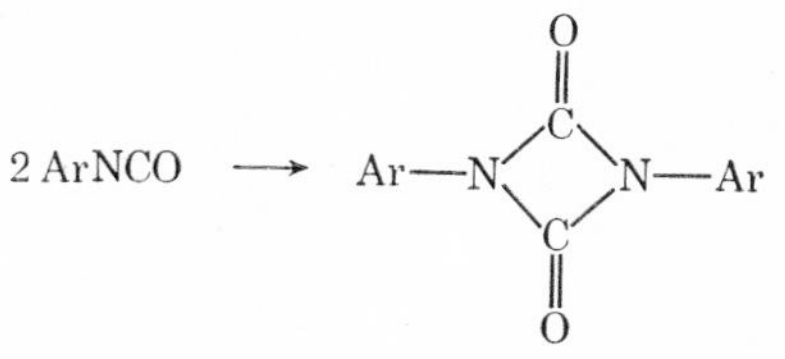

$$2\ ArNCO \longrightarrow Ar\!-\!N\overset{\displaystyle \overset{O}{\|}}{\underset{\underset{\displaystyle O}{\|}}{\overset{C}{\underset{C}{}}}}N\!-\!Ar \qquad\qquad (13)$$

mildly by tertiary amines such as pyridine. Some aromatic isocyanates such as 4,4′-diphenylmethane diisocyanate dimerize slowly on standing, even without catalyst. Both aliphatic isocyanates and aromatic isocyanates form *trimers* (eq. 14). Catalysts

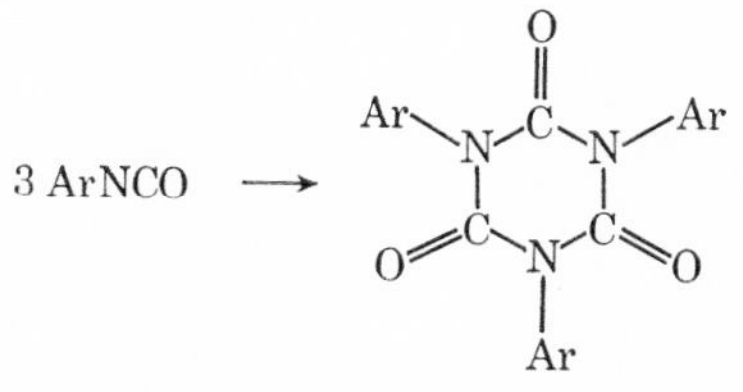

$$3\ ArNCO \longrightarrow \qquad\qquad (14)$$

such as triethylphosphine promote trimerization of aliphatic compounds. Other catalysts that have been used to induce trimerization of either aromatic or aliphatic isocyanates include calcium acetate, potassium acetate, sodium formate, sodium carbonate, sodium methoxide, triethylamine, oxalic acid, and a large number of soluble compounds of iron, sodium, potassium, etc. Isocyanate dimers at elevated temperatures react much as do isocyanates, sometimes by prior dissociation of the isocyanate, sometimes by direct reaction between dimer and active-hydrogen compound. Catalysts often produce effects similar to those with isocyanates; thus a dimer that may be present as an impurity in certain isocyanates may give nearly the same reaction as monomeric isocyanates.

Catalysts for the various reactions of isocyanates are summarized in Table 6. A number of catalysts are effective for more than one type of reaction.

Table 6. Summary of Catalysts[a] Promoting Various Isocyanate Reactions (8)

Amines	Alcohols	Water	Carboxylic acids	Ureas	Urethans	Phenols	Trimerization	Dimerization[b]
triethylenediamine	Bi cpds	triethylenediamine	Co cpds	Sn cpds	Pb cpds	tertiary amines	strong bases and basic salts	phosphines
Pb cpds	Pb cpds	tertiary amines	Fe cpds	triethylene-diamine	Co cpds	ZnCl₂	Pb cpds	tertiary amines
Sn cpds	Sn cpds	Sn cpds	Mn cpds	Zn cpds	Zn cpds	basic salts	Co cpds	(in high concn)
Co cpds	triethylene diamine		triethylene-diamine		Cu cpds	AlCl₃	Fe cpds	
Zn cpds	Ti cpds		tertiary amines		Mn cpds	acids	Cd cpds	
ureas	Fe cpds		CH₃COOK		Fe cpds		V cpds	
tertiary amines	Sb cpds				Cd cpds		tertiary amine plus alcohol	
carboxylic acids	U cpds				V cpds			
	Cd cpds							
	Co cpds							
	Th cpds							
	Al cpds							
	Hg cpds							
	Zn cpds							
	Ni cpds							
	tertiary amines							
	V cpds							
	Ce cpds							
	MgO							
	BaO							
	pyrones							
	lactams							
	acids							

[a] Catalysts that promote reaction with the active-hydrogen compound shown.
[b] For aromatic isocyanates only.

Formation of Polyurethans

The polycondensation reactions leading to the formation of polyurethans are influenced by a number of factors, prominent among which are the structure of the isocyanate, including its functionality and the type and location of substituents; the structure of the polyhydroxy compound; the solvent used and the dilution of the system; the presence of impurities, such as traces of acid in the isocyanate; and the temperature, particularly above 100°C. In addition, the reactions are complicated by the participation of reactant molecules (eg, alcohol and water) and products (eg, ureas and urethans) as catalysts. These factors will be discussed in detail in this section.

The uncatalyzed reaction of an isocyanate with a hydroxyl-containing compound proceeds as shown in equation 15 (see p. 63 for the resonance structures of the iso-

$$
\text{R'NCO} + \text{ROH} \rightleftharpoons \left[\begin{array}{c} \text{R'N}{=}\text{C}{-}\ddot{\text{O}}: \\ \uparrow \\ \text{H}{-}\text{O}{-}\text{R} \end{array}\right] \xrightarrow{\text{ROH}} \left[\begin{array}{c} \text{H}{-}\text{O}{-}\text{R} \\ | \\ \text{R'}{-}\ddot{\text{N}}{-}\text{C}{-}\ddot{\text{O}}: \\ | \\ \text{R}{-}\text{O}{-}\text{H} \end{array}\right] \rightarrow
$$

$$
\underset{\parallel}{\overset{\text{O}}{\text{R'NH}{-}\text{C}{-}\text{OR}}} + \text{ROH} \quad (15)
$$

cyanate group). A base-catalyzed reaction can be represented by equation 16 (21–23) and an acid catalyst could be expected to participate as shown in equation 17. In

$$
[\text{R}{-}\text{N}{=}\text{C}{=}\text{O} \leftrightarrow \overset{\oplus}{\text{R}{-}\text{N}}{=}\text{C}{-}\overset{\ominus}{\text{O}}] + :\text{B} \underset{k_2}{\overset{k_1}{\rightleftharpoons}} \left[\begin{array}{c} \text{R}{-}\text{N}{=}\text{C}{-}\text{O}^{\ominus} \\ \uparrow \\ \text{B}^{\oplus} \end{array}\right] \xrightarrow[k_3]{\text{R'OH}}
$$

$$
\left[\begin{array}{c} \text{H}{-}\overset{\oplus}{\text{O}}{-}\text{R'} \\ \overset{\ominus}{\text{R}}{-}\text{N}{-}\text{C}{-}\text{O}^{\ominus} \\ \uparrow \\ \text{B}^{\oplus} \end{array}\right] \rightarrow \underset{\parallel}{\overset{\text{O}}{\text{RNH}{-}\text{C}{-}\text{OR'}}} + :\text{B} \quad (16)
$$

$$
\text{R'NCO} + \text{HA} \rightarrow [\text{R'}{-}\text{N}{=}\text{C}{-}\ddot{\text{O}}: \rightarrow \text{H}\cdots\text{A}] \xrightarrow{\text{ROH}} \underset{\parallel}{\overset{\text{O}}{\text{R'NH}{-}\text{C}{-}\text{OR}}} + \text{HA} \quad (17)
$$

equations 15 and 16 the structure of the isocyanate and the active-hydrogen compound plays a major role in determining the rate of reaction. For example, an electron-withdrawing substituent on the isocyanate molecule increases the partial positive charge on the isocyanate carbon and moves the negative charge farther from the site of reaction (see structures (**5a**) and (**5b**)), thus making attack on the carbon by an electron donor

$$
\left[\begin{array}{cc} \overset{\text{O}}{\underset{\text{O}}{>}}\text{N}{-}\!\!\!\bigcirc\!\!\!-\text{NCO} & \leftrightarrow & \overset{\text{O}}{\underset{\ominus\text{O}}{>}}\text{N}{=}\!\!\!\bigcirc\!\!\!={\overset{\oplus}{\text{N}}}{-}\text{C}{=}\text{O} \end{array}\right]
$$

$$
\quad\quad\quad (\textbf{5a}) \quad\quad\quad\quad\quad\quad\quad\quad (\textbf{5b})
$$

Table 7. Relative Effects of Substituents on Activity of Substituted Phenyl Isocyanates with Alcohols[a]

| | Relative reactivity in system indicated | | | |
| | 2-Ethylhexanol, benzene, 28°C | | Methanol, dibutyl ether, $(C_2H_5)_3N$ catalyst, 20°C (Ref. 22) | 1-Butanol, toluene, $(C_2H_5)_3N$ catalyst (Ref. 25) |
Substituent	Ref. 24	Ref. 26		
p-SO$_2$	>50			
p-NO$_2$	>35	41	~130	
m-NO$_2$		33		
m-CF$_3$		10		
m-Cl	7	7.5		
m-Br		7.5		
m-NCO	6	5		6.7
p-NCO	6	4		4
p-Cl		3.5		
p-NHCOOR	2	1.5		1
p-C$_6$H$_5$		1.5		
m-NHCOOR	2	1.5		2
m-CH$_3$O		1.3		
none	1.0	1.0	1.0	1.0
p-n-C$_4$H$_9$		0.7		
m-CH$_3$	0.5	0.6		
p-CH$_3$	0.5	0.6	0.5	0.5
p-CH$_3$O		0.5	0.4	
o-CH$_3$	0.08			0.17–0.25
o-CH$_3$O	0.04			

easier, and hence giving a faster reaction. On the other hand, an electron-donating substituent reduces the partial positive charge on the isocyanate carbon, making attack at that point by an electron donor more difficult, thus giving a slower reaction. The effects of a number of substituents are indicated in Table 7.

The effect of substituents on the active-hydrogen compound is the opposite to that on the isocyanate, since the primary role of the active-hydrogen compound is that of an electron donor. For example, with amines, electronegative groups withdraw electrons, reducing the basicity of the nitrogen (structures (**6a**) and (**6b**)), making it a poorer electron donor. Conversely, electron-donating groups increase the basicity of the amino nitrogen.

(**6a**)   (**6b**)

In addition to the electronic effects of substituents, steric factors are also important. In aromatic compounds bulky groups in the ortho positions, or in aliphatic compounds branching or bulky substituents close to the site of reaction, retard the reaction. These steric factors not only affect the reactivity of both the isocyanate and the active-hydrogen compound, but also influence the effectiveness of a catalyst. Since a catalyst must approach the site of reaction as closely as a reactant itself, the ease of approach and extent to which the approach is possible will be influenced by steric relations between the catalyst and one or more of the reactants, usually the isocyanate.

More than one of the many reactions of the isocyanates may be occurring in a system (see the section on Reactions of isocyanates, p. 63). This possibility is most pronounced at elevated temperature and in the presence of catalysts. In some cases equilibria may be involved. The necessity of purifying all components of the reaction system under study cannot be overemphasized. A very wide range of compounds, both organic and inorganic, may catalyze one or more of the isocyanate reactions (see Table 6). Trace impurities in resins are particularly likely sources of catalysts. Another common oversight is the very small acid (HCl) content of the isocyanate itself. In many cases the acid will effectively neutralize a portion of a basic catalyst.

The choice of solvent may affect both the rate of an uncatalyzed reaction and the effectiveness of a catalyst. In general the solvents that readily complex with the active-hydrogen compound or catalyst, eg, by hydrogen bonding or dipole moment interaction, will provide a slower reaction than will the solvents that cannot so readily associate with reactant or catalyst. In extreme cases the solvents may even react with the isocyanate, as in the cases of dimethylformamide, dimethylacetamide, and dimethyl sulfoxide (27).

Whereas most kinetic studies have been made in dilute solution, most commercial applications utilize solvent-free systems. Since a change in a solvent may affect a rate or even relative rates in a series of reactants, one should not expect to find exactly the same kinetic results in a solvent-free system as in a dilute solution.

Diisocyanates. The reactions of diisocyanates are usually more complicated kinetically than are those of monoisocyanates. The initial reactivity of a diisocyanate is similar to that of a monoisocyanate substituted by an activating group, in this case the second isocyanate group. As soon as one isocyanate group has reacted with an alcohol, the remaining isocyanate group has a reactivity similar to that of a monoisocyanate substituted by a urethan group. As was shown in Table 7, a urethan group in the meta or para position has only a very mild activating effect, much less than an isocyanate group in the meta or para position. The reactivity of a diisocyanate having both isocyanate groups on one aromatic ring should decrease significantly as the reaction passes approx 50% completion.

This decrease in reactivity may be even greater if another substituent is present ortho to one isocyanate group. As shown in Table 7, an ortho alkyl substituent greatly decreases the reactivity of an isocyanate group.

In diisocyanates in which the two isocyanate groups are on different aromatic rings or are separated by aliphatic chains, the effect of one isocyanate or urethan group on a second isocyanate group is less pronounced. The effect becomes still less as the aromatic rings are separated farther and farther from each other; eg, by progressively longer aliphatic chains. The reactivities of a number of diisocyanates are indicated in Figure 1.

In spite of these complicating factors, the kinetics of diisocyanate reactions have been studied with great interest. This has arisen from the commercial importance of the diisocyanates in the productions of urethan foams, elastomers, coatings, adhesives, and other products.

Catalysts. The largest commercial applications of isocyanates utilize catalyzed reactions, especially the preparation of foams. For this reason the catalysis of the isocyanate–hydroxyl reaction has been the object of extensive research. In general, an

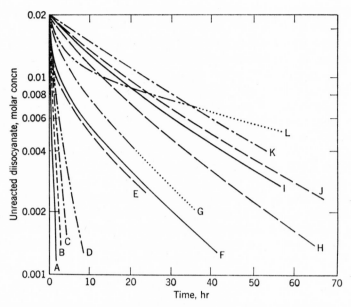

Fig. 1. Reactivity of aromatic diisocyanates with 2-ethylhexanol (A–K) and diethylene glycol adipate polyester (L), in benzene. Key:

A, 1-chloro-2,4-phenylene
   diisocyanate
B, *m*-phenylene diisocyanate
C, *p*-phenylene diisocyanate
D, methylenedi-*p*-phenylene
   diisocyanate
E, 2,4-tolylene diisocyanate
F, 2,4-/2,6-tolylene
   diisocyanate (60:40)

G, 2,6-tolylene diisocyanate
H, 3,3′-dimethyl-4,4′-biphenylene
   diisocyanate
I, methylenebis(2-methyl-*p*-phenylene)
   diisocyanate
J, 3,3′-dimethoxy-4,4′-biphenylene
   diisocyanate
K, 2,2′,5,5′-tetramethyl-4,4′-biphenylene
   diisocyanate
L, 2,4-/2,6-tolylene diisocyanate (80:20)
   plus polyester

increase in base strength of a catalyst, such as the tertiary amines, is accompanied by an increased catalytic strength, except when steric hindrance interferes (28). The effect of both acidic and basic catalysts has been studied, and two acids, hydrogen chloride and boron trifluoride etherate, were shown to be mild catalysts (29). When a large excess of alcohol was present the acids had little or no catalytic effect. Pyridine was a mild catalyst, even at high alcohol excess. A linear relation was found for the values of the experimentally observed rate constant and the pyridine concentration, as well as for the concentration of each acid catalyst when the alcohol concentration was low.

The choice of catalyst may affect the two groups in a diisocyanate differently (30). With 1,8-naphthylene diisocyanate, the isocyanate group in the 1 position was 1.5 times more reactive toward 1-butanol than was the 8-position group when copper naphthenate was used as a catalyst. With lead naphthenate the ratio was 5.8. It might be expected that the isocyanate group with less steric hindrance could coordinate with certain catalysts more easily than one with more steric hindrance, thus accounting for the difference.

Although most tertiary amine catalysts are effective in proportion to their base strength, an exception is triethylenediamine (1,4-diaza[2.2.2]bicyclooctane). This catalyst is much more powerful than would be predicted from its base strength (see Table 8) (31,32). The explanation may be the complete lack of steric hindrance in structure (7).

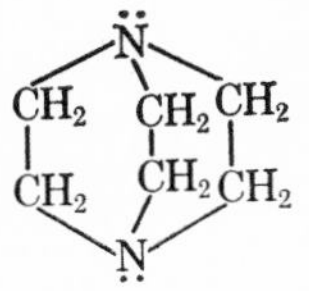

(7)

Although many compounds have catalytic activity for the isocyanate–hydroxyl reaction, none of the catalysts first used was powerful enough for satisfactory commercial production of poly(ether–urethan) foam by the one part, or "one-shot," process (see p. 538). Research directed toward such a process led to the recognition of many

Table 8. Reaction of Phenyl Isocyanate with 2-Ethylhexanol[a]

Catalyst		$k \times 10^4$, liter/(mole)(sec)	
	pK	In benzene	In dioxane
none		0.39	0.055
triethylamine	3.36	5.3	
triethylenediamine	5.40	21.7	6.0
N,N-dimethylpiperazine	5.71	4.3	
N-ethylmorpholine	6.49	1.6	

[a] Reactants, 0.072M; catalyst, 0.0014M; at 23°C.

metal compounds as powerful catalysts, in particular the organotin compounds, in commercial foam processes. The tremendously powerful effect of certain tin compounds, compared to some other catalysts, is shown in Table 9 (33,34). A synergistic

Table 9. Catalysis of Phenyl Isocyanate–Methanol Reaction in Dibutyl Ether[a]

Catalyst	Mole % catalyst used	Relative activity at 1.0 mole % catalyst
none		1
triethylamine	1.00	11
cobalt naphthenate	0.93	23
benzyltrimethylammonium hydroxide	0.47	60
stannous chloride	0.70	68
tetra-n-butyltin	1.01	82
stannic chloride	0.30	99
tri-n-butyltin acetate	0.10	500
n-butyltin trichloride	0.20	830
trimethyltin hydroxide	0.041	1,800
dimethyltin dichloride	0.020	2,100
di-n-butyltin dilaurate	0.0094	37,000

[a] At 30°C.

Table 10. Synergistic Catalysis of Phenyl Isocyanate–Butanol Reaction[a]

Catalyst	Catalyst concn, mole %	$k \times 10^4$, liter/(mole)(sec)
none		0.37
triethylamine	0.88	2.4
dibutyltin diacetate	0.00105	20
triethylamine plus dibutyltin diacetate	0.99	88
	0.00098	

[a] In dioxane at 70°C.

effect between an amine and an organotin catalyst has also been observed (Table 10) (33).

In a study in which a large number of metallic compounds were tested (by a simple screening test in which tolylene diisocyanate was reacted with a polyether triol) for catalytic activity in the isocyanate–hydroxyl reaction, a roughly descending order of catalytic activity was found as follows: bismuth, lead, tin, triethylenediamine, strong bases, titanium, iron, antimony, uranium, cadmium, cobalt, thorium, aluminum, mercury, zinc, nickel, trialkylamines, cerium, molybdenum, vanadium, copper, manganese, zirconium, and trialkylphosphines. Arsenic, boron, calcium, and barium compounds did not show any catalytic activity within the limits of the screening test used (35). However, a study of the relative strength of several catalysts for the reaction of phenyl isocyanate with poly(oxypropylene) glycol in toluene yielded somewhat different results, notably for diethylenetriamine and dibutyltin dilaurate (36). These differences in apparent effectiveness may be the result of the different reaction media used, and emphasize the danger of trying to make precise correlations between results from studies in dilute solution and reactions in essentially solvent-free systems.

A reaction mechanism for metal catalysts was proposed as shown in equation 18. This coordination effect, which proposes that the hydroxyl group enters on the metal

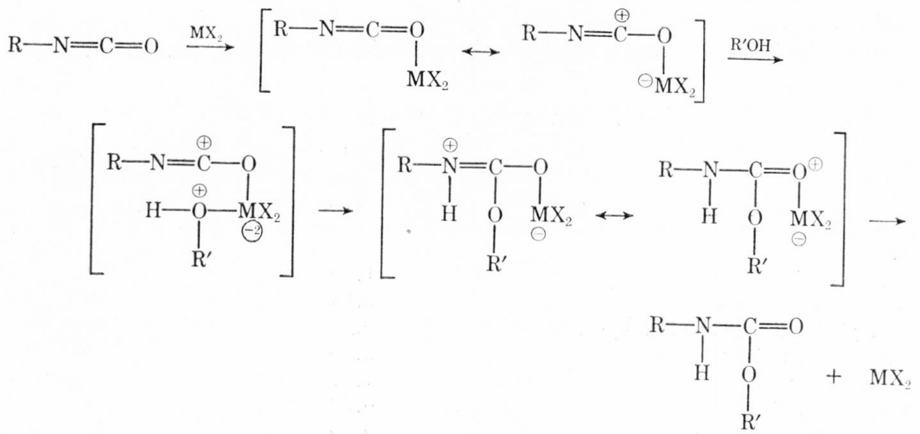

side of the complex and attaches to the metal in close proximity to the isocyanate-group nitrogen, can explain the remarkable catalytic activity of the metals. Obviously, the order of the metal coordination complex formation could be the reverse so that the hydroxy-compound complex forms first and the isocyanate second. Complexing of the metal with both isocyanate and alcohol appears to be a reasonable explanation of their much greater catalytic effect on the isocyanate–hydroxyl reaction than on the iso-cyanate water reaction.

Catalysis of the isocyanate–water reaction by tertiary amines is well known owing to its importance in polyurethan foams. Based on the rate of carbon dioxide evolution, the rate of the reaction generally increases as the base strength of the tertiary amine catalyst increases (see Fig. 2) (37).

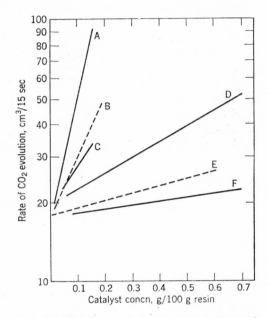

Fig. 2. Effect of catalyst concentration on the isocyanate–water reaction. Key: A, triethylene-diamine; B, tetramethyl-1,3-butanediamine; C, triethylamine; D, dibutyltin dilaurate; E, ethyl-morpholine; F, stannous octoate.

Although the choice of temperature, isocyanate, and active-hydrogen compound permits one to control isocyanate reactions to a large extent, the choice of catalyst often gives much broader control. Furthermore, the catalytic effect may be so powerful in many cases that it overrides the influence of changes in isocyanate structure or reaction medium. For these reasons many commercial uses of isocyanates include

Table 11. Effects of Catalysts on Relative Reactivities of Phenyl Isocyanate with Active-Hydrogen Compounds[a]

Catalyst	Relative reactivity with		
	1-Butanol	Water	Diphenylurea
none	1.0	1.1	2.2
N-methylmorpholine	40	25	10
triethylamine	86	47	4
tetramethyl-1,4-butanediamine	260	100	12
triethylenediamine	1,200	380	90
tributyltin acetate	80,000	14,000	8,000
dibutyltin diacetate	600,000	100,000	12,000

[a] In dioxane, at 70°C; 10 mole % catalyst.

catalysts to give the desired control over the reaction. Relative reactivities of several catalysts with various active-hydrogen compounds are given in Table 11 (34).

As mentioned above, steric hindrance in either reactant or catalyst can reduce the effect of a catalyst. A most notable example of the former is the failure of dialkylanilines to catalyze the reaction with alcohols. Catalyst effects may also be expected to be much more pronounced at low than at high temperatures.

Catalysis in Commercial Polymerizations. The effects of several catalysts on three types of commercially important reactions were reported in reference 38. The rise time of a prepolymer foam system was used as a guide to the rate of the isocyanate–water reaction. The curing of a castor oil prepolymer coating was observed; this was primarily a measure of the rate of the isocyanate–water reaction, but in the use of the cobalt naphthenate catalyst other reactions such as allophanate formation, trimerization, and free-radical type of reactions characteristic of a drying-oil cure may have contributed. Prepolymer gelation was also observed: gelation could result from isocyanate dimerization or trimerization, allophanate formation, or biuret formation if some water had been used in the prepolymer formation. Results are shown in Table 12.

Table 12. Relative Catalyst Activity

Catalyst	Foam, max rise time,[a] sec	Film drying time,[b] min	Gelation time,[c] days
triethylamine	255	76	7.5
diethylcyclohexylamine	315	90	12
dimethylethanolamine	570	105	27
methylmorpholine	930	245	30
cobalt naphthenate	3600	35	0.67

[a] Castor oil–TDI prepolymer foam system.

[b] Castor oil–MDI prepolymer solution, 0.001 mole catalyst/10 g prepolymer, poured on mercury and air dried.

[c] Castor oil–MDI prepolymer (10 g) in benzene (5 g); 0.002 mole catalyst.

Since bases and many soluble heavy metal compounds are very strong catalysts and acids are generally weak catalysts for only some of the isocyanate reactions, acids may often be used to stabilize isocyanate systems. In reality the acid must serve to neutralize a base or render inactive a metal which may be present, for example as a polyether polymerization catalyst or esterification catalyst.

The general relation between catalysis by inorganic bases, acid catalysis, and acid retardation or stabilization is shown in Figure 3 (39). The effects are shown for the isocyanate–hydroxyl, isocyanate–urethan, isocyanate–urea, and isocyanate trimerization reactions, since these were the most likely reactions in the prepolymer system studied.

Whereas simple acids such as hydrogen chloride are usually the best neutralizers of basic catalysts, heavy metal catalysts are often neutralized best by acidic compounds which may also function as chelating agents, eg, citric acid. Relatively little has been published on neutralization of heavy-metal catalysts, but it is generally known that hydrogen chloride or acid chlorides will often, though not always, neutralize a metal catalyst.

By the proper choice of temperature and catalyst one can often force reactions in the desired direction. For example, in the preparation of linear polymers one should

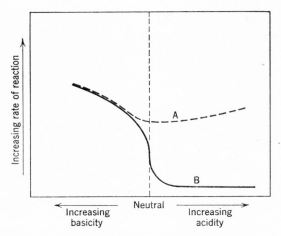

Fig. 3. General catalytic effects of strong acids and bases. Key: A, reaction with hydroxyl; B, reaction with urea and urethan; trimerization (37).

choose difunctional reactants, a low temperature so that biuret, allophanate, or trimerization reactions are not favored, and a catalyst that will promote only the chain-lengthening reaction.

As another example, if a basic catalyst that could promote allophanate, biuret, or trimer formation is present, the base should be neutralized with hydrogen chloride and the isocyanate–hydroxyl reaction can then be catalyzed with tertiary amine.

When a crosslinked polymer is desired it is usually obtained with best control by using a trifunctional reactant such as a triol, and catalyzing only the isocyanate–hydroxyl reaction. An added advantage of this technique is that the crosslink thus obtained has better thermal stability than does a biuret or allophanate crosslink.

Property–Structure Relationships

Although all polyurethans obviously contain urethan groups, they also contain a multiplicity of other groups, such as urea, ester, and ether, as well as aromatic rings, that affect the properties of the polymer. Furthermore, the reactants may be difunctional, trifunctional, or of even higher functionality. Because of the resulting wide range of possible structures, the study of property–structure relationships in polyurethans is highly complex.

General Considerations. The general principles of property–structure relationships in polymers may also be applied to polyurethans. These principles may be briefly summarized as follows:

1. *Molecular Weight.* Most properties of a polymer change as the molecular weight increases up to a limiting value and then remain relatively constant, as the molecular weight increases further. Properties that increase in value as the molecular weight increases include tensile strength, melting point, elongation, elasticity, and glass-transition temperature. Solubility generally decreases as the molecular weight approaches a limiting value. The polyurethans discussed in this article are generally of a sufficiently high molecular weight that they are little affected by this factor.

2. *Intermolecular Forces.* Intermolecular forces include hydrogen bonding, polarizability, dipole moments, and van der Waals forces. The bonds formed by these

forces are much weaker than primary chemical bonds and are affected by temperature and stress. The effectiveness of these forces is reduced by factors such as repulsion of like charges, bulky side chains or groups, poor geometric "fit" of attracting groups, moderate to high crosslink density, or the presence of plasticizers.

3. *Stiffness of Chain.* Chain units with limited rotation tend to stiffen polymer chains. Flexibility favors softness, low melting points, low glass-transition temperatures, and elasticity. Groups which restrict chain rotation tend to stiffen the polymer chain. Stiffening favors high melting point, hardness, and reduced elasticity. The ether group is very flexible, whereas aromatic rings tend to produce rigid polymers.

4. *Crystallization.* Crystallization of polymers is accomplished by strong intermolecular forces, close "fit" of polymer chains, linearity, and units in the chain which restrict rotation. An increase in crystallization generally leads to reduction in solubility, elasticity, elongation, and flexibility, and increase in tensile strength, melting point, and hardness.

5. *Crosslinking.* Increasing greatly the degree of crosslinking of a polymer increases the rigidity, softening point, and modulus of amorphous polymers, and reduces elongation and swelling by solvents. Largely crystalline polymers may be affected differently by small increases in crosslinking. A few crosslinks may reduce crystallinity by reducing chain orientation, changing a high-melting, hard, dense crystalline polymer to a more elastic, softer, amorphous material. Further crosslinking may then have the effects described above for amorphous polymers.

Effects of Various Structural Units. The relative contribution of groups commonly found in polyurethans to the intermolecular forces of the polymer is indicated by the molar cohesive energy (Table 13) (40). The melting point of polyureas

Table 13. Molar Cohesive Energy of Organic Groups

Group	Cohesive energy, kcal/mole
—CH$_2$— (hydrocarbon)	0.68
—O— (ether)	1.00
—COO— (ester)	2.90
—C$_6$H$_4$— (aromatic)	3.90
—CONH— (amide)	8.50
—OCONH— (urethan)	8.74

is higher than that of urethan or amide polymers with the same number of chain atoms in a repeating unit (41), indicating that the cohesive energy of the urea group is probably even greater than that of the urethan.

The relatively small effect of ester groups is evidently due to the flexible character of the C–O–C group in the ester, which tends to offset the moderately strong cohesive energy of the ester group itself. However, in poly(ester–urethans) or poly(ester–ureas), which contain strong hydrogen-donor groups, the ester group can be expected to participate much more in hydrogen bonding; thus, an increase in ester group concentration may be expected to increase the strength of the polymer aggregate. The effects of flexible ether and thioether groups on melting points of polyurethans are given in Table 14 (1). Rigid aromatic rings generally influence polymer properties in a manner opposite to that of the ether group, as illustrated by the melting points listed in Table 15.

Table 14. Influence of Ether Groups on Melting Points of Polyurethans[a]

Glycol	Polyurethan mp, °C
$HO(CH_2)_5OH$	151
$HO(CH_2)_2O(CH_2)_2OH$	120
$HO(CH_2)_2S(CH_2)_2OH$	129–134
$HO(CH_2)_9OH$	147
$HO(CH_2)_4O(CH_2)_4OH$	124
$HO(CH_2)_4S(CH_2)_4OH$	120–125

[a] Polymers from hexamethylene diisocyanate and glycol indicated.

Table 15. Effect of Aromatic Rings on Melting Points of Polyurethans

Urethan components		Mp, °C
Diisocyanate	Glycol	
$OCN(CH_2)_8NCO$	HOCH₂—⟨benzene⟩—CH₂OH	168
$OCN(CH_2)_8NCO$	$HO(CH_2)_6OH$	153
OCN—⟨benzene⟩—NCO	$HO(CH_2)_6OH$	230
$OCN(CH_2)_4NCO$	$HO(CH_2)_6OH$	180

Elastomers. Much of the study of property–structure relationships in polyurethans has been done on urethan elastomers. These are usually formed by reacting a diisocyanate, a linear long-chain diol, such as a polyester or polyether, and a low-molecular-weight chain extender such as a glycol or diamine.

Poly(ester–Urethans). Results of a study of the effects of isocyanate structure on poly(ester–urethans) are shown in Table 16 (42). Large and rigid aromatic rings, symmetrical structure, and absence of methyl substituents favor high modulus, tear strength, and hardness.

Table 16. Effect of Diisocyanate Structure on Physical Properties of Poly(ester–Urethan) Elastomers[a] (42)

Diiso-cyanate[b]	Tensile strength, psi	Elongation		Modulus at 300%, psi	Tear strength,[c] lb/in.	Hardness (Shore B)
		%	Set, %			
NDI	4300	500	85	3000	200	80
p-PDI	6400	600	25	2300	300	72
TDI	4600	600	1	350	150	40
MDI	7900	600	10	1600	270	61
DMDI	5300	500	0	600	40	47
DPDI	3500	700	10	300	90	56
TODI	4000	400	10	2300	180	70

[a] 0.1 equivalents poly(ethylene adipate), mol wt 2000; 0.32 equivalents diisocyanate; 0.2 equivalents 1,4-butanediol chain extender.

[b] NDI, 1,5-naphthylene diisocyanate; p-PDI, p-phenylene diisocyanate; TDI, 2,4-tolylene diisocyanate; MDI, 4,4′-diphenylmethane diisocyanate; DMDI, 3,3′-dimethyl-4,4′-diphenylmethane diisocyanate; DPDI, 4,4′-diphenylisopropylidene diisocyanate; TODI, 3,3′-dimethyl-4,4′-diphenyl diisocyanate.

[c] Split sample, Federal Test Method FTMS-601/M4221.

Table 17. Effect of Polyester Structure on Physical Properties of Cast Urethan Elastomers[a] (42)

Polyester	Tensile strength, psi	Elongation %	Set, %	Modulus at 300%, psi	Tear strength,[b] lb/in.	Hardness (Shore B)
poly(ethylene adipate)	6900	590	15	1550	240	60
poly(1,4-butylene adipate)	6000	510	15	1900	280	70
poly(1,5-pentylene adipate)	6300	450	10	1800	60	60
poly(1,3-butylene adipate)	3200	520	15	1100	100	58
poly(ethylene succinate)	6800	420	40	3200	200	75
poly(2,3-butylene succinate)	3500	380	105	c	520[d]	85

[a] Prepared from diphenylmethane diisocyanate, 1,4-butanediol extender, and the polyester indicated.

[b] Split sample, Federal Test Method FTMS-601/M4221.

[c] Exhibited cold drawing.

[d] Because of cold drawing this result is probably not a true tear value.

Greater changes in the properties of urethan elastomers result from varying the structure of the major component of the polymer, the polyester (Table 17) (42). Methyl substituents reduce tensile strength and 300% modulus, which are influenced more by the presence of side chains in the glycol than by ester group separation. Tear strength is apparently influenced by both the presence of methyl substituents and ester group separation since it is significantly lower in elastomers made both with poly(1,5-pentylene adipate) and with poly(1,3-butylene adipate) than in that made with poly(ethylene adipate). Effects on temperature characteristics of the polyester

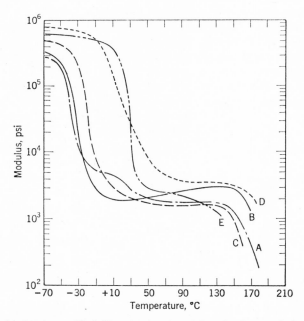

Fig. 4. Effect of polyester on Clash-Berg torsional modulus of polyester-derived polyurethan elastomers (42). System: polyester, diphenylmethane diisocyanate, 1,4-butanediol. Polyesters: A, poly(1,4-butylene adipate); B, poly(1,5-pentylene adipate); C, poly(ethylene adipate); D, poly-(2,3-butylene succinate); E, poly(ethylene succinate).

component are indicated in Figure 4. Stiffening at low temperatures varies widely and appears to be dependent upon the concentration of ester groups on the polyester; for example, the glass-transition temperatures of polymers prepared with poly(1,4-butylene adipate) and poly(1,5-pentylene adipate) are significantly lower than that of the poly(ethylene adipate) elastomer. Reducing the ester group concentration improves the low-temperature flexibility and reduces the tear strength. An increase of ester group concentration reduces flexibility at low temperatures, and at room temperatures results in higher hardness, higher modulus, and a marked increase in permanent elongation. These results may be attributed to increased van der Waals attractive forces.

Increasing the length of an aliphatic glycol chain extender through a homologous series has little effect on properties. An aromatic glycol results in higher tear modulus and hardness, changes that may be attributed to the greater rigidity of the aromatic group. The small effect of changes in glycol chain extender is undoubtedly also due to the relatively small amounts of extender used in comparison to the other reactants.

More drastic changes in the properties of the elastomers can be induced by varying the crosslink density. Normal crosslinking in the urethan elastomer is reported to occur by reaction of terminal isocyanate groups with urethan groups to form allophanate linkages. Chemical crosslinking can also be obtained and controlled by the substitution of a trifunctional hydroxy compound for the glycol extender. An initial decrease in modulus with increased crosslinking is contrary to results with hydrocarbon elastomers, where increased crosslinking usually results in increased modulus. In the case of poly(ester–urethan) elastomers it appears that increased chemical crosslinks reduce the ability of the chains to orient and, hence, the probability of forming hydrogen bonds and other intermolecular bonds. This phenomenon continues until the crosslink density is sufficiently high for the chemical crosslinks to begin to exert their own influence, at which point the modulus again increases. This observation tends to confirm the theory that a major portion of the strength of urethan elastomers is due to forces other than primary valence bonding.

The effect of crosslinking on thermal stability is indicated in Figure 5. When hydrogen bonding is greatest, as for curve A, the material is harder at room temperature. Its thermal stability is less than that of the polymers crosslinked with triol. Besides the fact that primary bonds are less readily disrupted by heat than are hydrogen bonds, the urethan linkages in the triol crosslink are probably more stable than the allophanate crosslink.

The effect of short branches in poly(ester–urethan) elastomers was studied, using polyesters prepared from a series of glycols, $RN(CH_2CH(CH_3)OH)_2$, and cured with tolylene diisocyanate (43). Although results were not conclusive, it was found that as R progressed from C_3 to C_6, the freezing temperature was lowered; however, when R was increased above C_6, the freezing temperature increased. Thus, the C_3 to C_6 branches may serve to hold chains apart, yet not be so large as to hinder rotation of polymer segments by their own bulk. On the other hand, the branch larger than C_6 may be so large that its own bulk is a hindrance to rotation, and the result is that T_g is raised.

The effect of isocyanate structure on the properties of poly(ester–urethans) with a diamine chain extender has also been investigated using prepolymers cured with 3,3'-dichloro-4,4'-diaminodiphenylmethane (44). These diamine-cured elastomers have higher modulus, higher tear strength, and frequently higher tensile strength and

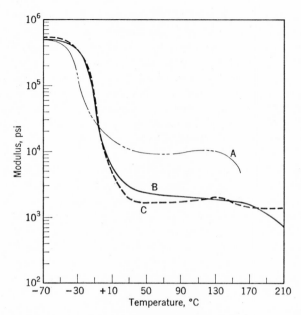

Fig. 5. The Clash-Berg torsional modulus of polyester-derived polyurethan elastomers (42). System: poly(ethylene adipate), diphenylmethane diisocyanate, Multrathane Extender XA, trimethylolpropane. Key: 0% trimethylolpropane in extender; B, 70% trimethylolpropane in extender; C, 100% trimethylolpropane in extender.

lower elongation than the corresponding glycol-cured elastomers. The presence of methyl substituents on the isocyanates appears to have somewhat less effect on properties than in the case of the glycol-cured elastomers.

Poly(ether–Urethans). Poly(ether–urethan) elastomers cured with aromatic diamines have structure–property relationships very similar to those of poly(ester–urethan) systems (45,46). Increases in crosslinking result in reduced modulus, elongation, compression set, and tear strength, as in the polyester elastomers. In comparing the effects of various aromatic diamines it is found that the ortho-methyl and chlorine groups in the amines do not cause a reduction of property values as do ortho methyl groups on the isocyanate groups in glycol-cured poly(ester–urethans).

Foams. Studies have also been made of relationships between the molecular structure and mechanical properties of polyurethan foams. Although the systems are more complex than those found in the elastomers, certain generalities can be made. Properties of foam prepared from poly(oxypropylene) glycols and triols with tolylene diisocyanate (80:20 isomer ratio) by the prepolymer technique are summarized in Table 18 (47). Tensile strength generally decreases as molecular weight per branch point increases although the exact relationship is obscured by the varying aromatic and urethan contents. Clash-Berg torsional stiffness curves are shown in Figure 6 for the foams prepared from polyether triols. The results show the combined effects of reducing the degree of crosslinking, aromatic content, and urethan content as noted in Table 18. Table 19 shows the results of an effort to separate the effects of M_c and of urethan and aromatic content. For foams with similar values of M_c (about 3375–3385) but higher aromatic and urethan contents, the modulus increased sharply at higher temperatures; of two foams with approximately equal urethan and aromatic

Table 18. Relationship between Structure and Physical Properties for Polyether-Based Polyurethan Foams (47)

	Triol	Triol	Triol	Triol	Triol	Triol	Triol and glycol	Triol and glycol
wt/branch pt	1650	2175	3375	3900	5175	6525[a]	10225	15185
aromatic, %	16.2	14.9	12.6	11.2	11.0	11.0	14.6	12.3
urethan, %	10.5	8.3	5.3	4.5	3.4	2.7	7.9	6.0
urea, %	7.9	7.6	6.6	6.6	6.7	6.6	6.6	7.0
density, lb/ft³	2.4	2.2	2.5	2.5	2.8	2.4	2.2	2.2
tensile strength, psi	30.0	21.0	18.0	20.0	22.0	15.0	24	21
tensile modulus at 100% elongation, %	100	16	12	10	7	6		
elongation, %		130	155	200	295	340	270	375
compression strength, psi								
25%	1.9	0.9	0.5	0.5	0.5	0.3	0.5	0.4
25%[b]	0.7	0.6	0.5	0.4	0.4	0.3	0.7	0.4
50%	2.8	1.3	0.7	0.7	0.7	0.4	0.7	0.4
75%	8.3	4.3	2.1	1.9	1.6	0.9	1.7	1.0
compression set (22 hr, 50% deflection, 70°C), %	7.0	3	4	3.0		14	3	6
rebound elasticity, %	16	15	44	49	42		25	
Yerzley resilience, % (at 30% deflection)	c	c	47	63	50		21	
point-load indentation,[d] sec	7100	130	1	1.0	1			
swell index, vol % (in dimethylacetamide)	145	170	237	240	350		390	

[a] Prepolymer prepared under different conditions.
[b] Measured after 1 min rest.
[c] Not applicable.
[d] Time required to recover from 90% indentation by a rod 1.13 in. in diameter held in indentation for 5 min.

Table 19. Effect of Structural Factors and Branching on Torsional Stiffness (Clash-Berg) (47)

wt/branch pt	3,375	3,385	10,225
aromatic (C_6H_3), %	12.6	14.7	14.6
urethan (—NHCOO—), %	5.3	8.1	7.9
temp, °C (at 50-psi modulus)	−37	−10	−35
temp, °C (at 100-psi modulus)	−40	−30	−45

contents, the one with the higher M_c value retained its flexibility at the lower temperature. Figure 7 shows the relationship between M_c (and, related to it, the urethan and aromatic contents), tensile modulus, elongation, and volume swell in dimethylacetamide.

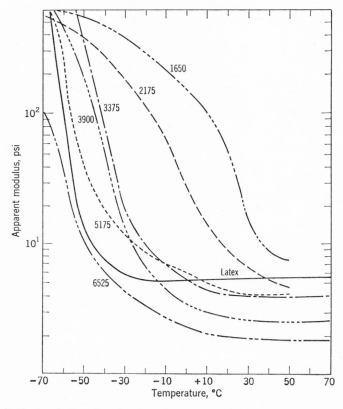

Fig. 6. Effect of calculated M_c values (as shown) and concurrent structural changes on Clash-Berg torsional modulus of polyether-derived polyurethan foams (47).

Flow, Creep, and Stress Relaxation. Flow, creep, and stress relaxation are important characteristics of most polymers, since they affect the permanence of the polymers' properties during their use. *Flow* is used here to indicate the permanent, nonrecoverable change in geometric relationship of polymer molecules as a result of stress or a change resulting from the permanent slippage of one or more molecules or segments past other molecules or segments. *Creep* is considered to be the reversible change in shape with time of the sample under a sustained load. *Stress relaxation* is

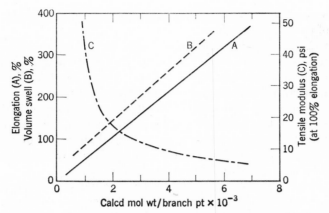

Fig. 7. Relation between calculated M_c and properties of polyether foams (48).

considered to be the reduction with time of the load required to give a sample a specified deformation.

Certain special features of polyurethans influence creep and stress relaxation. Several groups occurring in the structure of polyurethans include bonds that can break and reform as the result of stress. These include the biuret, allophanate, urea, and urethan groups. Breaking and reforming such bonds might require activation energies in the range of 20 to 50 kcal/mole. In studies of the stress relaxation of polyurethans, poly(ether–urethans) and poly(ester–urethans) give the same results; since the ether linkage is unlikely to break under the conditions used, it can be assumed that the ester link does not contribute to stress relaxation or creep (49,50). Rate studies point toward the allophanate and biuret groups as containing bonds easily broken at elevated temperatures. It has been shown that an equilibrium exists between urethan and allophanate with considerable disassociation at 106°C and above (51). Biurets react more readily with amines than do ureas. The relative instability of the allophanate and biuret groups contributes to their undesirability in the polymer. Urethan and urea groups are also known to disassociate, though less readily, at elevated temperatures.

It is also highly probable that the breaking of secondary bonds, hydrogen bonds, and bonds resulting from van der Waals' attractive forces, plays a significant role in creep and stress relaxation of urethan polymers. Since all urethan polymers contain a large percentage of groups participating in such attractions, a considerable part of the tensile and compression strength of these polymers may be assumed to be the result of such intermolecular attractions. Furthermore, such bonds are broken relatively easily, so that it is not surprising that an apparent softening of urethan foam occurs on flexing, and recovery of essentially the original strength occurs after an adequate rest period.

The compression set of polyurethans, and in particular, polyurethan foams, is an important measure of the usability of the polymer in many applications, such as cushioning, automobile seating, and packaging. The ability of a polymer to recover most of its original height after compression is a function of the ease (or difficulty) of forming primary chemical bonds during the compression period. In polyurethan foams this can be minimized by having as low a concentration as possible of biuret and allophanate groups and as few amine groups as possible, and by operating at a relatively low temperature. In addition, the formation of hydrogen bonds during compression will increase the compression set; this is a function of the sites available for such bonding and the fit of the polymer segment.

Several conditions influence the compression set of polyether-based polyurethan foams. Factors favoring bad set are: (a) large excess of water in foam formulation; (b) high amine catalyst concentration; (c) very active catalysts for the NCO/H₂O reaction; (d) slow curing (for example, room temperature for many prepolymer systems); and (e) high relative humidity while in the compressed state combined with recovery in the dry state. Factors favoring good set are: (a) mild catalyst for the NCO/H₂O reaction; (b) theoretical amount or only slight excess of water in the foam recipe; (c) low catalyst concentration; (d) fast cure; and (e) plasticizers permitting recovery of foam at 70°C after the compression period.

A similar explanation can be applied to the softening of urethan polymers during flexing. The extent of softening of the urethan polymer may be the direct function of the extent of the interchain hydrogen bond and van der Waals attractions. The greater the attractions the greater is the potential softening.

Foams

Urethan foams are normally prepared from diisocyanates and hydroxyl-terminated polyethers or polyesters. Linear or only slightly branched polymers are used to provide flexible foams, whereas more highly branched polymers produce rigid foams. Foaming is usually accomplished by including water in the system, the reaction between isocyanate and water providing carbon dioxide for foaming. For rigid foams, a low-boiling liquid such as trichlorofluoromethane has been used as a blowing agent. Appropriate catalysts and stabilizers control the foam formation and cure. See also Foamed plastics.

Although much foam production is carried out by suitably mixing all ingredients simultaneously (the "one-shot process"), the reactions involved may be more readily illustrated if they are considered in steps (the "prepolymer process"). The first reaction is that between excess diisocyanate and the hydroxyl-containing polymer (eq. 19a). This isocyanate-terminated prepolymer and excess diisocyanate are then foamed by reaction with water (eq. 19b). Thus the polymer contains the polyester or

$$4\,R(NCO)_2 + HO\text{---}OH \rightarrow OCN\text{---}R\text{---}NHCOO\text{---}OOCNH\text{---}R\text{---}NCO + 2\,R(NCO)_2 \quad (19a)$$
$$\mathbf{(8)}$$

$$\mathbf{(8)} \xrightarrow{\text{H}_2\text{O}} \text{---}OOCNHR\text{---}NHCONH\text{---}RNHCONHR\text{---}NHCONHR\text{---} + CO_2 \quad (19b)$$

polyether chain, urethan groups, aromatic rings, and many urea groups. At least trace amounts of free ---NH₂ end groups are present, and with the "one-shot" method traces of free ---OH end groups may also be present.

The balance of reactants may affect the polymer structure and the properties of the foam. For example, the density is usually controlled by regulating the water content; low-density foams require more water than higher-density foams. The isocyanate content is increased proportionately, so that low-density foams contain a higher percentage of aromatic rings and urea groups than do higher-density foams. For flexible foams at approximately 2.0 lb/ft³ density, about 25% of the isocyanate groups combine with hydroxyl groups of the resin, whereas about 75% react with water.

FLEXIBLE FOAMS

The principal raw materials employed in the preparation of flexible foams are diisocyanates, polyfunctional polyols, blowing agents, catalysts, and surfactants. Modifiers, fillers, plasticizers, etc, may also be used to impart special properties.

Although various *diisocyanates* have been used in preparing flexible foams, for all practical purposes tolylene diisocyanate (TDI) is used almost exclusively. The 80:20 mixture of 2,4- and 2,6-isomers is most commonly used commercially in the United States.

Polyesters (qv) were originally used as *polyols* in making flexible foam, and are still used in special applications. They are primarily adipate esters of diols and triols, the triol concentration chosen depending upon the degree of branching desired. However, *polyethers* constitute by far the largest group of hydroxyl-terminated foam intermediates. The bulk of polyethers now used are adducts of propylene oxide to diols, triols, or alcohols of higher functionality. When primary terminal hydroxyl groups are desired, the poly(oxypropylene) derivatives are capped with ethylene oxide. In many instances polyether blends are now used for slab stock or for molded goods with specific properties.

Water is used in all commercial flexible urethan foam systems to produce carbon dioxide as the blowing agent. *Fluorochemicals*, such as trichlorofluoromethane are frequently used in addition to water when it is desired to reduce further the foam density.

The most frequently used *catalysts* for flexible foam systems are combinations of tertiary amines and organometallic compounds. Both the foaming process and the properties of the resulting foam are influenced by the type and amount of catalyst used. Catalysts are discussed in detail in the section Polyurethan formation.

Surfactants in foaming formulations impart stability of the foam system and regulate the cell size. Many different classes of surfactants have been employed, usually organic nonionic and ionic surfactants and silicones. See also Surfactants.

Preparation. There are three principal methods of flexible-foam preparation: (a) prepolymer; (b) semi- or quasi-prepolymer; and (c) one-shot. In the prepolymer process a polyether or polyester is reacted with an excess of diisocyanate prior to foaming to yield the isocyanate-terminated prepolymer. Foaming occurs by reaction of the prepolymer with water to form urea linkages and carbon dioxide, which acts as the blowing agent. Catalysts and surfactants regulate foaming rate and cell size, respectively (see also the section on Isocyanate Reactions). In the semi- or quasi-prepolymer method, a diisocyanate is reacted with a portion of the polyether or polyester, producing a low-molecular-weight polymer of relatively low viscosity, dissolved in a large excess of diisocyanate. Foaming is brought about by reacting this semi- or quasi-prepolymer with the remainder of the polyol and water in the presence of catalysts and surfactants. This method is used extensively for the preparation of rigid and semirigid foams but is seldom employed for the manufacture of flexible foams. In the "one-shot" process the various foam components are mixed simultaneously. The process, originally used in the preparation of polyester-based foams, is now the most widely used method for the manufacture of polyether-based, flexible polyurethan slab stock, and is being increasingly used in molding applications.

The "one-shot" polyether and polyester systems differ considerably and will be dealt with separately. A foam-producing unit consists of an accurate metering (pumping) unit and an efficient mixing unit capable of handling two or more components. A metering pump with an accuracy of $\pm 0.5\%$ is adequate. Gear pumps work well on higher-viscosity components, whereas piston pumps or high-pressure fuel-injection pumps (Bosch) work well on low-viscosity materials. Variable-speed drives are a necessity. The mixing unit is extremely important since uniformity in blending is necessary for proper foam growth and properties. A continuous mixer, operating for long periods of time, is usually of the low-shear type. Intermittent units provide high shear and to a degree, are self-cleaning (a tapered housing is of value here). Outlet nozzles can vary in size, control, and mixing time.

In commercial continuous-slab production, the mixed reactants flow from the mixing head, which traverses at right angles to a continuously moving conveyor, so that overlapping parallel lines of liquid are laid down. Conveyors range from 50 to 100 ft in length, travel up to 20 ft/min, and are inclined by some 5 to 10 degrees to prevent liquid from falling into the rising foam. The higher conveyor rates are used on low-viscosity polyether systems because of their greater flow tendency. The foam slabs are then treated with steam or radiant heat to cure the surface. Cutting of standard lengths facilitates handling of the blocks, which are then conveyed to the curing ovens. Standard slab production ranges from 100 to 200 lb/min, blocks being up to 80 in. in width and 2–3 ft in thickness.

When polyesters are used, inherently stiffer foams that are less resistant to hydrolysis on aging, as compared to polyether foam, are produced. Although polyester foam has been largely replaced by the more versatile polyether products, special applications exist in the United States and broader use is found in Europe.

In one-shot polyester systems, the amount of tolylene diisocyanate necessary to develop good physical properties, and especially resistance to aging under conditions of high humidity, is at least 100% of the equivalent of the theoretical amount required for the hydroxyl groups present in the polyester and the water. In actual practice, however, a slight excess of 3–10% is preferred. The amount of water used largely determines the foam density: as the amount of water is increased with a corresponding increase in tolylene diisocyanate, the density decreases. If the water content is increased without an increase in the tolylene diisocyanate, foams may be obtained having coarse cells, harsh texture, as well as lower tensile and tear strengths and increased compression set. The addition of other blowing agents, such as fluorochemicals, will result in decreased density and in softening of the foam.

Tertiary amines are the most commonly used catalysts with polyester foams. Some of those which are commercially important are N-ethylmorpholine, N-cocomorpholine (derived from coconut amine), and Armeen DM16D (Armour Chemical Company). N-Ethylmorpholine can act as the sole catalyst whereas the other two are usually employed in combination with N-ethylmorpholine. Increasing catalyst concentration will, of course, increase the reactivity of systems, speed the cure, and tend to decrease the cell size, density, and compression modulus. However, an excess of catalyst may cause shrinkage and splits, and too little catalyst will also produce splits.

Emulsifiers or surfactants are used to promote more intimate mixing of the foam components, thus ensuring greater uniformity of the cell size. In addition, they may also serve to stabilize the foam and tend to reduce shrinkage and help to produce finer cells when used in the correct proportion. Typical emulsifiers are Witco 77-86 (Witco Chemical Company), Emulphor El-719 (GAF Corporation), and A3, A7, and A9 (Mobay Chemical Company). An insufficient amount of additive may lead to voids or splits owing to fast surface cure, whereas an excess will prolong the cure and promote shrinkage. Mineral oils in concentrations of 0.05–0.20 per hundred of resin (phr) may be utilized to increase the cell size and reduce shrinkage in certain systems.

Preparation of One-Shot Polyether-Based Foams. Commercially available polyether polyols range, generally, in molecular weight from 400 to 6000. The most widely used polyethers for flexible foams are triols having a molecular weight of about 3000 and having predominately secondary hydroxyl groups. Increasing the primary hydroxyl content by capping the polyethers with ethylene oxide increases the rate of cure and may increase slightly the compression modulus, tensile strength, tear strength, and elongation.

In polyether foams, 3–7 % excess of tolylene diisocyanate is used. If the excess is increased, the compression modulus goes up but there is also a tendency toward formation of harsh, boardy foams with a greater proportion of the large, enclosed cells. Tensile strength and resilience also increase whereas elongation decreases. When less than equivalent amounts of tolylene diisocyanate are used, the foam has little strength, poor resilience, may contain splits, and is difficult to reproduce.

Water in the formulation serves not only to furnish carbon dioxide to form the cells but also causes formation of primary amine, which further reacts with isocyanate to form urea linkages in the urethan polymer. This urea segment imparts higher load-bearing properties. As in the case of polyester foams, trichlorofluoromethane can also be used as an additional blowing agent to give added softness or to control density. Generally, as the amount of fluorochemical is increased, the silicone and tin catalyst content also has to be increased.

Since the majority of flexible foams are produced from polyols which contain secondary hydroxyl groups, a more active catalyst system is required than is found in the one-shot polyester foams. Organometallic compounds, such as *organic tin compounds*, in combination with tertiary amines are generally employed. Preferred catalysts for the one-shot polyether foam production are tertiary amines combined

Table 20. High-Density Foam Slab Formulations

Composition			
polyether triol, 3000 mol wt	100	100	100
silicone SF-1034	1.25	1.25	1.25
triethylenediamine	0.15	0.15	0.15
stannous octoate	0.25	0.25	0.25
water	2.25	2.25	2.25
TDI,[a] 80:20	32.2	35.3	38.3
TDI index	105	115	125
Properties			
density, lb/ft³	2.44	2.41	2.40
tensile strength, psi	17.5	16.9	14.9
elongation, %	248	163	107
tear strength, lb/in.	2.4	1.6	1.1
resilience, %	51	55	53
compression set, %			
(90% deflection, 22 hr, 70°C)	5.1	4.5	5.5
indentation load, lb/50 in.²			
2 in., 25% deflection	32.0	39.9	41.1
65% deflection	65.2	77.9	83.3
4 in., 25% deflection	40.9	50.0	54.1
65% deflection	85.0	101.6	112.6
sag factor (2 in.)	2.04	1.95	2.03

[a] Tolylene diisocyanate.

with organic tin compounds, such as stannous octoate. Tin compounds containing alkyl groups attached directly to the tin, such as dibutyltin dioctoate, promote oxidation (or degradation) of foam at temperatures above 120°C. Since the exotherm in the foam block usually exceeds 120°C, these catalysts cannot be used effectively unless an antioxidant is included. The tin catalyst promotes the isocyanate–hydroxyl reaction primarily, whereas the tertiary amine catalyst favors the isocyanate–water reaction. Thus, one can obtain a good balance between urethan formation and the blowing reaction.

Silicone surfactants are extremely important for the successful preparation of one-shot polyether flexible foams. Silicones initially used for one-shot polyether foams, and still the primary materials, are the alkylsilicone–polyoxyalkylene copolymers.

Table 21. Typical Medium-Density Slab Foam Formulation and Properties (52)

Composition	
polyether triol, 3000 mol wt	100
silicone L-520 or SF-1034	1.0
N-ethylmorpholine	0,4
triethylenediamine	0.05
stannous octoate	0.45
water	3.3
TDI, 80:20	42.5
TDI index	104
Properties	
density, lb/ft³	1.80
tensile strength, psi	19.9
elongation, %	258
tear resistance, lb/in.	3.3
resilience, %	48.5
compression set, % (method B,	
90% deflection, 22 hr, 70°C)	3.6
compression load, psi	
25% deflection	0.75
65% deflection	1.17
indentation load, lb	
25% deflection, no rest	36.2
1 min rest	32.2
65% deflection, no rest	59.9
1 min rest	54.7

They contain silicone–oxygen–carbon linkages and are susceptible to hydrolysis. The primary role of the silicone surfactant is to lower surface tension and provide cell-wall resilience. Resilient films prevent the collapse of the foam during rise and continue to stabilize it until the chain extension and crosslinking reactions have progressed sufficiently for the foam to be *self-supporting*. A secondary, but nevertheless important role of the silicone surfactant is cell-size control. The silicone is either metered separately, in combination with a polyol, or added as a water–amine–silicone mixture. It can also be incorporated in the fluorochemical. Formulations and properties of one-shot polyether foam are as varied as the many types of components that can be used to make them. Some typical formulations with properties are shown in Tables 20–24.

Table 22. Low-Density, Soft One-Shot Polyether–Urethan Foam Formulation (53)

polyether triol, 3000 mol wt	100
trichlorofluoromethane	17.7
stannous oleate	1.75
N-ethylmorpholine	0.5–1.0
silicone copolymer L-520	2.0
water	3.6
tolylene diisocyanate (TDI) (80:20)	42.5

Table 23. High-Load-Bearing Polyether Foam Formulation (54)

Composition

polyether triol, 3000 mol wt	85.0
Quadrol[a]	15.0
TDI, 80:20	47.5
water	2.3
silicone L-520 or SF-1034	1.5
N,N'-dimethylpiperazine	0.9
TDI index	100

Properties

density, lb/ft^3	2.3
tensile strength, psi	18.7
elongation, %	65
compression set, % (50% deflection, 22 hr, 70°C)	7.9
compression deflection, psi	
25%	1.31
50%	1.80
75%	4.80

[a] Trademark of Wyandotte Chemicals Corp. for amino-containing tetrol.

Molding. One of the rapidly growing areas of foam technology is molding of flexible and semiflexible foam to form, eg, such items as automotive safety pads, sunvisors, arm rests, and bucket seats. Molded parts based on tolylene diisocyanate can be produced either by the one-shot or the prepolymer method. Although the one-shot method was preferred in the past, today's production is largely by the prepolymer method because of better reproducibility.

In 1967, one-shot foaming systems based on polymeric isocyanates (4,4'-diphenylmethane diisocyanate analogs with a functionality between two and three) and new polyols with higher molecular weights (from 3000 to 6500 and capped with primary hydroxyl groups) were introduced for one-shot molding. This system has the following advantages:

1. A wide range of properties can be obtained, ranging from flexible to rigid foams, with a minimum of compounding in the formulation.

2. The polymeric semiflexible foams are essentially cold cured and require no external heat. The molds are generally heated to about 100°F at the beginning of a day's production and the heat of the reaction keeps the molds warm throughout the day. By contrast, tolylene diisocyanate-based materials must be oven cured and in many cases must even be heat cured after removal from the mold.

3. The vapor pressure of the polymeric isocyanates is considerably lower than that of tolylene diisocyanate, thus minimizing one of the manufacturing difficulties

Table 24. Self-Extinguishing Polyether Foam (55)

Formulation

polyether triol, 3000 mol wt	100
stannous octoate	0.35
dibutyltin dilaurate	0.05
water	3.5
triethylenediamine	0.09
silicone L-520	1.6
fluorocarbon-11	9.0
TDI	43.5
Firemaster T23P[a]	15
TDI index	103

Properties

density, lb/ft^3	1.5
tensile strength, psi	12
elongation, %	200
tear strength, lb/in.	2.4
compression load, psi	
25% deflection	0.35
50% deflection	0.43
rebound resilience, %	43
(Goodyear-Healey pendulum	
at 75% penetration)	
compression set, %	>10
(90% deflection, 22 hr, 70°C)	
extinguishing time, sec	3.2
distance burned, in.	1.1

[a] Michigan Chemicals Corp.

during the molding operation. On the other hand, tolylene diisocyanate-based foams generally have higher elongation than polymeric isocyanate-based materials and lower raw material costs. However, polymeric isocyanate-based materials have generally shorter processing cycles and do not require the prepolymer step, so that the total difference in cost is small.

RIGID FOAMS

One of the early areas of research in polyurethans was in the field of rigid polyurethan foams, starting around 1937 at Farbenfabriken Bayer under the direction of Otto Bayer. This research resulted in the development of a number of rigid polyurethan foams for use in lightweight, high-strength sandwich materials for aircraft construction and as insulation in submarines and tanks. These early foams were based on polyesters with a relatively high acid number. The foams were formed as a result of a reaction of an isocyanate with the free carboxyl and residual water in the polyester, causing the formation of carbon dioxide. The resultant foam was a high polymeric material and a hard, insoluble foam. This initial work, reported in the United States in the late 1940s, led to the development of machinery suitable for producing rigid foam in the early 1950s.

From the early limited use of rigid foam in the aircraft industry, the market has grown to include applications in insulation for refrigerators (both household and commercial), in construction, flotation, transportation, furniture, etc. The raw materials used in rigid foam usually consist of a polyfunctional isocyanate, a polyhydroxy material, a blowing agent, a catalyst, and a cell-size regulator.

Although the original rigid foams were based on the distilled purified tolylene diisocyanate (TDI), today's production is based primarily either on what is described as a "crude" (undistilled) TDI and on so-called "polymeric" isocyanates. This latter group of isocyanates consists of undistilled products derived from aniline–formaldehyde condensation; they are chemically related to diphenylmethane diisocyanate (MDI). Idealized structures of these materials are shown in (9) and (10).

(9)   (10)

The use of "polymeric" MDI offers several advantages, the most important being the relatively low vapor hazard associated with this product, since its vapor pressure is less than 5 torr at 200°C. In addition, "polomeric" MDI imparts improved heat and flame resistance in rigid foams when compared to TDI.

The most commonly used polyols are the polyethers, such as those based on propylene oxide adducts to various polyfunctional alcohols or amines such as glycerol, pentaerythritol, trimethylolpropane, sorbitol, alpha-methylglucoside, sucrose, and ethylenediamine. These polyethers have predominately secondary hydroxyl groups unless they have been deliberately terminated with ethylene oxide.

Polyesters have also been used as the polyol component for rigid foam. However, owing to their higher cost, they are gradually being replaced by the polyethers.

Halogenated hydrocarbons, such as trichlorofluoromethanes, are generally used as blowing agents in rigid foam. Water can also be used, utilizing its reaction with isocyanate to form carbon dioxide as the blowing agent. Other lower-boiling solvents, such as dichlorodifluoromethane, which is especially useful in the "frothing process," can also be used.

In the halogenated solvent-blowing systems, amine catalysts such as diaza[2.2.2]-bicyclooctane (triethylenediamine) or 1,2,4-trimethylpiperazine and tin compounds (such as dibutyltin dilaurate or stannous octoate) are used. In water-blown systems, the usual catalysts are tertiary amines, such as N-ethylmorpholine, triethylamine, or substituted ethanolamines.

The most common surfactants used in rigid foams are water-soluble silicones, such as polyoxyalkylene–polydimethylsiloxane block copolymers. These usually give extremely fine-celled uniform foams of high closed-cell content. For rigid foams used in construction, the need for flame retardance has been met by incorporating inorganic compounds (such as metal oxides, metal soaps, metal sulfates, phosphates, borates, titanium compounds, etc) or organic compounds containing phosphorus or chlorine.

Rigid foams can be made by either a semiprepolymer, prepolymer, or one-shot technique. The two methods most widely used at the present time are the semiprepolymer and one-shot methods. Most of the rigid foam is now produced either in the slab stock form, or by spraying or molding. In any of the three methods, a primary consideration must be the machinery used to produce the foam; it is similar to that used in producing flexible foam and consists of a metering unit and a mixing unit.

Continuous *slab stock* of rigid foams is produced either in the form of large buns, to minimize scrap loss, or as a composite laminate for uses such as roof decking. The slab stock is cut to the desired thickness and is then generally used by adhering it to another surface. Facings which have been used include plywood, gypsum board, cement, asbestos board, hardboard, or aluminum. The facing used in the continuous laminating procedure is generally an impregnated paper stock, although recently equipment has been developed to laminate continuously with metallic skins. Spraying is usually done on a job site with equipment which is lightweight, inexpensive, and compact.

An extremely important and rapidly growing technique of utilizing rigid foam is by molding. The size of parts being currently molded range from small decorative furniture items to panels as large as 4 ft × 18 ft with thicknesses ranging up to 4 in. A modified molding technique is used in what is called *foaming-in-place*, as exemplified by the void filling of masonry cavity walls.

Properties. Some properties common to both rigid polyether- and polyester-based polyurethan foams are: (a) excellent thermal insulation, especially when blown with fluorochemicals; (b) combination of high strength with light weight; (c) good

Table 25. Comparison of Properties of Various Insulating Materials (56)

	Cork	Polystyrene bead foam	Glass fibers	CO_2-blown rigid polyure-than	Solvent-blown rigid polyure-than
density, lb/ft³	6.5–7.5	1.2	0.5–12	2.0	2.0
compressive strength, psi	high	12	low	48	48
water absorption, lb/ft²	low	0.082–0.106	high	0.03	0.03
K factor, Btu/(hr)(ft²)(°F/in.)	0.26	0.25	0.23–0.30	0.25	0.11

Table 26. Properties of Typical Polyether-Based Foam Blown with Trichlorofluoromethane (57)

density (open blow), lb/ft³	2.0
density (molded), lb/ft³	2.5
closed cells, %	93
compressive strength (10% deflection), psi	32
tensile strength, psi	30
flexural strength, psi	55
moisture vapor transmission, perm	1.7
K factor, Btu/(hr)(ft²)(°F/in.)	
initial	0.110
aged at 60°C until equilibrium	0.160
dimensional stability, % swelling	
−32 to 93°C, linear	less than 5
95% rh, 66°C, linear	less than 5

Table 27. Variation of Properties of Rigid Polyether-Based Foams with Trichlorofluoromethane Content (56)

trichlorofluoromethane,						
% of total weight	41	25	19.5	15	12	4.0
density, lb/ft³	0.83	1.1	1.3	1.7	2.1	7.6
compressive strength						
at yield point, psi	5.73	12.50	23.4	34.4	48.4	200
deflection at yield						
point, %	12.7	9.0	9.0	7.5	9.0	
closed cells, %	85–90	85–90	85–90	85–90	85–90	85–90
K factor,						
Btu/(hr)(ft²)(°F/in.)	0.18	0.16	0.14	0.13	0.11	0.145

heat-resistance; (d) good energy-absorption, for sound deadening or vibration damping; and (e) excellent adhesion to wood, metal, glass, ceramic, and fibers. Properties of rigid polyurethan foams as compared with common insulating materials are shown in Table 25. Typical properties of a solvent-blown polyether foam are illustrated in Table 26. The properties of a specific rigid foam system can be varied by changing the density of the foam. Table 27 illustrates the wide range of properties that can be obtained. Most properties are dependent, at least to some degree, on the foam density.

APPLICATIONS

One of the large outlets for flexible polyurethan foam is for furniture cushions. Lighter weight, greater strength, and ease in fabrication as compared to latex foam are some of the deciding factors in its selection. In addition, the fabrication and application of slab stock foam is easier and faster than the use of animal hair or other filling materials. Improved molding techniques of flexible foam are responsible for its utilization in furniture with unusual shapes. In addition to flexible foam, molded rigid foam has in the past years made great inroads into the furniture industry. A current shortage of select hardwood and a scarcity of well-trained wood workers has forced the furniture industry to look for replacement materials. Among other candidate materials, rigid polyurethan foam has found utilization initially in decorative parts, mirror frames, chair shells, and the like. This is perhaps the most rapidly growing use of polyurethan foam today.

Nearly one million polyurethan foam mattresses are being produced in the United States per year. They have found acceptance because of their superior durability, freedom from odor, nonallergenic properties, ease of cleaning, resistance to drycleaning solvents, oils, and perspiration, and the fact that they are only one-fourth the weight of a comparable innerspring and one-half to one-third the weight of a latex foam mattress.

Another rapidly increasing use of flexible foam is in the automotive industry for seat cushioning, instrument-panel trim, safety pads, arm rests, floor mats, sunvisors, underlays, roof insulation, weather stripping, air filters, etc. It was estimated that the 1970 model cars would use an average of 19 lb of foam per automobile. One of the first uses of flexible polyurethan foam was in seating for aircraft where its light weight is of special importance.

In 1961, the use of flexible foam as a bonding material for fabric, primarily in the apparel industry, began. Polyester foam is used for this application, wherein the foam is bonded to the fabric by the flame-lamination technique. In this process, the surface of foam is heated until the surface layer fuses and becomes soft and tacky. It is then bonded under pressure to the fabric. Adhesives can also be used for bonding the foam to the fabric. A foam lining for the garments makes the fabric dimensionally stable and provides high insulating qualities. Other advantages include excellent hand and drape and outstanding crease and wrinkle resistance.

Another area of utility for flexible polyurethan foams is in carpet underlay; they provide cushioning, are nonskidding, and do not mat down. They also impart a luxurious feel, even to low-cost carpeting.

One of the major uses of rigid polyurethan foam is in home refrigerators. All major manufacturers are currently using rigid urethan foam as insulation in at least a portion of their total line. Because of the superior insulating characteristics of the fluorochemical-blown foams, the manufacturers can build refrigerators with thinner insulation and, therefore, larger inside capacity. All types of refrigerated trucks such as milk trucks, ice cream trucks, and trailers have been insulated with rigid polyurethan foams. Besides having particularly good insulating properties, rigid foams contribute to the structural strength of the body of the truck, have low moisture pickup, and can withstand solvents such as gasoline and temperatures up to 100°C.

The potentially biggest market for polyurethan foam insulation is in the building industry. The areas of utilization in this field include curtain-wall construction, preformed rigid panels, spray-applied wall construction, and roofing insulation (either sprayed or in preformed panels). The total market potential here includes residential homes and commercial and industrial buildings, such as large refrigerated warehouses.

The use of rigid polyurethan foam as flotation equipment is growing. Most boats built today utilize rigid foam in some manner to help support the boat in the water. Larger ships have used rigid polyurethan foam as void fillers; it is also used in lifeboats and refrigerator ships.

Elastomers

Solid polyurethan elastomers were developed in Europe in the early 1940s. The early products were what are called *millable gums* and were prepared from saturated polyesters, diisocyanates, and a chain extender such as water, glycol, amine, or amino alcohol. Other similar types of products have been introduced over the succeeding years with limited success owing primarily to their high cost.

During the early 1950s a method of liquid *casting* polyurethan elastomers was developed by Mueller (58). Similar systems were introduced in the United States during the mid-1950s, based both on polyesters and polyethers. During the late 1950s *thermoplastic polyurethans* were introduced (59,60). As with other polyurethans, a wide range of properties can be obtained with the polyurethan elastomers, eg, tensile strengths can range from 4000 to 8000 psi, and 300% moduli range from 1400 to 3200 psi. The hardness of polyurethan elastomers can be varied from a low of about 10 Shore A to about 75–80 Shore D. Over this range of hardness, the physical properties of the polyurethans are, in general, superior to those of the more conventional elastomeric compounds. Polyurethan elastomers also have remarkable resistance to most solvents, including gasoline, aliphatic hydrocarbons, and, to some degree, arodegree, aromatic hydrocarbons. See also Vol. 7, pp. 692–695.

Cast Elastomers. Although it is possible to prepare polyurethan elastomers by a one-shot technique, the commercial method today utilizes a prepolymer technique.

The prepolymer is formed by reacting a diisocyanate with hydroxyl-terminated polyester or polyether to form an isocyanate-terminated prepolymer (eq. 20). This

$$\text{OCN—R—NCO} + \text{HO—R'—OH} \rightarrow \text{OCN—R—}\overset{\text{H}}{\underset{}{\text{N}}}\text{—}\overset{\text{O}}{\underset{}{\text{C}}}\text{—OR'O—}\overset{\text{O}}{\underset{}{\text{C}}}\text{—}\overset{\text{H}}{\underset{}{\text{N}}}\text{—R—NCO} \qquad (20)$$

prepolymer is then further reacted (chain extended) with an active-hydrogen compound such as glycol, diamine, or trifunctional polyol (such as trimethylolpropane). *Chain extension* using the glycol takes place with the formation of urethan groups, as shown in equation 21. When diamines are used as the chain extender, substituted

$$\text{OCN—prepolymer—NCO} + \text{HO—R'—OH} \rightarrow \begin{array}{l} \text{OCN—prepolymer—}\overset{\text{H}}{\underset{}{\text{N}}}\text{—}\overset{\text{O}}{\underset{}{\text{C}}}\text{—O} \\ \hspace{5cm} \diagdown \\ \hspace{5.5cm} \text{R'} \quad (21) \\ \hspace{5cm} \diagup \\ \text{OCN—prepolymer—}\underset{\text{H}}{\overset{}{\text{N}}}\text{—}\underset{\text{O}}{\overset{}{\text{C}}}\text{—O} \end{array}$$

urea linkages are formed (eq. 22). In prepolymer formation an excess of isocyanate is used. In the chain-extension step, the active-hydrogen compound can be used in a ratio such that either an isocyanate-rich or hydroxyl-rich final product is obtained. Generally speaking, in the overall balance, an excess of isocyanate is used.

$$\text{OCN—prepolymer—NCO} + \text{H}_2\text{N—R'—NH}_2 \rightarrow \begin{array}{l} \text{OCN—prepolymer—}\overset{\text{H}}{\underset{}{\text{N}}}\text{—}\overset{\text{O}}{\underset{}{\text{C}}}\text{—}\overset{\text{H}}{\underset{}{\text{N}}} \\ \hspace{5.5cm} \diagdown \\ \hspace{6cm} \text{R'} \quad (22) \\ \hspace{5.5cm} \diagup \\ \text{OCN—prepolymer—}\underset{\text{H}}{\overset{}{\text{N}}}\text{—}\underset{\text{O}}{\overset{}{\text{C}}}\text{—}\underset{\text{H}}{\overset{}{\text{N}}} \end{array}$$

The above steps are usually carried out at an elevated temperature, approx 100°C. Following the chain-extension step, the liquid material solidifies and is removed from a mold and placed in a curing oven at an elevated temperature for varying periods of time.

Millable Gums. The polyurethan millable gums can be made in a manner similar to the prepolymer technique for cast elastomers or by a one-shot technique. A storage-stable polymer is first prepared, usually with a hydroxyl component in slight excess. This gum stock may be cured by incorporating a polyisocyanate and heating under pressure. If the gums contain certain functional groups, such as unsaturated or methylene groups, they may be cured by peroxide or sulfur-containing agents.

Thermoplastic Polyurethans. The preparation of this material is very similar to that of the millable gum. However, the product has a higher molecular weight and is much less soluble than the millable gum. The material is processed on conventional thermoplastic-forming equipment, such as by injection molding or extrusion.

Raw Materials. The polyol component of polyurethan elastomers, as in the case of foams, can be either a polyester or a polyether. In either case, the polyols used generally are *linear*. Although elastomers can be made from nearly any linear polyether, those based on poly(1,4-oxybutylene) glycols exhibit the best overall range of properties. Most polyester-based elastomers are prepared with at least a portion of the carboxylic acid being adipic acid. The glycol portion can be either ethylene glycol,

1,3-butylene glycol, 1,4-butylene glycol, or 1,2-propylene glycol. Higher-molecular-weight glycols can be used for specific requirements.

The isocyanates generally preferred for polyurethan elastomers are tolylene diisocyanate, 4,4'-diphenylmethane diisocyanate, and 1,5-naphthylene diisocyanate. When tolylene diisocyanate is used, best results are obtained with the 2,4 isomer alone. In addition, because of the lower reactivity of the tolylene diisocyanate molecule, it is necessary to use an aromatic diamine, such as methylenebis(o-chloroaniline) as the chain extender. Formulations utilizing 4,4'-diphenylmethane diisocyanate and 1,5-naphthylene diisocyanate are generally chain extended with simple diols or triols or combinations of both. With 4,4'-diphenylmethane diisocyanate, it is possible to use up to 15% of an aromatic diamine in the chain extender and still maintain reasonable processing characteristics. The polyhydroxy compounds commonly used with these systems are 1,4-butanediol, bis(2-hydroxyethyl) ether of hydroquinone, 1,2,6-hexanetriol, or trimethylolpropane.

Properties. It is possible to obtain a wide range of physical properties in polyurethan elastomers by merely changing the formulation or the raw materials of the system. This type of compounding freedom does not exist for conventional elastomers, for which changes in properties are achieved in a limited manner by blending with plasticizers, wax, reinforcing agents, and fillers (see Rubber compounding). Although different raw materials can be used in formulations, polyurethan elastomers exhibit certain similarities in properties. This can be seen in Table 28, where polyether- and polyester-derived cast elastomers, millable gums, and thermoplastic poly-

Table 28. Typical Properties of

	Hardness (Shore)	Polyether type[b,c]			
		A 10–40	A 45–75	A 88–98	D 68–75
	ASTM				
ultimate tensile strength, psi	D 412	250–425	600–4500	3900–5000	4000–8000[h]
ultimate elongation, psi	D 412	425–1000	430–700	200–480	120–270[h]
100% modulus, psi	D 412			750–1950	3000–3700
300% modulus, psi	D 412	40–225	150–1100	2100–4400	
permanent set, % elongation set at break	D 412				
compression set (method B, 22 hr at 70°C)	D 395	2–34	7–40		
tear strength (Graves), lb-in.	D 624	20	50–225	475–750	
resilience, % rebound				42–48	
Yerzley	D 945		65–80	75[i]	

[a] Each hardness range covers several commercial grades.

[b] Most of these liquid cast types can also be compression molded.

[c] Typified by properties given here for Adiprene L (Du Pont).

[d] Typical ranges given are for: (1) formulations based on Multrathrane (Mobay Chemical Co.); (2) Disogrin (Disogrin Industries, Inc.); (3) Daycollan (American Latex Products Corp., Division of Dayton Rubber Co.); (4) Elastacast (Acushnet Industries, Inc.); (5) Neothane (Goodyear Tire & Rubber Co.); (6) Vulkollan (Farbenfabriken Bayer A.G.).

ester-derived polyurethans are compared. The variation within a system is illustrated by Tables 29 and 30.

The hardness of polyurethan elastomers extends over a wide range from Shore A 10 to Shore D 75. Most commercial applications, however, require a range from Shore A 60 to 100. Elongations on the order of 600–800% can be obtained in this hardness range. Tensile strength of the polyurethan elastomers generally ranges upward from 4000 psi, although elongation generally decreases as the hardness of the polymer is increased. The load-bearing capability of polyurethan elastomers is greater than that of other elastomers of comparable hardness and elasticity, and is a function of the hardness of a material, as shown in Figure 8. The tear strength of polyurethan elastomers is exceptionally high (300–500 lb/in.) in comparison with general-purpose elastomers, such as natural rubber or SBR (50–110 lb/in.).

The *abrasion resistance* of polyurethan elastomers is excellent. Excessive heat buildup must be avoided, since the tear strength of polyurethans is a function of the internal temperature of the rubber; polyurethan elastomers are not recommended for use above temperatures of 88–100°C.

Uses. A large application is in tires and sheets, eg, for solid tires for industrial lift trucks, where the elastomer's ability to carry high loads and its excellent oil and tear resistance are especially valuable. In addition, the power consumption is reduced because less energy is absorbed in flexing of the tires. Friction drives and belts are other important applications; low abrasion, high coefficient of friction, and a low noise level are the major advantages. Because of the excellent resistance to most oils,

Polyurethan Elastomers[a] (61)

Polyester type[b,d]				Millable gum		Thermoplastic
A 60–73	A 78–83	A 88–92	A 93–98	A 62–66[e]	A 70–95[f]	A 85[g]
4000–6000	4500–8000	4000–6500	4000–5500	3150–5500	3600–4500	6200–8000
550–650	475–700	475–600	450–550	450–500	315–450	600
250–600	475–925	1000–1700	1200–1600			
625–1650	1400–2500	1600–3200	2100–3000	1700–2475	2750–3250	1200–1500
0–50	3–15	20–50	20–50			
1–25	20–25	20–25	20–25	18–21	14–47	87
280–320	450–525	550–600	550–700	260–380		430
45–50	50–60	45–50	45–50	69		
				56–57		78

[e] Black-reinforced; range covers Adiprene C (Du Pont) and Genthane S (General Tire and Rubber Co.).

[f] Typical range of properties obtainable with resin reinforcement in Genthane S (General Tire and Rubber Co.).

[g] Estane 5740X1 (B. F. Goodrich Chemical Co.); polyester type.

[h] Crosshead speed of 1 in./min; ASTM method is 20 in./min.

[i] 75% value is for A 89 durometer stock; at hardness of A 95, stock is too hard for reading.

Table 29. Miscellaneous Properties of Polyurethan Elastomers (62)

	Compound designation				
	A	B	C	D	E
Shore durometer hardness	55A	65A	75A	85A	92A
100% modulus, psi	200	300	450	700	1100
tensile strength, psi	2500	3000	4500	4500	5000
elongation at break, %	650	430	430	440	450
specific gravity	1.06	1.10	1.10	1.10	1.10
thermal conductivity, Btu/(hr)(ft²)(°F/in.)	1.181	1.040	1.010	0.952	0.917
linear coefficient of thermal expansion, in./in. per °F					
−32 to +32°F (−36 to 0°C)	1.22×10^{-4}	1.40×10^{-4}	1.35×10^{-4}	1.27×10^{-4}	1.43×10^{-4}
32–75°F (0–24°C)	0.77×10^{-4}	1.00×10^{-4}	1.05×10^{-4}	1.07×10^{-4}	1.01×10^{-4}
75–212°F (24–100°C)	0.95×10^{-4}	1.19×10^{-4}	1.15×10^{-4}	1.10×10^{-4}	0.95×10^{-4}
212–302°F (100–150°C)	0.98×10^{-4}	1.08×10^{-4}	1.02×10^{-4}	0.97×10^{-4}	0.90×10^{-4}
linear shrinkage, %	1.7	2.6	2.1	1.8	1.0
compression set, % (method B)	7	10	15	22	26
compression set, % (method A)					9
solenoid brittle temp, °C	below −68 flexed	below −68 flexed	below −68 flexed	below −68 flexed	below −68 flexed
impact resistance, ft-lb/in.					
abrasion index, Bureau of Standards, %	37	100	145	170	200
tear strength, Graves, lb/in.	150	175	225	400	550
tear strength, split, lb/in.	10	12	25	45	70

[a] Adiprene, E. I. du Pont de Nemours & Co. Inc.

Table 30. Standard and Softer Elastomers from an MDI-Polyester-Derived Prepolymer[a] (63)

prepolymer, parts by wt	100	100	100	100	100
prepolymer temp, °C	110	110	110	110	110
1,4-butanediol, parts by wt	6.5	4.8	4.1	3.5	2.8
diethylene glycol, parts by wt		2.0	2.8	3.5	4.3
Properties					
hardness, Shore A	80–85	70–75	68–73	65–70	60–65
tensile strength, psi	6500–8000	5500–7000	5000–7000	4500–6500	3500–5500
tensile modulus, psi					
100% elongation	700–900	400–550	350–500	350–400	250–350
200% elongation	1000–1250	600–750	550–700	450–550	400–500
300% elongation	1800–2000	900–1100	750–1000	650–850	500–600
elongation, %	550–650	530–630	550–650	600–700	600–700
elongation set, %	10–20	5–15	5–15	5–15	5–15
tear strength,[b] lb/in.	200–300	200–260	165–210	150–200	135–200
solenoid brittleness, °C	−50	−50	−50	−55	−60

[a] MDI = methylenedi-*p*-phenylene diisocyanate.

[b] Split-sample method.

urethan elastomers are being used as seals, wiper rings, and valve seats of various types. Many parts are used in the oil industry, especially in drilling.

Polyurethan elastomers also find utilization in shock-absorbing and vibration-dampening applications. These types of applications include flexible coupling connectors, shock absorbers for automobiles, pneumatic machinery, earth-moving equipment, and farm machinery.

Polyurethan elastomers are also used for conveyor rollers, drive sprockets, cable jacketing, and various types of gears.

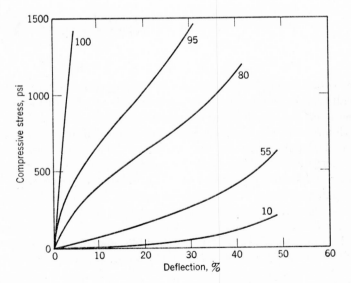

Fig. 8. Compression–deflection of cast polyether-derived polyurethane lastomers at various Shore A hardnesses (62).

Coatings

One of the earliest uses of polyurethans was in the field of coatings. It was found that by proper formulation of various resins and isocyanates, coatings could be produced that exhibit flexibility, impact resistance, toughness, and chemical and abrasion resistance.

The earliest, and still widely used, systems were *two-component* coatings, wherein an isocyanate reacts with an hydroxyl-bearing compound (eq. 23). By this reaction, polyurethans are formed when the starting materials are polyfunctional.

$$RNCO + HOR' \rightarrow RN\overset{H}{\underset{}{-}}\overset{O}{\underset{}{C}}-OR' \qquad (23)$$

The polyisocyanate used in this two-component system is derived from tolylene diisocyanate by reaction with a trihydric alcohol, eg, trimethylolpropane (eq. 24).

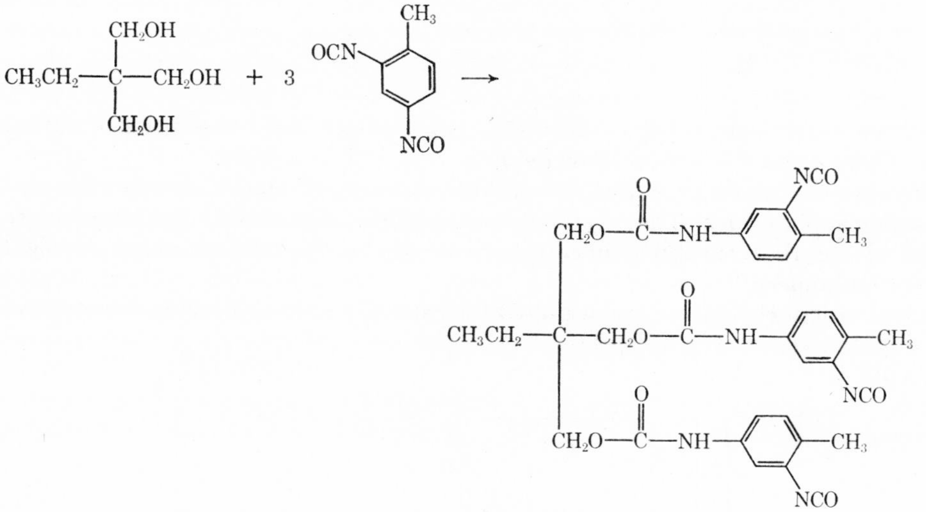

$$(24)$$

This product contains only trace amounts of tolylene diisocyanate and thus presents a reduced health hazard compared to the pure, volatile tolylene diisocyanate.

Polyfunctional polyesters based on adipic acid, phthalic anhydride, ethylene glycol, trimethylolpropane, etc, are generally used with the above polyisocyanate adduct, although branched polyethers can also be used. In general, the basic polyurethan structure–property relationships apply to polyurethan coatings. Since the polyisocyanate adduct is essentially the same in all formulations, the property variations are obtained through variations in the polyol structure. The most flexible, least chemically resistant, and softer coatings are obtained with linear or only slightly branched polyesters. As the degree of functionality (or branching) of the polyester increases, the coating film becomes tougher, harder, less flexible, and more resistant to chemicals. Abrasion resistance can be considered to be better with the harder films, when measured in the laboratory by such means as the Taber Abraser, but the overall characterization of abrasion resistance actually depends more on the end use and the substrate rather than only on chemical formulation.

In the late 1950s, *one-component polyurethan coatings* were developed (64). These systems are based mainly on stable isocyanate-terminated prepolymers obtained from tolylene diisocyanate and a polyfunctional polyether (eq. 25). Coatings such as these

$$(n + 1)\ OCN—R—NCO + n\ HO—R'—OH \rightarrow$$

$$OCN(RNH—\overset{\overset{\displaystyle O}{\|}}{C}—O—R'—O—\overset{\overset{\displaystyle O}{\|}}{C}—NH)_n RNCO \quad (25)$$

dry by utilizing the reaction of the free isocyanate groups with water or atmospheric moisture. They are said to be "moisture cured." This reaction proceeds through the unstable carbamic acid, with carbon dioxide being eliminated, to give primary amine groups, which then further react with isocyanate groups to form ureas (eqs. 26a and b).

$$R—N{=}C{=}O + H_2O \rightarrow R—NH_2 + CO_2 \uparrow \quad (26a)$$

$$R—NH_2 + R—N{=}C{=}O \rightarrow R—NH—\underset{\underset{\displaystyle O}{\|}}{C}—NH—R \quad (26b)$$

The same structure variables described in the two-component system above are valid with one-component coating systems. However, other polyisocyanates can be used in the preparation of the prepolymer. These include methylenedi-*p*-phenylene diisocyanate, triphenylmethane triisocyanate, and even the polyisocyanate adduct used in the two-component system. Polyethers are generally used as the resin portion although polyesters can also be used.

Another, and older type of one-component coating system, is the *"urethan oil"* or *"uralkyd,"* which makes use of isocyanate-modified drying oils. The urethan oils are the reaction products of diisocyanates (usually tolylene diisocyanate) with hydroxyl-containing drying-oil derivatives. Since they are generally formulated to have an NCO/OH ratio of one or lower, the main reaction responsible for final curing appears to be an oxidative polymerization through the double bonds of the drying-oil fatty acid portion of the polymer.

The urethan oils are prepared by alcoholysis of an unsaturated glyceride with a polyol such as trimethylolpropane (eq. 27). The ratio of glyceride to polyol controls the

$$2\ \underset{\overset{\displaystyle |}{CH_2OCOR}}{\overset{\overset{\displaystyle CH_2OCOR}{|}}{CHOCOR}} + R'(OH)_3 \rightarrow 2\ \underset{\overset{\displaystyle |}{CH_2OH}}{\overset{\overset{\displaystyle CH_2OCOR}{|}}{CHOCOR}} + R'\underset{\overset{\displaystyle |}{OH}}{\overset{\overset{\displaystyle OCOR}{|}}{—OCOR}} \quad (27)$$

average number of hydroxyl groups per molecule in the mixed alcoholysis products. The latter is then reacted with the diisocyanate, which serves to tie together the hydroxy esters, increasing the molecular weight and functionality (due to unsaturation) (eq. 28).

$$2\ \underset{\overset{\displaystyle |}{CH_2OH}}{\overset{\overset{\displaystyle CH_2OCOR}{|}}{CHOCOR}} + R'(NCO)_2 \rightarrow \underset{\overset{\displaystyle |}{CH_2—OCONH—R'—NHCOO—}}{\overset{\overset{\displaystyle CH_2OCOR}{|}}{CHOCOR}}\ \underset{\overset{\displaystyle |}{CH_2}}{\overset{\overset{\displaystyle CH_2OCOR}{|}}{CHOCOR}} \quad (28)$$

A third one-component coating system is the *"blocked" isocyanate,* which consists of a polyisocyanate reacted with a monofunctional hydroxy compound. The resultant product is unreactive towards hydroxyl compounds at room temperature but, at elevated temperatures, will function as an isocyanate. Typically, an adduct of tolylene diisocyanate and trimethylolpropane is first prepared in solution, followed by the addition of phenol to "block" the remaining isocyanate groups. The idealized structure of such a product is given by (**11**). Because of the high temperatures (usually

$$
\begin{array}{c}
\text{NHCOOC}_6\text{H}_5 \\
\text{CH}_2\text{OCONH}\!\!-\!\!\!\bigcirc\!\!\!-\!\text{CH}_3 \\[2ex]
\text{NHCOOC}_6\text{H}_5 \\
\text{CH}_3\text{CH}_2\!\!-\!\!\text{C}\!\!-\!\!\text{CH}_2\text{OCONH}\!\!-\!\!\!\bigcirc\!\!\!-\!\text{CH}_3 \\[2ex]
\text{CH}_2\text{OCONH}\!\!-\!\!\!\bigcirc\!\!\!-\!\text{CH}_3 \\
\text{NHCOOC}_6\text{H}_5
\end{array}
$$

(11)

150°C or higher) necessary to cause the reactivation of the isocyanate groups, these systems have found uses primarily as wire-coating enamels. However, they do find utility on any substrate that is relatively heat resistant.

Aliphatic Polyurethans. The urethan coating systems so far described were, until about 1960, based largely on tolylene diisocyanate. Although coating films prepared from these systems had good properties, they all had poor color stability when exposed to ultraviolet radiation. Additives improved their performance, but only delayed the color change. It had been known that, by using aliphatic diisocyanates in place of aromatic diisocyanates, such as tolylene diisocyanates, polyurethan coatings with outstanding light stability could be produced. Hexamethylene diisocyanate had been used for many years in experimental programs. However, owing to its high vapor pressure, with accompanying health hazards, and its high cost, very little use of this chemical was made until about 1960 when Farbenfabriken Bayer A.G. introduced a polyisocyanate of biuret structure based on hexamethylene diisocyanate (65). This compound is produced from the reaction of the diisocyanate and water (eq. 29). This polyisocyanate still retains the aliphatic charac-

$$
3\ \text{OCN(CH}_2)_6\text{NCO} + \text{H}_2\text{O} \xrightarrow{-\text{CO}_2}
\begin{array}{c}
\overset{\text{O}}{\overset{\|}{\text{C}}}\!-\!\overset{\text{H}}{\underset{}{\text{N}}}\text{(CH}_2)_6\text{NCO} \\
\text{OCN(CH}_2)_6\text{N} \\
\underset{\text{O}}{\overset{}{\text{C}}}\!-\!\underset{\text{H}}{\overset{}{\text{N}}}\text{(CH}_2)_6\text{NCO}
\end{array}
\qquad (29)
$$

teristics desirable for a nonyellowing coating and also has a very low vapor pressure, thus reducing the hazards of unmodified hexamethylene diisocyanate.

Since aliphatic isocyanates have a lower rate of reaction as compared to aromatic isocyanates and form softer resins when used in comparable formulations, it is necessary to be more selective in choosing coreactants for this adduct than when using the aromatic-based adducts. Thus, in order to obtain fast drying times, it is desirable to use polyols with only primary hydroxyl groups, since they are known to be more reactive to isocyanate reaction than secondary hydroxyl groups. In order to get films of sufficient hardness and chemical resistance, it has been found that phthalic polyesters with approx 8% hydroxyl groups and a softening point of 90–100°C are desirable. Polyether and epoxy resins are satisfactory for chemical resistance but are inadequate for chalking resistance.

Accelerators can be used to shorten the drying and curing times. Tertiary amines commonly used with aromatic diisocyanates are not effective with aliphatic isocyanates; however, certain metal compounds are effective, especially zinc octoate and zinc naphthenate. Dibutyltin dilaurate can be used instead of the zinc compounds but, because it drastically shortens the pot life, is used in concentrations of only 10% of the amount of zinc octoate that would have been required. The pot life is also influenced by temperature, concentration of the two-component coating, and nature of the solvent.

Coating films based on this hexamethylene diisocyanate adduct have been reported to have excellent color retention after being aged for 1600 hr at 90 and 120°C. Scarcely any changes were seen at 90°C and only slight changes in color were observed at 120°C.

Since the introduction of the polyisocyanate based on hexamethylene diisocyanate, several other aliphatic monomers have been offered, both commercially and experimentally. These include: methylcyclohexylene diisocyanate (**12**) (66); dimethylenedicyclohexyl diisocyanate (**13**) (66); lysine diisocyanate (**14**), where R is CH_3 or C_2H_5 (67); bis(2-isocyanatoethyl) fumarate (**15**) (68); bis(2-isocyanatoethyl) carbonate (**16**) (68); and "dimeryl" diisocyanate (**17**), where R is essentially a C_{36} hydrocarbon radical (69). With the possible exception of "dimeryl" diisocyanate, all of the above monomers can be used the same way as tolylene diisocyanates to form moisture-curing,

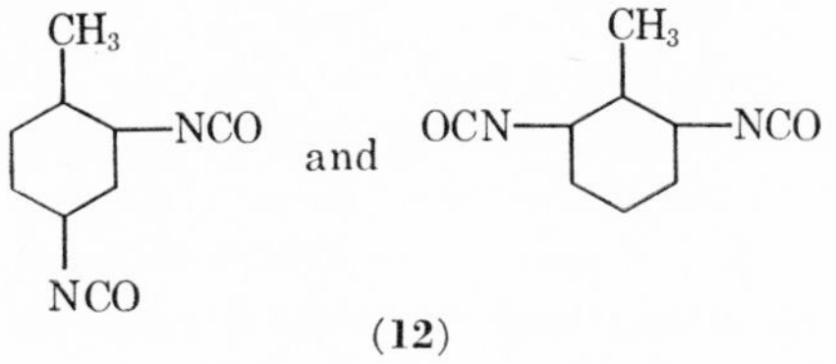

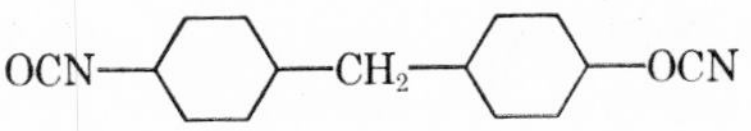

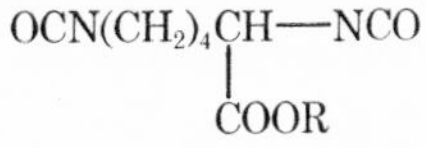

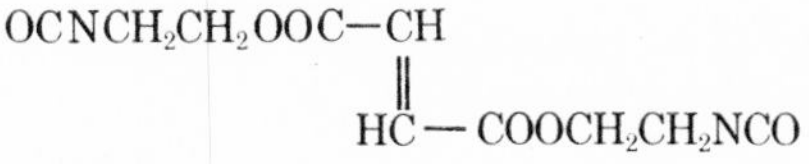

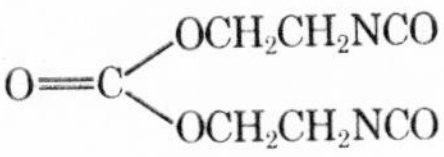

one-component systems or as reactants in urethan-oil preparation. However, because of their cost and the relatively small differences in the final products, they have not found utilization to any degree in urethan oils.

Methylenedicyclohexyl diisocyanate is used mainly in one-component moisture-cured prepolymers. In the preparation of the prepolymer, it is customary to use dibutyltin dilaurate as a catalyst in order to avoid a long processing time and high temperatures. With 0.1% dibutyltin dilaurate, such a prepolymer based on a low-equivalent-weight polyether can be prepared in one hour at 60–70°C. Curing time of a catalyzed prepolymer is good and the properties of the cured film are comparable to those of a tolylene diisocyanate-based moisture-cured prepolymer with the additional attribute of excellent color stability.

Lysine diisocyanate can also be used as a monomer in preparing moisture-cured prepolymers, but it has found greater utilization in the preparation of polyisocyanate adducts with polyfunctional polyols, such as trimethylolpropane. An idealized structure for such a compound is given by (**18**), where R is CH_3 or C_2H_5. This compound

$$
\begin{array}{c}
\text{COOR} \\
| \\
\text{CH}_2\text{—OCONH(CH}_2\text{)}_4\text{CH—NCO} \\
| \\
\text{CH}_3\text{CH}_2\text{—C—CH}_2\text{—OCONH(CH}_2\text{)}_4\text{CH—NCO} \\
| \qquad\qquad\qquad\qquad | \\
\text{CH}_2\text{—OCONH(CH}_2\text{)}_4\text{CH—NCO} \qquad \text{COOR} \\
| \\
\text{COOR}
\end{array}
$$

(18)

can then be used in two-component systems using the same type of polyols as described earlier for the comparable products made from tolylene diisocyanate.

By using modified polyamines with "dimeryl" diisocyanate in a two-component system, it is possible to make *polyurea coatings* with outstanding light stability. Because of the fairly rapid exothermic reaction of the two components, it is suggested that mixing be done at a solids content of 55%. For maximum pot life, a solids content of about 40% is desirable. The curing mechanism of the polyurea coating is given by equation 30. The reaction is relatively independent of ambient conditions

$$
O{=}C{=}N{-}R{-}N{=}C{=}O \;+\; \begin{array}{c} H \\ \diagdown \\ N{-}R'{-}N \\ \diagup \\ H \end{array} \begin{array}{c} H \\ \diagup \\ \diagdown \\ H \end{array} \;\rightarrow
$$

$$
\text{www-}(\text{NH—C—NH—R'—NH—C—NH—R})\text{-www} \overset{O}{\underset{}{\parallel}} \qquad \overset{O}{\underset{}{\parallel}} \qquad\qquad (30)
$$

although an accelerated cure can be obtained at elevated temperatures.

Bibliography

"Urethans" in *ECT* 1st ed, Vol. 14, pp. 473–480, by J. A. Garman, Fairfield Chemical Division, Food Machinery and Chemical Corp.; "Urethane Polymers" in *ECT* 1st ed., First Supplement Volume, pp. 888–908, by J. H. Saunders and E. E. Hardy, Mobay Chemical Co.

1. O. Bayer, *Angew. Chem.* **59,** 275 (1947).
2. O. Bayer, W. Siefkin, L. Orthner, and H. Schild (to I. G. Farbenindustrie), Ger. Pat. 728,981 (1942).
3. A. Wurtz, *Ann.* **71,** 326 (1849).
4. H. S. Rothrock (to E. I. du Pont de Nemours & Co., Inc.), U.S. Pat. 2,282,827 (1942).
5. O. Bayer, E. Muller, S. Petersen, H. F. Piepenbrink, and E. Windemuth, *Angew. Chem.* **62,** 57 (1959); *Rubber Chem. Technol.* **23,** 812 (1950).
6. A. Hochtlen, *Kunststoffe* **42,** 303 (1952).
7. C. M. Barringer, *Teracol 30–Polyalkylene Ether Glycol,* Bulletin HR-11, Du Pont, 1956.
8. J. H. Saunders, and K. C. Frisch, *Polyurethanes:* I. *Chemistry* and II. *Technology,* Vol. 16 in *High Polymers* Series, Interscience Publishers, a division of John Wiley & Sons, Inc., New York, 1962.
9. W. Siefken, *Ann.* **562,** 75 (1949).

10. S. Petersen and H. F. Piepenbrink, in E. Müller, ed., *Methoden der Organischen Chemie*, Vol. 8, Thieme-Verlag, Stuttgart, 1952.
11. J. H. Saunders and R. J. Slocombe, *Chem. Rev.* **43**, 203 (1948).
12. R. G. Arnold, J. A. Nelson, and J. J. Verbanc, *Chem. Rev.* **57**, 47 (1957).
13. S. Davis, J. M. McClellan, and K. C. Frisch, *Paper, Isocyanate Symp. Upper Midwest Section Plastics Engrs., Minneapolis, Minn., Oct. 1957.*
14. K. C. Frisch and S. Davis, *Paper, Am. Chem. Soc. Meet., Miami, Fla., April 1957.*
15. E. Müller, in E. Müller, ed., Houben-Weyl *Methoden der organischen Chemie*, Vol. 14, Thieme-Verlag, Stuttgart, 1963.
16. F. Hostettler (to Union Carbide Corp.), U.S. Pat. 2,933,477 (1960).
17. D. M. Young and F. Hostettler (to Union Carbide Corp.), U.S. Pat. 2,933,478 (1960).
18. T. L. Davis and J. M. Farnum, *J. Am. Chem. Soc.* **56**, 883 (1934).
19. R. Leuckart and M. Schmidt, *Ber.* **18**, 2339 (1885).
20. S. Petersen, *Ann.* **562**, 205 (1949).
21. J. W. Baker, M. M. Davies, and J. Gaunt, *J. Chem. Soc.* **1949**, 24.
22. J. W. Baker and J. B. Holdsworth, *J. Chem. Soc.* **1947**, 713.
23. J. W. Baker and J. Gaunt, *J. Chem. Soc.* **1949**, 9, 19, 27.
24. M. E. Bailey, V. Kirss, and R. G. Spaunburgh, *Ind. Eng. Chem.* **48**, 794 (1956).
25. J. Burkus and C. F. Eckert, *J. Am. Chem. Soc.* **80**, 4958 (1958).
26. M. Kaplan, *J. Chem. Eng. Data* **6**, 272 (1961).
27. W. R. Sorenson, *J. Org. Chem.* **24**, 987 (1959).
28. B. G. Aizner and K. C. Frisch, *Ind. Eng. Chem.* **51**, 715 (1959).
29. J. J. Tazuma and H. K. Latourette, *Paper, Am. Chem. Soc. Meet., Atlantic City, N.J., Sept. 1956.*
30. J. L. O'Brien and A. S. Pagano, *Paper, Delaware Regional Meet., Am. Chem. Soc., Feb. 5, 1958.*
31. A. Farkas and K. G. Flynn, *J. Am. Chem. Soc.* **82**, 642 (1960).
32. A. Farkas, G. A. Mills, W. E. Erner, and J. B. Maerker, *Ind. Eng. Chem.* **51**, 1299 (1959).
33. E. F. Cox and F. Hostettler, *Paper, Am. Chem. Soc. Meet., Boston, Mass., April 1959.*
34. F. Hostettler and E. F. Cox, *Ind. Eng. Chem.* **52**, 609 (1960).
35. J. W. Britain and P. G. Gemeinhardt, *J. Appl. Polymer Sci.* **4**, 207 (1960).
36. S. I. Axelrood, C. W. Hamilton, and K. C. Frisch, *Ind. Eng. Chem.* **53**, 889 (1961).
37. H. W. Wolf, Jr., *Catalyst Activity in One-Shot Urethane Foam*, Technical Bulletin, Du Pont, March 1960.
38. M. E. Bailey, A. Khawam, and G. C. Toone, *Paper, Am. Chem. Soc. Meet., Atlantic City, N.J., Sept. 1956.*
39. H. L. Heiss, F. P. Combs, P. G. Gemeinhardt, J. H. Saunders, and E. E. Hardy, *Ind. Eng. Chem.* **51**, 929 (1959).
40. C. W. Bunn, *J. Polymer Sci.* **16**, 323 (1955).
41. R. Hill and E. E. Walker, *J. Polymer Sci.* **3**, 609 (1948).
42. K. A. Pigott, B. F. Frye, K. R. Allen, S. Steingiser, W. C. Darr, J. H. Saunders, and E. E. Hardy, *J. Chem. Eng. Data* **5**, 391 (1960).
43. J. L. Boivin, *Can. J. Chem.* **36**, 1405 (1958).
44. C. F. Blaich, Jr., and A. J. Sampson, *Rubber Age* **89**, 263 (1961).
45. R. J. Athey, *Rubber Age* **85**, 77 (1959).
46. R. J. Athey, *Ind. Eng. Chem.* **88**, 611 (1960).
47. R. E. Bolin, J. F. Szabat, R. J. Cote, E. Peters, P. G. Gemeinhardt, A. S. Morecroft, E. E. Hardy, and J. H. Saunders, *J. Chem. Eng. Data* **4**, 261 (1959).
48. J. H. Saunders, *Rubber Chem. Technol.* **33**, 1259 (1960).
49. P. C. Colodny and A. V. Tobolsky, *J. Am. Chem. Soc.* **79**, 4320 (1957).
50. J. A. Offenbach and A. V. Tobolsky, *J. Colloid Sci.* **11**, 39 (1956).
51. I. C. Kogon, *J. Org. Chem.* **24**, 83 (1959).
52. S. Davis and W. Fijal, Wyandotte Chemicals Corp., private communication.
53. *TIB No. 35-F-12*, Technical Information Bulletin, Mobay Chemical Company, Nov. 25, 1959.
54. R. L. Sandridge, A. S. Morecroft, E. E. Hardy, and J. H. Saunders, *J. Chem. Eng. Data* **5**, 495 (1960).
55. *Flexible Urethane Foam Self-Extinguishing Properties*, National Aniline Research Notes RN-8, Allied Chemical Corp., 1961.
56. M. Kovarik, *Paper, Meet. Southeastern Section, Soc. Plastics Engrs., Atlanta, Ga., March 1961.*

57. L. R. LeBras, *SPE J.* **16,** 420 (1960).
58. E. Mueller, *Rubber Plastics Age* **39** (3), 195 (1958).
59. K. A. Pigott, J. W. Britain, W. Archer, B. F. Frye, R. J. Cote, and J. H. Saunders, *Ind. Eng. Chem. Prod. Res. Develop.* **1** (1), 28 (1962).
60. C. S. Schollenberger, H. Scott, and G. R. Moore, *Rubber World* **137,** 549 (1958).
61. M. W. Riley, *Mater. Design Eng.* **50** (4), 92 (1959).
62. J. G. DiPinto and S. D. McCready, *Engineering Properties of Urethane Elastomers*, Development Products Report No. 17, E. I. du Pont de Nemours & Co., Inc., Elastomer Chemicals Dept., April 1960.
63. K. A. Pigott, R. J. Cote, K. Ellegast, B. F. Frye, E. Muller, Wm. Archer, Jr., K. R. Allen, and J. H. Saunders, *Rubber Age* **91,** 629 (July 1962).
64. H. L. Heiss, J. H. Saunders, M. E. Morris, B. R. Davis, and E. E. Hardy, *Ind. Eng. Chem.* **46,** 1498 (1954).
65. *Products for Surface Coatings, Desmodur/Desmophen*, ed. 1.6, Farbenfabriken Bayer A. G., 1966, p. 14.
66. K. A. Pigott, E. R. Wells, and G. A. Hudson, *Paper, Polymer Conf. Series, Organic Coatings Technology, Wayne State University, June 1967.*
67. J. D. Garber, D. Wasserman, and R. A. Gasser (to Merck and Co.), Can. Pat. 756,115 (April 1967).
68. *Advance Technical Information, Aliphatic Isocyanates*, Technical Bulletin F-41195, Union Carbide Corp., Dec. 1964.
69. *Polyurea Coatings Based on DDI Brand Diisocyanate*, Technical Bulletin CDS 10-65, General Mills Corp., Sept. 1965.

General References

K. A. Pigott, "Polyurethans," in N. Bikales, ed., Mark-Gaylord Encyclopedia of Polymer Science and Technology, Vol. 11, Interscience Publishers, a division of John Wiley & Sons, Inc., New York, 1969, pp. 506–562.
P. F. Bruins, ed., *Polyurethane Technology*, Interscience Publishers, a division of John Wiley & Sons, Inc., New York, 1969.
J. M. Buist and H. Gudgeon, eds., *Advances in Polyurethane Technology*, Interscience Publishers, a division of John Wiley & Sons, Inc., New York, 1968.
Cellular Plastics, Publication 1462, National Academy of Sciences–National Research Council, Washington, D.C., 1967.
B. A. Dombrow, *Polyurethanes*, 2nd ed., Reinhold Publishing Corp., New York, 1965.
T. H. Ferrigno, *Rigid Plastic Foams*, 2nd ed., Reinhold Publishing Corp., New York, 1967.
J. H. Saunders and K. C. Frisch, *Polyurethanes: I. Chemistry* and II. *Technology*, Vol. 16 in *High Polymers* Series, Interscience Publishers, a division of John Wiley & Sons, Inc., New York, 1962–1964.
J. H. Saunders, "Polyurethane Elastomers," in J. P. Kennedy and E. G. M. Törnqvist, eds., *Polymer Chemistry of Synthetic Elastomers*, Interscience Publishers, a division of John Wiley & Sons, Inc., New York, 1969, pp. 727–765.

K. A. PIGOTT
Mobay Chemical Company

URIC ACID

Uric acid, 2,6,8($1H,3\dot{H},9H$) purinetrione, $C_5H_4N_4O_3$, exists as white or yellowish-white, odorless, and tasteless crystals or powder. It is a member of the purines and xanthines (see Vol. 1, p. 764) and exists in tautomeric forms, which probably are in equilibrium in solution (1). It has usually been represented as the trienol (lactim form, 2,6,8-trihydroxypurine) (**1a**) or the triketone (lactam form, 2,6,8-trioxypurine) (**1b**). The enol form accounts for the acidic properties. In this form, uric acid would be expected to be tribasic (1), but it forms only two series of salts, the acid or primary salts, $MHC_5H_2N_4O_3$, and the normal, neutral, or secondary salts, $M_2C_5H_2N_4O_3$, probably owing to the very weak dissociation constant of the third hydrogen ion. The ketone form (**1b**) accounts for its general lack of chemical reactivity, low solubility, etc (2). The structural formulas for uric acid may be written as shown in (**1a**) and (**1b**) or as a benzenoid structure as in (**1c**), shown as the "diolone" form or dibasic form as it is now sometimes written. See also Heterocyclic compounds, Vol. 10, p. 913.

lactim form (**1a**)   lactam form (**1b**)   (**1c**)

Von Scheele discovered uric acid in 1776 as a constituent of urine and bladder stones and named the substance "urine acid" as well as "bladder stone acid." Bergman made the same discovery at about the same time. Von Scheele described this substance in great detail, including its solubility and its behavior toward the various mineral acids, especially nitric acid. In 1793 Fourcroy described the properties of uric acid in great detail and established its relationship with urea (qv). Other workers contributed immeasurably to the chemistry of uric acid, for example, Prout with his discovery of ammonium acid purpurate as a color test for uric acid (murexide test) and Brugnatelli with his preparation of alloxan by the action of nitric acid on uric acid. Formula (**1b**) for uric acid was proposed by Medicus in 1875 and proved by Fischer as a result of his investigations into the preparation of the various methyl derivatives (2,3). The first successful syntheses of uric acid, by fusing glycine, NH_2CH_2COOH, with urea and by heating 3,3,3-trichlorolactic acid, $CCl_3CHOHCOOH$, with urea, were reported by Horbaczewski in the 1880s. Wöhler and Liebig, von Baeyer, Emil Fischer, Roosen, and others completed the monumental work on the properties and synthesis of uric acid.

Occurrence. Uric acid, usually in the form of its salts, is widespread in nature in both plants and animals. It is found in many seeds as well as other plant parts and is a constituent of the urine of all carnivorous animals, being formed from xanthine by the action of xanthine oxidase in the metabolism of purines and their derivatives, notably nucleic acids (qv). Uric acid is present in human urine to the extent of 0.3–2.0 g/day, varying with certain diets and with certain pathological conditions, such as arthritis and gout (sodium urate deposits in the joints in both diseases). In reptiles and birds, uric acid is the chief end product of the metabolism of proteins as well as of purines. Thus excreta of certain reptiles may contain as high as 75% of almost pure ammonium urate. Chicken manures contain 2–3% ammonium urate, and certain bird excreta (guano) contain 10–20%.

Physical and Chemical Properties

Uric acid decomposes above 250°C without melting, with the evolution of hydrogen cyanide. It is soluble in glycerol, sulfuric acid (without decomposition), solutions of lithium carbonate, sodium acetate, and sodium phosphate, and very soluble in alkalis. It is practically insoluble in boiling water (1:2000) and insoluble in alcohol, ether, and chloroform.

Of the two series of salts formed, the acid alkali metal salts are less soluble in water than the normal salts. The acid piperazine salt and the lysidine (methylglyoxalidine) salts are much more soluble in water than the metallic acid or normal salts.

Like other purines, uric acid absorbs in the ultraviolet region. The maxima are dependent on the pH.

Uric acid has a pK_a of 5.7. It is stable to hydrolysis, but this stability decreases with increasing N-substitution. Alkali solutions are rather unstable. On dry distillation of uric acid, Scheele obtained carbon, ammonia, carbon dioxide, and cyanuric acid.

Uric acid is not easily affected by reducing substances, with a few exceptions; for example, Fischer and Ach prepared 6-thiouramil, $NH.CO.NH.CO.CH(NH_2).CS$, by heating uric acid with an ammonium sulfide solution at 160°C. Electrolytic reduction in sulfuric acid solution below 8°C yields purone, dihydro-2,8-purinedione, $C_5H_8N_4O_2$, and at 14–17°C, isopurone, $C_5H_8N_4O_2$ (in which the imidazole ring has been broken), along with purone, tetrahydrouric acid, and other products.

Uric acid is easily oxidized by a variety of compounds to give a wide variety of products. In general, alkaline oxidation attacks the pyrimidine ring and acid oxidation the imidazole ring.

Scheme 1

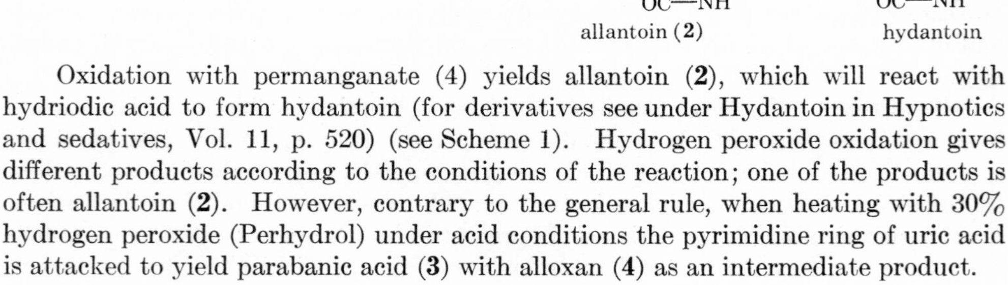

allantoin (2)   hydantoin

Oxidation with permanganate (4) yields allantoin (2), which will react with hydriodic acid to form hydantoin (for derivatives see under Hydantoin in Hypnotics and sedatives, Vol. 11, p. 520) (see Scheme 1). Hydrogen peroxide oxidation gives different products according to the conditions of the reaction; one of the products is often allantoin (2). However, contrary to the general rule, when heating with 30% hydrogen peroxide (Perhydrol) under acid conditions the pyrimidine ring of uric acid is attacked to yield parabanic acid (3) with alloxan (4) as an intermediate product.

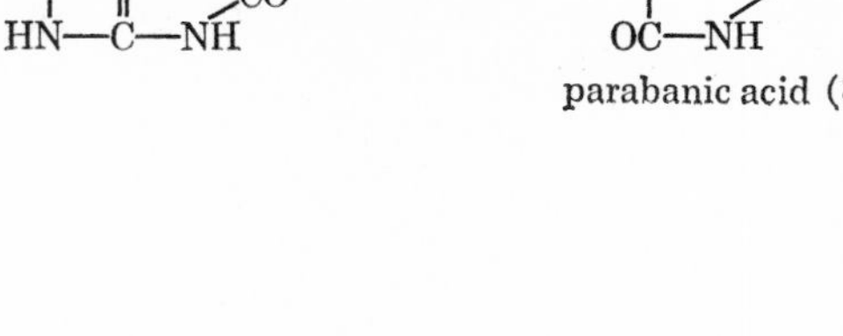

parabanic acid (3)

Oxidation with cold concentrated nitric acid or with chlorine (5) yields alloxan (4) and urea.

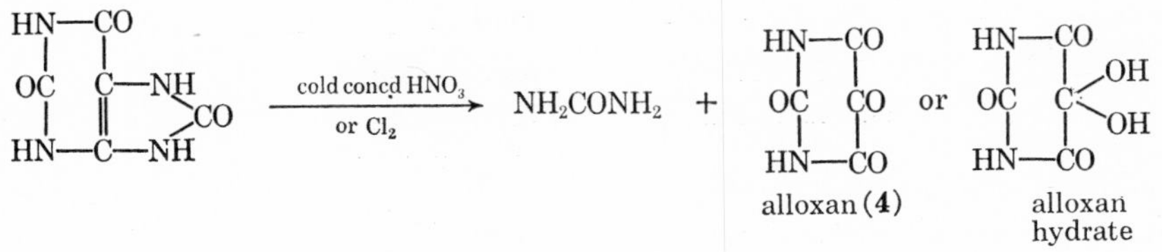

```
HN—CO
|    |
OC   C—NH            cold concd HNO₃
|    ‖     CO    ─────────────────────→   NH₂CONH₂   +
HN—C—NH               or Cl₂
```

```
HN—CO            HN—CO
|    |           |    |   OH
OC   CO   or    OC   C
|    |           |    |   OH
HN—CO            HN—CO
alloxan (4)      alloxan
                 hydrate
```

Heating with concentrated nitric acid (6) yields dialuric acid (5) and alloxan (4), both compounds on continued heating reacting further to give alloxantin (6). On addition of ammonia, the purple-red murexide (7) (acid ammonium purpurate) is obtained (7) (see Scheme 2). With warm, very dilute nitric acid, alloxantin (6) and urea are obtained directly.

Fischer and other investigators prepared various isomeric methyl (from mono to tetra) derivatives of uric acid (2,3). By heating uric acid with phosphorus oxychloride at 150–160°C, Fischer (8) was able to produce 2,6,8-trichloropurine and 2,6-dichloro-

Scheme 2

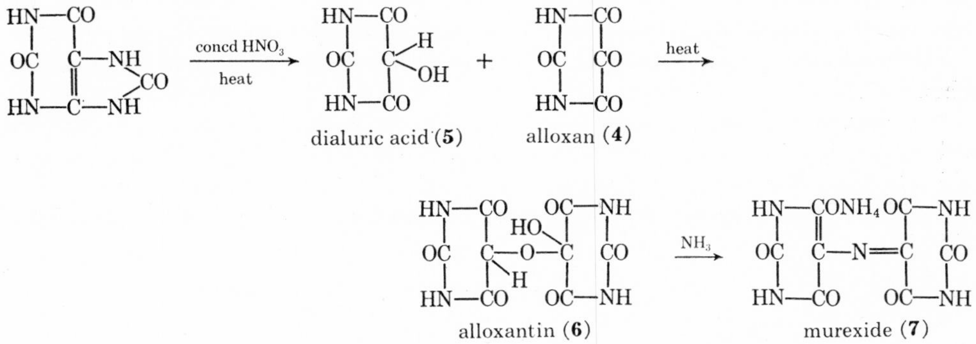

```
HN—CO                         HN—CO                  HN—CO
|    |        concd HNO₃       |    |   H             |    |
OC   C—NH    ─────────────→   OC   C             +   OC   CO      heat
|    ‖   CO      heat          |    |   OH            |    |     ─────→
HN—C—NH                        HN—CO                  HN—CO
                              dialuric acid (5)      alloxan (4)
```

```
HN—CO   OC—NH                       HN—CONH₄ OC—NH
|    |  HO |   |                     |    |       |   |
OC   C—O—C    CO        NH₃         OC   C—N═C    CO
|    | \   |   |       ─────→        |    |       |   |
HN—CO  H  OC—NH                      HN—CO    OC—NH
     alloxantin (6)                       murexide (7)
```

8-hydroxypurine. Heating uric acid with formamide until completely dissolved yields, upon cooling, xanthine. Theophylline can be prepared similarly from 1,3-dimethyluric acid and formamide (9).

Preparation and Manufacture

Uric acid is produced by extraction from natural sources or synthetically.

Extraction from Natural Sources. Uric acid can be obtained by extracting either reptilian excreta or bird guano with sodium hydroxide, precipitating with ammonium chloride, and decomposing the precipitate with hydrochloric acid. The guano may also be treated with calcium carbonate in a strongly alkaline solution, and the free acid precipitated with hydrochloric acid. Bird guano is used as the commercial source of the acid.

A slightly different procedure has been patented (10). Bird excrements are extracted with 8% aqueous sodium hydroxide and heated for 6–8 hr at 80–100°C with vigorous agitation. The solution is filtered and the crude uric acid precipitated with

Scheme 3

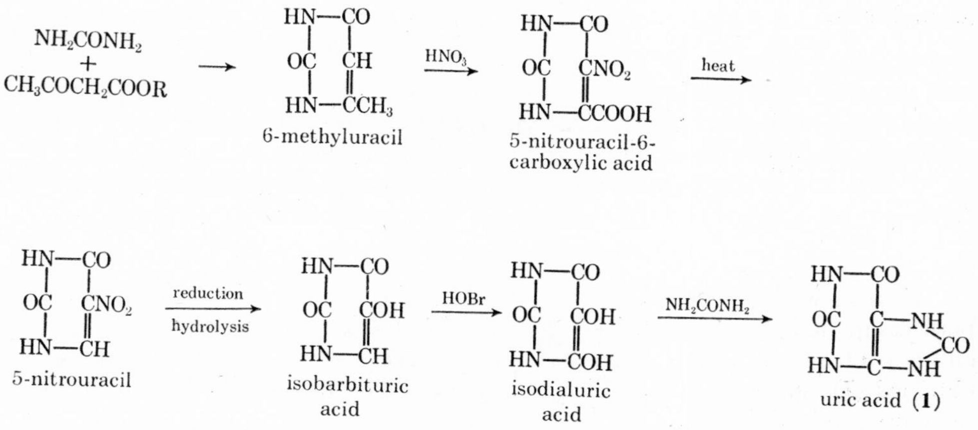

NH$_2$CONH$_2$
+
CH$_3$COCH$_2$COOR

$\longrightarrow$

HN—CO
| |
OC CH
| ‖
HN—CCH$_3$

6-methyluracil

$\xrightarrow{\text{HNO}_3}$

HN—CO
| |
OC CNO$_2$
| ‖
HN—CCOOH

5-nitrouracil-6-
carboxylic acid

$\xrightarrow{\text{heat}}$

HN—CO
| |
OC CNO$_2$
| ‖
HN—CH

5-nitrouracil

$\xrightarrow[\text{hydrolysis}]{\text{reduction}}$

HN—CO
| |
OC COH
| ‖
HN—CH

isobarbituric
acid

$\xrightarrow{\text{HOBr}}$

HN—CO
| |
OC COH
| ‖
HN—COH

isodialuric
acid

$\xrightarrow{\text{NH}_2\text{CONH}_2}$

HN—CO
| |
OC C—NH
| ‖   >CO
HN—C—NH

uric acid (**1**)

sulfuric acid. The crude acid is then redissolved in 5–6% sodium hydroxide and the solution evaporated to 20% of the original volume. Sodium urate crystallizes out of solution and is again dissolved in sodium hydroxide solution. The pure uric acid is precipitated with excess sulfuric acid.

Synthetic Methods. The thorough researches by Fischer and by Behrend and Roosen into the breakdown products (various stages of oxidation) of uric acid, and their possible combinations to produce uric acid, resulted in its successful synthesis. The Behrend and Roosen synthesis (11) starts with urea and an acetoacetic ester (see Scheme 3).

Fischer, continuing the work of von Baeyer, succeeded in converting pseudouric acid (**8**) to uric acid by concentrating a solution of pseudouric acid in 20% hydrochloric acid solution. The work of von Baeyer and Fischer is combined in Scheme 4 to show the total synthesis of uric acid from urea and malonic acid by way of barbituric acid (qv).

Scheme 4

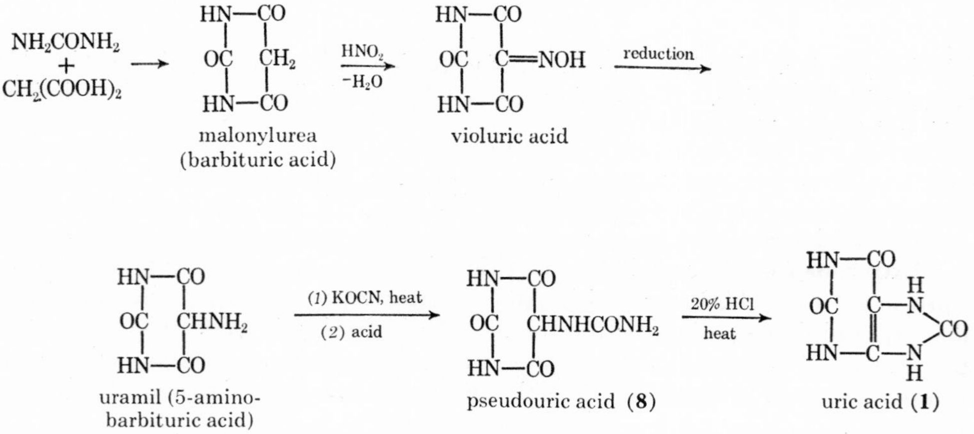

NH$_2$CONH$_2$
+
CH$_2$(COOH)$_2$

$\longrightarrow$

HN—CO
| |
OC CH$_2$
| |
HN—CO

malonylurea
(barbituric acid)

$\xrightarrow[-\text{H}_2\text{O}]{\text{HNO}_2}$

HN—CO
| |
OC C=NOH
| |
HN—CO

violuric acid

$\xrightarrow{\text{reduction}}$

HN—CO
| |
OC CHNH$_2$
| |
HN—CO

uramil (5-amino-
barbituric acid)

$\xrightarrow[\text{(2) acid}]{\text{(1) KOCN, heat}}$

HN—CO
| |
OC CHNHCONH$_2$
| |
HN—CO

pseudouric acid (**8**)

$\xrightarrow[\text{heat}]{20\% \text{ HCl}}$

HN—CO
| | H
OC C—N
| ‖   >CO
HN—C—N
         H

uric acid (**1**)

Cavalieri and co-workers (12) report overall yields of only about 27% with the combined Baeyer and Fischer method. However, with certain modifications, an overall yield of 55% based on the urea used was obtained.

In a more recent synthesis of uric acid (13), 5-sulfoaminouracil, 1,2,3,4-tetra-hydro-2,4-dioxo-5-pyrimidinesulfamic acid, $NH.CO.NH.CO.C(NHSO_3H):CH$, is heated with urea at 190–200°C for 0.5 hr. The melt is treated with hot water, yielding acid ammonium urate, which is dissolved in hot water containing the required amount of sodium hydroxide. The solution is clarified with carbon, and while still hot, acidified with hydrochloric acid to give the free uric acid.

Uric acid is available in three grades: cp, reagent, and technical (usually 90–95% pure). It is packaged in 1-lb bottles and 5- to 100-lb drums.

Analysis

Qualitative. *Murexide Test.* When uric acid is moistened with concentrated nitric acid and evaporated to dryness on a water bath, a purple-red color (resembling the dye from the sea snail *Murex*) develops on the addition of a few drops of ammonia. This test, however, is not restricted to uric acid, but will occur with other purines as well.

Quantitative. *In Urine.* (*1*) The Benedict and Francke method (14,15) utilizes the fact that solutions containing uric acid will develop a blue color, the intensity of which is proportional to the uric acid concentration, when arsenophosphotungstic acid (prepared from sodium tungstate, arsenic trioxide, phosphoric and hydrochloric acids) and sodium cyanide are added. This is a direct microchemical determination.

(*2*) In the Folin-Schaffer method (16), the uric acid is converted into ammonium urate, which is dissolved in hot sulfuric acid solution and then titrated with potassium permanganate solution.

In Blood. Both methods described above can be used to determine the uric acid concentration in the blood. A more recent method for the determination of uric acid in blood uses a lithium carbonate reagent (17).

In Fruit Products. Tilden has described a paper-chromatographic method (18).

Reagent Grade. The two methods described under urine can be used, or the nitrogen content can be determined by the Kjeldahl method.

Technical Grade. The Folin-Schaffer and Kjeldahl methods may be used.

Uses

Uric acid is used for the commercial preparation of allantoin, alloxan, alloxantin, parabanic acid, murexide, and other derivatives. Uric acid and its salts have been and are still being used on occasion in medicine. It has been used internally for edematous heart, pulmonary tuberculosis, paranoia, etc, and also externally in gout. A 4% ointment of ammonium urate has been used in the treatment of chronic eczema and also internally for coughs and grippe. Lithium acid urate is used as an antarthritic, and murexide is used as an organic indicator for the determination of calcium and other metal ions.

Because of its ability to absorb ultraviolet rays, uric acid is being extensively used in concentrations of 0.035–0.2% for stabilization of colors in shampoos and other cosmetic products which are affected by exposure to light. Uric acid is unique in that it will not affect the color, perfume, or pH of whatever preparation it is used in. It is compatible in both acid and alkaline medias as well as cationics, anionics, and nonionics.

Derivatives

SALTS

Ammonium acid urate, $NH_4HC_5H_2N_4O_3$, is a white crystalline powder, soluble in 1600 parts cold water and more so in boiling water. It is soluble in alkalis or concentrated sulfuric acid, and is prepared by interaction of ammonia with uric acid.

Calcium urate (calcium acid urate), $CaH_2(C_5H_2N_4O_3)_2.2H_2O$, is soluble in about 1500 parts cold water. It is prepared by precipitating a potassium or sodium urate solution with calcium chloride.

Lithium acid urate, $LiHC_5H_2N_4O_3$, is soluble in 380 parts cold or 39 parts boiling water. It is prepared by boiling uric acid with lithium carbonate.

Potassium urate (normal potassium urate), $K_2C_5H_2N_4O_3$, occurs in fine needles soluble in 44 parts cold water and 36 parts boiling water. It decomposes at high temperatures and is prepared in the same way as sodium urate.

Potassium acid urate, $KHC_5H_2N_4O_3$, exists as a white amorphous powder soluble in 800 parts cold and in 80 parts boiling water. It is prepared in the same way as sodium acid urate.

Sodium urate (normal sodium urate), $Na_2C_5H_2N_4O_3$, occurs as crystalline granules or a white crystalline powder soluble in about 77 parts cold and about 75 parts boiling water, and slightly soluble in alcohol. It decomposes at 150°C and is prepared by suspending uric acid in a cold dilute sodium hydroxide solution and concentrating under a vacuum.

Sodium acid urate, $NaHC_5H_2N_4O_3$, is a white powder soluble in 1100–1200 parts cold water and 125 parts boiling water. It is prepared by treating uric acid with a solution of sodium carbonate.

OTHER DERIVATIVES

Allantoin, 5-ureidohydantoin, glyoxyldiureide (**2**), $C_4H_6N_4O_3$, is a white odorless powder, stable in air; mp approx 228–235°C (decompn) (depending on method of determination). It is slowly soluble in about 200 parts cold water and about 25 parts boiling water. It dissolves, with decomposition, in solutions of alkali hydroxides. Allantoin occurs widely in plants and is excreted by almost all mammals, but not by anthropoid apes and man, as the end product of purine metabolism. It is formed by the action of the enzyme uricase on uric acid. It is produced synthetically in yields up to 75% by oxidizing uric acid with potassium permanganate in alkaline solution (4) (see p. 482). The commercial product, sold in drums, contains not less than 34.92% and not more than 35.45% of N, corresponding to not less than 98.5% of allantoin. It is used in ointments for medical and veterinary use, in wounds and ulcers to stimulate growth of healthy tissue, and also in hand lotions, creams, lipsticks, pomades, aftershave lotions, and suntan preparations for its healing effect.

Allantoin can be determined qualitatively by: (*1*) adding mercuric nitrate to a cooled, saturated, aqueous solution (a white precipitate forms which is soluble in excess mercuric nitrate); (*2*) boiling a small amount with 10% hydrochloric acid for 3–5 min and adding 1% phenylhydrazine hydrochloride solution (when cool, the addition of potassium ferricyanide and hydrochloric acid produces a cherry-red color); or (*3*) by mixing 2 ml of a 0.1% solution with 2 ml of ammoniacal copper tartrate solution and placing in a boiling water bath 15 min, then cooling and adding 2 ml of Folin acid molybdate solution (a brominated solution of sodium molybdate in a phosphoric–sulfuric–acetic acid medium). A blue color develops (19).

The following allantoin derivatives are extensively used in cosmetics and dermatologicals:

Aluminum Chlorhydroxy Allantoinate (Alcloxa) (20) (**9**). Used in antiperspirant

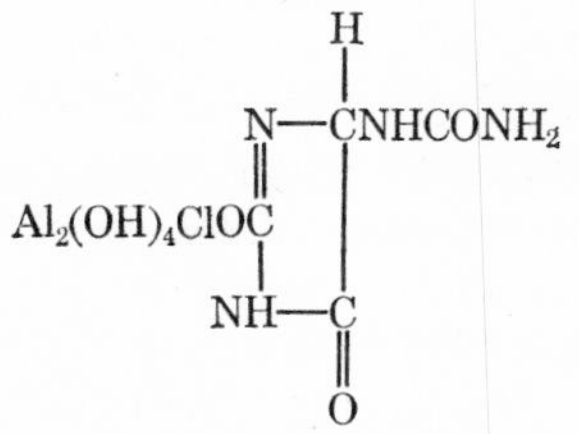

alcloxa (**9**)

deodorants, aftershave lotions, acne, and astringent preparations.

Aluminum Dihydroxy Allantoinate (Aldioxa) (20) (**10**). Used in diaper rash,

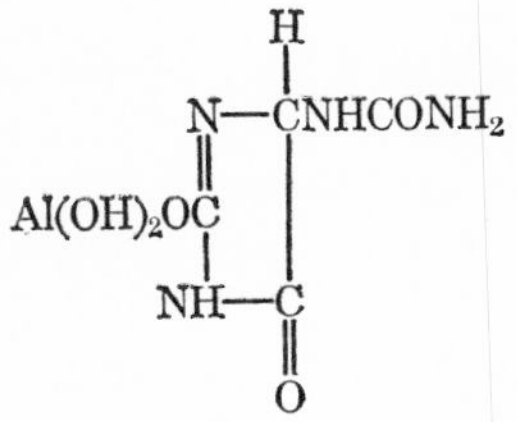

(**10**)

creams, lotions, compressed powders, talcums, etc. Also in the treatment of GI (gastrointestinal) ulcers.

Miscellaneous. Many other derivatives (21) have also been synthesized and are being widely used, primarily in cosmetics applications (22).

Alloxan, mesoxalylurea, 2,4,5,6(1H,3H)-pyrimidinetetrone (**4**), $C_4H_2N_2O_4$, exists as colorless to pink crystals easily soluble in water and in alcohol, slightly soluble in chloroform and in petroleum ether, and insoluble in ether. It is prepared by oxidizing uric acid with concentrated nitric acid or chlorine in the cold (see p. 482), by treating alloxantin (**6**) with fuming nitric acid (23), or by oxidizing barbituric acid with chromic acid (24). Alloxan monohydrate is available commercially for use in nutrition experiments and in organic syntheses, as of riboflavin (qv).

Alloxantin, uroxin, (**6**) forms a dihydrate (**9**), $C_8H_6N_4O_8 \cdot 2H_2O$, a white crystalline powder, which on exposure to air takes up ammonia, turning red. It is very soluble in hot water and slightly soluble in cold water, cold alcohol, and ether. It is prepared by the oxidation of uric acid with potassium chlorate followed by reduction with hydrogen sulfide (25). It is commercially available and is used in the synthesis of riboflavin.

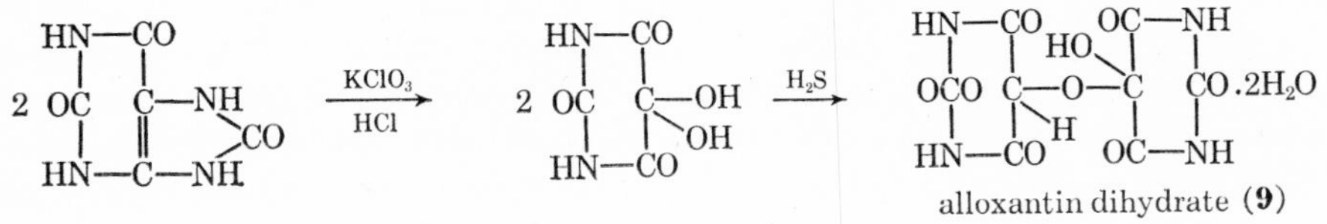

alloxantin dihydrate (**9**)

Parabanic acid, imidazolidinetrione oxalylurea, (**3**), $C_3H_2N_2O_3$, is a white crystalline powder; mp 243°C (decompn). It is soluble in 20 parts cold water and in alcohol. It is prepared by the oxidation of uric acid with 30% hydrogen peroxide (26), by the oxidation of alloxan, or by treating hydantoin with bromine and water.

Bibliography

"Uric Acid" in *ECT* 1st ed., Vol. 14, pp. 480–487, S. B. Mecca, Schuylkill Chemical Company.

1. A. P. Mathews, *Physiological Chemistry*, 5th ed., William Wood & Co., New York, 1930, p. 765.
2. *Thorpe's Dictionary of Applied Chemistry*, Vol. IX, 4th ed., Longmans, Green and Co., London, 1954, p. 802.
3. P. A. Levine and L. W. Bass, *Nucleic Acids*, Chem. Cat. Co., Monograph 56, 1931, pp. 88–91.
4. W. W. Hartman, E. W. Moffett, and J. B. Dickey, in A. Blatt, *Organic Syntheses*, Collective Vol. 2, John Wiley & Sons, Inc., New York, 1944, p. 21.
5. S. M. McElvain, *J. Am. Chem. Soc.* **57**, 1303 (1935).
6. Ref. 1, p. 767.
7. P. B. Hawk, B. L. Oser, and W. Summerson, *Practical Physiological Chemistry*, 12th ed., The Blakiston Co., Inc., div. McGraw-Hill Book Co., Inc., New York, 1947, p. 730.
8. Ref. 3, p. 92.
9. Ger. Pat. 864,870 (Jan. 29, 1953) (cl. 12p, 7₁₀), H. Bredereck and H. G. von Schuh.
10. U.S. Pat. 2,302,204 (1942), H. S. Gable and J. M. Bloodsworth.
11. Ref. 3, p. 93.
12. L. F. Cavalieri, E. V. Blair, and B. G. Brown, *J. Am. Chem. Soc.* **70**, 1240 (1948).
13. E. G. Fischer, W. P. Neumann, and J. Roch, *Chem. Ber.* **85**, 752 (1952).
14. S. R. Benedict and E. Francke, *J. Biol. Chem.* **52**, 387 (1922).
15. Ref. 1, p. 1172.
16. Ref. 1, p. 1173.
17. Lavagne-Mézière, *Anal. Biol. Clin. (Paris)* **9**, 340 (1951).
18. D. N. Tilden, *J. Assoc. Offic. Agr. Chemists* **34**, 498 (1951).
19. S. A. Katz, R. Turse, and S. B. Mecca, *J. Soc. Cos. Chem.* **15**, 303–310 (1964).
20. S. B. Mecca, *Proc. Sci. Sect. Toilet Goods Assoc.* **31** (May 1959).
21. I. I. Lubowe and S. B. Mecca, *Proc. Sci. Sect. Toilet Goods Assoc.* **42**, 6 (1964).
22. R. E. Faust and I. I. Lubowe, *Am. Perfumer* **83**, 29–32 (July 1968).
23. W. W. Hartman and O. E. Shepard, in L. I. Smith, *Organic Syntheses*, Vol. 23, John Wiley & Sons, Inc., New York, 1943, p. 3.
24. A. V. Holmgren and W. Wenner, in Arnold, *Organic Syntheses*, Vol. 32, John Wiley & Sons, Inc., New York, 1952, p. 6.
25. D. Nightingale, in L. I. Smith, *Organic Syntheses*, Vol. 23, John Wiley & Sons, Inc., New York, 1943, p. 6.
26. H. Blitz and G. Schiemann, *Chem. Ber.* **59**, 721 (1926).

S. B. Mecca
Schuylkill Chemical Co.

UV ABSORBERS

Ultraviolet radiation accelerates the physical and chemical deterioration processes of polymeric substrates and causes the fading of colorants, the yellowing of cellulosics, the photooxidation of polyolefins, the dehydrochlorination of poly(vinyl chloride), the embrittlement of coatings, erythema (sunburn), etc. Ultraviolet radiation is usually defined as electromagnetic radiation of wavelength between 4 and 400 nm. The sources of ultraviolet may be divided into two classes, natural and artificial. The most important natural source is the sun. The artificial sources comprise a wide variety of arcs and lamps. Although it is a relatively simple matter to match any limited region of the solar spectrum, no single type of arc or lamp is known which has exactly the same energy distribution as the sun's spectrum over the entire range. Solar simulators consist of some combination of arcs, incandescent lamps, and filters.

Approximately 10% of the sun's energy is in the ultraviolet. The solar ultraviolet intensity at the earth's surface depends on the thickness of the ozone layer in the upper atmosphere (ozone absorbs strongly between 200 and 300 nm), latitude, elevation

Table 1. 2-Hydroxybenzophenones

Trade name	Supplier	Composition
Uvinul 400	GAF	2,4-dihydroxybenzophenone
Inhibitor DHBP	Eastman	2,4-dihydroxybenzophenone
Rylex D	Du Pont	2,4-dihydroxybenzophenone
Cyasorb UV-9	Am Cy	2-hydroxy-4-methoxybenzophenone
Uvinul M-40	GAF	2-hydroxy-4-methoxybenzophenone
Cyasorb UV-531	Am Cy	2-hydroxy-4-octyloxybenzophenone
Picco UV-299	PICCO	2-hydroxy-4-octyloxybenzophenone
Carstab 700	Carlisle	2-hydroxy-4-octyloxybenzophenone
AM-300	Ferro	2-hydroxy-4-octyloxybenzophenone
Uvinul 410	GAF	4-decyloxy-2-hydroxybenzophenone
Mark 202A	Argus	4-decyloxy-2-hydroxybenzophenone
Inhibitor DOBP	Eastman	4-dodecyloxy-2-hydroxybenzophenone
Rylex H	du Pont	4-dodecyloxy-2-hydroxybenzophenone
Cyasorb UV-24	Am Cy	2,2'-dihydroxy-4-methoxybenzophenone
Cyasorb UV-207	Am Cy	2-hydroxy-4-methoxy-2'-carboxybenzophenone
Uvinul D-49	GAF	2,2'-dihydroxy-4,4'-dimethoxybenzophenone
Uvinul 490	GAF	mixture of Uvinul D-49 and other tetrasubstituted benzophenones
Uvinul D-50	GAF	2,2',4,4'-tetrahydroxybenzophenone
Uvinul MS-40	GAF	2-hydroxy-4-methoxy-5-sulfobenzophenone
Cyasorb UV-284	Am Cy	2-hydroxy-4-methoxy-5-sulfobenzophenone
Uvinul DS-49	GAF	sodium 2,2'-dihydroxy-4,4'-dimethoxy-5-sulfobenzophenone
DBR	Dow	2,4-dibenzoylresorcinol
HCB	Dow	5-chloro-2-hydroxybenzophenone
Permyl B-100	Ferro	substituted phenone

above sea level, atmospheric turbidity, time of day, and time of year. Practically all of the radiation of wavelengths shorter than 290 nm is absorbed in the earth's atmosphere. The shortest measured wavelength at the earth's surface is 286.3 nm at an elevation of 1600 m in the Alps. The total absorption depends on the amount of atmosphere, in addition to local conditions such as clouds, fog, dust, and smoke (1).

Desirable Features. Ultraviolet absorbers (commonly called light stabilizers) extend the useful life of irradiated polymers by absorbing ultraviolet radiation and/or an intermolecular energy process and harmlessly dissipating the energy (2). The ultraviolet absorber should preferentially absorb the pertinent ultraviolet components of sunlight and other ultraviolet sources such as fluorescent lights. Ideally it should have high absorptivity in the ultraviolet region of 290–400 nm and low or no absorptivity in the visible region to ensure low color. The ultraviolet absorber should be stable to the electromagnetic radiation and not be destroyed or used up in a relatively short time. It should be thermally stable to processing conditions and be unreactive and chemically inert with other additives present during fabrication and subsequent use. It must be compatible with the substrate and should have low volatility. Good color retention as well as good original color is desirable. A wash- and drycleaning-resistant absorber may be required for some applications. For food packages, Food and Drug Administration approval must be obtained. The absorber must not be a skin sensitizer if it is to be used in sunburn protective preparations.

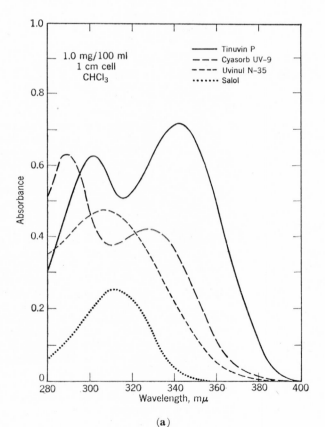

(a)

Factors to be considered in selecting an ultraviolet absorber include the substrate to be protected, the method and conditions (temperature, catalysts, time) of fabrication, the degree and type of protection desired, the size and shape of the finished product, the end use and environment during end use, and the presence of other additives such as antioxidants, antistatic agents, colorants, fillers, accelerators, hydrogen chloride scavengers, etc. The effectiveness of an ultraviolet absorber is dependent on its concentration and the thickness of the polymeric substrate. The effects of ultraviolet radiation are greatest at the surface of the polymer and it may be desirable to concentrate the ultraviolet absorber near the surface. The thinner the film or coating which contains the absorber, the greater is the concentration of absorber required.

Commercial Ultraviolet Absorbers. The most important commercial ultraviolet absorbers may be grouped into five types. They are the 2-hydroxybenzophenones, the 2-(2′-hydroxyphenyl)benzotriazoles, the salicylates, the aryl-substituted acrylates, and the p-aminobenzoates which are widely used in suntan lotions. In addition to miscellaneous absorbers, an important class of nonultraviolet-absorbing light stabilizers are the nickel chelates, complexes, and salts of aromatic systems. These are believed to provide protection by a mechanism which involves a direct transfer of the electronic energy from the excited state of the polymer molecule to the stabilizer

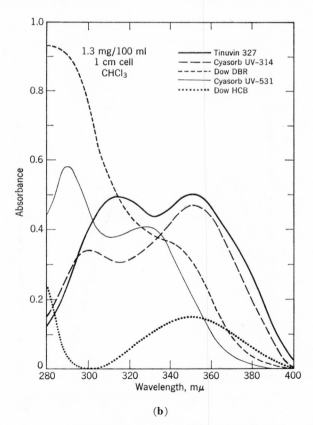

(**b**)

Fig. 1. Absorbance of ultraviolet absorbers used as stabilizers for polymeric materials at concentrations of (**a**) 1.0 mg/100 ml; (**b**) 1.3 mg/100 ml.

Table 2. 2-(2′-Hydroxyphenyl)benzotriazoles

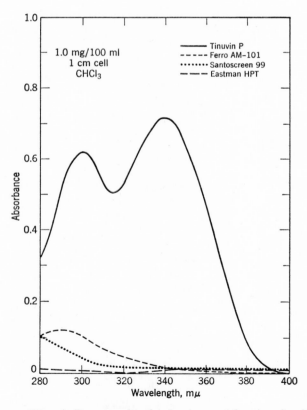

Trade name	Supplier	Composition
Tinuvin P	Geigy	2-(2′-hydroxy-5′-methylphenyl)benzotriazole
Tinuvin 320	Geigy	2-(3′,5′-dialkyl-2′-hydroxyphenyl)benzotriazole
Tinuvin 326	Geigy	2-(3′-*tert*-butyl-2′-hydroxy-5′-methylphenyl)-5-chlorobenzotriazole
Tinuvin 327	Geigy	2-(3′,5′-di-*tert*-butyl-2′-hydroxyphenyl)-5-chlorobenzotriazole
Tinuvin 328	Geigy	2-(3′,5′-dialkyl-2′-hydroxyphenyl)benzotriazole

molecule (3).   Another group of nonabsorbing compounds which become light stabilizers have the facility to photochemically rearrange or be converted to ultraviolet absorbing materials upon exposure to ultraviolet radiation.   An example is the conversion of the nonabsorbing *m*-hydroxylphenyl benzoate to the absorbing 2,4-dihydroxybenzophenone.

Fig. 2. The absorbance of Tinuvin P compared to the absorbance of three nonabsorbing light stabilizers.

Extraterrestrial environment requires ultraviolet absorbers with special properties. The absorbers must be stable to, and be able to absorb, the shorter wavelengths of ultraviolet radiation and have extremely low vapor pressures. Metal cyclopentadienyl derivatives (ferrocene type) provide some degree of extraterrestial protection (4).

Table 3. Substituted Acrylates

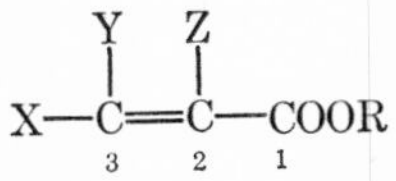

$$\underset{3}{X}-\underset{2}{\overset{\displaystyle Y}{\underset{|}{C}}}=\underset{1}{\overset{\displaystyle Z}{\underset{|}{C}}}-COOR$$

Trade name	Supplier	Composition
Uvinul N-35	GAF	ethyl 2-cyano-3,3-diphenylacrylate
Uvinul N-539	GAF	2-ethylhexyl 2-cyano-3,3-diphenylacrylate
UV-Absorber 317	Naftone	butyl 2-cyano-3-methyl-3-(*p*-methoxyphenyl)-acrylate
UV-Absorber 318	Naftone	methyl 2-cyano-3-methyl-3-(*p*-methoxyphenyl)-acrylate
UV-Absorber 340	Naftone	N-(β-cyano-β-carbomethoxyvinyl)-2-methylindoline
Cyasorb UV-1988	Am Cy	methyl 2-carbomethoxy-3-(*p*-methoxyphenyl)-acrylate

Table 4. Salicylates

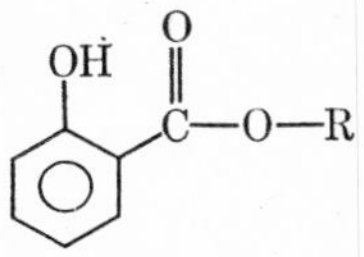

Trade name	Supplier	Composition
Salol	Dow, Heyden, Penick, Mallinckrodt	phenyl salicylate
TBS	Dow	p-*tert*-butylphenyl salicylate
Inhibitor OPS	Eastman	p-*tert*-octylphenyl salicylate
	Scher	dipropylene glycol salicylate
Sunarome W.M.O.	Felton	2-ethylhexyl salicylate

Table 5. Nickel Organic Stabilizers

Trade name	Supplier	Composition
AM-101	Ferro	nickel bis(octylphenol) sulfide
AM-103	Ferro	nickel organic complex
Cyasorb UV-1084	Am Cy	[2,2′-thio-bis(4-*tert*-octylphenolato)]-*n*-butylamine nickel
Irgastab 2001	Geigy	nickel bis[O-alkyl-(3,5-di-*tert*-butyl-4-hydroxy-benzyl)] phosphonate
Negopex A	ICI	nickel organic complex

Figures 1a and 1b show the absorbance of the type of ultraviolet absorbers used as stabilizers for polymeric materials. The 2-hydroxybenzophenones are represented by Cyasorb UV-9, Cyasorb UV-314, Cyasorb UV-531, Dow DBR, and Dow HCB; Tinuvin P and Tinuvin 327 are 2-(2'-hydroxyphenyl)benzotriazoles; Uvinul N-35 is an aryl-substituted acrylate and salol is phenyl salicylate.

Table 6. Miscellaneous Stabilizers

Trade name	Supplier	Composition
Inhibitor RMB	Eastman	*m*-hydroxyphenyl benzoate
Inhibitor HPT	Eastman	hexamethylphosphoric triamide
Busorb 30	Buchman	β-benzoyloxy-2'-hydroxychalcone
DIP	Nease	diisopropyl dixanthogen

Figure 2 compares the absorbance of Tinuvin P with three nonabsorbing light stabilizers. Ferro AM-101 is a nickel chelate, Santoscreen 99 is an ester which can rearrange to an ultraviolet-absorbing 2-hydroxybenzophenone, and Eastman HPT is hexamethylphosphoric triamide, $[(CH_3)_2N]_3PO$, which is offered as a light stabilizer.

Table 7. Key to Suppliers

Abbreviation	Supplier
Am Cy	American Cyanamid Company, Intermediates Department
Argus	Argus Chemical Corp., a subsidiary of Witco Chemical Co., Inc.
Buchman	Buchman Laboratories Inc.
Carlisle	Carlisle Chemical Works, Inc.
Dow	The Dow Chemical Company, Chemicals Department
Du Pont	E. I. du Pont de Nemours & Co., Inc.
Eastman	Eastman Chemical Products, Inc., Chemicals Division
Felton	Felton Chemical Company, Inc.
Ferro	Ferro Chemical Division, Ferro Corporation
GAF	GAF Corporation
Geigy	Geigy Industrial Chemicals, Division of Geigy Chemical Corp.
Heyden	Heyden Division, Tenneco Chemicals, Inc.
ICI	ICI America
Mallinckrodt	Mallinckrodt Chemical Works
Naftone	Naftone, Inc.
Nease	Nease Chemical Co., Inc.
Penick	S. B. Penick & Co.
PICCO	Pennsylvania Industrial Chemical Corporation
Scher	Scher Brothers

Tables 1 through 6 list the commercial ultraviolet absorbers by type, the trade names, the suppliers, and the chemical compositions. Table 7 shows the key to the suppliers.

Table 8. UV-Absorber Properties

UV Absorber	Maximum absorption[a]		Appearance	Melting point, °C
	λ, nm	a		
Uvinul 400	288-C	66.5	off-white	140–142
Inhibitor DHBP	323-C	43.0		
Rylex D				
Cyasorb UV-9	287-C	68.0	pale cream	63–64
Uvinul M-40	328-C	44.0		
Cyasorb UV-531	290-C	48.0	pale cream	48–49
Picco UV-299	330-C	32.0		
Carstab 700				
AM-300				
Uvinul 410	288-C	42.0	off-white	49–50
Mark 202A				
Inhibitor DOBP	288-C	40.0	pale yellow	43
Rylex H	325-C	28.0		
Cyasorb UV-24	285-C	46.0	pale yellow	68–70
	350-C	56.0		
Cyasorb UV-207	320-T	34.8	white	166–168
Uvinul D-49	288-C	45.0	yellow	130
	355-C	70.0		
Uvinul 490	288-C	46.0	yellow	80
	353-C	70.0		
Uvinul D-50	286-M	48.8	yellow	195
	345-M	58.0		
Uvinul MS-40	285-W	46.0	white	145
Cyasorb UV-284				
Uvinul DS-49	333-W	16.5	light yellow	>350
DBR	340-C	28.0	yellow	125–128
HCB	262-C	60.0	yellow	93–95
	350-C	21.0		
Tinuvin P	298-C	61.0	off-white	129–130
	340-C	70.0		
Tinuvin 320	305-C	50.0	off-white	152–154
	345-C	49.0		
Tinuvin 326	313-C	46.0	light yellow	140–141
	350-C	50.0		
Tinuvin 327	315-C	42.0	pale yellow	154–158
	352-C	47.0		
Tinuvin 328	300-C	45.0	off-white	81
	340-C	44.0		
Uvinul N-35	303-M	46.0	white	96–98
Uvinyl N-539	308-M	34.0	white	−10
UV-Absorber 317	321-C	45.5	pale yellow liquid	
UV-Absorber 318	321-C		pale yellow	65–85
UV-Absorber 340	338-C	129.6	yellow	98
Cyasorb UV-1988	315-C	95.9	white	58
Salol	310-C	24.0	white	41–43
TBS	311-C	17.0	white	62–64
Inhibitor OPS	311-C	16.0	white	72–74
AM-101	290-C	12.0	green	
Cyasorb UV-1084	296-C	13.5	pale green	258–261
Irgastab 2001	276.5-C	6.9	light tan	137–149
	283-C	7.1		

[a] C = chloroform, M = methanol, T = toluene, W = water.

The absorptivities, appearance, and melting points of the more important ultraviolet stabilizers are tabulated in Table 8, where the wavelength (λ) is given in nanometers, and the absorptivity is given by the formula (5)

$$a = \frac{A}{bc}$$

where a = absorptivity in liters per gram-centimeter
A = absorbance = $-\log_{10} T$
T = transmittance = amount of radiant power transmitted by the sample relative to the radiant power incident on the sample
b = sample path or cell length in centimeters
c = concentration in grams per liter

The main commercial uses for light stabilizers by type are the following:

2-Hydroxybenzophenones.   Polypropylene; polyethylene; poly(vinyl chloride); polyesters.

2-(2'-Hydroxyphenyl)benzotriazoles.   Polystyrene and copolymers; polypropylene; acrylics; polyesters; poly(vinyl chloride); polycarbonates.

Salicylates.   Acrylics; poly(vinylidene chloride copolymers); cellulosics; sunburn lotions.

Substituted Acrylates.   Poly(vinyl chloride); polystyrene.

Nickel-containing Stabilizers.   Polypropylene; polyethylene.

Bibliography

"Ultraviolet Absorbers" in *ECT* 1st ed., Suppl. 2, pp. 883–902, by George M. Gantz and Stiles M. Roberts, General Aniline & Film Corporation.

1. L. R. Koller, *Ultraviolet Radiation*, 2nd ed., John Wiley & Sons, Inc., New York, 1965, Chap. 4.
2. P. J. Briggs and J. F. McKellar, *J. Appl. Polymer Sci.* **12,** 1825 (1968).
3. M. Heskins and J. E. Guillet, *Macromolecules* **1,** 97 (1968).
4. *WADD (Wright Air Development Division) Tech. Rept. 61–108*, Air Research and Development Command, United States Air Force, Wright-Patterson Air Force Base, Ohio, May, 1961.
5. ASTM Designation E 131-68; ASTM Designation E 169-63.

David A. Gordon
Geigy Industrial Chemicals Division
of Geigy Chemical Corporation

V

VACUUM TECHNOLOGY

Vacuum technology comprises a variety of techniques for producing, measuring, and employing gas pressures less than atmospheric pressure. Each technique is limited to a certain pressure range, and within a given range the selection of the proper equipment and method of operation depends on the process requirements. The unit of pressure commonly employed is the torr defined so that one standard atmosphere $(1,013,250 \text{ dyn cm}^{-2})$ is exactly 760 torr. The torr, named in honor of E. Torricelli, has no abbreviation, but when used in conjunction with other abbreviated units it is written Torr. The torr has largely replaced the standard millimeter of mercury (abbreviated as mmHg) which is equal to 1.000,000,14 torr, but both of these units are incompatible with the standard International (SI) System of Units in which the recommended unit of pressure is the newton per square meter (abbreviated as $N \text{ m}^{-2}$). One torr equals exactly $(101,325/760) \text{ N m}^{-2}$. See also Units in Supplement Volume.

Because the range of pressures within the operating limits of a given pump or vacuum gage is usually several orders of magnitude, data on vacuum equipment and processes are frequently presented graphically on a logarithmic pressure scale. Many of the important applications of vacuum techniques involve pressures less than 1 torr, and pressures below 1 torr are expressed in terms of negative powers of ten. Pressures in the range from 1 torr to 10^{-3} torr are frequently expressed in millitorr (abbreviated mTorr). The pressure range from 760 to 25 torr may be referred to as low vacuum, the range from 25 to 10^{-3} torr as medium vacuum, and the range below 10^{-3} torr as high vacuum. The range below 10^{-9} torr is sometimes called ultrahigh vacuum.

Low- and Medium-Vacuum Techniques. Evaporation and distillation can be carried on at lower temperatures when the gas pressure is reduced. The use of lower temperatures reduces the thermal decomposition hazard and the rate of corrosion of process equipment. The vacuum required depends on the vapor pressure of the material and the temperature above which decomposition or corrosion is excessive. Many materials, which cannot be distilled at or above atmospheric pressure, can be distilled without decomposition in the range from 10^{-3} to 760 torr.

Vacuum pumps are sometimes used merely to create a pressure differential across material to be moved (eg, in filtration, lifting and circulation of liquids, and vacuum cleaners). In these applications little is gained by reducing the pressure below 5% of 1 atm (or 38 torr). Since the vapor pressure of water at room temperature is about 25 torr, the presence of water may frequently be tolerated in such applications. One of the most important applications of low-vacuum techniques is the evaporation of water from solutions or moist solids at temperatures below 100°C. Low-vacuum systems are therefore classified as wet or dry, depending on the presence or absence of water in the liquid phase. Wet systems can be pumped with wet vacuum pumps, such as piston pumps employing water as lubricant and sealant or water jet pumps.

Fig. 1. Applications of vacuum.

A: Mechanical pumps or steam ejectors. B: Booster pumps. C: Cryo, turbomolecular, sorption, ion or diffusion pumps.

Steam ejectors are also frequently used on wet systems. Dry mechanical vacuum pumps do not use water but in general employ some low-vapor-pressure liquid for lubrication of moving parts, sealing of valves, cooling, and for reducing the harmful space between piston and exhaust valve at the end of the discharge cycle.

In the vacuum range from 760 to 1 torr, pressures are measured by relatively simple mechanical or liquid-level gages, such as the Bourdon tube, bellows, or diaphragm gages, and the mercury U-tube manometer (see Pressure measurement). Leak hunting may be performed without the aid of expensive leak detectors by immersing the equipment in a water tank or coating suspected areas with a film of soap solution and watching for bubbles while raising the internal gas pressure slightly above atmospheric. However, halide- or helium-type leak detectors can be used to locate both large and small leaks quickly (see p. 137).

High-Vacuum Techniques. When the pressure falls below 1 torr, significant changes occur in the properties of the residual gas and vapors. The ability of air to transmit sound and to conduct heat begins to diminish rapidly with decreasing pressure, and high-voltage electric discharges change from an arc to a glow and finally black out as the pressure approaches 10^{-3} torr.

The molecular mean-free-path, or average distance a molecule can travel in a gas before colliding with another molecule, varies inversely as the pressure and becomes on the order of magnitude of the diameter of ordinary pipe lines when the pressure reaches 10^{-3} torr. As collisions between gas molecules become less frequent than collisions with the walls, the rate of flow of gas through a pipe ceases to depend on the internal viscosity of the gas and depends only on the temperature of the gas and the geometry of the pipe. When the mean-free-path exceeds the dimensions of the apparatus, the gas molecules move in straight paths from wall to wall, evaporating from each surface in a random direction, and are removed from the system only by a chance flight from some surface directly exposed to the mouth of the pump. This process is referred to as free molecular flow and requires relatively large-diameter pipes of minimum length to avoid throttling the pump.

At pressures below 10^{-3} torr, the chemical composition of the residual gas changes rapidly as the oxygen and nitrogen content becomes negligible, while water vapor, organic vapors, and other sorbed gases begin to evolve from the walls, gaskets, and process materials in the system. Modern high-vacuum equipment can produce and measure pressures in the range from 10^{-3} to 10^{-9} torr on an industrial scale when suitable techniques are observed in designing and fabricating the vacuum system (1–3). This equipment includes oil and mercury diffusion pumps, getter-ion pumps, cryopumps, and special mechanical pumps for high-vacuum service which will be described in the following sections. High-vacuum gages are relatively complicated electrical or mechanical devices requiring special operating and calibrating techniques to avoid large errors in measurement. Some of the common vacuum gages and the techniques for leak detection in high-vacuum systems will be described.

Several of the principal high-vacuum applications are arranged in chart form in Figure 1 according to the approximate pressure required in the vacuum system. High-vacuum techniques used in distillation and metallurgy are described briefly in the following. For other vacuum applications see Drying, Vol. 7, pp. 326–378 (particularly freeze drying); Filtration, Vol. 9, pp. 264–286; Film deposition techniques, Vol. 9, pp. 186–220; Electron tube materials, Vol. 8, pp. 1–23; Mass spectrometry, Vol. 13, pp. 87–99; Metallic coatings, Vol. 13, pp. 249–284; Electric furnaces, Vol. 10, pp. 252–278; Semiconductors (fabrication), Vol. 17, pp. 862–883.

Ultrahigh-Vacuum Techniques.   Techniques have been developed for producing and measuring pressures considerably below 10^{-9} torr for such purposes as the study of surfaces free of adsorbed gas or the obtaining of very pure gases (4,5).   By using systems constructed entirely of metal and glass, which can be baked in an oven at 400–500°C, and by cleaning up residual gases with getter-ion pumps, sorption pumps, or cryogenic pumps, pressures as low as 10^{-14} torr can be reached.   These techniques find application in the electronics industry, but at present they are of limited application in chemical technology.

Measurement of Low Pressures

The range of several common types of vacuum gages is illustrated in Figure 2.   The width of the band is a rough indication of the variation in precision of readings over the useful range of a given gage but is not intended to indicate the relative accuracy of different gages.   The ranges shown are somewhat arbitrary since modifications in design or operating conditions can shift the ranges up or down to some extent.   For illustrations of diaphragm, Bourdon, thermocouple, and ionization gages see Pressure measurement, Vol. 16, pp. 470–481.

Liquid-level gages, such as the U-tube manometer and the McLeod gage, measure the pressure directly, and their calibration can be determined from the geometry of the instrument and the density of the liquid.   These gages are constructed of glass and are easily broken and cumbersome to use, but they are important as one of the primary standards against which the other types of gages must be calibrated.   The other gages are more convenient because they are usually provided with dials or scales on which the pressure is indicated by a pointer, but in general they do not have the same calibration factor for all gases and are subject to electrical and mechanical defects (6).

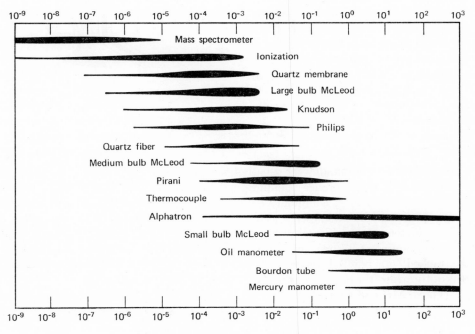

Fig. 2. Range of vacuum gages.

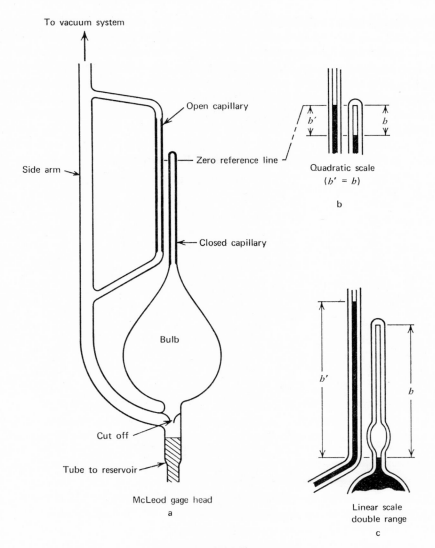

Fig. 3. McLeod gage.

The **mercury or oil manometer** usually has the form of a simple glass U-tube with means for providing a reference vacuum over the liquid in one arm. While n-butyl phthalate is the fluid most commonly employed in oil manometers, other nonviscous low-vapor-pressure oils of known specific gravity may be used. The sensitivity of these manometers can be increased by inclining the measuring arm at a small angle with the horizontal or by measuring the meniscus level with a vernier or micrometer head attached to a pointed vertical rod inside one arm. The range depends on the lengths of the arms, the density of the liquid, and the reference pressure.

A typical **McLeod gage** head is shown in Figure 3a. Mercury in a reservoir (not shown) is moved by the operator upward into the big glass bulb, trapping gas in the bulb as the mercury rises above the cut-off opening, and compressing the gas into a closed capillary attached to the top of the bulb. The mercury rises simultaneously through a side arm attached below the bulb into an open capillary located adjacent to the closed capillary and of equal bore until the mercury level in one of the capillaries reaches a reference line on a scale behind the capillaries. This reference line is usually located level with the top of the closed capillary (see Fig. 3a) but may be at any measured distance below the top of this capillary.

McLeod gages can be designed to cover, on one scale, a range of 3–4 cycles on the log scale of Figure 2 with the lower end of the useful range located at any point from 5×10^{-7}–1 torr. The double-range McLeod gage employs a second bulb of smaller volume located near the base of the closed capillary as shown in Figure 3c. By bringing the mercury meniscus to a zero reference line located between the bulbs, the pressure, p, in torr is indicated by the position of the meniscus in the open capillary on a linear scale according to the formula $p = kh \, h'/(1 - kh)$, where h is the distance (in mm) between the top of the closed capillary and the mercury meniscus in the closed capillary, h' is the difference in height (in mm) between the menisci in the open and closed capillaries, and k is a constant equal to the cross-sectional area (in cm^2) of the closed capillary divided by 10 times the combined volume (in cm^3) of the closed capillary and the bulb (or bulbs in the case of a double-range gage) to the cut-off. The second range is obtained by bringing the meniscus in the open capillary to a zero reference line at the top of the closed capillary (Fig. 3b) and noting the position of the meniscus in the closed capillary on a direct-reading scale. The pressure reading, p, on this scale, follows the quadratic relation $p = kh^2/(1 - kh)$, where h is the distance between the zero reference line and the scale reading and k is defined above. The quadratic scale gives higher sensitivity at low pressures than the linear scale. Correction of k for variation in density of mercury from room temperature to 0°C and the local value of gravitational acceleration is usually ignored since the error is less than 1%.

The McLeod gage cannot be used for gases or vapors that condense at the prevailing pressures and temperatures in the closed capillary. A solid carbon dioxide or liquid nitrogen trap must be included in the connecting line if the mercury vapor from the McLeod gage is to be kept out of the system or if vapors in the system are likely to contaminate the gage. The use of these traps is essential when calibrating other types of gage against the McLeod gage, but errors due to the pressure differential caused by the streaming of mercury vapor from the McLeod gage into the trap must be avoided by suitable techniques (6,7).

In the operation of the **Knudsen or radiometer gage,** a suspended vane system is rotated by directing gas molecules evaporated from a hot surface against one side of the vane while less energetic molecules coming from a cool surface impinge on the opposite side. While the calibration of the Knudsen gage varies only slightly with different gases and the gage has been found useful in chemical investigations, it is not commonly available commercially. The gage requires careful mounting, and vibration must be suppressed by the dampening action of magnets.

Various types of mass spectrometers in simplified form (residual gas analyzers) are often used to measure the partial pressure of gases, such as H_2, H_2O, CO, CO_2, N_2, O_2, and A, commonly found in vacuum systems. The gas molecules are first ionized by electron collision and then the positive ions are sorted out and collected according to their ratio of mass to electronic charge, the collector current being correlated with the partial pressure by previous calibration. Polyatomic molecules may dissociate into various fragments during this process, and a characteristic spectrum is obtained.

The **hot-filament ionization gage** can be calibrated for various gases but is not reliable with oxygen, hydrogen, carbon dioxide, and other active gases and vapors that are decomposed by chemical action or pyrolysis at the surface of the incandescent filament. The gage responds rapidly to changes in pressure, but the indicated pressure change is likely to be quite inaccurate at the lower pressures unless many precautions are taken. The error may be positive or negative, depending on the relative rates of cleanup of the gas by sorption and the outgassing of the walls and electrodes. Large errors may be introduced when the connecting pipe between the system and the gage head is long and narrow so that the flux of molecules into or out of the gage tube creates a pressure differential. Techniques for measuring the currents in ion gage tubes are explained in the operating instructions supplied with the electrical control circuits available from vacuum equipment manufacturers. The lower pressure limit of the gage is set primarily by the background currents due to photoelectrons emitted from the collector which intercepts soft x rays created by electron bombardment of the anode (6).

The principle of the **Philips gage,** also called the **cold-cathode ionization gage or Penning gage,** is the measurement of the current transmitted through a glow discharge in the gas between a wire loop anode and two disc-shaped cathodes located above and below the plane of the anode at potentials of about 2000 V in the presence of a magnetic field. The direction of the magnetic field is parallel to the axis of the loop so that the electrons travel in long spiral paths as they move toward the anode. This action increases the possibility of collision with the gas molecules present, produces greater sensitivity, and sustains the glow discharge at lower pressures where the electron mean-free-path greatly exceeds the distance between electrodes. The Philips gage has two advantages over the hot-filament ionization gage. Since the cathode is cold, there is nothing to burn out when the tube is accidentally exposed to high pressures. Also, the Philips gage can be used to cover the important range from 10^{-6} to 10^{-1} torr, whereas the ordinary ion gage cannot be used above 2×10^{-3} torr. However, the Philips gage is not as reliable as the ordinary ion gage because the parts of the tube are more difficult to outgas and the cleanup effect is intensified by the high voltages and longer electron trajectories. The gage should not be exposed to mercury vapor. The magnetron-type gages developed by Redhead and by Lafferty can read pressures as low as 10^{-13} torr (4,8).

The **Alphatron gage** uses the α particles from a radioactive source to ionize the gas. It has the widest range of the electrically operated gages, recent designs operating from 1000 torr to less than 10^{-4} torr. It has no filament to burn out and does not suffer from the cleanup effects present in other types of ion gages (6).

The **quartz fiber gage,** the **quartz membrane gage,** and all-glass gages of the Bourdon type are available for accurate measurements on extremely corrosive gases such as chlorine, but they are too fragile and tedious to operate for ordinary industrial use. Descriptions of these gages are given in references (3,6). The quartz fiber and quartz membrane gages operate on the principle of the damping action of the gas on the oscillations of a fiber or membrane which is set in motion by magnetically operated devices. The decay in the amplitude of the oscillations is usually observed through an optical system and can be correlated with the pressure.

The Bourdon gage has the same calibration factor for all gases, but the Pirani and thermocouple gages must be calibrated for the gas to be measured. The **Pirani and thermocouple gages** operate on the principle of the change in temperature of a heated filament (of tungsten, platinum, or other stable metal with large temperature coefficient of resistance) with changes in gas pressure at low pressures. The Pirani measures the change in filament temperature as a change in resistance of one branch of a Wheatstone bridge which may have a similar filament in a sealed-off tube in a parallel branch to compensate for bridge voltage fluctuations and ambient temperature effects on the zero balance. The thermocouple gage measures the temperature change by a thermocouple junction welded to the center of the filament. These gages are commonly used on vacuum stills and dehydration systems at pressures below one torr. It is not safe to operate a Pirani or thermocouple filament at temperatures above 250°C if organic vapors from the common greases, oils, and gasket materials are exposed in the system. Decomposition of these vapors on the filament usually leaves a dark deposit which seriously alters the sensitivity and the zero balance of a Pirani by increasing the emissivity of the surface. Pirani and thermocouple gages are excellent for indicating changes in pressure and the approximate value of the true pressure between 10 and 200 mTorr. However, they are not to be trusted in the 0–10 mTorr range above 200 mTorr unless calibrated by an experienced operator on a vapor-free system a few hours before using.

Production of Low Pressures

The pressure of gas in a chamber at a given temperature may be lowered by allowing the gas to escape into a vacuum pump or by exposing a surface that retains impinging gas molecules by sorption or chemical forces. The movement of the gas toward the pump or trapping surface may be accelerated by ionizing the molecules and applying an electric field. Pumps employing ionization or sorption may be attached to the chamber.

The pressure within a closed chamber of fixed volume containing a large percentage of condensable vapor can be greatly reduced by cooling the walls to a temperature at which the vapor pressure is negligible. This principle is employed in the steam turbine in conjunction with pumps to remove the residual air. In all pumps some cooling of the walls is necessary to remove heat of compression and the heat of condensation of any vapors to be condensed in the pump. In some pumping systems, refrigerated traps are used to condense vapors before reaching pumps whose efficiency might be impaired by condensation of the vapor within the pump.

The pump speed is defined as the volume of gas that flows into the pump in unit time at a constant gas pressure measured near the intake port. In high-vacuum technology speed is frequently expressed in liter/sec or in cubic feet per minute (which in the vacuum industry is commonly abbreviated as cfm). (1 liter/sec equals 2.12 cfm.)

Some high-vacuum pumps eject vapors or residual gases generated within the pump back into the system even while gas from the system enters the pump, and all pumps have a minimum pressure which they can produce under standard test conditions known as the ultimate pressure or ultimate vacuum which is dependent on the balance between the rate of flow of material in and out of the pump across the inlet opening.

The production of low pressures by chemical reactions, or electrical cleanup is, in general, limited to high-vacuum and ultrahigh-vacuum applications, such as the use of a barium getter in the production of radio receiving tubes (see Vol. 8, p. 20) and the use of titanium sublimation pumps on chambers for simulating the very low pressures in interplanetary space. Zeolite sorption traps cryogenically cooled may be used to lower the pressure from one atmosphere to less than 10^{-3} torr in small chambers without the aid of mechanical pumps, thus avoiding contamination of the chamber by vapors from the sealing and lubricating oils used in mechanical pumps (9).

Good vacuum technique consists largely in the use of materials of sufficiently low outgassing rate; the proper selection and matching of pumps, system, and pipe lines; and the design of flanges, gaskets, couplings, and welds so that leakage is reduced to a minimum and leak hunting is facilitated (10–12). However, even well designed systems sometimes fail to reach the required low pressure, and certain vacuum techniques must be followed to diagnose and correct the trouble. These techniques involve the use of measuring and control instruments such as vacuum gages, leak detectors, thermometers, liquid-level indicators, valves, and heater power controls. A skilled technician can also tell much about the operation of pumps by listening to the sounds emitted and feeling the temperature of the walls at certain points or noting the appearance of the pump fluid in pumps with glass walls or windows.

Vacuum Pumps

The first practical vacuum pumps were developed by Otto von Guericke and Robert Boyle during the years 1640–1670. They employed a moving piston fitting tightly in a cylindrical casing and having check valves that permitted the gas to flow only in the desired direction. These pumps were greatly improved by Fleuss (1890) who provided a quantity of oil in the exhaust end of the pump cylinder to reduce the harmful space at the end of the piston stroke so that less gas remained unexpelled through the exhaust valve. Various pumps employing liquid mercury as a moving piston or a check valve were developed by Geissler (1855), Toepler (1862), Sprengel (1865), Kaufmann (1905), Gaede (1905), and others. Gaede developed the oil-sealed rotary piston mechanical pump in 1906. Finally, vapor jet pumps employing high-velocity streams of vapor discharged from nozzles into ducts of such cross section that the gas cannot readily flow countercurrent to the stream were developed. Steam ejector pumps were invented by Parsons (1902) and Leblanc (1908). Mercury vapor diffusion pumps were invented by Gaede (1915), Langmuir (1916), and Crawford (1917). Oil vapor diffusion pumps were invented by Burch (1928), Hickman (1929), and others. The oil ejector pump was developed by Hickman and Kuipers in 1941, but is seldom used any more because of the availability of Roots-type blower pumps which cover the range from 10^{-4} to 20 torr.

Modern vacuum pumps are of ten general types: (1) reciprocating piston pumps, (2) rotary piston pumps, (3) centrifugal pumps, (4) liquid jet pumps, (5) rotary blower pumps (turbine wheel and two- or three-lobe Roots-type pumps), (6) molecular drag or turbomolecular pumps, (7) vapor jet pumps (ejector pumps and diffusion pumps), (8) getter-ion pumps, (9) sorption pumps, (10) cryogenic pumps (cryopumps). Some vacuum pumps, such as types (1) through (5) can be designed to discharge gases directly into the atmosphere, while others, such as types (6) and (7), require a backing pump or forepump to maintain a suitably low forepressure on the discharge side. The compression ratio produced by a given pumping stage is defined as the ratio of exhaust pressure to intake pressure at a given throughput when pumping a gas or a vapor that is not condensed within the pumping stage. Pumping stages of limited compression ratio are frequently combined in series within a common pump housing to give a multistage pump of broad operating range. The compression ratio for single-stage ejector pumps is about 10, and the compression ratio for individual stages of multistage diffusion pumps is about 4 at maximum throughput. Most oil diffusion pumps have four stages within a common pump casing, giving a total compression ratio of about 256 at maximum full-speed throughput.

Oil-sealed mechanical piston pumps can be operated with high compression ratios over a range from atmospheric pressure to about 10^{-3} torr. **Rotary piston pumps** employ a cylindrical housing (stator) and an eccentrically mounted rotor which may contain radial slots through which vanes can slide back and forth to keep contact with the stator wall, or the stator may contain a slot through which a vane slides with one edge maintaining contact with the rotor [see Fig. 15(b), page 738, Vol. 16]. **Rotary plunger pumps** have a cylindrical plunger (or piston) which is moved by an eccentric rotor in a sliding rotary motion with an oil seal against the walls of a cylindrical stator and which divides the stator into two compartments by means of an attached vane or blade which slides through a slot in a cylindrical bearing in the stator wall [see Fig. 15(c), page 738, Vol. 16. *Note*: The figure captions for Figs. 15(b) and (c) should be interchanged]. A spring-loaded valve on the discharge side of the stator prevents the return of gas as the rotary seal passes by the discharge opening. To assist the action of the valve and insure complete discharge of the gas some oil is discharged along with the gas, and means are provided for separating and recirculating this oil to the inside of the stator.

Centrifugal pumps do not employ a discharge valve but move the gas, or a mixture of gas and liquid, from the axis to the circumference of a housing by the propelling action of a rapidly rotating member provided with ducts, blades, or vanes. They are not efficient at low pressures. **Liquid jet pumps,** such as the water aspirator, entrain the gas along the boundary of a high-velocity jet of liquid directed by a nozzle into a Venturi tube. Water aspirators are commonly employed for suction in the chemical laboratory but are seldom used on an industrial scale because of the large flow of water required (about one cubic foot of water for each cubic foot of gas pumped). However,

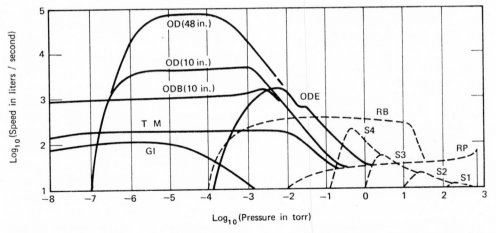

Fig. 4. Variation of pumping speed with inlet pressure for typical vacuum pumps. OD (48 in.) = oil diffusion pump with nominal 48 in. diameter inlet flange; OD (10 in.) = 10 in. oil diffusion pump; ODB (10 in.) = 10 in. oil diffusion pump with baffle; ODE = oil diffusion-ejector pump; TM = turbomolecular pump; GI = getter-ion pump (sputter type); RB = Roots type blower; RP = rotary piston oil-sealed mechanical pump; S1, S2, S3, S4 = matching stages in series of 4-stage steam ejector.

hydrosteam pumps are available in which the condensing water for one or more stages of steam ejector is recirculated through a water jet pump acting as forepump. Liquid mercury jet pumps of the Sprengel type entrain the gas between successive short columns of liquid formed by mercury falling into a narrow vertical tube. They are rarely used today except for handling small quantities of gases (such as tritium) which would react with organic liquids or water and which must be pumped from a vacuum system into storage tanks or reaction vessels at pressures above 10 torr. Automatic Toepler pumps (mercury piston pumps) may also be used for this purpose.

Table 1. Operating Ranges of Some Commercially Available Pumps

Type of pump	Operating range, torr
reciprocating piston:	
1-stage	$760 - 10$
2-stage	$760 - 1$
rotary piston oil-sealed:	
1-stage	$760 - 10^{-2}$
2-stage	$760 - 10^{-3}$
centrifugal multistage (dry)	$760 - 200$
liquid jet:	
mercury Sprengel	$760 - 10^{-3}$
water aspirator (18°C)	$760 - 15$
two-lobe rotary blower (Roots type)	$20 - 10^{-4}$
turbomolecular	$10^{-1} - 10^{-10}$
zeolite sorption (liquid nitrogen cooled)	$760 - 10^{-3}$
vapor jet pumps:	
steam ejector:	
1-stage	$760 - 100$
2-stage	$760 - 10$
3-stage	$760 - 1$
4-stage	$760 - 3 \times 10^{-1}$
5-stage	$760 - 5 \times 10^{-2}$
oil ejector (1-stage)	$2 - 10^{-2}$
diffusion-ejector	$2 - 10^{-4}$
mercury diffusion with trap:	
1-stage	$10^{-1} - {<}10^{-6}$
2-stage	$1 - {<}10^{-6}$
3-stage	$10 - {<}10^{-6}$
oil diffusion:	
1-stage	$10^{-1} - 5 \times 10^{-6}$
4-stage fractionating (untrapped)	$5 \times 10^{-1} - 10^{-9}$
4-stage fractionating (trapped)	$5 \times 10^{-1} - 10^{-12}$
getter-ion (sputter-ion)	$10^{-3} - 10^{-11}$
sublimation (titanium)	$10^{-4} - 10^{-11}$
cryopumps (20°K)	$10^{-2} - 10^{-10}$
cryosorption (15°K)	$10^{-2} - 10^{-12}$

Rotary blower pumps move the gas by the propelling action of one or more rapidly rotating members provided with lobes, blades or vanes revolving within a close-fitting housing without a discharge valve. The Roots type blower pump has a pair of lobed interengaging impellers, usually of figure eight cross-section, rotating in opposite directions. **Turbomolecular pumps** employ very high speed (3000–16,000 rpm) rotors with special turbine blades designed for pumping gas at pressures below 10^{-3} torr.

Vapor jet pumps employ a pump fluid which is heated in a boiler to generate the necessary vapor. The strength of the vapor jets depends on the boiler pressure, and each pumping stage has a limiting forepressure above which the pumping action is impaired by the gas in the forevacuum pushing the vapor jet away from the wall of the pump casing so that gas can leak back into the high vacuum. For diffusion pumps this limiting forepressure is approximately one-half of the take-hold pressure below which pumping action begins. The upper limit of the ranges shown in Table 1 for vapor jet pumps represents the take-hold pressure for the last stage of compression. The lower limit represents the ultimate pressure obtainable with typical pump fluids as measured on an untrapped gage of the type that reads total pressure of gas and vapors. When no traps are used, the ultimate pressure of oil and mercury diffusion pumps is determined primarily by the vapor pressure at room temperature of the pump fluid that condenses on the walls near the intake port.

Steam ejector pumps entrain the gas along the boundary regions of a jet of steam directed by a nozzle into a Venturi tube (diffuser) without condensation. Oil ejector pumps use a vapor jet of a stable organic compound directed into a Venturi tube (diffuser) which is cooled to condense the vapor striking the walls. Diffusion-ejector pumps combine one or two stages of diffusion pumping with one or two ejector stages, usually employing a relatively stable hydrocarbon or chlorinated hydrocarbon pump fluid.

Operation of Diffusion Pumps. A typical oil diffusion pump constructed of metal is shown in Figure 5. (The dashed lines show the modification necessary to convert to a fractionating type of pump.) The liquid in the bottom of the pump (or the boiler) is heated to temperatures in the range 150–250°C causing vapor to evaporate from the upper surface (bubbles do not form as in ordinary boiling) at a rate sufficient to create boiler pressures in the range 0.3–1.0 torr. The vapor passes upward inside the chimney and is accelerated to supersonic velocities as it expands through the nozzles. The nozzles direct the vapor downward, but the vapor diverges outward toward the cooled pump casing as shown by the arrows. Also, some vapor is scattered back toward the intake port. Gas molecules from the system can diffuse downward through this backstreaming vapor and penetrate into the denser forward-moving core of the vapor jet. Here they are driven at an acute angle toward the wall and on into the forevacuum. Because of the density and velocity of this forward-moving stream, it is almost impossible for gas molecules from the forevacuum to diffuse upstream through the vapor jet and escape into the high-vacuum system. There are always a small quantity of backstreaming vapor molecules which are scattered from the top jet through the mouth of the pump. These molecules condense on baffles placed above the pump and may then migrate slowly by reevaporation or creepage of oil films into the vacuum system unless stopped by cold traps and creepage barriers.

Metal diffusion pumps are now available commercially in sizes ranging from a casing diameter of 1 in. and a speed of a few liter/sec to a casing diameter of 52 in. with a speed of about 90,000 liter/sec. Air cooling can be used on the smaller pumps, but water cooling is to be preferred. Organic or silicone oils are preferred as pump fluids because in fractionating pumps they can produce ultimate pressures as low as 10^{-9} torr without cold traps, whereas mercury diffusion pumps always require a cold trap. Mercury vapor is toxic, and mercury vapor pumps are not recommended for vacuum systems in which foods or drugs are processed. Among the commonly used organic pump fluids are di(2-ethylhexyl) phthalate, petroleum fractions of average molecular weight between 300 and 400, the chlorinated biphenyls and the polyphenyl ethers (9). The organic diffusion-pump fluids will be decomposed when exposed to air pressures above a few torr without turning off the pump heaters. The boiler should therefore be allowed to cool to a safe temperature as recommended by the manufacturer before exposing to atmospheric pressure, or the pump should be kept under vacuum by using valves and a bypass system when admitting air to the process vessel.

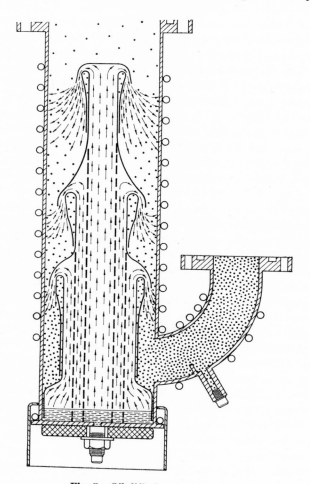

Fig. 5. Oil diffusion pump.

Getter-ion pumps trap the gas in layers of chemically active metal which are deposited (by continuous or intermittent evaporation or sputtering from a suitable source) on a collector surface which may be negatively charged and, in some cases, cryogenically cooled. Freshly deposited titanium metal layers will chemisorb oxygen, nitrogen, hydrogen, water vapor, and other common gases, while inert gases such as argon must first be ionized by an electric discharge and the ions then drawn by an electric field to the collector where they are buried by fresh layers of titanium before the neutralized ions can escape. **Sorption pumps** employ artificial zeolite or activated charcoal to adsorb gas on the large surface area provided by these porous materials, usually with the aid of liquid nitrogen cooling. **Cryopumps** employ liquid helium or mechanical refrigeration to cool exposed trapping surfaces to 20°K or below.

The types and average operating range of some commercially available pumps are listed in Table 1. The upper and lower limits of the operating range of each type of pump depend on the individual pump design, working fluid, and operating conditions. The values in Table 1 are, therefore, somewhat arbitrary. Actual performance curves for typical pumps are shown in Figure 4.

Baffles and Traps. It is frequently difficult to reduce the partial pressure of vapors by pumping alone to values at which contamination of the system, or the product being processed, is negligible. In this case, some type of zeolite sorption trap or re-frigerated baffle or trap must be installed (1). In all cases it is desirable to place a water-cooled baffle above the intake port of the diffusion pump in order to condense the backstreaming pump-fluid vapor and to return the fluid to the pump (12). If this is not done, the surfaces of the system directly exposed to the pump may become coated with a film of pump fluid, and the pump boiler may eventually be emptied by loss of fluid to the system. The baffle must be optically tight so that no vapor molecule can pass through by molecular flow without striking a cooled surface, but the passages through the baffle must be wide and short so that the resistance to gas flow is not too large.

Leak Detection. At pressures below 1 torr the rate of leakage through the walls of a system is substantially independent of the internal pressure. The partial pressure of air in the chamber due to a leak in the chamber will then be inversely proportional to the net pumping speed. In some applications it is not necessary to eliminate all of the small leaks in the chamber because suitable working pressures can be produced by employing high-speed pumps. However, in most applications there will be occasions when leaks occur which are too large to be tolerated.

If the system is properly designed, a leak can be readily located by using one of the following techniques (2,11): (1) helium gas is sprayed over the suspected leak area while the partial pressure of helium in the system is monitored with a mass spectrom-eter focused on helium; (2) Pirani, thermocouple, or various types of ionization gages to detect inleakage of a probe gas, such as acetone or hydrogen, for which the gage sensitivity is much different from the value for air; (3) palladium barrier ionization gage with hydrogen as probe gas which on entering a leak can diffuse through the hot palladium barrier into an evacuated and sealed ionization gage; (4) filling system with compressed air and coating with soap solution and looking for bubbles; (5) hot platinum-ceramic couple as halide-sensitive, positive-ion emitter with halogen com-pound as probe gas; (6) differential dual Pirani or ionization gage using a cold trap with one of the gage heads and a condensable probe gas which condenses in the trap to give a lower reading in the associated gage head (4); (7) rate of rise of pressure in sections isolated by valves, plugs, cover plates, etc; (8) high-voltage spark coil (Tesla coil) for glass vacuum systems with pin holes or cracks which will attract and hold the spark discharge when the tip of the metal core is brought near the leak; (9) ultraviolet absorption cell and organic probe gas which on penetrating the leak absorbs light from a hydrogen lamp and causes a deflection in the reading of a photocell (13). There are a number of other methods which involve subjecting the inside of the system to liquids or chemically active gases. These latter methods are not recommended be-cause the system becomes contaminated and must be thoroughly cleaned before a high vacuum can be produced.

The mass spectrometer type of leak detector, which employs helium for the probe gas, is the most sensitive and accurate instrument. Specially trained personnel are required for operation and maintenance, yet the time saved in finding leaks and avoiding shutdowns often justifies the use of this type of leak detector in industrial applications of large and complicated vacuum equipment. To use this method to best advantage, the process equipment should be designed with special double-gasketed flange joints for circulating helium, or the flanges should have a raised face so that the gaskets are easily accessible to a helium probe, and sufficient test connections should be provided at strategic locations for attaching the intake line to the leak detector. The location of these latter connections should be so chosen that the ratio of the system volume to the pumping speed is not more than 10 sec. After preliminary evacuation of the system with large pumps, the residual gases can be pumped directly into the leak detector using the small pumps within the latter. Or, the gases can be passed into an auxiliary pumping system consisting of a vapor jet pump and a mechanical pump with the leak detector connected to the backing space between the vapor pump and the mechanical pump. Helium from a standard cylinder with reducing valve is applied to the outside of the system under test by means of a long narrow metal tube with a fine orifice which directs a jet of helium at the points suspected of leaking. The jet may be moved along a weld or flange at the rate of about 1 in./sec. When the helium enters a leak, a surge will be observed in the reading of the output meter or a sound of increasing pitch and intensity will be heard on units equipped with an audioindicator.

The overall tightness of a system can be measured by the hood method. This consists in surrounding the equipment with a hood of rubber or plastic, or a gas-tight box, in which a fixed concentration of helium is maintained, and comparing the output meter deflection with that obtained when the helium mixture is allowed also to flow through a calibrated leak installed under the hood. The smallest leak rate that can be detected with the helium leak detector is about 10^{-11} ml/sec of air at STP. Large leaks require auxiliary pumping systems to lower the pressure to 0.1 torr or less, and the throttle valve on the leak detector is adjusted to maintain pressures of 10^{-4} torr or less in the mass spectrometer tube. Sampling probes are also available which draw helium mixed with air at atmospheric pressure into the leak detector when placed near a leak in a system filled with helium at 10–20 psi gage pressure.

When minute traces of certain organic vapors, especially those containing a halogen, come in contact with a hot platinum anode that has been sensitized by a source of alkali ions, the ability of the platinum to emit positive ions is greatly increased. There are two methods of using this phenomenon for leak detection. The system under test may be filled with gas containing a halogen compound, such as Freon refrigerant 12, at pressures above atmospheric and the gas escaping through a leak can be drawn into the detector by a small motor-driven fan (11). The sensitive platinum element can also be placed in a glass or metal envelope with a tubulation connected to the system, and the system evacuated while the halogen containing probe gas is played over the suspected areas on the outside of the system. Probe gas entering through a leak will diffuse into the detector head. In this latter method the voltages must be limited so that glow discharges are avoided. It is claimed that the sensitivity of these halide leak detectors is nearly equivalent to that of the helium leak detector. However, these halide leak detectors are not as satisfactory as the helium type for quantitative measurements of leak rate. In using Freon refrigerant 12 as the probe gas it must be remembered that this gas is about four times as heavy as air. Therefore, probing should be started at the bottom of the system and continued upward.

Design of Vacuum Systems

The design of vacuum systems requires a knowledge of the laws governing the flow of gases from the chamber through the pumps. For a given system geometry the flow behavior depends on the pressure and temperature of the gas, and four types of flow are distinguished by the names turbulent flow, viscous flow, transition flow (between viscous and molecular flow), and molecular flow. Poiseuille's law holds for the viscous flow of gases through cylindrical pipes (3). Turbulent flow is seldom encountered in actual vacuum practice except for short periods during which the mean pressure in the pipe is only slightly lower than atmospheric. If the pipeline is designed to be adequate in the viscous flow range, it will be found adequate in the turbulent range. The flow of air at room temperature becomes turbulent when the throughput exceeds $2 \times 10^5 D$ millitorr-liters per second, where D is the pipe diameter in centimeters. The laws of flow in the transition and molecular flow regions were developed by Knudsen, Smoluchowski, and Clausing (3).

Throughput is defined as the quantity of gas in pressure-volume units at a specified temperature (usually room temperature) flowing per unit time across a specified open cross section of a pump or pipe line (14). The measured speed of a pump for a given gas is then defined as the throughput of that gas, flowing from a standard test chamber across the pump inlet, divided by the pressure of that gas measured at a specified point in the test chamber near the pump inlet (15). Throughput is commonly expressed in torr-liter/sec or millitorr-cfm. The terms load and capacity are sometimes used to refer to the quantity of gas in mass units per unit time flowing through the pump at a given intake pressure. A load of 1 lb/hr of air at 25°C corresponds to a throughput of 80.5 torr-liter/sec.

The speed of exhaust, E, of a vacuum chamber is defined by the relation

$$E = -(dP/dt)(V/P) \tag{1}$$

where P is the total pressure in the chamber of volume V at time t. The speed of exhaust, E, approaches zero as the pressure, P, approaches the ultimate pressure, P_u, which the pump can produce in the chamber. The value of the system ultimate pressure, P_u, depends on the evolution of gases and vapors from the walls of the system including the chamber, the pump, and the connecting line. When P_u is nearly constant, the speed of exhaust is given by

$$E = S(1 - P_u/P) \tag{2}$$

where S is the speed or volumetric rate of flow of gas out of the chamber into the pipe line between chamber and pump, and where S is assumed to be nearly constant over the pressure range P_1 to P_u. Then substituting (2) in (1) and integrating gives a formula for the time of evacuation from P_1 to P_2 $(P_1 > P_2 > P_u)$:

$$t_2 - t_1 = 2.3(V/S) \log_{10} [(P_1 - P_u)/(P_2 - P_u)] \tag{3}$$

If the quantities S and P_u are not constant over a given pressure range, then the range may be broken into several intervals which are small enough that these quantities may be treated as constants and the above integration performed. The total evacuation time will be the sum of the times for the intervals.

The Net Speed Formula. The speed, S, for permanent gases at the outlet of the vacuum chamber will be less than the speed, S_0, at the inlet of the pump because of the resistance to the gas flow offered by the connecting line. The speed, S, is therefore sometimes referred to as the net speed. For a well-designed pipeline it is accepted practice that the net pumping speed S in the vacuum chamber shall be much less than 70% of the speed S_0 available at the mouth of the pump or at the entrance to the baffle (when the latter is not incorporated in the pump). The reduction in pumping speed depends on the resistance, or impedance, W, of the pipeline. The formula for computing the net pumping speed, S, is:

$$\frac{1}{S} = \frac{1}{S_0} + W \tag{4}$$

where S and S_0 are usually expressed in liter/sec and W in sec/liter. It is sometimes more convenient to use the reciprocal of the impedance, which is known as the conductance and is represented by $U = 1/W$. Then:

$$S = S_0 U/(S_0 + U) \tag{5}$$

Equations (4) and (5) are valid only when S_0 and U (or W) are suitably defined and measured (15,16).

Conductance of Circular Pipes. The conductance U is frequently expressed in liters per second, although to be strictly correct it should be considered as torr-liter/sec per torr pressure difference across the ends of the tube. Conductance problems are conveniently divided into two classes, the molecular flow type, for which the product of the mean pressure $\overline{P}$ (one-half the sum of the entrance and exit pressures) in millitorr of air at room temperature and the radius R of the tube in cm is ten or less, and the Knudsen-Poiseuille flow type, for which the $\overline{P}R$ product is greater than ten.

Figure 6 gives the molecular conductance, U_0, of air at 25°C through a circular pipe of diameter D cm and length L cm located between two large chambers in which the pressure is P_1 and P_2, respectively. The mean pressure will be given by:

$$\overline{P} = (P_1 + P_2)/2 \tag{6}$$

and the product $\overline{P}R$ should be computed (in millitorr-cm) to determine whether the flow is molecular or not. Figure 7 gives the relative conductance factor U/U_0, or ratio of true conductance to molecular conductance, for air at 25°C as a function of the $\overline{P}R$ product in millitorr-cm according to Knudsen's formula (3). To obtain the true conductance U, the molecular conductance U_0 as given by Figure 6 is multiplied by the relative conductance factor U/U_0 as given by Figure 7. To find this factor, locate the $\overline{P}R$ product on the horizontal axis. If the product is between 0.1 and 100, the value will be located at the bottom of the chart and curve A will be used. If the product is between 100 and 100,000, the top scale and curve B are used. For example, a $\overline{P}R$ product of 1000 corresponds to a factor of 30. The conductance U thus obtained represents the rate of flow in millitorr-liters per second per millitorr pressure drop at the given mean pressure. To find the actual quantity Q of gas flowing through the tube in millitorr-liter/sec, multiply this value of U by the difference in pressure at the two ends of the tube, thus:

$$Q = U(P_1 - P_2) \tag{7}$$

Figure 7 cannot be used when the flow becomes turbulent. The transition to turbulent flow occurs when the throughput exceeds 4×10^5 millitorr-liter/sec and for most sys-

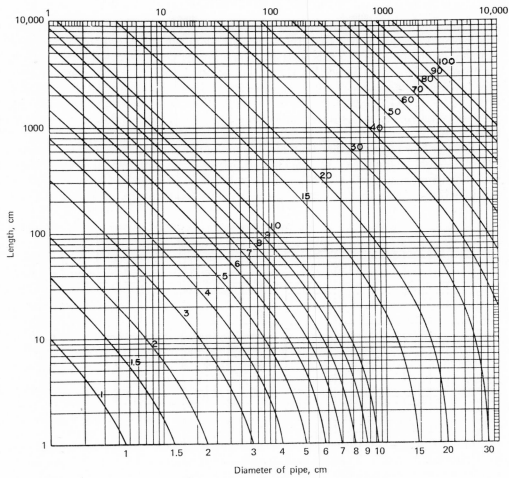

Fig. 6. Molecular conductance chart.

tems this is likely to occur in the region on Figure 7 where curve B is represented by a broken line.

The molecular flow of air at room temperature through long cylindrical tubes may also be estimated from the approximate formula:

$$U_0 = R^3/L \tag{8}$$

where U_0 will be given in liter/sec when the radius R and length L are expressed in mm. If the pressure is too high for molecular flow, then U_0 is computed from R^3/L or read from Figure 6 and the result is multiplied by the relative conductance factor as given by Figure 7.

If the pipeline is constructed of a series of long tubes of different diameters, the total molecular conductance U_0 in liter/sec for air at room temperature can be computed from the formula

$$\frac{1}{U_0} = \frac{L_1}{R_1{}^3} + \frac{L_2}{R_2{}^3} + \frac{L_3}{R_3{}^3} + \cdots \tag{9}$$

where L_1 is the length in mm and R_1 is the radius in mm of the first section, etc. If the length of any section is less than ten times the diameter and the diameter changes abruptly at the entrance to this section, then a correction term for the entrance or aperture impedance of this section should be included. To correct for the entrance impedance of any section of pipe of cross-sectional area A_n square centimeters, the term $(A_m - A_n)/11.7 \, A_m A_n$ is added to the right side of equation (9), where A_m is the cross-sectional area in cm² of the chamber from which the air enters the section in question.

For the flow of gases other than air, the molecular conductance, U_0, as given by Figure 6 should be multiplied by $(28.98/M)^{\frac{1}{2}}$, where M is the molecular weight of the gas. For temperatures other than 25°C the value of U_0 as given by Figure 6 should be multiplied by $(T/298.15)^{\frac{1}{2}}$, where T is the absolute temperature. Figure 7 for the

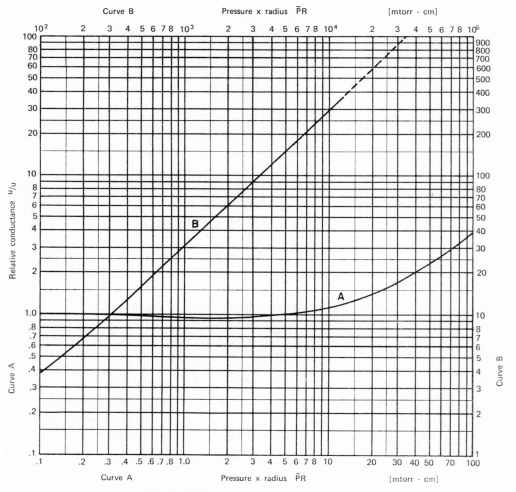

Fig. 7. Relative conductance factor chart.

relative conductance can be corrected for different gases and temperatures by multiplying the PR product by the factor $(1.83 \times 10^{-4}/\eta_i)(298M/29T)^{1/2}$, where η_t is the viscosity of the gas in poises, and locating this result on the horizontal axis instead of the original value of $\bar{P}R$.

The following method of changing the scale units permits Figure 6 to be used for tubes whose diameters are so small or so large that the point corresponding to the desired length does not fall on the chart. Multiply both the given length and diameter by 10, or some positive or negative power of ten, so that the point falls on the chart. Then the conductance read at the top of the chart must be divided by 100, or by the square of the power of ten chosen, to obtain the correct answer. For example, the conductance of a tube of 0.3 cm ID and 10 cm length is obtained by following the curve for a diameter of 3 cm to the intersection with the line for a length of 100 cm and reading 3 liter/sec from the scale at the top. The correct answer is then 3/100 or 0.03 liter/sec.

The resistance of a bend or elbow in the pipeline can usually be neglected for molecular flow, provided the length L is measured along the center line and the bend is not within two diameters of the end of the tube or another bend. For sharp bends located near the end of the tube or within two diameters of another bend, the effective length of the bend can be considered as its axial length plus one radius. Exact corrections for right-angle bends and other irregular shapes as calculated using Monte Carlo computer programs have been published (1). Formulas for the conductance of ducts of noncircular cross section will be found in references (11,17,18).

Materials of Construction. The vacuum chamber may be constructed of Pyrex brand (Corning Glass Works) borosilicate glass in the case of small chambers or bell jars, but industrial systems are usually constructed from steel (preferably stainless) or aluminum. For most of the materials exposed in high vacuum systems the outgassing rate can be assumed proportional to the exposed area, A_m, although for some very porous materials the rate is more proportional to the bulk or mass. Except for evaporation from pure liquid or solid phases, the outgassing rate normally decreases with time when the temperature is constant. For many industrial type vacuum systems it has been found that, after the initial evacuation to about 10^{-5} torr, the pressure, P, in the chamber decreases approximately according to

$$P = P_u + K_1 A_m/S_n t^\alpha \tag{10}$$

where P_u is the ultimate pressure, K_1 is a constant which may be considered equal to the average outgassing rate per unit area after one hour of pumping, A_m is the exposed area, S_n is the net pumping speed, t is the total pumping time (including the initial evacuation) in hours, and α is an exponent which is usually nearly constant for the first few hours. For rough calculations in typical metal vacuum systems it may be assumed that K_1 is on the order of 10^{-4} torr-liter/(sec) (ft^2) and $\alpha = 1$, but for systems containing large amounts of elastomers or plastics K_1 may be on the order of 10^{-3} torr-liter/(sec) (ft^2) and α may be closer to 0.5 (18,19).

Systems constructed entirely of metal and glass may be heated to temperatures as high as 500°C to accelerate the outgassing, the average rate increasing by approximately a factor of 10 for each 100°C increase in temperature. In order to reach 10^{-6} torr in 10 hr in an unconditioned vacuum chamber the net pumping speed (in liter/sec) should be 10–100 times the exposed area (in ft^2).

Assembly of Systems

The essential components of a typical large high-vacuum system employing vapor pumps are shown in Figure 8. The diffusion pump (A) must be located close to the vessel (B) and should always have a water- or refrigerant-cooled baffle (C) located over the pump mouth so that backstreaming vapor will be condensed and returned as liquid to the pump. The high-vacuum valve (D) and manifold connecting the vessel and the baffled pump should be of diameter equal to or larger than the mouth of the pump, and the length of the passage between the baffle and the vessel should preferably be not more than about three times the mean diameter of this passage. A space at least 6 in. high should be allowed below the diffusion pump boiler (E) for easy servicing of the heaters, draining the pump fluid, or removal of the pump from the system.

Selection of Pumps. The type of high-vacuum pump selected depends on the lowest pressure that must be attained (see Table 1). The size of the pump depends on the gas load during the process cycle and must be determined from performance curves similar to Figure 4 as supplied in the manufacturers' catalogs. As a general rule the net pumping speed in liter/sec at the entrance to the vessel should be of the same order of magnitude as the volume of the vessel in liters. (To convert volume in ft³ to liters multiply by 28.3, or roughly by 30.) For most chemical applications a water-cooled diffusion pump will be employed, and the cooling water should circulate through the coils from the high-vacuum toward the fore-vacuum side at a rate in cm³/min about equal to the heater input in watts. The maximum throughput of such pumps in milli-torr-liter/sec is approximately equal to twice the heater input in watts. Automatic safety controls are available to protect against failure of the cooling water. Efficient cooling is particularly important for mercury vapor pumps. Air cooling is possible with oil vapor pumps but is not recommended except on portable systems. For many physical applications in which the presence of organic vapors is undesirable, the system may be pumped down from atmospheric pressure to a few millitorr by cryogenically cooled zeolite sorption pumps and the pressure then reduced to the high-vacuum or

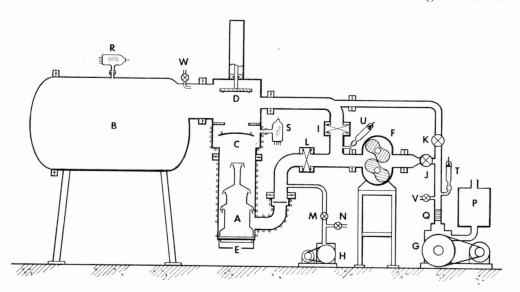

Fig. 8. Typical high-vacuum system.

ultrahigh-vacuum range by getter-ion pumps, titanium sublimation pumps, or cryo-pumps.

Since diffusion pumps will only operate against a forepressure (pressure on the exhaust side) of about 0.7 torr or less, the pressure range from atmospheric to the limiting forepressure for the diffusion pump must be covered by a suitable forepump or roughing pump. The oil-sealed rotary mechanical pump is the type most commonly used, but steam ejectors may be more efficient in some applications where high-pressure steam is available. The pressure range from 1 torr to 0.01 torr may be covered by either an oil vapor ejector or a mechanical pump. If a rotary piston or rotary plunger pump (G) is in use and it is found that too much time is being spent in this pressure range because of evolution of gas from the system, then an oil vapor ejector or a Roots-type blower (F) should be installed as shown in Figure 8.

The size of the mechanical pump required depends primarily on the time allowable for covering the range from atmospheric pressure to a pressure of 0.7 torr at which the typical diffusion pump begins to operate. This time can be estimated from the approximate formula:

$$t = 7V/S_f \tag{11}$$

where V is the total volume of the system and S_f is the average speed of the forepump (assumed constant over the whole pressure range). When V is given in ft^3 and S_f in cfm, then t will be given in min. When V is expressed in liters and S_f in liter/sec, t is obtained in sec. Thus, to pump a chamber of 10 ft^3 from 760 to 0.7 torr in 7 min requires at least a 10 cfm forepump. Since most mechanical pumps have an average speed about equal to $\frac{3}{4}$ of the rated speed or displacement at atmospheric pressure, a 14 cfm mechanical pump would be recommended in this case. Formulas and graphs for estimating pump-down time will also be found in the manufacturers' catalogs for the various mechanical pumps.

The size of the forepump selected may depend on the peak load as well as the pump-down time. Every diffusion pump has a maximum torr-liter/sec throughput determined by the jet design and operating conditions. However, if the speed of the forepump is not sufficient, the maximum torr-liter/sec throughput of the diffusion pump will be limited to the product of the forepump speed and the limiting forepressure for the diffusion pump. Since it is usually desirable to keep this product larger than the peak load of gas evolved from the vacuum system after the diffusion pump has begun to operate, the size of the forepump selected should be adequate to meet this peak-load condition. Frequently the size indicated by the peak-load condition is much smaller than the size required to meet the specified pump-down time. In this case, if the processing part of the cycle is much longer than the pump-down time, it is advisable to use two forepumps; a large one (G) for roughing down from atmospheric pressure and a smaller one (H) for holding the vapor pumps during the roughing period and backing them during the processing period.

Location of Valves, Traps, and Connecting Lines. Because of the long warm-up and cool-down times in some cases it is desirable to keep the diffusion pump hot by closing valves (D) and (L) and operating the holding pump (H) while evacuating the chamber to 0.5 torr or less with the forepump (G) and blower (F). Then valves (L) and (D) are opened and (I) is closed. When the time allowable for reaching 10^{-4} torr is more than 20 min, the roughing can be done through the diffusion pump, provided the boiler has been cooled to a safe temperature for exposure to atmospheric pressure.

The time required to cool the boiler to the safe temperature can be shortened by passing cold water through a separate cooling coil which surrounds the boiler, the water being blown out again by compressed air before again turning on the pump heaters. The use of this procedure obviates the need for high-vacuum valves, bypass lines, and holding pumps.

Provision for separating water or solvent vapors from the mechanical pump sealing oil reservoir (P) or for preventing condensation within the mechanical pump should be made whenever the system evolves large quantities of moisture or vapor. Most mechanical vacuum pumps are equipped with a gas-ballast valve which may be opened to admit air at a certain point in the compression cycle to avoid condensation of moisture within the pump (9). Modified rotary pumps are available in which hot oil or steam is circulated to maintain the pump above 100°C during operation so that any water load can be handled while still reaching an ultimate pressure below 0.1 torr (9,20). In extreme cases, a cold trap must be included in the forevacuum line to condense moisture or solvents which might contaminate the forepump. If the sealing oil becomes contaminated, it can sometimes be purified by centrifuging or filtration and evaporation, but it is usually cheaper to recharge the pump periodically with fresh oil.

Since the large mechanical pumps often have considerable vibration, flexible metal bellows (Q) are usually included in the forevacuum line. These bellows also aid in aligning the pipes during assembly. It is not necessary to have the rotary piston pump (G) or even the Roots-type pump (F) installed close to the vessel and the diffusion pump, since the resistance to gas flow of moderate lengths of pipe having the same diameter as the suction opening on these pumps is not a serious problem at pressures above 0.2 torr. In some installations it is desirable to locate the mechanical pumps at a distance for easy servicing without exposure of personnel to dangerous process equipment, or to reduce the noise level in the process room.

If the process involves a period of outgassing or there is a steady evolution of gas, the pressure that can be maintained is roughly inversely proportional to the net speed of the diffusion pump and connecting pipeline. When the outgassing load exceeds the maximum capacity of the diffusion pump, the jets break down and the pressure rises rapidly. If the heater is not shut off and the forepressure is not too high, the jet action may be resumed temporarily. However, the pressure oscillates over a wide range and the pumping action is unsteady. In some processes the period of outgassing is brief and the diffusion pump is allowed to run in the breakdown region until the pressure again falls to the point at which the process can begin. Roots blower pumps are extremely useful in shortening these periods. If the forepressure rises to more than 0.8 torr for more than five min, the heater of the diffusion pump should be shut off to avoid decomposition of the oil. Automatic devices for shutting off the heater are available.

Installation of Gages and Controls. Referring to the typical large vacuum system illustrated in Figure 8, a hot-filament ionization gage or a cold cathode ionization gage should be installed on the process vessel as shown at (R) and also on the baffle at (S) below the seat of the high-vacuum valve. If mercury vapor is present in the system, the gage may have to be trapped with liquid nitrogen or with solid carbon dioxide and acetone. The rate of rise of pressure on gage (R) when valves (D), (I), and (K) are closed will indicate the leak rate or outgassing rate and give information about the presence of vapors.

A Pirani, thermocouple, Alphatron, or other gage reading in the 1–1000 millitorr range, should be installed in the forevacuum line at (T). A Bourdon or diaphragm gage can be installed on the vessel or in the forevacuum line if the progress of evacuation from atmosphere down to a few torr must be watched. The condition of the forepump can always be checked by closing valves (J) and (K) and observing the gage at (T). A Pirani or thermocouple gage can be installed at (U) when information on the interpump pressure is of interest (for example, when pumping hydrogen or any gas at interpump pressures near the limiting forepressure for the diffusion pump).

The valve at (V) is used to vent the forepump while holding the rest of the system under vacuum by closing (J) and (K). Similarly, the holding pump (H) may be vented to atmosphere by closing valve (M) and opening (N). This should be done when shutting off the motor on the forepump or holding pump for a long period since the sealing oil in some types of mechanical pump may otherwise leak back into the vacuum system. The openings at (N) or (V) can also be connected to a leak detector.

The process vessel should be vented through a valve located on the vessel as shown at (W). The blast of inrushing air should be directed away from the process material in the chamber. The time, t, in min to vent a vacuum chamber of volume V ft³ from pressures less than 10^{-3} torr to atmospheric pressure through a standard globe valve of nominal diameter D in. is given approximately by:

$$t = V/100D^2 \tag{12}$$

A thermal switch may be installed on the cooling coils of the diffusion pump to protect against failure of cooling water or may be located in the boiler to prevent overheating due to a forepressure above the critical value. A pressure-operated safety switch can be installed in the forevacuum line to protect the diffusion pump against excessive rise in forepressure during operation.

Applications

Vacuum Distillation. Most organic compounds will not tolerate prolonged heating at temperatures greater than 250°C without excessive decomposition. Compounds of molecular weight greater than 150 will in general have vapor pressures less than 760 torr at temperatures below 250°C and must therefore be evaporated at reduced pressure by vacuum or steam distillation techniques. Steam distillation can only be used with materials that are immiscible and unreactive with water. Low pressure is also useful in the fractional distillation of mixtures that are azeotropic at atmospheric pressures. In some cases it is more economical to use low pressures in fractional distillation because less heat is required to achieve a given separation (21).

Vacuum-distillation techniques are classified by the type of still employed. Molecular stills are designed so that the distance between the evaporating surface and the condenser is not large compared with the mean-free-path of the vapor molecules; they operate at pressures of 0.01 torr or less (22,23). The material being distilled (distilland) is usually spread out in a thin film over the surface of the heating unit and caused to flow by gravity (falling-film still) or centrifugal force (centrifugal still) over the hot surface into a reservoir for the residue. Falling-film stills and rotary drum stills are usually provided with rotary or fixed blades which spread the oil in a more uniform thin film by a wiping action (wiped-film stills). Blades with slots in the edge cut at an

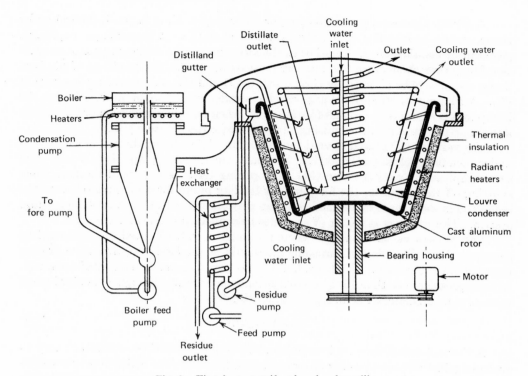

Fig. 9. Five-foot centrifugal molecular still.

angle to the plane of rotation serve to accelerate or slow down the rate of fall of the film, depending on the desired transit time through various zones of distillation. When no obstruction is interposed between evaporator and condenser, the distillation is substantially projective, that is, very few evaporated molecules return to the distilland. In short-path rotating drum stills the distilland is spread in a thin film on a horizontal roller and the residue is scraped off before the surface again dips into the distilland (24). Wiper blades or brushes cause turbulence in the oil film, which improves the distillation rate by avoiding surface depletion of the more volatile components.

The open-path or short-path pot still is not generally regarded as a molecular still, but it does employ a short, wide rectification column with a very low pressure drop from pot to condenser. Stirring of the distilland in the pot is desirable since ebullition does not occur in high-vacuum distillation. The equilibrium pot still employs a long, narrow column, or a wide column with stationary packing, or a concentric rotating condenser which flings the condensate back to the heated wall of the column. The fractionating efficiency is high, but the pressure drop and thermal decomposition hazard are also higher than for short-path stills (21).

For large-scale fractional distillation or rectification at pressures in the moderate-vacuum region (760 to 1 torr), conventional bubble-cap plate towers and tubular boilers may be used. To avoid a pressure drop of more than 1 or 2 torr per tray, the slot submergence, or liquid head above the top of the slots in the bubble cap, must be maintained at a suitably low value.

Figure 9 shows a typical commercial molecular still of the centrifugal type. The evaporator is a conical aluminum casting, with a diameter of 155 cm at the top and

110 cm at the bottom, which rotates at about 400 rpm.   This rotor is heated by radiation from surrounding electrical resistance heaters which in turn are backed by a layer of reflective insulation.   The distilland is fed to the bottom of the conical inside wall of the rotor through a pipe pointing in the direction of rotation and having a nozzle which projects the liquid at the peripheral speed of the rotor.   The liquid is lifted up the distilling surface by the centrifugal force and forms a film about 0.05 mm thick. The residue is flung into a gutter at the top of the still from which it flows through a heat exchanger to a gear pump which discharges the oil at atmospheric pressure, or the residue may be fed directly from the gutter to the rotor of another still in series.   In the latter case the conical rotor serves as a lift pump.

The vapors travel a short distance to a condenser, sometimes formed of vertical leaves or vanes closely spaced to form a louvre system.   Each leaf is provided with three gutters to separate the condensate into top, middle, and bottom fractions.   By circulating warm water through attached tubes, the temperature of the condenser is adjusted to collect about 95% of the wanted constituents while the unwanted semicondensable constituents and permanent gases are rejected and pass through to the center of the still where a separate cold condenser collects some of the semicondensables.   Other vapors are condensed on the dome and on a baffle placed at the entrance to the pumps, and the remainder pass on into the pumps.

The pumping system may consist of three or four steam ejectors in series acting as forepump for an oil vapor ejector and a diffusion pump in series.   The latter pump has a speed of about 5000 liters per second in the range from 10^{-3} to 10^{-2} torr.

The rate of feed on distilland is on the order of 500 kg/hr.   Care is taken to adjust the feed so that the upper edge of the rotor does not run dry.   If complete distillation is required, 90–95% of the residue may be recirculated.   A nonvolatile carrier oil may be added to the feed at about the same rate as the bleed-off until sufficient recycling residue is built up to keep the rotor wet.

The 5 ft centrifugal still with unobstructed path between rotor and condenser gives separations of 0.80–0.95 theoretical molecular plate.   A theoretical molecular plate is defined as the maximum separation achieved when the condenser is indefinitely close to the evaporator, and distilland is circulated so rapidly that the evaporating surface has substantially the same composition as the main bulk of the distilland, and the vapor molecules emerge at such a low density that there are relatively few collisions in flight.   The separation of an ideal binary mixture under these conditions would be proportional to the ratio of the partial pressure divided by the square root of the molecular weight for the two constituents.   The ratio of partial pressures often varies widely with the absolute rate of distillation which adds a further complication to the assessment of one theoretical molecular plate.   The separation can be increased by adding a fractionating barrier between evaporator and condenser.   This may take the form of a wire-mesh screen so formed that condensate drips on the rotor.   Various schemes for improving separation by grouping banks of stills in cascade have been described (22,25).   Centrifugal molecular stills are available commercially in all sizes from a 5 in. to a 5 ft rotor.   Wiped-film stills are also available in sizes from 2 in. to 8 ft inside diameter.

The decomposition hazard (26), which may be estimated by multiplying the average pressure in the still by the time of exposure, is lowest for centrifugal molecular stills, but increases with their size.   The hazard is one order of magnitude higher for wiped-film stills, two orders of magnitude higher for simple falling-film stills, and many thousands of times higher for column stills (22,27). The centrifugal still is particularly adapted for thermally unstable liquids of high viscosity and requiring temperatures of 100–280°C for distillation in a high vacuum. It has been successfully applied to the commercial distillation of oil-soluble vitamins, sterols, animal fats, monoglycerides, and high-molecular-weight plasticizers.

Vacuum Metallurgy. Vacuum techniques are used in metallurgy for the following applications (28–35): (a) melting and casting of special alloys; (b) degassing molten steel; (c) distillation and purification of zinc, alkali metals, lead, and tin; (d) sintering of refractory metal powders and cermets; (e) production and purification of barium, strontium, calcium, magnesium, and uranium; (f) purification of refractory metals by arc melting; (g) welding and brazing; (h) bright annealing; (i) analysis of the gas content of metals; (j) zone refining of metals and alloys; (k) growing single crystals of silicon and of germanium; and (l) preparation of pyrophoric metal powders.

Various methods of heating the metal in a special vacuum chamber, called a vacuum furnace, have been employed depending on the size of the metal charge, the required temperature, and the reactivity of the metal with available crucible materials; see also Furnaces, Vol. 10, pp. 252–294. For small and medium quantities and moderate temperatures (up to 1200°C) tube and retort furnaces heated externally by electrical resistance heaters, gas burners, or induction coils are used for heat treating solid metal parts, production of magnesium, and vacuum fusion gas analysis of small metal samples. For higher temperatures and the melting of large quantities of metals in crucibles, water-cooled induction heating coils are wrapped around the crucible inside a large vacuum chamber with water-cooled walls (vacuum induction furnace). For small quantities of metals which react with available crucible materials the metal may be suspended by levitation in an inhomogeneous ac electromagnetic field and melted by eddy current heating, but this technique is seldom employed. For high temperatures and metals that react with available crucible materials when molten, the metal may be melted in an electric arc furnace or an electron bombardment furnace. For high temperature heat treatment of solid metal parts, sintering or melting of metals in crucibles in moderate quantities, electric resistance heating units may be placed around the crucible inside the vacuum chamber and shielded from the furnace walls by several radiation shields of refractory metal sheet (usually molybdenum) (vacuum resistance furnaces).

The advantages of using low pressures when heat treating or melting metals are obviously related to the protection of the hot metal from reaction with atmospheric or ambient gases and the removal of evolved gases and volatile impurities thus favoring reactions which form gaseous products such as the deoxidation of steel with carbon by forming carbon monoxide. It is also possible to use protective atmospheres of inert gases, but it should be noted that commercially pure argon is only 99.98% pure while a vacuum of 10^{-3} torr is equivalent in residual gas content to argon of 99.99987% purity (35). Metals can be distilled or sublimed in a vacuum at a faster rate at reduced temperatures and condensed to denser deposits. Removal of dissolved gases from metals by vacuum melting results in greater strength, somewhat improved corrosion resistance, more cohesive crystal structure (less embrittlement), longer life for steel bearings, springs, and fine drawn wire, freedom from flakes, and improvement in hot-workability. Removal of C, O, and S from transformer steel results in lower watt losses. Vacuum-melted copper has a higher conductivity, higher density, and greater ductility than OFHC copper (see Vol 6, p. 127). High-purity vacuum-melted germanium and silicon are used in solid-state electronic devices.   Almost all of the world's production of lead is subjected to vacuum distillation (see Vol. 12, p. 234) to remove zinc which has been added to precipitate other elements present.   Pure uranium and zirconium are produced by vacuum techniques for use in nuclear reactors.

Vacuum melting of metals began in 1903–1912 when W. von Bolton of the Siemens Company in Germany arc-melted tantalum for use in electric lamps.   W. Rohn began vacuum melting nickel-chromium alloys with resistance furnaces in 1917 and

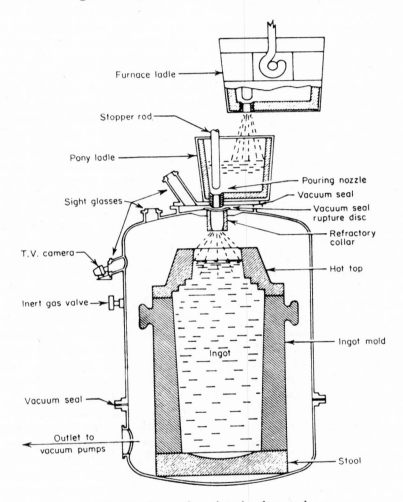

Fig. 10.   Stream degassing of molten steel.

during the period 1923–1940 developed large vacuum induction furnaces at the Heraeus Vakuumschmelze Company in Germany.   Large vacuum arc furnaces were introduced in 1956 in the United Kingdom and the United States for melting titanium and special steel alloys for the aero-engine industry.   Large-scale electron bombardment furnaces were first described by H. Smith in 1957 (30).   The size of induction and arc melting furnaces has steadily increased until now ingots up to 60 in. in diameter and weighing 50 tons may be produced (34).

The vacuum degassing of steel in large quantities began during the period 1950–1956 in Germany and in the Soviet Union and was introduced into the United States in 1956 to overcome flaking of large ingots used in forging the rotors for electric power generators. The elimination of flaking is largely due to the reduction of the hydrogen content, but oxygen, nitrogen, carbon, and sulfur may also be partially removed with beneficial results. The degassing is performed by transferring molten steel from a large ladle into an evacuated container under conditions such that the liquid inside the vacuum is broken up into droplets or violently agitated to expose a large surface from which the gas can escape. Figure 10 shows a stream degassing furnace in which the molten steel emerges from a pouring nozzle in the form of a spray which is directed by a refractory collar into an ingot mold. The pressure during pouring varies between 0.4 and 0.8 torr. After the ingot is poured, the tank is back-filled with inert gas and returned to atmospheric pressure to accelerate the cooling of the ingot. Various other methods have been used in which the molten steel is sucked from a large (20–250 ton) ladle through a tube into an evacuated vessel and then returned periodically to the ladle through the same tube (Dortmund-Hörder method) or continuously through a separate return tube (Ruhrstahl-Heraeus method) (34,35) (see also Vol. 18, p. 739). About 3 million tons of steel per year is vacuum degassed in the United States.

Furnace Design. Externally heated vacuum retorts of nickel-chrome steel were used in the manufacture of magnesium by the Pidgeon process during World War II, and are still used in the production of calcium. Retorts built from iron cannot be heated above 500°C without excessive scaling of surfaces exposed to the atmosphere. Iron sprayed with a coat of stainless steel can be used up to 700°C. Iron sheathed in Inconel (International Nickel Co.) nickel–chromium–iron can be used up to 1200°C, but above 800°C mild steel begins to creep under moderate pressures. Above 1100°C both mild steel and stainless steel show pronounced creeping. Although the retorts used in the Pidgeon process had walls more than 1 in. thick, they gradually became deformed. Iron retorts enclosed in a vacuum chamber can be used up to temperatures of 1200°C, but heating cannot be accomplished from the outside with gas or oil burners.

Fused quartz, mullite, alundum, zircon, sillimanite, and pure refractory oxides (see Refractories) are used for small tubular retorts heated externally by electrical resistance ovens or by induction heating coils. Furnaces for melting and casting of metals use vertical crucibles of graphite, alumina, magnesia, zirconia, beryllia, or thoria enclosed in a large metal vacuum chamber. Heating is accomplished by an arc, induction coils, or resistance windings located inside the vacuum tank and operated at sufficiently low voltages to avoid glow discharges or uncontrolled arcing.

External heaters made from Nichrome (Driver-Harris Co.) chromium–nickel–iron may be used up to 1000°C and Kanthal (C. O. Jelliff Mfg. Corp.) resistance alloy up to 1200°C. Tungsten or molybdenum resistance heating elements protected by inert gas or vacuum can be used above 1200°C but become brittle at high temperature. Tubular tantalum heaters have been used inside vacuum furnaces up to 2500°C. Graphite resistance heating units can be employed inside vacuum furnaces operating up to 1600°C, above which carbon evaporates at a rate that may cause trouble. Silicon and aluminum vapors attack graphite heating elements or crucibles to form carbides, but graphite can be used in calcium reduction furnaces. Graphite reacts with most refractories and oxides above 1300°C. Silicon carbide resistance heating units are not recommended above 1400°C in vacuum. Tungsten, molybdenum, and tantalum heating units are attacked by oxygen and water vapor. Resistance heating units are always surrounded by insulation or radiation shields.

High-frequency (1–20 kHz) induction heating is used for high-temperature furnaces (1000–2000°C). For melts up to 5 lb, a radio-frequency generator may be used. For larger melting furnaces, motor generators with frequencies of 9.6, 4.2, or 3.0 kHz are employed, and for very large furnaces a frequency of 0.96 kHz or less is recommended. The lower frequencies give greater stirring of the melt. The power required depends on the melting time and size of the charge. For pilot-plant furnaces with melting times of about 30 min, approximately 1 kW of generator capacity per pound of steel is required (36). Large furnaces for melting 1000 lb or more of ferrous or nickel base alloys require about 0.3–0.5 kW generator output per pound of melt for melting times of 1–2 hr. The metal is held in a crucible surrounded by several turns of water-cooled copper coil through which is passed the high-frequency current. The coil must be kept sufficiently far from structural metal parts of the furnace to avoid undue power losses. A coil voltage of about 200 V is preferred to avoid arcing and glow discharge, but very large furnaces (over 200 kW) usually employ 400 V. Completely insulated coils may be operated at 800 V in vacuum. The crucible is usually filled to about two-thirds of its volume. Frequently constituents must be added to the batch under vacuum. In small furnaces, it is sometimes advisable to reduce the temperature of the crucible below the melting point since the added material may release quantities of gas causing splattering. In large furnaces, this precaution may not be necessary if the material is added slowly. The additions are usually dropped from buckets into a chute. Devices for emptying the buckets by external controls have been described by Nisbet (37). Additional ingredients may also be suspended above the crucible or chute on a wire which can be fused by an electric current.

The melt must be poured into water-cooled molds, since allowing the melt to solidify in the crucible requires a long time and usually results in large pipes (see Vol. 18, p. 755) and nonuniformity. The melt may be tapped from the bottom of a crucible provided with a hole sealed by a stopper rod. Alternatively the bottom of the crucible may have a short pipe sealed by a fusible plug which can be melted by an electric heater surrounding the pipe. Glass viewing ports are usually required so that the operation can be controlled from the outside. The glass must be protected by retractable metal shields to avoid being obscured by material volatilized and spattered from the melt.

Ceramic or sand molds outgas too much and water-cooled copper or steel molds are therefore commonly used. Casting of ingots must be done slowly to allow the liberated gas to be pumped away and to avoid long pipes, but for investment casting the total pouring time should be about one or two seconds. Water cooling rates approximate one gallon per minute for each pound of melt (except for metals like beryllium that produce brittle ingots if cooled too rapidly). Bottom disks of ceramic may be used to prevent excessive cooling of the first run of metal. Colloidal graphite applied to the walls of the mold and baked on will facilitate the release of the ingot from a chilled metal mold.

Vacuum arc furnaces are used for metals with a very high melting point, such as molybdenum, tungsten, and tantalum, or for melting metals, such as zirconium, that react in the molten state with available crucible materials. The ingot is formed in a water-cooled crucible or mold, usually made from copper, under such conditions that only a cup or cone-shaped pool of metal is molten at the top of the ingot and contact between the molten metal and the mold is limited to a narrow ring which quickly solidifies as the ingot grows. The metal to be melted is usually formed into a long narrow rod, which becomes the negative electrode, by compacting sponge, granules, and perhaps some scrap in a hydraulic press and sintering or welding sections as required. This rod is mounted vertically inside an elongated vacuum chamber either by means of pairs of rolls which grip and move the rod to maintain the required arc length above the molten pool or by means of a smooth support rod attached to the top of the rough electrode and sliding through a vacuum seal at the top of the furnace in response to a motor-driven differential gear system to maintain the proper arc length (29,30). The arc is struck between the movable (negative) electrode and an initial disk of the same metal in the bottom of the crucible or mold. The pressure must be either less than 0.05 torr or else kept above the point at which glow discharges occur by admitting an inert gas, such as argon. Usually, the pressure is maintained in the range from 10^{-3} to 10^{-1} torr by Roots-type pumps or oil diffusion-ejector pumps, and the arc is confined to the vaporized metal after striking. The consumable electrode melts quite rapidly as the arc plays over the tip and drops of metal fall into the molten pool which is also heated by the arc and stirred by thermal convection currents and the wandering of the arc, stirring sometimes being aided by the action of a dc coil placed around the outside of the mold near the level of the pool. The gap between the electrode and pool must be less than that to the wall of the mold to avoid arcs which might burn through the mold.

The consumable electrode may be formed continuously by compacting and sintering powder in a mold at the top of the furnace. The consumable electrode may also consist of a previously formed ingot which is being remelted into a larger diameter ingot mold. In small furnaces it may sometimes not be convenient to form a consumable electrode (consutrode) and the metal is added as powder or granules directly to the molten pool from a feeder while the arc is formed between the pool and a permanent movable electrode rod made of tungsten, molybdenum, or graphite, usually in the presence of argon, or a mixture of argon and helium at a pressure between 5 and 200 torr. However, nonconsumable electrodes contaminate the melt and are seldom used.

The open circuit voltage before striking the arc may be as high as 100 V dc, but during melting the arc voltage drops to between 20 and 45 V, depending on the metal being melted (34). The arc current in large consumable electrode furnaces is on the order of 1000 A per in. of mold diameter. Ingot diameters vary from a few inches to 60 in. Toward the end of the melting cycle the arc current may be gradually reduced to allow the melted zone to flatten so that the shrinkage cavity in the cooled ingot will be a minimum, a procedure known as hot topping. For high-quality homogeneous cast ingots the metal may be vacuum arc melted twice or first vacuum induction melted and then arc melted.

Because of the danger of explosions due to burning of the water-cooled mold by a wandering arc or collapse of a mold due to failure to supply a continuous water flow, the operator of large industrial furnaces views the arc through a teleoptic system in a control room behind an explosion-proof wall. However, the danger of explosions is practically eliminated in modern computer-controlled installations with large Roots-type pumps to handle gas bursts and power supplies designed to quench parasitic arcs and closely control the arc length.

Electron beam or electron bombardment furnaces are used for the melting, evaporation, welding, and heat treating of refractory metals such as molybdenum, niobium, tantalum, and tungsten and of reactive metals, such as beryllium, titanium, and zirconium where products of very high purity are required (30,33,34). Laboratory furnaces range in power up to 10 kW, and industrial furnaces up to 1200 kW power have been constructed which can melt the common metals at rates of more than one ton per hour. Electron beams may also be used for machining or the drilling of fine holes. Pressures of 10^{-4} torr or less must be maintained at the electron source, and relatively expensive, very high speed pumping systems are required to handle the normal gas loads. Operating voltages for melting furnaces range from 10 to 30 kV. Welding and machining systems may use voltages up to 150 kV where special precautions must be taken against x rays. The electron trajectories are usually curved by means of electric or magnetic fields so that the electron gun source can be shielded from vaporized metal and target splatter. By passing the beam through orifices across which a pressure differential can be maintained by auxiliary pumping equipment, the gun can be protected from excessive ion bombardment and the effects of gas bursts from the target. The feed stock for melting furnaces can take the form of powder, pellets, scrap, or bars inserted through a vacuum lock. The advancing surface of the feed stock is bombarded by the electrons directed at such an angle that the molten material will drip into a water-cooled ingot mold. A molten pool is maintained at a constant level at the top of the ingot by electron bombardment while the solidified ingot is continuously drawn downward through the mold side walls (38).

Chemical Reactions at Low Pressures. Inorganic chemical reactions at room temperature are very slow in high-vacuum systems because of the absence of water and other ionizing solvents that are too volatile to remain in the condensed phase. However, at elevated temperatures reactions may occur at appreciable rates in a vacuum system. For most inorganic compounds and metals, temperatures above 800°C are required for melting or dissociation. Above 1300°C so many substances react that it is difficult to find materials for the structural parts of the vacuum system that will not be corroded or cause contamination. Many high-temperature reactions are facilitated by the use of low pressures because surface films of oxides and nitrides are avoided and volatile products are pumped or distilled away from the reaction mixture. For example, beryl ore reacts with aluminum or magnesium fluoride at 1000°C in a vacuum to produce beryllium fluoride, which is evaporated in the process (39).

Reactions to be carried on in molten baths require a large exposed surface with vigorous stirring of the bath so that the gaseous products can escape readily. The hydrostatic pressure of the liquid a few millimeters below the surface may be greater than the partial pressure of the reaction product at the prevailing temperature. For some typical reactions used in vacuum metallurgy see Magnesium and magnesium alloys, Vol. 12, p. 675; Calcium and calcium alloys, Vol. 3, p. 919.

Bibliography

"Vacuum Technique" in *ECT* 1st ed., Vol. 14, pp. 503–536, by B. B. Dayton, Consolidated Vacuum Corporation.

1. C. M. Van Atta, *Vacuum Science and Engineering*, McGraw-Hill Book Co., Inc., New York, 1965.
2. A. Guthrie, *Vacuum Technology*, John Wiley & Sons, Inc., New York, 1963.
3. S Dushman, in J. M. Lafferty, ed., *Scientific Foundations of Vacuum Technique*, 2nd ed., John Wiley & Sons, Inc., New York, 1962.
4. P. A. Redhead, J. P. Hobson, and E. V. Kornelsen, *The Physical Basis of Ultrahigh Vacuum*, Chapman and Hall Ltd., London, 1968.
5. R. W. Roberts and T. A. Vanderslice, *Ultrahigh Vacuum and its Applications*, Prentice-Hall, Inc., Englewood Cliffs, N. J., 1963.
6. J. H. Leck, *Pressure Measurement in Vacuum Systems*, Chapman and Hall Ltd., London, 1964.
7. H. Ishii and K. Nakayama, *Eighth National Vacuum Symposium Transactions*, Pergamon Press, New York, 1962, pp. 519–524.
8. J. M. Lafferty, *Proc. 4th Intern. Vac. Congress*, Inst. of Physics and Physical Soc., London, 1969, pp. 647–650.
9. B. D. Power, *High Vacuum Pumping Equipment*, Reinhold Publishing Corp., New York, 1966.
10. H. A. Steinherz, *Handbook of High Vacuum Engineering*, Reinhold Publishing Corp., New York, 1963.
11. M. Pirani and J. Yarwood, *Principles of Vacuum Engineering*, Reinhold Publishing Corp., New York, 1961.
12. D. J. Santeler, D. H. Holkeboer, D. W. Jones, and F. Pagano, *Vacuum Engineering*, Boston Technical Publishers, Cambridge, Mass., 1967.
13. J. Romand, V. Schwetzoff, and B. Vodar, *Vide* **6**, 1046 (1951).
14. American Vacuum Society, *Glossary of Terms Used in Vacuum Technology*, Pergamon Press, New York, 1958.
15. B. B. Dayton, *Vacuum* **15**, 53–57 (1965).
16. B. B. Dayton, *Ind. Eng. Chem.* **40**, 795–803 (1948).
17. A. Guthrie and R. K. Wakerling, *Vacuum Equipment and Techniques*, McGraw-Hill Book Co., Inc., New York, 1949.
18. A. H. Beck, ed., *Handbook of Vacuum Physics*, The Macmillan Co., New York, 1964, Vol. 1, part 1, pp. 1–32.
19. B. B. Dayton, Transactions Sixth National Vacuum Symposium, Pergamon Press, Oxford, 1960, pp. 101–119.
20. R. V. D. Strong and C. Francksen, *J. Vac. Sci. and Techn.* **6**, 263–265 (1969).
21. E. S. Perry and A. Weissberger, eds., *Technique of Organic Chemistry*, Interscience Publishers, a division of John Wiley & Sons, Inc., New York, 1965, Vol. IV, Distillation.
22. P. Ridgway Watt, *Molecular Stills*, Reinhold Publishing Corp., New York, 1963.
23. G. Burrows, *Molecular Distillation*, Oxford University Press, Oxford, 1960.
24. E. Holland-Merten, *Handbuch der Vacuum Technik*, Wilhelm Knapp, Halle, 1950.
25. K. C. D. Hickman, *Ind. Eng. Chem.* **39**, 686 (1947).
26. K. C. D. Hickman and N. D. Embree, *Ind. Eng. Chem.* **40**, 135 (1948).
27. K. C. D. Hickman, *Ind. Eng. Chem.* **40**, 16 (1948).
28. J. M. Blocher, ed., *Vacuum Metallurgy, 1954 Symposium*. The Electrochemical Society, New York, 1955.
29. W. E. Kuhn, ed., *Arcs in Inert Atmospheres and Vacuum, 1956 Symposium*, John Wiley & Sons, Inc., New York, 1956.
30. R. F. Bunshah, ed., *Vacuum Metallurgy*, Reinhold Publishing Corp., New York, 1958.
31. R. F. Bunshah, ed., *Transactions of the Vacuum Metallurgy Conference 1959*, New York University Press, New York, 1960.
32. J. Wesley Cable, *Vacuum Processes in Metalworking*, Reinhold Publishing Corp., New York, 1960.
33. R. F. Bunshah, ed., *Transactions of the Vacuum Metallurgy Conference 1960*, Interscience Publishers, a division of John Wiley & Sons, Inc., New York, 1961.

34. American Vacuum Society, *Annual Transactions of the Vacuum Metallurgy Conference*, American Vacuum Society, New York, 1963–1969.

35. J. A. Belk, *Vacuum Techniques in Metallurgy*, The Macmillan Co., New York, 1963.

36. E. D. Malcolm, *J. Sci. Instr.* **28**, Suppl. No. 1, 63 (1951).

37. J. D. Nisbet, *Iron Age* **159** (25), 56–59 (1947); **161** (12), 79 (1948).

38. R. Bakish, ed., *Introduction to Electron Beam Technology*, John Wiley & Sons, Inc., New York, 1962.

39. W. J. Kroll, *Vacuum* **1**, 163–184 (1951).

B. B. DAYTON
The Bendix Corporation

VACUUM TUBES.   See Electron tube materials.

VALDET.   See Surfactants.

VALERIAN.   See Oils, essential, Vol. 14, p. 214.

VALONIA.   See Leather, Vol. 12, p. 318.

VALVES.   See System valving, under Piping systems, Vol. 15, p. 651.

VANADIUM AND VANADIUM ALLOYS

Vanadium, symbol V, at. no. 23 and at. wt 50.942, is a member of group VA of the periodic system and of the first transition series.   It is a gray-colored metal which crystallizes in the body-centered cubic form and, in the pure state, is very soft and ductile.   Because of its high melting point, it is referred to as a refractory metal, along with niobium, tantalum, chromium, molybdenum, and tungsten.   The principal use of vanadium is as an alloying addition to iron and steel, particularly in high-strength steels and, to a lesser extent, in tool steels and castings.   It is also an important beta stabilizer for titanium alloys.   Recent interest in vanadium-base alloys as claddings for fuel elements in advanced nuclear reactors has stimulated research on the metallurgy of this relatively unknown element.

Vanadium was first discovered in 1801 by Manuel del Rio while he was examining a lead ore obtained from Zimapán, Mexico.   He reported that this ore contained a new element and, because of the red color imparted to its salts on heating, he named it "erythronium."   Later when this discovery was challenged, he accepted the explanation that it was an impure form of chromium, probably lead chromate.   The actual identification of the element vanadium did not occur until 1830 when N. G. Sefström isolated the new element from cast iron processed from an ore from mines near Taberg, Sweden. He gave it the name vanadium after Vanadis, the Norse goddess of beauty.   Shortly after Sefström's discovery, Friedrick Wöhler showed that vanadium was identical to the erythronium which del Rio had found several years earlier.

Physical Properties

Vanadium is a soft, ductile metal in pure form but it is hardened and embrittled by the presence of oxygen, nitrogen, and hydrogen (1,2).   It is a poor conductor of heat as compared with most metals, its conductivity being lower than that of copper, for example, by a factor greater than ten.

Table 1. Physical Properties of Vanadium Metal (3–5)

Property	Value
crystal structure	body-centered cubic
lattice constant, Å	3.026
density, g/cm³	6.11
melting point, °C	1900 ± 25
boiling point, °C	3000
vapor pressure, at 1393–1609°C	$R \ln P = 121.95 \times 10^3\, T^{-1} - 5.123 \times 10^{-4}\, T + 36.29$ where R = gas constant, cal/mole/deg C; P = pressure, atm; T = absolute temperature, °K
specific heat, at 20–100°C, cal/g	0.120
latent heat of fusion, kcal/mole	4
latent heat of vaporization, kcal/mole	106
enthalpy, at 25°C, kcal/mole	1.26
entropy, at 25°C, cal/mole/deg C	7.05
thermal conductivity, at 100°C, cal/(sec)(cm²)(deg C/cm)	0.074
electrical resistance, at 20°C, $\mu\Omega$-cm	24.8
temperature coefficient of resistance, at 0–100°C, $\mu\Omega$-cm/deg C	0.0034
magnetic susceptibility, cgs units	1.4×10^{-6}
superconductivity transition, °K	5.13
coefficient of linear thermal expansion, per deg C	
at 20–720°C (x ray)	$9.7 \pm 0.3 \times 10^{-6}$
at 200–1000°C (dilatometer)	8.95×10^{-6}
thermal expansion at 23–100°C, μin./in./deg C	8.3
recrystallization temperature, °C	800–1000
modulus of elasticity, psi	$18–19 \times 10^6$
shear modulus, psi	6.73×10^6
Poisson ratio	0.36
thermal neutron absorption, barns/atom	5.00 ± 0.01
capture cross section for fast (1 MeV) neutrons, mbarns/atom	3

The most important physical properties of vanadium are listed in Table 1. The values included are taken from the most recent scientific literature. The purity of vanadium has been improving each year, making this compilation subject to continuous modification or revision.

Chemical Properties

Vanadium exhibits positive oxidation states of 2, 3, 4, and 5. When heated in air at different temperatures, it oxidizes to a brownish-black trioxide, a blue-black tetroxide, or a reddish-orange pentoxide. It reacts readily with chlorine at fairly low temperatures (180°C) to form VCl_4 and with carbon and nitrogen at high temperatures to form VC and VN, respectively. The pure metal in massive form is relatively inert toward oxygen, nitrogen, and hydrogen at room temperature.

Vanadium is resistant to attack by hydrochloric or dilute sulfuric acid and to alkali solutions. It is also quite resistant to corrosion by sea water but is reactive toward nitric, hydrofluoric, or concentrated sulfuric acids. Galvanic corrosion tests run in simulated seawater indicate that vanadium is anodic with respect to stainless steel and copper, but cathodic to aluminum and magnesium. Vanadium exhibits good corrosion resistance to liquid metals such as bismuth and low-oxygen sodium.

Occurrence

Vanadium is widely distributed throughout the earth but in rather low abundance, ranking twenty-second among the elements of the earth's crust. It is estimated that the lithosphere contains about 0.07% vanadium with few deposits containing more than 1 or 2%. Vanadium is found in uranium-bearing minerals of Colorado; in the copper, lead, and zinc vanadates of Africa; and with certain phosphatic shales and phosphate rocks in the western United States. It is a constituent of titaniferous magnetites, which are widely distributed, with large deposits in the U.S.S.R., South Africa, Finland, China, eastern and western United States, and Australia. At one time the largest and most important vanadium deposits were the sulfide and vanadate ores from the Peruvian Andes, but these are now depleted.

Trace amounts of vanadium have been found in meteorites and sea water, and it has been identified in the spectrum of many stars including our own sun. The occurrence of vanadium in oak and beech trees and some forms of aquatic sea life indicates its biological importance.

There are over sixty-five known vanadium-bearing minerals, some of the more important of which are listed in Table 2. Five of these (patronite, bravoite, sulvanite,

Table 2. Important Minerals of Vanadium

Mineral	Color	Formula	Location
patronite	greenish-black	$V_2S + nS$	Peru
bravoite	brass colored	$(Fe,Ni,V)S_2$	Peru
sulvanite	bronze yellow	$3Cu_2S.V_2S_5$	Australia, U.S. (Utah)
davidite	black	titanate of Fe, U, V, Cr, and rare earths	Australia
roscoelite	brown	$2K_2O.2Al_2O_3(Mg,Fe)O.$ $3V_2O_5.10SiO_2.4H_2O$	U.S. (Colorado and Utah)
carnotite	yellow	$K_2O.2U_2O_3.V_2O_5.3H_2O$	southwest U.S.
vanadinite	reddish-brown	$Pb_5(VO_4)_3Cl$	Mexico, U.S., Argentina
descloizite	cherry red	$4(Cu,Pb,Zn)O.V_2O_5.H_2O$	Africa, Mexico, U.S.
cuprodescloizite	greenish-brown	$5(Cu,Pb)O.(V,As)_2O_5.2H_2O$	Africa
vanadiferous phosphate rock		$Ca_5(PO_4)_3(F,Cl,OH)$ VO_4 ions replace some of the PO_4 ions	U.S. (Montana)
titaniferous magnetite		$FeO.TiO_2-FeO.(Fe,V)O_3$	U.S.S.R., China, Finland, Africa

davidite, and roscoelite) are classified as primary minerals, whereas all of the others are secondary products formed in the oxidizing zone of the upper lithosphere. Until recently the principal source of vanadium was the complex sulfide ore, patronite, found in Mina Ragra, Peru. With the gradual depletion of this deposit, the carnotite and roscoelite ores found in the sandstones of the Colorado Plateau have taken on greater importance. For many years these carnotite–roscoelite ores have been the major source of vanadium as well as its important coproduct, uranium.

The metallic vanadates of lead, copper, and zinc which are found in South-West Africa and Zambia are also a large resource of vanadium-bearing ores as are the phosphatic shales and rocks of the Phosphoria formation in Idaho and Wyoming. For several years vanadium salts have been obtained as a by-product of the phosphoric acid and fertilizer industries. Large reserves of vanadium have recently been discovered in Arkansas and the titaniferous magnetite ores will probably come into greater prominence on the world production scene. Certain petroleum crude oils, especially those from South America, contain varying amounts of vanadium compounds. These accrue as fly ash or boiler residues upon combustion of the crude oils and they may be reclaimed.

Manufacture

ORE PROCESSING

Vanadium is recovered domestically as a principal mine product, as a coproduct or by-product from uranium–vanadium ores, and from ferrophosphorus as a by-product in the production of elemental phosphorus. In Canada it is recovered from crude oil residues and in the Republic of South Africa, as a by-product of titaniferous magnetite. Whatever the source, however, the first stage in ore processing is the production of an oxide concentrate.

The principal vanadium-bearing ores are generally crushed, ground, screened, and mixed with a sodium salt such as $NaCl$ or Na_2CO_3. This mixture is roasted at about 850°C and the oxides are converted to a water-soluble sodium metavanadate, $NaVO_3$. The vanadium is extracted by leaching with water and precipitated as sodium hexavanadate, $Na_4V_6O_{17}$ (referred to as red cake) by the addition of sulfuric acid to adjust the pH to between 2 and 3. This is then fused at 700°C to yield a dense black product that is sold as technical-grade vanadium pentoxide. This product contains a minimum of 86% V_2O_5 and 6–10% Na_2O.

The red cake may be further purified by dissolving it in an aqueous solution of Na_2CO_3. The iron, aluminum, and silicon impurities are precipitated from the solution by pH adjustment. Ammonium metavanadate is then precipitated by the addition of NH_4Cl and calcined to give vanadium pentoxide of greater than 99.8% purity.

Vanadium and uranium are extracted from carnotite by direct leaching of the raw ore with sulfuric acid. An alternative method is to give the ore an initial roast followed by successive leaching with H_2O and dilute HCl or H_2SO_4. In some cases the first leach is made with a Na_2CO_3 solution. The uranium and vanadium are then separated from the pregnant liquor by liquid–liquid extraction techniques involving careful control of the oxidation states and pH during extraction and stripping.

A process that is presently being used on a very large scale by Highveld Steel and Vanadium Corporation in the Republic of South Africa is the recovery of high-vanadium slags from titaniferous magnetites (6). The ore, containing about 1.75% V_2O_5, is partially prereduced with coal in large rotary kilns. The hot ore is then fed to an enclosed, submerged-arc electric smelting furnace which produces a slag (containing substantial amounts of titania) and pig iron containing most of the vanadium which is in the ore. After tapping from the furnace and separating off the waste slag, the molten pig iron is blown with oxygen to form a slag containing up to 25% V_2O_5. The slag is separated from the metal and may then be used as a high-grade raw material in the usual roast-leach process described in the preceding section.

PRODUCTION OF FERROVANADIUM

The steel industry accounts for over three-fourths of the world's consumption of vanadium as an additive to steel. It is added in the steel-making process as a ferrovanadium alloy which is produced commercially by the reduction of vanadium ore, slag, or technical-grade oxide with carbon, ferrosilicon, or aluminum. The product grades, which may vary from 35 to 80% vanadium, are classified according to their vanadium content. The consumer use and grade desired dictate the choice of reductant.

Carbon Reduction. The production of ferrovanadium by reduction with carbon has been supplanted by other methods in recent years. An important development has been the use of vanadium carbide as a replacement for ferrovanadium as the vanadium additive in steel making. A product containing about 85% vanadium, 12% carbon, and 2% iron, marketed by the Union Carbide Corporation as Carvan, is produced by the solid-state reduction of vanadium oxide with carbon in a vacuum furnace.

Silicon Reduction. The preparation of ferrovanadium by the reduction of vanadium concentrates with ferrosilicon (see Silicon, metallurgical) has been used, but not extensively. It involves a two-stage process in which technical-grade vanadium pentoxide, ferrosilicon, lime, and fluorspar are heated in an electric furnace to reduce the oxide, yielding an iron alloy containing about 30% vanadium, but undesirable amounts of silicon. The silicon content of the alloy is then decreased by the addition of more V_2O_5 and lime to effect the extraction of most of the silicon into the slag phase. An alternative process involves the formation of a vanadium–silicon alloy by the reaction of V_2O_5, silica, and coke in the presence of a flux in an arc furnace. The primary metal is then reacted with V_2O_5 to produce ferrovanadium.

A silicon process has been developed recently by the Foote Mineral Company (7). The method is currently being used commercially to produce tonnage quantities of ferrovanadium. A vanadium silicide alloy containing less than 20% silicon is produced in a submerged-arc electric furnace by reaction of vanadium-bearing slags with silica, flux, and a carbonaceous reducer, followed by refinement with vanadium oxide. This is then reacted with a molten vanadiferous slag in the presence of lime to produce a ferrovanadium alloy called Solvan containing approx 28% vanadium, 3.5% silicon, 3.8% manganese, 2.8% chromium, 1.25% nickel, 0.1% carbon, and the remainder iron. A unique feature of this process is its applicability to the pyrometallurgical process of vanadium-bearing slags of the type described in the preceding section.

Aluminum Reduction. The aluminothermic process for preparing a ferrovanadium alloy differs from either of the methods described above in that this reaction is highly exothermic. A mixture of technical-grade vanadium oxide, aluminum, iron scrap, and a flux are charged into an electric furnace and the reaction between aluminum and vanadium pentoxide is initiated by the arc. The temperature of the reaction is controlled by adjusting the size of the particles and the feed rate of the charge, or by using partially reduced material, or by replacing some of the aluminum with a milder reductant such as calcium carbide, silicon, or carbon. Ferrovanadium containing as much as 80% vanadium is produced in this way.

Ferrovanadium may also be prepared by the well-known thermite reaction in which vanadium and iron oxides are co-reduced by aluminum granules in a magnesite-lined steel vessel or in a water-cooled copper crucible (8). The reaction is initiated by use of a barium peroxide–aluminum ignition charge. This method is also used to prepare vanadium–aluminum master alloys for the titanium industry.

Preparation of Pure Vanadium

Vanadium metal can be prepared either by the reduction of vanadium chloride with hydrogen or magnesium or by the reduction of vanadium oxide with calcium, aluminum, or carbon. The oldest and most commonly used method for producing vanadium metal on a commercial scale is the reduction of V_2O_5 with calcium. Recently a two-step process involving the aluminothermic reduction of vanadium oxide, combined with electron-beam melting, has been developed. This method has the capability of producing a purer grade of vanadium metal, of the quality required for the nuclear reactor program.

Calcium Reduction. High-purity vanadium pentoxide is reduced with calcium to produce vanadium metal of about 99.5% purity. The exothermic reaction is carried out adiabatically in a sealed vessel or "bomb." In the original Marden and Rich (9) process calcium chloride was added as a flux for the CaO slag. The vanadium metal was recovered in the form of droplets or beads. McKechnie and Seybolt (10) were able to obtain a massive ingot or regulus by replacing the calcium chloride flux with iodine. This reaction became the basis of the first large-scale commercial process for producing vanadium. The reaction is initiated either by preheating the charged bomb or by internal heating with a fuse wire embedded in the charge. Calcium iodide formed by reaction of calcium with iodine serves both as a flux and a thermal booster. Thus sufficient heat is generated by the combined reactions to yield liquid metal and slag products. The resulting metal contains about 0.2% carbon, 0.02–0.08% oxygen, 0.01–0.05% nitrogen, and 0.002–0.01% hydrogen. Two factors that contribute to the relative inefficiency of this process are the rather low metal yields (75–80%) and the amount of calcium reductant required (50–60% excess over stoichiometry).

Vanadium powder can be prepared by substituting V_2O_3 for the V_2O_5 as the vanadium source. The heat generated during the reduction of the trioxide is considerably less than for the pentoxide so that only solid products are obtained. The powder is recovered from the product by leaching away the slag with dilute acid.

Aluminothermic Process. In the development of the liquid-metal fast-breeder reactor, vanadium has been considered for use as a fuel-element cladding material. Difficulty was encountered in the fabrication of alloys prepared from the calcium-reduced metal, a factor attributable to the high interstitial impurity content. In order to meet the more stringent purity requirements for this application an aluminothermic process was developed by the U.S. Atomic Energy Commission (11). In this process vanadium pentoxide is reacted with high-purity aluminum in a bomb to form a massive vanadium–aluminum alloy. The alloyed aluminum and dissolved oxygen are subsequently removed in a high-temperature, high-vacuum processing step to yield metal of greater than 99.9% purity. A more detailed description of the process follows.

Purified V_2O_5 powder and high-purity aluminum granules are charged into an alumina-lined steel crucible. The vessel is flushed with an inert gas in order to minimize atmospheric contamination and then sealed. The reaction is initiated by a vanadium fuse wire. Sufficient heat is generated by the chemical reaction to produce a molten alloy of vanadium containing 15 wt % aluminum and a fused aluminum

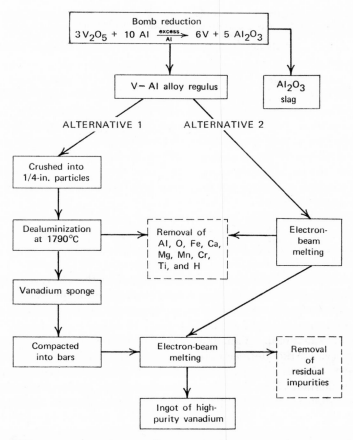

Fig. 1. Flow diagram for aluminothermic process showing alternative methods of aluminum removal from alloy regulus.

oxide slag. The liquid alloy separates from the alumina slag and settles to the bottom of the crucible as a massive product. The feasibility of carrying this reaction out in a water-cooled copper crucible, thus eliminating the alumina liner which is a source of some contamination, has been demonstrated recently.

The vanadium alloy is then purified and consolidated by one of two alternative procedures. These are seen from the flow diagram of the entire aluminothermic reduction process presented in Figure 1. In one procedure the brittle alloy is crushed and heated in a vacuum at 1790°C to sublime out most of the aluminum, oxygen, and other impurities. The presence of the aluminum has been shown to facilitate removal of the oxygen, which is the feature that makes this process superior to the calcium process. Further purification and consolidation of the metal is accomplished by electron-beam melting of pressed compacts of the vanadium sponge.

The alternative procedure involves direct electron-beam melting of the vanadium–aluminum alloy regulus. Two melting steps are required in order to reach the desired levels of aluminum and oxygen in the final ingot. Chemical analyses of two ingots of vanadium metal prepared from the identical vanadium–aluminum alloy and processed by the two methods described above are presented in Table 3. Comparable purities

Table 3. Vanadium Prepared by Dealuminization with Vacuum Heating Compared to That Prepared by Direct Electron-Beam Melting of Vanadium Alloy Containing 15 wt % Aluminum

Impurity	By dealuminization, ppm		By direct melting, ppm	
	Sponge	Electron-beam-melted ingot	First melt	Second melt
C	100	100	100	100
O	50	60	200	150
N	40	45	40	50
H	< 1	< 1	< 1	< 1
Al	> 1000	300	∼1000	100
Ca	< 10	< 10	< 10	< 10
Cu	< 20	< 20	< 20	< 20
Fe	70	70	60	60
Mg	< 20	< 20	< 20	< 20
Mn	< 20	< 20	< 20	< 20
Ni	60	60	60	60
Si	300	300	300	300

are obtained by these procedures. Vanadium metal in the alloy, sponge, and ingot forms representing different processing stages on an experimental scale are shown in Figure 2. Ingots weighing up to 500 lb have been prepared by this process, employing direct electron-beam melting of the alloy.

Refining of Vanadium. Vanadium can be purified by one of three methods: iodide refining (van Arkel-deBoer process), electrolytic refining in a fused salt, and electrotransport. Metal of greater than 99.95% purity has been prepared by the iodide refining method (12). In this process an impure grade of vanadium metal is reacted with iodine at 800–900°C to form vanadium diiodide and the volatilized

Fig. 2. V–Al alloy regulus, compacted sponge for melt stock, and high-purity vanadium ingot.

iodide is thermally decomposed and deposited on a hot filament at about 1300°C. The refining step is carried out in an evacuated and sealed tube. The major impurities removed in the process are the gaseous elements and those metals that form stable, nonvolatile iodides. Vanadium metal containing 5 ppm nitrogen, 150 ppm carbon, and 50 ppm oxygen has been prepared in this way.

An electrolytic process for purifying "crude" vanadium has been developed at the U.S. Bureau of Mines (13). It involves the cathodic deposition of vanadium from an electrolyte consisting of a solution of VCl_2 in a fused KCl–LiCl eutectic. The vanadium content of the mixture is between 2 and 5% and the operating temperature of the cell is between 650 and 675°C. Metal crystals or flakes of up to 99.995% in purity have been obtained by this method.

The highest-purity vanadium reported to date has been purified by an electro-transport technique (14). A high-density current was passed through a small rod of electrolytically refined metal, heating it to 1700–1850°C. Under these conditions interstitial solute atoms such as carbon, oxygen, and nitrogen migrate to the negative end of the bar, resulting in a high degree of purification along the remainder of the rod. Small amounts of vanadium containing less than 10 ppm of carbon, oxygen, and nitrogen and having a resistance ratio, $R_{300°K}/R_{4.2°K}$, of greater than 1100 have been prepared by this technique.

Economic Aspects

The United States has dominated world vanadium production since World War II. Prior to 1956 the U.S. was the chief importer of vanadium ores, but since that time it has become self-sufficient and is now a net exporter. This reflects the U.S. Atomic Energy Commission's need for uranium ores whereby uranium–vanadium ores are processed for the uranium content and large quantities of the vanadium by-product stockpiled. Although the U.S. continues to lead in vanadium ore production, there has been a significant increase in the supply from other countries in recent years. Canada began production in 1965 and the U.S.S.R. announced plans to double production by 1967. Australia may also become a large producer. Production figures for vanadium ores and concentrates are given for the period of 1960–1968 in Table 4.

Table 4. World Production of Vanadium in Ores and Concentrates by Countries, short tons (15–18)

Year	United States	Finland	Republic of South Africa	South-West Africa	Norway[a]	Others[b]	Total
1960	4971	625	656	838			7090
1962	5211	629	1393	1019			8252
1964	4362	1084	1282	1102	740	3	8573
1966	5166	1069	1711	1353	730		10029
1967	4963	1292	2100	1500	740		10595
1968	6483	1321	2498	1323	927		12562

[a] Total is of listed figures only; no undisclosed data included.
[b] Includes Argentina and Mexico.

The 1968 price of technical-grade vanadium pentoxide was in the range of $0.90–1.15/lb of contained V_2O_5. Based on contained vanadium, Carvan sold at $2.46/lb and ferrovanadium at $2.90/lb. The 1969 price of calcium-reduced vanadium of 99.3% purity is $33/lb. The aluminum-reduced metal is quoted at $35 to over $200/

lb, depending on form and quantity. Electrolytically refined granules of 99.8% purity sell for an estimated $30–40/lb while metal of 99.99% purity is priced at $2000/lb.

Uses

The most important use of vanadium is as an alloying element in the steel industry where it is added to produce grain refinement and hardenability in steels. Vanadium is a strong carbide former so that carbide particles form in the steel, thus restricting the movement of grain boundaries during heat treatment. This produces a fine-grained steel which exhibits greater toughness and impact resistance than a coarse-grained steel and which is more resistant to cracking during quenching. In addition the carbide dispersion confers wear resistance, weldability, and good high-temperature strength. Vanadium steels are used in dies or taps because of their depth-hardening characteristics and for cutting tools because of their wear resistance. They are also used as constructional steel in both light and heavy sections, for heavy iron and steel castings, forged parts such as shafts and turbine motors, automobile parts like gears and axles, and for springs and ball bearings. Vanadium is an important component of ferrous alloys used in jet aircraft engines and turbine blades where high-temperature creep resistance is a basic requirement.

The principal application of vanadium in nonferrous alloys is the titanium "6–4" alloy (6% Al–4% V) which is becoming increasingly important in supersonic aircraft where strength-to-weight ratio is a primary consideration. Vanadium and aluminum impart high-temperature strength to titanium, a property that is essential in jet engines, high-speed air frames, and rocket motor cases. Vanadium foil can be used as a bonding material in the cladding of titanium to steel. Vanadium is added to copper-base alloys to control the gas content and microstructure. Small amounts of vanadium are added to aluminum alloys to be used in pistons of internal combustion engines to enhance the alloys' strength and reduce the thermal expansion coefficient. Because of its low capture cross section for fast neutrons as well as its resistance to corrosion by liquid sodium and good high-temperature creep strength, vanadium alloys are receiving considerable attention as a fuel-element cladding for fast-breeder reactors.

Bibliography

"Vanadium and Vanadium Alloys," in *ECT* 1st ed., Vol. 14, p. 583, by Jerome Strauss, Vanadium Corporation of America.

1. S. A. Bradford and O. N. Carlson, *Am. Soc. Metals Trans. Quart.* **55**, 493 (1962).
2. R. W. Thompson and O. N. Carlson, *J. Less-Common Metals* **9**, 354 (1965).
3. C. A. Hampel, ed., *The Encyclopedia of the Chemical Elements*, Reinhold Publishing Corp., New York, 1968, p. 790.
4. C. A. Hampel, ed., *Rare Metals Handbook*, 2nd ed., Reinhold Publishing Corp., New York, 1961, p. 634.
5. American Society for Metals, *Metals Handbook*, Vol. 1, 8th ed., Reinhold Publishing Corp., New York, 1961, p. 1227.
6. T. J. McLeer, Foote Mineral Co., Exton, Pa., personal communication, July, 1969.
7. U.S. Pat. 3,420,659 (Oct. 11, 1967), H. W. Rathmann and R. T. C. Rasmussen (to Foote Mineral Co.).
8. F. H. Perfect, *Trans. Met. Soc. AIME* **239**, 1282 (1967).
9. J. W. Marden and M. N. Rich, *Ind. Eng. Chem.* **19**, 786 (1927).
10. R. K. McKechnie and A. U. Seybolt, *J. Electrochem. Soc.* **97**, 311 (1950).
11. O. N. Carlson, F. A. Schmidt, and W. E. Krupp, *J. Metals* **18**, 320 (1966).

12. O. N. Carlson and C. V. Owen, *J. Electrochem. Soc.* **108,** 88 (1961).

13. T. A. Sullivan, *J. Metals* **17,** 45 (1965).

14. F. A. Schmidt and J. C. Warner, *J. Less-Common Metals* **13,** 493 (1967).

15. *Mineral Facts and Problems, U.S. Bur. Mines Bull.* **630,** 1965, p. 1044.

16. *Economic Analysis of the Vanadium Industries, U.S. Dept. of Comm. Doc. PB-176 471*, June 1967, p. 70.

17. G. L. DeHuff, "Vanadium," in *Minerals Yearbook*, U.S. Bureau of Mines, 1967, p. 1179.

18. G. L. DeHuff, "Vanadium," Preprint from *Minerals Yearbook*, U.S. Bur. Mines, 1968.

General References

R. Rostoker, *The Metallurgy of Vanadium*, John Wiley & Sons, Inc., New York, 1958.

T. E. Dietz and J. W. Wilson, *Behavior and Properties of Refractory Metals*, Stanford University Press, Stanford, Cal., 1965.

M. E. Weeks, *Discovery of the Elements*, 6th ed., Mack Printing Company, Easton, Pa., 1956.

C. A. Hampel, ed., *Rare Metals Handbook*, 2nd ed., Reinhold Publishing Corp., New York, 1961, p. 634.

C. A. Hampel, ed., *The Encyclopedia of the Chemical Elements*, Reinhold Publishing Corp., New York, 1968, p. 790.

Economic Analysis of the Vanadium Industries, U.S. Dept. of Comm. Doc. PB-176 471, June, 1967.

O. N. Carlson
E. R. Stevens
Ames Laboratory of the U.S.
Atomic Energy Commission

VANADIUM COMPOUNDS

Nomenclature. Largely because of the numerous oxidation states shown by vanadium, considerable confusion has arisen in the past over the naming of vanadium compounds. This is particularly noticeable with the indiscriminate use of terms such as vanadite and vanadyl. Thus "vanadyl" is used for compounds containing V=O bonds, and both $VOCl_2$ and $VOCl_3$ have been named vanadyl chloride. To lessen these problems, this article is using the nomenclature generally accepted by leading scientific journals; in particular it denotes the formal oxidation state in the Stock manner by means of Roman numerals. These formalities tend to produce turgid prose, so from time to time more trivial variants may be used provided they do not introduce ambiguities. Table 1 lists examples illustrating the nomenclature.

Introduction. Vanadium, with the ground state electronic configuration, $3d^3 4s^2$, is a typical transition element, and accordingly shows a considerable number of oxidation states, ranging from the expected group state of V down to II, all the intermediate states being well characterized. These oxidation states are found in aqueous

Table 1. Nomenclature of Vanadium Compounds

Compound	Approved scientific name	Acceptable name
$VOCl_3$	vanadium(V) oxytrichloride	vanadium oxytrichloride
VCl_4	vanadium(IV) chloride	vanadium tetrachloride
VCl_3	vanadium(III) chloride	vanadium trichloride
VO_2	vanadium(IV) oxide	vanadium dioxide
Na_3VO_4	trisodium tetraoxyvana-date(V)	sodium orthovanadate

solutions, as well as in solid compounds. There is also considerable evidence for formal oxidation states of 0 and -1 in solid compounds, such as the carbonyl and certain complexes.

In the group-oxidation state (V), vanadium is diamagnetic and gives compounds that are usually colorless or pale yellow; thus the colors result from charge-transfer bands occurring in the ultraviolet region of the spectrum and "tailing" into the visible region. With compounds of vanadium in lower oxidation states, one or more $3d$ electrons are present, and these are generally unpaired and give rise to paramagnetic and colored species. Table 2 lists some such simple compounds with appropriate information.

Table 2. Magnetism and Spectra of Vanadium Compounds

Compound	Oxidation state	No. of d electrons	Magnetic moment, B.M.[a]	Color
$VOCl_3$	V	0	0	yellow
$VOSO_4.5H_2O$	IV	1	1.73	blue
$(NH_4)V(SO_4)_2.12H_2O$	III	2	2.80	blue
$VSO_4.6H_2O$	II	3	3.74	violet

[a] Bohr magnetons.

It must be appreciated, of course, that the magnetic moments and colors of compounds of a particular oxidation state depend on the precise nature and geometrical arrangement of the atoms associated with the vanadium. If, for instance, the compound is not magnetically dilute, then spin-pairing may occur between neighboring vanadium atoms and lead to low moments or even diamagnetism. All compounds of vanadium with unpaired electrons are colored, but it must be emphasized that the absorption spectra can be quite complex so that no oxidation state is characterized by a particular color. This can be illustrated by comparing vanadium(IV) oxy salts, which are generally blue or blue-green, with vanadium(IV) chloride, which is deep-red in color. For a more detailed discussion of these topics, see references 1, 2, and 3.

Several general observations may be made concerning the dependence of the chemical properties of vanadium compounds, eg acidity, upon the formal oxidation state of the metal. Thus, as the oxidation state decreases, so the basic character of the oxide increases; V_2O_5 is acidic, VO_2 amphoteric, and the remaining oxides basic. Vanadium(V) is found normally only in anionic form, as in the vanadates, although the ion $(VO_2)^+$ is known in strongly acidic solutions. The quadrivalent element, on the other hand, readily yields both oxyanions and oxycations, eg $(VO_4)^{4-}$ and $(VO)^{2+}$, respectively. In the lower oxidation states the simple aquated ions $(V(H_2O)_6)^{3+}$ and $(V(H_2O)_6)^{2+}$ are well established, and a number of salts of oxyacids (eg sulfates) have been characterized (4).

Quinquevalent vanadium compounds are reduced fairly readily, the oxidation state of the product depending upon the reducing agent. Thus, sulfur dioxide or mercury in acid solution brings about reduction to vanadium(IV), but a more vigorous reducing agent, such as zinc amalgam, yields divalent vanadium. The lower oxidation state compounds are good reducing agents and must be kept in an inert atmosphere.

Analysis. The best known qualitative test for vanadium is the red coloration produced by the addition of hydrogen peroxide to a sulfuric acid solution of a vanadate. A related complex is produced by titanium, but in this case the color is discharged by the addition of fluoride or phosphate ions. The borax bead test for vanadium gives a pale yellow bead in an oxidizing flame and a green bead in the reducing flame.

On the quantitative side there is a wide choice of methods ranging from atomic absorption spectroscopy to conventional chemical procedures; the latter can be divided into colorimetric, volumetric, and gravimetric methods (5). The best established colorimetric method is based on the hydrogen peroxide complex, although the second method, which makes use of the brownish-yellow phosphovanadotungstate complex, is more accurate and sensitive.

Volumetric methods involve the reduction of vanadium to either the quadrivalent state (with sulfur dioxide or mercury) or the divalent state (with zinc amalgam), followed by oxidation with standard potassium permanganate or ceric sulfate solution. The reduction to divalent vanadium subsequently gives a much bigger titration value, but this method is less easy to use because of the ease with which vanadium(II) solutions are oxidized. The sulfur dioxide reduction has the advantage that it can be repeated after the titration with potassium permanganate.

Gravimetric methods are not very satisfactory for vanadium. In the absence of other metals one useful technique is the quantitative oxidation of the vanadium compound to vanadium(V) oxide.

Interstitial and Intermetallic Compounds

Vanadium resembles its neighboring elements (Ti, Zr, Hf, Nb, Ta, Mo, and W) in a tendency to take up atoms of nonmetals into its lattice, the uptake being accompanied by a change in the packing pattern of the metal atoms to a cubic close-packed arrangement. This arrangement is characterized by the presence of so-called octahedral and tetrahedral holes in the lattice, which are surrounded by six and four metal atoms, respectively, there being one octahedral and two tetrahedral holes per metal atom. The octahedral holes can take up a relatively large atom, such as carbon or nitrogen, so that carbides and nitrides of compositions MC and MN can be formed, each having the NaCl structure. The tetrahedral holes, which are much smaller, can accommodate hydrogen atoms, leading us to expect the formation of a hydride, VH_2. These carbides, nitrides, and hydrides are referred to as *interstitial compounds*. Their composition is determined by geometrical packing considerations rather than by the conventional concepts of valence. Since not all the possible vacant sites need to be filled the observed compositions often fall below the theoretical and, with VN for instance, the homogeneity range is $VN_{(0.71-1.00)}$.

These carbides, nitrides, and hydrides are genuine interstitial compounds in that the nonmetal atoms are not linked and metal–metal distances remain short and similar to those in the metals themselves. Thus, the V—V distance of 2.92 Å in VN may be compared to 2.63 Å in the metal. (The comparable Ti—Ti distances for TiN and Ti metal are 2.99 and 2.93 Å, respectively.)

Borides are also known, but the boron atoms are linked together to fill spaces between the close-packed vanadium layers, each boron atom having six surrounding vanadium atoms in a trigonal prism arrangement. In VB the boron atoms form a simple zig-zag chain but in VB_2 they link in a layer of hexagons (rather like the carbon atoms in graphite).

Other nonmetal atoms give rise to related compounds, ie silicides, germanides, phosphides, arsenides, selenides, and tellurides, but since these atoms are much bigger than carbon and nitrogen they cannot be considered purely as simple interstitial compounds but rather as intermediates between interstitial and intermetallic compounds.

All the compounds are very hard crystalline solids (9–10 on Mohs' scale) with high melting points, and considerable resistance to chemical attack. These characteristics suggest possible uses as high-temperature refractories and bearing materials, although there are problems of fabrication, especially since the compounds are very brittle. This can be overcome by "alloying" with metals; one patent (6) has suggested that VC and other carbides can be flame-sprayed onto steels and nickel alloys to give a hard, dense, wear-resistant refractory coating. Mixtures of borides, carbides, and nitrides have also been converted by powder metallurgy techniques into suitably shaped ceramic materials. With increasing needs for such materials, the interstitial compounds of the early transition metal compounds in general, and of vanadium in particular, have a promising future. They can be made by a variety of high-temperature techniques, the simplest being the direct reaction of the elements in an electric furnace.

Vanadium Borides. (See Boron Compounds, Vol. 3, p. 673.) Each of the two borides, VB and VB_2, is very hard and indifferent to attack by common acids. The melting point of VB_2 is 2400°C. For the properties of borides and the related silicides and phosphides, see reference 7.

Vanadium Carbides. (See Carbides.) In addition to VC, a second carbide, V_2C, has been characterized. In this compound the metal assumes a hexagonal close-packed structure but only one half of the octahedral holes are occupied by carbon atoms; the V—V distance is 2.84 Å. VC (mp, 2830°C; sp gr, 5.3) is gray-brown and resists attack by H_2O, HCl gas, Cl_2, and S below red heat, and by the mineral acids (except HNO_3) at room temperature.

Vanadium Silicides. The vanadium–silicon system (7) is quite complicated and a number of compositions have been reported (8) in addition to the well-established compound VSi_2. The silicides can be made from elemental silicon and either vanadium or vanadium(V) oxide at a temperature of around 1200°C. They are resistant to chemical attack, VSi_2 being more resistant than V_2Si with respect to halogens or HCl. They, like the borides, have considerable potential as ceramic materials because they have a better thermal shock resistance than the oxide ceramics.

Vanadium Nitride. (See Nitrides.) VN (mp, 2050°C; sp gr, 5.63) can be obtained as a gray-brown powder by direct combination of the elements, or as a yellow-brown crystalline solid by the reduction of VCl_4 with H_2 in the presence of N_2. The crystalline solid obtained contains some interstitial hydrogen. The nitride is fairly inert to mineral acid attack but reacts slowly with molten NaOH or $KHSO_4$. It becomes superconducting at low temperature (as do other similar nitrides) with a transition temperature around 8°K.

Vanadium Phosphides and Arsenides. Both V_3P and VP have been prepared as black solids, and V_2P and VP_2 have also been reported (7). The phosphides are more readily attacked by acids than the nitride. VP has the NiAs structure. The gray-colored arsenides VAs (sp gr, 6.28) and V_2As (sp gr, 6.39) are also known (9).

Vanadium Selenides and Tellurides. VSe, VSe_2, and VTe are known but appear to be of no great industrial importance at the moment.

Vanadium Hydride (10) (see Vol. 11, p. 208). Hydrogen is readily taken up by vanadium metal with a notable change of the metal lattice. Up to a composition of $VH_{0.05}$ the lattice remains body-centered cubic and merely expands slightly, but a body-centered tetragonal phase then takes over up to a composition of $VH_{0.90}$. At this stage the face-centered cubic form takes over. There is some doubt about the maximum hydrogen content, although experiments have shown that up to $VH_{1.80}$ can be prepared. It seems likely that the hydrogen is present as discrete atoms in the "tetrahedral holes" of the metal lattice, so that a maximum hydrogen content would correspond to $VH_{2.0}$. The precise nature of the hydrogen in the lattice is uncertain; it is there as atoms rather than as molecules, but there is insufficient evidence to decide if it is hydridic or protonic.

Vanadium Oxides and Vanadates

The oxides V_2O_5, VO_2, and V_2O_3 are well characterized and an oxide approximating in composition to VO is also known, but only V_2O_5 has any commercial significance.

Vanadium(V) oxide (vanadium pentoxide), V_2O_5, is a yellowish-red powder (sp gr, 3.36). When allowed to cool from the molten state (mp, 658°C) it forms long red needles. In the laboratory it can be prepared as a brick-red powder by slightly acidifying an alkaline aqueous solution of ammonium vanadate (NH_4VO_3). It is also the final product of heating any vanadium compound in air. This forms the basis of industrial production where ammonium vanadate is carefully ignited, the operation being carried out in stages to lessen the chance of lower oxides being produced.

Vanadium(V) oxide forms an irregular layer structure in which each vanadium atom has an environment of five oxygen atoms in a distorted trigonal-bipyramidal arrangement.

In its chemical behavior and general stability V_2O_5 is intermediate in character between the highest oxides of titanium, TiO_2, and chromium, CrO_3. Thus, while V_2O_5 is not quite as stable to heat as TiO_2, it is much more stable than CrO_3; similarly, V_2O_5 is more acidic and a better oxidizing agent than TiO_2 but less so than CrO_3. V_2O_5 is fairly reactive and yields oxyhalides with halogenating agents, such as BBr_3 which gives $VOBr_2$. With fluorine VOF_3 is obtained under ordinary conditions, but at higher temperature and pressure (10–50 atm, 200–475°C) VF_5 is the product. The oxide is quite readily reduced, especially at elevated temperature, by H_2 and CO, and by a considerable number of metals, eg alkali metals, alkaline earth metals, Al, Zn, Cd, etc. Electrochemical reduction at 900°C of a solution of V_2O_5 in molten $CaCl_2$ yields vanadium metal. Aqueous HCl also brings about reduction, chlorine gas being liberated.

The oxide has a slight solubility in water (around $0.1 \text{ g}/100 \text{ g H}_2O$), the extent depending upon the history of the specimen; samples recrystallized from the molten state are much less soluble than precipitated specimens, as might be expected. Colloidal solutions of vanadium(V) oxide can be formed by slowly pouring molten V_2O_5 into water. In these sols the oxide is present as rodlike particles and the sols show stream birefringence.

Being amphoteric, V_2O_5 dissolves in both acids and alkalis, but the species produced depend upon the pH of the solution and the vanadium concentration. The solution chemistry of vanadium, and of related niobium and tantalum species, which has been studied by many investigators, has been reviewed by Pope and Dale (11). In very alkaline solutions (pH > 13) simple mononuclear vanadate anions are present, while in strongly acid solutions the predominant species is the dioxyvanadium(V) cation $(VO_2)^+$. Between these pH limits more complex vanadate(V) species are present. Thus, the acidification of an alkaline solution containing a simple vanadate anion, $(VO_4)^{3-}$, leads to very rapid protonation, followed by a much slower condensation, as shown below.

$$(VO_4)^{3-} + H^+ \rightleftharpoons (HVO_4)^{2-}$$
$$2\,(HVO_4)^{2-} \rightleftharpoons (V_2O_7)^{4-} + H_2O$$

As the solution is made more acidic so the condensation becomes more complex. Tri- and tetravanadates have been identified in solutions of pH 6.5–8.2, the trivanadate species, $(V_3O_9)^{3-}$, being formed at low concentrations of vanadium, and the tetravanadate, $(V_4O_{12})^{4-}$, at high concentrations. The $(V_{10}O_{28})^{6-}$ ion and its protonated forms, $(HV_{10}O_{28})^{5-}$ and $(H_2V_{10}O_{28})^{4-}$, appear to be present in more acidic solutions (pH 2–7).

Vanadates(V) have been fully characterized by x-ray methods and a number of types identified. The stoichiometry of these compounds has led to the use of the following nomenclature, commonly applied to the phosphates: $M_3^IVO_4$, orthovanadates; $M_4^IV_2O_7$, pyrovanadates; and M^IVO_3, metavanadates.

The orthovanadates contain discrete tetrahedral $(VO_4)^{3-}$ anions in which the V—O bond lengths all approximate to 1.66 Å. The pyrovanadates, which are isomorphous with analogous pyrophosphates, also contain discrete ions, $(V_2O_7)^{4-}$, with two VO_4 tetrahedra sharing a corner. The position in the metavanadates is a little less simple, since the structures depend upon whether the salt is hydrated or not. In the anhydrous salts KVO_3 and NH_4VO_3 the vanadium atoms are four-coordinate, with VO_4 tetrahedra linking through two oxygen atoms; the V—O distances are 1.81 Å for bonds to the bridging oxygen atoms and 1.66 Å to terminal oxygen atoms. In the hydrated form, $KVO_3 \cdot H_2O$, the vanadium atoms are five-coordinate, there now being three shared oxygen atoms per vanadium (V—O = 1.95 Å) and two terminal oxygen atoms (V—O = 1.66 Å), the environment of the vanadium atoms being trigonal bipyramidal, as in V_2O_5. Salts of the $(V_{10}O_{28})^{6-}$ anion have also been examined, and here the vanadium atoms have a distorted octahedral configuration, the ten units being linked through edge sharing to give a rectangular sheet.

Vanadium(IV) oxide (vanadium dioxide), VO_2, is a blue-black solid (mp, 1967°C; sp gr, 4.34) with a distorted rutile (TiO_2) structure. It can be prepared by reduction of V_2O_5 at melting-point temperature with a variety of materials, including oxalic acid, sugar, carbon, and sulfur. It is slightly hygroscopic and slowly oxidizes in air. It is amphoteric, and dissolves in acids to give oxyvanadium(IV), $(VO)^{2+}$, salts, and in hot alkalis to yield vanadates(IV) (hypovanadates, vanadites) that have not been well characterized.

Vanadium(III) oxide (vanadium sesquioxide), V_2O_3, is a black solid (sp gr, 4.87), which is formed by the reduction of the pentoxide with carbon or hydrogen. It has the corundum (Al_2O_3) structure. It is slowly oxidized in air, and more rapidly by chlorine, which gives $VOCl_3$ and V_2O_5. It is insoluble in water and does not react readily with acids and therefore is not a convenient source of vanadium(III) salts.

Vanadium(II) oxide, VO, is a gray-black powder (sp gr, 5.6) with a metallic luster. It is a nonstoichiometric compound with a defect NaCl structure in which the oxygen content may vary from 45 to 55%. It has a good electrical conductivity of a metallic nature and it seems that V—V bonding may be present. It is best prepared from V_2O_5 by reduction with hydrogen at 1700°C. It is insoluble in water and alkalis, but dissolves in acids to give vanadium(II) salts.

INDUSTRIAL USES OF VANADIUM(V) OXIDE AND THE VANADATES

The oxide and the vanadates are by far the most important chemical compounds of vanadium in the commercial world. V_2O_5 is widely used as a catalyst for a variety of reactions, particularly those involving oxidation, of both inorganic and organic compounds. One major use is in the oxidation of sulfur dioxide to sulfur trioxide in the contact process for the manufacture of sulfuric acid. V_2O_5 is much cheaper than the alternative, platinum, and it is much less susceptible to catalyst poisoning. A very extensive range of organic oxidations are catalyzed by V_2O_5. Typical industrial uses are in the oxidation of many hydrocarbons (eg naphthalene to phthalic anhydride) and in reactions involving oxidation and ammonolysis (eg propene to acrylonitrile or o-xylene to phthalic anhydride). Closely allied to these uses is the recent development of afterburners in automobiles, where devices based on V_2O_5 and Al_2O_3 have been developed to oxidize the hydrocarbons in exhaust gases. These afterburners are said to facilitate conversion of 91% of the hydrocarbons in the gases and to have a lifetime of around 12,000 miles. With the current concern for smog elimination this use is likely to increase greatly. See Automobile exhaust control in Vol. 2 and in Supplement Volume.

Vanadium(IV) oxide is also used as a catalyst for various reactions and recently has been introduced as a resistor material; for the latter purpose a mixture of V_2O_5 and various metal oxides is reduced with NH_3 at 300–400°C. Semiconducting materials can also be made by fusing V_2O_5 with Na_2O, to give black substances containing both quinquevalent and quadrivalent vanadium. The composition range is $Na_yV_2O_5$, where $0.15 < y < 0.33$.

There are various other uses of oxyvanadium compounds, but none of these employ very large quantities. V_2O_5 is added to glass to eliminate wavelengths below 3590 Å, the addition of as little as 0.02% V being sufficient to remove the high-energy ultraviolet radiation that injures the eye and causes fading of fabrics. This type of glass is used especially in store windows. In larger amounts V_2O_5 produces a yellowish-green color in glass, the tone depending on the composition of the glass. Glasses whose compositions involve phosphates and vanadates are of some interest to the electronics industry because of their characteristic low electrical resistivities (16). Small amounts of vanadium oxides are also used in ceramic pigments.

The oxide and metavanadates are used in the manufacture of printing inks, where they catalyze the formation of resinous black pigments from tar oils; some of the modern quick-drying inks depend for their performance on the addition of ammonium metavanadate. Small amounts of V_2O_5 are used in the textile-printing trade, where they help the oxidation of aniline to give black dyes of great intensity and fastness.

In 1965 the price of vanadium(V) oxide in the United States was $1.15/lb.

Halides and Oxyhalides

Table 3 lists the known halides and oxyhalides for each oxidation state. It is interesting to note that the only pentahalide is the fluoride. The nonexistence of the

Table 3. Vanadium Halides and Oxyhalides

Oxidation state	Halide		Oxyhalide	
	formula	appearance[a]	formula	appearance[a]
V	VF_5	white	VOF_3	pale yellow
			$VOCl_3$	yellow liquid
			$VOBr_3$	deep-red liquid
			VO_2F	brown
			VO_2Cl	orange
IV	VF_4	green	VOF_2	yellow
	VCl_4	red-brown liquid	$VOCl_2$	green
	VBr_4[b]	magenta	$VOBr_2$	yellow-brown
III	VF_3	green		
	VCl_3	red-violet	$VOCl$	brown
	VBr_3	gray	$VOBr$	violet
	VI_3	dark brown		
II	VF_2	blue		
	VCl_2	pale green		
	VBr_2	orange-brown		
	VI_2	red		

[a] At room temperature; solids unless otherwise specified.
[b] Decomposes at $-23°C$.

pentachloride is not a consequence of the larger size of the chlorine atoms because ions such as $(VCl_6)^{2-}$ are well known; the simplest explanation is a thermodynamic one related to the weakness of the F—F bond and the comparative strength of the Cl—Cl bond. Even the tetrachloride is rather unstable thermally and the tetrabromide decomposes above $-23°C$. Not all the halides are of great significance and therefore only the more important ones appear in the following section. Clark's book (3) gives an excellent account and should be consulted for further details.

The halides and oxyhalides are excellent sources of other vanadium compounds; they are used commercially mainly for catalytic purposes, particularly in the field of polymerization of olefins, etc. $VOCl_3$, VCl_4, and VCl_3 have all been used, for instance, as Ziegler-Natta-type catalysts for the copolymerization of ethylene and propylene.

Vanadium(V) fluoride is a white solid which melts at 19.5°C to a pale-yellow viscous liquid (sp gr, 2.46). It is best prepared by the fluorination of the powdered metal at 300°C under rigorously anhydrous conditions with the fluorine diluted with nitrogen. In the solid state the structure is one of endless chains of octahedrally coordinated vanadium atoms linked through cis-bridging fluorine atoms (12); the viscous nature of the liquid state suggests the retention of much of this linkage.

Vanadium(V) oxytrichloride is a pale-yellow liquid (bp, 127°C; sp gr, 2.459) with a tetrahedral monomeric structure, which is readily prepared by the chlorination of V_2O_5 mixed with charcoal at red heat. It is readily hydrolyzed, forms coordination compounds with simple donor molecules such as ethers, but is reduced when allowed to react with sulfur-containing ligands and molecules such as trimethylamine. It is completely miscible with many hydrocarbons and nonpolar metal halides such as $TiCl_4$, and it dissolves sulfur.

Vanadium(IV) chloride (vanadium tetrachloride), VCl_4, is a dark red-brown liquid (bp, 152°C; sp gr, 1.82). It is best prepared by the chlorination of the metal powder at 300°C, and freed from dissolved chlorine by repeated freezing and evacuation. Care has to be taken to avoid contamination by $VOCl_3$. VCl_4 has a simple tetrahedral monomeric structure. It is readily hydrolyzed, dissolves in nonpolar solvents, but forms addition compounds with donor solvents such as pyridine; many such molecules reduce the chloride and produce compounds of trivalent vanadium (eg $(CH_3)_3N$ gives $VCl_3 \cdot 2N(CH_3)_3$). VCl_4 is barely thermodynamically stable and dissociates into VCl_3 and Cl_2. This decomposition takes place slowly at room temperature and more rapidly with increasing temperature. Decomposition is also induced both catalytically and photochemically and therefore there are difficulties in transporting and handling this chemical on an industrial scale.

Vanadium(III) chloride (vanadium trichloride), VCl_3, is a violet solid (sp gr, 2.82), which is easily hydrolyzed. It can be prepared in a pure state by the thermal decomposition of the tetrachloride, or by heating V_2O_5 with sulfur monochloride under reflux for 8 hr. VCl_3 cannot be purified by sublimation because it decomposes around 400°C into VCl_4 and Cl_2. It has a structure closely related to that of iron(III) chloride, with the vanadium atoms having an octahedral environment of shared chlorine atoms. The chloride is insoluble in nonpolar solvents but dissolves readily in donor solvents, such as acetonitrile, with the formation of coordination compounds. The tribromide closely resembles the trichloride.

Vanadium(II) chloride (vanadium dichloride), VCl_2, is a green crystalline solid (sp gr, 3.09), which sublimes without decomposition in an inert atmosphere around 900°C. It has the cadmium iodide structure and is best made by disproportionation of the trichloride at 800°C in a stream of nitrogen. It dissolves in water and alcohol to give a blue-purple solution which acts as reducing agent.

Vanadium Sulfates

Quinquevalent vanadium does not form simple salts, such as sulfates, and the quadrivalent salts are solely based on the $(VO)^{2+}$ ion. Sulfates of vanadium(III) and (II) are both known, besides several other rather ill-defined salts of organic acids, such as acetates and benzoates. Many complex anionic species are known for all the oxidation states (II–V), but these are discussed briefly under the section dealing with coordination compounds.

Vanadium(IV) oxysulfate (vanadyl sulfate), $VOSO_4 \cdot 5H_2O$, is a sky-blue solid, which is readily soluble in water and sparingly soluble in alcohol. It may be obtained by the reduction of V_2O_5 with SO_2 in sulfuric acid. Other hydrates can be made and an anhydrous green form can also be prepared, which is virtually insoluble in water. Both forms contain double $V{=}O$ bonds, with the vanadium atom in an octahedral environment of oxygen atoms; in the hydrated form four molecules of water and one oxygen of the sulfate group make up the environment of oxygen atoms around vanadium.

Vanadium(III) sulfate, $V_2(SO_4)_3$, can be prepared in both anhydrous and hydrated forms. The structures of these are not fully understood but are likely to be complex, especially since the anhydrous form is insoluble in both water and sulfuric acid. The sulfate is a powerful reducing agent which bleaches indigo and vegetable dyes.

Vanadium(II) sulfate, $VSO_4 \cdot 6H_2O$, is a light red-violet crystalline powder produced by the electrolytic reduction of $VOSO_4$. The powder is readily oxidized in air, and dissolves in water to give a red-violet solution.

Organometallic Compounds of Vanadium

Twenty years ago this area of vanadium chemistry could have been dismissed in a sentence since it was virtually an unknown field. In recent years organometallic compounds of the transition metals have received enormous attention, and, while the number of characterized vanadium compounds is not high, our knowledge of this region of chemistry is substantial (13).

By organometallic compounds we refer to compounds in which there is direct bonding between metal and carbon atoms, although the definition is by no means restricted to conventional σ-bonded metal alkyls and aryls. Indeed while simple alkyls are known for titanium, eg $(CH_3)_nTiCl_{(4-n)}$ ($n = 1-4$), the vanadium analogs have not been reported and appear to be too unstable to be isolated. Conventional preparative techniques, such as those based on reactions between vanadium halides and Grignard reagents, lead only to decomposition products of the alkyl groups. Compounds with O-bonded alkyl groups may well be involved as intermediates in Ziegler-type polymerizations of olefins, but they are likely to be isolated as stable species at room temperature only if other stabilizing groups are also bonded to vanadium (see under Cyclopentadienyl compounds and under Coordination compounds), or additional π-bonding is present as in the carbonyl.

A breakthrough came in the 1950s with the discovery of the cyclopentadienyl compounds, the first being the iron derivative, ferrocene, $\pi(C_5H_5)_2Fe$. Similar derivatives have now been made of many other transition metals including vanadium. All $\pi(C_5H_5)_2M$ compounds are characterized by having a peculiar sandwich structure, in which the metal atom is between two parallel planar cyclopentadienyl rings, with unusual bonding involving the π electrons of the ring system and the metal d orbitals.

Other unsaturated hydrocarbons are now known to bond to vanadium and these compounds are likely to be investigated fully within the next few years.

The main impetus for the investigation of organovanadium compounds came from the knowledge that similar compounds of titanium were the active catalytic species in the Ziegler-Natta stereospecific polymerization of olefines. So far the vanadium compounds appear to be less effective than the titanium analogs and they suffer from the considerable disadvantage of being toxic (see under Toxicology) and thus an undesirable component of domestic plastics, even at a low level of concentration.

Vanadium(0) Carbonyl. Vanadium hexacarbonyl, $V(CO)_6$, forms blue-green pyrophoric crystals that can be sublimed. It is made from VCl_3 and CO by treatment with sodium at high pressure (250 atm). The compound is unique among metal carbonyls in that it is paramagnetic with one unpaired electron. It might have been expected that the molecule would be dimeric with a metal–metal bond (as is $Mn(CO)_5$), but this is not the case. Steric reasons account for this, with the loss of some d_π-p_π bonding a further contributing factor. Various carbonyl derivatives have been prepared including several salts of the anion $(V(CO)_6)^-$; CO groups can be replaced by tertiary phosphines to give compounds such as $V(CO)_4(PEt_3)_2$.

Cyclopentadienyl and Similar Compounds. Vanadocene, $\pi(C_5H_5)_2V$, the analog of ferrocene (see Vol. 6, p. 697), can be prepared as violet-black crystals (mp, 168°C) by the reaction of vanadium(IV) chloride with sodium cyclopentadiene, NaC_5H_5. By varying the precise experimental conditions, the two chloro compounds, $\pi(C_5H_5)_2$-VCl and $\pi(C_5H_5)_2VCl_2$, can be obtained. The monochloro compound forms indigo-blue crystals (mp, 206°C), while the dichloro derivative is light green in color and melts with decomposition above 250°C. Vanadocene itself is very soluble in organic solvents, but this solubility decreases along the series $\pi(C_5H_5)_2V$, $\pi(C_5H_5)_2VCl$, and $\pi(C_5H_5)_2VCl_2$. The dichloro compound is insoluble in solvents of the light petroleum type, but dissolves in ethanol and water.

The chlorine atom in $\pi(C_5H_5)_2VCl$ can be replaced by a phenyl or methyl group by reaction with phenyllithium or methyllithium, to give compounds which are sensitive to air but otherwise stable in the solid state. Alternative arene groups, such as benzene and substituted benzenes, can be bonded to vanadium instead of cyclopentadiene; for example, dibenzenevanadium, $\pi(C_6H_6)_2V$, has been prepared as reddish-brown crystals (mp, 278°C) by the reduction of VCl_4 by aluminum in benzene in the presence of aluminum chloride.

Allyl and Acetylene Derivatives. These have not been characterized very well but unquestionably do exist. Thus, the reaction of $\pi(C_5H_5)_2VCl$ with allyl magnesium chloride in tetrahydrofuran solution is said to give the allyl derivative $\pi(C_5H_5)_2V(C_3H_5)$ and the pure π allyl complex $V(C_3H_5)_3$ has been reported. $\pi(C_5H_5)_2VCl$ also reacts with dimethyl acetylenedicarboxylate in benzene to give the air-sensitive, dark-green substance $\pi(C_5H_5)_2VC_2(CO_2CH_3)_2$, in which the vanadium atom forms a typical "bent-bond" linkage with the acetylene fragment, as shown below:

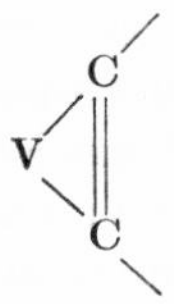

Coordination Compounds of Vanadium

In this area the studies on the chemistry of vanadium compounds have shown the biggest jump forward over the past ten years, although much of the interest has been rather academic (eg theoretical implications of magnetic and spectra measurements), and very few industrial applications have been proposed so far. Because of the limited use of these compounds and in view of an excellent review by Nicholls (14), only a relatively brief survey is given here.

The term coordination compound, and the alternative term complex compound, apply to compounds in which the number of bonds formed by the central metal atom differs from the valence of the metal. Thus, typical coordination compounds of trivalent vanadium would be $(Ven_3)Cl_3$, $VCl_3 \cdot 3py$ (en = ethylenediamine, py = pyridine), and $(N(C_2H_5)_4)_3VCl_6$, in each of which the vanadium atom forms six bonds rather than three, the compounds containing cationic, neutral (uncharged), and anionic complex vanadium species, respectively.

As with neighboring transition elements, the stereochemistry of these compounds is largely concerned with six-coordination, in which vanadium has a pseudooctahedral structure apart from a few trigonal-prismatic compounds formed by chelating bidentate sulfur ligands. Coordination number four occurs only with $(VCl_4)^-$ in conjunction with a tetraphenylarsonium cation. Lower coordination numbers have not been reported. Coordination number five is now well established, particularly for complexes of trivalent vanadium, such as $VCl_3 \cdot 2N(CH_3)_3$, and for oxyvanadium(IV) species (see below). The high coordination number of eight has been found for vanadium in the complex $VCl_4 \cdot 2diars$ (diars = o-phenylene-bis-dimethylarsine).

Coordination Compounds of Vanadium(V). These are anionic in character and besides the vanadates, which can be regarded formally as complex, include a number of oxyacid salts of the types $M_3^I(VO_2(C_2O_4)_2)$ and $M^I(VO_2SO_4)$. Complex halogen anions, such as $(VF_6)^-$ and $(VOCl_4)^-$, are also known. All these compounds are diamagnetic and colorless or pale-yellow.

Coordination Compounds of Vanadium(IV). These fall into two categories, the adducts based on VCl_4 and VF_4, and those related to the oxyvanadium(IV) species $(VO)^{2+}$; the latter are much better known. The adducts include the hexahalogeno anions $(VF_6)^{2-}$ and $(VCl_6)^{2-}$ and a series of neutral complexes of the type $VCl_4.2L$, where L (ligand) includes acetonitrile, pyridine, and tetrahydrofuran. With some ligands, such as thioethers, reduction to trivalent vanadium readily takes place.

The chlorocompounds are extremely sensitive to moisture and readily hydrolyzed to oxyvanadium(IV) derivatives. In fact, the whole chemistry of vanadium(IV) is dominated by the tendency to form compounds based on $(VO)^{2+}$; this is in contrast to the analogous titanium systems. Many complexes containing the VO grouping have been prepared either as neutral or anionic species. Two examples of neutral complexes are, $VOCl_2.2N(CH_3)_3$ and $VO(acac)_2$ ($acac = CH_3COCHCOCH_3$); both are soluble and monomeric in nonpolar solvents. The trimethylamine adduct is especially interesting in that the bulky amine groups evidently force a transtrigonal bipyramidal five-coordinate structure on the vanadium atom, in which the oxygen, vanadium, and chlorine atoms form a trigonal plane. There is a considerable tendency for complexes, such as $VO(acac)_2$, to take up a further molecule of a donor ligand (eg pyridine) to achieve the more symmetrical six-coordinate arrangement. Oxyvanadium(IV) anions are formed with the halogens and many carboxylate groups. All of the oxyvanadium(IV) complexes are very similar in color, either blue or green, because their electronic spectra characterize the VO grouping and relatively little is contributed to the energy diagram by the other ligands. Some oxyvanadium(IV) complexes, especially those soluble in organic solvents, have found use in homogeneous catalyst systems for rubber synthesis.

Coordination Compounds of Vanadium(III). Trivalent vanadium is sufficiently basic for straightforward V^{3+} ions to be formed, stabilized, of course, by the attachment of donor molecules, such as water and urea. Most of the complexes are neutral, however, or anionic. In the neutral complexes, the majority of the complexes are based on the trihalides, and many compounds of the type $VX_3.3L$ have now been prepared. This stoichiometry has been achieved with such ligands as pyridine, acetonitrile, and cyclic ethers, but with more bulky ligands (eg $N(CH_3)_3$ or α picoline) the complexes have the formula $VX_3.2L$; they have a trans-trigonalbipyramidal structure.

Salts of $(VA_6)^{3-}$ anions have been prepared in which the cation is F, Cl, CN, NCS, or NCSe, and of various oxyanions, such as $(V(C_2O_4)_3)^{3-}$. No major industrial uses have been proposed up until now.

Coordination Compounds of Vanadium(II). These have not been well characterized. They include simple cations, eg $(V(H_2O)_6)^{2+}$, neutral complexes, eg $VCl_2.4$-pyridine, and anions, eg $(V(CN)_6)^{4-}$. They are good reducing agents and must be handled carefully to avoid oxidation by the atmosphere.

Coordination Compounds of Vanadium(0). The curious compound $V(bipyridyl)_3$ has been prepared by the reduction of an alcoholic solution of $(V(bipyr)_3)I_2$ by magnesium. In this, and the analogous o-phenanthroline complex, the vanadium is formally in a zero oxidation state, but there is almost certainly a delocalization of the vanadium electrons onto the ligand ring systems.

Toxicology and Biological Significance of Vanadium Compounds

Vanadium compounds are toxic to both humans and animals and should be considered as industrial hazards of the same order as selenium and tellurium compounds. At the industrial level, the problem has shown up particularly in connection with the ash of oil-fired boilers which may contain appreciable amounts of vanadium(V) oxide. Some workers engaged in the handling of such ash have suffered nose bleeding and considerable irritation of the respiratory tract.

Faulkner Hudson (15) has surveyed the general problems of the health hazards of vanadium compounds and pointed out that the respiratory center is affected and accompanied by constriction of the peripheral arteries of the viscera, hyperperistalsis, and enteritis. Other effects may include fatty degeneration of the liver and damage to the convoluted tubules of the kidneys. The toxicity depends upon the compound and the method by which it is administered. Thus, vanadium compounds in general, and V_2O_5 in particular, can be absorbed quickly from the lungs into the bloodstream, while bowel absorption takes place much less readily. Much of the vanadium taken by mouth is excreted and there appears to be no cumulative effect, provided the daily tolerance dose is not exceeded. When inhaled, the chief damage is to the respiratory tract, and there appears to be no fibrotic changes or specific chronic lesions in the lungs. Experiments with rabbits have shown that when V_2O_5 (as sodium metavanadate) is injected subcutaneously, the lethal dose is around 10 mg per kg body weight. For human beings, about 30 mg of V_2O_5 (as sodium metavanadate), given intravenously, is quoted as the probable fatal dose for a man of 70 kg body weight. The hexavanadate and vanadium(IV) oxysulfate are appreciably less toxic, perhaps by a factor of five or ten.

Vanadium compounds have a significant effect on the body metabolism, particularly with reference to sulfur and cholesterol. The effect on the sulfur metabolism shows up markedly in the keratinous tissues, there being a decreased cystine content of hair and nails, for instance. The cholesterol metabolism is profoundly affected, the synthesis of cholesterol by the body being reduced considerably when vanadium is present; this has possible significance in the treatment of cardiovascular disease.

Bibliography

"Vanadium Compounds" in *ETC* 1st ed., Vol. 14, pp. 594–602, by H. E. Dunn and C. M. Cosman, Vanadium Corporation of America.

1. J. Lewis and R. G. Wilkins, *Modern Coordination Chemistry*, Interscience Publishers, Inc., New York, 1960.
2. C. J. Ballhausen, *Introduction to Ligand Field Theory*, McGraw-Hill Book Co., Inc., New York, 1962.
3. R. J. H. Clark, *The Chemistry of Titanium and Vanadium*, Elsevier Publishing Co., Amsterdam, 1968.
4. L. F. Larkworthy, J. M. Murphy, K. C. Patel, and D. J. Phillips, *J. Chem. Soc.* **1968A**, 2936.
5. G. Charlot and D. Bézier, *Quantitative Inorganic Chemistry;* translated by R. C. Murray, Methuen and Co., Ltd., London, 1957.
6. Fr. Pat. 1,447,081 (July 22, 1966), F. J. Dittrich (to Metro Inc.). *Chem. Abstr.* **66**, 68627 (1967).
7. B. Aronsson, T. Lundström, and S. Rundavist, *Borides, Silicides, and Phosphides*, Methuen and Co., Ltd., London, 1965.
8. J. Hallais, J. Sénateur, and R. Fruchart, *Compt. Rend.* **264C**, 1947 (1967).
9. H. Boller and H. Notworthy, *Monatsh. Chem.* **97**, 1053 (1966).

10. A. J. Maeland, *J. Phys. Chem.* **68**, 2197 (1964).
11. M. T. Pope and B. W. Dale, *Quart. Rev. (London)* **22**, 527 (1968).
12. A. J. Edwards and G. R. Jones, *J. Chem. Soc.* **1969A**, 1651.
13. M. L. H. Green, *Organometallic Compounds*, Vol. 2, 3rd ed., Methuen and Co., Ltd., London, 1968.
14. D. Nicholls, *Coord. Chem. Rev.* **1**, 379 (1966).
15. T. G. Faulkner Hudson, *Vanadium Toxicology and Biological Significance*, Elsevier Publishing Co., Amsterdam, 1964.
16. Brit. Pat. 1,058,186 (Feb. 8, 1967) (to General Electric Co.). *Chem. Abstr.* **66**, 98102 (1967).

G. W. A. Fowles
University of Reading
Reading, England

VANILLIN

Vanillin is the common or trivial name for 4-hydroxy-3-methoxybenzaldehyde. (It was formerly called protocatechualdehyde-3-methyl ether and vanillaldehyde.)

It crystallizes from water, water–alcohol, and organic solvents in the form of monoclinic, prismatic needles (1,2). Vanillin is found widely in the plant kingdom in trace amounts and its commercial sources have usually been plant related. Today it is derived principally from lignin which is present in large amounts as a waste material in spent sulfite liquors from paper-pulp manufacture. Vanillin is the most popular flavor in the United States in terms of tons used in food production. It is also used in increasing quantities in the perfume, pharmaceutical, and metal-plating industries. See also Flavors and spices; Perfumes.

Physical and Chemical Properties

The physical properties of vanillin are listed in Table 1. For crystallographic data, see reference 3. Data (4,5) have been compiled on the absorption spectra of

Table 1. Physical Properties of Vanillin

Property	Value
melting point, °C	81–83
boiling point, °C	
at 760 mm Hg	284 (dec)
at 1 mm Hg	127
density, d_4^{20}, g/ml	1.056
heat of soln, in H_2O at infinite	
dilution, cal/mole	5.2
heat of neutralization,[a] cal/mole	9.26
heat of combustion,[b] cal/mole	914.7

[a] When one mole of vanillin in 30 liters of water is neutralized with 0.2N NaOH.

[b] Constant pressure and volume.

vanillin in the vapor state and in the following solvents: ethyl alcohol; ethyl alcohol and sodium ethoxide; ethyl alcohol and sodium hydroxide; and hexane.

Vanillin is slightly soluble in water (1 part and 5 parts per 100 parts by wt at 14 and 75°C, respectively); moderately soluble in glycerol (approx 4.5% by wt at 25°C); as well as freely soluble in propylene glycol (approx 23% by wt at 25°C), ethyl alcohol (50, 70, and 95%), chloroform, ether, carbon disulfide, glacial acetic acid, pyridine, and caustic solutions.

REACTIONS

Three different types of reactions are possible with vanillin: those of the aldehyde group, the phenolic hydroxyl, and the aromatic nucleus. The aldehyde group undergoes certain typical aldehyde condensation reactions that allow various substitutions for the aldehyde group. The aldehyde group may also be partially or completely reduced. However, being a *p*-hydroxybenzaldehyde, vanillin does not undergo some very common aromatic aldehyde reactions. These include the Cannizzaro reaction, the benzoin condensation, and oxidation with Fehling solution to the corresponding acid, vanillic acid. If the hydroxyl group in vanillin is protected, oxidation to vanillic acid derivatives readily occurs. As a phenol, vanillin forms esters and ethers, and the nucleus is easily substituted by halogen and nitro groups. In comparison with most other aldehydes, vanillin is notable for its stability.

Condensation. With acetone in the presence of alkali, vanillin forms vanillyl-idenacetone, which can be reduced to zingerone (mp, 41°C), the chief flavoring agent of ginger.

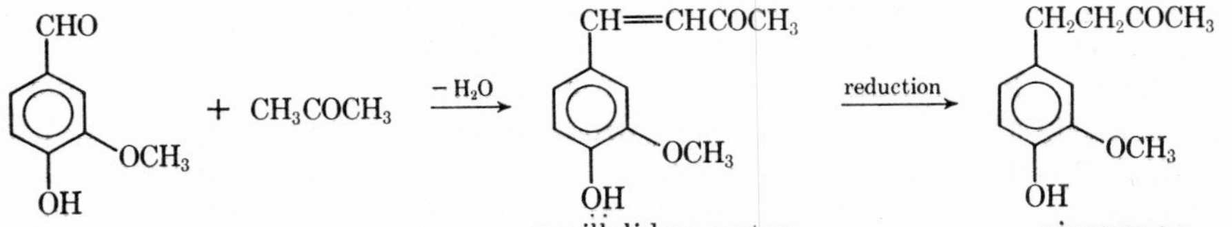

vanillylidene acetone   zingerone

Condensation of vanillin with hydroxylamine yields vanillin oxime, which is reduced with sodium amalgam and acetic acid to vanillylamine.

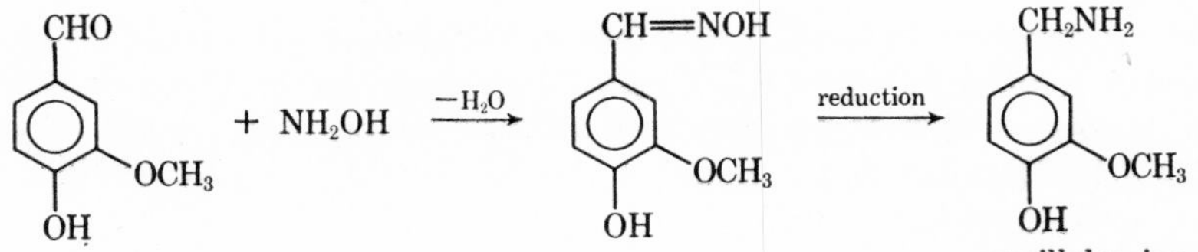

vanillylamine

The amides derived from this compound are pungent materials. Capsaicin (*trans*-8-methyl-*N*-vanillyl-6-nonamide), $CH_3O(OH)C_6H_3CH_2NHCO(CH_2)_4CH=CHCH(CH_3)_2$, for example, has a potency one thousand times as great as that of zingerone. With phenylhydrazines, vanillin forms hydrazones. With heptaldehyde in dilute alcoholic potassium hydroxide, vanillin yields α-amyl-3-methoxy-4-hydroxycinnamaldehyde, which has a stronger aroma than jasminal, a fraction of oil of jasmine.

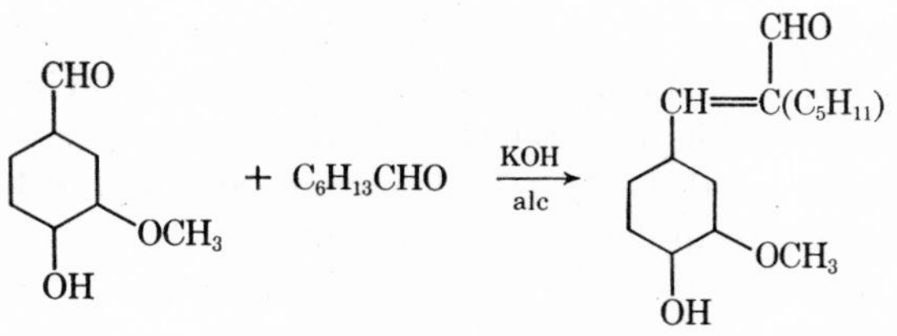

CHO + $C_6H_{13}CHO$ $\xrightarrow[\text{alc}]{\text{KOH}}$ [structure with CHO, CH=C(C_5H_{11}), OCH_3, OH]

In a Perkin condensation, Tiemann prepared the acetate of ferulic acid, $CH_3O(CH_3COO)C_6H_3CH=CHCOOH$, by prolonged boiling of vanillin with sodium acetate and acetic anhydride. Vanillylidene cyanohydrin, $CH_3O(OH)C_6H_3CH(OH)CN$, can be made by the action of potassium cyanide on a sodium bisulfite solution of vanillin. Vanillylidenenitromethane, $CH_3O(OH)C_6H_3CH=CHNO_2$, is prepared from nitromethane by using methylamine hydrochloride and sodium carbonate or ethylamine as the condensing agent.

Reduction. Catalytic hydrogenation using platinum oxide in alcohol, nickel, or palladium–barium in glacial acetic acid as catalyst, or the action of sodium amalgam in water, yields vanillyl alcohol (mp, 115°C).

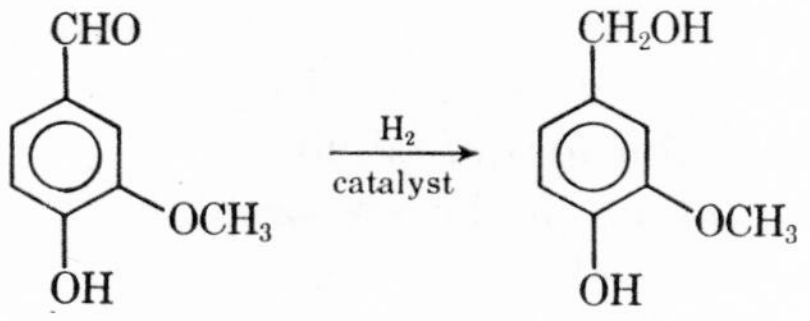

[structure: CHO, OCH_3, OH] $\xrightarrow[\text{catalyst}]{H_2}$ [structure: CH_2OH, OCH_3, OH]

Creosol, $CH_3O(OH)C_6H_3CH_3$, may also be produced with the palladium–barium catalyst, and hydrovanilloin, $CH_3O(OH)C_6H_3CH(OH)CH(OH)C_6H_3(OH)(OCH_3)$, is an additional product of the sodium amalgam reduction.

Oxidation. For optimum conditions for obtaining high yields of vanillic acid, $CH_3O(OH)C_6H_3COOH$, see p. 192. Dehydrodivanillin (5,5'-bivanillin), $OHC(CH_3O)(OH)C_6H_2C_6H_2(OH)(CH_3O)CHO$, is obtained in the presence of iron in almost theoretical yields on treatment of vanillin with sodium peroxydisulfate. Optimum yields of methoxyhydroquinone, $CH_3O(OH)C_6H_3OH$, which has antioxidant properties, are obtained by the oxidation of vanillin with alkaline hydrogen peroxide.

Ethers. Methylation with methyl sulfate and sodium hydroxide yields veratraldehyde, $(CH_3O)_2C_6H_3CHO$. Ethyl, propyl, isopropyl, and benzyl ethers have also been prepared.

Esters. Treatment of vanillin with acetic anhydride in alkaline solution produces vanillin acetate ($-OH \rightarrow -OCOCH_3$); prolonged heating with acetic anhydride, with or without catalysts such as concentrated sulfuric acid, gives the triacetate, $CH_3O(CH_3COO)C_6H_3CH(OOCCH_3)_2$. Benzoyl chloride and alkali give vanillin benzoate ($-OH \rightarrow -OCOC_6H_5$) (4).

Substitution. Generally substitution occurs in the 5-position when vanillin is halogenated or nitrated. If the hydroxyl group is blocked by esterification or etherification, substitution takes place in the 2- and 6-positions. Bromination of vanillin acetate gives high yields of a 6-bromo derivative, while nitration of the acetate gives high yields of the 2-nitro derivative, and small amounts of a 6-nitro derivative. Raiford and others (6) have prepared a variety of chlorine-substituted products of vanillin, including the monochloro derivatives with substitution in the 2-, 5-, and 6-positions.

Occurrence

Vanillin occurs widely in the plant kingdom in woody tissues, in resins, and in the cambium layer of coniferous trees. It occurs likewise in hardwoods, having been found along with syringaldehyde in the forty-six varieties studied by Pearl (7). The best known natural source of vanillin is the vanilla plant, a tropical orchid, which produces fruits or seed pods 6–10 in. long and 0.25 in. thick. These so-called vanilla "beans" of commerce are cultivated in Mexico, a native habitat, in Malagasy and the nearby islands, and in Comores, Réunion, Mauritius, and Seychelles (vanilla beans known to commerce as the "Bourbons" are produced in the latter). Islands of the French West Indies, especially Guadeloupe and Basse Terre, produce fine vanilla beans. Java vanilla beans of Indonesia are a favorite source of vanilla; and also those from Tahiti. The vanillons, or wild beans, are found in Brazil and other countries of South America (8). Though widespread, vanillin is normally found in trace amounts; even in vanilla beans its concentration is limited to less than 5%.

Preparation and Manufacture

From Vanilla Pods. The history of vanilla is lost in antiquity. Perhaps it was associated first with religious rites (9). About 1520, Hernando Cortez, Spanish conqueror of Mexico, was served a drink of chocolate and vanilla by the Aztec chief Montezuma. He was so pleased with the flavor that he introduced it into Europe. From that time until today it has been a favorite food flavor, derived exclusively from the unripened seed pods of *Vanilla planifolio* and *Vanilla fragrans* until the last quarter of the nineteenth century. Gobley is cited as the first to isolate vanillin from the vanilla pod in 1858 (2,10).

Vanilla "bean" culture is still the basis of a flourishing industry. Most of the vanillin, which is the dominant flavor characteristic of vanilla, is formed in the fruit pods during the curing process; it is not present in significant amounts when the green pod is harvested. Climatic conditions, timing of harvest, killing and sweating of the pods are some of the factors determining vanillin content of the pod, as well as overall quality of the final vanilla extract. It is believed that β glucosidase acts on gluco-vanillin with the formation of vanillin and sugar as curing of the pods progresses. The vanilla that is marketed is normally an alcohol extract of the cured crushed vanilla pods.

From Coniferin. Vanilla was prepared by Tiemann and Haarmann (11,12) in 1874 from the glucoside coniferin of which coniferyl alcohol is the aglucone. Coniferin was obtained by extracting the cambium layer of coniferous trees. It was oxidized by dichromate to convert it to glucovanillin, the latter compound being hydrolyzed to vanillin with enzymes or mineral acid.

$$CH_3OO(C_6H_{11}O_5)C_6H_3CHCHCH_2OH \xrightarrow{CrO_3}$$
coniferin

$$CH_3OO(C_6H_{11}O_5)C_6H_3CHO \xrightarrow[\text{or enzyme}]{\text{acid}}$$
glucovanillin

$$CH_3O(OH)C_6H_3CHO$$
vanillin

This method is of historical and academic interest only. However, Haarmann patented his method (13), one of the earliest vanillin patents.

From Eugenol. The oil of cloves contains 85–95% of eugenol (see Phenolic ethers) which provided a source material of historical and practical significance. When eugenol is treated with strong alkali it rearranges by migration of the double bond to form isoeugenol. Oxidation produces vanillin.

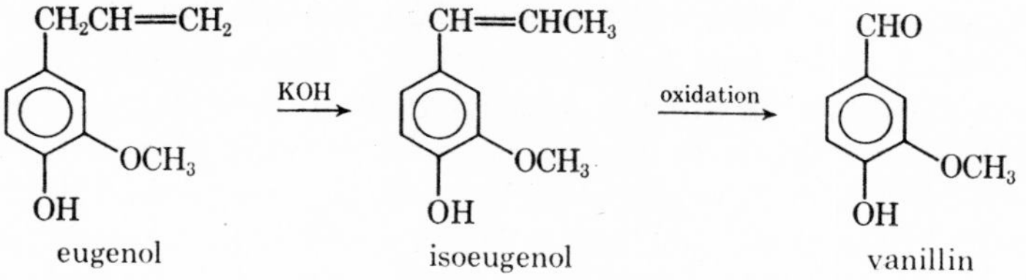

eugenol   isoeugenol   vanillin

During the last quarter of the nineteenth century and the first quarter of the twentieth century eugenol was the most popular commercial source of synthetic vanillin production. One of the accepted ways was to treat oil of cloves with potassium hydroxide at 200°C. The eugenol was thereby transformed to isoeugenol. The reaction mixture was acidified with sulfuric acid and extracted with benzene. The isoeugenol from the benzene extract was acetylated to protect the phenolic group and oxidation to vanillin was accomplished with ozone, potassium permanganate, or potassium dichromate (14). Nitrobenzene was also recommended (15). Variations of the process are possible, many of which have been patented (16–19).

From Guaiacol. Guaiacol (o-methoxyphenol, sometimes called pyrocatechol monomethylether) was first obtained from wood tar or coal tar in the destructive distillation of wood and coal. (See Phenolic ethers.) It was also made from o-dichlorobenzene, a by-product in the production of p-dichlorobenzene. The ortho compound was treated with alkali to form 1,2-benzenediol (catechol) which upon methylation gave a good yield of guaiacol. Starting with guaiacol numerous vanillin syntheses are possible.

1.   Reimer-Tiemann's reaction for the introduction of an aldehyde group into the benzene ring can be used. Resin formation is one disadvantage of this method.

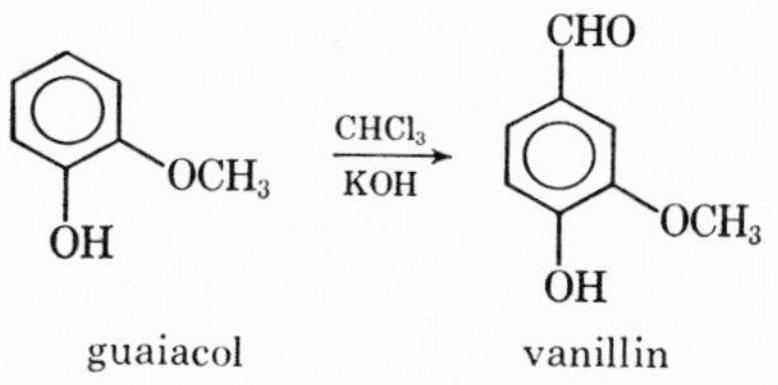

guaiacol   vanillin

2.   A procedure using guaiacol and hydrocyanic acid takes advantage of the Gatterman synthesis. In this case unwanted vanillin isomers are formed as by-products.

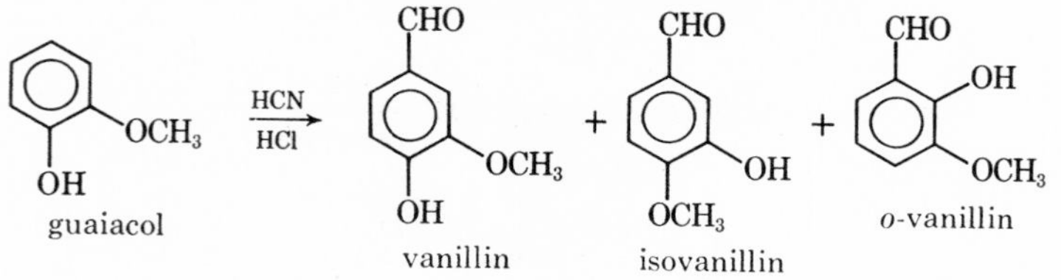

guaiacol   vanillin   isovanillin   o-vanillin

3.   Mottern (12) reported good yields by causing guaiacol acetate to undergo a Fries rearrangement. The resulting apocynin (3-methoxy-4-hydroxyacetophenone)

was then oxidized with nitrobenzene to vanilloylformic acid, which, in turn, was decarboxylated, without resinification, in a solution of dimethyl-*p*-toluidine.

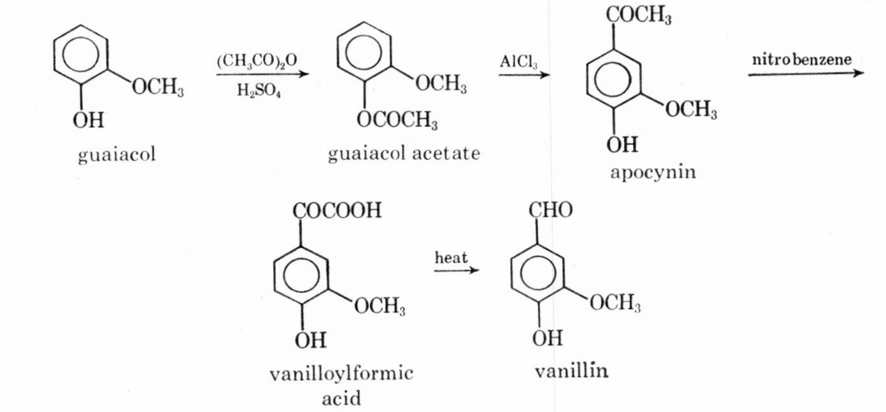

4. A modified Sandmeyer reaction can be applied to guaiacol to prepare vanillin. In this case guaiacol is treated with formaldehyde and *p*-nitrosodimethylaniline by heating on a water bath for several hours in methanol solution, while bubbling in gaseous hydrochloric acid. The products are vanillin and *p*-aminodimethylaniline (20,21).

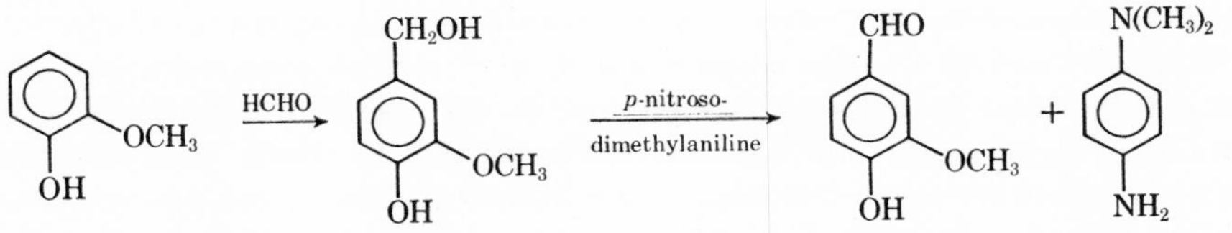

5. Other methods are known. In most if not all cases, unwanted by-products or unsatisfactory yields result (22,23).

From Safrole. Safrole is obtained from camphor oil. When treated with alkali it is converted to isosafrole by migration of the double bond. Isosafrole can be oxidized to give piperonal—called heliotropine in the perfume industry—which, on treatment with PCl$_5$, gives protocatechualdehyde (3,4-dihydroxybenzaldehyde). Methylation with dimethylsulfate in alkali yields vanillin and also isovanillin. There are several patents covering this process (24–26), which is illustrated below.

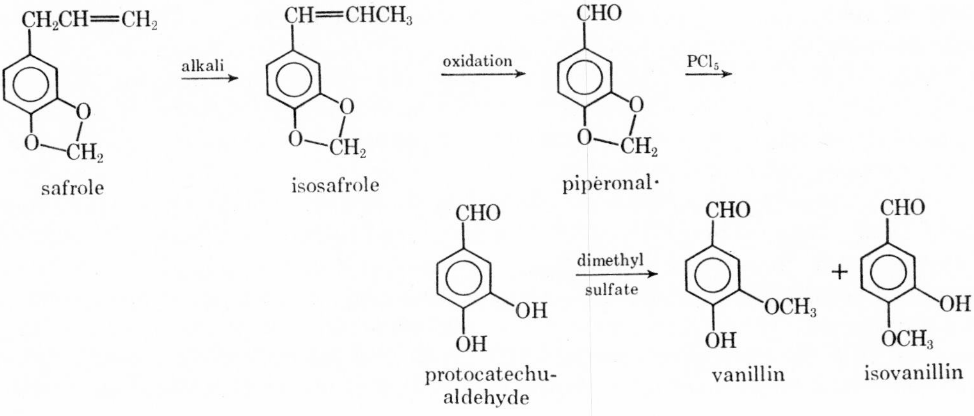

VANILLIN FROM LIGNIN

An anonymous statement appeared in a technical journal published in Germany in 1875 that is of particular interest because it forecasts present vanillin technology, and because it precedes by a quarter century the generally acknowledged discovery. The anonymous investigator described experiments performed on waste sulfite liquor from new chemical pulping methods for coniferous wood chips. His investigations of the waste liquors containing lignin (qv) proved that vanillin was present, noticable by its highly aromatic fragrance. His notes went on to state that he was convinced that vanillin could be made from papermill waste liquors in a technically feasible and profitable way, provided the development of a production process were fully explored (27). The next record of producing vanillin from lignin is by V. Grafe, who in 1903 heated dry waste sulfite liquor solids with lime to 180°C, noting in his journal that he smelled vanillin (28,29). He is credited by most authorities as the first to prepare vanillin from lignin.

Systematic investigations for obtaining vanillin from lignin began about 1927, two general methods being used: hydrolysis, by alkali or alkaline earth hydroxides at elevated temperature and atmospheric or higher pressure; and oxidation of lignin-containing liquors with air, metal oxides, or nitrobenzene, in an alkaline medium, either with or without catalysts.

Hydrolysis. In 1927 and 1928 Kurschner (30–33) and also Toppel (34) carried out a systematic study of alkaline hydrolysis of waste sulfite liquor. Kurschner and Schramek (35) continued the work. They found that maximum yields of vanillin were obtained when the alkaline liquor was refluxed for 20 hr with sodium hydroxide. They found that calcium oxide with sodium carbonate or sodium sulfate also produced vanillin but in lesser yields. The earliest attempts at determining yields were handicapped by inadequate analytical methods which permitted interference from non-vanillin constituents. In 1931 Honig and Ruziczka (36) improved the procedure by a superior method of the isolation and assay of vanillin. After the reaction had taken place they acidified the mixture with sulfuric acid, extracted it with trichloroethylene and extracted the latter with aqueous sodium sulfite. The sodium bisulfite addition compound was decomposed by acidifying; and heating, and extracted with ether; the ether was evaporated; and the vanillin in the residue was precipitated with m-nitrobenzoic hydrazide and weighed as the vanillin hydrazone. Yields of 0.97–2.39 g/liter were found by using this method. Combining Kurschner's method of hydrolysis with Honig's method of isolation and estimation of vanillin, Tomlinson and Hibbert in 1936 (37) obtained 2.6 g of vanillin per liter of waste sulfite liquor. They refined the definition of yield further by basing it on lignin, reporting 5.9% of the lignin content as a typical yield of vanillin. They reported that the yield per unit of lignin increased with an increase in sulfur content of the lignosulfonate starting material and this was confirmed by Hagglund (38). There was considerable variation in vanillin yields from one waste sulfite liquor source to another.

At the same time that lignin was being investigated as a vanillin source in Europe and Canada, Howard in the United States discovered a simple and practical means of isolating lignin from waste sulfite liquor by precipitation (39–41). The process he described accomplished practically complete separation of the liquor into the following three parts: (1) an inorganic product, calcium sulfite, which was filtered off and recycled to make wood chip cooking acid; (2) an organic precipitate, calcium ligno-sulfonate, containing most of the lignin and a convenient starting point for vanillin manufacture; and (*3*) a liquid carbohydrate fraction. Howard's selective precipitation of calcium sulfite and calcium lignosulfonate was accomplished by stepwise additions of lime to pH 9.0 and 12.0, respectively. In 1936, Salvo Chemical Corporation in cooperation with Marathon Paper Mills Company of Rothschild, Wis., began manufacturing vanillin by alkaline hydrolysis of lignosulfonates, basing its process on the patents of Howard and others (42–44). A year later, Howard-Smith Chemicals Ltd., of Cornwall, Ont., began commercial vanillin production by alkaline hydrolysis of waste sulfite liquor (45,46). These were the only two manufacturers of vanillin using lignin as a source material up to this time. Hydrolysis, ie splitting vanillin from the complex lignin molecule, was accomplished by heating with sodium hydroxide under pressure at about 160°C. Isolation and purification of vanillin were accomplished in about the same way by both companies, and similar to the scheme shown in Fig. 1 for the oxidation processes (see p. 188). The processes differed, however, in the following essentials: lignin was concentrated by precipitation (41) from the waste sulfite liquor at Salvo, while at Howard-Smith whole waste sulfite liquor was used, after concentration by evaporation; at Salvo an alcohol solvent was used to extract vanillin from the alkaline reaction mixture, while at Howard-Smith the reaction mixture was acidified and extracted with benzene; at Salvo partially reacted lignin residues were returned to Marathon for recovery of lignin products, while at Howard-Smith the effluent was concentrated and kilned for the recovery of sodium carbonate after removal of calcium salts. Vanillin yields of 3–6% of the lignin present in the waste sulfite liquor were obtained.

Oxidation. Freudenberg worked with oxidizing agents, oxygen carriers, and catalysts in the 1940s and 1950s, oxidizing wood meal, as well as waste sulfite liquor (47–51). He found substantially increased yields of vanillin as compared with alkaline hydrolysis. Oxidation with air or nitrobenzene gave yields of 10 and 20–30%, respectively. Using catalysts like cobaltic hydroxide (see Vol. 5, p. 741) and cupric hydroxide (see Vol. 6, p. 275) enabled him to increase the yields still further. Throughout the world the many aspects of increasing the yield with oxidizing agents and oxygen carriers have been investigated. Nitrobenzene is not a useful oxidizing agent, the main difficulty being in balancing the sale of nitrobenzene reduction products with the sale of vanillin. More successful has been the controlled air oxidation of lignosulfonates, each of the major producers in North America using some modification of this method at the present time. References 39–56 describe some of the techniques of Salvo Chemical Corporation which revised its process from alkaline hydrolysis to air oxidation in 1945; for the methods of Monsanto Chemical Company, see references 57–59; the major patents referring to the processing procedures of the Ontario Paper Co., Ltd., are described in references 60–67. These three companies are leaders in vanillin production, each in its own way. Salvo Chemical Co., subsidiary of Sterling Drug, Inc., in cooperation with Marathon Paper Mills Co. (now Marathon Rothschild Division of American Can Co.) of Rothschild, Wis., pioneered in the commercial production of lignin vanillin, having been in the business continuously since 1936 (Sterwin Chemicals, Inc., a division of Sterling Drug, Inc. is closely associated with lignin vanillin because it sells 100% of Salvo's output). Monsanto of St. Louis, Mo., has one of the earliest and longest production records of vanillin manufacture—from 1904 to the present time—first from eugenol in 1904, then from guaiacol in 1929 and, starting in 1953, from lignin purchased from Puget Sound Pulp and Timber Co. at Bellingham, Wash., (now Puget Sound Division of Georgia Pacific Co.). Ontario

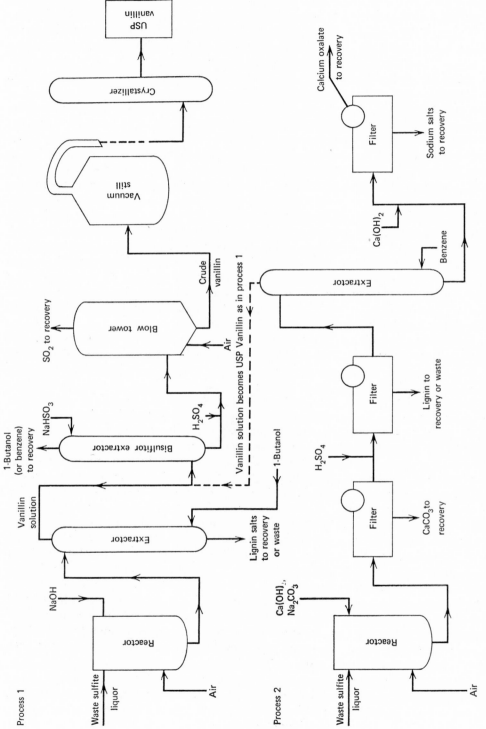

Fig. 1. General schemes for the production of vanillin.

Paper Co., Ltd., of Thorold, Ont., Canada, is the world's largest producer, having started manufacturing lignin vanillin about the same time as Monsanto (68,69).

Although each manufacturer has distinguishing process variations, the general schemes for producing vanillin are illustrated by two flow diagrams, Process 1 and Process 2, representing Salvo and Monsanto, and Ontario, respectively (see Fig. 1.).

Although all the world's lignin found in spent papermill wastes is not available for vanillin production, the potential of usable sulfite mill waste is staggering. For example, a 300 ton-per-day pulp mill produces at least 75 tons of lignin (25% of the pulp weight). If a 6% vanillin yield is assumed, this average mill has a daily potential of 4.5 tons of vanillin or 2,700,000 lb per 300-day year. When consideration is given to a world production figure of 60,000,000 tons of pulp for 1964 and a projected figure of 130,000,000 tons for 1975 (70) some idea of the abundance of vanillin raw material can be realized. Economically feasible processes for the utilization of lignin for the manufacture of vanillin were delayed until 1936 for two very logical reasons: first, technical handling difficulties were considerable; and, second, a feasible scheme was lacking for utilizing the more than 90% lignin residue left over from vanillin manufacture, without which vanillin manufacture was not economically sound. Technical difficulties were finally overcome, as partly revealed in the patent literature; other procedures are well-guarded know-how of the respective companies. Handling of the 90+% organic matter remaining in the effluent was mastered by Salvo in its mutually beneficial relationship with Marathon Rothschild Div. of American Can Co., to which all vanillin plant effluent is returned for the manufacture of lignin products. The lignosulfonate products sold in million-pound quantities per year by Marathon include dispersants for dyes, carbon black and boiler waters, and oil-well drilling muds. It also sells millions of pounds as binding chemicals for feedstuff pellets, ceramics and linoleum mastics, and as chelating material for industrial cleaners and dyes. At the Puget Sound plant of Georgia Pacific Co. a dispersant for use in cement making and soil amendments is produced from lignosulfonates. How much is derived from Monsanto's vanillin plant effluent is unknown. Figure 1 shows the treatment of waste sulfite liquor at Ontario Paper Company, Ltd. In all cases lignin utilization, and/or chemical recovery, are an integral part of the economics. See also Lignin.

During the depolymerization and oxidation of lignin compounds a score of other organic compounds are produced, such as acetovanillone, $CH_3O(OH)C_6H_3COCH_3$ (the largest component), p-hydroxybenzaldehyde, p-$(OH)C_6H_4CHO$, 5-formylvanillin, $CH_3O(OH)C_6H_2(CHO)_2$, and syringic aldehyde, $(CH_3O)_2(OH)C_6H_2CHO$. The "vanillin solution" from the solvent extractor is purified by treatment with bisulfite, the crude vanillin is then distilled and, upon multiple crystallization, USP-grade vanillin of very high purity and excellent flavor is obtained. Only small quantities of the other compounds are salvaged for commerce.

The methyl ether of vanillin, *veratraldehyde*, is manufactured and marketed by both Salvo and Monsanto. Salvo has an extensive market in 3-ethoxy-4-hydroxy-benzaldehyde, or "*ethylvanillin*," (the correct common name is *bourbonal*), a misnomer used in the perfume and flavor trade. It is made from vanillin in a patented process (71) in which vanillin is first methylated in the 4-position, the 3-methoxyl group removed, the resulting isovanillin ethylated, and the mixed ether hydrolyzed to produce ethyl vanillin (bourbonal). Monsanto sells bourbonal it manufactures from sources other than lignin vanillin.

Lignin vanillin is produced not only in North America, but probably also in the Peoples Republic of China (73) and in Europe by the Aktieselskapet Borregaard at Sarpsborg, Norway (licensed by Monsanto) and the Wloclavek plant in Poland (72). Interest in the production of vanillin is increasing throughout the world and a great deal of literature in the field is published in Russia and Japan especially.

Economic Aspects

Vanillin derived from eugenol and similar raw materials has always been expensive and is still selling in the range of $6.75/lb.

Vanillin derived from guaiacol is no longer available, having been replaced by lignin vanillin which is more economically obtained from waste sulfite liquor. Lignin vanillin domestically produced has sold in the price range of $3.00–3.25/lb in the 1960s.

Synthetic vanillin must compete, to some extent, with natural vanilla extracted from the vanilla pod (see Vol. 9, p. 375). Although the flavor of the natural product is largely due to its 2–3% content of vanillin, resins and other ingredients of the pod alter the flavor to some extent. The synthetic product is always of consistent flavor, whereas the natural product varies to some degree from batch to batch. Marketing of vanillin for flavor uses is complicated by these differences in extracts from natural sources and the preference of users for vanillin from a particular raw material or a specific blend.

In 1951 U.S. production was 999,000 lb at a unit value of $2.93/lb. It is estimated that the combined North American production capacity of Salvo Chemical Co. in Wisconsin, Monsanto on the West Coast, and Ontario Paper Co., Ltd., in Ontario, Canada, is on the order of 6.5 million lb. World consumption for 1970 is estimated at 8.8 million lb.

Grades and Specifications

Vanillin is available in both USP and FCC (Food Chemicals Codex) grades. The latter grade is a relatively new designation prescribing minimum requirements of purity for an appropriate grade of food chemicals. Both USP and FCC specify that vanillin should contain no less than 97% and no more than $100 \pm 3\%$ of (CH_3O)-$(OH)C_6H_3CHO$, calculated on the dried basis. The USP grade should lose no more than 1%, and the FCC grade no more than 0.5% of its weight on drying for 4 hr over silica gel; and they should leave no more than 0.05% residue on ignition. Vanillin should be stored in tight, light-resistant containers. It slowly oxidizes on exposure to moist air.

Vanillin is also available in reagent grade (99.9% pure) and in technical grade. The latter is 98+% pure and contains foreign odors and colors or related organic material from lignin that make this grade less desirable for flavoring. Actually, USP- and FCC-grade vanillin, obtained from the major producers, is of reagent-grade purity.

Analysis

Identification. Vanillin can be identified by the following tests: (1) When a solution of ferric chloride is added to a cold saturated vanillin solution, a blue color appears which changes to brown upon heating to 20°C for a few min. On cooling, a white to pearly gray precipitate of silky needles forms. This is dehydrodivanillin,

$(CH_3O(OH)C_6H_2CHO)_2$. (2) Vanillin can be completely extracted from its ether solution by saturated aqueous sodium bisulfite from which it can be precipitated by mineral acids. (3) It can be identified by the white to slightly yellow precipitate formed by adding lead acetate to a cold water solution of vanillin. The precipitate is soluble in acetic acid and sparingly soluble in hot water (74).

Determination. Various techniques are used for the analysis of vanillin, such as colorimetric, gravimetric, spectrophotometric, and chromatographic (thin-layer, paper, and gas–liquid) methods. (Polarographic, titrimetric, and oximetric procedures have been proposed but are not widely used.) The divisions are not rigid because often two or more methods are combined.

The Folin-Denis colorimetric method (75) detects about 0.2 mg of vanillin and depends on a color complex developed with phosphotungstic and phosphomolybdic acids. A slight modification was the colorimetric procedure accepted by AOAC for a number of years (76). Subsequently AOAC recommended a combination spectrophotometric–column chromatographic method for the analysis of vanillin in the presence of "ethyl vanillin" (bourbonal) and coumarin (77). In this procedure the mixture is separated on a silica gel column, using an isooctane–chloroform solvent to develop the separations. Elution volumes for each constituent are predetermined on a known solution and the unknown mixture developed and read on an ultraviolet spectrophotometer at 270 and 325 nm. The latest USP (74) also prescribes ultraviolet spectrophotometry for identifying and assaying vanillin. A rapid micromethod by Stone and Blundell (78) combines paper chromatography with spectrophotometry. Their method is especially applicable for microwork. Vanillin, p-hydroxybenzaldehyde, and syringaldehyde are separated by paper chromatography and the compounds eluted from the developed strip, with readings carried out in an alkaline alcohol medium at the respective maxima (352 nm for vanillin, 368 nm for syringaldehyde, and 335 nm for p-hydroxybenzaldehyde). Heide and Lemmens (79) described a paper-chromatographic separation of vanillin, "ethyl vanillin," veratraldehyde, heliotropin, vanitrope, and coumarin, with spectrophotometric determination of the separated compounds. The developer used was a mixture of benzene, petroleum ether, and methanol. A similar procedure is reported by Mitchell (80). Newcombe and Reid (81) claim an interesting separation of acetovanillone and acetoveratrone from vanillin and veratraldehyde—separations which have proved especially troublesome—using a bisulfite technique.

The gravimetric method of Iddles et al. (82), in which vanillin is precipitated as the 2-4-dinitrophenylhydrazone, suffers from the disadvantage that all water-soluble aldehydes and ketones are precipitated. A more specific agent for gravimetric determination is m-nitrobenzoic acid hydrazide, although this reagent also precipitates such aldehydes as veratraldehyde and syringic aldehyde.

The manufacture of vanillin, whether from eugenol, guaiacol, or lignin, is accompanied by the simultaneous formation of one or more closely related or similar compounds that interfere with the analysis of vanillin by many of the usual procedures. Vanilla extract and compounded flavors containing vanillin have been tedious to analyze for a similar reason. The separation, identification, and assay of mixtures in one operation became possible with the advent of gas–liquid chromatography (GLC). A number of food and flavor laboratories began to explore the possibilities of this procedure in the late 1950s. Wilkins Instruments and Research Co. was one of the first to publish a method (83). At the present time most of the flavor producers and compounders have their own GLC techniques and a number of good procedures have been published (84–90). Prabucki and Lenz did some fundamental work (91). Pepper et al. (92) describe a method used on lignin oxidation products. Johansen described a method and conditions for separating, identifying, and assaying vanillin and ten other vanillin-related compounds on Carbowax 20M (Union Carbide Corp.) or 60/80 mesh acid-washed Chromosorb G (Johns-Manville Corp.) (93). GLC is always changing as new materials for coating and packing columns are found. Currently one laboratory prefers a capillary column coated with phenyldiethanolamine succinate or Carbowax 20M TPA (94).

Uses

Until recent years about 85% of the vanillin production was used as a flavoring agent in ice cream, candies, cookies, puddings, cake mixes, gelatin desserts, soft drinks, etc. The balance was used in deodorants, perfumes, and odor fixatives, and as a masking agent in pharmaceutical and vitamin preparations.

As of 1970, the greatest use for technical vanillin is as chemical intermediate in the production of pharmaceutical products, surpassing the quantities used for flavoring purposes. Vanillin has been used in the synthesis of papaverine, L-dihydroxyphenylalanine and other potentially useful derivatives. These include certain hydrazones of vanillin with a plant-killing action similar to that of 2,4-D (95) and the zinc salt of dithiovanillic acid (made by the reaction of vanillin and ammonium polysulfide in alcoholic hydrochloric acid), which is a vulcanization inhibitor (96). 5-Hydroxymercurivanillin, 5-acetoxymercurivanillin, and 5-chloromercurivanililn have been prepared and found to have disinfectant properties (97). Although the 5-hydroxy compound possesses a toxicity toward microorganisms comparable with sodium pentachlorophenate and ethylmercuric phosphate, its toxicity toward fish is much less. It is therefore of potential interest as a slime-control agent in mills where wastes are discharged into fishing streams.

Other potential uses for vanillin include the prevention of foaming in lubricating oils (98); the preparation of syntans (see Vol. 5, p. 325) for tanning (99); as a brightener in zinc plating baths (100); as an aid to the oxidation of linseed oil (101); and as a solubilizing agent for riboflavin (102).

Derivatives and Related Compounds

Ethyl vanillin (3-ethoxy-4-hydroxybenzaldehyde, bourbonal, vanillal), C_2H_4O-$(OH)C_6H_3CHO$ (mp, 77–78°C), forms white crystalline needles completely soluble in ethyl alcohol. Ethyl vanillin is a synthetic flavoring and aromatic agent similar to vanillin, but much more intense. It is sold under such trade names as Ethavan (Monsanto Co.) and Vanaldol (Fries Bros.), and was priced (early 1969) at $5.75/lb in 100-lb drums. It is made by a patented process (71) from lignin vanillin (see p. 189).

Vanillic acid (4-hydroxy-4-methoxybenzoic acid), $CH_3O(OH)C_6H_3COOH$, forms colorless needles from water (mp, 207°C) and sublimes on heating. It is soluble to the extent of 0.12 and 2.5 g in 100 ml water at 14 and 100°C, respectively. Vanillic acid is very soluble in ethyl alcohol and ethyl ether.

Controlled potassium hydroxide fusion of vanillin gives high yields of vanillic acid (103), and vanillin is oxidized almost quantitatively in caustic soda to the acid by silver oxide (104), as shown below.

Oxidations of vanillin have also been effected with mercuric and auric oxides (105). A good yield of vanillic acid may also be obtained by the treatment of vanillin oxime with acetic anhydride followed by hydrolysis (106). In addition to processes for making vanillic acid from vanillin, several methods of producing vanillic acid in high yield directly from isolated lignosulfonates and waste sulfite liquor have been developed (107,108).

Vanillic acid esters are easily prepared in high yield from vanillic acid and the appropriate alcohol. These esters have been used as nontoxic food preservatives, slime-control agents, disinfectants, in pharmaceuticals, and in suncreams.

Ethyl vanillate, $CH_3O(OH)C_6H_3COOC_2H_5$, forms colorless needles (mp, 44°C; bp, 293°C); it is insoluble in water, very soluble in ethyl alcohol and ether, and soluble in alkalis. Ethyl vanillate is less toxic than sodium benzoate when administered in oil. During World War II, the U.S. Food and Drug Administration considered the use of ethyl vanillate as a preservative in amounts up to 0.10%, when such use would permit the delivery of acceptable food products to the Armed Forces. Ethyl vanillate has remarkable preservative properties in foods, such as salt fish and fresh fruit (109).

Acetovanillone (4-hydroxy-3-methoxyacetophenone, apocynin), $CH_3COC_6H_3$-$(OH)OCH_3$, crystallizes from water as colorless prisms (mp, 115°C; bp, 295–300°C); it is very soluble in hot water; 7.7 g dissolve in 100 g of ethyl alcohol at 9°C; it is soluble in ether and benzene and insoluble in petroleum ether. For many chemical purposes, crude vanillin, containing a small amount of acetovanillone as an impurity, is available commercially and may be substituted for pure vanillin.

Veratraldehyde, $(CH_3O)_2C_6H_3CHO$, is made by methylating vanillin, using dimethyl sulfate and sodium hydroxide. It is used as a metal brightener in the plating industry and as a perfume.

Bibliography

"Vanillin" in *ECT* 1st ed., Vol. 14, pp. 603–611, by D. M. C. Reilly, Food Machinery and Chemical Corp. (Chemical Div.).

1. M. B. Jacobs, *Am. Perfumer Essent. Oil Rev.* **57**, 45–47 (July, 1952).
2. F. K. Beilstein, *Handbuch der organischen Chemie,* 4th ed., Vol. 8, Springer Verlag, Berlin, 1925, p. 247 ff.
3. W. C. McCrone, *Anal. Chem.* **22**, 500 (1950).
4. A. Lioxin, *Combined Phenol and Aldehyde,* Ontario Paper Co., Ltd., Thorold, Ont., Can., 1953, 16 pp.
5. Data, Salvo Chemical Co., Subsidiary of Sterling Drug Inc., 1950.
6. L. C. Raiford and J. G. Lichty, *J. Am. Chem. Soc.* **52**, 4576 (1930).
7. I. A. Pearl, *Tappi* **41**, 621–624 (1958).
8. T. Clendenning, *Mfg. Confectioner* **31**, 30–31 (May, 1950).
9. D. E. Lakritz, *Drug Cosmetic Ind.* **65** (6), 646, 647, 720–724 (1949).
10. F. A. Arana, "Vanilla Curing and Its Chemistry," *U. S. Dept. Agr. Agr. Infor. Bull.* **42** (1944).
11. Ph. Traynard, *Assoc. Tech. Ind. Papetière, Bull.* **1956** (5), 136–138. (A review of methods for obtaining vanillin, in French).
12. H. O. Mottern, *J. Am. Chem. Soc.* **56**, 2107–2108 (1934).
13. U.S. Pat. 151,119 (May 19, 1874), W. Haarmann.

14. U.S. Pat. 1,643,804 (Sept. 27, 1927), R. H. Bots.
15. U.S. Pat. 1,643,805 (Sept. 27, 1927), R. H. Bots.
16. U.S. Pat. 192,542 (June 26, 1877), F. Tiemann.
17. U.S. Pat. 487,204 (Nov. 29, 1892), F. Ach (to Boehringer und Söhne).
18. U.S. Pat. 560,494 (May 19, 1896), W. Haarmann (to Haarmann and Riemer).
19. U.S. Pat. 1,939,491 (Dec. 12, 1933), F. Elger (to Hoffmann-LaRoche Inc.).
20. U.S. Pat. 1,345,649 (July 6, 1920), A. Weiss (to Société Chimique des Usines du Rhône).
21. R. N. Shreve, *Chemical Process Industries*, McGraw-Hill Book Co., Inc., New York, 1945, pp. 567–568.
22. U.S. Pat. 2,640,083 (May 26, 1953), J. Kamlet (to Mathieson Chemical Corp.).
23. U.S. Pat. 1,878,061 (Sept. 20, 1932), A. W. Titherley and D. P. Hudson (to Monsanto Chemical Co.).
24. U.S. Pat. 1,792,716 (Feb. 17, 1931), F. E. Stockebach.
25. U.S. Pat. 1,792,717 (Feb. 17, 1931), F. E. Stockebach.
26. U.S. Pat. 1,812,132 (June 30, 1931), F. Boedecker (to J. D. Riedel Aktiengesellschaft).
27. Anonymous, *Dinglers Polytech. J.* **216** (4), 372 (1875). (In German.)
28. L. E. Wise, *Wood Chemistry*, Reinhold Publishing Corp., New York, 1944, pp. 321–322.
29. V. Grafe, *Monatsh.* **25,** 987 (1904).
30. K. Kürschner, *J. Prakt. Chem.* **118,** 238 (1928).
31. K. Kürschner, *Angew. Chem.* **40,** 231 (1927).
32. *Ibid.,* **41,** 1106 (1928).
33. K. Kürschner, *Cellulosechemie* **19,** 136 (1941).
34. O. Toppel, *Papier* **15,** 81–93 (1961); *Chem. Abstr.* **55,** 19238e (1961).
35. K. Kürschner and W. Schramek, *Tech. Chem. Papier Zellstoff-Fabr.* **29,** 35 (1932).
36. M. Honig and W. Ruziczka, *Z. Angew. Chem.* **44,** 845 (1931).
37. G. H. Tomlinson, Jr., and H. Hibbert, *J. Am. Chem. Soc.* **58,** 340–364 (1936).
38. E. Hagglund and L. C. Bratt, *Svensk Papperstid.* **39,** 347 (1936).
39. U.S. Pat. 1,551,882 (Sept. 1, 1925), G. C. Howard.
40. U.S. Pat. Re. 18,268 (Dec. 1, 1931), G. C. Howard (to Guy C. Howard Co.). Reissue of U.S. Pat. 1,699,845 (Jan 22, 1929).
41. U.S. Pat. 1,856,558 (May 3, 1932), G. C. Howard.
42. U.S. Pat. 2,057,117 (Oct. 13, 1936), L. R. Sandborn, J. R. Salvesen, and G. C. Howard (to Marathon Paper Mills Co.).
43. U.S. Pat. 2,104,701 (Jan. 4, 1938), L. T. Sandborn (to Marathon Paper Mills Co.).
44. U.S. Pat. 2,399,607 (Apr. 30, 1946), R. Servis (to Marathon Paper Mills Co.).
45. U.S. Pat. 2,069,185 (Jan. 26, 1937), H. Hibbert and G. Tomlinson, Jr. (to Howard Smith Chemicals Ltd.).
46. H. Frank, *Riechstoffindustrie* **13** (10), 229–231 (1938).
47. Ger. Pat. 945,327 (July 12, 1956), K. Freudenberg, W. Lautsch, and H. Brenek (to Zellstoffabrik Waldhof).
48. Ger. Pat. 947,365 (Aug. 16, 1956), K. Freudenberg, W. Lautsch, and H. Brenek (to Zellstoffabrik Walhof).
49. Ger. Pat. 947,402 (Sept. 20, 1956), K. Freudenberg, W. Lautsch, and H. Brenek (to Zellstoffabrik Walhof).
50. U.S. Alien Property Custodian, Serial No. 318,386 (Apr. 20, 1943), K. Freudenberg, W. Lautsch, and H. Brenek.
51. Zyamunt Kin, *Przeglad Papier.* **9,** 236–241 (1953).
52. U.S. Pat. 2,434,626 (Jan. 13, 1948), J. R. Salvesen, D. L. Brink, D. G. Diddams, and P. Owzarski (to Marathon Paper Mills Co.).
53. U.S. Pat. 2,598,311 (May 27, 1952), E. W. Schoeffel (to Salvo Chemical Corp.).
54. U.S. Pat. 3,049,566 (Aug. 14, 1962), E. W. Schoffel (to American Can Co.).
55. U.S. Pat. Appl. 780,592 (Dec. 2, 1968), D. G. Diddams and N. Renaud.
56. U.S. Pat. Appl. 789,599 (Jan. 7, 1969), W. B. Gitchel, D. G. Diddams, and C. A. Hoffman.
57. U.S. Pat. 2,506,540 (May 2, 1950), C. C. Bryan (to Monsanto Chemical Co.).
58. U.S. Pat. 2,692,291 (Oct. 19, 1954), C. C. Bryan (to Monsanto Chemical Co.).
59. U.S. Pat. 2,721,221 (Oct. 18, 1955), C. C. Bryan (to Monsanto Chemical Co.).
60. U.S. Pat. 2,516,827 (July 25, 1950), H. B. Marshall and C. A. Sankey (to The Ontario Paper Co., Ltd.).

61. U.S. Pat. 2,544,999 (Mar. 31, 1951), H. B. Marshall and C. A. Sankey (to The Ontario Paper Co., Ltd.).
62. U.S. Pat. 2,576,752 (Nov. 27, 1951), J. H. Fischer and H. B. Marshall (to The Ontario Paper Co., Ltd.).
63. U.S. Pat. 2,576,753 (Nov. 27, 1951), J. H. Fischer and H. B. Marshall (to The Ontario Paper Co., Ltd.).
64. U.S. Pat. 2,576,754 (Nov. 27, 1951), J. H. Fischer and H. B. Marshall (to The Ontario Paper Co., Ltd.).
65. U.S. Pat. 2,686,120 (Aug. 10, 1954), H. B. Marshall and C. A. Sankey (to The Ontario Paper Co., Ltd.).
66. U.S. Pat. 3,054,659 (Sept. 18, 1962), D. Craig and C. D. Logan (to The Ontario Paper Co., Ltd.).
67. U.S. Pat. 3,054,825 (Sept. 18, 1962), D. Craig and C. D. Logan (to The Ontario Paper Co., Ltd.).
68. *Chem. Week* **70** (5), 33–34 (Feb. 2, 1952).
69. *Chem. Week* **89** (11), 103–104 (Sept. 16, 1961).
70. K. W. Britt, *Handbook of Pulp and Paper Technology*, Reinhold Publishing Corp., New York, 1964, p. 13.
71. U.S. Pat. 3,367,972 (Feb. 6, 1968), W. B. Gitchel, E. M. Pogainis, and E. W. Schoeffel (to Sterling Drug Inc.).
72. H. Wawrzyniak, *Przeglad Papier.* **22** (9), 292–296 (1966); *Chem. Abstr.* **66**, 56826d (1967).
73. W. X. Clark, Sterwin Chemicals Inc., personal communication to author, 1969.
74. *The Pharmacopeia of the USA, XVII*, Mack Publishing Co., Easton, Pa., 1965, p. 748.
75. O. Folin and W. Denis, *J. Ind. Eng. Chem.* **4**, 670–672 (1912).
76. H. A. Lepper, *Official Methods of Analysis of the Association of Official Agricultural Chemists*, 7th ed., AOAC, Washington, D.C., 1950, p. 306.
77. W. Horwitz, *Official Methods of Analysis of the Association of Official Agricultural Chemists*, 10th ed., AOAC, Washington, D.C., 1965, pp. 287–289.
78. J. E. Stone and M. J. Blundell, *Anal. Chem.* **23**, 771–774 (1951).
79. R. Heide and J. F. Lemmens, *Am. Perfumer Essent. Oil Rev.* **59**, 21–23 (1954).
80. L. C. Mitchell, *J. Assoc. Offic. Agr. Chemists* **36** (4), 1123 (1953).
81. A. G. Newcombe and S. G. Reid, *Nature* **172**, 455–456 (1953).
82. H. A. Iddles, A. W. Low, B. D. Rosen, and R. T. Hart, *Ind. Eng. Chem. Anal. Ed.* **11**, 102–103 (1939).
83. *Research Notes*, Wilkens Instruments and Research Co., Walnut Creek, Calif. (Spring, 1961).
84. F. J. Feeny, *J. Assoc. Offic. Agr. Chemists* **47**, 555 (1964).
85. G. E. Martin, F. J. Feeny, and F. T. Scaringelli, *J. Assoc. Offic. Agr. Chemists* **47**, 561 (1964).
86. J. Fitelson, *J. Assoc. Offic. Agr. Chemists* **47**, 1161 (1964).
87. D. N. Smith, *J. Assoc. Offic. Agr. Chemists* **48**, 509 (1965).
88. J. Fitelson, *J. Assoc. Offic. Agr. Chemists* **48**, 911 (1965).
89. *Ibid.*, 913.
90. *Ibid.*, **49**, 566 (1966).
91. A. L. Prabucki and F. Lenz, *Helv. Chim. Acta* **45**, 2012–2014 (1962) ((in German); *Chem. Abstr.* **58**, 8395c (1963).
92. J. M. Pepper, M. Manolopoulo, and R. Burton, *Can. J. Chem.* **40**, 1976–1980 (1962); *Chem. Abstr.* **57**, 15393i (1962).
93. N. G. Johansen, *J. Gas Chromatog.* **3** (6), 202–203 (June, 1965).
94. J. Barr and F. Priebe, Salvo Chemical Corporation, personal communication to author, 1969.
95. H. S. Reed, J. Dufrenoy, J. R. Parikh, and J. R. Oneto, *Compt. Rend.* **230**, 2317–2318 (1950).
96. G. Bruni and T. G. Levi, *Atti Acad. Lincei* **32**, i, 5–8 (1923); *Chem. Zentr.* **1923**, III, 1642.
97. U.S. Pat. 2,489,380 (Nov. 29, 1949), H. F. Lewis and I. A. Pearl (to Sulphite Products Corp.).
98. U.S. Pat. 2,528,465 (Oct. 31, 1950), V. N. Borsoff and J. O. Clayton (to California Research Corp.).
99. U.S. Pat. 2,564,022 (Aug. 14, 1951), J. Miglarese (to U.S. Leather Co.).
100. W. Eckardt, *Oberflachentech.* **18**, 170–171 (1941); *Chem. Zentr.* **1942**, I, 1936.
101. H. Hadert, *Farbe Lack* **56**, 349–351 (1950).
102. U.S. Pat. 2,449,640 (Sept. 21, 1948), J. Charney (to Wyeth, Inc.).
103. I. A. Pearl, *J. Am. Chem. Soc.* **68**, 2180–2181 (1946).

104. *Ibid.*, 429–432, 1100–1101 (1946).

105. U.S. Pat. 2,414,119 (Jan. 14, 1947), I. A. Pearl (to Cola G. Parker as trustee).

106. L. C. Raiford and D. J. Potter, *J. Am. Chem. Soc.* **55**, 1682–1685 (1933).

107. U.S. Pat. 2,431,419 (Nov. 25, 1947), I. A. Pearl (to Sulphite Products Corp.).

108. U.S. Pat. 2,433,227 (Dec. 23, 1947), H. F. Lewis and I. A. Pearl (to Sulphite Products Corp.).

109. I. A. Pearl, *Food Inds.* **17**, 1173 (1945); I. A. Pearl and F. J. McCoy, *Food Inds.* **17**, 1458 (1945).

Donald G. Diddams
Sterling Drug Inc.

Jack K. Krum
The R. T. French Co.

VAPOR–LIQUID EQUILIBRIA

The equilibrium of a vapor in contact with a liquid is a special case of the general phenomena of phase equilibria. In this discussion the topic will be treated from a basic thermodynamic approach; see also Phase rule; Thermodynamics.

The celebrated phase rule of Gibbs (1b) is one of the crowning achievements of scientific deduction. Both its derivation and simple statement constitute the foundation of any study of vapor–liquid equilibria. The statement of the phase rule; namely:

$$f = n - r + 2 \tag{1}$$

where f = degrees of freedom, n = number of independent components and r = number of distinct phases, is deceptively simple, since it is based on the culmination of an intricate series of deductions stemming from the first and second laws of thermodynamics. For each distinct phase in equilibrium Gibbs derived his famous equation of existence which relates the intensive variables of state (temperature, pressure and the chemical potentials of each component) for all the phases. For equilibrium, Gibbs proved that each of these intensive variables must be the same for all the phases. The final result is therefore a set of equations equal to the number of phases (r) containing 2 more variables than the number of components (n). Therefore, the degrees of freedom are simply the difference between the number of variables and equations. To make proper application of the phase rule, the concepts of independent components and distinct phases must be fully understood, especially in systems undergoing chemical change (2). In this discussion, the topic will be limited to systems in which there is *no* chemical reaction taking place.

When limited to a single component, the vapor–liquid equilibrium reduces to a monovariant system ($f = 1 - 2 + 2 = 1$), in which case only the pressure or temperature can be varied at will. Normally, this relationship is expressed as a vapor-pressure equation in which the pressure is given as a function of temperature. For systems containing more than one component, and in which the number of phases is one more than the number of components, the system will again be monovariant. Thus for a binary system to be monovariant, there will be three phases in equilibrium; and since the vapor is always a single phase, the vapor–liquid equilibrium would consist of two distinct liquids in equilibrium with the vapor phase. Systems with more than one component are usually divided into two classes: binary systems and multicomponent systems (more than two components). For the purpose of convenience and continuity the discussion will be treated in three separate sections in the following order:

1. Single-component systems
2. Binary systems
3. Multicomponent systems

Vapor Pressure of a Single Substance

For a single-component system the vapor–liquid equilibrium reduces to a vapor-pressure relationship in which the two intensive variables of state, namely pressure and temperature are interrelated. This relationship is a stable equilibrium over the region between the triple point and the critical point as shown in Figure 1 for water, and is very much the same for all liquids. The triple point is defined as that system containing all three phases in equilibrium; in this case vapor, liquid and solid. According to the phase rule, this point is invariant ($f = 1 - 3 + 2 = 0$) and can exist in equi-

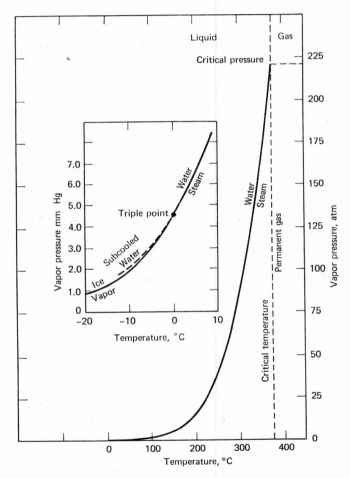

Fig. 1. Typical vapor-pressure curve for water from its freezing point to the critical point with insert showing all three phases in the region of the triple point.

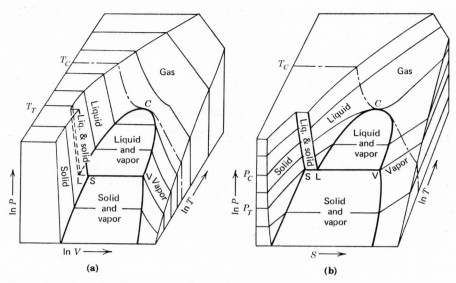

Fig. 2. Phase diagrams for water at constant mass showing all three phases in stable equilibrium from below the triple point to above the critical point: (**a**) P–V–T diagram with isotherms; (**b**) P–S–T diagrams with isobars. Logarithmic coordinates are used for P, V and T as suggested by Gibbs to yield straight lines in the region approaching an ideal gas (9).

librium only at a single pressure and temperature. It is possible to have vapor and liquid in equilibrium below the triple point, but the equilibrium is defined as metastable, since the slightest disturbance will result in the precipitation of the stable solid phase. It is very analogous to the condition of a supersaturated solution. The upper limit is defined by the critical temperature; above this temperature both the vapor phase (steam) and the liquid phase (water) become a so-called "permanent gas," since no amount of compression will reduce the gas to the liquid state. The vapor pressure corresponding to the critical temperature is known as the critical pressure, but it is important to realize that the temperature is the controlling factor concerning the phase change and for this reason is usually treated as the independent variable.

It should be noted that Figure 1 is in effect a projection of a three dimensional phase diagram into the P–T plane. Including the third variable, volume V (or sometimes entropy S) would give diagrams as shown in Figure 2. In the case of water, the unusual phenomenon of expansion during solidification shows up as a reversal of the liquid–solid surface in Figure 2a. For most substances a contraction occurs during solidification so that the liquid–solid surface appears the same in the volume diagram (Fig. 2a) as in the entropy diagram (Fig. 2b). The triple point also appears as a line perpendicular to the P–T plane in both parts of the diagram.

From a macroscopic point of view the vapor–liquid equilibrium appears to be a stable stationary condition; however from the microscopic or molecular scale it consists of an active interchange between vapor and liquid molecules at the interface between the phases. In other words, there is in effect a kinetic equilibrium or steady-state condition, in which the rate of evaporation of liquid molecules is exactly equal to the rate of condensation of vapor molecules. By combining the Langmuir equation of mass rate of evaporation with the Clausius-Clapeyron equation (defined below), Othmer (3) was able to show that for the highest rates of evaporation encountered in engineering practice, the equilibrium between vapor and liquid is maintained for all practical purposes. This is because the net rate of vaporization is a minute fraction of the dynamic rate of interchange between liquid and vapor molecules across the interface at equilibrium.

Equations of State for Vapors and Gases. In a thermodynamic analysis of the equilibrium between a vapor and its liquid, a reference state is required from which the fundamental functions of fugacity f and Gibbs free energy G may be determined. The fugacity function is a measure of the chemical potential or tendency for a chemical species to change phases. For a vapor that behaves as an ideal gas, the fugacity becomes exactly equal to the vapor pressure. For real vapors and gases the difference between the actual and ideal volumetric behavior is a measure of the fugacity. Thus, it is apparent at this point that the definition of the ideal gas is fundamental. The simple equation of state for the ideal gas is given as follows:

$$PV = NRT \qquad\qquad (2)$$

where P = pressure, V = volume, N = number of moles, R = gas constant, T = absolute temperature.

In this definition the units chosen for the constant R must be consistent with those selected for the remaining variables. The mole corresponds to a mass of gas numerically equal to its molecular weight; when expressed in grams, it contains 6.03×10^{23} molecules and occupies 22.414 liters at standard conditions (273.16°K and 1 atmosphere pressure). The corresponding value of $R = 0.08205$ l. atm per °K

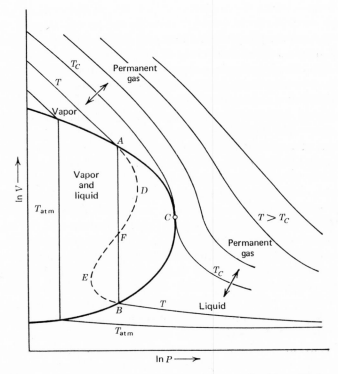

Fig. 3. Isothermal volume chart at constant mass for the vapor–liquid equilibrium phase region (9).

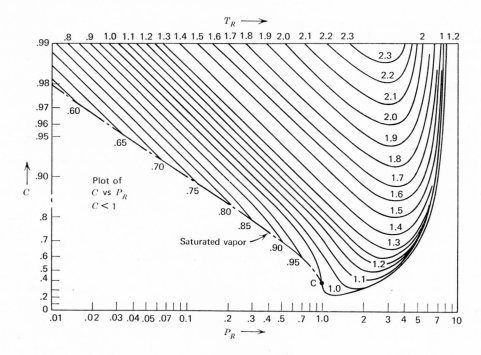

Fig. 4a

and gm-mole. As a consequence of the first and second laws of thermodynamics, the absolute temperature as defined in terms of the pressure and volume of an ideal gas is identical to the thermodynamic temperature as defined by the second law. By allowing the pressure of a given mass of real gas to approach zero, while its volume approaches infinity, at a constant temperature, it is possible to define the absolute temperature scale in terms of the observable pressure and volume at two distinct temperatures, such as the freezing point and boiling point of water. If the temperature of the triple point of water is given an assumed value, the absolute temperature scale can be defined with much greater accuracy by measurements of pressure and volume at this single point. (This in fact has been approved by the International Temperature Scale Conference). The ideal gas may also be defined from the kinetic or statistical mechanical approach in which the pressure of the gas is related to the kinetic energy of translation of the molecular assembly (4).

Many analytical expressions for an equation of state of a real gas have been proposed (5,6) but none can ever be exact. The P–V–T relationships for a real gas including its vapor and liquid states may be plotted as shown in Figure 3, which is in effect a projection of Figure 2 into the P–V plane showing the isothermals for the vapor and liquid phases. The dotted line $ADFEB$ is used to differentiate the metastable and unstable phases from the mixture of stable vapor and liquid phases indicated by the solid line AFB. It is for this reason that the graph has been plotted as V vs P rather than the conventional form of P vs V. For a stable phase, as shown by Gibbs, $(d \ln V/d \ln P)_T < 0$; thus the extrapolated vapor AD and the extrapolated liquid BE are metastable, but the region shown by EFD is unstable.

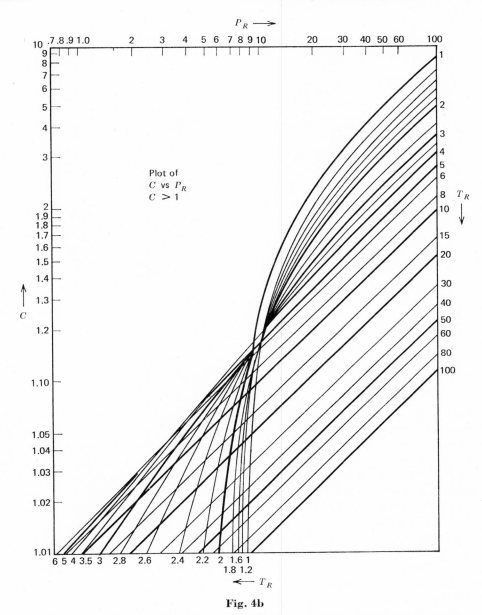

Fig. 4b

Fig. 4. Generalized compressibility chart plotted as a function of reduced pressure ($P_R = P/P_C$) with reduced temperature ($T_R = T/T_C$) as parameter. The compressibility scale is plotted as (**a**) $\log (1 - C)$ when C is less than unity and (**b**) $\log (C - 1)$ when C is greater the unity. These coordinates yield straight lines in the ideal gas region (9).

On the critical isotherm, which passes through C (the critical point) an inflection point occurs so that at least the first and second derivatives must reduce to zero; namely: $(dP/dV)_T = 0$ and $(d^2P/dV^2)_T = 0$ at $P = P_c$ and $T = T_c$ (critical point).

These conditions have been useful in deriving analytical equations of state for real gases, but never with the necessary degree of accuracy over the complete region

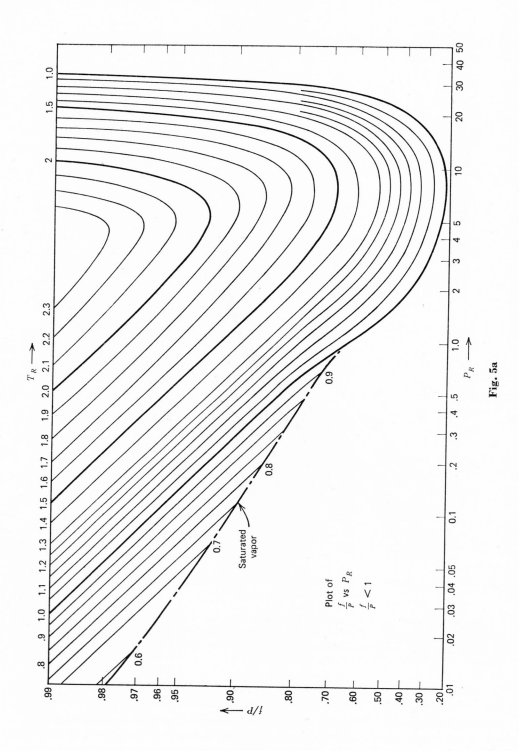

$T_R \longrightarrow$

$P_R \longrightarrow$

$\longleftarrow f/P$

Plot of
$\frac{f}{P}$ vs P_R
$\frac{f}{P} < 1$

Saturated
vapor

Fig. 5a

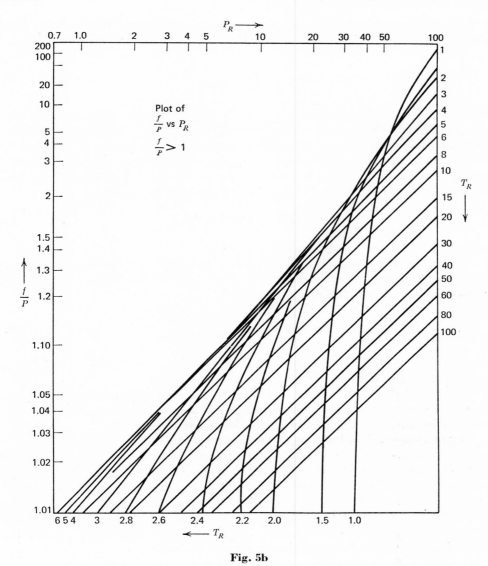

Fig. 5b

Fig. 5. Generalized fugacity–pressure ratios plotted as in Figure 4 with values (**a**) less than unity and (**b**) those greater than unity (9).

of phase equilibria. To express the deviations of a real gas from an ideal one in an exact way the compressibility factor is used, thus:

$$C = \frac{PV}{NRT} \tag{3}$$

These factors may be tabulated or plotted for real gases from experimental data. Generalized correlations based upon corresponding states have been devised for predicting the compressibility of real gases when the critical conditions are known (7,8).

A typical plot is shown in Figure 4. From the known deviations between the real and ideal gas the fugacity (f) can be related to the compressibility factor as follows:

$$\ln \frac{f}{P} = \int_0^P \frac{C - 1}{P} \, dP \tag{4}$$

As $P \to 0$, $(C - 1)/P \to$ finite value so that the integral is finite. The integration may be performed graphically or by computer to obtain values with an accuracy consistent with the experimental data. Alternately, generalized correlations of f/P may be obtained from those of compressibilities based on corresponding states (9). The plot derived from Figure 4 is shown in Figure 5. The thermodynamic relationship between fugacity and free energy is usually stated in differential form; namely,

$$dG = NRTd \ln f \tag{5}$$

where G = Gibbs free energy function.

When the reference state (indicated by a superscript °) is selected as that condition where the pressure of the real gas vanishes (that is, when the gas becomes ideal) the differential form may be integrated to give:

$$(\Delta G°)_T^D = NRT \ln \frac{f}{P} \tag{6}$$

where $(\Delta G°)_T^D = (\Delta G°)_T - (\Delta G°)_T^I =$ (difference function), $(\Delta G°)_T = NRT \ln f/f°$ (real gas) and $(\Delta G°)_T^I = NRT \ln P/P°$ (ideal gas).

It is noted that both the real and ideal free energy functions become infinite as the reference state of zero pressure and fugacity is approached. However, the difference function $(\Delta G°)_T^D$ remains finite in accordance with equation 4.

Vapor Pressure as a Function of Temperature. Having established the thermodynamic reference state for the pure vapor phase, it is now possible to relate it to the liquid phase when a vapor–liquid equilibrium has been established for the single component system. Again we make use of the fundamental relationship for a homogeneous phase in equilibrium first established by Gibbs (1a); and the fact that the temperature, pressure and chemical potential of each component must be the same for all phases in equilibrium (1b). For the single component vapor–liquid equilibrium, this leads to the thermodynamically exact Clapeyron equation:

$$\frac{dP}{dT} = \frac{\Delta S}{\Delta V} = \frac{\Delta H}{T \Delta V} \tag{7}$$

where $\Delta S = S_V - S_L =$ entropy change on vaporization; $\Delta H = H_V - H_L =$ enthalpy change on vaporization; $\Delta V = V_V - V_L =$ volume change on vaporization. The vaporization process occurs at constant P and T; subscript V = vapor and L = liquid.

We note that the enthalpy change ΔH is for the same mass of substance as for the change in volume ΔV, which for convenience is taken as one mole. The enthalpy change is identical to the latent heat of vaporization L, thus:

$$L = \Delta H \tag{8}$$

When the vapor phase obeys the ideal gas law and the volume of liquid is negligibly small compared to the same mass of vapor, equation 7 reduces to the famous Clausius-Clapeyron equation:

$$\frac{dP}{dT} = \frac{LP}{RT^2} \tag{9}$$

This may then be integrated assuming that L is constant; thus:

$$\log P = B - \frac{A}{T} \tag{10}$$

where

$$A = \frac{L}{2.303\,R}$$

and

$$B = \text{constant of integration}$$

One would expect that equation 10 would apply only over a limited range of temperature and pressure where the assumptions made would be valid. The fact is that this equation can be applied with reasonably fair accuracy over the complete range of vapor pressure from the triple point to the critical point when A and B are evaluated from a plot of $\log P$ vs $1/T$. The constant slope A may also be evaluated with a fair degree of accuracy from the latent heat L in the vicinity of the boiling point. What makes the Clausius–Clapeyron equation so remarkable is that although the latent heat becomes zero at the critical point, the assumption of constant L appears to be valid. The answer to this paradox lies in the exact Clapeyron equation given above, for as the critical point is approached both ΔH and ΔV approach zero and their ratio remains finite.

Using the principle of a reference substance, it is possible to correlate the vapor-pressure data of practically all known liquids (see Data-interpretation and correlation Vol. 6, pp. 714–18). Combining the above principle with that of corresponding states, the correlation can be further extended using critical constants (see Vol. 6, pp. 718–21). Many equations have been proposed for the correlation of vapor-pressure data and an excellent comparison has been made (10) among a substantial number of these. More recently, two correlations using the reference substance principle have been presented; in the former reference (11), correlation has been made with latent heat, and in the latter (12) with vapor volume, in which double log functions are used ($\log \log P$ and $\log \log T$).

Binary Systems

Up to now, the vapor-liquid equilibrium discussed has been the simple case of a single component; nevertheless, a very important one. When there is more than one component, the variables of composition enter. Although the binary system is a special case of a multicomponent system, it is treated separately and for good reason. Just as the single-component system serves as the boundary conditions for a binary system, so in turn do the binaries serve as the boundary conditions for a multicomponent system. In applying the phase rule, it is immediately apparent that a binary system has in general 2 degrees of freedom ($f = 2 - 2 + 2 = 2$). Having three independent variables; namely temperature, pressure and composition, this means that three independent relationships are possible between any two of the variables taken while the third is considered constant:

1. P and T at constant composition
2. P and composition at constant T
3. T and composition at constant P

In general, both the vapor and liquid contain varying amounts of each component, and are normally defined by the following symbols:

$$x_1 = \text{mole fraction of component 1 in the liquid phase}$$
$$x_2 = 1 - x_1 = \text{mole fraction of component 2 in the liquid phase}$$
$$y_1 = \text{mole fraction of component 1 in the vapor phase}$$
$$y_2 = 1 - y_1 = \text{mole fraction of component 2 in the vapor phase}$$

Because of the phase rule constraints, relationships 2 and 3 above imply that only one independent relationship between P and composition (or T and composition) is possible. Thus if P is expressed as a function of x, then the corresponding function between P and y is a necessary consequence and not independent, and similarly for the relationship between y and x. In fact, a number of methods have been developed in which the relationship between y and x at constant T is calculated from experimental measurements of the total pressure P as a function of x at constant T. The exact thermodynamic equations for the above relationships is again a consequence of the Gibbs equation of existence for a homogeneous phase (1a).

A generalized Clapeyron equation (see eq 7) relates the pressure to temperature at either constant x or y; at constant x

$$\left(\frac{dP}{dT}\right)_x = \frac{\Delta H_y}{T\Delta V_y} \tag{11}$$

where ΔH_y = enthalpy change on vaporization; ΔV_y = volume change on vaporization. The subscript y indicates that the liquid phase is taken at the composition of the equilibrium vapor; ie, the vaporization process is for a liquid having mole fractions y_1 and y_2 rather than x_1 and x_2.

Similarly at constant y,

$$\left(\frac{dP}{dT}\right)_y = \frac{\Delta H_x}{T\Delta V_x} \tag{12}$$

where the subscript x indicates that the vapor phase is taken at the composition of the equilibrium liquid.

At constant T,

$$\left(\frac{dP}{d\Delta \ln f}\right)_T = \frac{RT(y_1 - x_1)}{\Delta V} \tag{13}$$

where $\Delta \ln f = \ln f_1 - \ln f_2$; f_1 = fugacity of component 1; f_2 = fugacity of component 2.

At constant P,

$$\left(\frac{dT}{d\Delta \ln f}\right)_P = -\frac{RT^2(y_1 - x_1)}{\Delta H} \tag{14}$$

Equations 11–14 are exact and may be reduced to simplified form by making the same assumptions in reducing the exact Clapeyron equation to the approximate Clausius–Clapeyron equation, thereby, making them more readily useful when only a paucity of experimental data are available; thus:

$$\left[\frac{d \ln P}{d\left(\frac{1}{T}\right)}\right]_x = -\frac{\Delta H_y}{R} \tag{15}$$

or

$$\left[\frac{d \ln P}{d\left(\frac{1}{T}\right)}\right]_y = -\frac{\Delta H_x}{R} \tag{16}$$

$$\left(\frac{d \ln P}{dy_1}\right)_T = \frac{(y_1 - x_1)}{y_1 y_2} \tag{17}$$

and

$$\left[\frac{d\left(\frac{1}{T}\right)}{dy_1}\right]_P = \frac{R(y_1 - x_1)}{\Delta H \; y_1 \; y_2} \tag{18}$$

In equations 15–18, the assumption of an ideal gas for the vapor phase made it possible to substitute the partial pressure for the fugacity of each component:

For an ideal vapor phase,

$$f_1 = Py_1 = p_1 \tag{19}$$

and similarly for component 2.

Equations 15 and 16 are identical to the Clausius–Clapeyron equation for a single component in which the composition is automatically constant. Equation 17 was first introduced by van der Waals (9a) and clearly shows that the relationship between y and x is implicit if P is known as a function of y at constant T. Equation 18 is the analogous van der Waals relationship at constant P, and also clearly shows that the relationship between y and x is implicit if T is known as a function of y at constant P. The factor $R/\Delta H$ may be treated as a constant in equation 18, unless the temperature range is unusually large, in which case it must be treated as a function of y and T.

Relative Volatility Function. In treating the experimental data of vapor–liquid equilibria, the relative volatility function has been found most useful. In its most general form, it is defined as the volatility of one component relative to another in a multicomponent system, the volatility being defined simply as the ratio of vapor to liquid composition. Thus, for a binary system, the relative volatility function appears as follows:

$$\alpha_{12} = \frac{y_1 x_2}{x_1 y_2} \tag{20}$$

Normally component 1 is selected as the more volatile so that the relative volatility function is greater than unity. However, in the case of systems with components of nearly equal volatility, but with large deviations from ideality, a reversal may take place in relative volatility; so that over one range of composition, component 1 may be the more volatile, while over another, component 2 is more volatile. If such is the case, then a constant boiling mixture occurs at the crossover point where the relative volatility is unity. The system at that point is called an *azeotrope* which literally means to boil unchanged.

Relative volatility has a thermodynamic significance and can be defined in terms of fugacity. For convenience the relative fugacity (known as activity) referred to liquid composition is the function most useful in this connection. It is known as the activity coefficient; thus:

$$\gamma_1 = \frac{a_1}{x_1} = \frac{f_1}{f_1^\circ x_1} = \text{activity coefficient}$$

where

$$a_1 = \frac{f_1}{f_1^\circ} = \text{activity} \tag{21}$$

The standard state indicated by the superscript ($^\circ$) is normally taken as the pure component at the same temperature and pressure as that of the system, especially in the case of vapor–liquid equilibria. In addition, the activity coefficient is taken relative to mole fraction as being most convenient, rather than molality or mole ratio as sometimes used in physical chemical studies. Other standard states and representations of composition have been employed in the definition of activity coefficient, and although these find useful application in the study of solutions (with nonvolatile solutes), they could lead to considerable confusion, which indeed they have. The symmetrical simplicity of using the pure components for the standard states and mole fractions for composition, has ultimately proven its superiority by virtue of its universal acceptance in phase equilibria. The thermodynamically exact relationship between relative volatility and activity coefficient can be written as follows:

$$\alpha_{12} = \frac{\gamma_1 p_1^\circ \nu_2 \nu_1^\circ}{\gamma_2 p_2^\circ \nu_1 \nu_2^\circ} \tag{22}$$

where $\nu_1 = f_1/p_1$, $\nu_1^\circ = f_1^\circ/p_1^\circ$ and $p_1^\circ = $ vapor pressure of pure component 1 (similarly for p_2°).

The new symbol for ratio of fugacity to partial pressure is introduced here for convenience. For the pure components, this ratio has already appeared in equation 4, and its method of evaluation described. For mixtures (which includes binary systems) two approaches to a generalized solution of these factors have been proposed, the first involves the use of an equation of state for gas mixtures (13) and the other employs generalized factors for these mixtures (8,14). Without the use of these generalized methods, the amount of experimental data required for the exact representation of a multi-component system would be staggering. For an ideal system (in both phases) the activity coefficients and fugacity–pressure ratios reduce to unity, so that the ideal relative volatility becomes:

$$\alpha_{12}{}^I = \frac{p_1^\circ}{p_2^\circ} \tag{23}$$

which is simply the ratio of the vapor pressures of the pure components. The deviation function for relative volatility is defined as the ratio of the function to its ideal value; namely:

$$\alpha_{12}{}^D = \frac{\alpha_{12}}{\alpha_{12}{}^I} = \frac{\gamma_1 \nu_2 \nu_1{}^{\circ}}{\gamma_2 \nu_1 \nu_2{}^{\circ}} \tag{24}$$

At moderate pressures and temperatures where the usual assumptions of ideal gas for the vapor phase and negligible volume of liquid in comparison to the vapor are valid, this reduces to:

$$\alpha_{12}{}^D = \frac{\gamma_1}{\gamma_2} \tag{25}$$

Simply stated, when the vapor phase is ideal, the deviation in relative volatility is a measure of the non-ideality in the liquid phase. Since the activity coefficients in the binary system are dependent (see eq 26), it is possible to express each in terms of the relative volatility deviation. The so-called "Gibbs–Duhem" equation, which is nothing more than a reduction of Gibbs equation of existence (1a) to constant pressure and temperature, relate the activity coefficients as follows:

$$x_1 \, d \log \gamma_1 + x_2 \, d \log \gamma_2 = 0 \tag{26}$$

Combining equations 25 and 26 and integrating from the standard states (9a):

$$\log \gamma_1 = \int_0^{x_2} x_2 \, d \log \alpha_{12}{}^D$$

and

$$\log \gamma_2 = - \int_0^{x_1} x_1 \, d \log \alpha_{12}{}^D \tag{27}$$

For the convenient representation of vapor–liquid equilibria a dimensionless function of free energy has been proposed:

$$Q = \frac{\Delta G^D}{2.303 \, RT} \tag{28}$$

where ΔG^D = difference function for free energy of mixing (see eq 6) and $Q = x_1 \log \gamma_1 + x_2 \log \gamma_2$.

This function is especially useful in expressing the vapor–liquid equilibria of multicomponent systems in terms of the composite binaries as discussed below. A combination of the two parts of equation 27 results in a test of the thermodynamic consistency of vapor–liquid equilibrium data:

$$\int_0^1 \log \alpha_{12}{}^D \, dx_1 = \int_0^1 \log \frac{\gamma_1}{\gamma_2} \, dx_1 = 0 \tag{29}$$

An example of this test is shown in Figure 6.

This test has been extended to multicomponent systems (15), used to evaluate effects of vapor non-ideality and enthalpy of mixing (16) and for systems with incomplete data (17). For the data to be consistent, the algebraic area under the curve must vanish. It must be noted that the test is rigorous only if the data are reduced to

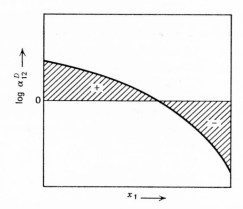

Fig. 6. Thermodynamic area test for the relative volatility deviation. Note that the (+) area is equal to the (−) area when the data are consistent (9).

constant temperature and pressure and that all factors in equation 24 are accounted for. Non-ideality in the vapor phase may be omitted only if negligible. In addition, because of the use of molar quantities, the correct molecular weight for each component must be used. This is the molecular weight in the reference vapor phase (ie, the pure component vapor at the same temperature and pressure of the system in equilibrium). For associating vapors, such as acetic and formic acid, the true molecular weight is a relatively complicated function of temperature and partial pressure, which may be evaluated from vapor density measurements. The Gibbs equation, from which the phase rule has been deduced, requires that the same molecular weight be applied to all phases in order that the chemical potential per unit mass of each component be the same in each phase. This means that α_{12}^D remains independent of molecular weight and that correction need only be applied to x_1. Thus, any errors of inconsistency due to association would appear as a horizontal translation of the curve away from the associating component. On the other hand, equation 24 indicates that any non-ideality in the vapor phase due to the pure components or to mixing would cause a vertical translation of the α_{12}^D curve, which had been calculated by assuming ideality in the vapor phase. For constant temperature systems in which the pressure varies, a volumetric correction must be made when the pressure is high enough to make it significant. This is usually negligible at pressures below 10 atm. However, for constant pressure systems in which the temperature varies, especially with high relative volatilities and high enthalpies of mixing, an enthalpic correction must be made to obtain a valid consistency test. In fact the discrepancy indicated by the test may be used to evaluate the effects mentioned above (18).

Analytical Solution of the Gibbs-Duhem Equation. For purposes of analysis and prediction of vapor–liquid equilibria, an analytical solution of the Gibbs-Duhem equation has been sought since the turn of the century. The oldest and most famous are the Margules and van Laar equations. Margules proposed his analytical solution back in 1895 and to this day it is probably the most useful in spite of the many new solutions proposed. For a discussion of some of these newer proposals reference is made to articles in the literature (19,20).

The equations of van Laar date back to 1910 only 15 years after Margules proposed his equations; and have been found to be nearly as useful. They do have one shortcoming in that they cannot correctly describe a system in which a maximum or minimum occurs in the activity coefficient function when plotted against composition. To solve the Gibbs–Duhem differential equation, Margules used a simple power series and obtained the following when limited to the first two terms of the series:

$$\log \gamma_1 = A_{12}x_2^2 + B_{12}x_2^3$$

and

$$\log \gamma_2 = A_{21}x_1^2 + B_{21}x_1^3 \tag{30}$$

where

$$A_{21} = A_{12} + 3/2\, B_{12} \text{ and } B_{21} = -B_{12}.$$

Equations 30 have been written in symmetrical form involving four constants, A_{12}, B_{12}, A_{21} and B_{21}, although only two of these are independent, because of the advantages in extending the equations to multicomponent systems. On the other hand van Laar used a simple hyperbolic function obtaining

$$\log \gamma_1 = \frac{A_{12}}{(1 + B_{12}\, x_1/x_2)^2} \text{ and } \log \gamma_2 = \frac{A_{21}}{(1 + B_{21}\, x_2/x_1)^2} \tag{31}$$

where

$$A_{21} = \frac{A_{12}}{B_{12}} \text{ and } B_{21} = \frac{1}{B_{12}}.$$

Again the symmetrical notation has been used, which has the additional advantage that the equation for the second component can be obtained by interchanging the subscripts for the first component. The constants in the van Laar equations have been written with the same symbols as in the Margules for convenience only and are in general quite different.

Because the analytical solutions of the Gibbs–Duhem equation are essentially mathematical with very little physical significance, although some of the more recent proposals do have a semi-theoretical basis, the following discussion will be limited to the Margules equations. In this way more of the thermodynamic principles can be covered with a minimum of mathematics. A typical plot of the activity coefficients as a function of liquid composition is shown in Figure 7 in which a maximum and minimum appear in the respective curves. The Gibbs–Duhem equation requires that the maximum and minimum occur at the same composition and that the area under each curve be equal.

The two constant Margules equations can be used to obtain analytical expressions for the relative volatility deviation and the dimensionless Gibbs free energy function (Q of equation 28) as follows:

$$\log \alpha_{12}{}^D = a_{12}\, (1/2 - x_1) + b_{12}\, (1/6 - x_1 + x_1^2) \tag{32}$$

where $a_{12} = 2A_{12} + 3/2\, B_{12}$ and $b_{12} = 3/2\, B_{12}$ and conversely $A_{12} = 1/2\, a_{12}$

$- 1/2\, b_{12}$ and $B_{12} = 2/3\, b_{12}$

$$Q = x_1x_2(A_{12}x_2 + A_{21}x_1 + B_{12}x_2^2 + B_{21}x_1^2) \tag{33}$$

(the constants coming from equations 30).

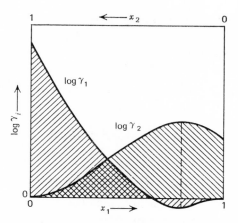

Fig. 7. Activity coefficients as a function of liquid composition for a binary system at constant temperature and pressure (9).

Equations 32 and 33 are especially useful in predicting binary vapor–liquid equilibria and can readily be extended to multicomponent systems. They are also useful in expressing the change of activity coefficient with temperature and pressure to evaluate partial enthalpies and volumes of mixing as discussed below.

Change of Activity Coefficient with Temperature and Pressure. Since the activity coefficient is a measure of the Gibbs free energy function, its change with temperature and pressure is a consequence of the general Gibbs–Duhem equation in which pressure and temperature are permitted to vary. The final results may be written as follows:

Change with P at constant T and x

$$\left(\frac{d \log \gamma_1}{dP}\right)_{T,x} = \frac{v_1 - v_1{}^\circ}{2.303\,RT} \tag{34}$$

and similarly for component 2 where v_1 = partial molar volume of component 1 in the mixture; $v_1{}^\circ$ = molar volume of pure component 1.

Change with T at constant P and x

$$\left(\frac{d \log \gamma_1}{d\,1/T}\right)_{P,x} = \frac{h_1 - h_1{}^\circ}{2.303\,R} \tag{35}$$

where h_1 = partial molar enthalpy of component 1 in the mixture and $h_1{}^\circ$ = molar enthalpy of pure component 1.

Equation 35 is quite similar to the Clapeyron equation in which the slope of a fugacity function with temperature is measured by a latent heat function. In general this type of equation is a specific example of the well-known Gibbs–Helmholtz equation in which the rate of change of the Gibbs free energy function with respect to temperature at constant composition is a measure of the enthalpy. It is one of the most important deductions of thermodynamics since it permits one to deduce heat quantities from equilibrium measurements of state and conversely to deduce equilibrium conditions from thermal measurements alone.

One of the significant applications of reference plotting (see Data-interpretation and correlation, Vol. 6, p. 728, Fig. 8) is to the plotting of activity coefficients as a

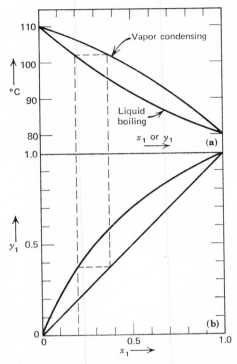

Fig. 8. Benzene(1)–toluene(2) mixtures: (**a**) vapor–liquid curves as functions of temperature; (**b**) vapor compositions as a function of liquid compositions. The horizontal dotted line is for the specific system at 102°C.

function of temperature at constant composition by using the vapor pressure of a reference substance as the temperature coordinate. The slopes of the lines of constant composition are measures of the ratios of partial enthalpies of mixing to the latent heat of vaporization of the reference substance taken at the same temperature.

Equation 34 is normally not used for equilibria at moderate pressures; however, in the analysis of high pressure equilibria, the partial volumes of mixing become quite significant and are essential to a consistent analysis (21).

Types of Binary Systems. The first classification of binary systems may be made into ideal and non-ideal. Although the ideal behavior of gases and gas mixtures is a common phenomenon, that of liquid mixtures is relatively scarce, at least as far as engineering practice is concerned. In fact it is quite surprising that any ideal behavior of liquid binary systems should exist at all. Nevertheless, binary mixtures of close members of a homologous series of relatively simple compounds, such as straight chain hydrocarbons, and some aromatic hydrocarbons, form ideal binaries. A classical example of an ideal vapor–liquid binary system is given by benzene and toluene. The vapor–liquid equilibria at one atmosphere pressure are shown in Figure 8. Because this is an ideal system, both activity coefficients reduce to unity as does the relative volatility deviation.

The second general classification is between systems that have positive deviations from ideality as distinguished from those with negative deviations. Those with positive deviations have activity coefficients greater than unity and conversely those with negative deviations have coefficients less than unity. Each of these classifications are further differentiated by the complexity of the activity coefficient functions. In terms of relative volatility deviation, the simplest function is a straight line when $\log \alpha_{12}{}^D$ (where component 1 is the more volatile) is plotted against x_2, a positive slope indicating positive deviations and conversely a negative slope indicating negative deviations. This is equivalent to a single constant in the Margules equations ($B_{12} = 0$) so that the activity coefficient curves for each component are symmetrical mirror images. In addition, thermodynamic consistency requires that $\alpha_{12}{}^D$ has a value of unity at the midpoint composition. The binary system in Figure 9 is an excellent example of the above case with positive deviations. When the degree of non-ideality

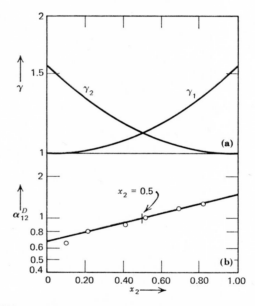

Fig. 9. Carbon disulfide(1)–chloroform(2) mixtures: (**a**) logarithmic plot of activity coefficients. The lines cross at the midpoint composition. (**b**) Logarithmic plot of relative volatility deviation. The straight line passes through the midpoint composition at $\log \alpha_{12}{}^D = 0$, thus satisfying the thermodynamic consistency test of Figure 6.

increases, the complexity of the activity coefficients requires that at least two constants be employed in the Margules (or other integrated form of the Gibbs-Duhem equation) relationships. Fortunately, two constants are sufficient for practically all systems at moderate pressure especially for engineering applications (22). Even when the activity coefficient curves exhibit maxima and minima two constants are sufficient as seen in Figure 10. In this system, the activity coefficient of the less volatile component becomes negative in the region of the minimum point, so that it is possible to encounter both positive and negative deviations in a system which shows complex deviation from ideality.

For systems with sufficiently high deviation from ideality and components that are relatively close in volatility (similar boiling points), an *azeotrope* or constant boiling mixture (CBM) will form. At this point the vapor and liquid compositions are equal and the relative volatility reduces to unity. Under these conditions the relative volatility deviation equals the reciprocal of the ideal relative volatility; thus, at the azeotrope (see equations 23 and 24):

$$\alpha_{12} = 1 \quad \text{and} \quad \alpha_{12}{}^{D} = \frac{1}{\alpha_{12}{}^{I}} = \frac{p_2{}^{\circ}}{p_1{}^{\circ}} \tag{36}$$

When the deviations from ideality are positive, the azeotrope corresponds to a minimum boiling point for an isobaric system (or a maximum pressure for an isothermal system). Conversely, when the deviations are negative, the azeotrope corresponds to a maximum boiling point at constant pressure (or a minimum pressure at constant temperature). Several remarkable deductions can be made concerning

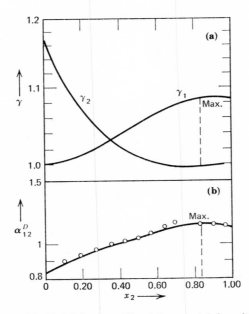

Fig. 10. Carbon tetrachloride(1)–heptane(2) mixtures: (**a**) logarithmic plot of activity coefficients, showing a maximum for component 1 and a minimum for component 2 at $x_2 = 0.83$; (**b**) logarithmic plot of relative volatility deviation shows a maximum at the same composition.

homogeneous (single liquid phase) azeotropic systems from thermodynamics. To illustrate these deductions, the phase diagram for such a system is shown in Figure 11. At the azeotrope it is necessary that a minimum or maximum with respect to liquid composition coincide respectively with a minimum or maximum with respect to vapor composition, ie:

$$\left(\frac{dT}{dx_1}\right)_P = \left(\frac{dT}{dy_1}\right)_P = 0 \tag{37}$$

and

$$\left(\frac{dP}{dx_1}\right)_T = \left(\frac{dP}{dy_1}\right)_T = 0 \tag{38}$$

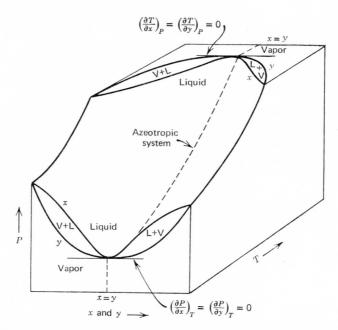

$$\left(\frac{\partial T}{\partial x}\right)_P = \left(\frac{\partial T}{\partial y}\right)_P = 0$$

$$\left(\frac{\partial P}{\partial x}\right)_T = \left(\frac{\partial P}{\partial y}\right)_T = 0$$

Fig. 11. Phase diagram for a binary vapor–liquid azeotropic (homogeneous) system (9).

The condition of maximum or minimum is determined by the second derivatives at the point. Thermodynamically, the relationship is as follows:

$$\left(\frac{d^2P}{dx^2}\right)_T = -\left(\frac{dP}{dT}\right)_x \left(\frac{d^2T}{dx^2}\right)_P \tag{39}$$

But the pressure–temperature derivative for a vapor–liquid system is always positive (see equation 7 in which ΔS and ΔV are always positive), ie:

$$\left(\frac{dP}{dT}\right)_x > 0 \tag{40}$$

By virtue of the exact thermodynamic relationships given in equations 39 and 40, the second derivatives in equation 39 must be opposite in sign, proving that a minimum in the isothermal curve must be associated with a maximum in the isobaric curve and vice versa at the azeotropic composition. Examples of homogeneous azeotropic systems of both kinds are shown in Figure 12 (minimum boiling point) and Figure 13 (maximum boiling point). See also Azeotropy.

Equation 36 may be used to predict the change of azeotropic composition with temperature and pressure. Over relatively short ranges, the $\alpha_{12}{}^D$ curve may be assumed to remain constant and the azeotropic compositions predicted by using the appropriate value of $\alpha_{21}{}^I$ as performed in Figure 14. Over larger ranges, the change in $\alpha_{12}{}^D$ may be accounted for by the use of partial enthalpies of mixing (equation 35 for both components).

As the deviation from ideality continues to increase, it is possible to reach a point where the single liquid phase separates into two distinct liquids. The critical point at which this occurs is quite analogous to the critical point for pure gases where the

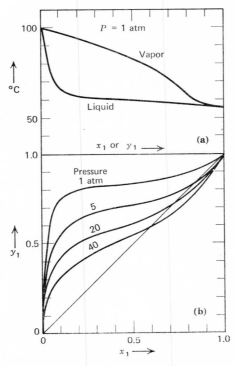

Fig. 12. Acetone(1)–water(2) mixtures. Boiling point curves at atmosphere (**a**). Several x,y curves are shown at different pressures (**b**). The homogeneous azeotrope does not appear until the pressure reaches about 2.5 atm.

"permanent gas" first separates into liquid and vapor phases. From the Gibbs relationships for stable phases, at constant P and T, one may write:

$$\left(\frac{dp_1}{dx_1}\right) > 0 \text{ and } \left(\frac{dp_2}{dx_2}\right) > 0 \tag{41}$$

For convenience equation 41 has been written assuming that the partial pressures are sufficiently close to the fugacities (the latter are necessary only at high pressures). In Figure 15 these partial pressures are plotted for a binary system exhibiting two distinct liquid phases. The dotted line $ADFEB$ represents the homogeneous equilibrium, while the solid line AB represents the mixture of the two liquids. For the regions of AD and EB the equilibrium is metastable; however, for DFE it is unstable and will spontaneously begin to separate into two liquid phases. An example of a system with partial immiscibility is shown in Figure 16. According to the phase rule when two liquids are in equilibrium with the vapor, a binary system is monovariant. Therefore, among the three variables of temperature, pressure and composition only one can be independent. Thus, at constant pressure, the temperature and composition of the phases are fixed. For the system in Fig. 16, the temperature of the three-phase mixture is 92°C, the composition of the vapor is 0.73, the butanol-rich liquid is 0.53 and the water-rich liquid is 0.97 (expressed as mole fraction water). Because of the presence of two liquids, the binary system is a constant boiling mixture defined as a heterogeneous azeotrope. At the azeotrope the two vapor curves intersect in a point of

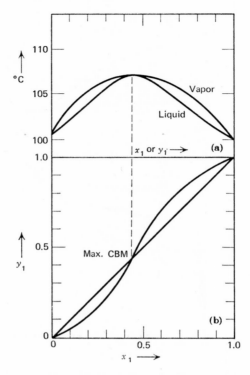

Fig. 13. Water(1)–formic acid(2) mixtures. The maximum boiling point of 107°C occurs at the azeotropic composition of 0.43 (**a**). The second derivative in the y_1 vs x_1 curve is negative at the CBM in contradistinction to a minimum boiling point in which it is positive (**b**).

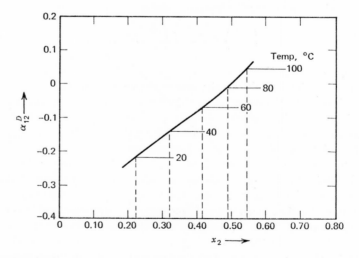

Fig. 14. Logarithmic plot of relative volatility deviation of ethyl acetate (1) with respect to ethyl alcohol (2). The variation of the azeotropic composition with temperature is obtained by intersecting the curve with the value of the ideal relative volatility at the corresponding temperature.

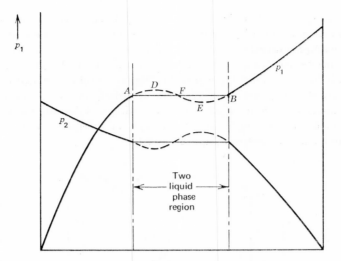

Fig. 15. Isothermal partial pressures from a binary system with partial immiscibility (9).

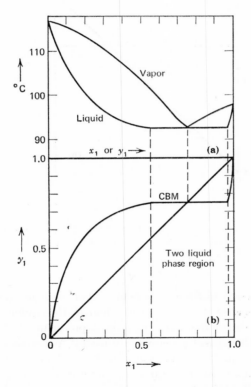

Fig. 16. Water(1)–1, butanol(2) mixtures. The region of two liquid phases lies between the two vertical dotted lines. The heterogeneous azeotrope is indicated by CBM (**b**) (9).

discontinuity having two derivatives, one for each curve. When the two liquids are completely immiscible (for all practical purposes) the vapor–liquid equilibria becomes analogous to a simple eutectic system in a liquid–solid binary system as seen in Figure 17. Heterogeneous azeotropes are quite important in the steam distillation of immiscible organic liquids. They have also been found useful in separating homogeneous azeotropic mixtures. Because the heterogeneous azeotropes are formed by extremely high positive deviations from ideality they are always of the minimum boiling point type. Those of the maximum boiling point type, of thermodynamic necessity, must always be homogeneous (compare Figs. 12 and 13).

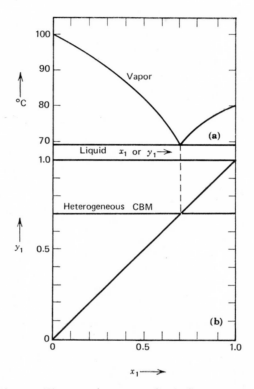

Fig. 17. Benzene(1)–water(2) system is an example of a heterogeneous azeotrope with complete immiscibility. The azeotrope is completely analogous to the eutectic point in a liquid–solid binary system consisting of a single liquid and two pure solids.

Correlation and Prediction of Binary Systems. The correlation of vapor–liquid equilibria is more of an exercise in mathematics than in the application of thermodynamics. Nevertheless, thermodynamics can be a powerful tool in shaping the forms upon which the mathematical processes are used. Although many integrated forms of the Gibbs-Duhem equation have been found useful as discussed above, two-parameter equations have been found adequate for practically all engineering applications. In fitting the experimental data to any of these two-parameter sets of equations, it has been shown advantageous to employ statistical weighting factors (22). When the data are expressed in the form of relative volatility deviations (see equation 32) the two independent parameters may be evaluated statistically by using the weighting factors illustrated in Figure 18, which is a plot of the following equation:

$$W = 64 \left[\frac{x_i x_j}{1 + \alpha_{ji}(x_j + \alpha_{ij}x_i)^2} \right]^2 \qquad (42)$$

where component i is the more volatile.

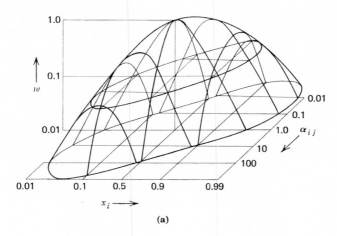

(a)

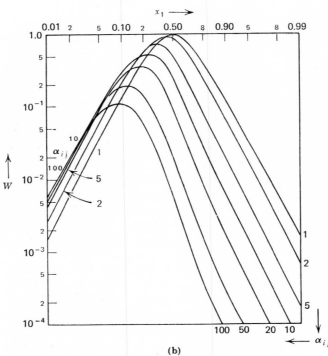

(b)

Fig. 18. Plot of weighting factor which shows characteristic distribution about the maximum point. Component i is the more volatile. (a) Three dimensional plot; (b) projection onto the relative volatility-composition plane (22).

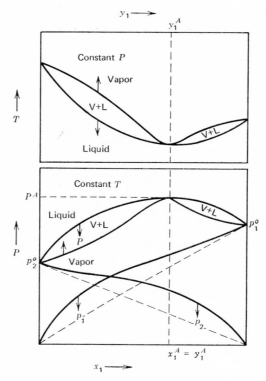

Fig. 19. Vapor–liquid equilibria of a binary system with a homogeneous azeotrope (9).

When conditions no longer permit the usual assumptions of ideal vapor phase and negligible liquid volume, such as vapor–liquid equilibria at low temperatures and high pressures, more complex methods of correlation are required such as those recommended for hydrocarbon mixtures (23).

The prediction of binary systems is even less subject to thermodynamic principles and utilizes more of statistical mechanical and molecular theory. However, there are still a number of cases where prediction can be made without resorting to molecular theory. In the case of a homogeneous azeotropic system, it is possible to obtain the composition of the azeotrope by non-equilibrium measurement, for example by fractional distillation. Once this composition is determined the complete vapor–liquid equilibrium curve may be predicted by solving an integrated form of the Gibbs-Duhem equation containing two parameters. For example, the two independent parameters of equation 30 are calculated as follows:

$$A_{12} = \frac{3 - 2x_1{}^A}{(x_2{}^A)^2} \log \left(\frac{P^A}{p_1{}^\circ}\right) - \frac{2x_2{}^A}{(x_1{}^A)^2} \log \left(\frac{P^A}{p_2{}^\circ}\right)$$

and

$$B_{12} = \frac{2}{(x_1{}^A)^2} \log \left(\frac{P^A}{p_2{}^\circ}\right) - \frac{2}{(x_2{}^A)^2} \log \left(\frac{P^A}{p_1{}^\circ}\right) \tag{43}$$

the superscript A referring to the azeotrope.

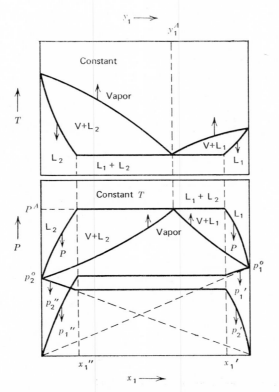

Fig. 20. Vapor–liquid equilibria of a binary system with partial immiscibility showing the typical heterogeneous azeotrope (9).

The phase diagrams for a typical minimum boiling point (maximum vapor pressure) homogeneous azeotropic system are shown in Figure 19. In the case of a heterogeneous azeotropic system with partial immiscibility, the two parameters may be evaluated from a knowledge of the mutual solubilities in each liquid phase. When the solubilities are small the calculations are direct; however, when they are relatively large a method of successive approximations is necessary. Thus, as the mutual solubilities become sufficiently small:

$$A_{12} \rightarrow \log \left(\frac{x_2''}{x_2'}\right)^2 \left(\frac{x_1''}{x_1'}\right)$$

and

$$B_{12} \rightarrow 2 \log \left(\frac{x_1' x_2'}{x_1'' x_2''}\right) \tag{44}$$

the superscript ′ referring to the liquid phase rich in component 1 and ″ to the phase rich in component 2.

Equation 44 is quite accurate when the solubilities x_2' and x_1'' are less than 0.1. For larger values, a second approximation will usually be sufficient. As the mutual solubilities approach 0.5 more approximations are needed and may be continued until

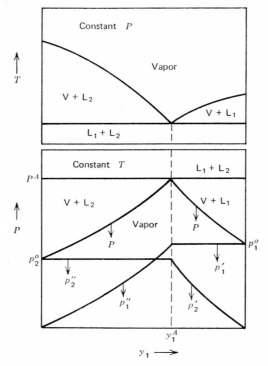

Fig. 21. Vapor–liquid equilibria of a binary system with total immiscibility (9).

the desired degree of accuracy is obtained. When the accurate activity coefficients are obtained the azeotropic composition and total pressure may be calculated as follows:

$$y_1{}^A = \frac{\gamma_1{}'p_1{}^\circ x_1{}'}{P^A}$$

and

$$P^A = \gamma_1{}'p_1{}^\circ x_1{}' + \gamma_2{}''p_2{}^\circ x_2{}'' \tag{45}$$

In the first approximation, each solvent-rich component is assumed to obey Raoult's law; namely: $\gamma_1{}' = \gamma_2{}'' = 1$. An example of the above system is shown in Figure 20. The phase diagrams clearly show that the two vapor curves intersect in a point of discontinuity as pointed out in Figure 16. The typical partial-pressure curves appear as in Figure 15. The dotted lines are shown for comparison and indicate the partial pressures for an ideal system in which both components obey Raoult's law over the complete range of composition. As each pure component is approached, the partial pressure for the solvent liquid does in fact approach Raoult's law.

For complete immiscibility, the mutual solubilities reduce to zero and the phase relationships are greatly simplified as seen in Figure 21. The two liquids consist of the pure components and the vapor compositions are calculated directly from the vapor pressures of the pure components and the total pressure; thus, at the azeotrope:

$$y_1{}^A = \frac{p_1{}^\circ}{P^A}$$

and

$$P^A = p_1^\circ + p_2^\circ \tag{46}$$

For vapor compositions greater than that of the azeotrope:

$$y_1 = \frac{p_1^\circ}{P}$$

and

$$p_2' = P - p_1^\circ \ (p_2'' = p_2^\circ) \tag{47}$$

Similarly for vapor compositions less than that of the azeotrope:

$$y_1 = 1 - y_2 = 1 - \frac{p_2^\circ}{P}$$

and

$$p_1'' = P - p_2^\circ \ (p_1' = p_1^\circ) \tag{48}$$

To predict the vapor–liquid equilibria of binary systems from data taken for the pure components only, requires the application of principles beyond thermodynamics (except for the rather trivial case of complete immiscibility). Statistical mechanical and molecular theory have been quite useful in predicting as well as correlating vapor–liquid equilibria. By utilizing vapor pressure data, parachors, critical temperatures and critical pressures of the pure components it has been shown possible to predict the two parameters in equation 32 by means of field factors grouped according to classes of compounds (22). When the effects of pressure can no longer be neglected, predictions can be made by employing molecular shape factors in combination with the principle of corresponding states (24). Using a reference fluid with accurately known properties in the range of the vapor–liquid equilibria involved, excellent predictions have been made for aliphatic hydrocarbon binaries employing the above method.

Although vapor–liquid equilibria where, in addition, chemical reaction takes place have been omitted from this discussion, except for the case of the association of vapor components, another exception should be made in regard to the recent application of treating non-ideality in the liquid phase in terms of complex formation as well as physical forces alone (25). The method does become somewhat more involved from a mathematical point of view, but in certain applicable systems it does enable a better correlation or prediction of data with a smaller number of parameters. For example, if component A and B interact to form a complex AB, then the equilibrium constant for the reaction can be stated in thermodynamic terms as:

$$K = \frac{a_{AB}}{a_A a_B} = \frac{\gamma_{AB} z_{AB}}{\gamma_A z_A \gamma_B z_B} \tag{49}$$

where the subscripts refer to the complex and pure components and z refers to the true mole fractions of the same molecular species.

The apparent activity coefficients for each component can then be stated as follows:

$$\gamma_2 = \frac{z_A \gamma_A}{x_2} \text{ and } \gamma_1 = \frac{z_B \gamma_B}{x_1} \tag{50}$$

in which each activity coefficient is a function of only two parameters as follows:

$$\log \gamma_2 = F(K,\beta,x_2) \tag{51}$$

and similarly for component 1, where β = van Laar parameter for physical interaction only.

The functional relationship may be stated in terms of regular solution theory, thereby reducing the number of constants (25).

Multicomponent Systems

In the light of the phase rule, it is clearly apparent that as the number of components increases in a vapor–liquid equilibrium so do the degrees of freedom in equal number; thus for two phases:

$$f = n - 2 + 2 = n \tag{52}$$

To arrive at a monovariant system, so as to express the pressure as a function of temperature, it is therefore necessary to maintain constant composition. Under these conditions equation 11 for the binary may be extended to a multicomponent as follows:

$$\left(\frac{dP}{dT}\right)_{x_1,x_2\cdots x_n} = \frac{\Delta H_y}{T\Delta V_y} \tag{53}$$

where $\Delta H_y = H_V - \sum_{i=1}^{n} y_i h_i;\ \Delta V_y = V_V - \sum_{i=1}^{n} y_i v_i;\ h_i$ = partial liquid enthalpy of component i and v_i = partial liquid volume of component i.

It should be noted that the respective enthalpy and volume changes correspond to a change in vaporization of liquid having the equilibrium vapor composition, in which the partial quantities refer to the liquid phase (26). Equations similar to that of 12, 13 and 14 can be written for the multicomponent system. In addition, by assuming ideal gas behavior for the vapor phase and neglecting the volume of the liquid in comparison to the vapor, equations similar to 15, 16, 17 and 18 may also be written for the multicomponent system. In those equations in which composition is a variable at either constant temperature or pressure, the derivatives with respect to the mole fraction of component i are taken with $n - 2$ mole fractions remaining constant. Because the sum of the mole fractions must be unity it is not physically possible to vary the single mole fraction x_i without allowing another mole fraction such as x_j to vary simultaneously. For a two-phase monovariant system at constant temperature or pressure, it is thus necessary to hold all but two of the mole fractions of the n components constant and allow the remaining two to vary simultaneously, but dependently so that their sum is always constant.

The correlation and prediction of vapor–liquid equilibria for multicomponent systems is based upon setting up a free energy function (Q of equation 28) in terms of all the composition variables at a given P and T. Then by a successive treatment of allowing only two composition variables to change at a time, each of the binary interactions (in terms of $\log \alpha_{ij}{}^D = \log \gamma_i - \log \gamma_j$) are evaluated. From these, the activity coefficients for each component may be obtained using the relationship between the partial and total functions; namely:

$$\log \gamma_i = Q + \sum_{j=1}^{n} x_j \log \alpha_{ij}{}^D \tag{54}$$

where

$$Q = \sum_{i=1}^{n} x_i \log \gamma_i$$

and

$$\log \alpha_{ij}{}^{D} = \frac{\partial Q}{\partial x_i} - \frac{\partial Q}{\partial x_j} \tag{55}$$

The partial derivatives are purely mathematical in nature and have no physical meaning.

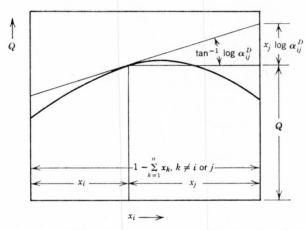

Fig. 22. Free energy function Q plotted as a function of x_i (with x_j as the other variable) for a multicomponent system at constant P, T, x_1, $x_2 \ldots x_k \ldots x_n$ where $k \neq i$ or j (9).

Equation 54 may be solved analytically providing the function Q can be expressed explicitly in terms of the mole fractions. Otherwise, it is necessary to resort to graphical or iterative methods of calculation. A generalized tangent slope method for the graphical solution of a multicomponent system is shown in Figure 22. For a ternary system, the above generalized method reduces to a tangent plane, in which the three partial quantities (activity coefficients) at a given composition are given by the intersection of the tangent plane at the Q surface with each of the ordinates at the three pure components. The three-dimensional plot of the above ternary is shown in Figure 23.

Prediction in Multicomponent Systems. The prediction of multicomponent vapor–liquid equilibria can be made by setting up the free energy function for the complete system from those of the constituent binaries. In the absence of any additional data, the assumption can be made that the chemical potential or free energy function for each component in the mixture is additive. Using two-parameter equations similar to equation 33 for each constituent binary, the multicomponent function would appear as follows:

$$Q = \sum_{i=1}^{n} \sum_{j=1}^{n} x_i x_j (A_{ij} x_j + A_{ji} x_i + B_{ij} x_j{}^2 + B_{ji} x_i{}^2) \tag{56}$$

where $i < j$.

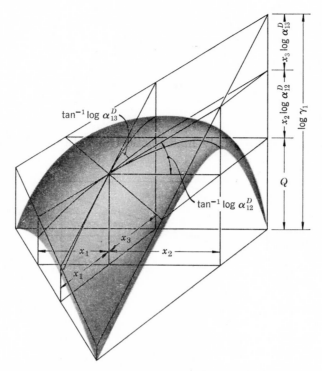

Fig. 23. Free energy function plot for a ternary system showing the surface Q as a function of x_1, x_2 and x_3 at constant P and T. The tangent plane corresponding to a specific composition (x_1, x_2 and x_3) is shown intersecting the ordinates erected at the pure components (9).

In equation 56 the constants for each binary are known, thus defining the free energy function for the multicomponent system of n components. The activity coefficient for any component in the system can then be calculated by solving equation 54 utilizing equation 55, and obtaining:

$$\log \gamma_k = \sum_{i=1}^{n} \left(A_{ki} x_i^2 + 2A_{ik} x_i x_k - 2 \sum_{j=1}^{n} A_{ij} x_i x_j^2 \right)$$
$$+ \sum_{i=1}^{n} \left(B_{ki} x_i^3 + 3B_{ik} x_i x_k^2 - 3 \sum_{j=1}^{n} B_{ij} x_i x_j^3 \right) \quad (57)$$

where: $i \neq k$, $i \neq j$ and $k = 1$ to n.

For a ternary system, the activity coefficient for component 1 would appear as follows in terms of independent parameters only (see equation 30):

$$\log \gamma_1 = A_{12} x_2^2 + A_{13} x_3^2 - 2A_{23} x_2 x_3 + 2(A_{12} + A_{13} + A_{23}) x_1 x_2 x_3 + B_{12} x_2^3$$
$$+ B_{13} x_3^3 - 3B_{23} x_2 x_3^2 + 3(2B_{12} + B_{13} - B_{23}) x_1 x_2^2 x_3 + 3(B_{12} + 2B_{13} + B_{23}) x_1 x_2 x_3^2 \quad (58)$$

The activity coefficients for components 2 and 3 can be obtained from equation 58 by a transformation of components according to the sequence $1 \to 2 \to 3 \to 1$.

In the event that multicomponent data are available, it is possible to use equations 56 and 57 for correlation purposes, in which the constants are either checked or calculated from the data. In some cases it may be desirable to increase the number of

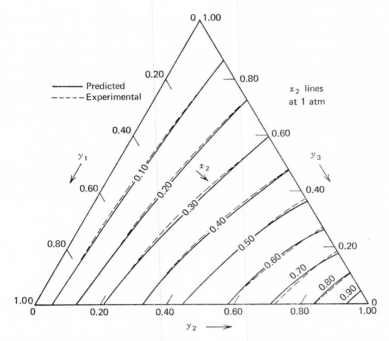

Fig. 24. Vapor–liquid equilibria for the ternary system acetone(1)–chloroform(2)–benzene(3) at 1 atm pressure. Only the values for chloroform liquid mole fractions are shown on the familiar triangular coordinate plot.

parameters for certain of the binaries and add additional multicomponent terms to the free energy function. In general, because of the scarcity of multicomponent data, prediction from the binary systems is the only alternative. Such a prediction for a ternary system is shown in Figure 24 and compared with the experimental data. This is a constant-pressure system and requires a method of successive approximation to calculate the data from the predicted activity coefficients. First a reasonable estimate is made of the temperature, then the partial pressures are calculated and summed as follows:

$$P = \sum_{i=1}^{n} \gamma_i p_i{}^{\circ} x_i \tag{59}$$

This total pressure is now compared with the pressure of the system and a closer approximation made to the temperature. The procedure is continued until the desirable degree of agreement is obtained. The method is easily adaptable to a computer program. Having obtained the correct partial pressures, the vapor composition for each component is calculated from:

$$y_i = \frac{\gamma_i p_i{}^{\circ} x_i}{P} \tag{60}$$

Naturally, in the case of a constant temperature system equations 59 and 60 give a direct solution after calculating the activity coefficients for each component from equation 57.

When conditions no longer permit the assumption of ideal gas behavior in the vapor phase, negligible liquid volume, and significant change in the free energy function with respect to pressure at constant temperature or with temperature at constant pressure, fugacities must replace the partial pressures, and volumetric changes in the liquid phase at constant temperature or enthalpic changes at constant pressure must be considered. Procedures for making these calculations will be found in the literature (21,27). The equation of equilibrium under the above conditions at constant temperature may be written as:

$$\ln f_i = \ln \gamma_i x_i f_i^{\circ} + \frac{v_i P}{RT} \tag{61}$$

where f_i = fugacity of component i in each phase = $\nu_i y_i P$; ν_i = fugacity coefficient and v_i = partial molar volume.
The vapor-phase compositions are then given by:

$$y_i = \frac{f_1}{P \nu_i} \tag{62}$$

where

$$P = \sum_{i=1}^{n} \frac{f_i}{\nu_i}$$

The above calculations lend themselves to computer operation usually employing iterative procedures, in which the final criterion is that $\Sigma\, y_i$ equal unity within a reasonably selected small tolerance.

Experimental Methods

The methods employed for the experimental determination of vapor–liquid equilibria remain essentially the same as those developed in the last 60 years. In the following discussion it is intended to cite examples of the latest forms of these methods without attempting to give a complete review of the literature. As will be noted in the description of these methods, many sophisticated forms of apparatus have been introduced and have greatly improved both the convenience and the accuracy of measurements under a wide range of difficult conditions.

Static Method. Because of the many practical difficulties associated with this simple method, very few applications have appeared in the literature. However, with the introduction of improved instruments, this method still finds useful application (28). The static method for the accurate determination of total pressure described in the above reference is illustrated in Figure 25. If the total pressure can be measured with sufficient accuracy, it is not necessary simultaneously to measure the vapor composition to obtain accurate vapor–liquid equilibrium data. However the apparatus shown in Figure 25 can easily be adapted to isolate an equilibrium vapor sample for analysis if desired, as shown in the dynamic steady state methods described below. The total pressure apparatus is essentially one in which the carefully isolated and degassed liquid mixture is allowed to reach equilibrium with its vapor at constant temperature. The approach to equilibrium is hastened by magnetic stirring. The pure components are added to the equilibrium cell from a degassing still for each component, by intermittently cooling the equilibrium cell with liquid nitrogen. The

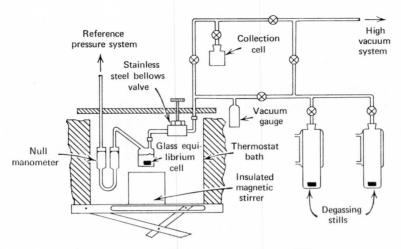

Fig. 25. Total pressure apparatus (28).

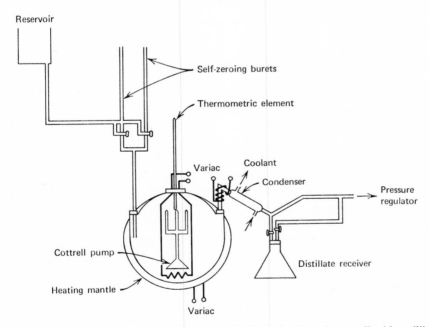

Fig. 26. Schematic diagram of apparatus for the rapid determination of vapor–liquid equilibria (29).

total pressure is accurately determined by a reference pressure system using a null manometer to equalize the reference system with that of the system. Pressures can be measured to within plus or minus 0.04 torr accuracy, and extended if necessary to 0.001 torr with more precise gages.

Dynamic Methods, Unsteady State. The objections to the static method because of its difficulty in achieving equilibrium without contamination can be overcome with a dynamic method. The oldest method is due to Brown which dates back to 1879 and employs a *continuous distillation at constant pressure*. By avoiding superheating, entrainment and rectification, the instantaneous equilibrium conditions are determined

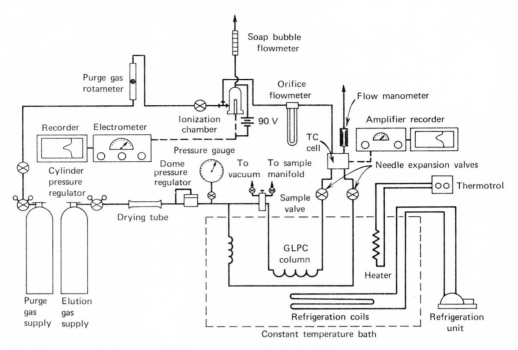

Fig. 27. Schematic diagram of high pressure gas–liquid partition chromatography apparatus (30).

by differentiation and material balancing of the data for the condensate composition. The apparatus shown in Figure 26 has the advantage in that vapor–liquid equilibrium data can be obtained in a relatively short time (29). To improve accuracy, the boiling point data are first determined by a dynamic steady state method as described below, in which the boiling vapors are returned to the pot by a reflux condenser.

Almost as old is the method Linebarger suggested back in 1895 which involves a continuous evaporation or transpiration from the equilibrium mixture. A modern rendition of this method is shown in Figure 27 employing a highly complicated gas–liquid partition chromatography (GLPC) apparatus (30,31). In effect the GLPC column resembles a rectifying distillation column with many theoretical plates sufficient to separate the components completely and determine the amounts of each quantitatively. The equilibrium data may be derived by either the use of theoretical plate theory or by rate theory. Using radioactively tagged molecules it is possible to demonstrate the congruence of the above two theories in regard to the final equilibrium data. Linebarger used an inert gas as the transpiration agent which is justified at ordinary pressures. However, at the high pressures used in the GLPC apparatus an inert gas would affect the equilibrium. Thus, the elution gas used in the GLPC apparatus is one or more of the components of the system. An inert gas is used for purging the system and does not interfere with the equilibrium. Compositions are measured by means of a differential thermal conductivity cell. The tagged radioactive molecules are measured by an electrometer and ionization chamber.

Dynamic Methods, Steady State. The advantages of the steady-state dynamic method of vapor–liquid equilibrium measurements have made this by far the most popular. By maintaining a constant recirculation of the equilibrium phases at steady

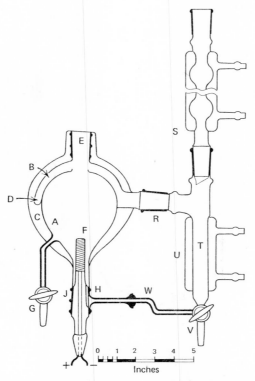

Fig. 28. The Othmer still with an integral jacketing arrangement to minimize heat losses from the pot: A, still body; B, vapor jacket; C, vacuum jacket; D, drain trough; E, thermometer part; F, glass immersion heater; G, liquid sample stopcock; H, groove in heater joint; J, heater joint; R, vapor line joint; S, condenser; T, condensate receiver; U, condensate cooler; V, condensate sample stopcock; W, capillary return line (32).

state conditions of pressure, temperature and composition accurate and convenient measurements are possible. The use of a hold-up trap to insure sufficient quantities of the equilibrium vapor-phase in liquid form for analysis dates back to Yamaguchi early in this century, in 1913. His rather cumbersome apparatus that required manual adjustment of pressure to achieve uniform boiling at constant temperature was improved by the introduction of a remarkably simplified compact glass apparatus by Othmer in 1928. The original Othmer still which operates at isobaric conditions continues to be the standard laboratory apparatus to this day. In spite of the many improvements he and his associates have made over the original model, the latest modification shown in Figure 28 remains the same simple compact isobaric equilibrium still with a few modern embellishments to improve adiabatic conditions (without the use of auxiliary heat) and ease of assembly of the basic components (32). Equilibrium between the liquid phase in the still body and the condensed vapor phase in the trap is reached when the thermometer maintains a constant reading. At these conditions a steady state of boiling is reached in which the boiling vapors from the still pot are equal in composition and quantity to the condensate returning through the capillary. Active boiling of the liquid in the pot is sufficient to achieve uniform composition without resorting to mechanical agitation. As previously mentioned, Othmer (3)

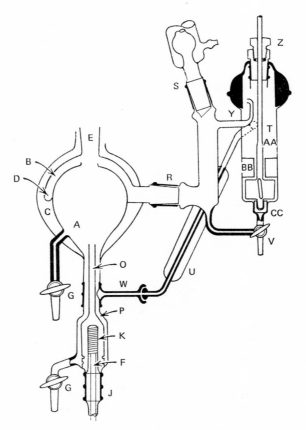

Fig. 29. Othmer still for partially miscible systems. In addition to parts listed in Figure 28: K, boiling chamber; O, uptake tube; P, annular recycle to boiler; Y, vent to receiver; Z, PTFE gland for stirrer; AA, stirrer; BB, baffles; CC, PTFE-bearing stirrer; (32).

clearly showed that Langmuir's equation of mass rate of evaporation leads to the conclusion that the equilibrium between boiling liquid, and vapor in the pot is maintained under the highest rates of evaporation encountered in engineering practice. The limiting conditions in the Othmer still are established by the extent of superheating and entrainment created by the rate of boiling. Because of the nearly perfect adiabatic jacketing, active boiling far below the point of significant superheating or entrainment is possible with a minimum heat input.

Several modifications of the above Othmer still are included in the reference cited; however, the most significant is that shown in Figure 29, which permits data to be obtained on partially miscible systems in either the liquid or condensed-vapor phase or both. Vigorous mechanical agitation of the condensed-vapor phase in the trap insures that the equilibrium composition is returned to the still pot. The thermosyphon is sufficient for adequately mixing the two liquid phases in the pot.

In the attempt to eliminate the inhomogeneity in composition caused by the return of the condensate to the pot of the Othmer still, Gillespie added a Cottrell pump and disengagement chamber in 1946. However, several modifications since then have appeared (33,34,35) in order to better achieve the original purposes of

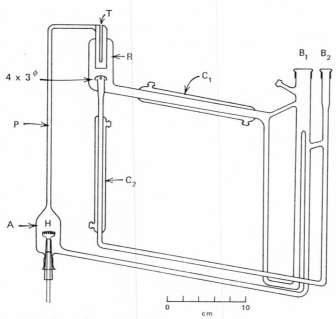

Fig. 30. Modified Gillespie still: A, boiler; B_1, liquid sampling opening; B_2, vapor condensate sampling opening; C_1, condenser; C_2, cooler; H, internal heater; P, Cottrell pump; R, disengagement chamber; T, thermometer well (35).

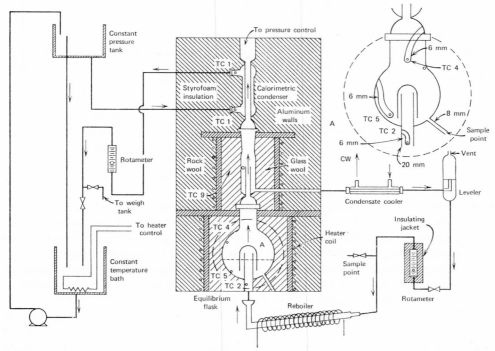

Fig. 31. Equilibrium still with flash reboiler and auxiliary equipment for enthalpy measurement. All parts in contact with the system are fabricated of glass and PTFE to reduce contamination (36).

Gillespie. One of the more recent modifications is shown in Figure 30 (35). The boiling mixture enters the disengagement chamber from the Cottrell pump where the equilibrium vapor and liquid phases are separated. The vapor phase is condensed and collected in an overflow trap. The corresponding liquid is cooled and collected in a second overflow trap. Both traps overflow into the same return to the boiler where the cycle of recirculation is repeated by the Cottrell pump. Although this arrangement reduces the sensitivity of the apparatus to the rate of boiling and is capable of obtaining fairly consistent data, previous modifications utilize more complex devices to improve attainment of true equilibrium. One employs a packed separation column to smooth out the approach to true equilibrium and a magnetically stirred

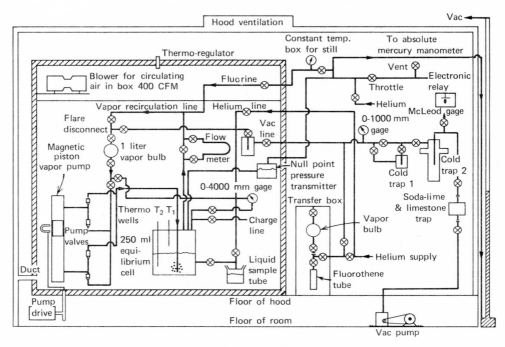

Fig. 32. Layout of isothermal recirculation vapor–liquid equilibrium still using a magnetic vapor pump (39).

chamber to thoroughly mix the liquid returning to the boiler (34). The other employs a magnetic stirrer directly in the boiler (33). Even with the latest improvements, it is still debatable whether the complications introduced by Gillespie's modification of the Othmer still is beneficial or detrimental to the attainment of true equilibrium.

Another attempt to remove the above mentioned inhomogeneity was proposed by Jones and co-workers in 1943; the idea being to vaporize the returning condensate so that it in effect has the equilibrium phase conditions as it enters the boiler. In this way the equilibrium vapor merely bubbles through the liquid in the pot on its way to the condenser. Unfortunately, to make effective use of this method, the control of heat to the flash boiler must be quite exact and as a consequence the apparatus is not as simple as originally conceived. For example in Figure 31 an intricate application is made of the flash reboiler (36). So precise is the means of heat control that the apparatus is also used for calorimetric purposes to measure simultaneously the latent heat of vaporization in conjunction with the vapor–liquid equilibria. The flash reboiler principle has recently been used in a high-pressure recirculating still (37). Built entirely of stainless steel with PTFE seats in all the valves it resembles an earlier high-pressure Othmer still (38). The operating conditions can be varied from −50 to +500°F and from atmospheric to 1,200 psia. It was found more convenient to operate at constant temperature by controlling the heat to the head tank, and easily maintain constant circulation with a constant heat input to the still vaporizer. A sophisticated

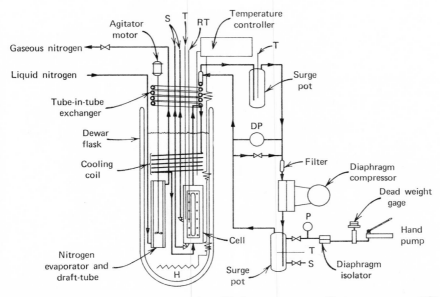

Fig. 33. Low-temperature vapor–liquid equilibrium apparatus: P, pressure gage; DP, differential pressure gage; T, thermocouple recorder; RT, resistance thermometer; S, sample connection; H, electric heater (37).

differential temperature controller insures adiabatic conditions in the still body which is surrounded by vapors from a secondary boiler. The heat to this boiler is controlled so as to maintain the same temperature inside and outside the still.

Up to now circulation of the fluids in the dynamic steady state stills has been accomplished by heating to the boiling point. Mechanical methods have been used for agitation only. In the more recent highly complicated apparatus, circulation is achieved by pumping the vapor phase, liquid phase or both. Using pumps with negligible pressure drops in comparison to the total pressure of the system, highly precise measurements can be made of vapor–liquid equilibria under extremely adverse conditions. A recirculation still employing a magnetic piston vapor pump is shown in Figure 32 (39). Made of special materials to withstand the highly corrosive action of halogen compounds, it is a static system converted to a steady state dynamic one by means of the recirculating vapor pump. A high vacuum system is used to remove any contamination. Pressure is measured with a null point diaphragm transmitter continuously balanced by a reference helium pressure system. To insure that the fluorides would not break down to simpler compounds, a small amount of elemental fluorine was added to the mixture.

A still for operation under cryogenic temperatures ($-300°$F) and high pressures (3,000 psia) utilizes a vapor diaphragm pump (Fig. 33) (37). A very similar apparatus capable of operation down to $-340°$F at pressures as high as 8,000 psia employs an electromagnetic recirculating vapor pump (40). The low temperatures are obtained with liquid nitrogen in a Dewar flask. Constant temperature is controlled with a regulated heat supply to the liquid-nitrogen bath.

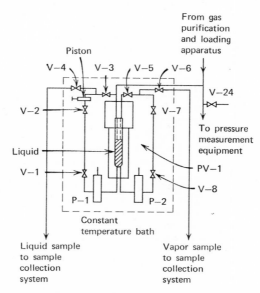

Fig. 34. Flow diagram of dynamic vapor–liquid equilibrium system employing rotary vane high-pressure pumps for circulating both phases. Pump P-1 circulates liquid from the bottom of the cell to the top and P-2 circulates vapor from the top to the bottom where it bubbles through the liquid (41).

A highly elaborate system employing recirculating pumps for both phases and designed for pressures as high as 18,000 psia is shown in Figure 34 (41). In spite of the advanced instrumentation and exotic materials employed, in principle, it is probably the simplest dynamic steady state apparatus consistent with obtaining the most precise data. Circulation of both phases greatly hastens the attainment of equilibrium and allows for the removal of samples without disturbing the equilibrium.

Nomenclature

A	=	constant in Clausius-Clapeyron equation
A_{12}, A_{21}	=	constants in Margules or van Laar equations
A_{ij}, A_{ji}	=	constants in equation of Q and γ_k for a multicomponent system
a_i	=	activity of component i
a_{12}	=	constant in relative volatility deviation equation
B	=	constant in Clausius-Clapeyron equation
B_{12}, B_{21}	=	constants in Margules or van Laar equations
B_{ij}, B_{ji}	=	constants in equations of Q and γ_k for a multicomponent system
b_{12}	=	constant in relative volatility deviation equation

C	= compressibility
f	= fugacity, also degrees of freedom
f_i	= fugacity of component i
G	= Gibbs free energy
H	= enthalpy
h_i	= partial molar enthalpy of component i
K	= equilibrium constant for chemical reaction
L	= latent heat of vaporization
N	= total number of moles
n	= number of components
P	= total pressure
p_i	= partial pressure of component i
Q	= dimensionless free energy function
R	= gas constant
r	= number of phases
S	= entropy
T	= temperature
V	= volume
v_i	= partial molar volume of component i
W	= statistical weighting factor
x_i	= mole fraction of component i in liquid phase
y_i	= mole fraction of component i in vapor phase
z	= true mole fraction of molecular species (with subscript A, B or AB)

Greek letters

α_{ij}	= relative volatility of component i with respect to component j
β	= van Laar parameter of physical interaction
γ_i	= activity coefficient of component i
ν_i	= fugacity-pressure ratio of component i

Subscripts

A	= the molecular species A
AB	= the complex species AB
B	= the molecular species B
i	= component i (1 to n)
j	= component j (1 to n)
k	= component k (1 to n)
L	= the liquid
T	= at constant temperature
V	= the vapor
x	= at constant liquid composition or taken at the composition of the equilibrium liquid
y	= at constant vapor composition or taken at the composition of the equilibrium vapor

Superscripts

A	= the azeotrope
D	= difference function
I	= ideal function
$'$	= liquid phase rich in component 1 in a binary system
$''$	= liquid phase rich in component 2 in a binary system
$\circ$	= reference or standard state

Operators

d	= differential of
Δ	= increment of
$F(\ )$	= function of variables included in the parentheses
Σ	= summation of
$\int$	= integral of
∂	= partial derivative of

Bibliography

"Vapor-Liquid Equilibria" in *ECT* 1st ed., Vol. 14, pp. 613–645, by D. F. Othmer, Polytechnic Institute of Brooklyn and R. Gilmont, The Emil Greiner Co.

1. J. W. Gibbs, *Collected Works*, Yale Univ. Press, New Haven, 1948, Vol. 1; a. p. 88, b. p. 96.
2. V. J. Lee, *J. Chem. Educ.* **44**, 164 (1967).
3. D. F. Othmer, *Ind. Eng. Chem.* **21**, 576 (1929).
4. R. Hulme, *J. Chem. Educ.* **34**, 459 (1957).
5. K. K. Shah and G. Thodos, *Ind. Eng. Chem.* **57** (3), 30 (1965).
6. J. J. Martin, *Ind. Eng. Chem.* **59** (12), 34 (1967).
7. D. S. Viswanath and G. J. Su, *A. I. Ch. E. J.* **11**, 202 (1965).
8. T. W. Leland, Jr. and P. S. Chappelear, *Ind. Eng. Chem.* **60** (7), 15 (1968).
9. R. Gilmont, *Thermodynamic Principles for Chemical Engineers*, Prentice-Hall, Inc., Englewood Cliffs, N. J., 1959; a. p. 230; b. 236.
10. D. G. Miller, *Ind. Eng. Chem.* **56** (3), 46 (1964).
11. D. F. Othmer and H. N. Huang, *Ind. Eng. Chem.* **57** (10), 42 (1965).
12. D. F. Othmer and E. S. Yu, *Ind. Eng. Chem.* **60** (1), 22; (4), 39 (1968).
13. D. F. Othmer and H. T. Chen, *A. I. Ch. E. J.* **12**, 488 (1966).
14. J. S. Busch and L. N. Canjar, *A. I. Ch. E. J.* **7**, 343 (1961).
15. L. C. Tao, *Ind. Eng. Chem. Fundamentals* **1**, 119 (1962).
16. L. C. Tao, *Ind. Eng. Chem.* **56** (2), 36 (1964).
17. F. D. Stevenson and V. E. Sater, *A. I. Ch. E. J.* **12**, 586 (1966).
18. D. Jaques, *J. Chem. Educ.* **42**, 651 (1965).
19. H. Renon and J. M. Prausnitz, *A. I. Ch. E. J.* **14**, 135 (1968).
20. I. A. Wiehe and E. B. Bagley, *Ind. Eng. Chem. Fundamentals* **6**, 209 (1967).
21. P. L. Chueh and J. M. Prausnitz, *Ind. Eng. Chem.* **60** (3), 34 (1968).
22. R. Gilmont, D. Zudkevitch and D. F. Othmer, *Ind. Eng. Chem.* **53**, 223 (1961).
23. H. L. Chang, P. S. Chappelear and R. Kobayashi, *A. I. Ch. E. J.* **14**, 318 (1968).
24. J. W. Leach, P. S. Chappelear and T. W. Leland, *A. I. Ch. E. J.* **14**, 568 (1968).
25. H. G. Harris and J. M. Prausnitz, *Ind. Eng. Chem. Fundamentals* **8**, 180 (1969).
26. L. C. Tao, *A. I. Ch. E. J.* **15**, 362, 460 (1969).
27. R. V. Orye and J. M. Prausnitz, *Ind. Eng. Chem.* **57** (5), 18 (1965).
28. H. G. Harris and J. M. Prausnitz, *A. I. Ch. E. J.* **14**, 737 (1968).
29. R. S. Ramalho, F. M. Tiller, W. J. James and D. W. Bunch, *Ind. Eng. Chem.* **53**, 895 (1961).
30. K. T. Koonce and R. Kobayashi, *J. Chem. Eng. Data* **9**, 494 (1964).
31. K. T. Koonce, H. A. Deans and R. Kobayashi, *A. I. Ch. E. J.* **11**, 259 (1965).
32. D. F. Othmer, R. Gilmont and J. J. Conti, *Ind. Eng. Chem.* **52**, 625 (1960).
33. A. Rose and W. R. Supina, *J. Chem. Eng. Data* **6**, 173 (1961).
34. S. Yerazunis, J. D. Plowright and F. M. Smola, *A. I. Ch. E. J.* **10**, 660 (1964).
35. I. Nagata, *J. Chem. Eng. Data* **10**, 106 (1965).
36. C. H. Bloom, C. W. Clump and A. H. Koeckert, *Ind. Eng. Chem.* **53**, 829 (1961).
37. H. Hipkin, *A. I. Ch. E. J.* **12**, 484 (1966).
38. D. F. Othmer, S. J. Silvis and A. Spiel, *Ind. Eng. Chem.* **44**, 1864 (1952).
39. R. C. Liimatainen and B. S. Swanson, *A. I. Ch. E. J.* **10**, 860 (1964).
40. F. P. Stein, C. J. Sterner and J. M. Geist, *Chem. Eng. Progr.* **58** (11), 70 (1962).
41. N. K. Muirbrook and J. M. Prausnitz, *A. I. Ch. E. J.* **11**, 1092 (1965).

ROGER GILMONT
Roger Gilmont Instruments, Inc.
and Polytechnic Institute of Brooklyn

VARAMIDE, VARCONIC, VARIQUOT, VARISOFT. See Surfactants.

VARNISHES. See Coatings, industrial; Paint.

VAT DYES

Vat dyes are colored, water-insoluble organic compounds that possess the property of forming water-soluble alkali metal salts in an alkaline reducing bath, and whose solutions have affinity for cellulosic fibers. The chief classes of vat dyes are anthraquinone dyes (Vol. 2, p. 517), and indigoid dyes (qv); the sulfur dyes (qv) are also sometimes included. See also Vol. 7, pp. 554, 575. For pigments from vat dyes see Vol. 15, p. 583.

VERIPON. See Surfactants.

VERMICULITE. See Insulation, thermal, Vol. 11, p. 832; Micas, Vol. 13, p. 399.

VERMILION. See Pigments (Inorganic), Vol. 15, p. 525.

VETERINARY DRUGS

Animals make an important contribution to the health, comfort, and pleasure of mankind in the world. Livestock for food or labor, consisting of cattle, sheep, pigs, goats, water buffalo, horses, donkeys, mules, and camels, number over three billion, and there is an equal or greater number of poultry. House pets may number 50 million throughout the world. The health of all these animals is of great importance to their owners, and to others who depend on their availability. How they prosper depends on many factors. Some, such as environmental conditions, are outside the purview of the veterinarian. Others, such as disease, breeding, nutrition, and growth, are susceptible to his control. Infectious diseases are limiting factors in the efficient production of livestock, and of great concern to owners of animal pets. Many chemical and biological tools are at the disposal of the veterinarian to control the causative agents of such diseases and alleviate their symptoms. They may be included in the diet on a continuous basis to prevent specific diseases; they may be used orally or by injection to treat diseases. Some are used to improve the rate of gain, milk or egg production, or feed efficiency. Hormones may be used to regulate reproduction in the interest of more efficient meat production. Other agents serve as sedatives, tranquilizers, or analgesics.

Some veterinary medicines are naturally occurring substances which are extracted from plants or animal tissues. Synthetic processes have resulted in the replacement of naturally occurring drugs in some cases where they can be produced more cheaply and with more effective control of purity and potency. Antibiotics are examples of natural substances which, as in the case of penicillins, have been made more useful by the development of chemical—as opposed to biological—production methods.

Biologicals (bacterins and vaccines) are not medicines by a strict definition of the term. They are produced by growth and modification of disease-producing organisms so that they cannot cause disease but, when injected or otherwise applied, stimulate an animal to produce antibodies which protect it from a specific disease. Some viral diseases of animals for which vaccines are available are listed in Table 1, and bacterial diseases treated with vaccines are shown in Table 2.

All purposes for which veterinary medicines are used may involve application of the material in a number of ways. Medication or vaccination may be by injections given intramuscularly, intradermally, subcutaneously, intraperitoneally, or topically.

Table 1.   Some Viral Animal Diseases for Which Vaccines are Available

Disease	Animal	Disease	Animal
avian encephalomyelitis	poultry	fowl pox	poultry
bluetongue	sheep	hog cholera	swine
bovine virus diarrhea	cattle	infectious bovine	
canine distemper	dog	rhinotracheitis (IBR)	cattle
contagious ecthyma	sheep	infectious bronchitis	poultry
encephalitis,		infectious canine	
Eastern and Western	horses	hepatitis	dogs
equinine influenza	horses	infectious	
equinine		laryngotracheitis	poultry
rhinopneumonitis	horses	Newcastle disease	poultry
feline distemper	cats	pigeon pox	poultry
foot and mouth		rabies	cattle, cats, dogs
disease	cattle	warts	cattle, dogs

Table 2.   Some Bacterial Diseases for Which Vaccines are Available

Disease	Bacterial organism	Animals affected	Vaccine material
anthrax	*Bacillus anthracis*	cattle, sheep, horses, mules	attenuated spore
blackleg	*Clostridium chauvoei*	cattle, goats	anaculture
botulism	*Clostridium botulinum*	cattle, sheep, goats	toxoid
brucellosis			
(Bang's disease)[a]	*Brucella abortus*	cattle	lyophilized vaccine
enterotoxemia	*Clostridium perfringens*	cattle, sheep	anaculture or antitoxin
erysipelas		poultry (mainly turkeys),	
	Erysipelothrix insidiosa	sheep, swine, cattle	bacterin
fowl cholera	*Pasteurella multocida*	chickens, turkeys	bacterin
hemorrhagic	*Pasteurella multocida* plus		
septicemia	*Parainfluenza viruses*	cattle, sheep	bacterin
hemoglobinuria	*Clostridium haemolyticum*	cattle, sheep, goats	anaculture
leptospirosis	*Leptospira pomona*	cattle, swine	chick embryo vaccine
leptospirosis	*Leptospira canicola* and		
	L. icterohemorrhagiae	dogs	bacterin
paratuberculosis	*Mycobacterium*		
(Johne's disease)	*paratuberculosis*	sheep, cattle	bacterin
salmonellosis	*Salmonella abortivoequina*	cattle, sheep, goats	bacterin
staphylococcal			bacterin; bacteriophage
infections	*Staphylococci*	cattle	lysate
strangles	*Streptococcus equi*	horses	bacterin
tetanus	*Clostridium tetani*	horses, sheep	toxoid

[a] Essentially eradicated in the United States.

Oral administration can be by the use of tablets or pills, or by inclusion of the material in the feed or drinking water of animals.   In addition to these routes of administration, certain poultry vaccines are given intranasally or intraocularly, while others may be brushed into the vent or feather follicles or applied to the wing web with a needle.

Animal Diseases.   Many of the degenerative disorders and nutritional diseases found in animals are closely related to their counterparts found in humans, and respond to the same type of therapy.   This is much less true for the infectious diseases which, being caused by great varieties of bacteria, viruses, protozoa, helminths, and other organisms, often manifest themselves in unique ways.   Many of these are still referred to by traditional, often colorful, names which describe the symptoms rather than the cause. Some examples are foot and mouth disease of cattle, blackhead of poultry, strangles of horses, and lumpy jaw of cattle.

One example worth special mention is bovine mastitis, a disorder of commercial importance to the dairy industry. Dairy farmers lose millions of dollars annually as a result of these infections of the mammary gland due primarily to species of streptococci and staphylococci, and sometimes to *Pseudomonas, Aerobacter*, yeasts, molds, and other organisms. The result may be reduced or complete loss of milk production and reduced productive life of cows in herds because of permanent damage to udders. Although an occasional infection becomes systemic requiring the use of drugs by intramuscular or intravenous injection, most infections are localized in the udder and are treated by infusions with antibacterials via the teat canal. Because inflammation of the udder is often associated with infections, anti-inflammatory agents, such as hydrocortisone, are frequently incorporated into treatment formulations not only to reduce inflammation per se, but also to facilitate dispersion of the antibacterials to all sites of infection. The selection of vehicles for the drugs is as important as the selection of the drugs themselves. Vehicles must be nonirritating to udder tissues, they must be able to carry the active ingredients either in solution or suspension so that they are uniformly distributed and they themselves must mix readily with milk to allow for extensive distribution in the gland. They must hold the drugs in the udder long enough to have the desired effect, but they and the drugs must also be excreted within 96 hr after administration, the maximum time permitted by the Food and Drug Administration, which recognizes that farmers are not likely to delay marketing milk from treated cows longer than that time.

The drugs used in veterinary medicine are in many cases the same as are used for humans. They are described and structural formulas given, in a number of articles in other volumes to which cross reference is made. A few structural formulas are given here for important drugs which are used exclusively, or mainly, in veterinary medicine.

Antibacterials

The agents used against bacterial infections are mostly antibiotics. See Antibiotics; Chloramphenicol; Macrolide antibiotics; Penicillin; Polypeptide antibiotics; Polyene antibiotics; Streptomycin and related antibiotics; Tetracycline antibiotics.

In addition to their use as actual curative agents, antibiotics are often added continuously to the feed of animals. This, by preventing the onset of disease, enables them to grow and produce meat, milk, and eggs at optimum efficiency. Although there has been concern that this practice might lead to the development of drug-resistant strains with adverse effects in animals or man, such has not occurred. In fact it has been shown that resistant organisms thus formed are likely to be less virulent than the unmodified strain. (H. Jarolmen and G. Kemp, "Association of Increased Recipient Ability for R Factors and Reduced Virulence Among Variants of *Salmonella choleraesuis* var. *kunzendorf*," *J. Bacteriol.* **97** (2), 962–963 (1969).) The antibiotics used most commonly in this manner are penicillin, streptomycin, chlortetracycline, oxytetracycline, bacitracin, tylosin, and oleandomycin.

The penicillins (for example, procaine penicillin G) are effective against Gram-positive bacteria. They are used alone or in combination with other drugs for the control of streptococcal bovine mastitis (by intramammary infusion), infectious cystitis and pyelonephritis of cattle, calf pneumonia, and other bacterial infections.

They are also used for treating localized infections of the skin and eyes, topically and systemically. As noted above, penicillin is also fed to animals to promote growth. Feeding of penicillin can also be useful in reducing the possibility of bloat in ruminants. Some of the semisynthetic penicillins (eg ampicillin) are effective against Gram-negative bacteria.

Streptomycin is active primarily against Gram-negative organisms. It is used in the treatment of bovine mastitis (*E. coli, A. aerogenes*), actinomycosis of cattle, calf pneumonia, sterility due to vibriosis, and, by infusion in combination with penicillin, metritis. It has been used for cystitis in horses, infectious nephritis and infectious dysentery in dogs, leptospirosis in swine and dogs, infectious sinusitis in turkeys, and chronic respiratory disease of chickens. It can be given orally, or parenterally; doses must be frequently repeated. Since it is not absorbed from the intestinal tract, the oral route is used only for intestinal infections. Although it is active in promoting growth when fed to swine and chickens, it is not widely used in this way.

Dihydrostreptomycin, which is derived from streptomycin by reduction of an aldehyde to an alcohol, has similar uses. Both of these compounds can, in certain circumstances, produce deafness.

Neomycin (see Vol. 19, p. 41) is effective against Gram-positive and Gram-negative bacteria, especially *Pseudomonas* and *Proteus*. It is used orally only for the treatment of intestinal bacterial infections because it is not absorbed, and topically for treating infections of ear, eye, and skin, in cattle, lambs, and dogs, often in combination with other drugs, such as penicillin and cortisone.

Chlortetracycline is active against Gram-positive and Gram-negative bacteria, as well as rickettsiae. Among the diseases against which it is effective are anaplasmosis, metritis, foot rot, shipping fever, and calf scours, strangles in horses, enteric infections in swine, and synovitis in poultry and swine. The continuous feeding of low levels of chlortetracycline is used to promote growth in poultry, swine, and cattle. Recently a combination of chlortetracycline, sulfamethazine, and penicillin has proven to be especially useful in raising pigs. Chlortetracycline is readily absorbed on oral administration; it is also administered by injection and intramammary infusion.

Oxytetracycline, is used to treat mastitis (by intramammary infusion), metritis, foot rot, shipping fever, calf scours in cattle, anthrax and strangles in horses, diarrhea and enterotoxemia in sheep, necrotic enteritis in swine, and infectious synovitis of poultry. It is also employed as a growth stimulant for pigs and poultry.

Tetracycline, like chlortetracycline and oxytetracycline, possesses broad-spectrum antibiotic properties and has many applications in the treatment of endometritis, pneumonia, and strangles in horses, foot rot and calf scours and various respiratory diseases in cattle, as well as in dogs and cats.

Bacitracin (see Vol. 16, p. 315) is effective primarily against Gram-positive organisms. It is not well absorbed via oral administration but is effective against diarrhea in dogs and swine dysentery. It has also been used for bovine mastitis (*Streptococcus agalactiae*), but its effectiveness is lower than some other antibiotics. It is also used topically on infected wounds, otitis externa, and conjunctivitis, and has been used for infectious keratitis in cattle. It possesses activity as a growth factor for swine and poultry.

Tylosin (see Vol. 12, p. 647) is effective principally against Gram-positive bacteria but certain Gram-negatives, spirochetes, mycoplasmas, and vibrios are also susceptible. It has been used in intramammary infusions against bovine mastitis (staphylococcal and streptococcal), and for vibrionic dysentery in swine, and infectious sinuitis due to *Mycoplasma gallisepticum* in turkeys. It has also been used to improve growth and feed efficiency in swine and poultry.

Oleandomycin (see Vol. 12, p. 643) is used as a mastitis treatment, and in feed to improve growth of swine and poultry.

Antibiotics which are used to treat infections, once clinical disease has appeared, include the above plus chloramphenicol, novobiocin, erythromycin, neomycin, and polymyxin B.

Chloramphenicol is a broad spectrum antibiotic active against Gram-positive and Gram-negative bacteria, and rickettsiae. It has been used effectively against pneumonia, pasteurellosis, enteric conditions, infectious cystitis, kidney infections, and typhoid in dogs, cats, and horses in the United States.

Novobiocin is an antibiotic obtained from *Streptomyces spheroides* (C. H. Shunk et al., *J. Am. Chem. Soc.* **78**, 1770 (1956)). It is used primarily against *Staphylococcus aureus* and other Gram-positive organisms especially when the disease organism is resistant to other antibiotics. The major use is in turkey and mink feeds.

Erythromycin (see Vol. 12, p. 641) is active mainly against Gram-positive bacteria, and has been employed in dogs, either as the base or a salt, to treat secondary bacterial infections associated with distemper. It has also been used for treatment of pneumonia, pharyngitis, and urinary tract infections of horses, cattle, swine, and sheep, and mycoplasma infections of poultry.

Polymyxin B Sulfate (see Vol. 16, p. 319) is useful against Gram-negative bacteria, especially *Pseudomonas aeruginosa*. Although originally recommended only for topical use, it has been used parenterally in combination with neomycin for shipping fever in cattle, strangles in horses, and tonsilitis in dogs.

Sulfonamides. The sulfa drugs (see Vol. 19, p. 261) possess a wide spectrum of antibacterial activity, being variously effective against coccidiosis, staphylococci,

Table 3. Veterinary Uses of Sulfa Compounds

Sulfa compound	Coccidiosis (poultry)	Intestinal infections (swine, cattle, pets)	Mastitis (cows)	Respiratory diseases (cows, horses, sheep, pets)	Systemic infections	Urinary infections
phthalylsulfacetamide		+				
phthalylsulfathiazole		+				
succinylsulfathiazole		+				
sulfabromomethazine			+	+	+	
sulfadiazine				+	+	+
sulfadimethoxine		+		+		
sulfaethoxypyridazine				+	+	
sulfaguanidine	+					
sulfamerazine	+		+	+	+	
sulfamethazine	+	+	+	+	+	
sulfanilamide[a]				+	+	
sulfapyridine[b]				+	+	
sulfaquinoxaline	+	+			+	
sulfathiazole	+			+	+	+
sulfisoxazole					+	+

[a] Also used in topical applications.

[b] Also used in treating vaginitis.

streptococci, *Pasteurella, Proteus,* and *E. coli,* as well as other organisms. The selection of a particular sulfa compound for a particular disease is usually determined by its rate of absorption and/or rate of excretion. The great variety of applications is shown in Table 3.

In addition some sulfa compounds, such as sulfamethazine, are used in animal feeds as part of a growth-promotant combination (see Chlortetracycline above).

Nitrofurans. The nitrofurans (qv) are effective against Gram-positive and Gram-negative bacterial. Activity against protozoa, fungi, and large viruses has been reported.

Furazolidone is useful in poultry diseases, such as fowl typhoid, pullorum disease, and synovitis, as well as for increasing feed efficiency; it is also useful in swine diseases, such as bacterial enteritis.

Nitrofurantoin controls urinary-tract pathogens including *Proteus* and *Aerobacter,* and epizootic tracheobronchitis (kennel cough) in dogs.

Nitrofurazone is used against bacterial infections as a topical agent, as well as by oral administration in drinking water, especially against enteritis such as infectious necrotic enteritis of swine. It is also used in controlling cecal and intestinal coccidiosis in poultry by administration in the feed.

2-Methoxymethyl-5-nitrofuran (nitrofurfuryl methyl ether), is useful against a number of dermatophytes, such as ringworm, when applied topically. It is not used on cats.

Other Antibacterial Agents. Methenamine (hexamethylenetetramine) (see also Formaldehyde), is used as an oral urinary antiseptic in horses, dogs, and cats.

Chlorquinaldol, 5,7-dichloro-2-methyl-8-quinolinol, is used as an antiseptic in wounds and eczema, often in combination in an ointment with hydrocortisone.

Sodium hydrogen *p*-aminobenzenearsonate (see Vol. 2, p. 729) is used against strongyle infestation and trypanosomiasis in horses, and against enteritis in swine. It was one of the first chemotherapeutic agents applied for the treatment of syphilis in humans.

Anticoccidials and Antihistomonal Agents

Coccidiosis is a major protozoal disease affecting primarily poultry. Nearly all of the birds around the world raised for meat receive an anticoccidial drug sometime during their lives. Anticoccidial drugs are used almost exclusively in poultry feeds, the major use being in the feeding of broilers over the entire life of the birds to prevent disease outbreaks. When there is a clinical infection therapeutic levels are used in feed or drinking water. As a result of these practices mortality losses due to coccidiosis have been virtually eliminated in the United States.

Antibiotics (chlortetracycline, oxytetracycline, chloramphenicol, etc) are used for treatment of coccidiosis, but most are effective against only one or two of the nine prevalent species causing coccidiosis. An antibiotic, monensin, having broad-spectrum anticoccidial activity was reported (A. Agtarap et al., *J. Am. Chem. Soc.* **89** (22) 5737–5739 (1965)); in early 1970 it had not been registered for commerical use in the United States.

Sulfa drugs, particularly sulfaquinoxaline, are also useful. They are frequently used in combinations with other drugs. The nitrofuran drug, nitrofurazone, is also used as an anticoccidial.

A number of highly effective, broad spectrum drugs or combinations of drugs have been developed. The first of these was nitrophenide (**1**), which was shown in 1949 to be

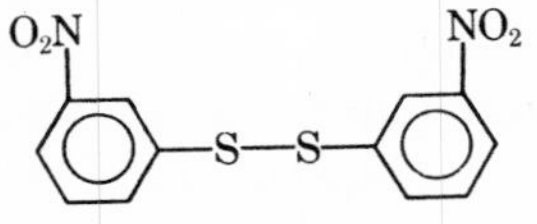

nitrophenide (**1**)

effective against cecal and intestinal coccidiosis in chickens. It has been largely replaced by newer drugs, such as amprolium (**2**), amprolium in combination with ethopabate (**3**), or combinations of roxarsone (**4**), nitromide (**5**), sulfanitran (**6**), and alkomide (**7**), shown in Scheme 1. Amprolium, is an effective broad-spectrum coccidiostat; in combination with ethopabate it is marketed as Amprol Plus (Merck & Co., Inc.).

Scheme 1.

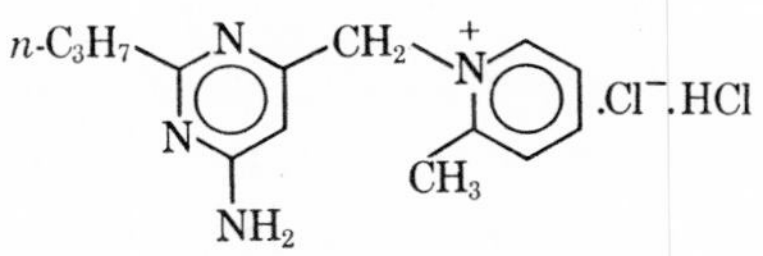

amprolium (**2**) 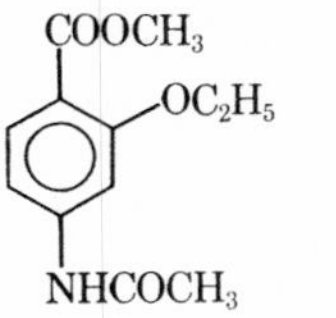   ethopabate (**3**) 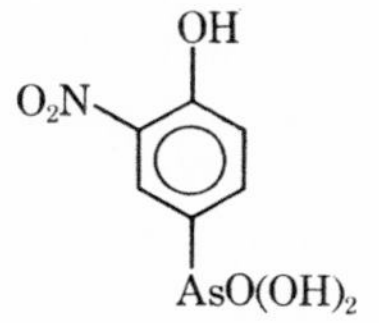   roxarsone (**4**)

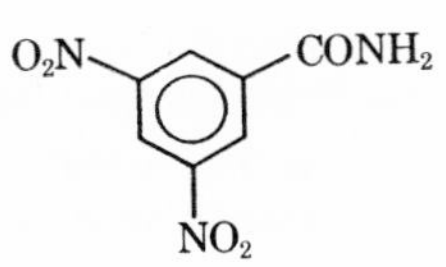

nitromide (**5**) 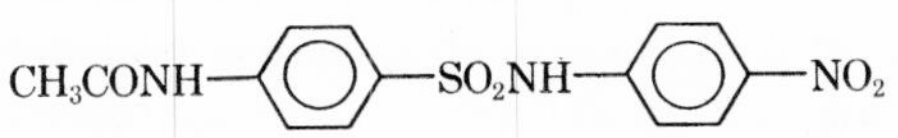   sulfanitran (**6**) 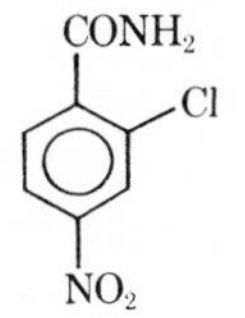   alkomide (**7**)

Unistat (Salsbury Laboratories) is a combination of antimicrobial agents which has proven very effective in the control of coccidiosis. It contains nitromide and sulfanitran. Unistat-3 (Salsbury Laboratories) contains roxarsone and the ingredients of Unistat. Novastat (Salsbury Laboratories) is a combination of alkomide and sulfanitran.

Zoalene (Dow Chemical Co.) (**8**) is a coccidiostat of proven utility. Clopidol (**9**) is a new compound which has been gaining acceptance in poultry markets around the world.

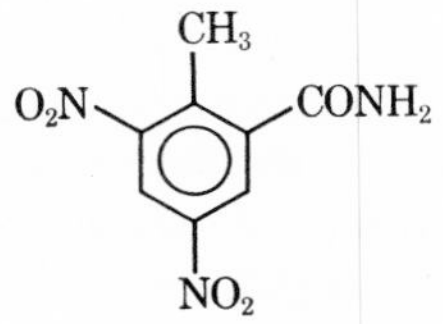

Zoalene (**8**) 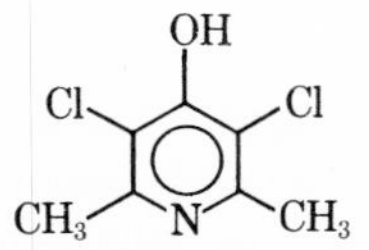   clopidol (**9**)

Nicarbazin, a complex of 4,4'-dinitrocarbanilide and 2-hydroxy-4,6-dimethylpyrimidine, is a highly effective coccidiostat for poultry; however, reports that it affects the color of egg shells and produces discolored yolks have limited its application to broilers rather than laying hens.

Four of the newer anticoccidial drugs introduced within the last few years are substituted quinolones. The structures of these compounds, buquinolate, methyl benzoquate, decoquinate, and cyproquinate, are shown in Scheme 2.

Scheme 2.

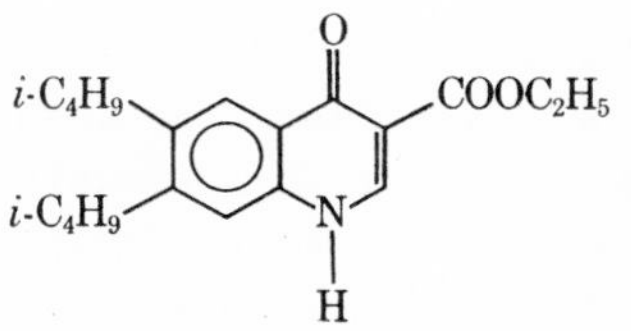

buquinolate (**10**)

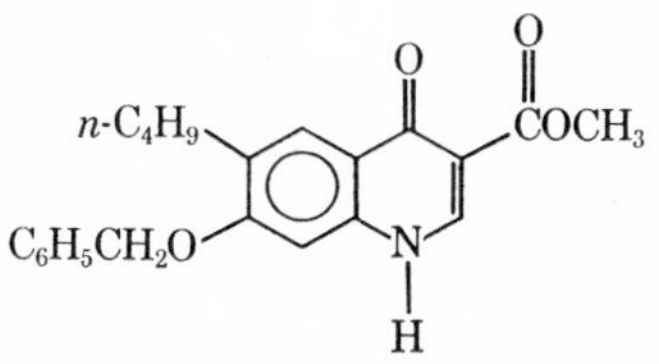

methyl benzoquate (**11**)

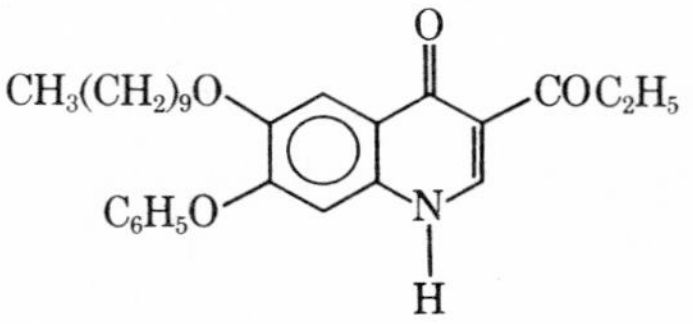

deccoquinate (**12**)

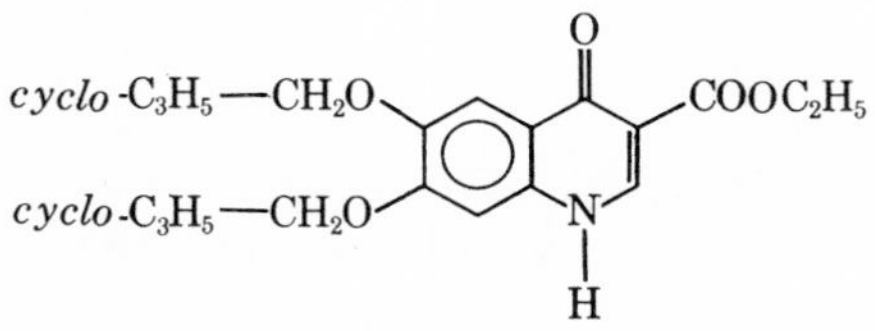

cyproquinate (**13**)

Coccidiosis is also a serious disease in cattle and sheep. Prevention has been most successful with sulfaquinoxaline (see Sulfonamides), sulfaguanidine, and nitrofurazone (see Nitrofurans). Treatment relies heavily on sulfa drugs.

Protozoa cause other diseases in all species of domestic animals and pets. In each species the parasite may attack specific systems or organs, such as the liver and ceca of turkeys (*Histomonas meleagridis*), or the parasites may produce lesions throughout the body, as in toxoplasmosis of dogs.

Nithiazide (**14**), dimetridazole (**15**), 4-nitrophenylarsonic acid, 2-amino-5-nitro-thiazole, and its *N*-acetyl derivative, and *p*-ureidobenzenearsonic acid are among the drugs used for the control of histomoniasis (the disease known as "blackhead" because of the head being cyanotic) in turkeys and chickens.

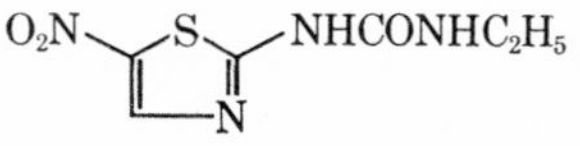

nithiazide (**14**)

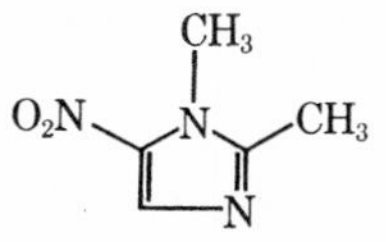

dimetridazole (**15**)

Fungicides

Fungal infections generally involve the skin, hair, and nails. Therapy involves long-term application of solutions, creams, or powders and thorough cleaning of the affected areas, since the dead tissue may prevent medication from reaching deep-seated infections. The recent development of oral products, such as griseofulvin, which carry medication to the infection site via the blood stream (ie act systemically) has materially aided the treatment of these conditions. Several of the available agents are polyene antibiotics (qv).

Amphotericin B is useful in histoplasmosis, coccidioidomycosis, and nocardiosis. Nystatin prevents or treats fungal infections in poultry and swine and is also used to treat mastitis and skin conditions caused by yeasts in cattle and other animals. Griseofulvin (16) is a systemic fungicide administered orally or topically. It is effective against dermatophytes which cause ringworm.

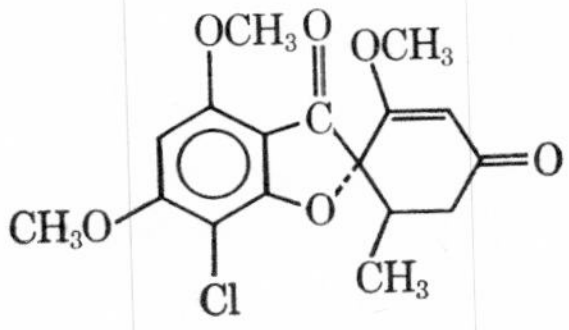

griseofulvin (16)

Undecylenic acid, $CH_3(CH_2)_9COOH$, usually in combination with zinc undecylenate, is applied topically for ringworm and other fungal infections of the skin.

Anthelmintics

Worms of various types and sizes can affect the intestinal tract, the respiratory tract and the circulatory system. Damage is caused by interference with proper nutrition or respiration, actual tissue injury (especially from migrating larvae), loss of blood, occlusion of major blood vessels or combinations of these effects. The important types of worms, drawn from the phyla of nemathelminthes (round worms) and platy-helminthes (flat worms), vary among geographical areas and among the species attacked. Cattle, sheep, poultry, pigs, horses, dogs, and cats all suffer from such infections.

Effective anthelmintics may be given orally or by various parenteral routes. The death of parasites in the circulatory system may result in blockage of vessels, and absorption of dead parasites elsewhere in the body may lead to anaphylactic reactions. These chemicals should therefore be administered with caution and only by people trained in their use.

There are many kinds of anthelmintics used in veterinary practice, including inorganic compounds, natural products, halogenated hydrocarbons, organophosphates, and phenol derivatives. See Parasitic infections, chemotherapy. Of special interest at the present time are six highly active broad, spectrum anthelmintics, shown in Scheme 3:

Scheme 3.

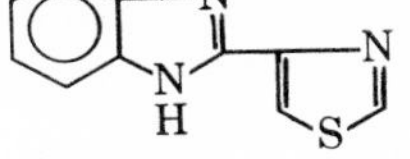

thiabendazole (17)

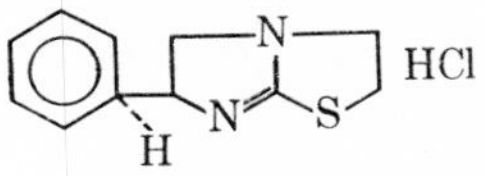

levamisole (18)

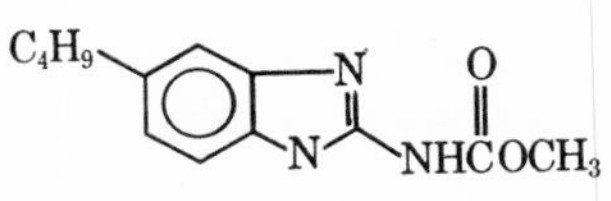

parbendazole (19)

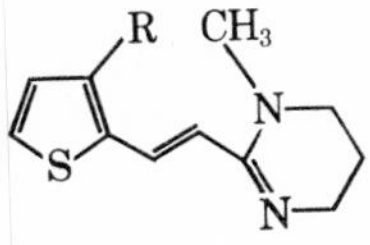

pyrantel (R = H) (20)

morantel (R = CH$_3$) (21)

thiabendazole (**17**), dl-tetramisole and its purified l-isomer-levamisole (**18**), parbendazole (**19**), pyrantel (**20**), and morantel (**21**). These are finding, or are expected to find, extensive application to cattle and sheep.

Insecticides and Acaricides

External parasites cause heavy agricultural losses by reducing weight gains and feed efficiency of food-producing animals, by damaging wool and hides, and by actually causing deaths. Pet animals may also suffer damage to skin and hair coat, and severe infestations can cause death.   The principal insect pests are fleas, screwworms, warble flies, nose bot flies, and face flies. The acarid pests are ticks and mites. In addition to their direct effects, insects and ticks carry diseases; eg ticks spread tick pyemia in lambs, cattle tick fever in cattle, and tularemia in man, rodents, and sheep.

Insecticides and acaricides are generally applied to animals as dusts, dips, liquid "pour-ons," or in backrubbers. Newer methods of treatment include orally administered tablets or liquids which act by means of their presence in the animal's blood circulation, and medicated collars which, by controlled release of the compound as a vapor, are effective over an extended period. Insect control is also achieved by treating the environment (barns, pens, etc) with suitable insecticides. Insecticides and acaricides are used extensively in veterinary medicine (see Insecticides).   The compounds in widest use belong to the following classes: arsenicals, sulfur compounds, carbamates, chlorinated hydrocarbons, pyrethrins, rotenone, nicotine, organophosphates, benzyl benzoate, cresol solutions, and carbon disulfide. Among the compounds most recently introduced for control of ectoparasites are the following organophosphates, shown in Scheme 4: chlorfenvinphos (**22**), Dursban (The Dow Chemical Co.) (**23**), famphur (**24**), and cythioate (**25**).

Scheme 4.

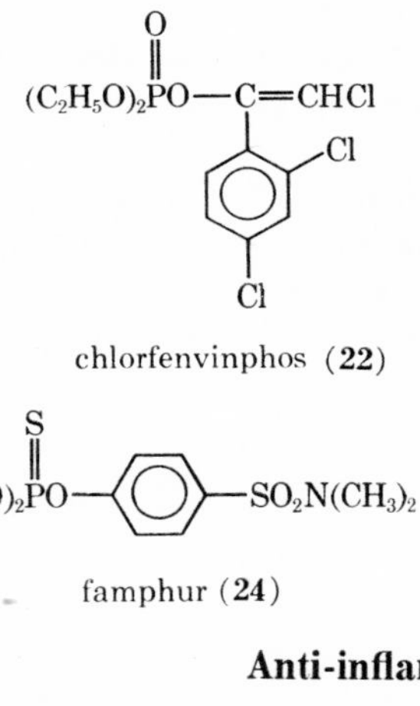

chlorfenvinphos (**22**)

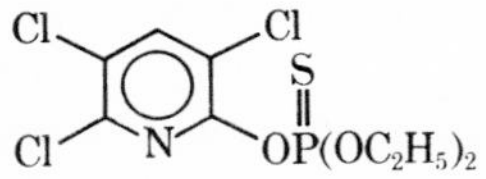

Dursban (**23**)

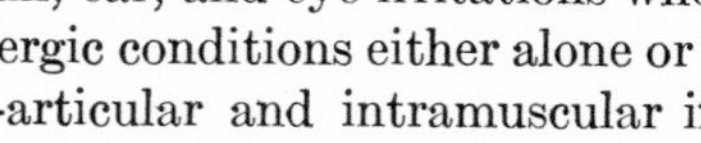

famphur (**24**)

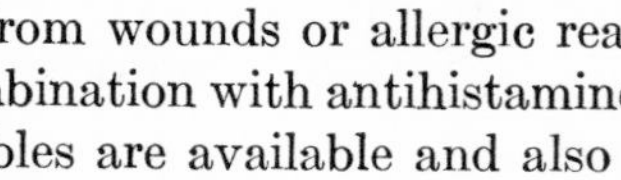

cythioate (**25**)

Anti-inflammatory Agents

These compounds have proven to be of value in arthritic disorders, in the treatment of skin, ear, and eye irritations whether from wounds or allergic reactions, and general allergic conditions either alone or in combination with antihistamines.

Intra-articular and intramuscular injectables are available and also tablets for either primary or supportive therapy. Topical ointments, creams, and solutions are used for eye, ear, and skin problems.

Hydrocortisone acetate, cortisone acetate, prednisolone *t*-butyl acetate, and tri-amcinolone acetonide are widely used members of the glucocorticoids (see Vol. 11, p. 77). Intramuscular injection in connection with an injection of glucose is used in treating ketosis of cattle, an impairment of carbohydrate metabolism often resulting from stress accompanying pregnancy and parturition. Similar treatment of ketosis in sheep has produced conflicting results. These agents are usually also used in combination with antibiotics in treating conjunctivitis and pink-eye (infectious oph-thalmia) in pets, cattle, sheep, and goats, and generally for treating inflammations of the skin and joints.

Agents for Control of Growth and Reproduction

Rate of growth, sexual maturity, and in fact all major bodily functions and processes are regulated by hormones secreted by glands within the body. In order for the body to function properly certain hormones must be produced in proper amounts since they in turn regulate the production of other hormones. Deficiencies or excesses must be treated with either injectable or oral hormones in order to restore the natural balance.

Certain sex hormones which also regulate growth are used to improve rates of gain and feed efficiency of domestic animals. Among these are progesterone, testoster-one, and estradiol. Synthetic hormones are often used in veterinary practice. The most important example is diethylstilbestrol which is similar to estradiol in action but much cheaper. When fed rations containing diethylstilbestrol cattle gain faster with a greater feed efficiency. Norethandrolone acts similarly. See also Hormones.

Artificial insemination has been widely used to improve dairy stock and thereby increase milk production. If estrus could be regulated at will in beef cattle, artificial insemination could become practical in improving meat production. Progesterone is used to delay estrus, testosterone to enhance masculinity or repress female charac-teristics, and estradiol is used to enhance female characteristics and induce estrus.

Diethylstilbestrol is widely used to promote increased weight gains in young cattle. In addition it has high estrogenic activity and receives some use on this account against certain disorders, such as anestrus and hypertrophy of the prostate. Norethandrolone is a 17-α-ethyl-19-nor analog of testosterone and a potent anabolic steroid and androgen. Stanozolol, 17-β-hydroxy-17-α-methylandrostano-[3,2-c]pyr-azole, is an anabolic steroid for dogs and cats.

Among the new compounds for estrus regulation in cattle, sheep, dogs, etc, are chlormadinone (**26**), medroxyprogesterone acetate (**27**), alphasone acetophenide (**28**), and melengestrol acetate (**29**), shown in Scheme 5.

Scheme 5.

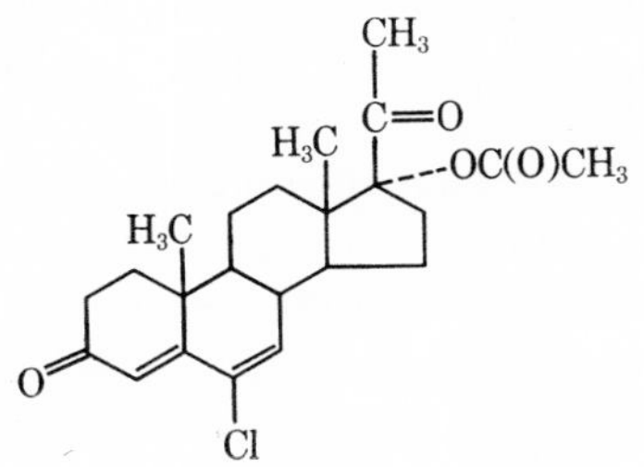

chlormadinone (CAP) (**26**) 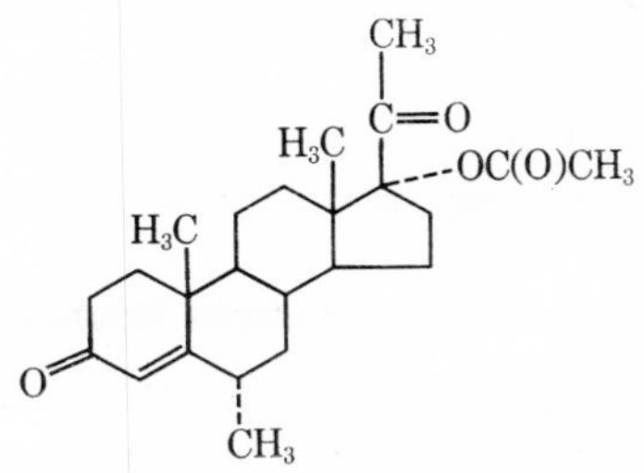   medroxyprogesterone acetate (MAP) (**27**)

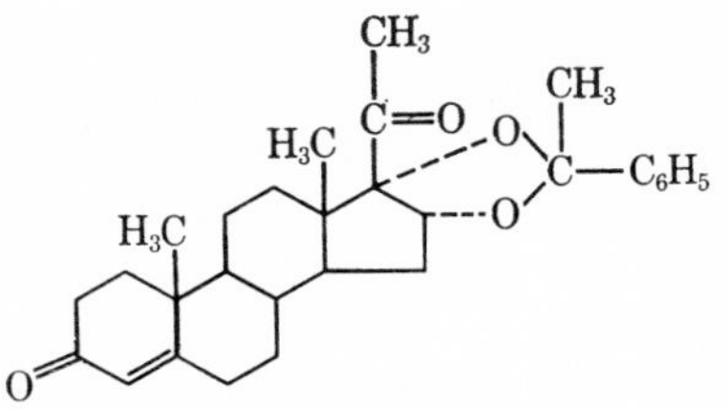

alphasone acetophenide (DHPA) (28)

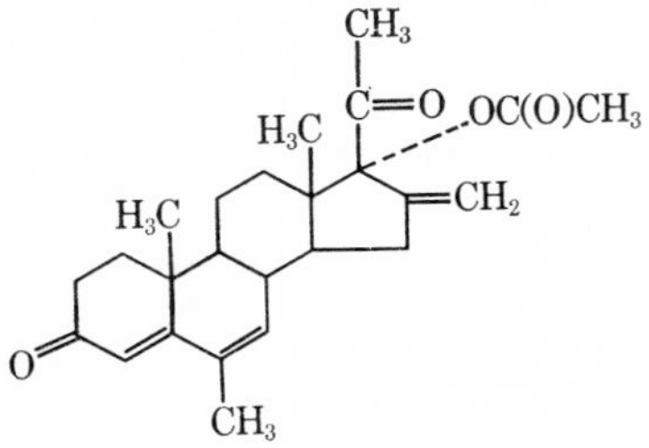

melengestrol acetate (MGA) (29)

Miscellaneous

Certain classes of compounds, although not used extensively, or which have been covered adequately elsewhere in these volumes, are deserving of mention here. Among the latter are anesthetics (qv) which are generally the same as those used in human medicine. Tranquilizers form another category (see Psychopharmacological agents).

Cardiovascular agents (qv) such as digitoxin, amyl nitrite, and bismuth subnitrate are used, much as they are in humans, for the management of appropriate conditions in animals.

Diuretics (qv) are used to reduce udder edema in parturient cattle and to relieve congestions in various cardiac and respiratory problems of small animals and horses and certain other less common conditions. The most frequently used compounds are acetazolamide, hydrochlorothiazide, and furosemide.

Autonomic blocking agents, such as atropine and scopolamine, are used as antispasmodics and in conditions affecting the eye. Atropine is recommended as the treatment for poisoning by organic phosphates, the active ingredients of various insecticides.

Parasympathomimetic agents are employed to stimulate intestinal motility for primary effects (constipation) and for secondary effects (to remove parasites). An example of a drug for the latter use is arecoline. Other agents are physostigmine and pilocarpine.

Central nervous system stimulants are used to counteract depression as a result of traumatic injury, anesthetic overdoses, debilitating diseases and general lassitude, and lack of appetite. Some drugs have action on specific systems, such as respiratory stimulants, while others may have a general effect on the body, exerting a "tonic" effect.

Apomorphine is used as an emetic. Its greatest value is associated with the early treatment of poisonings where the vomition of toxic materials minimizes the toxic effects and reduces the dose of specific antidote required.

Stomachics are used in cases of indigestion and as appetite stimulants to facilitate recovery from surgery or debilitating diseases and generally include mixtures of chemicals. The most commonly used compounds are strychnine (as nux vomica), arsenic, dried rumen contents, iron, copper, cobalt, and aromatic spirits of ammonia.

Government Regulation of Veterinary Medicine

Section 512 of the Federal Food, Drug and Cosmetic Act requires that the Food and Drug Administration of the Department of Health, Education and Welfare review the safety and efficacy of veterinary medicines before they may be offered for commercial sale.   In the case of pesticides used in animals, certain requirements of the U.S. Department of Agriculture must also be met.   Each application must include information not only with respect to the use of the drug itself but also manufacturing, labeling, and other facts related to the proposed product.   A brief abstract of the requirements follows:

1. Safety and efficacy.

a. Detailed reports on studies made on laboratory animals.

b. Reports on all clinical tests.   These must have been conducted by "more than one independent, competent investigator" and at least some of the investigations should have been performed within the United States.

c. If the drug is a combination of previously investigated or marketed drugs, an adequate summary of preexisting information must be supplied.

2. Components of the drug.   All substances used in the synthesis, extraction, or any other method of preparation of the drug itself and the finished dosage form must be identified by the established name or complete chemical name, including structural formulas.   Components of the product, whether active or not, must be identified quantitatively in the batch formula and the final dosage form.

3. A full description of the methods used and controls required for the manufacture, processing and packing of the drug, including controls to "determine and preserve the identity, strength, quality, and purity of the drug."   Each person who will perform any part of these operations must be identified and must sign a statement which describes the methods, facilities, and controls in his part of the operation.

4. Instructions to be placed on the label, including any special precautions, must be approved.

5. Analytical controls of adequate accuracy and sensitivity and other controls with respect to labeling, storage, and inventory control must be described.

When the Food and Drug Administration concludes that the drug performs as claimed, that its use does not result in hazards to the health of either animals which are treated or humans consuming edible tissues or products of those animals, formal approval is published in the weekly *Federal Register*.   Such notice authorizes the sale of a drug subject to the various control procedures outlined above.   The Food and Drug Administration constantly monitors the commercial use of approved products, however, and should it determine that unanticipated hazards exist, either through drug usage under certain unusual conditions or through misuse of the drug, it has the power to require changes in the instructions for use, or, following specified procedures, to withdraw its approval for the sale of the drug.

Bibliography

O. H. Siegmund, et al., *The Merck Veterinary Manual*, 3rd ed., Merck & Co., Inc., Rahway, N. J., 1967.

H. M. Costello, *Modern Veterinary Practice Red Book*, American Veterinary Publications, Inc., Santa Barbara, Calif., 1969.

H. Marsh, *Newsom's Sheep Diseases*, 2nd ed., The Williams and Wilkins Co., Baltimore, 1958.

H. E. Biester and L. H. Schwarte, *Diseases of Poultry*, 5th ed., The Iowa State University Press, Ames, 1965.

L. M. Jones, *Veterinary Pharmacology and Therapeutics*, 2nd ed., Iowa State College Press, Ames, 1957.

L. S. Goodman and A. Gilman, *The Pharmacological Basis of Therapeutics*, 2nd ed., The Macmillan Co., New York, 1960.

H. J. Milks, *Practical Veterinary Pharmacology, Materia Medica and Therapeutics*, 6th ed., Alex Eger, Inc., Chicago, Ill., 1949.

M. G. Fincher, et al., *Diseases of Cattle*, American Veterinary Publications, Wheaton, Ill., 1956.

H. P. Hoskins, *Canine Medicine*, 2nd ed., American Veterinary Publications, Wheaton, Ill., 1959.

H. W. Dunne, *Diseases of Swine*, 2nd ed., Iowa State College Press, Ames, 1964.

D. H. Udall, *The Practice of Veterinary Medicine*, 5th ed., D. H. Udall, Ithaca, N.Y., 1947.

W. A. Hagan and D. W. Bruner, *The Infectious Diseases of Domestic Animals*, 3rd ed., Comstock Publishing Associates, Ithaca, N.Y., 1957.

Feed Additive Compendium, The Miller Publishing Co., Minneapolis, Minn., 1969.

The Merck Index, 8th ed., Merck & Co., Inc., Rahway, N. J., 1968.

D. C. Blood and J. A. Henderson, *Veterinary Medicine*, The Williams and Wilkins Co., Baltimore, Md., 1960

A. LOUIS SHOR AND
RICHARD J. MAGEE
American Cyanamid Co.

VETIVER.   See Oils, essential, Vol. 14, p. 214.

VILLAVECCHIA TEST.   See Fats and fatty oils, Vol. 8, p. 794.

VINEGAR

Vinegar is a condiment prepared by two separate microbial processes: First, an alcoholic fermentation of various raw materials, containing either naturally occurring or converted fermentable sugars, by species of yeasts of the genus *Saccharomyces*, and second, the so-called oxidative fermentation of alcohol, so produced by species of bacteria of the genus *Acetobacter*. (See Enzymes; Ethanoic acid; Fermentation.)

Although cull apples and apple peels and cores are the most important economic raw material in the United States, grapes and grape products, barley malt, honey, and various cull fruits (fresh and dried) may be used for the production of table vinegar. Citrus vinegar has been produced in California, Florida, and in other citrus-growing regions (1). In Russia and other countries, whey also has been used as a base for vinegar production (2), and products such as palm juice have been converted into vinegar (3). Industrial alcohol is the most important source of vinegar for use in mayonnaise, tomato products, prepared mustard, pickles, sauces, and other products. Specially flavored vinegars are prepared by infusion of cider or wine vinegar with herbs and spices, such as garlic, eschalots, tarragon, etc.

There is no standard of identity for vinegar under the Food, Drug and Cosmetic Act of 1938. Under the Federal Food and Drugs Act of 1906, in the latest effective service and regulatory announcement of 1933, the following six types of vinegar are defined (4):

"1. Vinegar, cider vinegar, apple vinegar, is the product made by the alcoholic and subsequent acetous fermentations of the juice of apples, and contains, in 100 cubic centimeters (20°C), not less than 4 grams of acetic acid.

"2. Wine vinegar, grape vinegar, is the product made by the alcoholic and subsequent acetous fermentations of the juice of grapes, and contains, in 100 cubic centimeters (20°C), not less than 4 grams of acetic acid.

"3. Malt vinegar is the product made by the alcoholic and subsequent acetous fermentations, without distillation, of an infusion of barley malt or cereals whose starch has been converted by malt, and contains, in 100 cubic centimeters (20°C), not less than 4 grams of acetic acid.

"4. Sugar vinegar is the product made by the alcoholic and subsequent acetous fermentations of sugar syrup, molasses, or refiners' syrup, and contains, in 100 cubic centimeters (20°C), not less than 4 grams of acetic acid.

"5. Glucose vinegar is the product made by the alcoholic and subsequent acetous fermentations of a solution of glucose, is dextrorotatory, and contains, in 100 cubic centimeters (20°C), not less than 4 grams of acetic acid.

"6. Spirit vinegar, distilled vinegar, grain vinegar, is the product made by the acetous fermentation of dilute distilled alcohol, and contains, in 100 cubic centimeters (20°C), not less than 4 grams of acetic acid."

The composition of vinegar depends upon the following factors: the nature of raw material that is converted to vinegar stock; the materials added to facilitate alcoholic fermentation, (such as nitrogenous, phosphate, and potassium compounds added to products deficient in these—for example, honey—and sulfurous acid or sulfites added to control the alcoholic fermentation); the materials added to facilitate acetification (usually growth and activity stimulants for *Acetobacter*, such as yeast hydrolyzates, vitamin-B concentrates, and phosphates); the extent of alcoholic and subsequent acetic fermentation; and the method of acetification. Only small amounts of acetoin, 3-hydroxy-2-butanone, are formed in submerged fermentation but appreciable amounts occur in generator vinegars (5). There are considerable differences in the rate and extent of change in nonvolatile organic acids in cider vinegar between surface and submerged acetification (6).

The early standards for vinegars, particularly cider vinegar, prohibited the use of fruit waste accumulating in other industries, such as apple peels and cores from applesauce and dried-apple production. To prevent the use of these products and adulteration of a fruit vinegar, such as cider vinegar, with vinegars from another source or addition of acetic acid or water, the early definition of vinegar (7) specified a minimum of apple-solids and apple-ash content. The early standard specifying the presence of not less than 1.6 g of apple-solids and not less than 0.25 g of apple-ash was so impractical that it actually compelled the manufacturer of high-grade cider vinegar to adulterate his product to meet the requirements. Present standards are more realistic

Table 1. Composition of Cider Vinegar

Constituent	Average	Max	Min
total acid as acetic, %	4.94	7.96	3.29
total solids, %	2.54	4.52	1.37
nonsugar solids, %	1.90	2.89	1.26
reducing sugars in solids, %	19.6	45.0	5.6
total ash, %	0.367	0.52	0.20
alkalinity of water-soluble ash, ml $0.1N$ acid[a]	35.7	56.0	21.5
ash in nonsugar solids, %	18.8	26.5	11.2
soluble phosphoric acid, mg P_2O_5[a]	17.3	39.9	6.7
insoluble phosphoric acid, mg P_2O_5[a]	29.3	64.2	15.1
alcohol by vol, %	0.35	2.0	0.03
glycerol, %	0.30	0.46	0.23
polarization (direct), °V[b]	−1.46	−3.6	−0.2
polarization (invert), °V[b]	−1.69	−3.1	0.0

[a] Ash from 100 ml vinegar.

[b] One degree Ventzke scale, 1°V, = 0.34657° angular rotation, sodium D light.

and allow the manufacture of cider vinegar not only from the juice of whole apples but also from the juice of sound apple peels and cores. For the development of standards, see references 7–10a.

The most widely quoted data (8) on the composition of cider vinegar are given in Table 1 (11). In comparison with cider vinegar, the average solids content of wine vinegar is 1.89%, malt vinegar 2.70%, and distilled vinegar 0.35%. The ash content for these vinegars is 0.27, 0.34, and 0.04%, respectively; the P_2O_5 content, in comparison with 0.035% for cider vinegar, is 0.053, 0.105, and 0.0%, respectively, for the others.

Recently published data on the composition of malt vinegar are cited by Morgan and Voelcker (12), Nagarathnamma et al. (13), and on the composition of wine vinegar by Antoni and Celi (14), and Mecca and Davi (15).

In addition to acetic acid, vinegars contain variable amounts of nonvolatile organic acids, including chiefly malic and citric acids with smaller amounts of succinic and lactic acids; glycerol; unfermented and unfermentable sugars; unoxidized alcohol and acetaldehyde; acetoin; and phosphates, chlorides, and sulfates of potassium, calcium, and other cations. Formic acid may be present in distilled vinegar.

The official analysis of vinegars (16) includes the determination of the total solids content, total and water-soluble and insoluble-ash content, alkalinity of soluble ash, total and soluble and insoluble phosphate, total acid content, volatile acid content, total reducing substances before and after inversion, nonvolatile and volatile reducing substances, alcohol, glycerol, coloring matter, polarization, sulfate, heavy metals, dextrin, preservatives, and permanganate oxidation number. From these analyses, interpreted in terms of the known variability of authentic vinegars, it is possible to determine the type and extent of adulteration. The permanganate oxidation number is particularly useful for differentiating between vinegar and commercial acetic acid. The value for the nonvolatile reducing substances is useful in calculating nonsugar solids.

Improved differentiation between natural fermented vinegars, acetified distilled vinegar, and artificial vinegar is now possible by the application of paper-partition chromatography, gas–liquid chromatography, and ultraviolet absorption spectroscopy. Acetoin has been used for a long time to distinguish natural from artificial vinegar (17–20). Diacetyl was not detected in distilled vinegar (21) but was found by gas chromatography (20,22,23). The determination of cider vinegar in wine vinegar is based on malic acid (24) and the presence of acetone was used to detect vinegars made from alcohol (25). The presence of amino acids, detected and determined by chromatography, was used to distinguish fermented from synthetic vinegars (26–28). Ultraviolet absorption was employed by Grote et al. (29) and Webb and Galetto (30). The latter report volatile acid, total tartaric acid content, and absorbance at 280 nm for fifty-eight samples of authentic California red wine vinegars produced commercially by submerged fermentation.

The permanganate or oxidation number, usually reported as ml of $0.01N$ permanganate required to oxidize the distillate from a 100-ml sample in 30 min, is a measure of volatile reducing substances, chiefly acetoin, formed during fermentation and not present in synthetic acetic acid. The volatile reducing substances expressed as invert sugar range from 0.14 to 0.34 g/100 ml in cider vinegar. The permanganate number, as a measure of the content of volatile carbonyl compounds, unsaturated organic acids, and secondary alcohols, is discussed in detail by Saruno (31) and more recently by Llaguno (32). Data on the acetoin content of vinegar are given by Doro and Sadini (33), and others (18–20, 22,23,34).

Mecca (35) found the presence of monomethylguanidine acetic acid anhydride characteristic of fermentation vinegars. Subsequently (36) he introduced a new iodine number—based on a colorimetric method for determining monomethylguanidine acetic acid—for differentiating between fermentation and artificial vinegar and this was then applied to commercial vinegars (37).

The volatile components of the vapors of natural and distilled vinegars were investigated by gas–liquid chromatography (GLC) (22,23,38). The determination of sorbitol and mannitol by thin-layer chromatography (39) or by GLC (40) was proposed to distinguish fermented from artificial vinegars.

The strength of vinegar, ie the concentration of acetic acid, was expressed historically in terms of the grains of sodium bicarbonate neutralized by one fluid oz wine measure of vinegar, and in the early vinegar tests this was measured by the volume of carbon dioxide gas evolved from excess of sodium bicarbonate treated with a measured volume of vinegar. The original grain strength multiplied by 0.1565 gave the concentration of acid as acetic acid in g/100 ml. Although the term "grain strength" is still commonly used to express the concentration of the acetic acid in a vinegar, one-grain vinegar is defined now as containing 0.1 g of acetic acid in 100 ml at 20°C. The present grain strength is thus ten times the acetic acid content. A vinegar meeting the legal minimum of 4 g acetic acid per 100 ml would be of 40-grain strength in comparison with the old standard of 25.6-grain strength.

Manufacture

ALCOHOLIC FERMENTATION

Acetic acid is formed in two stages from sugar-containing media, such as fruit juices, or starches converted to sugars by enzymic or acid hydrolysis. In the first stage the sugars are fermented to alcohol through the agency of enzymes of yeast. In this alcoholic fermentation sugar is converted chiefly into ethyl alcohol and carbon dioxide, while part of the sugar is assimilated by the yeast cells, and part is transformed into glycerol, acetaldehyde, lactic acid, etc. The succinic acid that accumulates in the alcoholic fermentation is derived largely from carbon dioxide and oxalacetic acid, and the 2,3-butylene glycol (2,3-butanediol) from acetaldehyde; the higher alcohols are derived from the amino acids. According to the Gay-Lussac equation (41)

$$C_6H_{12}O_6 \rightarrow 2\ CO_2 + 2\ C_2H_5OH$$

1 g of glucose should yield 0.5114 g of alcohol. The yield of alcohol is usually only about 85–90% of the stoichiometrical yield, but experiments with five strains of compressed baker's yeast and five strains of brewer's yeast and pure D-glucose gave fermentation efficiencies of 88–94%. The fermentation efficiency, ie the ratio of the amount of alcohol obtained from the weight of sugar fermented to the theoretical yield, was found to be greater with proliferating yeasts than with resting cells, and actively fermenting yeast gave a slight but significant increase in fermentation efficiency. To obtain optimum yields of alcohol and to prevent undesirable changes in quality as a result of contamination with undesirable species of yeast and bacteria, the fermentation is conducted at a temperature of 75–80°F. A selected strain of *Saccharomyces cerevisiae*, in the presence of the minimum amount of sulfur dioxide, is used as a starter. The raw material is prepared for alcoholic fermentation by the conversion of starches

into fermentable sugars and extraction of these sugars; expression of juice from products such as fresh fruits; and dilution of honeys, syrups, or dried fruits with water. The fermentable sugar content of the product is adjusted to not less than 8% nor over 20%. Where necessary, potassium and ammonium phosphates are added, as, for example, with honey, which is low in these originally and whose content of these is further reduced by the dilution necessary for fermentation. Ammonium and potassium phosphates are not required in the fermentation of crushed grapes and grape juice or with apple juice. Apple juice usually ranges from 10 to 15% in sugar content, but it may be as low as 6.5%, particularly when expressed from peels and cores. Apple juice of low sugar content should be blended with juice of higher sugar content to obtain vinegar of 50–60-grain strength. Concentrated apple juice may be used as blending stock for juices low in sugar content and also for the production of higher-strength cider vinegar. To facilitate the fermentation of apple juice concentrated by vacuum evaporation, it is necessary to add ammonium phosphate, particularly at lower fermentation temperatures.

The alcoholic fermentation is conducted usually in two stages. The primary stage normally lasts from 3 to 7 days, during which the bulk of the sugar is converted into alcohol, and this is followed by the slower and less active fermentation, which extends over a period of several weeks. The primary fermentation is usually conducted in open fermentation vats with cooling coils or external coolers to facilitate the control of the temperature and to follow the course of fermentation by periodic determination of Brix degree. (This is a commonly used measure of specific gravity by hydrometry and refers to the percentage of sucrose by wt for pure sucrose solutions or to the concentration of a sucrose solution having the same specific gravity as the liquid tested.) The secondary fermentation is conducted in closed fermentors to prevent spoiling by aerobic yeasts and acetic acid bacteria. Sulfur dioxide as liquid sulfur dioxide or as sodium or potassium metabisulfite is added to the apple juice or other sugar-containing medium at the start of the fermentation and before the addition of the yeast starter to inhibit the growth and activity of undesirable yeasts and bacteria. An initial level of 50–100 mg of sulfur dioxide per liter is sufficient. Although it is possible to convert fermented fruit juices containing up to a total of 100 ppm of sulfur dioxide, it is better to keep the total and free sulfur dioxide content to 50 ppm or less. When the alcoholic fermentation is completed, the yeast cells and other suspended material settle and the clarified wine is separated from the sediment by racking (drawing off). This stock may then be stored for additional aging or used directly for acetification.

Usually only apple juice is fermented by vinegar producers; grape and other fruit wine is purchased from wineries, beer and fermented malt from breweries or yeast plants, and distilled alcohol from distilleries. Fermented apple juice, and wine used for vinegar stock, are usually denatured by addition of sufficient strong vinegar to bring the acetic acid content to 1% in order to facilitate storage and to avoid payment of internal-revenue wine tax. Alcohol for vinegar stock is also denatured with acetic acid. The added acetic acid also serves to protect the vinegar stock against spoilage by lactic acid bacteria if it is to be stored. Small quantities of sulfur dioxide may also be used for this purpose if the final sulfur dioxide content after fermentation is low enough to permit this without exceeding the concentration toxic to acetic acid bacteria used in converting this stock into vinegar. The vinegar stock should be low in reducing sugar content, preferably below 0.3%.

ACETIC FERMENTATION

The conversion of alcohol to acetic acid, according to the overall reaction,

$$C_2H_5OH + O_2 \rightarrow CH_3COOH + H_2O$$

is catalyzed by enzymes secreted by species of *Acetobacter*. It requires oxygen and, like the alcoholic fermentation, is exothermic. Theoretically 1 g of alcohol should yield 1.304 g of acetic acid, but in practice 1 g of acetic acid is obtained, and only under exceptional circumstances is a yield of 1.1 g attained. Thus, although a fermentation efficiency of 90% can be attained in the alcoholic fermentation (33,61), a fermentation efficiency of 77–84% is attained in the acetification process with fruit vinegars. With distilled vinegars higher efficiencies can be attained industrially. Thus, Allgeier et al. (42), in pilot-plant vinegar generators, obtained fermentation efficiencies on the basis of alcohol in the charge (denatured alcohol containing ethyl acetate), varying from 79.8 to 90.3%, when the water used in dilution of alcohol was varied. With the best water supply used, fermentation efficiencies of 85–90% were obtained. If the conversion of sugar into alcohol is 90% of the theoretical and the conversion of alcohol into acetic acid is 80% of the theoretical, apple juice used for vinegar production must contain not less than 8.32 g of fermentable hexose per 100 ml to yield a vinegar meeting the legal limit of 4 g of acetic acid per 100 ml at 20°C.

The vinegar stock for the production of 40-grain-strength table cider vinegar is adjusted to an alcohol content of not less than 3 g of alcohol and 1 g of acetic acid per 100 ml. For the production of distilled vinegar of 100-grain strength the stock is adjusted to about 10.5% alcohol by vol and 1% acetic acid. Allgeier et al. (42), using 190-proof alcohol diluted with water to 9 g of alcohol per 100 ml of fluid, obtained vinegar of 9.9–10.6 g of acetic acid per 100 ml. High-strength apple and wine vinegars can be made from stock of 10% alcohol and 1% acetic acid content by the generator process. The apple-vinegar stock is adjusted either by dilution with water or blending of several stocks of high and low alcohol content. Wine is diluted with water to the desired alcohol strength and adjusted as to acetic acid content by the addition of strong vinegar. To meet the present standards of identity the acetic acid content of a particular vinegar stock must be derived from vinegar prepared previously from similar stock; that is, cider vinegar cannot be added to wine or wine vinegar to apple cider stock. The distilled vinegar stock must be fortified with added growth stimulants to promote acetification, and, when necessary, similar preparations may be added to cider- or wine-vinegar stock.

Until recently the growth stimulants required by *Acetobacter* were not known, and several empirically developed nutrient mixtures distributed under trade names, such as Aceto-pep (Industrial Alcohol Co.), were used at the rate of 2–3 lb/100 gal of vinegar stock. These nutrients usually contain inorganic salts and malt or yeast extract. Allgeier et al. (42) reported that a filtered fermented solution of a diluted malt concentrate added at the rate of 1% by vol was quite satisfactory for the acetic fermentation of solutions of denatured alcohol. The nutritional requirements of several species of *Acetobacter* are now known (43–52), but this knowledge has not yet been applied to the selection of the best growth and activity stimulants in vinegar generator production. It has been applied, however, in experimental studies of submerged fermentation. Rao and Stokes (49) reported that, although *Acetobacter suboxydans* and *A. melanogenum* cannot grow in chemically defined media with ethyl alcohol as the sole source of carbon, they develop when small amounts of yeast autolyzate, liver extract.

or peptone are added to the medium. The growth-promoting activity of these biological materials is believed to be due to the reducing sugars and related substances which they contain. Small amounts of glucose, fructose, mannitol, and glycerol were substituted successfully for yeast autolyzate.

In laboratory generators charged with beechwood shavings and inoculated with *Acetobacter xylinum*, glucose, but not maltose, sucrose or invert sugar, is required for maximum acetification (52a). The concentrations of glucose and other nutrients for optimum acetification, per liter, are as follows: glucose, 0.5–0.7 g; ammonium, 20 mg; phosphate, 120 mg; magnesium, 5 mg; potassium, 15 mg. In commercial generators the addition of glucose does not increase the rate of acetification but it does increase the yield of acetic acid and reduces carbon dioxide formation. The requirement of the other nutrients was confirmed.

A number of varieties of *Acetobacter* are described in the literature, and the particular species and strain possessing the characteristics desired for vinegar production in the quick vinegar process now generally used have been isolated. Pure cultures of the several species of *Acetobacter* that have been isolated previously from vinegar do not have the desired characteristics, and vinegar generators as yet are not inoculated commercially with pure cultures. A suitable quantity of unpasteurized vinegar obtained from a generator in satisfactory production is used to start a new generator.

Owing to difficulties encountered in culturing *Acetobacter* from operating generators, the bacteria actually active in acetic acid production have not been isolated and the changes in bacterial flora and bacterial activity during vinegar production under industrial conditions are still unknown. Shimwell (51) succeeded in isolating a pure culture of the "true working bacteria" in a vinegar acetifier and described its morphological, cultural, and physiological characteristics. He considers this strain as belonging to the "mesoxydans" group of Frateur's classification (52b). The pure culture was used successfully in making spirit, malt, and wine vinegars, and was found to retain its acid tolerance and other desirable properties on prolonged subcultivation. The pure culture, unlike the mixed flora of unknown nature, gave a faster rate of acetification and a more uniform product, resulting in the elimination of undesirable contaminants.

The dehydrogenation of ethyl alcohol to acetic acid by the endocellular alcohol dehydrogenase of *Acetobacter* is greatly influenced in rate and extent by the species and strain of predominant bacteria and by the environmental conditions under which this conversion is allowed to occur. The concentration of bacterial cells relative to the volume of vinegar stock being converted, the method of controlling the diffusional operations involved in the mass transfer of the vinegar stock to the bacterial surface and of removal of the vinegar formed, the method of supplying the concentration of oxygen required for this oxidation, the temperature at which the oxidation occurs, and the composition of the vinegar stock, all markedly affect the rate, extent, and efficiency of acetification.

At the present time most commercial vinegar is produced by the generator process, in which the vinegar stock is pumped into specially devised distributors from which it is uniformly sprayed over the surface of an inert porous medium or packing material at the surface of which it is oxidized by the bacterial enzymes. The stock flows from the packing by gravity into receiving tanks from which it is recycled. The rate and extent of acetification in these generators are controlled by the rate of flow of the vinegar in the recycling process, the type and concentration of nutrients present or added, the overall size of the generator (alcohol conversion usually is slower in smaller generators), the extent of aeration, and the temperature. The slower acetification which occurs in the Pasteur modification of the Orleans process, in which a specially constructed wooden grating is placed at the surface of the liquid in partly filled barrels or in shallow tanks to support the film of vinegar bacteria, is still used in Europe for wine-vinegar production; it is not used extensively in the United States, even for wine-vinegar production. At one time it was believed that the generator process could not be used for the production of red wine vinegar because of excessive oxidation of anthocyanin pigments, but by proper control of temperature and aeration this difficulty has been overcome.

The design and operation of vinegar generators have been based largely on empirical observations, rather than on scientific studies of the mass-transfer operations involved in the conversion of alcohol into acetic acid, the factors influencing the efficiency of the oxidation process, and the production of vinegar of improved aroma and stability. The manufacture of vinegar is very ancient in origin and very little change in vinegar production occurred (53) from the first complete description of vinegar production by Olivier de Serres in 1616 until the first development of the quick (generator) process by Schützenbach in 1823. Schützenbach provided ventilating holes near the bottom of the large tank used to hold the stock, which was distributed over the packing material for the purpose of obtaining a greater supply of air. Gravity flow was used for the liquid trickling down the packing material and for the air flowing up the tank. Bitting (53) presented a synopsis of the U.S. patents issued on various modifications of the generator process from 1849 to 1929. No important advance was made in the generator design, however, until Frings introduced his airtight generator in 1932 (54–64). In this generator cooling coils are provided in the collection chamber at the bottom for temperature control, and the oxygen supply is controlled by a damper located at a vent pipe at the top of the tank and by the number and size of air inlets at the lower part of the generator. The rate of flow of the stock through the generator is controlled by the rate of pumping of the vinegar stock from the collection chamber to the sprayer or distributing arm at the top of the generator packing. This generator has since been improved by installation of air blowers to control the supply and distribution of air, and to give better distribution of stock over the entire volume of packing and improved temperature control. Automatic control of air and liquid flow and temperature was introduced in 1939 (62). Aside from patents, few data were published on the operation of vinegar generators until the early 1950s (42,64). Hromatka and Ebner (64) investigated the oxygen requirements of *Acetobacter* on beechwood shavings and found that the oxygen uptake was linear with time at more than 12% of oxygen in the gas phase, but the rate of oxygen uptake dropped at lower partial pressure and reached zero at 4.5% of oxygen in the gas phase. Pure oxygen was not found to be harmful in generator fermentation but was harmful in submerged fermentation.

Frings Generator. The Frings generator consists of an airtight tank usually 14 ft in diam and 15 ft high. Inside the tank near the bottom is a wooden grating and near the top is the sprayer or distributing arm. At the top a 4-in. vent, with a damper connected to a tile exhaust or a vapor–liquid separator, serves to control the air supply. Air enters the generator through ten air intakes, containing air filters, located around the tank near the level of the wooden grate. Tubular cooling coils are located at the bottom of the tank in the collection chamber, and these are connected by a centrifugal pump to a feed line supplying the sprayer. The generator is packed with beechwood shavings, supported by the wooden grating, up to about 1.5 ft from the top of the generator. The vinegar stock or mix is allowed to circulate repeatedly through the beechwood shavings until vinegar of the desired strength has been obtained. The cycle time in such a generator for distilled vinegar is usually 3–5 days in a good operation for a mix containing 9 g of alcohol per 100 ml of feed. In the Frings-type pilot-plant vinegar generator, used by Allgeier et al. (42), the cycle time was deliberately increased to 7 days by decreasing nutrients slightly in order to allow one operator to carry out the experimental work. The rate of liquid flow was 12 gal/(hr)(100 ft³) of packing, and the rate of air flow was 55 ft³/(hr)(100 ft³) of packing when the operating temperature was approximately 85°F. The feed temperature was adjusted to 28°C; the average temperature at the upper level of the generator was 29°C and at the lower end 30–32°C. Under these conditions the rate of production of acetic acid was 11 lb/ (24 hr)(100 ft³). In a large Frings generator a 2500-gal batch of mix containing 10.5% ethyl alcohol, 1% acetic acid, and 7 lb of Aceto-pep requires 3–5 days to reach 10.5% acetic acid. In comparison the production of cider vinegar is assumed to be 0.2–0.3 gal/(24hr)(ft³) of packing of 6% vinegar.

Theoretically 1 gal of vinegar stock containing 11% alcohol by vol would require about 28 ft³ of air for oxidation of alcohol to acetic acid and would generate about 3300 Btu. In a closed generator, such as the Frings, the heat evolved in this oxidation would rapidly raise the temperature considerably above the desirable range of 80–90°F. Actually much more than the stoichiometric amount of air is required in the process and an adequate supply of cooling water must be supplied for the required temperature control.

In recirculating generators, even when inoculated with the desired flora by the use of carefully selected raw vinegar, difficulty is experienced with wine or cider vinegar due to sliming caused by the development of the so-called "mother of vinegar," actually exocellular bacterial cellulose produced by *A. xylinum* which encapsulates the bacterial cells.

Plugging of Frings-type generators may result also due to slime formation occurring when the nutrients are not balanced properly. This sliming occurs more frequently when beechwood shavings are used as packing, but even with coke packing sliming occurs in 6–12 months of use, necessitating repacking. Slime formation is considerably decreased as the acetic acid content of the finished vinegar is increased, so that, with cider or wine vinegar stock containing sufficient alcohol to produce 10% vinegar, sliming occurs less frequently. Distilled-vinegar generators operate indefinitely without being plugged with cellulose slime formed by *Acetobacter* when maintained in proper balance with nutrients under optimum conditions for normal operation. In the generation of lower-strength cider or wine vinegars the generators are packed with coke, rattan bundles, or corn cobs. When the generator becomes so plugged as to be uneconomical to operate, the packing is removed and thoroughly cleaned.

Beechwood shavings, properly curled, provide a good surface for growth and activity of *Acetobacter* species and are sufficiently inert and porous to allow for good operation of generators. They are, however, more expensive and more difficult to clean when plugged with slime.

Although beechwood shavings are the preferred packing for modern generators of the circulating type, the type of packing has varied considerably. Often the most readily available or cheapest packing obtainable in a given district has been used, such as coke, pumice, rattan, grape and other twigs, and corn cobs. Crushed coke, size graded to remove fine particles and dust, is a cheaper substitute. It is more inert and easily cleaned, but it is difficult to remove the acid-soluble iron and copper salts. Metallic contamination is undesirable not only because it renders the vinegar more susceptible to formation of hazes and deposits but also because it affects the acetification process. Allgeier et al. (42) reported that lead, copper, iron, zinc, and tin, when present in the vinegar at levels of 10, 15, 50, 100, and over 100 ppm, respectively, were toxic to vinegar bacteria and caused decrease in efficiency of conversion of ethyl alcohol to acetic acid, accompanied in most cases by a corresponding rise in residual alcohol.

Allgeier et al. (42) reported comparative data on the performance of generators packed with beechwood shavings, coke, rattan, and Berl saddles. Surface area and impurities in coke were found to be limiting factors. Removal of impurities by acid washing, increasing the surface by using smaller pieces ($\frac{1}{4}$ to $\frac{1}{2}$ in. in diam), and proper addition of nutrients to facilitate growth and adherence of bacteria to the surface, increased the operating efficiency of the coke-packed generator from 76 to 98% of that obtained with beechwood shavings. A higher concentration and a different type of nutrient were required with coke than with beechwood packing. The coke-packed generator responded markedly to a yeast autolyzate–steep water mixture. With specially prepared rattan, yeast autolyzate also produced marked improvement in operating efficiency. The nutrient requirements were found to change with the type of packing, special nutrients being required by packings deficient in extent of surface or by packings with surfaces not readily coated with bacteria.

Submerged Fermentation. After a period of about 130 years the packed quick vinegar generator is now being replaced by the submerged fermentation process. This is an obvious application of the successful biological production of antibiotics during and following World War II. In the literature credit is given to Hromatka and Ebner (64) and Haeseler (65) for introducing submerged acetification of vinegar stock. Actually one of the first, if not the first, report of successful acetification by this means was that of Fowler and Subramanian (66) who used a modified activated sludge process designed for biological oxidation of domestic sewage and report the existence of at least one commercial plant using their process. At present three methods of submerged acetification are available. The "acetator" developed on the basis of extensive investigations of Hromatka and associates in Austria (64, 67–71); the "cavitator" based on a device used for biological oxidation of domestic sewage described by Cohee and Steffen (72), patented by Burgoon et al. (73), and modified for continuous-flow production by Mayer (74); and the so-called "aspirator" developed by Richardson (75). These three processes differ in the method of introducing and distributing air into the vinegar stock. In the "acetator," described in some detail by Allgeier and Hildebrandt (76) and Beaman (77) a rotor and a stator are used to draw in and intimately mix air with the vinegar stock; the "cavitator," also described in references 63 and 76, is a draft tube used to mix the incoming air with the substrate. In the Richardson "aspirator" the mixture of vinegar bacteria and alcoholic substrate is aerated by pumping it under pressure through a venturi-tube ejector (78). Hromatka and his colleagues (64, 67–71) investigated the submerged fermentation acetification of red wines and synthetic bases (79–83). Lopez et al. (84) investigated the nutritional requirements of certain *Acetobacter* species under submerged fermentation conditions and studied factors influencing the initiation of submerged fermentation and the rate of acetification. They used a modification of the medium developed by Litsky and Tepper (50). In their experimental fermenters operated at 76–82°F with 20 liters of air introduced per hour, *Acetobacter acetigenum* and *pasteurianum* were superior to other species tested. The optimum pH range for acetification was 3.9 to 5.0. Higher acetic acid production was found when the basal medium in the fermenters was recharged daily than when it was continuously diluted. The bacterial count at peak acid levels was 1–100 × 10⁶ microorganisms per ml. Richardson (85) investigated the optimum conditions necessary for the production of vinegar from waste pineapple juice. Vinegars containing up to 7% acetic acid could be produced in less than 24 hr with a conversion efficiency greater than 90% in both laboratory and pilot-plant equipment. Davé and Vaughn (78) studied the factors influencing the production of 10% vinegar from pineapple brandy denatured with added acetic acid in 5-gal carboys fitted with sintered-glass air spargers and in a pilot-scale Richardson aspirator. The effects of aeration, temperature, and nutrient composition were investigated in some detail. The optimum temperature was found to be 80°F; optimum acetification occurred at an oxygen solution rate of 60 mM per hour per liter. The best medium mixture contained corn sugar, diammonium hydrogen phosphate, autolyzed yeast, citric acid, powdered whey, and potassium chloride. It was difficult to obtain a vinegar of 10 g acetic acid per 100 ml regularly, although this was obtained in runs with stock containing 3–8% ethanol and 4.7–9.0 g acetic acid per 100 ml.

Commercially, difficulty is experienced in completing the acetification of vinegar stock in submerged fermentation. When the concentration of alcohol in the fermenting wash is lower than 0.25–0.30% the rate of acetification becomes significantly lower. The vinegar made by the submerged process has a characteristic flavor and aroma, different from that obtained by the generator process. The product, however, has met with favorable customer acceptance. Vinegar produced by submerged fermentation is always cloudy due to suspended bacteria and insoluble components of the mash. This necessitates effective filtration in plant operation.

Concentrated Vinegar. A product of up to 100-grain strength may be readily obtained in the production of distilled vinegar by suitable adjustment of the alcohol and acid content of denatured vinegar stock. Wine vinegar of this strength also can be produced by the adjustment of wine stock and cider vinegar may be obtained by concentrating apple juice before fermentation. It is also possible to concentrate vinegar by freeze concentration (86,87). This is believed to produce a concentrated vinegar of improved flavor and stability.

Vinegar Eels. Vinegar eels (*Anguilla aceti*), which may infest vinegar, are usually considered undesirable and are generally removed from the finished vinegar by pasteurization and filtration. These "eels" are nematodes which are usually about ¹⁄₁₆ in. long. They can be readily observed with a hand lens. Zalkan and Fabian (88), however, have observed that many industrial vinegar generators are infested with these eels with no apparent detriment to the operation. They investigated the effect of vinegar eels on acid production in vinegar pilot-plants and observed higher acetic acid production and greater efficiencies of conversion in generators containing vinegar eels. This effect was particularly evident when the diluted alcohol was fortified with phosphates or with phosphate and nitrate or ammonium chloride.

They suggested that the vinegar eels act as scavengers to keep the beechwood shavings free of dead vinegar bacteria and also to supply more readily available nutrients to the bacteria.

PREPARATION FOR MARKET

The freshly made vinegar is usually harsh in flavor and odor, unstable, and not brilliantly clear. It is prepared for marketing by aging, and is filtered and clarified, blended and stabilized, and bottled. Vinegar for manufacturing use is shipped in wooden barrels, casks, or tank cars and trucks.

Aging. Cider, wine, malt, and distilled vinegar to be bottled for retail distribution as condiment vinegar, is pumped from the generator receiving tank into storage containers for aging. During aging and storage the vinegar should be protected against excessive aeration or exposure to air, as this favors growth of acetic acid bacteria, loss of acetic acid by oxidation, and formation of mother of vinegar (a thick mucilageneous layer of acetobacter cells imbedded in exocellularly synthesized bacterial cellulose).

Distilled vinegar, particularly when used for manufacturing, is usually not aged, and is marketed as produced. The generator-produced distilled vinegar may be used as a source of acetic acid (qv) for industrial use, as well as for food vinegar uses. Clean, sound cooperage, treated to prevent absorption of off-flavors, is used for storage, which may extend for several weeks or months. Distilled vinegar to be used in manufacture is stored only for clarification treatment. The storage period required to produce a mellow smooth vinegar varies with the source of vinegar stock, the conditions of acetification, and the size and type of storage containers. During storage residual ethyl alcohol is esterified by acetic acid and other organic acids. Other chemical changes, still ill defined, also take place. The storage containers must be kept well filled to avoid oxidation of acetic acid by vinegar bacteria to carbon dioxide and water, which can occur with vinegars of 6% acetic acid content and lower. The final flavor obtained is determined by the type of vinegar stock used and the conditions of acetification and storage. The flavor of distilled vinegar is markedly influenced by the type of denaturant added to distilled high-proof alcohol (42) and the type of water used in diluting the alcohol (42,77). Allgeier et al. (42) obtained distilled vinegars of best aroma from denatured alcohol containing specially selected still fractions. The flavor of wine vinegar can be improved by the addition of small amounts of aged dry wine. The herb-and-spice-flavored vinegars are carefully compounded and aged to obtain more desirable and more stable flavors.

Filtration and Clarification. The aged vinegars or the freshly prepared vinegars are filtered clear either before or after clarification and the removal of metal. Cider and particularly wine vinegars usually contain excessive amounts of iron and copper impurities picked up largely from the packing material in the generator (particularly from coke) and from contact with metal surfaces during fermentation and subsequent handling. High iron content leads to formation of suspended ferric phosphate and ferric tannate colloids and to unsightly deposits. High copper content leads to formation of insoluble copper colloids. The formation of metallic hazes does not occur in the absence of proteins and other colloids; they may be removed from vinegar by fining, ie adding an agent such as bentonite or sodium alginate before filtration. The concentration of iron and copper is reduced by treatment with potassium ferrocyanide (blue-fining) or with specially prepared adsorption compounds of potassium ferrocyanide. Cufex (Scott Laboratories, Inc.) is a fining agent prepared from ferrous ferrocyanide, potassium ferrocyanide, citrate, and bentonite. When properly used so that no residual ferrocyanide remains, blue-fining is quite satisfactory. Bentonite and other fining agents may be used either before or after blue-fining. The clarified and metalfree vinegar must be protected against contamination with heavy metals during the final filtration and filling operations. Filtration is also used in reducing the bacterial population and in removing vinegar eels if present.

Sterilization and Packaging. Pasteurization, addition of small amounts of sulfites, and sterile filtration are used to protect the clarified and filtered vinegar against spoilage by growth of cellulose-synthesizing acetic acid bacteria. The temperature and time of pasteurization vary with the extent of infection and conditions of filling. For filling into small containers, the vinegar is brought to 150–160°F, filled hot, sealed, and allowed to cool to room temperature after inversion to sterilize the neck and cap. Vinegar may be pasteurized in bulk by heating to 150°F, holding for 30 min, and cooling to 100°F, before filling into large containers. Sulfur dioxide may be used to inhibit growth of vinegar bacteria and also to stabilize the vinegar against oxidative deterioration in color and flavor, but the quantities added should be small (usually not over 50 ppm) to avoid effect on aroma and flavor. Contamination with metals (particularly iron and copper) and excessive aeration should be avoided in the final filtration and filling operations. Hard rubber, plastic, stainless steel, and glass materials are used in the fabrication of pumps, fillers, and closing machines, and only chemically inert materials are used in fabricating containers and closures.

Economic Aspects

Very little information on the production and use of vinegar in the United States is available. The early data published in the *Census of Manufactures* by the Bureau of the Census, U.S. Dept. of Commerce, give only the value of the total vinegar produced at the plant and the value added in manufacture.

The *Census of Manufactures 1963*, reported that the production of vinegar and cider in 1963 was at a relative level of 106, as compared with 100 in 1954 and the unit value increased from 100 (1954) to 124. Fermented-vinegar production in consumer-size containers of 1 gal or less amounted to 35,500,000 gal in 1963, valued at over $21,000,000; production in commercial-size containers of over 1 gal was 35,000,000 gal, valued at $6,500,000. Distilled-vinegar production amounted to 6,000,000 gal in consumer-size containers ($6,000,000) and 53,000,000 gal in commercial-size containers ($8,600,000). Vinegar stocks were given as 11,600,000 gal ($2,000,000) and vinegar and cider, not specified, as 13,000,000 gal ($4,000,000). Internal Revenue statistics indicate the production of about 622,000 gal of wine vinegar during the period of July 1 to December 31, 1968. During the period of January 1, 1967 to June 30, 1968 a production of 1,300,000 gal of wine vinegar is given.

The U.S. pack of vinegar in glass containers was 10.0 million av in 1956–1960, 12.3 million in 1961–1965 and 14 million in 1965, and 1967, according to statistical data summarized in the statistical review and yearbook number of *Canner/Packer*. The annual production of vinegar is estimated to be 8–9 million gal in Great Britain (64).

Bibliography

"Vinegar" in *ECT* 1st ed., Vol. 14, pp. 675–686, by M. A. Joslyn, University of California.

1. R. R. McNary and M. H. Dougherty, *Florida Univ. Agr. Expt. Sta. (Gainesville) Bull.* **622** (1960).
2. U.S.S.R. Pat. 188,449 (Nov. 1, 1966), L. A. Erzinkyan, M. S. Pakhlevanyan, E. A. Muradyan, L. M. Charyan, and A. B. Akopova; *Chem. Abstr.* **67**, 3981 (1967).
3. U. P. Varma, *Bihar Agr. Coll. Mag.* **10**, 35 (1959).
4. *Service and Regulatory Announcements, Food and Drug Administration*, No. 2, 4th revision, U.S. Dept. of Agriculture, Washington, D.C., 1933, p. 19.
5. H. Hadora and W. Beetschen, *Mitt. Gebiete Lebensm. Hyg.* **56** (1), 46 (1965).
6. T. Takeuchi, S. Furukawa, and R. Veda, *Hakko Kogaku Zasshi* **46** (4), 288, 292 (1968); *Chem. Abstr.* **69**, 4755 (1968).
7. L. M. Tolman, *J. Assoc. Offic. Agr. Chemists* **22**, 30 (1939). B. G. Hartman and L. M. Tolman, *Ind. Eng. Chem.* **9**, 759 (1917).
8. R. O. Brooks, *Critical Studies in the Legal Chemistry of Foods*, Reinhold Publishing Corp., New York, 1927.
9. C. A. Mitchell, *Vinegar: Its Manufacture and Examination*, 2nd ed., Charles Griffin & Co. Ltd., London, 1926.
10. A. D. Herrick, *Food Regulation and Compliance*, Revere Publishing Co., New York, 1944, pp. 177, 178, 218, 427, 428, 431, 433, 441.
10a. R. A. Robinson, *The Legal History of Vinegar*, Vinegar Brewers' Foundation, London, England.
11. R. W. Balcom, *U.S. Dept. Agr. Bur. Agr. Chem. Eng. Bull.* **132**, 96 (1910).
12. R. H. Morgan and E. Voelcker, *Food Process. Packaging* **32**, 207 (1963).
13. M. Nagarathnamma and G. S. Sidappa, *Indian Food Packer* **18** (2), 8 (1964); M. Nagarathnamma, C. Dwarakanoth, and J. S. Prathi, *Indian Food Packer* **18**, 4 (1964).
14. I. A. Antoni and G. Celi, *Ann. Sta. Chim. Agrar. Sper. Roma, Ser. III, Pubbl.* **232–240** (1965).
15. F. Mecca and B. Davi, *Bull. Lab. Chim. Prov.* **17**, 781 (1966).
16. *Official Methods of Analysis*, Association of Official Agricultural Chemists, Washington, D.C., 1965, pp. 480–483.
17. E. Rentschler, *Deut. Essigind.* **46**, 94 (1942).
18. V. Curzel, *Riv. Viticolt Enol. (Coneagleani)* **8**, 93 (1955).
19. O. T. Rotini and C. Galoppini, *Ann. Sper. Agrar. (Rome)* **11**, 1355 (1957).
20. A. Cesari, A. M. Cusmano, and L. Bomforti, *Boll. Lab. Chim. Provinciali (Bologna)* **15**, 444 (1966).
21. E. Toth, *Z. Lebensm. Untersuch. Forsch.* **82**, 439 (1941).
22. H. Suomalainen and J. Kangasperko, *Z. Lebensm. Untersuch. Forsch.* **120**, 353 (1963).
23. L. W. Aurand, J. A. Singleton, and J. Etchells, *J. Food Sci.* **31**, 172 (1966).
24. P. Armandola, *Riv. Viticolt Enol (Conegleani)* **7**, 256 (1954).
25. I. A. Pontin, *Anales Direc. Nacl. Quim. (Buenos Aires)* **6** (12), 83 (1953).
26. H. Schanderl and T. Staudenmayer, *Z. Lebensm. Untersuch. Forsch.* **104**, 26 (1956).
27. J. Bourgeois, *Branntweinwirtschaft* **79**, 250 (1957); *Mitt. Gebiete Lebensm. Hyg.* **48**, 217 (1957).
28. K. G. Bergner and H. R. Petri, *Z. Lebensm. Untersuch. Forsch.* **111**, 319, 494 (1960).
29. B. Grote, G. Lemke, and B. Westbank, *Z. Lebensm. Untersuch. Forsch.* **100**, 265 (1959).
30. A. D. Webb and W. Galetto, *Am. J. Enol. Viticult.* **16**, 79 (1965).
31. R. Saruno, *Ferment. Technol. (Japan)* **32**, 90, 128 (1954).
32. C. Llaguno, *Rev. Agroquim. Technol. Alimentos* **6**, 496 (1965).
33. B. Doro and V. Sadini, *Boll. Lab. Chim. Provinciali (Bologna)* **4**, 85 (1953).
34. O. T. Rotini and C. Galoppino, *Ann. Sper. Agrar. (Rome)* **11**, 1355 (1957).
35. F. Mecca, *Chim. Ind. (Milan)* **40**, 20 (1958).
36. F. Mecca, *Boll. Lab. Chim. Provinciali (Bologna)* **15**, 578 (1964).
37. G. Sereri and B. Doro, *Boll. Lab. Chim. Provinciali (Bologna)* **17**, 217 (1966).
38. G. Yamaguchi and K. Arima, *Kagaku Seibutsu* **6**, 351 (1968); *Chem. Abstr.* **69**, 7054 (1968).
39. G. Valentinis, *Ind. Aliment. Agr. (Paris)* **7**, 73 (1968).
40. H. K. Hundley, *J. Assoc. Offic. Anal. Chem.* **51**, 1272 (1968).

41. M. G. Blair and W. Pigman, *Arch. Biochem. Biophys.* **42**, 278 (1953).
42. R. J. Allgeier, R. T. Wisthoff, and F. M. Hildebrandt, *Ind. Eng. Chem.* **44**, 669 (1952); **45**, 489 (1953); **46**, 2023 (1954).
43. J. O. Lampen, L. A. Underkofler, and W. H. Peterson, *J. Biol. Chem.* **146**, 277 (1942).
44. L. A. Underkofler, A. C. Bantz, and W. H. Peterson, *J. Bacteriol.* **45**, 183 (1943).
45. A. N. Hall, G. A. Thomas, K. S. Tiwari, and T. K. Walker, *Arch. Biochem.* **46**, 485 (1953).
46. J. L. Stokes and A. Larsen, *J. Bacteriol.* **49**, 495 (1945).
47. C. Rainbow and G. W. Mitson, *J. Gen. Microbiol.* **9**, 371 (1953).
48. I. O. Foda and R. H. Vaughn, *J. Bacteriol.* **65**, 79 (1953).
49. M. R. R. Rao and J. L. Stokes, *J. Bacteriol.* **65**, 405 (1953); **66**, 634 (1954).
50. W. Litsky and B. S. Tepper, *Growth* **17**, 193 (1953).
51. J. L. Shimwell, *J. Inst. Brewing* **60**, 136 (1954).
52. A. N. Hall, I. Husan, K. S. Tiwari, and T. K. Walker, *J. Appl. Bacteriol.* **19**, 31 (1956).
52a. *Food Engr.* **27** (2), 175 (1955).
52b. J. Frateur, "Essai sur la systématique des acetobacter," *La Céllule (Belgium)*, **53**, 287–392 (1950).
53. A. W. Bittig, *Fruit Prod. J.* **8** (7), 25; (8), 16; (9), 13; (10), 18; (11), 16; (12), 18 (1928); *ibid.* **9**, 19, 25, 59, 86 (1929).
54. C. A. Mitchell, *Vinegar: Its Manufacture and Examination*, 2nd ed., Griffin, London, 1926.
55. H. Wüsteinfeld, *Lehrbuch der Essigfabrikation*, Verlag Paul Parey, Berlin, 1930.
56. A. E. Hansen, *Food Inds.* **7**, 277 (1935).
57. A. Steinmetz, *Fruit Prod. J.* **14**, 299, 313, 328 (1935).
58. F. M. Hildebrandt, *Food Inds.* **13** (8), 47 (1941).
59. L. Worrell, in M. B. Jacobs, ed., *The Chemistry and Technology of Food and Food Products*, Interscience Publishers, Inc., New York, 1951, pp. 1728–1730.
60. R. H. Vaughn, in L. A. Underkofler and R. J. Hickey, eds., *Industrial Fermentations*, Vol. 1, Chemical Publishing Co., New York, 1954, pp. 498–535.
61. S. C. Prescott and C. G. Dunn (revised by C. G. Dunn), *Industrial Microbiology*, McGraw-Hill Book Co., Inc., New York, 1959, pp. 428–473.
62. W. V. Cruess, *Commercial Fruit and Vegetable Products*, McGraw-Hill Book Co., Inc., New York, 1958, pp. 681–707.
63. A. H. Rose, *Industrial Microbiology*, Butterworths, Washington, D.C., 1961, pp. 167–171. L. E. Casida, Jr., *Industrial Microbiology*, John Wiley & Sons, Inc., New York, 1968, pp. 333–344.
64. O. Hromotka and H. Ebner, *Enzymologia* **13**, 369 (1949).
65. G. Haeseler, *Branntweinwirtschaft* **75**, 17 (1949).
66. G. J. Fowler and V. Subramaniam, *J. Indian Inst. Sci.* **6**, 147 (1923).
67. U.S. Pat. 2,707,683 (1955), O. Hromatka and H. Ebner.
68. O. Hromotka and H. Ebner, *Enzymologia* **13**, 369 (1949); **14**, 96 (1950); **15**, 57 (1951).
69. O. Hromotka, G. Kastner, and H. Ebner, *Enzymologia* **15**, 134, 337 (1951).
70. O. Hromotka and W. Polesofsky, *Enzymologia* **24**, 341, 372 (1962); **25**, 37, 52 (1963).
71. O. Hromotka, G. Kastner, T. Gruber, and H. Gsur, *Enzymologia* **25**, 65, 73 (1963).
72. R. F. Cohee and G. Steffen, *Food Eng.* **31**, 58 (1959).
73. U.S. Pat. 2,966,345 (1960), D. W. Burgoon, E. J. Ciabattari, and C. Yeomans.
74. U.S. Pat. 2,997,424 (1961), E. Mayer. E. Mayer, *Food Technol.* **17**, 582 (1963).
75. U.S. Pat. 2,913,343 (1959), A. C. Richardson.
76. R. J. Allgeier and F. M. Hildebrandt, *Advan. Appl. Microbiol.* **2**, 163–182 (1960).
77. R. G. Beaman, in H. J. Peppler, ed., *Microbiology Technology*, Reinhold Publishing Corp., New York, 1967, pp. 344–376.
78. B. A. Davé and R. H. Vaughn, *Am. J. Enol. Viticult.* **20**, 56 (1969).
79. J. Kostal, *Kvasny Prumysl.* **2**, 256 (1956); *Chem. Abstr.* **52**, 1543 (1958).
80. L. G. Notkina, *Vopr. Pishchevoi Brodil'noi Mikrobiol.* **1956**, 136; *Chem. Abstr.* **53**, 20683 (1959).
81. H. Thom, *Kvasny Prumysl.* **8**, 159 (1962); *Chem. Abstr.* **57**, 14290 (1962).
82. V. E. Plyaka, *Tr. Peterogof. Biol. Inst. Leningrad Gas. Univ.* **19**, 88 (1965); *Chem. Abstr.* **67**, 969 (1967).
83. M. Purgal, *Przem. Ferment. Rolny.* **10**, 299 (1966); *Chem. Abstr.* **67**, 970 (1967).
84. A. Lopez, L. W. Johnson, and C. B. Wood, *Appl. Microbiol.* **9**, 425 (1961).

85. K. C. Richardson, *Biotechnol. Bioeng.* **9**, 171 (1967).
86. F. K. Lawler, *Food Eng.* **33**, 68, 82 (1961).
87. U.S. Pat. 2,559,204 (1951), E. P. Wentzelberger.
88. R. C. Zalkan and F. W. Fabian, *Food Technol.* **7**, 453 (1953).

MAYNARD A. JOSLYN
University of California

VINYL CHLORIDE. See Vol. 5, p. 171.

VINYL ETHER, $(CH_2=CH)_2O$. See Anesthetics, Vol. 2, p. 399.

VINYL FIBERS. See Vinyon and related fibers.

VINYLIDENE POLYMERS

POLY(VINYLIDENE FLUORIDE) ELASTOMERS

Vinylidene fluoride monomer, $CH_2=CF_2$, and its homopolymers were dealt with under Fluorine compounds, organic, Vol. 9, p. 840. This article deals with elastomeric copolymers containing vinylidene fluoride. These are manufactured in the United States by E. I. du Pont de Nemours and Co., Inc. under the trademark Viton fluoroelastomer and by Minnesota Mining and Manufacturing Company under the trademarks Fluorel and Kel-F. In Italy, Montecatini Edison S.p.a. manufactures a fluorochemical rubber called Tecnoflon. Fluorochemical rubbers, presumably copolymers containing vinylidene fluoride, are manufactured also in the Soviet Union and by Daikin Cogyo Co., Ltd., in Japan. All these products are high priced relative to general purpose elastomers. Viton, for example, costs $10.00/lb. In addition, its specific gravity is high (1.82–1.86) so that the materials cost of a product made of Viton is a significant factor in its sales price. Its use in a highly competitive industry, such as the automotive industry, is a testimonial to its premium performance or "value in use."

Manufacture

The elastomeric copolymers of vinylidene fluoride with hexafluoropropylene, chlorotrifluoroethylene, or pentafluoropropylene are prepared in free-radical polymerization systems, eg with peroxide or azo-type catalysts, or by irradiation. Patents to du Pont (1,2), the 3M Company through the M. W. Kellogg Company (3), and to Montecatini (4), all indicate that an emulsion polymerization process can be effectively used in the synthesis of copolymers. These processes are operated under several hundred lb pressure at temperatures between 40 and 100°C with a peroxy initiator, such as an organic peroxide, hydrogen peroxide, a peroxydisulfate or a peroxydisulfate–bisulfite initiation system. Fairly careful control of composition is required to obtain optimum elastomeric properties. Polymers containing a very high proportion of vinylidene fluoride are classified as plastics rather than elastomers. Polymerization to low-molecular-weight fluid elastomers can be carried out with sources of free radicals soluble in an organic polymer solvent, which also acts as a chain-transfer agent (5).

The fluorocarbon elastomers currently of greatest commercial importance and which are discussed here, are dipolymers of vinylidene fluoride and hexafluoropropylene in the approximate weight ratio of 60:40 and a terpolymer of vinylidene fluoride, hexafluoropropylene, and tetrafluoroethylene. They contain $-CH_2CF_2-$ units and $-CF(CF_3)CF_2-$ units. Viton A, A-35, AHV, B, B-50, and E-60, as well as Fluorel 2140 and 2160, all fit these descriptions. The following discussion pertains only to these types.

Processing. Vinylidene fluoride-based elastomers are available in a wide variety of viscosity grades. The low Mooney viscosity of Viton A-35 (ML-10 at 100°C of about 35) and the high value for Viton AHV (ML-10 at 121°C of over 150) exemplify the range. Processing characteristics are closely related to viscosity. Viton differs from many other elastomers in that, for a given viscosity, it has relatively low "nerve," ie a short-lived visco-elastic memory. Viscosity breakdown does not occur to any large extent on mastication.

Most of the fluoroelastomers described above may be mixed on open, two-roll mills or in internal mixers in much the same manner as other elastomers. Their compounds likewise are processed on equipment conventional to the rubber industry. With minor consideration to compounding, they are readily molded, extruded, or calendered. Extrudates of the higher-viscosity compounds tend to have a rough, sandpaperlike finish which is best overcome by using a cool, long-land die. Water-jacketed aluminum dies have been used successfully. In fact, with the exception of molding, most processing operations are best carried out at temperatures lower than those used with other elastomers.

Fluoroelastomer compounds are virtually devoid of tack. Building operations (eg, roll building) entail, therefore, the use of a solvent such as methyl ethyl ketone to wet the plies before laminating.

Curing. The best vulcanizate properties are achieved by a two-step curing process. The first step entails the application of both heat and pressure, eg molding in a press or fabric wrapping, followed by curing in a pressurized open-steam autoclave. The second step is called a postcure and is carried out in an air oven. For thick sections, the temperature of the oven usually must be raised gradually or stepwise, to prevent fissuring of the part. A typical curing process would take place for 10–30 min in a press at 325–350°F, depending on the size of the molding, followed by 24 hr at 450°F in an oven. Depending again on the size of the part, the oven postcure may require a 4–8-hr rise to curing temperature. Postcuring is required for vulcanizates to have good heat and compression-set resistance. For solvent, oil, and chemical resistance alone, postcuring is not required.

The following theory of the curing of Viton fluoroelastomers at least partially explains why these conditions are required.

Curing Chemistry. Both free-radical and ionic processes have been recommended for the curing of vinylidene fluoride copolymer elastomers. The free-radical crosslinking process, comparatively little used, involves the thermal decomposition of relatively large amounts, up to 5%, of a source of energy-rich radicals, such as benzoyl peroxide (6) or difluorodiazine (7). It has been envisioned that, as the initiator decomposes, the free radicals formed abstract hydrogen from the methylene groups along the polymer chain, and the resultant polymer radicals interact, directly or through the intermediacy of radical traps, to form the crosslink. Magnesium oxide, as an acid acceptor, and carbon-black filler are usually included in the recipe. A postcure of the molded articles is usually required to produce satisfactory vulcanizate properties. The good resistance of the vinylidene fluoride polymers to attack by free radicals makes this curing chemistry relatively inefficient; radiation curing, presumably by radical processes, has also not been practiced extensively.

The commonly used curing chemistry for the vinylidene fluoride copolymer elastomer is believed to involve initially the removal of hydrogen fluoride from the polymer chain by reaction of the polymer with base (8–11). The resulting highly polarized —CH=CF— units react with each other to form crosslinks, if no other reagent is present, but react with suitable bisnucleophiles (diamines, bisphenols, bisthiols) much more readily. A variety of bases have been used for the dehydrofluorination reaction, including primary, secondary, and tertiary amines, tetraalkyl guanidines, and strongly alkaline metal alkoxides and hydroxides. In various studies the initial formation of the hydrogen fluoride reaction product with most of these bases has been established.

These bases convert the vinylidene fluoride polymers to a crosslinked condition relatively slowly. Better vulcanizate properties are achieved if the recipe contains a bisnucleophile capable of reacting with two of the —CF=CH— units postulated to have been created by the dehydrofluorination described above. Bisprimary amines (which act as their own bases) (11), bisthiols (12), and bisphenols (13), have been utilized.

With all of these systems it is advantageous to provide a metal oxide acid acceptor to remove hydrogen fluoride from the system and to deactivate any source of easily available fluoride ion. Magnesium, calcium, and lead oxides are preferred, perhaps as a consequence of the insolubility and stability of the corresponding fluorides.

The function of the postcure step, usually necessary to produce optimum vulcanizate properties, is the removal of volatile byproducts, in many cases water, from the molded articles, rather than to establish additional crosslinks (10). The enhanced physical properties seen after postcure may also reflect improved polymer-filler interaction as a result of the volatization of moisture.

The crosslinking with hexamethylenediamine can be represented as follows (11):

$$
\begin{array}{c}
\text{CF}_3 \\
| \\
+(\text{CH}=\text{CFCF}_2\text{CF})_n
\end{array}
\qquad
\begin{array}{c}
\text{CF}_3 \\
| \\
+(\text{CH}_2\text{CFCF}_2\text{CF})_n \\
| \\
\text{NH} \\
| \\
(\text{CH}_2)_6 \\
| \\
\text{NH} \\
| \\
+(\text{CH}_2\text{CFCF}_2\text{CF})_n \\
| \\
\text{CF}_3
\end{array}
$$

$$
+\quad
\begin{array}{c}
\text{NH}_2 \\
| \\
(\text{CH}_2)_6 \\
| \\
\text{NH}_2
\end{array}
\qquad\longrightarrow
$$

$$
\begin{array}{c}
+(\text{CH}=\text{CFCF}_2\text{CF})_n \\
| \\
\text{CF}_3
\end{array}
$$

$$
\longrightarrow
\begin{array}{c}
\text{CF}_3 \\
| \\
+(\text{CH}_2\text{CCF}_2\text{CF})_n \\
\| \\
\text{N} \\
| \\
(\text{CH}_2)_6 \quad + \; 2\ \text{HF} \\
| \\
\text{N} \\
\| \\
+(\text{CH}_2\text{CCF}_2\text{CF})_n \\
| \\
\text{CF}_3
\end{array}
$$

When bisphenols are used as the nucleophiles instead of the bisprimary amine, the crosslink is believed to contain ether linkages in the place of the Schiff base structure shown.

Compounding. Practically all curing systems include a metal oxide. Magnesia is by far the most commonly used but litharge (PbO), calcium oxide, and zinc oxide, together with dibasic lead phosphite are used occasionally for special properties. Amounts vary from 4 to 20 phr.

Either of two different types of cross-linking agent are currently used commercially. The older is based on diamines and includes Diak (Du Pont) No. 1, hexamethylene diaminemonocarbamate; Diak No. 31, N,N'-dicinnamylidene-1,6 hexanediamine ($C_6H_5CH\!\!=\!\!CHCH\!\!=\!\!N(CH_2)_3)_2$; and Diak No. 4 (an alicyclic amine salt). Amounts vary from 1 to 3 phr. The second and more recently developed type is an aromatic dihydroxy compound of which Diak No. 5, hydroquinone, is an example. Aromatic dihydroxy compounds require an alkaline activator, such as calcium hydroxide and/or an organic base to effect a crosslink. The aromatic dihydroxy–alkaline activator curing system provides an excellent balance between processing safety (Mooney scorch resistance, see Vol. 17, p. 608) and rapid cure rate in the press. More important, it affords a means for obtaining vulcanizates with excellent resistance to compression set.

Filler Loading. One of the unique and unusual processing features in compounding fluoroelastomers is the limitation in the amount and type of filler or reinforcing agent that can be used. As discussed subsequently, limitations on the permissible amount and type of plasticizer prohibit loading with large amounts, especially of highly reinforcing fillers, for they would cause processing to become intolerably difficult and the vulcanizate hardness and stiffness excessive for all but a few end-use applications. Generally, a loading of a soft carbon black is used, such as 20–40 phr of MT Carbon black. A fibrous calcium or magnesium silicate in the same amount makes a good nonblack filler. Without filler, tensile strength is low.

Plasticizer. Choice of a plasticizer is limited by the fact that most of those that are useful in other elastomers are too volatile above 400°F and, therefore, are lost during the oven postcure or during subsequent service at high temperature. Furthermore, because fluoroelastomer vulcanizates are very solvent resistant, the polymer and its vulcanizates have limited compatibility with most fluids used as plasticizers. Fluorosilicone oils and certain high-molecular-weight aromatic hydrocarbons are the most useful. These can be used to aid processing or to reduce vulcanizate hardness. No plasticizers that improve low-temperature properties and are practical to use, have so far been found.

Processing Aid. Viton LM, a low-molecular-weight vinylidene fluoride hexafluoropropylene copolymer, may be used to lower the viscosity of uncured compounds for ease of processing. Generally, 10–20 phr suffice. It is retained during postcuring but the product may lose some weight as a result of loss of volatiles during high-temperature service. A slight reduction in state of cure also results from the use of Viton LM.

Pigmentation. Viton fluoroelastomers may be compounded to give bright, color-stable vulcanizates if the dihydroxy curing system is used. Hitherto, color was limited to dark shades because the diamine curing agents discolored during postcure or subsequent heat service. With nonblack-filler loading, commonly used inorganic pigments and titanium dioxide offer a wide range of colors.

Properties of Vulcanizates

Fluoroelastomers in general are best known and used for their inherent resistance to the deteriorating influences of heat, oils, solvents, and chemicals alone or in combination. The retention of mechanical properties after prolonged exposure to temperatures of 400°F and above is remarkable.

In reading the discussion that follows, it must be borne in mind that compounding can enhance certain properties, while compromising others to some degree. Unless the changes in compounding are drastic, however, the properties described are typical of most products made from fluoroelastomers.

General-purpose fluoroelastomer vulcanizates have a Durometer A hardness (see Vol. 10, p. 818) of about 70. Formulations can be provided over the range of 50 to 95. Diamine-cured vulcanizates drop in hardness 15–20 points as the temperature is raised from 75 to 400°F. Dihydroxy-cured vulcanizates do not change in hardness over this temperature range.

Tensile strength values of 2000 psi at 75°F are typical. At higher temperatures, strength is less, eg 1000 psi at 200°F or 300 psi at 500°F. Elongation at break varies greatly with the formulation and state of cure but generally falls within the range of 125–300% at 75°F, dropping off, like the modulus, as the temperature is increased.

Fluoroelastomers have excellent resistance to degradation by heat. Vulcanizates remain usefully elastic almost indefinitely when exposed to laboratory air-oven aging up to 400°F or to intermittent test exposures up to 500°F. Continuous-service limits are generally considered to be as follows:

Continuous-service limits, hr	Temperature, °F
3000	450
1000	500
240	550
48	600

The solvent resistance of fluoroelastomers is better than that of other commercially available elastomers with respect to aliphatic and aromatic hydrocarbons, chlorinated solvents, oils, fuels, lubricants, and most mineral acids. It is not suitable for use with ketones, certain esters and ethers, some amines, hot anhydrous hydrofluoric acid or chlorosulfonic acid, and a few proprietary fluids.

The brittle temperature (ASTM D-746) varies with the thickness of the specimen. Typical values are the following: $-40°F$ for 0.075 in. thick and $-55°F$ for 0.025 in. thick. The glass-transition temperature, T_g, as measured by the Clash-Berg Stiffness Test (ASTM D-1043) is about 0°F. Temperature retraction (ASTM D-1329), T_{10}, is $-5°F$.

Using the dihydroxy curing system, very low compression-set values are attainable over a wide temperature range, particular with polymers recently developed specifically for this property. ASTM D-395 Method B values for a Viton E-60 compound are shown in Table 1 for pellets (0.5 in. × 1.129 in.).

Fluoroelastomers are essentially ozone proof. Aging in oxygen under pressure and at high temperature has virtually no effect.

For steam resistance special compounding is required but, even so, the fluoroelastomers discussed herein are not particularly good in steam service. Resistance to

Table 1. Compression-Set Values for a Viton E-60 Compound

Time, hr	Temperature, °F	Compression set, %
70	75	10
22	400	9
70	400	19
168	400	29
336	400	36

both β and γ radiation is about in the middle for elastomers in general. Gas permeability is low, being about equivalent to that of butyl rubber. Fungus does not grow readily on typical vulcanizates. Typical values for dc resistivity are on the order of 2×10^{13} Ω-cm, for specific inductive capacity around 15, for power factor about 5%, and for dielectric strength 500 V/mil. Vulcanizates of Viton fluoroelastomer showed little or no change in physical properties or appearance after ten years of outdoor exposure in southern Florida. Resilience is not particularly high, running about 50% at room temperature for a 70 Durometer A hardness vulcanizate as measured by ASTM D-945. Vulcanizates of the fluoroelastomers under discussion have good tear strength and excellent abrasion resistance compared to those of most general-purpose elastomers.

An amendment to the Federal Food, Drug and Cosmetic Act provides for the use of vulcanizates of Viton A, Viton AHV, and Viton B fluoroelastomers, containing magnesia and Diak No. 1 or Diak No. 4 in specified maximum amounts in the formulation of rubber articles intended for repeated food-contact use.

Uses

The commercial use of copolymers of vinylidene fluoride and hexafluoropropylene, as well as those containing tetrafluoroethylene, are governed largely by the excellent resistance to heat, fluids, and compression set of their vulcanizates. Elastomeric seals for industrial, aerospace and automotive equipment constitute the largest markets. These are mostly O-rings of all sizes, flat or lathe-cut gaskets, and, to a lesser extent, lip-type rotating or reciprocating shaft seals. Coated fabrics for diaphragms and sheet goods account for considerable consumption. Hose linings, tubing and industrial gloves for chemical service, rolls for hot or corrosive service, and extruded goods for such purposes as autoclave and oven seals constitute examples of the varied end uses to which fluoroelastomers are put.

Bibliography

NOTE: Much information can be obtained from the trade bulletins of the manufacturing companies.

1. U.S. Pat. 3,051,677 (1962), D. R. Rexford (to E. I. du Pont de Nemours and Co., Inc.).
2. U.S. Pat. 2,968,649 (1961), J. R. Pailthorp and H. E. Schroeder (to E. I. du Pont de Nemours and Co., Inc.).
3. U.S. Pat. 2,689,241 (1954), A. L. Dittman, A. J. Passino, and J. M. Wrightson (to The M. W. Kellogg Co.).
4. U.S. Pat. 3,331,823 (1967), D. Sianesi, G. C. Bernardi, and A. Reggio (to Montecatini Società Generale per l'Industrià Mineraria e Chimica).
5. U.S. Pat. 3,069,401 (1962), G. A. Gallagher (to E. I. du Pont de Nemours and Co., Inc.).
6. A. S. Novikov, F. A. Galil-Ogly, and N. S. Gilinskaya, *Kauchuk i Rezina* **21** (2), 4–10 (1962).
7. J. F. Smith and J. R. Albin, *Ind. Eng. Chem. Prod. Res. Develop.* **2** (4), 284–286 (1963).
8. K. L. Paciorek, L. C. Mitchell, and C. T. Lenk, *J. Polymer Sci.* **45**, 405–413 (1960).

9. J. F. Smith, *Rubber World* **142** (3), 102–107 (1960).
10. J. F. Smith and G. T. Perkins, *Rubber Plastics Age* **41** (11), 1362 (1960).
11. J. F. Smith, *Proc. Intern. Rubber Conf., Amer. Chem. Soc.*, Washington, D.C., 1959, pp. 575–581.
12. J. F. Smith, *Rubber World* **148** (2), 263–266 (1959).
13. U.S. Pat. 3,142,660 (1964), R. P. Conger (to U.S. Rubber Co.).

D. C. Thompson and A. L. Barney
E. I. du Pont de Nemours & Co., Inc.

POLY(VINYLIDENE CHLORIDE)

The two outstanding characteristics of vinylidene chloride (VDC) polymers are thermal instability and impermeability. The present commercial success of these materials is due to the fact that the instability problem was overcome so that their valuable properties could be exploited. The techniques, ie, copolymerization and plasticization, were developed by Ralph Wiley and co-workers at The Dow Chemical Company during the period of 1932–1939. The commercialization of these polymers under the trade name Saran began in 1939.

The homopolymer, saran A, though it has valuable properties, has not been used to any extent because of the difficulty in fabricating it. A great many copolymers were prepared, but only three types are of commercial interest: vinylidene chloride–vinyl chloride copolymers, saran B; vinylidene chloride–alkyl acrylate copolymers, saran C; and vinylidene chloride–acrylonitrile copolymers, saran F.

In the United States saran is now a generic term for high VDC-content polymers. In other countries it is still a trademark of The Dow Chemical Co. Unfortunately, the practice of using "poly(vinylidene chloride)" and "saran" synonymously developed even in the very early days. As a consequence, many of the materials identified in the literature as poly(vinylidene chloride) were actually copolymers of unknown composition.

The chemistry and properties of vinylidene chloride polymers are discussed in Schildknecht (1). A more detailed article will appear in the Encyclopedia of Polymer Science and Technology (Wiley-Interscience).

Vinylidene Chloride

Properties. Pure vinylidene chloride (1,1-dichloroethylene) is a colorless, mobile liquid with a characteristic "sweet" odor. Its properties are summarized in Table 1.

Vinylidene chloride is soluble in most polar and nonpolar organic solvents. It forms an azeotrope with 6% methanol (2). Distillation of the azeotrope, followed by extraction of the methanol with water, is a method of purification.

Table 1. Physical Properties of Vinylidene Chloride Monomer

Property	Value
boiling point at 760 mm Hg, °C	31.56
freezing point, °C	−122.5
flash point, °F	
Tag open-cup	3
Tag closed-cup	−2
autoignition temperature, °F	1031
density	
at −20°C	1.2902
at 0°C	1.2517
at 20°C	1.2132
refractive index, n_D^{20}	1.42468
specific heat, cal/g	0.275
heat of combustion, kcal/mole	261.93
heat of polymerization, kcal/mole	18.0
latent heat of vaporization	
at boiling point, cal/mole	6257
viscosity at 20°C, cP	0.3302
explosive mixture with air, % by vol	
lower limit	6.5
upper limit	15.5
dielectric constant at 16°C	4.67
solubility of VDC in H_2O at 25°C, wt %	0.25
solubility of H_2O in VDC at 25°C, wt %	0.035
vapor pressure, mm Hg	$\log P_{mm} = 0,98200 - 1104.29/(t + 237.697)$

MANUFACTURE

Vinylidene chloride monomer may be conveniently prepared in the laboratory by the reaction of 1,1,2-trichloroethane (see Vol. 5, p. 157) with aqueous alkali:

$$2\ CH_2ClCHCl_2 + Ca(OH)_2 \xrightarrow{90°C} 2\ CH_2{=}CCl_2 + CaCl_2 + 2\ H_2O$$

Other methods are based on bromochloroethane, trichloroethyl acetate, tetrachloroethane, and catalytic cracking of trichloroethane (3). Vinylidene chloride is prepared commercially by the dehydrochlorination of 1,1,2-trichloroethane by lime or caustic used in slight excess (2–10%) (4). A liquid-phase reaction, operated continuously at 98–99°C, yields approximately 90%. Performance with caustic is better than with lime. Vinylidene chloride is purified by washing with water, drying, and fractional distillation. An inhibitor is usually added at this point. Two commercial grades are available: one with 200 ppm of the monomethyl ether of hydroquinone, MEHQ, and the other with 0.6–0.8 wt % phenol. A typical analysis is shown in Table 2. Many inhibitors for the polymerization of vinylidene chloride have been described in patents, but the two listed above are most often used. For many polymerizations, the low level of MEHQ does not need to be removed; it can be overcome by polymerization initiators. But phenol, which is used at a higher level, must be removed, either by alkali extraction or distillation. Vinylidene chloride with inhibitor removed should be refrigerated at −10°C under a nitrogen atmosphere in the dark and in a nickel- or baked phenolic-lined storage tank. If not used within one day, it should be reinhibited.

Table 2. Typical Analysis of Vinylidene Chloride (5)

Constituent, ppm[a]	Value
vinylidene chloride, [b] wt %	99.7
vinyl chloride	850
cis-1,2-dichloroethylene	500
trans-1,2-dichloroethylene	1500
1,1-dichloroethane	10
ethylene dichloride	10
1,1,1-trichloroethane	150
trichloroethylene	10
inhibitor	
MEHQ grade	200
phenol grade, wt%	0.6–0.8

[a] Except where otherwise indicated.
[b] Excluding inhibitor.

Vinylidene chloride is produced commercially in the United States by The Dow Chemical Company, PPG Industries, and Vulcan Materials Co; in England, by Imperial Chemical Industries, Ltd.; in Germany, by Badische Anilin- und Soda-Fabrik and Chemische Werke-Huels; in Japan, by Asahi-Dow Chemical Company and Kureha Chemical Company.

Safety (5). Vinylidene chloride is highly volatile and, when free of decomposition products, has a mild "sweet" odor. Its warning properties are ordinarily inadequate to prevent excessive exposure. Inhalation of vapor presents a definite hazard which is readily controlled by observance of precautions commonly taken in the chemical industry. A single exposure for a few minutes to a high concentration of vinylidene chloride vapor (such as 4000 ppm) rapidly produces a "drunkenness" which may progress to unconsciousness on prolonged exposure. However, prompt and complete recovery from the anesthetic effects takes place when the exposure is of short duration. A single, prolonged exposure and repeated short-term exposures can be dangerous, even when the concentrations of vapor are too low to cause an anesthetic effect. They may produce organic injury to the kidneys and liver. For repeated exposures (8 hr per day, five days a week) the vapor concentration of vinylidene chloride should be below a time-weighted average of 25 ppm. Single exposure of about 1000 ppm for up to one hr or 200 ppm for up to 8 hr will probably not result in injury.

If a person should be affected or overcome from breathing vinylidene chloride vapors, he should be removed to fresh air at once, be made to rest, kept warm, and receive immediate medical attention.

Precautions should be taken to prevent skin contact with vinylidene chloride. The liquid is irritating to the skin after only a few minutes of contact. The inhibitor (MEHQ or phenol) may be partly responsible for this irritation. There appears to be little likelihood of systemic injury from skin absorption of the amounts of phenol to be encountered in handling inhibited vinylidene chloride. If contact does occur, the affected skin area should be washed with soap and water.

Inhibited vinylidene chloride is moderately irritating to the eyes. Contact can be expected to cause pain and conjunctival irritation and, possibly, some transient corneal injury and iritis; permanent damage is not likely. Chemical safety goggles should be worn when handling vinylidene chloride.

Vinylidene chloride vapor is flammable at concentrations between 6.5% and 15.5% by volume in air.

Uninhibited vinylidene chloride, in the presence of air or oxygen, forms a complex peroxide compound at temperatures as low as −40°C, which is violently explosive. Decomposition products of vinylidene chloride peroxides are formaldehyde, phosgene, and hydrochloric acid. A sharp acrid odor indicates oxygen exposure and probable presence of peroxides. This can be confirmed by the liberation of iodine from a slightly acidified dilute potassium iodide solution. Any formation of insoluble polymer may also indicate peroxide formation. The peroxide adsorbs on the precipitated polymer and any separation of the polymer may result in an explosive composition. Any dry composition containing more than about 15% peroxide detonates from a slight mechanical shock or from heat. Vinylidene chloride which contains peroxides may be purified by washing several times, either with 10% sodium hydroxide at 25°C, or with a fresh 5% sodium bisulfite solution. Any residues found in vinylidene chloride containers should be handled with great care, and the peroxides destroyed with water at room temperature.

Copper, aluminum, and their alloys should not be used in handling vinylidene chloride. Under the proper conditions, copper can react with acetylenic impurities to form copper acetylides, and aluminum with the vinylidene chloride to form aluminum chloralkyls. Both of these compounds are extremely reactive and potentially hazardous.

Polymerization

Vinylidene chloride polymerizes by both ionic and free-radical reactions. Processes based on the latter are, by far, the more common. For a review of the literature on free-radical polymerization of VDC, see reference 6. VDC is of average reactivity when compared to other unsaturated monomers. The chlorine substituents stabilize radicals in the intermediate state of an addition reaction. Since they are also strongly electron withdrawing, they polarize the double bond, making it susceptible to anionic attack. For the same reason, a carbonium ion intermediate would not be favored.

The 1,1-disubstitution causes significant steric interactions in the polymer (7), as evident from the heat of polymerization (see Table 1). When corrected for the heat of fusion, it is significantly less than the theoretical value of 20 kcal/mole for the process of converting a double bond to two single bonds. The steric strain is apparently not significant in the addition step since VDC polymerizes easily. Nor is it sufficient to favor depolymerization; the estimated ceiling temperature for PVDC is about 400°C.

Homopolymerization. The free-radical polymerization of VDC has been carried out in slurry, suspension, or emulsion. Solution polymerization in a medium that dissolves both monomer and polymer has not been studied. Slurry polymerizations are normally used only at the laboratory level. They can be carried out in mass or in common solvents, such as benzene. PVDC is insoluble in these media and separates from the liquid phase as a crystalline powder. The heterogeneity of the reaction makes stirring and heat transfer difficult; as a consequence, these reactions cannot be easily controlled on a large scale. Aqueous emulsion or suspension reactions are preferred for any large-scale operation. Slurry reactions are usually initiated by the thermal decomposition of organic peroxides or azo compounds. Purely thermal initiation can

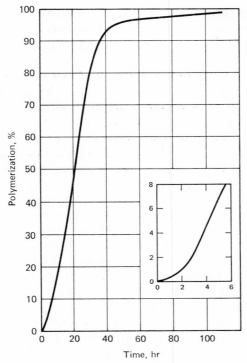

Fig. 1. Mass polymerization of vinylidene chloride at 45°C with 0.5% benzoyl peroxide as initiator (11).

occur, but rates are very slow (8). The spontaneous polymerization of VDC, so often observed when the monomer is stored at room temperature, is due to peroxides formed by the reaction of VDC with oxygen. Very pure monomer does not polymerize under these conditions. Irradiation also induces polymerization of VDC. Sources include both uv (9) and γ rays (10).

The heterogeneous nature of the mass polymerization of VDC is apparent from the rapid development of turbidity in the reaction medium following initiation. The turbidity is due to the presence of minute PVDC crystals. As the reaction progresses, the crystalline phase grows at the expense of the liquid. Eventually, a point is reached where the liquid slurry solidifies into a solid mass. A typical conversion-time curve is shown in Figure 1 for a benzoyl peroxide catalyzed mass polymerization (11). The reaction can be divided into three stages. The first stage is characterized by a rapidly increasing rate. The rate levels off in the second stage to a fairly constant value. This is often called the steady-state region. Throughout the first two stages, monomer concentration remains constant because the polymer separates into another phase. The third stage shows a gradual decrease in rate to zero due to depletion of monomer supply. Since the mass solidifies while monomer is still present (usually at conversions below 20%) further polymerization generates void space. The final solid, therefore, is opaque and quite porous. A similar pattern of behavior is observed when vinylidene chloride is polymerized in solvents such as benzene that do not dissolve or swell the polymer. In this case, of course, the reaction mixture may not solidify if the monomer concentration is low.

Heterogeneous Polymerization. This is characteristic of a number of monomers, including vinyl chloride and acrylonitrile. A completely satisfactory mechanism for these reactions has not yet evolved. In the case of VDC, this is at least partly due to a lack of experimental data. Only two kinetic studies have been reported: in the work of Burnett and Melville (9), uv initiation in mass and in cyclohexane was studied; Bengough and Norrish (12) investigated the reaction initiated by benzoyl peroxide. In neither case was the investigation broad enough to pin down the mechanism. There is little doubt that the solid polymer phase is the key factor. Until recently, the structure of the solid phase was ignored, but experiments by Cotman et al. (13) now show this to be a significant factor. These authors have made a thorough comparison of the various theories proposed before 1967. They conclude that none are adequate; they suggest, pessimistically, that the problem is probably unresolvable because of the complexity of the solid phase.

Emulsion and suspension reactions are doubly heterogeneous. The polymer is insoluble in the monomer, and both are insoluble in water. Suspension reactions are similar in behavior to slurry reactions. Oil-soluble initiators are employed, so the monomer–polymer droplet is like a small mass reaction. Emulsion polymerizations are more complex. Since the monomer is insoluble in the polymer particle, the simple Smith-Ewart theory should not apply (14). It is most likely that polymerization occurs on the particle surface.

Redox initiator systems are normally used in the emulsion polymerization of VDC in order to get high rates at low temperatures. Reactions must be carried out below about 80°C to prevent degradation of the polymer. PVDC in emulsion is also attacked by aqueous base. Therefore, reactions should be carried out at low pH.

The instability of PVDC is one of the reasons that ionic initiation of VDC polymerization has not been used extensively. Many of the common catalysts either react with the polymer or catalyze its degradation. For example, butyllithium polymerizes VDC via an anionic mechanism; but the product is a low-molecular-weight, discolored polymer with a low chlorine content (15). Cationic polymerization of VDC seems unlikely in view of its structure (16). Some available data, however, suggest the possibility. In the low-temperature radiation-induced copolymerization of VDC with isobutylene, reactivity ratios varied markedly with temperature, indicating a change from a free-radical mechanism (17). Coordination complex catalysts may also polymerize VDC by a nonradical mechanism. Again, this speculation is based on copolymerization studies.

Copolymerization. The importance of VDC as a monomer is due to its ability to copolymerize with other vinyl monomers. It is most easily copolymerized with acrylates, but it also reacts, though more slowly, with monomers such as styrene that form highly resonance-stabilized radicals. Reactivity ratios with various monomers are shown in Table 3. See also Vol. 16, p. 232. Many other copolymers have been prepared using monomers for which the reactivity ratios are not known. The commercially important copolymers include those with vinyl chloride (VC), acrylonitrile, or various alkylacrylates, but many commercial saran polymers contain three or more components, VDC being the major one. Usually, one component is introduced to improve the processability or solubility of the polymer. The others are added to modify specific end-use properties. Most of these compositions have been described in the patent literature. A partial listing of terpolymers is given in Table 4. A tetrapolymer

Table 3. The Reactivity of Vinylidene Chloride with Important Monomers (18)

Monomer	$r_1{}^a$	r_2
vinylidene chloride		
styrene	0.14	2.0
vinyl chloride	3.2	0.3
acrylonitrile	0.37	0.91
methyl acrylate	1.0	1.0
methyl methacrylate	0.24	2.53
vinyl acetate	6	0.1

^{a} VDC.

Table 4. Terpolymers^a of Vinylidene Chloride (19)

Second monomer	Third monomer
acrylates or acrylonitrile	isopropenyl acetate
acrylic or methacrylic acid	vinyl or acrylate ester
acrylonitrile	butadiene
	methyl methacrylate
	α-methylstyrene
	vinylidene chloride
	5-vinyl-2-picoline
α-alkylacrylates	vinyl chloride
alkyl maleates	vinyl esters
allyl chloride	diallyl fumarate
1,3-butadiene	chloroprene
butadiene	ethyl acrylate
	isobutylene
	methyl methacrylate
	α-methylstyrene
	styrene
	vinyl acetate
	vinyl chloride
	vinyl compounds
N-(2-formamidoethyl)acrylamide	vinyl chloride
isobutylene	"polymerizable substances"
1-chloro-1-bromoethylene	vinylidene bromide
glycidyl methacrylate	(2-methacryloyloxyethyl)-diethylammonium methyl sulfate
methyl acrylate	trichloroethylene
	vinyl chloride
styrene	vinyl chloride
vinyl halides	acrylic or methacrylic compound

^{a} Limited to polymers containing at least 50% VDC.

has been prepared with acrylic acid, acrylonitrile (VCN), and methyl methacrylate (MMA).

Mass copolymerizations yielding high VDC-content copolymers are normally heterogeneous. Two of the most important pairs, VDC/VC and VDC/VCN, are heterogeneous over the entire composition range. In both cases, at either composition extreme, the product separates initially in a powdery form, but for intermediate compositions, the reaction mixture may only gel. Copolymers in this composition range are swollen, but not completely dissolved, by the monomer mixture at normal polymerization temperatures. Copolymers containing more than 10–15 mole % acrylate are normally soluble. These reactions are therefore homogeneous, and if carried to completion, yield clear solid castings of the copolymer. Most copolymerizations can be carried out in solution because of the greater solubility of the copolymers in common solvents.

During copolymerization, one monomer may add to the copolymer more rapidly than the other. Except for the unusual case of equal reactivity ratios, therefore, batch reactions carried to completion yield polymers of broad composition distribution. More often than not, this is an undesirable result.

VDC copolymerizes randomly with methyl acrylate and *nearly* so with other acrylates. Very severe composition drift occurs, however, in copolymerizations with VC or methacrylates. Several methods have been developed to produce homogeneous copolymers regardless of reactivity ratio. These methods are applicable, mainly, to emulsion and suspension where adequate stirring can be maintained. (See below under Commercial methods.) Rates of copolymerization of VDC with small amounts of a second monomer are normally lower than its rate of homopolymerization. The kinetics of the copolymerization of VDC and VC have been studied (20).

Copolymers of VDC can also be prepared by methods other than free-radical polymerization. The possible cationic copolymerization by irradiation of VDC and isobutylene at low temperatures was already mentioned (see p. 280). Irradiation of VDC and styrene at low temperatures, however, generates a conventional radical reaction so this is not a general method (10). Copolymers have been prepared using various kinds of organometallic catalysts (21–24). The butyllithium reactions are clearly anionic, based on known chemistry and observed reactivity ratios (22). The mechanism by which more complex catalyst systems operate is not well established (23,24).

The systems described above yield normal copolymers. Ziegler-type catalysts can also be used to make block copolymers of VDC (25). This method is of recent origin and has not yet been developed extensively. Graft copolymers containing VDC can normally be prepared by free-radical methods but have not been thoroughly investigated. Irradiation of VDC vapor in the presence of a polymeric substrate produces a graft (26). Polymerization of VDC in the presence of vinyl acetate copolymers using chemical initiators also yields grafted products (27). Vinyl chloride can be grafted to a VDC–olefin copolymer in the same way (28). None of these materials have been well characterized.

COMMERCIAL METHODS OF POLYMERIZATION

Processes that are essentially modifications of laboratory methods allowing operation on a larger scale are used for commercial preparation of vinylidene chloride polymers. The nature of the end use dictates the polymer characteristics and, to some extent, the method of manufacture. Emulsion polymerization and suspension polymerization are the preferred industrial processes. Either process is carried out in a closed, stirred reactor, glass-lined and jacketed for heating and cooling. The reactor must be purged of oxygen and the water and monomer must be free of metallic impurities to avoid an adverse effect on the thermal stability of the polymer.

Emulsion polymerization is used commercially to make vinylidene chloride copolymers; they are utilized in two ways. In some applications, the latex which results is used directly (usually with additional stabilizing ingredients) as a coating vehicle to apply polymer to various substrates; in others, the polymer is first isolated from the latex before using. In applications where the polymer is not used in latex form, the emulsion process is used because it gives advantages over alternative methods that outweigh the disadvantages. In those cases, the polymer is recovered in dry powder form, usually by coagulating the latex with electrolyte, followed by washing and drying.

The principal advantages of emulsion polymerization as a process for preparing vinylidene chloride polymers are the following: (1) High-molecular-weight polymers can be produced with reasonable reaction times (especially pertinent to copolymers with vinyl chloride). The initiation and propagation steps can be controlled more independently than is the case for the suspension process. (2) Monomer can be added during the polymerization to maintain copolymer-composition control.

The disadvantages of emulsion polymerization essentially result from the relatively high concentration of additives in the recipe. The water-soluble initiators, activators, and surface-active agents generally cause the polymer to have greater water sensitivity, poorer electrical properties, and poorer heat and light stability. Recovery of the polymer by coagulation, washing, and drying improves, to some extent, these properties over those of the polymer deposited in latex form.

A typical recipe for batch emulsion polymerization is shown in Table 5 (29).

Table 5. Typical Recipe for Batch Emulsion Polymerization (29)

Ingredient	Parts by wt
vinylidene chloride	78
vinyl chloride	22
water	180
potassium peroxysulfate	0.22
sodium bisulfite	0.11
aerosol[a] MA(80%)	3.58
(dihexyl sodium sulfosuccinate)	
nitric acid (69%)	0.07

[a] Registered trademark, American Cyanamid Co.

A reaction time of 7–8 hr at 30°C is required for 95–98% conversion. A latex is produced with an average particle diameter of 1000–1500 Å. In addition to these ingredients, other modifying ingredients may be present, such as other colloidal protective agents (eg gelatin or carboxymethylcellulose), initiator activators (redox types), chelates, plasticizers, stabilizers, and chain-transfer agents.

Surfactants used commercially are generally anionic emulsifiers, alone or in combination with nonionic types. Representative anionic emulsifiers are the sodium alkylarene sulfonates, the alkyl esters of sodium sulfosuccinic acid, and the sodium salts of fatty alcohol sulfates. Nonionic emulsifiers are of the ethoxylated alkylphenol type.

Free-radical sources other than peroxysulfates may be used, such as hydrogen peroxide, organic hydroperoxides, peroxyborates, and peroxycarbonates. Many of these are used in "redox" pairs where an activator promotes the decomposition of the peroxy compound. Examples are peroxysulfate or perchlorate activated with bisulfite, hydrogen peroxide with metallic ions, and organic hydroperoxides with sodium formaldehydesulfoxylate (see Vol. 19. p. 422). The use of activators causes the decomposition of initiator to occur at lower reaction temperatures, which allows the preparation of higher-molecular-weight polymer within reasonable reaction times. This is an advantage, particularly for copolymers of vinylidene chloride with vinyl chloride. The use of oil-soluble initiators has been reported, but they are usually effective only when activated by a water-soluble activator or reducing agent.

Suspension polymerization is used commercially for vinylidene chloride polymers that are utilized as molding and extrusion resins. The principal advantage of the suspension process over the emulsion process is the use of fewer ingredients that can detract from the polymer properties. Stability is improved and water sensitivity is decreased. Extended reaction times and the practical difficulty of preparing higher-molecular-weight polymers are disadvantages of the suspension process vs the emulsion process, particularly with copolymers with vinyl chloride.

A typical recipe for suspension polymerization (30) is shown in Table 6.

Table 6. Recipe for Suspension Polymerization

Ingredient	Parts by wt
vinylidene chloride	85
vinyl chloride	15
deionized water	200
400 cP methylhydroxpropylcellulose[a]	0.05
lauroyl peroxide	0.3

[a] See Vol. 4, p. 650.

At a reaction temperature of 60°C, the polymerization proceeds to 85–90% conversion in 30–60 hr. Unreacted monomer is removed by vacuum pumps, condensed, and reused after processing. The polymer is obtained in the form of small (30–100 mesh) beads, that are dewatered by centrifugation and then dried in a flash dryer or fluid-bed dryer. Suspension polymerization employs monomer-soluble initiators and polymerization occurs inside suspended monomer droplets which are formed by the shearing action of the agitator and are prevented from coalescence by the protective colloid. It is important that the initiator be uniformly dissolved in the monomer before the formation of droplets. Unequal distribution of initiator causes some droplets to polymerize faster than others, leading to monomer diffusion from slow droplets to fast droplets. The fast polymerizing droplets form polymer beads that are dense, hard, and glassy, and extremely difficult to fabricate, due to their inability to accept stabilizers and plasticizers. Common protective colloids which prevent droplet coalescence and control particle size are poly(vinyl alcohol), gelatin, and methylcellulose. Organic peroxides, peroxycarbonates, and azo compounds are used as initiators for vinylidene chloride suspension polymerization. This process does not compensate for composition drift. Constant-composition processes have been designed using either emulsion or suspension reactions. It is more difficult to design controlled-composition processes using suspension methods.

In one approach (31), the less reactive component is removed continuously from the reaction in order to keep the unreacted monomer composition constant. This method has been used effectively in a case such as VDC–VC copolymerization where the slower reacting component is a gas and can be bled off during the reaction to maintain constant pressure. In many other cases, there is no practical way of removing the slow reacting component.

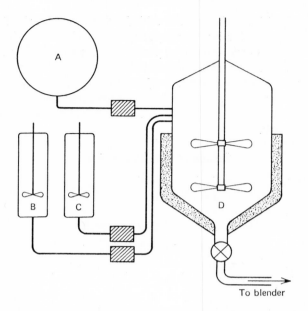

Fig. 2. Apparatus for continuous-addition emulsion polymerization of a VDC–acrylate mixture (32).

The method of emulsion polymerization by continuous addition is quite general. In this process, one or more components are metered continuously into the reaction. If the system is properly balanced, a steady state is reached in which a copolymer of uniform composition is produced. Woodford has described a process of this type in some detail (32). It can be used for the copolymerization of VDC with a variety of monomers. A flow diagram of the apparatus is shown in Figure 2. A typical recipe is shown in Table 7.

Table 7. Recipe for Emulsion Polymerization by Continuous Addition

Ingredient	Parts by wt
vinylidene chloride	468
comonomer	52
soap	
Tergitol[a] NP.35	12
sodium lauryl sulfate 25%	12
initiator	
ammonium peroxysulfate	10
sodium metabisulfite,[b] $Na_2S_2O_5$, 5%	10
water	436

[a] Registered trade name of Union Carbide for their nonionic wetting agent.
[b] See Vol. 19, p. 421.

The monomers are charged to the weigh tank A which is kept under a nitrogen blanket. The soaps, initiator, and part of the water are charged to tank B; the reducing agent and some water, to tank C. The remaining water is charged to the reactor D and the system sealed and purged. The temperature is raised to 40°C and one tenth of the monomer and initiator charges are added. Then one tenth of the activator is pumped in. When the reaction is underway, as indicated by an exotherm and pressure drop, feeds of A, B, and C are started at programmed rates, slowly at first and then gradually increasing. The emulsion is maintained at constant temperature during the run by pumping cooling water through the jacket. When all components are in and the exotherm begins to subside, an end shot of initiator and reducing agent is added to finish off the reaction. This process produces a latex suitable for coating paper, plastic film, and other substrates.

Structure and Properties

Chain Structure. The chemical composition of poly(vinylidene chloride) has been confirmed by various techniques including elemental analysis, x-ray diffraction analysis, infrared, raman and NMR spectroscopy, and degradation studies. The polymer chain is made up of vinylidene chloride units added head to tail.

$$-CH_2CCl_2CH_2CCl_2CH_2CCl_2-$$

Since the repeat unit is symmetrical, no possibility exists for stereo isomerism. Variations in structure can come about only by head to head addition, branching, or degradation reactions that do not cause chain scission. This includes such reactions as thermal dehydrochlorination, which creates double bonds in the structure to give

$$-CH_2CCl_2CH=CClCH_2CCl_2-$$

or a variety of ill-defined oxidation and hydrolysis reactions generating carbonyl groups.

The infrared spectra of PVDC often show traces of both unsaturation and carbonyl groups. The slightly yellow tinge of many of these polymers comes from the same source; the pure polymer is colorless. Elemental analyses for chlorine are normally slightly lower than theoretical (73.2%).

The high crystallinity of PVDC indicates that significant amounts of head to head addition, or branching, cannot be present. This has been confirmed by an examination of NMR spectra (33). There is only one peak in the spectrum due to the methylene hydrogens. Either branching or another mode of addition would produce nonequivalent hydrogens and a more complicated spectrum. However, the method would not be able to detect small amounts of such structures. The infrared spectra can also be interpreted in terms of simple head to tail structure (34).

While the chemical composition of the PVDC chain is well established, almost nothing is known about its size or size distribution. No direct measurements of molecular weights have ever been reported. Viscosity studies indicate, however, that the chain is unbranched and can be made with degrees of polymerization from 100 to well over 10,000 ([η] from 0.01 to >2.0 dl/g) (35). The interpretation of viscosity is based mainly on comparison with copolymers. The latter can be studied by conventional methods. Both molecular weights and distributions are typical of what would be expected from a free-radical polymerization.

Crystal Structure. The crystal structure of PVDC has not been completely solved. Several unit cells have been proposed. The best representation is probably that of Okuda (36). The unit cell contains four monomer units with two monomer units per repeat distance. The calculated density is higher than the experimental values which range from 1.80 to 1.94 g/ml at 25°C, depending on the sample. This is usually the case with crystalline polymers because samples of 100% crystallinity cannot normally be obtained. A direct calculation of the polymer density from volume changes during polymerization yields a value of 1.97 g/ml (37). If this value is correct, the unit-cell densities may be on the low side.

The repeat distance along the chain axis, 4.68 Å, is significantly less than that calculated for a planar zigzag structure. Therefore, the polymer must be in some other conformation; a number of conformations have been proposed, but there is no agreement as yet which is correct.

Morphology and Transitions. The highly crystalline particles of PVDC precipitating during polymerization are aggregates of thin lamellar crystals (38). The substructures are 50–100 Å thick and 100 or more times larger in the other dimensions. In some respects, they resemble the lamellar crystals grown from dilute solution (36). Not much is known about either type. Both appear to be built up of folded chains oriented normal to the lamellar surface.

Melting points of "as polymerized" powders are high, ranging from 198 to 205°C when measured by DTA (see Thermal analysis) or hot-stage microscopy. A lower melting point is often observed at high heating rates indicating that the powders may be annealing during the measurement. "As polymerized" PVDC does not show a well defined glass-transition temperature, T_g, due to its high crystallinity. But a sample can be melted at 210°C and quenched rapidly to an amorphous state at −20°C. The amorphous polymer shows a glass transition at −17°C by dilatometry (39). Values ranging from −19 to −11°C are observed, depending both on methods of measurement and on sample preparation.

Once melted, PVDC does not regain its "as polymerized" morphology when subsequently crystallized. The polymer recrystallizes in spherulitic habit. Spherulites observed between crossed polaroids show the usual maltese cross and are positively birefringent (40). The size and number of spherulites can be controlled. Quenching and low-temperature annealing generate many small nuclei which, on heating, grow rapidly into small spherulites. Slow crystallization at higher temperatures produces fewer, but much larger spherulites (41). The melting point, T_m, and degree of crystallinity of recrystallized PVDC also depend on crystallization conditions. T_m increases with crystallization temperature, but the "as polymerized" value cannot be achieved. There is no reason to believe that even these values are the true melting point of PVDC. It may be as high as 220°C. But slow, high-temperature recrystallization and annealing experiments are not feasible due to the thermal instability of the polymer (42), so this point is not resolved. Other transitions in PVDC have been observed by dynamic mechanical methods. Properties of PVDC are given in Table 8.

The properties of PVDC are usually modified by copolymerization. Copolymers of high VDC content have lower melting points and higher glass-transition temperatures than PVDC itself. Copolymers containing more than about 15 mole % acrylate or methacrylate, MA, are amorphous. Substantially more acrylonitrile (∼25%) or vinyl chloride (∼45%) are required to destroy crystallinity completely.

The effect of different types of comonomers on T_m is shown in Figure 3. The VDC/MA copolymers obey Flory's melting-point-depression theory and give a reasonable value for the heat of fusion. The VDC/VC system does not (36). This is probably caused by the fact that VC units can enter into the PVDC crystal structure as defects. Consequently, VC is less effective in lowering T_m.

The glass-transition temperatures of saran copolymers have been studied extensively (32,39,43–45). The effect of various comonomers on the glass transition tem-

Table 8. Properties of PVDC

Property	Preferred value	Range
melting point, °C	202	198–205
glass transition, T_g, °C	−17	−19 to −11
α transition,[a] °C	80	
density at 25°C, g/ml		
amorphous at 25°C	1.775	1.67–1.775
unit cell	1.96	1.949–1.96
cryst		1.80–1.97
refractive index, cryst, n_D^{25}	1.63	
heat of fusion, cal/mole	1500	1100–1900

[a] Between T_m and T_g.

perature is shown in Figure 4. In every case T_g increases with the comonomer content at low comonomer levels, even in cases where the T_g of the other homopolymer is lower. In some cases, a maximum T_g is observed at intermediate compositions. In others, where the T_g of the homopolymer is much higher than the T_g of PVDC, the glass temperatures of the copolymers increase over the entire composition range. There are differences in behavior even here. T_g increases most rapidly at low VCN levels, but changes the slowest at low VC levels. This suggests that polar interactions affect the former; but the increase in T_g in the VDC/VC copolymers may be simply due to loss of chain symmetry.

Because of these effects, the temperature range in which copolymers can crystallize is drastically narrowed. Crystallization induction times are prolonged and subsequent crystallization takes place at a low rate over a long period of time. Plasticization, which lowers T_g, decreases crystallization induction times significantly.

Copolymers with lower glass-transition temperatures tend to crystallize more rapidly also (46). Crystallization curves were obtained for 10 mole % acrylate copolymers of varying side-chain length. Among the acrylate copolymers the butyl

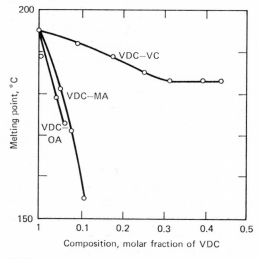

Fig. 3. Melting points vs compositions of various VDC copolymers.

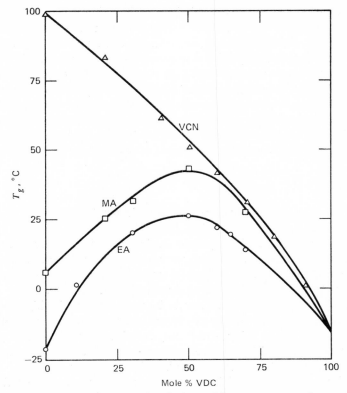

Fig. 4. Effect of comonomer structure on the glass-transition temperature of saran copolymers (45).

acrylate has a T_g of 8°C, the octyl acrylate, OA, −3°C, and the octadecyl, −16°C. The rates of crystallization of these copolymers are inversely related to the glass-transition temperatures. Apparently, the long alkyl side chains act as internal plasticizers, lowering the melt viscosity of the copolymer even though the acrylate group acts to raise it.

The more common crystalline copolymers show their maximum rates of crystallization in the range of 80–120°C. In many cases, these have broad composition distributions containing both fractions of high VDC content which crystallize rapidly, and other fractions that do not crystallize at all. PVDC itself probably crystallizes at a maximum rate at 140–150°C, but the process is difficult to follow because of severe polymer degradation. The copolymers may remain amorphous for a considerable period of time if quenched to room temperature. The induction time before the onset of crystallization depends on both the type and amount of comonomer; PVDC itself crystallizes within minutes at 25°C.

Orientation or mechanical working accelerate crystallization and have a pronounced effect on morphology. Okuda (36) has examined uniaxially oriented filaments in detail. He finds the crystals to be oriented along the fiber axis. The long period, by small angle scattering, is 76 Å and decreases with comonomer content. The fiber is 43% crystalline and has a melting point of 195°C with an average crystal thickness of 45 Å. The crystal size is not greatly affected by comonomer content, but both crystallinity and melting point decrease.

Copolymerization also affects morphology under other crystallization conditions. Copolymers in the form of cast or molded sheets are much more transparent due to the small spherulite size.   In extreme cases, crystallinity cannot even be detected optically; but its effect on mechanical properties is pronounced.   Before crystallization, films are soft and rubbery with low modulus and high elongation.   After crystallization, they are leathery and tough with a higher modulus and lower elongation.

Significant amounts of comonomer also reduce the ability of the polymer to form lamellar crystals from solution.   In some cases, the polymer merely gels the solution as it precipitates rather than forming distinct crystals.   At somewhat higher VDC content, it may precipitate out but in the form of aggregated, ill-defined particles and clusters.

Solubility and Solution Properties

PVDC, like many high-melting polymers, does not dissolve in most common solvents at ambient temperatures.   Copolymers, particularly those of low crystallinity, are much more soluble.   But still, one of the outstanding characteristics of saran polymers is resistance to a wide range of solvents and chemical reagents.

The insolubility of PVDC is due less to its polarity than its high melting point. It dissolves readily in a wide variety of solvents at temperatures above 130°C (42). The polarity of the polymer is important only in mixtures with specific polar aprotic solvents; many solvents of this general class are able to solvate PVDC strongly enough to depress the melting point more than 100°C.

Solubility is normally correlated with cohesive energy densities or solubility parameters.   Burrell has reported a value of 12.2 for the solubility parameter of PVDC (47). This is much higher than a value of 10.1 ± 0.3, estimated from solubility studies in nonpolar solvents, or of 1025 as calculated from Small's relationship.   The latter value is more likely to be correct.   In any case, the use of the solubility-parameter scheme for polar crystalline polymers like PVDC is of limited value.   (See also Solubility parameter in Supplement Volume.)

A typical nonpolar solvent of matching solubility parameter is tetrahydronaphthalene.   The lowest temperature at which PVDC dissolves in this solvent is 140°C. Specific solvents, on the other hand, can dissolve PVDC at much lower temperatures. A list of both good nonpolar solvents and specific solvents is given in Table 9.   The relative solvent activity is characterized by the temperature at which a 1% mixture of polymer in solvent becomes homogeneous when heated rapidly.

PVDC also dissolves readily in certain solvent mixtures (48).   One component must be a sulfoxide or amide of the type described above.   Effective cosolvents are less polar and have cyclic structures.   They include both aliphatic and aromatic hydrocarbons, ethers, sulfides, and ketones.   Acidic or hydrogen bonding solvents have an opposite effect, rendering the polar aprotic component less effective.   Both hydrocarbons and hydrogen bonding solvents are nonsolvents for PVDC.

As polymerized, PVDC is not in its most stable state; annealing and/or recrystallization can raise the temperature at which it dissolves.   Low crystallinity polymers can be dissolved at a lower temperature to form metastable solutions.   On standing at the solution temperature, however, they gel or become turbid, indicating precipitation.

Table 9. Solvents for Polyvinylidene Chloride

Solvents	$T_m,^a$ °C
nonpolar	
1,3-dibromopropane	126
bromobenzene	129
1-chloronaphthalene	134
2-methylnaphthalene	134
o-dichlorobenzene	135
polar aprotic	
hexamethylphosphortriamide	−7.2
tetramethylene sulfoxide, TMSO	28
N-acetylpiperidine	34
N-methylpyrrolidone	42
N-formylhexamethyleneimine	44
trimethylene sulfide	74
N-n-butylpyrrolidone	75
isopropyl sulfoxide	79
N-formylpiperidine	80
N-acetylpyrrolidine	86
tetrahydrothiophene	87
N,N-dimethylacetamide	87
cyclooctanone	90
cycloheptanone	96
n-butyl sulfoxide	98

^{a} Temperature at which a 1% mixture of polymer in solvent becomes homogeneous.

Copolymers with a high enough vinylidene chloride content to be quite crystalline, behave much like PVDC. They are more soluble, however, because of their lower melting points. The solubility of amorphous copolymers is much higher. The selection of solvents, in either case, varies somewhat with the type of comonomer. Some of the more common types are listed in Table 10. Solvents that dissolve PVDC also dissolve the copolymers, but at lower temperatures.

Table 10. Common Solvents for Saran Copolymers

Solvent	Copolymer type	Temperature range, °C
tetrahydrofuran	all	<60
methyl ethyl ketone	low crystallinity	<80
1,4-dioxane	all	50–100
cyclohexanone	all	50–100
cyclopentanone	all	50–100
ethyl acetate	low crystallinity	<80
chlorobenzene	all	100–130
dichlorobenzene	all	100–140
dimethylformamide	high VCN	<100

Solution properties of PVDC homopolymers have not been studied to any extent. The few data that are available have been obtained at high temperatures. On the other hand, a rather complete investigation of the dilute solution properties of several acrylate copolymers has been reported (49). The authors find the behavior to be typical of flexible backbone, vinyl-type polymers. The length of the acrylate ester side chain had little effect on properties.

Molecular weights of PVDC have not been measured directly, but a Mark-Houwink relationship in trichlorobenzene has been estimated from kinetics studies as follows: $[\eta] = 0.45 \times 10^{-4} M_w^{0.7}$ ($[\eta]$ in 100 ml/g). This is not too different from the relationships obtained for copolymers in methyl ethyl ketone. A 15% EA copolymer gave the value $[\eta] = 2.88 \times 10^{-4} M_w^{0.6}$.

The relationship for PVDC in good solvents, such as TMSO at 25°C, is estimated to have a K (coefficient) somewhere between the two, and an a (exponent) of 0.7, or $[\eta] \cong 1 \times 10^{-4} M_w^{0.7}$.

In many cases, the molecular weights of PVDC and saran copolymers are characterized by the absolute viscosity of a 2% solution in o-dichlorobenzene at 140°C. The exact correlation between this number and molecular weight is not known.

Mechanical Properties

Due to the difficulty of fabricating PVDC into suitable test specimens, very few direct measurements of its mechanical properties have been made. In many cases, however, the properties of copolymers have been studied as a function of composition and the properties of PVDC can be estimated by extrapolation. Table 11 lists some

Table 11. Mechanical Properties of PVDC

Property	Range
tensile strength, psi	
unoriented	5000–10,000
oriented	30,000–60,000
elongation, %	
unoriented	10–20
oriented	15–40
softening range (heat distortion), °C	100–150
flow temperature, °C	>185
brittle temperature, °C	−10 to 10
impact strength, (ft)(lb)/in.	0.5–1.0

properties characteristic of high VDC-content unplasticized copolymers. The actual performance of a given specimen is very sensitive to morphology, including the amount and kind of crystallinity, as well as orientation. Tensile strength increases with crystallinity while toughness and elongation drop. Orientation, however, improves all three properties. Table 12 shows the effect of orientation on properties of VDC/VC copolymer.

The dynamic mechanical properties of VDC/VC copolymers have been studied in detail. The incorporation of VC units in the polymer results in a drop in dynamic modulus due to the reduction in crystallinity. At the same time, however, the glass-transition temperature is raised; the softening effect observed at room temperature, therefore, is accompanied by increased brittleness at lower temperatures. Saran B copolymers are normally plasticized in order to avoid this difficulty. Small amounts of plasticizer (2–10%) can depress T_g significantly without loss of strength at room temperature.

Table 12. Effect of Stretch Ratio Upon Tensile Strength and Elongation[a] (50)

Stretch ratio	Tensile strength, psi	Elongation, %
2.50:1	34,080	23.2
2.75:1	33,960	21.7
3.00:1	43,960	26.3
3.25:1	38,900	33.1
3.50:1	45,820	19.2
3.75:1	47,830	21.8
4.00:1	46,380	19.7
4.19:1	45,520	16.2

[a] Average of five determinations, using the Instron test at 2 in./min.

At higher levels of VC, the T_g of the copolymer is above room temperature and the modulus rises again. A minimum in modulus (or maximum in softness) is normally observed in copolymers where T_g may be above room temperature. A thermomechanical analysis of both VDC/VCN and VDC/MMA copolymer systems shows a minimum in softening point at 79.4 and 68.1 mole % VDC, respectively (51).

In cases where the copolymers have substantially lower glass-transition temperatures, the modulus decreases with increasing comonomer content. This is due to a drop both in crystallinity and in glass-transition temperature. The loss in modulus in these systems is, therefore, accompanied by an improvement in low-temperature performance. At low acrylate levels (<10%), however, T_g increases with comonomer content. The brittle points in this range may, therefore, be higher than that of PVDC.

The long side chains of the acrylate ester group can apparently act as internal plasticizers. Substitution of a carboxyl group on the polymer chain, in itself, increases brittleness. A more polar substituent, such as an N-alkyl amide group, is even less desirable. Copolymers of VDC with N-alkyl acrylamide are more brittle than the corresponding acrylates even when the side chains are long (52). Side-chain crystallization may be a contributing factor.

Barrier Properties. Vinylidene chloride polymers are, in comparison with other polymers, impermeable to a wide variety of gases and liquids. Without doubt, this is a consequence of the combination of high density and high crystallinity in the polymer. An increase in either tends to reduce permeability. But a more subtle factor may be the symmetry of the polymer structure. Lasoski (53) has shown that both polyisobutylene and PVDC have unusually low permeabilities to water, compared to their monosubstituted counterparts polypropylene and PVC. The values shown in Table 13 include estimates for the completely amorphous polymers. The estimated value for

Table 13. Comparison of the Permeabilities of Various Polymers to Water Vapor (18,53,54)

Polymer	Density, g/ml		Permeability[a]	
	amorphous	crystalline	amorphous	crystalline
ethylene	0.85	1.00	200–220	10–40
propylene	0.85	0.94	420	
isobutylene	0.915	0.94	90	
vinylchloride	1.41	1.52	~300	90–115
vinylidene chloride	1.77	1.96	~ 30	4–6

[a] In g/(hr) (100 in.²) at 53 mm Hg pressure differential and 39.5°C for a film 1 mil thick.

Table 14. Permeability of Saran Film to Various Gases (18)

Gas	Temperature, °C	Activation energy, E_p, kcal/mole	Permeability,[a] $\bar{P} \times 10^{10}$
N_2	30	16.8	0.00094
O_2	30	15.9	0.00053
CO_2	30	12.3	0.03
He	34		0.31
H_2O	25	11.0	0.5
H_2S	30	17.8	0.03

[a] In $cm^3(STP)/(sec)(cm^2)(cm\ Hg/cm)$.

highly crystalline PVDC was obtained by extrapolating data on copolymers. Table 14 gives the permeability of a saran film and the activation energy for a variety of gases. Permeability is affected by both kind and amounts of comonomer, as well as crystallinity. In contrast to mechanical behavior, however, orientation seems to have only a minor effect.

It is observed that a more polar comonomer, such as VCN, increases the water-vapor transmission more than VC, other factors being constant. For the same reason, VCN copolymers are more resistant to low CED penetrants of low cohesive energy density.

Comonomers that lower T_g and increase the free volume in the amorphous phase increase permeability. This seems to be the way in which the higher acrylates act, though they undoubtedly influence solubility as well. Plasticizers increase permeability for similar reasons.

Degradation Chemistry. PVC has a marked tendency to eliminate HCl. The reaction is normally described as a two-step process:

1. Formation of a conjugated polyene

$$+CH_2CCl_2+_n \xrightarrow{\text{fast}} +CH=CCl+_n + n\ HCl$$

2. Carbonization

$$+CH=CCl+_n \xrightarrow{\text{slow}} 2n\ C + n\ HCl$$

While this simplified description serves to illustrate the main features of the reaction, it should by no means be accepted as a mechanism. In fact, PVDC can degrade by several distinct mechanisms. Degradation can be effected by heat, uv radiation, ionizing radiation (x rays, γ rays), alkaline reagents and catalytic metals or salts. The common feature of these reactions is that chlorine is removed from the polymer either as Cl^- or HCl, depending on the medium. But the carbonaceous residue remaining after various extents of decomposition is very much dependent on the method. In addition, most PVDC degradation reactions are heterogeneous, the polymer being present as a solid phase. In such cases, the course of the reaction is also influenced by the polymer morphology and the interaction between phases.

Thermal degradation has been studied widely (55–61). PVDC begins to decompose at about 125°C. In the very early stages of the thermal decomposition (<1%) PVDC begins to discolor and becomes insoluble. The melting point drops and bands corresponding to unsaturation appear in the ir spectrum. As the reaction proceeds conjugated double bonds are detected. The polymer becomes infusible and the PVDC crystal structure is destroyed. An increasing concentration of free radicals can be detected by ESR. If further reaction is brought about by raising the temperature, aromatic structures are formed. Finally, at very high temperature, graphitization of the carbonaceous residue takes place.

PVDC does not appear to degrade at a measurable rate in the dark at temperatures below 100°C. When exposed to uv or sunlight, however, it discolors. HCl is eliminated in the process and crosslinking takes place. Oster et al. (62,63) made a careful study of the photodegradation reaction.

Unlike uv, higher energy irradiation (γ rays) of PVDC causes chain scission. Copolymers of VDC and VC undergo both crosslinking and chain scission (PVC itself crosslinks) but the net result is dependent on polymer morphology, as well as copolymer composition (64).

The alkaline decomposition of PVDC is quite clearly an ionic reaction. But details of the mechanism are not known. The ultimate products are carbon and chloride ion. At intermediate stages conjugated polyene structures and acetylenic bonds are formed. The reaction is very fast, even at liquid-ammonia temperature (-33°C) (65). The reaction can be effected by various strong bases including alcoholic KOH, metal alkyls, metal alkoxides, metal amides in liquid ammonia, active metals, LiCl in dimethyl formamide, etc. The reaction is normally heterogeneous because PVDC is not soluble in the usual solvents for such reactions. The rate is, therefore, probably controlled by surface area. Aqueous bases have a limited effect on PVDC primarily because the polymer is scarcely wet or swollen by water. It is decomposed by hot concentrated caustic over a period of time, however. Weak bases, such as ammonia, amines, or polar aprotic solvents, also accelerate the decomposition of PVDC. The products from these reactions do not appear to have a simple polyene structure. In many cases, substitution reactions leading to bound nitrogen can occur competitively (66). These reagents, in addition, can also swell the polymer. Pyridine, for example, would be a fairly good solvent for saran if it did not attack the polymer chemically. In a nonsolvent mixture, however, pyridine does not penetrate into the polymer phase (67). Studies on single crystals indicate that it removes HCl only from the surface. Both kinetic studies and product characterization suggest that the reaction of two units in each chain fold can easily take place; further reaction is greatly retarded either by inability of the pyridine to diffuse into the crystal, or by steric factors.

Various metal salts such as $FeCl_3$, $AlCl_3$, and $ZnCl_2$ catalyze the thermal decomposition of PVDC (55). This problem is of great practical importance. Saran polymers, when heated, release HCl. If they are in contact with a metal surface, the metal chloride forms and catalyzes further decomposition. The reaction is thereby greatly accelerated. As a consequence, attempts to extrude unstabilized saran in conventional steel equipment lead to almost explosive decomposition. Very little is known about the mechanism of the reactions; the major industrial emphasis has been on preventative measures. Metal parts intended to be used with saran are fabricated from acid-resistant alloys or nickel. Nickel salts are much less active as catalysts. In addition, the polymers are usually stabilized with some type of metal-ion scavenger.

Stabilization. The art of stabilizing saran-type polymers is highly developed. While not much is understood about mechanisms in detail, some general principles have been established. The ideal stabilizer system should (*1*) absorb, or combine with, hydrochloric acid gas in an irreversible manner under the conditions of use, but not have such strong affinity as to strip HCl from the polymer chain; (*2*) act as a selective ultraviolet absorber to reduce the total ultraviolet energy absorbed in the polymer; (3) contain a reactive dienophilic molecule capable of destroying the discoloration by reacting with, and breaking up, the color-producing conjugated polyene sequences; (4) possess antioxidant activity in order to prolong the induction period of the oxidation process and prevent the formation of carbonyl groups and other chlorine activating structures resulting from oxidation of polymer molecules; and (5) have the ability to chelate metals, such as iron, and prevent the formation of metallic chloride which acts as a catalyst for polymer degradation.

Acid acceptors are of three general types: alkaline earth and heavy metal salts of weak acids, such as barium, cadmium, or lead fatty acid salts; epoxy compounds, such as epoxidized soybean oil or glycidyl ethers and esters; and organotin compounds, such as salts of carboxylic acids and organotin mercaptides.

Effective light stabilizers have a chemical configuration that leads to hydrogen bonding and ring "chelation," showing exceptional resonance stability with very great ultraviolet absorption properties. The principal compounds of commercial interest are derivatives of salicylic acid, resorcylic acid, benzophenone, and benzotriazole. Examples of dienophiles that have been used are maleic anhydride and dibasic lead maleate.

Antioxidants fall, generally, into the two following classes: those that react with a free radical to stop a radical chain and those that reduce hydroperoxides to alcohols. Phenolic antioxidants, such as Ionol (2,6-ditertiarybutyl-4-methylphenol) and substituted bisphenols, fall into the first class. The second class is exemplified by organic sulfur compounds and organic phosphites. The phosphites have the ability to chelate metals as do the Versenes (registered trademark, The Dow Chemical Co., EDTA) and citric acid or citrates.

The performance of organic phosphites to function as antioxidants and as chelating agents illustrates the dual ability of many of the stabilizing compounds. It is common practice to use a combination of stabilizing compounds to achieve optimum benefit. Synergism of stabilizing action is frequently observed.

An excellent list of patents on stabilization of vinyl chloride and vinylidene chloride polymers and copolymers has been tabulated by Chevassus and de Broutelles (68). This list illustrates the wide variety of materials that possess stabilization activity.

Processing and Uses

Molding Resins (69). Vinylidene chloride copolymer resins are fabricated by injection, compression, and transfer molding. The fairly sharp crystalline melting temperature of the resins makes the injection-molding process unique. Whereas, with other thermoplastics cold molds hasten the cooling of molded parts and shorten the molding cycle, with vinylidene chloride copolymers a cold mold supercools the resin and produces soft, flexible, amorphous pieces. Rapid hardening is achieved by heat treatment, using heated molds, to induce crystallinity. With heated molds, the molded pieces can be ejected from the mold at temperatures as high as 100°C in a strainfree, dimensionally stable form. Very rapid cycles can be realized with heavy sections. Sink marks are not usually a problem since the outer skin of the molded piece cools quickly below the crystalline melting temperature and crystallization takes place rapidly. Sufficient material must be injected into the mold or internal voids are likely to develop.

The range of molding temperatures is rather narrow due to the crystalline nature of the resin and thermal sensitivity. All crystallites must be melted to obtain low polymer-melt viscosity, but prolonged or excessive heating must be avoided to prevent dehydrochlorination.

Thermal degradation is a problem, even when the resin is formulated with the very best thermal stabilizers. Molding equipment is designed to alleviate this problem by having all passages through the heating cylinder streamlined to prevent plastic buildup. Any plastic that remains in the heating cylinder for longer than a few minutes decomposes, thereby releasing hydrochloric acid gas and forming carbon. The carbon may build up or break off and contaminate molded parts with occlusions. It is especially important that an injection-molding heating cylinder not be shut down when loaded with vinylidene chloride molten resin. The cylinder must be purged with a more stable resin, such as polystyrene.

The metal parts of the injection molder (liner, torpedo, nozzle) that contact the hot molten resin must be of the noncatalytic type to prevent an accelerated decomposition of the polymer. In addition, they must be resistant to corrosion by acid gas. Iron, copper, and zinc are catalytic to the decomposition and cannot be used as such or in many of their alloys. Magnesium is noncatalytic but is subject to corrosive attack, as is chromium, when used as plating. Nickel alloys, such as Duranickel and Hastelloy B, are recommended as the material of construction for injection-molded metal parts. These, along with pure nickel, are noncatalytic and corrosion-resistant; however, pure nickel is rather soft and not recommended.

The injection mold need not be made of noncatalytic metals; any high-grade tool steel may be used, since the plastic is cooling in the mold and undergoing little decomposition. However, the mold requires good venting to allow the passage of small amounts of acid gas, as well as air. Vents tend to be clogged by corrosion and have to be cleaned periodically.

Vinylidene chloride copolymer resins can be compression molded by conventional heating and cooling of the mold, as is done with other thermoplastics. Hot plastic from an extruder or dielectrically heated preforms may be used. The preferred method is dielectric heating of compressed powder preforms, where gentle heating can be applied to a preform practically free of air. A 5-kW generator at about 27-Mc frequency heats approximately one lb of resin per minute in a six-in. diam. preform. The heated preform has a powder skin which should be kneaded into the hot plastic to yield a smooth compression-molded surface. The mold can have a noncatalytic surface provided by chrome plating, although slight etching may occur. The mold should be well cored to allow an adequate supply of water for quick cooling. The plastic may be molded in conventional transfer molds at temperatures of 150–200°F. Dielectrically heated preforms, unkneaded, may be used and material remaining in the transfer cylinder may be collected for reheating to make a full cylinder charge. Practically no sprue scrap is generated to be reground.

Molded parts of vinylidene chloride copolymer plastics are used to satisfy the industrial requirements of chemical resistance and extended service life. They are used in such items as gasoline filters, valves, pipe fittings, containers, and chemical process equipment. Complex articles are constructed from molded parts by welding, where the welds are practically as strong as the other portions of the molded part. Welding is preferred over cementing because of the chemical resistance of the plastic. The weld is strong due to the high fluidity of the hot plastic, which, in turn, is due to the sharp-melting crystalline character. Hot-air welding at 400–500°F is an excellent method; in addition, hot-plate welding, frictional welding, or radiant heating may be used. Molded parts have good physical properties, but have lower tensile strength than films or fibers, since crystallization is random in molded parts and developed by orientation in films and fibers. Table 15 shows physical properties of a typical molded vinylidene chloride plastic.

Table 15. Properties of Molded Saran 281

Property	Value
tensile strength, psi	3500–4500
elongation at yield point, %	15–25
modulus of elasticity tension, psi	$0.7–2.0 \times 10^5$
impact strength, $\frac{1}{2}$ in. $\times$ $\frac{1}{2}$ in.	
notched bar, Izod, ft-lb/in.	2–8
flexural strength, psi	15,000–17,000
compressive strength at yield point, psi	7500–8300
resistance to heat (continuous), °F	170
resistance to heat (intermittent), °F	220
specific heat, cal/(g) (°C)	0.32
thermal expansion per °F	8.78×10^{-5}
thermal conductivity, cal/(sec)(cm²)(°C/cm)	2.2×10^{-4}
hardness, Rockwell M	50–65
specific gravity, g/ml	1.68–1.75
refractive index n_D	1.61
power factor, 60, 10^3, and 10^6 Hz	0.03–15
breakdown voltage, 60 Hz instantaneous, V/mil	
0.125 in. thickness	500
0.020 in. thickness	1500
0.001 in. thickness	300
volume resistivity,[a] Ω-cm	$10^{14}–10^{16}$
water absorption,[b] %	>0.1
burning rate	self-extinguishing
mold shrinkage,[c] in./in.	0.005–0.030
injection molding temperature, °F	300–350
injection molding pressure, psi	10,000–30,000
compression molding temperature, °F	250–350
compression molding pressure, psi	250–5000

[a] Relative humidity of 50%, at 25°C.
[b] ASTM 570-40T.
[c] Injection molded.

Extrusion Resins (50,69). Extrusion of vinylidene chloride copolymers, which is the major fabrication technique for filaments, films, rods, and tubing or pipe, requires the same concerns for thermal degradation, streamlined flow, and noncatalytic materials of construction that were described above for injection-molding resins. The plastic leaves the extrusion die in a completely amorphous condition and is maintained in this state by quenchcooling in a water bath at about 50°F, inhibiting the recrystallization. In this state, the plastic is soft, weak, and pliable. If it is allowed to remain at room temperature, it hardens gradually and recrystallizes partially at a slow rate with a random crystal arrangement. Heat treatment can be used to recrystallize at controlled rates.

Crystal orientation is developed in the supercooled extrudate by plastic deformation (stretching) and heat treatment. In the manufacture of filaments, the stretching produces orientation in a single direction, developing unidirectional properties of high tensile strength, flexibility, long fatigue life, and good elasticity. The filaments are taken from the supercooling tank, wrapped several times around smooth takeoff rolls, and then wrapped several times around orienting rolls which have a linear speed about four times that of the takeoff rolls. The difference in roll speeds produces the mechanical stretching and causes orientation of crystallites along the longitudinal axis during partial recrystallization. Heat treatment may be used during or after stretching to affect the degree of crystallization and control the physical properties of the oriented filaments.

A variation of the orientation process is used to produce vinylidene chloride copolymer films. The plastic is extruded into tube form, which is supercooled and subsequently biaxially oriented in a continuous process known as the bubble process. The supercooled tube is flattened and passed through two sets of pinch rolls arranged so that the second set of rolls is traveling faster than the first set. Between the two sets of rolls, air is injected into the tube to create a bubble which is entrapped by the pinch rolls. The entrapped air bubble remains stationary, while the extruded tube is oriented as it passes around the bubble. Orientation is produced in the transverse, as well as the longitudinal direction, creating excellent tensile strength, elongation, and flexibility in the film. The commercial procedure has been described by Stephenson (70).

Extruded monofilaments in diameters of 5–15 mils have been widely used in the textile field as furniture and automobile upholstery, drapery fabric, outdoor furniture, venetian-blind tape, filter cloths, etc. Chemically resistant tubing and pipe liners are extruded where the pipe liner is inserted into an oversize steel pipe which is swaged to size, and lengths connected by flanged joints and vinylidene chloride copolymer gaskets. Pipe fittings are lined with injection-molded liners.

The biaxially oriented extruded films are used in packaging applications where their excellent resistance to water vapor and most gases makes them an ideal transparent barrier. Being highly oriented, these films exhibit some shrinkage when exposed to higher than normal temperatures. Preshrinking or heat-setting may be performed to minimize residual shrink, or the shrinkage may be used to advantage in the heat-shrinking of overwraps on packaged items. The films are used in tube form or as flat film for overwraps or conversion to film bags on modified bagmaking machinery. The electronic or dielectric seal is the most satisfactory type for sealing the film to itself, although hot-plate sealers or cement-type seals produce an airtight seal on overwraps without the material actually being fused.

Lacquer Resins (71). Vinylidene chloride polymers have several properties which are valuable in the coatings industry. These are excellent resistance to gas and moisture-vapor transmission, good resistance to attack by solvents and by fats and oils, high strength, and the ability to be heat sealed. These characteristics derive from the highly crystalline nature of the very high vinylidene chloride composition of the polymer, which ranges from about 80 to more than 90% by wt. Minor constituents in these copolymers are generally from the following group: vinyl chloride, alkyl acrylates, alkyl methacrylates, acrylonitrile, methacrylonitrile, and vinyl acetate. Small concentrations of vinyl carboxylic acids, such as acrylic acid, methacrylic acid, or itaconic acid, are sometimes included to enhance adhesion of the polymer to the substrate. The ability to crystallize and the extent of crystallization are reduced with increasing concentration of the comonomers; some polymers of commercial use fail to crystallize at all.

Acetone, methyl ethyl ketone, methyl isobutyl ketone, dimethylformamide, ethyl acetate, and tetrahydrofuran are solvents for vinylidene chloride polymers used in lacquer coatings. Methyl ethyl ketone and tetrahydrofuran are the most commonly used solvents. Toluene is used as a diluent for either one. Lacquers prepared at 10–20% polymer solids by weight in a solvent blend of 2 parts ketone and 1 part toluene have a viscosity in the range of 200–1000 cps. Lacquers can be prepared with polymers of very high vinylidene chloride content and stored at room temperature in the tetrahydrofuran–toluene. Methyl ethyl ketone lacquers must be prepared and maintained at 60–70°C or the lacquer forms a solid gel. It is critical in the manufacture of polymers for lacquer application to maintain a fairly narrow compositional distribution in the polymer in order to achieve good dissolution properties.

The lacquers are applied commercially by roller coating, dip and doctor (see Coated fabrics), knife coater, and spraying. Spraying is useful only with lower-viscosity lacquers and solvent balance is important to avoid webbing from the spray gun. Solvent removal is difficult from heavy coatings and multiple coatings are recommended where a heavy film is desired, allowing sufficient time between coats to avoid lifting of the previous coat by the solvent. In the machine coating of flexible substrata (paper, plastic films), the solvent is removed by infrared heating or forced-air drying at temperatures of 90–140°C. Temperatures in the range of 60–95°C promote the recrystallization of the polymer, after the solvent has been removed. Failure to recrystallize the polymer leaves a soft, amorphous coating that blocks or adheres between concentric layers in a rewound roll. A recrystallized coating can be rewound without blocking. Handling properties of the coated film are improved by adding small amounts of wax as a slip agent and of talc or silica as an antiblock agent to the lacquer system. The concentration of additives is kept low to prevent any serious detraction from the vapor-transmission properties of the vinylidene chloride polymer coating. For this reason, plasticizers are seldom, if ever, used.

A primary use of vinylidene chloride polymer lacquers is the coating of films made from regenerated cellulose (72) or of board or paper with polyamide, polyester, polyethylene, polypropylene, poly(vinyl chloride), and polyethylene to impart resistance to fats, oils, oxygen, and water vapor. These are used, mainly, in the packaging of foodstuffs, where the additional features of inertness, lack of odor or taste, and nontoxicity are required. Vinylidene chloride polymers have been used extensively as interior coatings for ship-tanks (73), railroad tank cars, and fuel storage tanks (74), and for coating of steel piles (75) and structures. The excellent chemical resistance and good adhesion have resulted in excellent long-term performance of the coating. Brushing and spraying are suitable methods of application.

The excellent adhesion to primed films of polyester, combined with good dielectric properties and good surface properties, make the vinylidene chloride polymers very suitable as binders for iron oxide pigmented coatings of magnetic tapes (76). They have shown very good performance in audiotapes, videotapes, and computer tapes. See Magnetic tape.

Saran Latex (77–79). Vinylidene chloride polymers are often made in emulsion, but normally they are isolated, dried, and used as conventional resins. Beginning in the 1960s, stable latexes have been prepared which can be used directly for coatings and other applications. They have the desirable characteristic of forming an extremely water-resistant coating from an aqueous medium.

Not surprisingly, the major applications for these materials are as barrier coatings on paper products and, more recently, on plastic films. While barrier is the primary property sought, the heat-seal characteristics of saran coatings are equally valuable in many applications. Saran latexes have also been used in combination with cement to make high-strength mortars and concretes. They are also used as binders for paints and nonwoven fabrics; these latexes yield composites that are both water resistant and nonflammable.

PVDC latexes can be easily prepared by the same methods, but have few uses because they do not form films. Copolymers of high VDC content are film forming when freshly prepared, but soon crystallize and lose this desirable characteristic. Since crystallinity in the final product is very often desirable, in barrier coatings for example, a major developmental problem has been to prevent crystallization in the latex during storage but to induce rapid crystallization of the polymer after coating. This has been accomplished by using the proper combination of comonomers with VDC.

Table 16. Typical Latex Properties

Properties	Value
total solids, %	60–62
viscosity, at 25°C, cP	25
pH	4.5–5.0
color	white-cream
particle size, μ	0.25[a]
specific gravity, g/ml	1.38
mechanical stability	excellent
storage stability	excellent
chemical stability	not stable to di- or trivalent ions

[a] Approximately.

Table 17. Polymer Film Properties

Properties	Value
water-vapor transmission, g/(24 hr)(100 in.²) for 0.5 mil	0.37
grease resistance	excellent
scorability and fold resistance	moderate
oxygen permeability, ml(STP)/(sec)(cm²)(cmHg/mm)	0.5
heat sealability[a]	excellent
light stability	fair
density, g/ml	1.60
color	water white
clarity	excellent
gloss	excellent
odor	none

[a] Face to face.

Most latexes are made with varying amounts of acrylates, methacrylates, acrylonitrile, and vinyl chloride, the total amount of comonomer ranging from about 7 to 20%. The properties of a typical latex used for paper coating are listed in Table 16.

Latexes are usually formulated with antiblock, slip, and wetting agents. They can be deposited in conventional coating processes, such as with a doctor blade (see Coated fabrics). Coating speeds in excess of 1000 ft/min can be attained. The latex coating can be dried in forced-air or radiant-heat ovens. Two coats are normally applied to reduce pinholing. A precoat is often used on porous substrates to reduce the quantity of the more expensive saran latex needed to cover (80). The properties of a typical coating are listed in Table 17.

Bibliography

1. C. E. Schildknecht, *Vinyl and Related Polymers*, John Wiley & Sons, Inc., New York, 1952, Chap. 8.
2. U.S. Pat. 2,293,317 (1942), F. L. Taylor and L. H. Horsley (to The Dow Chemical Co.).
3. U.S. Pat. 2,238,020 (1947), A. W. Hanson and W. C. Goggin (to The Dow Chemical Co.).
4. P. W. Sherwood, *Ind. Eng. Chem.* **54**, 29 (1962).
5. P W. Sherwood, *Ind. Eng. Chem.* **54** (12), 29–33 (1962).
6. G. Talamini and E. Peggion, in G. E. Ham, ed., *Vinyl Polymerization*, Vol. 1, Part 1, Marcel Dekker, New York, 1967, Chap. 5.
7. P. J. Flory, *Principles of Polymer Chemistry*, Cornell University Press, Ithaca, New York, 1953, Chap. VI.
8. C. E. Bawn, T. P. Hobin, and W. J. McGarry, *J. Chim. Phys.* **56**, 791 (1959).
9. J. D. Burnett and H. W. Melville, *Trans. Faraday Soc.* **46**, 976 (1950).
10. W. J. Burlant and D. H. Green, *J. Polymer Sci.* **31**, 227 (1958).
11. R. C. Reinhardt, *Ind. Eng. Chem.* **35**, 422 (1943).
12. W. I. Bengough and R. G. W. Norrish, *Proc. Royal Soc. Ser. A* **218**, 149 (1953).
13. J. D. Cotman, M. F. Gonzalez, and G. C. Claver, *J. Polymer Sci. A-1* **5**, 1137 (1967).
14. B. M. E. Van der Hoff, *Advan. Chem. Ser.* **34**, 6 (1962).
15. A. Konishi, *Bull. Soc. Chem. Japan* **35**, 197 (1962).
16. *Ibid.*, 193 (1962).
17. A. P. Sheinker et al., *Dokl. Akad. Nauk SSSR* **124**, 632 (1959).
18. J. Brandrup and E. H. Immergut, eds., *Polymer Handbook*, Interscience Publishers, a div. of John Wiley & Sons, Inc., New York, 1966.
19. J. F. Gabbett and W. Mayo Smith, in G. E. Ham, ed., *Copolymerization*, John Wiley & Sons, Inc., New York, 1964, Chap. 10.
20. W. I. Bengough and R. G. W. Norrish, *Proc. Royal Soc. Ser. A* **218**, 155 (1953).
21. N. Yamazaki, K. Sasaki, T. Nisiimura, and S. Kambara, *ACS Polymer Preprints* **5**, 667 (1964).
22. A. Konishi, *Bull. Chem. Soc. Japan* **35**, 395 (1962).
23. B. L. Erusalimskii, I. G. Krasnosel'skaya, V. V. Mazurek, and V. G. Gasan-Zade, *Dokl. Akad. Nauk SSSR* **169**, 114 (1966).
24. Brit. Pat. 1,119,746 (1967), to Chisso Corp.
25. U.S. Pat. 3,453,346 (1969), H. J. Hagemeyer and M. B. Edwards (to Eastman Kodak Co.).
26. A. V. Vlasov, L. G. Tokareva, D. Ya. Tsvankin, B. L. Tsetlin, and M. V. Shablygin, *Dokl. Akad. Nauk SSSR* **161**, 857 (1965).
27. Can. Pat. 798,905 (1968), R. Buning and W. Pungs (to Dynamit Nobel Corp.).
28. U.S. Pat. 3,366,709 (1968), M. Baer (to Monsanto Co.).
29. U.S. Pat. 3,033,812 (1962), P. K. Isacs and A. Trafimow (to W. R. Grace & Co.).
30. U.S. Pat. 2,968,651 (1961), L. C. Friedrich, Jr., J. W. Peters, and M. R. Rector (to The Dow Chemical Co.).
31. U.S. Pat. 2,482,771 (1944), J. Heerema (to The Dow Chemical Co.).
32. D. W. Woodford, *Chem. Ind.* (*London*) **1966**, 316. U.S. Pat. 3,291,769 (1966), to Scott Bader & Co.
33. T. Fisher, J. B. Kinsinger, and C. W. Wilson, *Polymer Letters* **5**, 285 (1967).

34. S. Krimm, *Fortschr. Hochpolymer. Forsch.* **2**, 51 (1960).

35. R. A. Wessling, unpublished results, The Dow Chemical Co.

36. K. Okuda, *J. Polymer Sci. A* **2**, 1749 (1964).

37. E. J. Arlman and W. M. Wagner, *Trans. Faraday Soc.* **49**, 832 (1953).

38. R. A. Wessling, unpublished results, The Dow Chemical Co.

39. R. F. Boyer and R. S. Spencer, *J. Appl. Phys.* **15**, 398 (1944).

40. G. Schuur, *Some Aspects of the Crystallization of High Polymers*, Rubber—Stichting, Delft, 1955, p. 18.

41. R. M. Wiley, unpublished results, The Dow Chemical Co.

42. R. A. Wessling, paper submitted to *J. Applied Polymer Sci.* (1970).

43. K-H Illers, *Kolloid Z.* **190**, 16 (1963).

44. E. F. Jordan, W. E. Palm, L. P. Witnauer, and W. S. Port, *Ind. Eng. Chem.* **49**, 1695 (1957).

45. E. Powell, M. J. Clay, and B. J. Sauntson, *J. Appl. Polymer Sci.* **12**, 1765 (1968).

46. G. R. Riser and L. P. Witnauer, *ACS Polymer Preprints* **2**, 218 (1961).

47. H. Burrell, *Official Digest* 726 (1955).

48. R. A. Wessling, to be published (1970).

49. M. Asahina, M. Sato, and T. Kobayashi, *Bull. Chem. Soc. Japan* **35**, 630 (1962).

50. E. D. Serdensky, "Polyvinylidene Chloride Fibers," in H. Mark, E. Cernia, and S. M. Atlas, eds., *Man-Made Fibers*, John Wiley & Sons, Inc., New York, 1968, p. 319.

51. G. S. Kolesnikov, L. S. Fedorova, B. L. Tsetlin, and N. V. Klimentova, *Izv. Akad. Nauk SSSR, Odt. Khim. Nauk* **1959**, 731–735.

52. E. F. Jordan, G. R. Riser, B. Artymyshyn, W. E. Parker, J. W. Pensabene, and A. N. Wrigley, *J. Appl. Polymer Sci.* **13**, 1777 (1969).

53. S. W. Lasoski, *J. Appl. Polymer Sci.* **4**, 118 (1960).

54. S. W. Lasoski and W. H. Cobbs, *J. Polymer Sci.* **36**, 21 (1959).

55. L. A. Matheson and R. F. Boyer, *Ind. Eng. Chem.* **44**, 867 (1952).

56. C. B. Havens, in *Polymer Degradation Mechanisms*, NBS No. 525, U.S. Govt. Printing Office, Washington, D. C., 1953, pp. 107–122.

57. J. R. Dacey and D. A. Cadenhead, *Proc. 4th Conf. Carbon*, Buffalo University Press, 1960, pp. 315–319.

58. J. R. Dacey and R. G. Barradas, *Can. J. Chem.* **41**, 180 (1963).

59. F. H. Winslow, W. O. Baker, and W. A. Yager, *Proc. 1st and 2nd Conf. Carbon*, Buffalo University Press, pp. 93–102, 1956.

60. G. M. Burnett, R. A. Haldon, and J. N. Hay, *European Polymer J.* **3**, 449 (1967).

61. A. Bailey and D. H. Everett, *J. Polymer Sci. A-2* **7**, 87 (1969).

62. G. Oster, G. K. Oster, and M. Kryszewski, *J. Polymer Sci.* **57**, 937 (1962).

63. M. Kryszewski and M. Mucha, *Bull. Acad. Polon. Sci. Ser. Sci. Chim.* **13**, 53 (1965).

64. D. E. Harmer and J. A. Raab, *J. Polymer Sci.* **55**, 821 (1961).

65. E. Tsuchida, C-N. Shih, I. Shinohara, and S. Kambara, *J. Polymer Sci. A* **2**, 3347 (1964).

66. T. Nakagawa, *Kogyo Kagaku Zasshi* **71**, 1272 (1968).

67. I. R. Harrison and E. Baer, *J. Colloid Interface Sci.* **31**, 176 (1969).

68. F. Chevassus and R. de Broutelles, *The Stabilization of Polyvinyl Chloride*, St. Martin's Press, New York, 1963.

69. W. C. Goggin and R. D. Lowry, *Ind. Eng. Chem.* **34**, 327 (1942).

70. U.S. Pat. 2,452,080 (1948), W. T. Stephenson (to The Dow Chemical Co.).

71. *Saran F Resin*, Tech. Bull. Dow Chemical-Europe, 1969.

72. U.S. Pat. 2,462,185 (1949), P. M. Hauser (to E. I. du Pont de Nemours & Co., Inc.).

73. W. W. Cranmer, *Corrosion* **8** (6), 195–204 (1952).

74. J. E. Cowling, I. J. Eggert, and A. L. Alexander, *Ind. Eng. Chem.* **46**, 1977–1985 (1954).

75. R. L. Alumbaugh, *Mater. Protec.* **3** (7), 34–36, 39–45 (1964).

76. U.S. Pat. 3,144,352 (1964), (to Ampex Corporation).

77. L. J. Wood, *Mod. Packaging* **33**, 125 (1960).

78. R. F. Avery, *Tappi* **45**, 356 (1962).

79. A. D. Jordan, *Tappi* **45**, 865 (1962).

80. E. A. Chirokas, *Tappi* **50**, 59A (1967).

R. Wessling and F. G. Edwards
The Dow Chemical Company

VINYL POLYMERS

POLY(VINYL ACETALS)

Poly(vinyl acetal)-type resins are made from poly(vinyl acetate) by hydrolysis followed by condensation with an aldehyde (1). These reactions can be carried out either simultaneously, sequentially, or semisequentially. In the simultaneous, or one-stage process, hydrolysis and acetal formation are carried out at the same time. In the sequential, or two-stage process, the poly(vinyl acetate) is first hydrolyzed and this product is then reacted to form the acetal resin. In the semisequential process the aldehyde is added after the hydrolysis has progressed to a certain point, but without separation of the hydrolysis product.

A typical poly(vinyl acetal) resin can be represented by the general formula

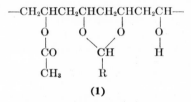

(1)

showing that to a polyvinyl chain there are attached some acyl groups, some acetal groups, and some hydroxyl groups (2). The properties of poly(vinyl acetal) resins are dependent on several factors, such as: (1) the average viscosity of the poly(vinyl acetate) used. (2) the extent of hydrolysis; ie, the percentage of acetyl groups which have been removed from the poly(vinyl ester) and replaced by (OH) groups. (3) the aldehyde used. (4) the percentage acetal reaction; that is, the percentage of the (OH) groups present or potentially present which are reacted to form acetal groups. (5) stabilization.

Any poly(vinyl acetate) can be used. Increasing the viscosity of the polymer increases the softening point, flow temperature, impact strength, hardness, and viscosity in solution of the poly(vinyl acetal) resin.

The hydrolysis of the poly(vinyl acetate) can vary to any extent from very slight to virtually complete. Increasing the percent hydrolysis raises the softening point, tensile strength, etc, and has a marked effect on the form stability of the resulting poly(vinyl acetal) resins. With some aldehydes an increase in percentage hydrolysis is accompanied by a marked change in the solubility of the resins in ordinary organic solvents.

Any aldehyde may be used in the acetal reaction. As might be expected, when the first member of a series of aldehydes is used the acetal resins produced tend to have properties somewhat different from those made from higher aldehydes. As a rule, increasing the length of the acetal side chain tends to give more flexible resins. Two or more aldehydes, or compounds containing a reactive carbonyl group, $>C=O$, or combinations of these with aldehydes may be used. Differences in reaction rate frequently necessitate special processing techniques.

The percentage acetal can be varied from very little to essentially complete and this is accompanied by a wide variation in the properties of the acetal resins. For example, in one series of acetal resins with increasing percentage acetalization, a wide range of properties is possible from water soluble materials to water impervious resins soluble in nonpolar solvents such as benzene. With increased acetalization accompanied by decreased ester content, the heat distortion point and flow temperature are raised; deformation under load is lowered, percent elongation and impact strength are increased (3).

The poly(vinyl acetal) resins may be stabilized so that they are color stable and essentially free from crosslinking when subjected to ordinary forming or molding temperatures. A modification in the regular stabilizing technique is used where crosslinking with a reactive resin is desirable. The chemical and patent literature contain many examples of the application of these principles in the general reaction of poly(vinyl alcohol) and aldehydes (4–9).

Nomenclature. It is to be remembered that the name "acetal" is somewhat peculiar. Acetals, as a group, are gem diethers from *any* aldehyde, $HRC(OR'')_2$. If the aldehyde was formaldehyde, one speaks of formals; and propionaldehyde and butyraldehyde lead to propionals and butyrals, respectively. But what if acetaldehyde was used? Then the product is an *acetal* in a narrow sense, as well as belonging to the wider group of acetals.

Poly(vinyl formal), poly(vinyl acetal), and poly(vinyl butyral) are commercially available. The grades of poly(vinyl formal) are denoted by two numerals, whereof the first is the approximate viscosity of the parent poly(vinyl acetate) (86 g in 1000 ml solution in benzene, measured in cP at 20°C). The second numeral indicates the extent of hydrolysis, that is, the percentage of the original acyl groups that have been removed whether or not the resulting hydroxyl group is acetalized. Commercial

Table 1. Trademarks for Poly(vinyl acetals) and Related Products

Trademark	Manufacturer
Alvar	Monsanto Company
Butvar	Monsanto Company
Formetic	Monsanto Company
Formvar	Monsanto Company
Mowital B	Farbwerke Hoechst A.-G.
Mowital F	Farbwerke Hoechst A.-G.
Mowital O	Farbwerke Hoechst A.-G.
Pioloform B	Dr. Alex. Wacker, GmbH, Munich
Pioloform F	Dr. Alex. Wacker, GmbH, Munich
Revyl A	Nobel-Bozel S.A., Paris
Revyl B	Nobel-Bozel S.A., Paris
Revyl F	Nobel-Bozel S.A., Paris
Rhovinal A	Soc. Rhône-Poulenc, Paris
Rhovinal B	Soc. Rhône-Poulenc, Paris
Rhovinal F	Soc. Rhône-Poulenc, Paris
Vinal	Pittsburgh Plate Glass
Vinylite XYHL	Union Carbide Corporation
Vinylite XYSG	Union Carbide Corporation
Vulcaprene A	I.C.I. Ltd., London

poly(vinyl acetal) resins are usually made from poly(vinyl acetates) having grade viscosities of 7–60.

Analysis. Residual acetate groups are estimated by saponification with excess alkali, and hydroxyl groups by acetylation by means of acetic anhydride in pyridine. Acetal groups can be estimated by measuring the amount of free aldehyde released by distilling the resin with 50% aqueous phosphoric acid, but the results are generally somewhat low (10).

Hazards. The commercial acetal resins themselves are nontoxic and burn slowly, but the hazards of other components present in the compounding should be taken into consideration.

Poly(vinyl formal)

Poly(vinyl formal) [(1), R = H] (Formvar, Revyl F, Mowital F, Rhovinal F, Pioloform F) is available as a white or pale yellow, free-flowing powder. Annual U.S. production for 1969 is estimated at 5–6 million lb. Price per lb is about $0.90.

Commercial poly(vinyl formal) is manufactured by a simultaneous process in acetic acid solution. Poly(vinyl acetate) of the proper viscosity is dissolved in a mixture of acetic acid, water, and formaldehyde. Sulfuric acid catalyst is added, the mixture maintained at an elevated temperature, usually 75–85°C, until the desired reaction has been completed. The sulfuric acid catalyst is then neutralized and the resins precipitated as porous granules by blending with water with rapid agitation. After washing and further stabilizing if necessary, the resin is centrifuged and dried (11–13).

By varying the charge composition and reaction conditions, an almost endless series of poly(vinyl formal) resins with wide variations in properties may be obtained. The bulk density and grain structure can be controlled by precipitation conditions, more dilute conditions giving a finer, more porous grain. As in most organic reactions, longer times at lower temperatures give lighter colors, a fact utilized along with stabilization in the production of resins for molding.

Table 2 lists the principal types of poly(vinyl formal) resins now available in the U.S. "E" indicates a material most suitable for wire enamel; "S," a thermally stabilized material. Electrical properties are given in references 14, 15.

Examination of the properties of the typical formals shown in Table 2 shows that, in general, strength, rigidity, and characteristic temperatures become lower as the proportion of acetate groups increases. Changes in molecular weight have little effect on properties below the inflection temperature. Flow temperature, on the other hand, is dependent on molecular weight.

Increasing acetate group content lowers the inflection temperature, with the resultant greater flow or deformation under load. The heat distortion temperature is independent of molecular weight between a degree of polymerization of about 350 to >1000 when the percent acetate content is held constant. A reduction of elongation and impact strength results from increasing the acetate group content. Water absorption increases with hydroxyl group content. In comparison with other resins, poly(vinyl formal) uniquely combines rigidity, toughness, and dimensional stability although excelled by another resin in each property separately.

Commercial poly(vinyl formals) with % hydrolysis in the 95% range are soluble in acetic acid, aniline, chloroform, cresylic acid, dimethyl sulfoxide, dioxane, ethylene dichloride, furfural, methyl benzoate, 1-nitropropane, pyridine, and certain mixtures

Table 2. General Properties of Poly(vinyl formal)

	ASTM test method	Formvar type					
		15/95E	15/95S	7/95E	7/95S	12/85S	7/70
		500[a]	500[a]	350[a]	350[a]	430[a]	350[a]
poly(vinyl alcohol), %[b]		5–6	7–9	5.0–6.5	7–9	5–7	5–7
poly(vinyl acetate), %[c]		9.5–13	9.5–13	9.5–13	9.5–13	20–27	40–50
viscosity, cP[d]		40–60	40–60	15–20	15–22	18–22	8–10
stability, min 1 hr at °C		120	150	120	150	120	120
specific gravity		1.2	1.2	1.2	1.2	1.2	1.2
refractive index, n_D^{25}	D542–42	1.5	1.5	1.5	1.5	1.5	1.5
flow temperature, °C	D569–48	160–170	160–170	140–145	140–145	145–150	
heat distortion point, °C	D648–49T	88–93	88–93	88–93	88–93	75–80	50–60
inflection temperature, °C	D1043–49T	103–108	100–105	103–108	100–105	90–95	75–80
apparent modulus of elasticity, psi $\times$ 10⁵	D1043–49T	3.5–6.0	3.5–6.0	3.5–6.0	3.5–6.0	4–6	4–6
	D638–41T		2.5–3.0				
tensile strength, 10⁴ psi	D638–41T	1.0–1.1	1.0–1.1	1.0–1.1	1.0–1.1	1.0–1.1	1.0–1.1
elongation, %	D638–41T	7–20	10–50	7–20	10–50	4–5	3–4
deformation at 50°C, 4000-psi load, %	D621–48T	0.2–0.4	0.2–0.4	0.2–0.4	0.2–0.4	4–6	very high
deformation at 70°C, 4000-psi load, %	D621–48T	1–3	1–3	1–3	1–3	very high	very high
impact strength, Izod, notched, ½ $\times$ ½ in., ft-lb	D256–43T	1.2–2.0	1.2–2.0	1.0–1.4	1.0–1.4	0.5–0.7	0.4–0.6
impact strength, Izod, notched, ½ $\times$ ⅛ in., ft-lb	D256–43T				1.3–2.0		0.4–0.6
impact strength, Izod, unnotched, ft-lb/in.	D256–43T	>60	>60	>60	>60		
water absorption, %	D570–40T	0.75	1.1	0.8	1.1	1.5	1.5

[a] Equal to one half the number of carbon atoms in the main chain.

[b] By acetylation.   [c] By saponification.   [d] 5 g resin in 100 ml ethylene dichloride at 20°C.

of aliphatic alcohols with aromatic hydrocarbons such as 30 parts ethanol to 70 parts toluene. The C_1–C_4 alcohols mixed with benzene, toluene, or xylene (depending on the volatility and toxicity requirements) will form solvents. The list of solvents is much greater for resins of high acetate content. Poly(vinyl formal) containing more than 10% poly(vinyl alcohol) is soluble only in liquids which are completely miscible in water.

The thermal degradation of poly(vinyl formal) is largely a surface atmospheric oxidation. A molded disc 10 mm thick of type S resin when heated 1 hr at 200°C showed only slight discoloration, about 1% increase in viscosity and 0.7% insolubles. The same resin in the unmolded state, when heated at 160°C for 1 hr, became discolored and was largely insoluble. Oxidation changes are greatly reduced by the use of p-tert-amylphenol (see Vol. 1, p. 907) as a stabilizer. Oxidative degradation at 150°C or higher splits off formaldehyde and water (16).

The extent of crosslinking in the poly(vinyl formals), when the reaction was carried out in aqueous solution using varying concentrations of formaldehyde and hydrochloric acid, has been studied recently (50).

Applications. The first and most extensive use of poly(vinyl formal) (15/95E and 7/95E) is the manufacture of a tough, heat-resistant wire enamel. An alkylphenolic resin is combined with it in a solvent mixture of cresylic acid or furfural and naphtha (17–20). After application to the wire the coating is cured to a crosslinked material harder than its components and virtually incapable of flow.

The great toughness, adhesion, flexibility, and abrasion resistance of this type of coating assure maintaining dielectric strength by retaining integrity of the coating during winding, treatment, and service. Resistance to heat shock, solvents, greases, and oils is outstanding (11).

Insulation requirements led to the demand for improved poly(vinyl formal) based wire coatings (21). An insulating coating possessing the general properties of poly(vinyl formal) phenolic coatings which could be dip-tinned at 350°–370°C instead of the 450–500°C required by standard Formvar-phenolic coatings was developed by replacing part or all of the polyester in a polyester–urethan coating by Formvar. Coatings containing about 25% Formvar dip-tinned at 350°C have adequate flexibility, resistance to heat shock and abrasion, and solvent resistance. Replacing a small portion of the phenolic resin in the standard Formvar-phenolic insulation with a blocked polyfunctional isocyanate gives an insulated wire with a cut-through temperature of 320+°C rather than the usual 200–220°C value for normal Formvar–phenolic insulated wire.

The introduction of refrigerant 22, $CHClF_2$ (see Vol. 17, p. 306) for air-conditioning applications led to the development and production of a magnet wire which was more resistant to refrigerant 22 than the standard poly(vinyl formal)–phenolic coated wire. Modifying the curing system for the poly(vinyl formal) resin by means of blocked isocyanates modified by thermosetting phenolic and melamine resins gave enamels called Formetic. Formetic enameled wire has excellent resistance to refrigerant 22, exceptionally high cut-through resistance as well as the outstanding balance of properties of standard Formvar–phenolic enameled wire.

Varnish-wire enamel systems capable of giving insulated systems which will give in excess of 20,000 hr operation at 130°C were developed (25–28). Varnishes containing epoxy resins and suitable curing agents give this type of thermal performance with poly(vinyl formal)–phenolic magnet wire (29).

Specialty applications were met by multicoat magnet wire: an overcoat of poly(vinyl butyral) over poly(vinyl formal)–phenolic, applied in the wire enameling oven, forms a heat-bondable magnet wire commonly used for tv yoke coils. The overcoating of standard poly(vinyl formal)–phenolic magnet wire with nylon gives a product with excellent windability characteristics necessary for tightly packed windings as used in armature windings for motors and generators.

Multicoat magnet wires were developed to obtain improved thermal performance. The extra coatings may be applied under or over the poly(vinyl formal). In some cases the end properties were additive: in others, the combination gave unexpectedly improved results. Thus, it was known that poly(vinyl formal)-based magnet wires impregnated with polyester gave systems that could be used at 130°C; it was found that the combination of a base coat of Formetic (roughly $\frac{2}{3}$ the thickness of the coating) with a top coat of polyester ($\frac{1}{3}$ of the thickness) gave wire with 20,000 hr life at temperatures in excess of 150°C. Over and above this, it was unexpectedly found that the combination showed better dielectric life at 300°C than wire coated with polyester alone (35).

The combination of a top coat of poly(vinyl formal)–phenolic enamel over a base coat of a 220°C coating such as a polyimide (Du Pont ML) shows very high burn-out resistance and 20,000 hr life at 160–190°C (30,33). When Formetic is employed as the top coat, the coating possesses excellent resistance to fluorinated refrigerants. In the above combination the ML coating makes up approximately ⅓ of the coating thickness and the combination wires can be coated by standard die-coating techniques with a split applicator and oven temperature gradient readily attainable in standard ovens. Results from a new overload test more severe than that normally used showed the 220°C polyimide provides the greatest resistance to burn-out, followed closely by the multicoat wire consisting of a thin base of polyimide coated with poly(vinyl formal)–phenolic enamel (32).

Wire enamels made using catalyst-free phenolic resins in the standard poly(vinyl formal)–phenolic enamel permit higher application rates and give wire with improved shelf life (34).

The addition of organometallic compounds has been suggested to improve poly-(vinyl formal) wire enamels (35).

Adhesives and Coatings. The versatility of poly(vinyl acetals) is attributed to their terpolymer constitution: acetate and hydroxyl groups as well as acetal. Poly(vinyl formal), due to its limited solubility in common solvents, stiffness, and high softening point is generally used as a component in a two-polymer adhesive system. In a single-component system it was first used in the bonding of mica. Bonding of solenoid coils and similar uses in the electrical industry are among its chief uses in the adhesive field. More recently the coacetal, poly(vinyl formal-ethylal) containing a minor amount (less than 0.1%) of tricresyl phosphate showed excellent adhesion to aluminum foil under thermal aging at 150–200°C (36). (Note: the expression "ethylal" is taken from the Russian source; it seems to provide a useful alternative to "acetal" in the narrow sense.)

The first synthetic resin adhesive in the structural bonding of metals was a two-polymer system: granular poly(vinyl formal) and a phenolic resin. Granular poly-(vinyl formal) placed between surfaces coated with a liquid phenolic resin provides, after the application of heat and pressure, stronger metal-to-metal bonds than riveting or welding, as well as savings of weight and labor in aircraft production (37). This adhesive process is known as the Redux process and is widely used for gluing metal–to–metal, metal-to-wood, metal-to-thermoset resins, etc.

Homogeneous solutions of the phenolic and formal are usually used in structural adhesive work. Film adhesives, either self-supporting or on a carrier base, are readily made from these solutions and are especially useful, for instance, in bonding the "skin" to the metal honeycomb in sandwich construction.

The very high shear strength values at temperatures up to 250°F, coupled with high peel strength at low temperatures and outstanding creep resistance, make the formal–phenolic combination very useful in many exacting adhesive applications. Fatigue strength is very high (38). Durability throughout a wide range of climates has proved to be satisfactory in airplane construction, where it is highly important (39,40).

Formal–phenolic adhesives are used extensively in printed circuit bonding. Here, dielectric properties, peel strength and blister resistance are most important. Test methods for this type of product have been standardized by the National Electrical Manufacturers' Association and the American Society for Testing Materials.

Epoxy resins are showing promise in the two-polymer adhesive system, often as a modifying resin rather than a replacement for the phenolic resin. Melamine–formaldehyde resins are being investigated (41).

Considerable recent work is being done on adhesive film based primarily on poly(vinyl formal)–phenolics (42–44).

Poly(vinyl formal), when used in combination with other resin types such as phenolics, epoxies, isocyanates, melamines, etc improves coating uniformity, minimizes cratering, improves adhesion and increases coating toughness and flexibility. Application of these coatings can be made by brush, dip, spray, fluidized bed, etc. They are used as drum and can linings as well as a wide variety of coated metallic substrates. Curable coatings may be formulated to meet the extractability requirements of the Food and Drug Administration.

Wash primers, or metal conditioners capable of minimizing corrosion, possessing excellent resistance to solvents, hot lubricants and hydraulic fluids, yet furnishing outstanding adhesion between the metallic substrate and a wide variety of topcoats, including the newer epoxy and acrylic coatings, are formulated from poly(vinyl formal), a phenolic resin and passivating agents.

Other Applications. Molding compositions containing poly(vinyl formal) are satisfying demanding applications. Molded articles exhibit excellent toughness, rigidity, dimensional stability, and high tensile strength. By formulation with certain rubbers, impact strengths may be raised very markedly (45,46).

Poly(vinyl formal) may be plasticized to form tough, elastomeric compositions suitable for use as self-sealing gasoline tanks and gasoline drop tanks. Gasoline resistant diaphragms and leatherlike cloths are made from cured mixtures of poly(vinyl formal) and a diisocyanate modified polyester amide. (Vulcaprene A.) The articles have a high abrasion and flex resistance and are nonthermoplastic (47). Poly(vinyl formal) household sponges (Ivalon) are tough, resilient, and soft (5,48).

A water solution of poly(vinyl alcohol) may be foamed with air in the presence of a wetting agent, inorganic acid, and formaldehyde to produce a very strong foamed poly(vinyl formal) which can be used as a substitute for balsa wood.

A mixture of poly(vinyl formal) and poly(vinyl acetal), or the coacetal, is used in the preparation of abrasion-resistant coatings which contain organic polymers and reactive monomers (49).

Fibers of poly(vinyl alcohol) spun from water solution and treated with formaldehyde, as developed in Japan (Vinylon, Kuralon, Kuravilon, Kaneviyon) have good strength, abrasion resistance and heat retention. See under Poly(vinyl alcohol).

Poly(vinyl acetal)

Poly(vinyl acetal) [(1), R = CH₃], although the name is used to designate the general class of acetal resins, specifically indicates the resins made with acetaldehyde, (Alvar, Revyl A, Rhovinal A). They are available as pale-yellow granules, soluble in a wide range of solvents: alcohols, ketones, esters, chlorohydrocarbons, benzene, toluene, dioxane, pyridine, and acetic acid.

Alvar was the first poly(vinyl acetal) to be produced commercially. Limited production was started in Canada by SCL in 1935 as a replacement for shellac. (See Phonograph record composition, Vol. 15, p. 227.) Production peaked in the early 1950s, then gradually fell off due to competition from less expensive resins. Commercial poly(vinyl acetal) is manufactured in Canada (SCL) by a simultaneous process in an ester–alcohol solvent. Poly(vinyl acetate) of the proper viscosity is dissolved in a n-butyl alcohol–n-butyl acetate mixture; water, paraldehyde (trimer of acetaldehyde; see Vol. 1, p. 81) and sulfuric acid are added and the whole kept at elevated temperature, usually 70–72°C, until the desired reaction has been completed. After neutralizing the sulfuric acid with sodium acetate, the alcohol–ester solvent is removed by steam distillation and washing, and the product is shredded and dried. Standard grades are 7/70, 13/80, and 15/70 (51).

In Germany (Farbwerke-Hoechst) a sequential process was used. Poly(vinyl acetate) of the desired viscosity was dissolved in a methanol–methyl acetate mixture and heated at 50–55°C for about 5 min using sodium hydroxide in methanol as catalyst. The solvent was removed by distillation and water added to produce an 8% solution of poly(vinyl alcohol). After filtration, excess aldehyde was added along with 2% of hydrochloric acid. The reaction temperature was 10°C ± 2°, with a short finish period at 40°C. The granular product was washed free from salts and finally used as an interlayer in safety glass.

Applications. Poly(vinyl acetal)'s ease of compression molding (212–265°F) and property of retaining its molded form, coupled with high resistance to abrasion made it outstanding in the filled type of phonograph record. Early uses included injection-molded (340–400°F) articles such as shoe heels, printing plates, toothbrush handles, combs, and novelties. It was particularly interesting in molded electrical parts where carbon tracking on flash-over could not be tolerated.

Spirit soluble lacquers of poly(vinyl acetals) give hard, tough, abrasion resistant coatings which can be applied by brush, spray or dip (52,53). High gloss, good abrasion resistance, ease of modification by the use of plasticizers or extending resins, along with solubility in cheap solvents, make them interesting as shellac substitutes, paper or fabric coatings and grease resistant coatings. Plasticized or modified acetal solutions give high-strength adhesives for laminated products, abrasion wheel bonding, insulation tapes and washable bandages. They increase the transparency and grease resistance of wrapping paper and fabrics. Impregnating dopes are used in treating or sizing fabrics, felt and paper.

Poly(vinyl butyral) [(1), R = CH₃CH₂CH₂] (Vinylite XYHL, Vinylite XYSG, Butvar, Revyl B, Rhovinal B, Mowital B, Pioloform B Vinal) is available as a white, free-flowing powder. Annual U.S. production for 1969 is estimated at 55–60 million pounds. Price per pound, about $1.05.

Commercial poly(vinyl butyral) is manufactured by a sequential process in an alcohol–ester solvent. Poly(vinyl acetate) of the proper viscosity is dissolved in an alcohol–ester solution and hydrolyzed with a sulfuric acid catalyst. The poly(vinyl alcohol) is separated, washed with solvent and suspended in an alcohol–ester solvent. Sulfuric acid and the proper amount of butyraldehyde are added and reaction carried out at moderate temperature with adequate agitation. After neutralizing, the product is precipitated by water, carefully washed, and dried (4,8,10).

The solvents are usually either methyl or ethyl alcohol–ester mixtures, and an alkaline catalyst is sometimes preferred in the hydrolysis step. By varying the poly(vinyl acetate) and the catalyst in the hydrolysis, and the charge composition and reaction conditions in the final step, a series of poly(vinyl butyral) resins with wide variations in properties may be obtained. The bulk density, grain structure, solvent and plasticizer absorption rate can be controlled by precipitation conditions.

Commercial production of one or more poly(vinyl butyrals) has been undertaken in Europe, Japan, and Russia, but the bulk of the production is in the United States. Examples of some of the commercial resins are given in Table 3.

Table 3. Partial List of Commercial Poly(vinyl butyral) Resins

Trademark		Manufacturer	poly(vinyl alcohol),[a] wt %	poly(vinyl acetate),[b] wt %	poly(vinyl butyral), wt %	Average degree of polymerization[c]
Butvar	B-98	Monsanto Co.	19	1.5	79.5	500
Butvar	B-90	Monsanto Co.	19	0.5	80.5	650
Butvar	B-76	Monsanto Co.	11	1.5	87.5	750
Butvar	B-73	Monsanto Co.	19	1.5	79.5	1000
Butvar	B-72A	Monsanto Co.	19	1.5	79.5	3500
Mowital	B-30H	Farbwerke Hoechst A.-G.	18	2.5	79.5	500
Mowital	B-30T	Farbwerke Hoechst A.-G.	24	2.5	73.5	500
Mowital	B-60H	Farbwerke Hoechst A.-G.	18	1.5	80.5	1000
Pioloform BL-18		Wacker-Chemie	19	3.0	78.0	200
Pioloform BW		Wacker-Chemie	23	3.0	74.0	1000
Pioloform BS		Wacker-Chemie	23	3.0	74.0	2000
Vinylite XYHL		Union Carbide	19	0.5	80.5	650
Vinylite XYSG		Union Carbide	19	0.5	80.5	2000

[a] By acetylation.   [b] By saponification.   [c] Equal to one-half the number of carbon atoms in the main chain.

The list of poly(vinyl butyral) resins listed in Table 3 is not intended to be complete, but is presented to show the variations in composition available in typical commercial resins (2,54). As the poly(vinyl acetate) content is maintained at a low level it exerts relatively little influence on properties of the commercial poly(vinyl butyral) resins. On the other hand, the poly(vinyl alcohol) content varies from about 11% to over 23% and it is this variation, coupled with the viscosity of the parent poly(vinyl acetate), that determines the properties of the poly(vinyl butyral) resins.

Unplasticized poly(vinyl butyral) sheeting has a specific gravity of 1.07–1.10, a heat distortion point of 50–60°C, a tensile strength (at break) of 4600–8000 psi, an elongation (at break) of 70–110%, an impact strength (Charpy, notched) of 0.7–1.1 ft-lb/in., and a dielectric strength (step by step) of 370–400 V/mil (55).

Poly(vinyl butyrals) are generally soluble in esters, alcohols, glycol ethers, ketones, and have limited solubility in chlorohydrocarbons. The lower-hydroxyl-content resins are soluble in a wider variety of organic solvents as compared to those with higher hydroxyl content. A notable exception is the insolubility of the low-hydroxyl-content resin in methanol. The presence of both hydroxyl and butyral groups permits solution in mixtures of alcohol and aromatic solvents.

Poly(vinyl butyrals) have high resistance to aliphatic hydrocarbons, mineral, animal, and vegetable oils with the exceptions of castor and blown oils. They withstand strong alkalis but are subject to attack by strong acids. When they are used as components of cured coatings their stability to acids as well as solvents is greatly improved. Crosslinking of poly(vinyl butyral) is carried out by reacting with various thermosetting resins such as phenolics, epoxies, diisocyanates, and melamines (8,56,57).

Availability of secondary hydroxyl groups in poly(vinyl butyral) for condensation of this kind is an important consideration in many applications. Incorporation of even a small amount of poly(vinyl butyral) into thermosetting compositions will markedly improve toughness, flexibility and adhesion of the cured coating.

Applications. The original and largest use of poly(vinyl butyral) containing 18–20% poly(vinyl alcohol) groups is as the chief component in the interlayer for automotive safety glass (laminated glass). The laminate consists of two layers of glass permanently bonded together under heat and pressure with a plasticized poly(vinyl butyral) interlayer. The excellent transparency, adhesion to glass, and impact resistance, coupled with moisture resistance, stability to light and heat, toughness and flexibility over a wide temperature range make this material outstanding (8).

Conventional safety glass interlayer normally contains controlled amounts of salts some of which is due to the use of a slight excess of alkali-metal or alkaline earth-metal salts over that necessary to neutralize the acid catalyst used in making the resin. The influence of this addition of certain salts to poly(vinyl butyral) interlayers has been studied intensively (57–61) and meaningful improvements in the impact performance of the resulting laminates have ensued. The improvement in the safety glass is believed to be a complex interaction between the glass, the interlayer, and the added salt. Various methods of controlling the adhesion characteristic of the interlayer in safety glass, including conditioning with water vapor, etc have been studied (62,63).

Other uses for the plasticized poly(vinyl butyral) sheeting are in high-load-supporting laminates (64), as the structural layer in architectural glass, between poly-(methyl methacrylate) sheets. As a lacquer, it is used to coat window glass to render it nonshattering.

Poly(vinyl butyral) resins are used in coating textiles to render fabrics waterproof and stain-resistant without noticeably affecting the appearance, feel, drape, and color (56). As a rule, the plasticized poly(vinyl butyral) resin is compounded with crosslinking resins such as urea–formaldehyde, phenol–formaldehyde, isocyanate, or other resins. Pigments may also be incorporated. The coating may be applied by calendering or knife-coating. Curing conditions are not critical, varying from 1 hr at 250°F to 5 min at 350°F, and are influenced by the crosslinking of the resin, the characteristics of the substrate and the equipment used. Cotton, wool, silk, nylon, viscose rayon, and other synthetic fabrics can be successfully coated for both indoor and outdoor use.

Butvar Dispersion BR (65) is also used in the textile industry. It gives tough, light, transparent films that develop full strength properties when dried at room temperature. The properties of the coated fabrics, eg, flexibility, smoothness, bonding ability, etc, are the same as those produced from solvent-based compounds. The dispersion has high solids, no solvent cost, and no fire hazards. It can be applied by dipping, padding, spraying, etc.

In protective coating for metals the best known poly(vinyl butyral) application is in wash primers, or metal conditioners. These protective coatings with poly(vinyl butyral), chromate pigment, solvent, and phosphoric acid as a passivating agent, have excellent adhesion to clean metal surfaces, and air-dry rapidly to leave a uniform protective film to which most paints and coatings will adhere (66,67). Wash primers are used on metal structures such as ships, airplanes, storage tanks, bridges, dam locks, guard rails, trucks, and house trailers. Weldable corrosion preventive primers for iron or steel are based on poly(vinyl butyral) composition which contains a phenolic resin and a rust inhibitor (68,69).

Poly(vinyl butyral) is used in metal coating applications in combination with other resin types such as phenolics, epoxies, melamines, and isocyanates. Its use improves coating uniformity and adhesion, and increases coating toughness and flexibility. Application can be made by brush, spray, dip, etc, and end uses include drum and can linings (70). Curable coatings of this type can be formulated to meet the extractability requirements of the Food and Drug Administration.

Poly(vinyl butyral) is widely used as a component of wash coats and sealers in wood finishing. Its use markedly improves the holdout, intercoat adhesion, moisture resistance, flexibility, and toughness. Combinations involving nitrocellulose and shellac are used under commonly used topcoats. Poly(vinyl butyral) greatly improves the holdout of polyester or polyurethan coatings as well as protecting the wood substrate against color changes due to light.

Combined with phenolic resins, poly(vinyl butyral) gives a superior knot sealer due to the excellent barrier qualities to bleeding of the terpenaceous matter from knots, rosin ducts, etc.

Structural Adhesives. Structural adhesives originally were developed for use in the aircraft industry to replace rivets, welding, etc. Refinements in the resins and their formulation led to their use in bonding brake linings, in the electrical and electronic industries, in printed circuits, in honeycomb coil construction, and in the architectural field for the manufacture of interior and exterior curtain walls. Structural adhesives are based on combinations of poly(vinyl butyral) and thermosetting resins, usually phenolics. Epoxy and epoxy–phenolic combinations with poly(vinyl butyral) are used in structural adhesive formulations (71). The ratio of poly(vinyl butyral) to other components has ranged from 10:1 to 10:20 although 10:5 seems to be the best ratio for a compromise of properties (72). Reducing the ratio of phenolic increases the flexibility and peel strength in the adhesive. The addition of small amounts of suitable plasticizer to an adhesive system combining a poly(vinyl butyral) with a thermosetting resin increases the flexibility and impact resistance of the bond with only a minor drop in high-temperature shear.

Structural adhesives based on poly(vinyl butyral) can be applied as solution, unsupported film, supported film on paper or cloth; or as two phases, liquid and solid (73). In airplane skin work the area to be adhered is first painted with the liquid phenolic, then the finely divided poly(vinyl butyral) is added by dusting or some similar method. The sandwich is then closed and the bond made by the application of heat or pressure.

Poly(vinyl butyral) forms an excellent base for hot-melt formulations, particularly where difficult-to-bond surfaces are involved. Their chief advantages are: no solvent release problem, ease of handling below their melt temperature, quick initial tack, and little shrinkage on hardening. These characteristics indicate their use in packaging, bookbinding, labeling, and adhesives for metal foil, plastic film, etc where absence of released solvent is very advantageous.

Other Applications. Poly(vinyl butyral) as an additive to brake fluids gives improved operating properties (74). Weftless tapes made from parallel strands of twisted paper and coated with a dispersion of plasticized poly(vinyl butyral) are moisture resistant, abrasion resistant and have great transverse strength (75). Poly(vinyl butyral) is used in the coating composition for combustible cartridge cases (76). Plasticized poly(vinyl butyral) is the binder in pastes and putties for sealing the seams between structural glass and resin sheets (77). Polyethylene–poly(vinyl butyral) blends give greater crack resistances than polyethylene alone (78,79). A self-adhesive light polarizing film is prepared from poly(vinyl butyral) compositions (80). Complex metal shapes are coated with poly(vinyl butyral) compounds by an electrostatic powder spraying technique (81).

Other Poly(vinyl acetals)

Poly(vinyl alcohol) has been condensed with other aldehydes as well as mixtures of aldehydes, forming coacetals; and with ketones, forming ketals (11). The ketal made from poly(vinyl alcohol) and cyclohexanone (Mowital O) was used for coating electric wires. This same ketal was later produced in powder form (82). Ethylene–poly(vinyl acetate) copolymers have been hydrolyzed and reacted with aldehydes and ketones, yielding acetals and/or ketals (83). The ketal derived from the use of cyclohexanone gave clear films which adhered well to glass and was used in the production of moldings, insulators and the interlayer in safety glass.

Bibliography

1. U.S. Pat. 2,036,092 (March 31, 1936), G. O. Morrison, F. W. Skirrow, and K. G. Blaikie (to Canadian Electro Products), Reissue 20,430 (June 29, 1937).
2. *Technical Bulletin No. 6070*, p. 4, Monsanto Company, St. Louis, Mo.
3. A. F. Fitzhugh and R. N. Crozier, *J. Polymer Sci.* **8**, 225–241 (1952); **9**, 96 (1952).
4. U.S. Pat. 2,162,678 (June 13, 1939), H. F. Robertson (to Carbide and Carbon Chemicals Corporation).
5. U.S. Pat. 2,609,347 (Sept. 2, 1952), C. L. Wilson.
6. U.S. Pat. 2,917,482 (Dec. 15, 1959), E. Lavin, A. F. Price, and R. N. Crozier (to Shawinigan Resins Corp.).
7. Ger. Pat. 947,114 (Aug. 9, 1956), W. Starck and W. Langbein (to Farbwerke Hoechst).
8. C. E. Schildknecht, *Vinyl and Related Polymers*, John Wiley & Sons, Inc., New York, 1952, pp. 323–385.
9. T. Motoyana and S. Okamura, *Chem. High Polymers (Japan)* **8**, 321 (1958).
10. N. Platzer, *Mod. Plastics* **28** (10), 128, 142, 144, 146 (1951).
11. A. F. Fitzhugh, E. Lavin, and G. O. Morrison, "Manufacture and Uses of Polyvinyl Formal," *J. Electrochem. Soc.* **100** (8), (Aug. 1953).
12. U.S. Pat. 2,258,410 (Oct. 7, 1941), J. Dahle (to Monsanto Chemical Co.).
13. U.S. Pat. 2,282,037 (May 5, 1942), J. Dahle (to Monsanto Chemical Co.).
14. *Table of Dielectric Materials*, **2**, National Defense Research Commission, Washington, D.C.
15. T. D. Callinan, "Thermal Evaluation of Formvar Resin," *N.R.L. Report C-3397*, Naval Research Laboratory, Washington, D.C. (Dec. 27, 1948).
16. H. C. Beachell, P. Fortis, and J. Hucks, *J. Polymer Sci.* **7**, 353–376 (1951).
17. U.S. Pat. 2,114,877 (April 19, 1938), R. W. Hall (to General Electric Company).
18. U.S. Pat. 2,154,057 (April 11, 1939), R. H. Thielking (to Schenectady Varnish Co.).
19. U.S. Pat. 2,307,588 (Jan. 5, 1943), E. H. Jackson and R. W. Hall (to General Electric Co.).
20. W. Patnode, E. J. Flynn, and J. A. Weh, *Ind. Eng. Chem.* **31**, 1063 (1939).
21. E. Lavin, A. H. Markhart, and R. W. Ross, "Recent Developments in Magnet Wire Based on Formvar," *Insulation* **8** (4), 25–29 (April 1962).
22. U.S. Pat. 3,067,063 (Dec. 4, 1962), R. W. Hall, E. L. Smith, and G. D. Hilker (half to Phelps Dodge Copper Products Corp., and half to Shawinigan Resins Corp.).
23. U.S. Pat. 3,069,379 (Dec. 18, 1962), E. Lavin, A. F. Fitzhugh, and R. N. Crozier (half to Shawinigan Resins Corp. and half to Phelps Dodge Copper Products Corp.).
24. U.S. Pat. 3,104,326 (Sept. 17, 1963), E. Lavin, A. H. Markhart, and R. E. Kass (to Shawinigan Resins Corp.).
25. R. B. Young and J. R. Learn, *High Temperature Varnish-Wire Enamel Systems*, National Conference on the Application of Electrical Insulation, Cleveland (1958).

26. A. F. Fitzhugh, E. Lavin, and A. H. Markhart, *New 130°C Varnished Formvar-Based Magnet Wire Systems*, National Conference on the Application of Electrical Insulation, Cleveland (1958).
27. C. T. Straka, "Thermal Stability of Class F Insulating Varnishes," *Insulation*, **7** (9), 17 (Sept. 1961).
28. E. W. Daszewski, *Thermal Endurance of Epoxy Encapsulated Magnet Wire*, National Conference of Electrical Insulation, Washington, D.C. (1959).
29. H. Lee, *Compatibility of Magnet Wire Insulation and Epoxy Encapsulating Resins*, National Conference on the Application of Electrical Insulation, Washington, D.C. (1959).
30. AIEE No. 57 "Test Procedures for Evaluation of the Thermal Stability of Enameled Wire in Air" (Jan. 1959).
31. R. V. Carmer and E. W. Daszewski, *Overload Resistance of Film Insulated Magnet Wires*, National Conference on the Application of Electrical Insulation, Chicago (1960).
32. C. F. Hunt, A. F. Fitzhugh, and A. H. Markhart, "Overload Characteristics of Enameled Magnet Wire," *Electro-Technology* **60** (3), 131–135 (Sept. 1962).
33. Belg. Pat. 615,937 (Oct. 3, 1962), E. Lavin, A. H. Markhart, and C. F. Hunt (to Shawinigan Resins Corp.).
34. U.S. Pat. 3,141,005 (July 14, 1964), J. E. Noll (to Westinghouse Electric Co.).
35. Fr. Pat. 1,483,395 (June 1, 1967), R. G. Flowers and C. A. Winter (to General Electric Co.).
36. A. I. Kislov, N. G. Drozdov, S. V. Yakubovich, and A. V. Uvarov, *Lakokrasochnye Materialy i ikh Primenenie* **6,** 32–36 (1967).
37. Brit. Pat. 577,823 (1942), (to Aero Research Ltd., now Ciba (A.R.L) Ltd.).
38. *Technical Note No. 144*, Ciba (A.R.L.) Ltd., Dec. 1954.
39. "The Tropical Durability of Metal Adhesives," *R.A.F. Technical Note No. Chem. 1349*, Feb. 1959.
40. H. W. Eichner, "Environmental Exposure of Adhesive-Bonded Metal Lapjoints," *W.A.D.C. Technical Report 59*, 564, Part I (1960).
41. R. J. Schliekelmann, *Adhesion* **10** (9), 343–350 (1966).
42. Ger. (East) Pat. 40,127 (July 15, 1965), K. Thinius and L. Dimter.
43. Ger. (East) Pat. 40,516 (Mar. 20, 1966), L. Dimter.
44. Ger. Pat. 1,239,046 (April 20, 1967), G. Koerner (to The Goldschmidt A.-G.).
45. U.S. Pat. 2,684,352 (July 20, 1954), C. F. Fisk (to United States Rubber Co.).
46. U.S. Pat. 2,775,572 (Dec. 25, 1956), C. F. Fisk (to United States Rubber Co.).
47. D. A. Harper, W. F. Smith, and H. G. White, *Papers of Proc. 2nd Rubber Technical Conference, London,* **1948**.
48. Brit. Pat. 973,951 (Nov. 4, 1964), (to Kalle, A.-G.).
49. U.S. Pat. 3,324,055 (June 6, 1967), C. L. Wilson and B. L. Orofino (to Alvin M. Marks and Mortimer M. Marks).
50. Shuzi Matsuzawa, Tomosaku Imoto, and Masaki Okazaki, "Crosslinking of Poly(Vinyl Alcohol) with Formal Formation" (Shinshu Univ., Ueda, Japan), *Kobunshi Kagaku* **25**(273), 25–30(275), 173–176.
51. T. P. G. Shaw and C. Monfet, *Alvar Bulletin*, Shawinigan Chemicals Limited (1952).
52. *Ind. Eng. Chem.* **3,** 72–76 (1931).
53. T. P. Sager, *J. Res. Nat. Bur. Stand.* **13,** 884 (1934).
54. E. Lavin and J. S. Snelgrove, "Polyvinyl Acetal Adhesives" in *Handbook of Adhesives*, Reinhold Publishing Corp., New York, 1962, pp. 383–399.
55. "Properties of Butvar and Formvar Resins," *Pub. No. 6130*, Monsanto Co.
56. P. S. Plumb, *Ind. Eng. Chem.* **36,** 1035–1038 (1944).
57. R. Houwink and G. Salomon, *Adhesion and Adhesives, Vol. 1*, American Elsevier Publishing Co., New York, 1965, pp. 318–326.
58. U.S. Pat. 3,231,461 (Jan. 25, 1966), P. T. Mattimoe (to Libbey-Owens-Ford Glass Co.).
59. U.S. Pat. 3,249,487 (May 3, 1966), F. T. Buckley and J. S. Nelson (to Monsanto Co.).
60. U.S. Pat. 3,262,837 (July 26, 1966), E. Lavin, G. E. Mont, and A. F. Price (to Monsanto Company).
61. U.S. Pat. 3,271,235 (Sept. 6, 1966), E. Lavin, G. E. Mont, and A. F. Price (to Monsanto Company).
62. Brit. Pat. 1,093,864 (Dec. 6, 1967) (to E. I. du Pont de Nemours & Co.).

63. U.S. Pat. 3,372,074 (Mar. 5, 1968), A. Rocher and H. Rhety (to Rhône-Poulenc).
64. U.S. Pat. 3,388,033 (June 1, 1968), F. T. Buckley, G. R. Macon, and R. F. Riek (to Monsanto Co.).
65. U.S. Pat. 2,611,755 (Sept. 23, 1952), W. H. Bromley, Jr. (to Shawinigan Resins Corp.).
66. "Bakelite Corp. Vinylite Vinyl Butyral Resins," New York (1949), *Chem. Age* (London) **62,** 889–890 (1950).
67. U.S. Pat. 2,525,107 (Oct. 10, 1950), L. R. Whiting and P. F. Wanger, Jr. (to U.S.A. Secretary of the Navy).
68. U.S. Pat. 3,325,432 (June 13, 1967), M. D. Kellert and R. V. DeShay (to Monsanto Co.).
69. Brit. Pat. 1,093,200 (Nov. 29, 1967), W. Borkenhagen, R. Pohlmann, and W. Bauer (to VEB Lack- und Lackkunstharzfabrik).
70. Netherlands Appln. 6,608,087 (March 30, 1967), American Can Co.
71. U.S. Pat. 2,920,990 (Jan. 12, 1960), J. L. Been and M. M. Grover (to Rubber and Asbestos Corp.).
72. W. Whitney and S. C. Herman, *Adhesive Age* **34** (1), 22–25 (Jan. 1960).
73. U.S. Pat. 2,499,134 (Feb. 28, 1950), N. A. deBruyne (to Rohm and Haas).
74. U.S.S.R. Pat. 161,858 (April 1964), R. I. Arunov and V. P. Barannik.
75. U.S. Pat. 3,126,312 (March 24, 1964), R. F. Nickerson (to Shawinigan Resins Corp.).
76. U.S. Pat. 3,282,146 (Nov. 1, 1966) and U.S. Pat. 3,293,056 (Dec. 20, 1966), W. S. Baker (to U.S. Dept. of the Army).
77. U.S.S.R. Pat. 183,310 (June 17, 1966) and U.S.S.R. Pat. 184,382 (July 21, 1966), V. A. Adintsov, S. E. Titov, T. M. Andrianova, and N. N. Matveev (to Volgoorgtekhstroi Trust).
78. Japan Pat. 1485 (Jan. 19, 1968), M. Hasebe (to Furukawa Electric Co., Ltd.).
79. U.S. Pat. 3,382,298 (May 7, 1968), H. R. Larsen and R. S. Zalkowitz (to Union Carbide Canada Ltd.).
80. U.S. Pat. 3,300,436 (Jan. 24, 1967), A. M. Marks and M. M. Marks.
81. U.S. Pat. 3,377,183 (April 9, 1968), W. C. Hurt, Jr., and R. P. Boehm (to General Electric Co.).
82. Japan Pat. 8743 (July 19, 1962), S. Meguro, T. Ozawa, S. Sukekawa, and K. Masaki (to Kanegafuchi Spinning Co., Ltd.).
83. Ger. Pat. 1,239,475 (April 27, 1967), D. Hardt and H. Bartl (to Farbenfabriken Bayer A.-G.).

GEORGE O. MORRISON
Technical Consultant

POLY(VINYL ACETATE)

Vinyl acetate, $CH_2{=}CHOCOCH_3$, is a colorless, flammable liquid having an odor whose immediate aspect is pleasant but which quickly becomes sharp and irritating. Its chief use is as a monomer for making poly(vinyl acetate) and vinyl acetate copolymers, which are widely used in water-based paints, adhesives, and several other coating or binding applications not requiring service at extreme temperatures. Poly(vinyl acetate) also serves as the precursor for poly(vinyl alcohol) and the poly(vinyl acetal) resins. Vinyl acetate is also copolymerized, as the minor constituent, with vinyl chloride and with ethylene to form polymers of commercial interest, and with acrylonitrile for acrylic fibers.

Vinyl acetate was first reported in 1912 in a patent describing the preparation of ethylidene diacetate from acetylene and acetic acid; the vinyl acetate was a minor by-product (1). Industrial interest in vinyl acetate monomer and poly(vinyl acetate) developed by 1925 and processes for their production were perfected (2). World production of vinyl acetate has increased steadily and rapidly since about 1950. In 1968, 708 million pounds of monomer were produced in the United States (3) and comparably large amounts in Japan and Europe.

Vinyl Acetate Monomer

Physical Properties.   Some physical properties of vinyl acetate are given in Table 1. The values of vapor pressure, heat of vaporization, vapor heat-capacity, liquid heat-capacity, liquid density, vapor viscosity, liquid viscosity, surface tension, vapor thermal conductivity, and liquid thermal conductivity over ranges of temperature have been calculated and graphed (9).

Table 1.   Physical Properties of Vinyl Acetate

Property	Value	Reference
boiling point, °C	72.7°	(4–7)
melting point, °C	−100°, −93°	(4–7)
specific gravity, 20/20	0.9338	(4–7)
refractive index, n_D^{20}	1.3952	(4–7)
viscosity 20°C, cP	0.42	(4–7)
flash point, Tag open cup, °F	15–23	(4–7)
heat of vaporization at 72°C, cal/g	90.6	(4–7)
heat of combustion, Kcal/g	5.75	(4–7)
heat of polymerization, Kcal/mole	21.3	(6, 8)
critical temperature, °C, estd	252	(9)
critical pressure, psia, estd	609	(9)
critical density, g/ml, estd	0.324	(9)

Vinyl acetate is completely miscible with organic liquids but not with water. At 20°C, a saturated solution of vinyl acetate in water contains 2.0–2.4% by weight vinyl acetate, while a saturated solution of water in vinyl acetate contains 0.9–1.0% water (4–7). At 50°C, the solubility of vinyl acetate in water is 0.1% more than at 20°C, but the solubility of water in VAc has doubled to about 2%.

Azeotropes containing vinyl acetate are listed in Table 2.

Table 2.   Vinyl Acetate Azeotropes (10)

Second component	Azeotropic boiling point, °C	Vinyl acetate, wt %
water	66.0	92.7
methanol	58.9	63.4
2-propanol	70.8	77.6
cyclohexane	67.4	61.3
heptane	72.0	83.5

The infrared and nuclear magnetic resonance spectra of vinyl acetate are shown in Figure 1.

Pure vinyl acetate does not absorb significantly in the ultraviolet at wavelengths longer than 250 nm (ethanol solvent) or 253nm (hexane solvent), ie, log ϵ = 0 at the points given, and rapidly cuts off at shorter wavelengths (11,12).

Chemical Properties.   The most important chemical reaction undergone by vinyl acetate is free-radical polymerization. The scientific as distinguished from the technological aspects of the subject have been reviewed (13). The reaction is shown in equation 1.

$$CH_2{=}CH{-}OCOCH_3 \rightarrow {+}CH_2{-}\underset{\overset{|}{OCOCH_3}}{CH}{+}_n \tag{1}$$

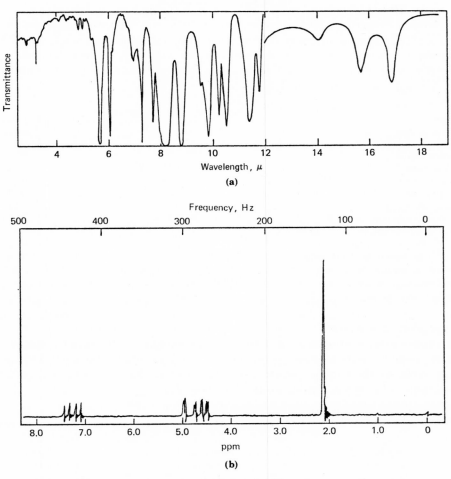

Fig. 1. Spectra of vinyl acetate: (**a**) infrared; (**b**) nuclear magnetic resonance.

Polymerization may be initiated by organic and inorganic peroxides, azo compounds, redox systems (which may include organometallic components), light, and high-energy radiation. There have been a few reports of polymerization by anionic and cationic mechanisms, but being so rare, they may be questionable (57). Polymerization is inhibited by or strongly retarded by the usual aromatic hydroxyl, nitro, or amine compounds, and also by oxygen, quinone, crotonaldehyde, copper salts, sulfur, conjugated polyolefins and ene-ynes. Recently compiled tabulations of quantitative information such as polymerization-rate constants, chain-transfer constants, and activation energies for the polymerization reactions of vinyl acetate may be found in the *Polymer Handbook* (14) as well as in reference (13). Poor agreement among the findings of different investigators is often encountered.

Vinyl acetate has been polymerized in bulk, suspension, solution and emulsion. It copolymerizes readily with some monomers but not with others. Several reactivity ratios (see Vol. 16, p. 233) are presented in Table 3. The Q and e values are 0.026 and -0.22, respectively.

Table 3. Reactivity Ratios in Vinyl Acetate Copolymerizations (13,14)

Comonomer	r_{VAc}	r_M	Temperature, °C
acrylic acid	0.1	2	70
acrylonitrile	0.07	6	70
chloroprene	0.01	50	65
diallyl phthalate	0.72	2.0	
diethyl fumarate	0.01	0.44	60
ethylene	1	1	90–150
ethyl vinyl ether	3	0	60
isobutyl methacrylate	0.025	30	60
isopropenyl acetate	1	1	75
maleic anhydride	0.07	0.01	
styrene	0.01	55	60
vinyl chloride	0.6	1.4	40
N-vinylpyrrolidone	0.2	3.3	50

Other chemical reactions undergone by vinyl acetate are those which may be expected from an ester and from a compound containing a double bond; that the alcohol portion of the ester is α-unsaturated does, however, lend the molecule characteristics not found in ordinary esters. Thus vinyl acetate is hydrolyzed with acidic and basic catalysis to form acetic acid; the expected alcohol, vinyl alcohol, is unstable and acetaldehyde is found.

Chlorine or bromine adds readily to give the 1,2-dihaloethyl acetate which may be distilled under vacuum without decomposition. Hydrogen chloride and/or hydrogen bromide adds readily and quantitatively at low temperatures to give the 1-haloethyl acetate (15). Some 2-haloethyl acetate forms at higher temperatures. With ozone, an ozonide is formed which is explosive when dry.

Vinyl acetate can be used to prepare other vinyl esters by transferring its vinyl group to another organic acid; a mercuric salt is necessary as catalyst (16, 17). Numerous vinyl carboxylates have been prepared by this method, shown in equation 2.

$$CH_3CO\text{—}OCH\text{=}CH_2 + RCOOH \xrightarrow[H^+]{Hg^{2+}} RCOOCH\text{=}CH_2 + CH_3COOH \qquad (2)$$

Acetic acid will add to the double bond, with acid catalysis, to give ethylidene diacetate, $CH_3CH(OCOCH_3)_2$. Oxidation of vinyl acetate with palladium acetate at 60°C produces 1,4-diacetoxy-1,3-butadiene accompanied by significant amounts of 1,1,4-triacetoxy-2-butene; 1,1,4,4-tetraacetoxybutane, and 1-acetoxy-1,3-butadiene (18). By use of diiodomethane in combination with a zinc–copper couple, methylene may be added to vinyl acetate to give cyclopropyl acetate (19). Polyhaloalkanes, of which CCl_4 and CCl_3Br are examples, react with vinyl acetate by a free-radical mechanism to yield 1:1 adducts or telomers having the fragments of the polyhalomethane as end groups (20); the structure of the trichlorobromomethane product is CCl_3CH_2-$CHBrOCOCH_3$. Numerous other molecules which can participate in free-radical chains similarly add across the double bond or are active in telomerization with vinyl acetate.

Vinyl acetate undergoes the oxo reaction to form acetoxypropionaldehydes. It is active in the Diels-Alder reaction to form six-membered rings, or four-membered rings with fluoroolefins. Thiols add across the double bond with Lewis acid catalysis as well as by free-radical reaction. Acetylacetone has been obtained from the reaction of vinyl acetate with acetyl chloride in the presence of aluminum trichloride (7).

MANUFACTURE

There are three chemically distinct processes for the commercial production of vinyl acetate. The oldest and still most extensively used method is the *addition of acetic acid to acetylene* (eq. 3).

$$CH_3COOH + HC\equiv CH \rightarrow CH_3COOCH=CH_2 \tag{3}$$

The reaction may be conducted in the liquid phase or in the vapor phase, but industrial practice is now all in the vapor phase.

The newest method, and one which is gaining adherents rapidly, is by *oxidative addition of acetic acid to ethylene* in the presence of a palladium catalyst (eq. 4). Liquid-phase and vapor-phase processes are in operation commercially.

$$CH_3COOH + H_2C=CH_2 + \tfrac{1}{2}O_2 \rightarrow CH_3COOCH=CH_2 + H_2O \tag{4}$$

The third method of commercial synthesis is a *two-step process* in which acetic anhydride is combined with acetaldehyde to form ethylidene diacetate which is then pyrolyzed to vinyl acetate and acetic acid (eqs. 5 and 6).

$$CH_3CHO + (CH_3CO)_2O \rightarrow CH_3CH(OCOCH_3)_2 \tag{5}$$

$$CH_3CH(OCOCH_3)_2 \rightarrow CH_2=CHOCOCH_3 + CH_3COOH \tag{6}$$

Acetic Acid to Acetylene. The greatest portion of the world's vinyl acetate production at present (1970) uses this reaction but the situation will be changed as the large expansion in monomer capacity, based on ethylene, now underway is completed in the near future (54). See Ethylene; Olefins; both in Supplement Vol. Estimated 1968 United States installed capacity for acetylene-based vinyl acetate was 710 million pounds per year, derived from the facilities of six companies. These companies and their estimated 1969 capacities are listed in Table 4 (21,54).

Table 4. United States Acetylene-Based Vinyl Acetate Production Capacity

Company	Million lb/yr	Plant location	Process
Air Reduction	95	Calvert City, Kentucky	Wacker
Borden	150	Geismar, Louisiana	Borden-Blaw-Knox
duPont	75	Niagara Falls, N.Y.	Wacker
Monsanto	80	Texas City, Texas	Monsanto-Scientific Design
National Starch	50	Long Mott, Texas	Wacker
Union Carbide	260	South Charleston, W. Va.	Wacker
		Texas City, Texas	Wacker

Acetylene-based capacity in Japan in 1968 was estimated at 710 million pounds per year (22); Western European overall capacity, most of which is acetylene-based, was more than 530 million pounds per year (23).

The *catalytic vapor-phase* process operates at temperatures of 180–210°C with a feed having a molar ratio of acetylene to acetic acid of 4 or 5 to 1 (24). The catalyst is zinc acetate supported on about three times its weight of activated carbon. A flow diagram is shown in Figure 2.

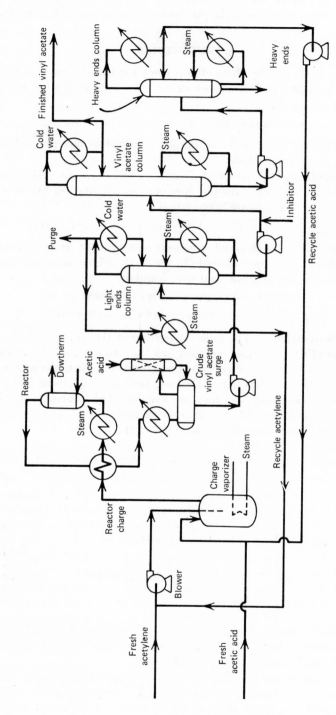

Fig. 2. Acetic acid to acetylene vapor-phase vinyl acetate synthetic process. Courtesy *Petroleum Refiner* (24).

During operation of the process, acetylene is sparged through acetic acid at a temperature which will give the desired ratio of the reactants in the exiting vapor. Typically, operating conditions of 70–80°C at 4–5 psig are used. The vaporized charge is preheated to 180–210°C before being passed through the reactor in which the catalyst is contained. The preheat temperature is dependent on the age of the catalyst, the higher temperatures being used toward the end of the catalyst life. The reactor may consist of a parallel array of 2-inch diameter, 12-foot long tubes contained in a shell through which a heat-exchange fluid is circulated. The reaction operates at 2–3 psig and a space velocity of 300–400 hr^{-1} to give acetic acid conversions of 80% or more. Gas temperatures increase 5–10°C on passage through the reactor. The overall yields are 92–98% on acetylene and 95–99% on acetic acid. Catalyst life is a few months. The vinyl acetate product is purified by distillation in a series of columns which also separate the unreacted acetylene and acetic acid for recycle, and the principal by-products, acetaldehyde and ethylidene diacetate. Stainless steel is extensively used because of the corrosive operating conditions; contact of cuprous metals with monomer after distillation is avoided because of the potent inhibiting action of copper on polymerization but copper stillpots are used to prevent polymerization during distillation.

Cadmium salts have also been used as the catalyst for the vapor-phase acetic acid–acetylene reaction, and other supports, such as silica gel or activated alumina.

The *liquid-phase process* for the acetic acid–acetylene route to vinyl acetate employs a mercuric salt catalyst which is precipitated in the reaction medium, acetic acid, by sulfuric acid, oleum, or phosphoric acid (15). Acetylene gas is blown into the reactor and the vinyl acetate product is entrained by unreacted acetylene and carried to the separation section. It is necessary to avoid long residence of vinyl acetate in the reaction vessel because it would react further to form ethylidene diacetate. Ethylidene diacetate is formed anyway in 10–30% yields. Reaction temperatures of 30–75°C have been employed, and batch or continuous operation is possible. The process is obsolete commercially.

Acetic Acid to Ethylene. This reaction, catalyzed by palladium salts, is a recent development in industrial organic processes. It followed closely the discovery of the palladium-catalyzed synthesis of acetaldehyde from ethylene and oxygen in an aqueous medium (see Vol. 1, p. 86). Producers have estimated that the use of ethylene instead of acetylene for vinyl acetate synthesis would lead to a 20% reduction in raw material costs. As a result, much of the capacity for monomer which will be installed in the future and is being built at present (1970) will use an ethylene feed (25,54).

As with acetylene, vinyl acetate processes using both liquid-phase and vapor-phase ethylene feed are known. Liquid-phase processes have been developed by Hoechst (Germany), ICI (England), and Nippon Gosei (Japan); the ICI process has been operated on a commercial scale in England and the United States for several years. The U.S. plant, operated by Celanese, has a capacity of 100 million pounds of vinyl acetate per year. Vapor-phase processes have been developed by Bayer and Hoechst cooperatively (Germany) and National Distillers and Chemicals, USI Chemicals (USA). While at present no plants are in operation on a commercial scale, confidence in the process evidently exists since USI Chemicals is building a 300 million lb/yr plant in Texas to be on stream in 1970, Celanese is building a 200 million lb/yr plant, and both the USI and Bayer processes have been licensed to six companies in Japan with the new plants to have an aggregate capacity of 512 million lb/yr.

In operation of the *liquid-phase process* (see Fig. 3), a mixture of ethylene and oxygen at 30 atm pressure is fed to the single-stage reactor which contains acetic acid, water, and catalyst at 100–130°C. The products, vinyl acetate and acetaldehyde, are separated from the exiting gas stream by a series of distillation columns (26). Acetic acid and water are also fed to the reactor; the proportion of water in the catalytic solution determines the ratio of acetaldehyde to vinyl acetate formed. The acetaldehyde can be converted to acetic acid which is fed back to the vinyl acetate process thus providing a method for producing monomer from ethylene and oxygen.

A product ratio of 1.14 moles of acetaldehyde to moles of vinyl acetate is optimum for such production and is within the limits of variability available by control of the water feed.

The catalyst solution contains soluble palladium salts, at a concentration of 30–50 mg Pd^{2+}/l and soluble copper salts at a concentration of 3–6 g Cu^{2+}/l. Chloride ion is also necessary; its presence in the form of HCl led to serious corrosion problems downstream from the reactor when the liquid-phase processes were first placed in operation, but use of proper materials of construction, ie, titanium, resin–graphite composites, and ceramics, has overcome this problem (25).

The ethylene–oxygen feed is maintained outside the ignition limits of the mixture, ie, at 5.5% oxygen or less, and this limits the conversion of ethylene per pass to 2–3%. The unreacted gas, however, carries the product out of the reactor (which is a bubble column) and is cleaned up for recycle without much reduction of pressure. Too little oxygen leads to precipitation of cuprous chloride in the reactor.

Overall yields are 90% based on ethylene and 95% based on acetic acid. Capital costs are about 50% higher than required by the acetylene vapor-phase process and energy consumption is higher but so long as the current large price difference between ethylene and acetylene exists, this route may be advantageous economically. The main chemical reactions occurring in this process are shown by equations 7–10.

$$CH_2{=}CH_2 + CH_3COOH + PdCl_2 \rightarrow CH_2{=}CH{-}OCOCH_3 + Pd + 2HCl \qquad (7)$$

$$CH_2{=}CH_2 + H_2O + PdCl_2 \rightarrow CH_3CHO + Pd + 2HCl \qquad (8)$$

$$Pd + 2CuCl_2 \rightarrow PdCl_2 + 2CuCl \qquad (9)$$

$$2CuCl + 2HCl + \tfrac{1}{2}O_2 \rightarrow 2CuCl_2 + H_2O \qquad (10)$$

The reaction of equation 7 occurs in a molecular complex of Pd^{2+} with acetate and ethylene where acetate displaces hydride which is simultaneously transferred to the central palladium atom (27). Equations 9 and 10 show the in situ regeneration of the catalytic system by oxidation of the precipitated palladium metal by cupric ion, and oxidation of the cuprous ion by oxygen. The overall change is shown by equation 4.

The *vapor-phase process* differs from that described above in that very little acetaldehyde is formed. Ethylene is converted to vinyl acetate in 91–94% yields while 1% acetaldehyde is formed. The extensive licensing by Japanese interests of the vapor-phase processes follows from the pattern of use of PVAc in Japan: about 80% is converted to poly(vinyl alcohol). The acetic acid is then available for recycling into the monomer synthesis, as shown in Figure 4.

In operation, a gaseous mixture of acetic acid, ethylene and oxygen is blown over the catalyst in a tubular reactor and the exit stream, containing vinyl acetate, unreacted starting materials, water and small amounts of acetaldehyde, carbon dioxide and other by-products, is separated by a combination of scrubbers and distillation stages. Ethylene at 5–10 atm gage is saturated with acetic acid at about 120°C and

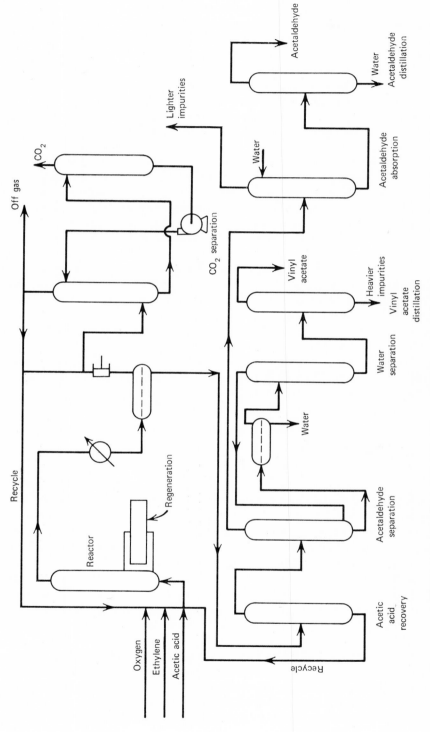

Fig. 3. Acetic acid to ethylene liquid-phase process. Courtesy *Hydrocarbon Processing and Petroleum Refiner* (26).

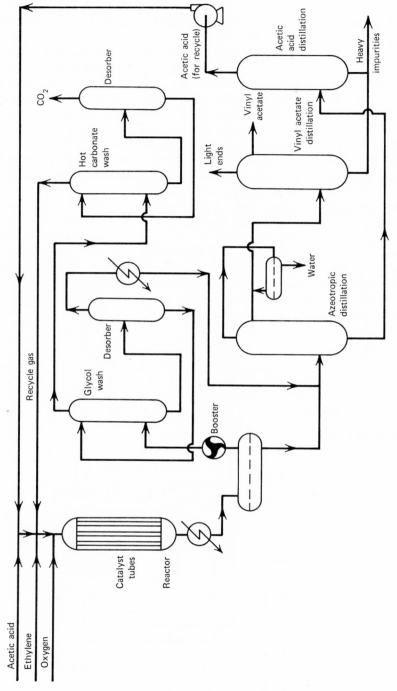

Fig. 4. Acetic acid to ethylene vapor-phase process. Courtesy *Hydrocarbon Processing and Petroleum Refiner* (26).

preheated before entering the reactor. Oxygen is added just before the reactor; the amount that can be added is limited by the explosive limit; at a pressure of 8 kg/cm² gage, 7 volume percent oxygen is the boundary, limiting the attainable conversions of ethylene and acetic acid. The gases are reacted over a catalyst of palladium metal, 0.1 to 2%, on an inert support at 175–200°C. The catalyst is held in a multi-tube reactor and the heat of reaction is used to generate steam for other process needs. Oxygen is 60–70% reacted per pass, acetic acid 20% and ethylene 10%, with space–time yields of over 200 grams of vinyl acetate per liter of catalyst per hour. About 10% of the reacted ethylene is converted to carbon dioxide. When the gases exit from the reactor they are cooled and the liquid and gaseous phases separated. The gaseous portion is scrubbed with propylene glycol to separate vinyl acetate and with hot carbonate solution to remove CO_2, and is then recycled. The combined liquids, ie, initial condensate and desorbed vinyl acetate, are distilled; acetic acid is recycled. In contrast to the liquid-phase ethylene-based process, corrosion is not a problem and the normal materials required for hot acetic acid are used in construction.

Two-Step Process. The one remaining commercial process for vinyl acetate is operated by the Celanese Corporation at Pampa, Texas (28). See Figure 5. Acetaldehyde and acetic anhydride are reacted in the presence of a catalyst at an elevated temperature to form ethylidene diacetate. This product is fed to a cracking tower where pyrolysis occurs and vinyl acetate and acetic acid are formed. Separation and purification are carried out in a number of distillation steps. This process, with a plant capacity of 65 million lb/yr, has been in operation since 1953; in the late 1960s numerous predictions of its imminent shut-down were made.

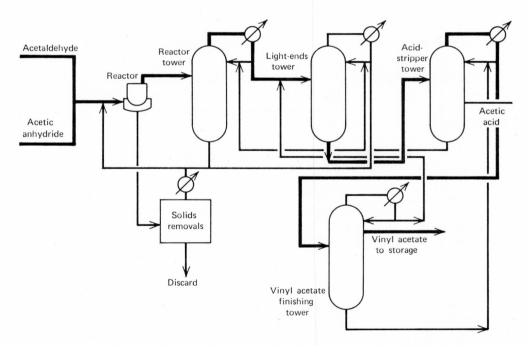

Fig. 5. Process for vinyl acetate via ethylidene diacetate. Courtesy *Hydrocarbon Processing and Petroleum Refiner* (28).

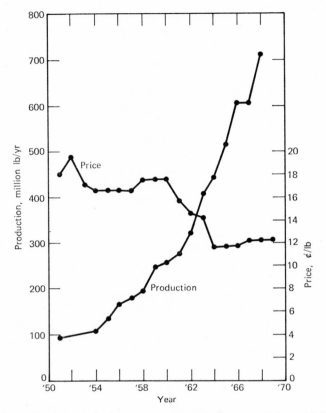

Fig. 6. Production and prices of vinyl acetate monomer. Courtesy U.S. Tariff Commission (30).

Other Possibilities. In addition to the industrial processes discussed above, vinyl acetate may be made in excellent yields by the reaction of vinyl chloride and sodium acetate in solution at 50–75°C in the presence of catalytic amounts of palladium chloride (29) (eq. 11).

$$CH_2{=}CHCl + CH_3CO_2^- \xrightarrow{\ PdCl_2\ } CH_2{=}CHOCOCH_3 + Cl^- \tag{11}$$

ECONOMIC ASPECTS

Vinyl acetate monomer production in the United States has increased at an overall annual rate of 12% since 1951. In 1968, 708 million pounds were produced, 735 million pounds in Japan, and approximately 600 million pounds in Western Europe (Fig. 6). The current rate of increase in monomer production in Japan is greater than that of the United States, while in Western Europe it is less (3,22,23). The U.S. list price of monomer in 1970 was 12 cents per pound, railroad tank-car quantities, delivered; monomer was being sold at 10¢/lb under contracts. Prices have tended to fall as production increased but not in a direct reciprocal relationship (Fig. 6). From a high of 19–20¢/lb in 1952, prices fell to 16.5¢/lb in 1954–1957, rose to 17.5¢/lb during 1958–1960, and after that declined to 11.5¢/lb by 1964. About a billion pounds of additional monomer capacity worldwide was under construction in 1969 or had been announced (54).

SPECIFICATIONS AND STANDARDS

Vinyl acetate monomer is supplied in three grades which differ in the amount of inhibitor they contain but otherwise have identical specifications. A low-hydroquinone grade, containing 3–7 ppm hydroquinone, is preferred where use is expected to be within two months of delivery. For monomer stored up to four months before polymerization, the grade containing 12–17 ppm hydroquinone is used. When indefinite storage is anticipated, a grade containing 200–300 ppm diphenylamine is supplied. The diphenylamine must be removed by distillation before the monomer is used for polymerization; it is not necessary to distill before using the hydroquinone-inhibited grades. Typical manufacturers' specifications are:

Vinyl acetate, %	99.8, min
Boiling point, °C	72.3–73.0
Acidity as acetic acid, %	0.007, max
Carbonyls as acetaldehyde, %	0.013, max
Water, %	0.04, max
Color, APHA system	0–5
Suspended matter	None

Vinyl acetate is commonly stored in carbon steel tanks. Baked-phenolic-coated steel, aluminum, glasslined and stainless steel are also suitable materials of construction. Copper and its alloys cause discoloration as well as inhibition of polymerization if brought into solution, and should not be used unless the monomer is distilled before use. The flammable limits of vinyl acetate vapors in air are 2.6–13.4% by volume. At normal storage temperatures, an explosive vapor–air mixture can exist in the vapor space of a storage tank, unless air is excluded. Thus, nitrogen blanketing should be considered. All tanks and lines should be electrically bonded and grounded, and explosion proof equipment, Underwriters Class 1, Group D, should be used.

Tests indicate that inhibited vinyl acetate has good storage stability at normal temperatures (below 100°F). For above-ground tanks, water-spray cooling or exterior white paint is sometimes used to minimize storage temperatures during hot summer weather. Unloading and storage precautions are the normal ones employed when handling a volatile, flammable liquid having a low flash point and autoignition temperature.

Vinyl acetate requires a red precautionary label for shipment. It should not be stored without inhibitor, or transferred between vessels by use of air pressure. It may be obtained in tank cars, tank trucks, and drums. Additional information concerning the safe handling of vinyl acetate is provided in the Manufacturing Chemists' Association *Chemical Safety Data Sheet SD-75*, and *Manual Sheet TC-4*.

ANALYTICAL AND TEST METHODS

Gas chromatography is an excellent method for determining vinyl acetate and its volatile impurities simultaneously. Using wet chemical techniques, vinyl acetate is assayed by adding excess bromine to an aliquot, followed by excess potassium iodide and titration with standard sodium thiosulfate. Acidity is determined by direct titration in methanol solution with standard caustic, aldehydes by the addition of excess sodium bisulfite followed by titration with standard iodine solution; and water by the Karl Fischer method (see Aquametry). Hydroquinone and diphenylamine may be determined by standard titration techniques, or spectrophotometrically in the ultraviolet region, after evaporation of vinyl acetate. Several companies describe in their product brochures empirical procedures, for determining the polymerization activity of the monomer. In these tests, a given amount of monomer is polymerized under standard conditions and one measures the rate of temperature increase or volume shrinkage, and the induction period.

ASTM specifications and test procedures have been established. They have numbers D 2190, D 2191, D 2193, and D 2083.

HEALTH AND SAFETY

Vinyl acetate is only moderately toxic by mouth or by absorption through the skin. It is not a severe irritant on contact with skin or eyes; however, as with most organic chemicals, prolonged or repeated contact with the skin should be avoided. The vapor is not especially irritating to the nose and throat, but good industrial hygiene practices should be followed and adequate ventilation provided. Repeated or prolonged exposure should be avoided.

When tested on rats the LD_{50} for oral ingestion was found to be 2.9 g/kg body weight, and for absorption through the skin, LD_{50} was more the 5 ml/kg in 24 hours. Breathing a vapor concentration of 1000 ppm in air was not fatal to any animals in 4 hours of exposure. At 4000 ppm, half of the test animals died within two hours.

First aid procedures to be followed in the event of over-exposure to vinyl acetate are:

(1) *Ingestion*—induce vomiting.

(2) *Inhalation*—provide fresh air; keep victim warm and quiet. Apply artificial respiration if necessary.

(3) *Skin contact*—wash with water.

(4) *Eye contact*—flush with water for 15 minutes.

USES

Polymerization is the only major use for vinyl acetate. The breakdown following is for the United States. Products which retain their character as poly(vinyl acetate) absorb 52–60% of monomer production; these will be discussed further in the following section on poly(vinyl acetate). Most of this use is in emulsion polymerization processes. Poly(vinyl alcohol) requires 13–15% of the monomer produced, the poly(vinyl acetals) about 15%, and vinyl chloride copolymers about 8%. Ethylene–vinyl acetate copolymers are a small but growing factor (3,30). Additional quantities are used in polymeric lube oil additives and in acrylonitrile copolymers for acrylic fibers. In Japan, 80% of the vinyl acetate produced goes into poly(vinyl alcohol) which is used there as a fiber.

Vinyl Acetate Polymers

POLYMERIZATION

Vinyl acetate has been polymerized industrially in bulk, in solution, in suspension, and in emulsion (31). Most, perhaps 90%, of the material identified as poly(vinyl acetate) or copolymers which are predominantly vinyl acetate are made by emulsion techniques. The patent literature is rich in descriptions of this technology, and recipes and procedures are readily available in the monomer brochures of producing companies, as well as in the open literature (4,7,13,31).

The emulsions are milk-white liquids containing about 55% poly(vinyl acetate), the balance being water, with small quantities of wetting agents and protective colloids. Their use eliminates the need for expensive, flammable, odorous, or toxic solvents, and for the recovery of such solvents. They are easy to apply, and the equipment is easy to clean with water, if done promptly. Emulsions also offer the advantage of high solids-content with fluidity, since the viscosity of the emulsion is independent of the molecular weight of the resin.

An emulsion recipe, in general, contains monomer, water, protective colloid and/or surfactant, initiator, buffer and may also contain a molecular-weight regulator. The monomer may consist of 30–60% of the charge but most commercially available emulsions contain about 55% solids. Several monomers are copolymerized commercially with vinyl acetate in emulsion polymerization and numerous others have been copolymerized on the laboratory scale. Among the comonomers most commonly employed industrially in emulsion copolymerization with vinyl acetate are ethylene; dibutyl maleate; dibutyl fumarate; di-2-ethylhexyl maleate and fumarate; ethyl, butyl, and 2-ethylhexyl acrylates; and vinyl laurate, caprate and Versatate. See Fatty acids (trialkylacetic). Vinyl hydrogen maleate and vinyl hydrogen fumarate have also been used as monomers, as well as acrylic acid and sodium ethylenesulfonate, when it was desired to incorporate an acidic or ionic group on the polymer. The neutral comonomers are added primarily to decrease the brittle temperature of the polymer below commonly encountered ambient temperatures since many uses of poly(vinylacetate) require some degree of flexibility in service. The monomers which contain acidic groups are added primarily to make the copolymer soluble in basic media such as aqueous ammonia. Copolymerization, compared to softening (which may also be accomplished by addition of plasticizers such as dibutyl phthalate, tricresyl phosphate, etc to the preformed polymer), gives a polymer which is innately and permanently plasticized. Plasticization, whether internal (copolymerization) or external (additives), is also extremely important for proper performance at the time of application. The ease of coalescence and the wetting characteristics of the polymer emulsion particles are related to their softness and the chemical nature of the plasticizer. The most efficient plasticizing comonomer, on a weight or price basis is ethylene (32); copolymerization of vinyl acetate with ethylene is a recent technological innovation and one of growing importance.

Many different combinations of surfactant and protective colloid are used in emulsion polymerizations of vinyl acetate. The properties of the emulsion and the polymeric film formed from it depends to a large extent on the identity and quantity of emulsifiers. These properties include average value and distribution of particle size; stability of the emulsion under conditions of mechanical shear, change of temperature, compounding, and mere passage of time; characteristics of the coalesced resin after application, such as smoothness, opacity, water-resistance, etc, as well as characteristics of the emulsion as it is being used such as flow properties and setting time (33).

Poly(vinyl acetate) emulsions may be made with a surfactant alone or with protective colloid alone but the usual practice is to use a combination of the two. Up to 3% emulsifiers may be included in the recipe normally, but when water-sensitivity or wet tack of the film is desired (as in some adhesives), more may be included. The most commonly used surfactants (qv) are the well-known anionic sulfates and sulfonates, but cationic emulsifiers and nonionics are also suitable. Indeed, some emulsion compounding formulas require the use of cationic or nonionic surfactants if the formulation is to be stable. The most commonly used protective colloids are poly(vinyl alcohol) and hydroxyethylcellulose, but again, there are many others, natural and synthetic, which are usable if not actually preferable for a given application.

In general, the greater the quantity of emulsifiers in a recipe, the smaller will be the particle size of the emulsion. At very low levels of emulsifier, eg 0.1%, the polymer does not form a creamy dispersion which stays indefinitely suspended in the aqueous phase but is in the form of small beads which settle and may be easily separated by filtration. This suspension or pearl polymerization process has been used to prepare polymers for adhesive and coating applications as well as for conversion to poly(vinyl alcohol), and products in bead form are available from several commercial suppliers of poly(vinyl acetate) resins. At the higher emulsifier levels, 1% and above, used for emulsion polymerization, the polymer forms tiny particles which do not settle but which remain indefinitely suspended. Particle sizes resulting from high surface-active agent to low protective-colloid ratios may range from 0.005 to 1 micron; such emulsions have lower solids and lower viscosities. The poly(vinyl acetate) emulsions of commerce are made, for the most part, with higher ratios of protective colloids in the recipes and usually contain about 56% solids. The average particle-size ranges from 0.2 to 10 microns and the viscosity of the emulsions from 4 to 50 poise. These latter compositions are occasionally described as "stable dispersions" rather than "true emulsions" which designation is reserved for the former compositions, but this distinction is generally ignored in commercial characterization (4,7,34–38). The term "latex" is also used to denote these products, particularly those with small, ie, $< 0.2 \mu$ particle size.

The initiators or catalysts used in vinyl acetate polymerizations are the familiar free-radical types such as hydrogen peroxide, peroxysulfates, benzoyl peroxide, and redox combinations. Emulsion polymerizations are conducted with water-soluble catalysts; benzoyl peroxide has been employed in emulsion polymerizations in conjunction with water-soluble catalysts, especially where monomer has been added continuously during the reaction. Suspension polymerizations, on the other hand, are run with monomer-soluble initiators predominantly.

Buffers are frequently added to emulsion recipes and serve two main purposes. The rates of decomposition of some initiators are affected by pH and the buffer is added to stabilize those rates since decomposition of catalyst frequently causes changes in pH. The rate of hydrolysis of vinyl acetate (and some other monomers) is also pH-sensitive and it is desirable to minimize hydrolysis of monomer since it produces acetic acid (which can affect the catalyst) and acetaldehyde (which may lower the molecular weight of the polymer undesirably). Emulsion recipes are usually buffered on the acid side at pH 4–5 with phosphate or acetate for example, but buffering at neutral pH with bicarbonate also gives excellent results. The pH of most commercially available emulsions is 4 to 6.

Often a chain transfer agent is added to vinyl acetate polymerizations, whether emulsion, suspension, solution, or bulk, to control the polymer molecular weight at a desired value. Aldehydes, thiols, carbon tetrachloride, etc have been added; when polymerization in solution is carried out, the solvent, of course, acts as a chain transfer agent and, depending on its transfer constant, has an effect on the molecular weight of the product. The rate of polymerization is also affected by the solvent but not in the same way as the degree of polymerization: the reactivity of the solvent-derived radical plays an important part here. Chain transfer constants for solvents in vinyl acetate polymerizations have been tabulated (14). Some emulsion procedures call for the recipe to include a quantity of preformed poly(vinyl acetate) emulsion, and sometimes antifoamers must be added.

A polymerization may be carried out simply by charging all ingredients to the reactor, heating to reflux and stirring until the reaction is over. This simple procedure is seldom followed. Typically, only a portion of the monomer and catalyst are initially charged and the remainder added during the course of the reaction. Better control of the rate of polymerization may be maintained in this fashion and this is particularly important in industrial-scale operations where large quantities of material are involved and heat-dissipation capacity may be limited. Continuous monomer addition in emulsion polymerization also leads to a smaller particle size and a more stable dispersion. On the molecular level, delayed monomer feed leads to more branches in the polymer chain. As a result, the rate of monomer addition has some effect on final film properties. Copolymerizations usually must be conducted with continuous monomer feed in order to obtain homogeneous polymer compositions. Emulsifiers, too, may be added in increments. Another common practice, after the polymerization has been run at a given temperature, is to raise the temperature and finish the reaction in that fashion, in order to ensure the absence of unreacted monomer. Industrially, polymerizations are carried out to over 99% conversion and thus there is no requirement to strip unreacted monomer.

Semicontinuous emulsion polymerization processes are known in which materials are added continuously to a first kettle and partially polymerized, then run over into a second reactor where, at a higher temperature and with additional catalyst, the reaction is concluded.

Large-scale bulk polymerizations are difficult to conduct because of the increase in viscosity as polymer forms and the consequent difficulty in removing heat. Low-molecular-weight polymers have been made in this fashion, however, and continuous processes are known (39).

All the processes discussed above are operated at atmospheric pressure in conventional glasslined or stainless steel kettles or reactors. The ethylene–vinyl acetate copolymer (EVA) processes must of necessity be operated under high pressure (32). The low vinyl acetate EVA copolymers, containing 10–40% vinyl acetate, are made in processes similar to those used to make low-density polyethylene where pressures are usually 15,000 psi and higher. A medium, ie 45%, vinyl acetate copolymer with rubber like properties is made by solution polymerization in t-butyl alcohol at 5000 psi. The 70–95% vinyl acetate emulsion copolymers are made in emulsion processes under ethylene pressures of 300 to 750 psi.

At the molecular scale, vinyl acetate polymerizations are understood as free-radical polymerizations generally, but are characterized in particular by a relatively large amount of chain transfer (13,40). This high reactivity of the poly(vinyl acetate) growing chain radical is attributed to its low degree of resonance stabilization. The high reactivity of the vinyl acetate radical also contributes to the high rate constant for propagation in vinyl acetate polymerization compared to styrene, the acrylates and the methacrylates. Chain transfer to monomer, as well as to other small molecules, leads to lower-molecular-weight products, but when polymerization occurs in the relative absence of monomer and other transfer agents such as solvent, chain transfer to polymer becomes more important. As a result, toward the end of batch-suspension or emulsion polymerizations branched polymer chains tend to form. In suspension and emulsion processes where monomer is fed continuously, the products tend to be more branched than when polymerizations are conducted in the presence of a plentiful supply of monomer. Chain transfer also takes place with the emulsifying agents, leading to their permanent incorporation into the product.

Investigation has shown that chain transfer to polymer occurs predominantly on the acetate methyl group in preference to the chain backbone; one estimate of the magnitude of the predominance was 40-fold (58,59). To study chain transfer to polymer, after determining the molecular weight, one hydrolyzes the polymer to poly(vinyl alcohol) and then reacetylates it to reform poly(vinyl acetate). The branches whose starting points were located on acetate are no longer present after this procedure, and the decrease in molecular weight observed can be related to the original branching. The number of branches per molecule of poly(vinyl acetate) polymerized at 60°C was calculated to be about 3 at 80% conversion; it rises rapidly thereafter and is about 15 at 95% conversion and 1–2×10^4 number average degree of polymerization.

Chain transfer to monomer is an extremely important factor controlling the molecular weight (12). Several determinations of C_m, the transfer constant to monomer (the ratio of rate constants of the transfer reaction and the propagation step k_{tr}/k_p), show that C_m goes from 1×10^4 to 3×10^4 as the temperature varies from 0°C to 75°C (13,42,53).

An assessment of the "best" values of the rate constants for propagation and termination has been made, with the following results (43);

$$\text{Propagation:} \quad k_p = 3.2 \times 10^7 \exp\,(-3150/T) \quad \text{l/(mol)(sec)}$$
$$\text{Termination:} \quad k_t = 3.7 \times 10^9 \exp\,(-1600/T) \quad \text{l/(mol)(sec)}$$

At 60°C, for example, $k_p = 2300$, and $k_t = 2.9 \times 10^7$.

An estimate of kinetic chain lifetime, that is, the time from initiation to termination by reaction with another radical, was 1–2 seconds at 50°C and 4% per hour rate of polymerization (12). If there are five chain transfer steps in the course of the kinetic chain, then a poly(vinyl acetate) molecule forms in 0.2 to 0.4 seconds. Faster rates of conversion give shorter kinetic chain lifetimes, in inverse proportion, but increased percent of conversion leads to longer chain lifetimes. At 75% conversion and 60°C, the radical lifetime is about 10 seconds.

Vinyl acetate polymerizes chiefly in the usual head-to-tail fashion found in vinyl polymerization, but some of the monomers do orient themselves head-to-head and tail-to-tail as the chain grows. The fraction of head-to-head addition increases with temperature. Flory found 1.15 mole % head-to-head structure at 15°C and 1.86 mole % at 110°C (41).

Because of its considerable industrial importance, as well as its intrinsic interest, emulsion polymerization of vinyl acetate has been extensively studied (13,44). The Smith-Ewart theory, which describes emulsion polymerization of several other monomers well, does not fit the behavior of vinyl acetate. The reason for this is the substantial water-solubility of vinyl acetate monomer; as a result, much initiation of polymerization occurs at dissolved monomer in the water phase as well as in the micelles. Since even low polymers are water-insoluble, phase separation soon occurs when chain growth begins at truly dissolved monomer. The rate of polymerization has been found to be proportional to the 0.7 to 1.0 power of the initiator concentration; the Smith-Ewart theory predicts $R_p = K[I]^{0.4}$. The rate of polymerization shows small or virtually no dependence on emulsifier concentration (depending on the study), whereas the Smith-Ewart theory predicts the rate to be proportional to the 0.6 power of the emulsifier concentration. In vinyl acetate polymerizations, the molecular weights of the products increase with the extent of conversion and the ratio of weight to number average degree of polymerization also changes, becoming larger at higher conversions (44,45). The dilute solution viscosity of poly(vinyl acetate) may be related to the molecular weight by using the following equations (46).

$$[\eta] = 1.02 \times 10^{-2} \ \bar{M}_w^{0.72} \ \text{(acetone solvent, 30°C)}$$
$$[\eta] = 3.14 \times 10^{-2} \ \bar{M}_w^{0.60} \ \text{(methanol solvent, 30°C)}$$

These equations apply to linear polymers of $\bar{M}_w/\bar{M}_n = 2.0$.

Currently, the best guide to the information available in the complex field of vinyl acetate polymerization is Lindemann's review (13).

PROPERTIES

Poly(vinyl acetates) vary with increasing values of molecular weight from viscous liquids through low-melting solids to tough, horny materials. They are neutral, water-white to straw-colored, tasteless, odorless, and nontoxic. The resins have no sharply defined melting point but become softer as the temperature is raised. They are soluble in organic solvents such as esters, ketones, aromatics, halogenated hydrocarbons, carboxylic acids, etc, but insoluble in the lower alcohols (excluding methanol which is a solvent), glycols, water and the very nonpolar liquids such as ether, carbon disulfide, aliphatic hydrocarbons, oils and fats. Alcohols such as ethyl, propyl and butyl containing 5–10% water will dissolve poly(vinyl acetate); butyl alcohol or xylene, which only swell the polymer at normal temperatures, will dissolve it when heated. Some physical properties are listed in Table 5.

Table 5. Physical Properties of Poly(vinyl acetate)

Property	Value
density, g/cc at 20°C	1.19
Index of refraction, n_D^{20}	1.467
water absorption, % in 24 hr	2–3
transition temperature, second order, °C	30–34
thermal coefficient of linear expansion per °C	8.6×10^{-5}
thermal conductivity, cal/(sec)(cm²)(°C/cm)	38×10^{-5}
specific heat, cal/gm	0.389
dielectric constant,[a] ϵ'	
at 30°C, 60 Hz	3.3
60°C, 60 Hz	6.5
25°C, 1 kHz	3.15
25°C, 100 kHz	3.09
25°C, 1 MHz	3.06
25°C, 10 MHz	3.02
loss factor,[a] ϵ''	
at 30°C, 60 Hz	0.1–0.06
60°C, 60 Hz	1.8
25°C, 1 kHz	0.025
25°C, 100 kHz	0.028
25°C, 1 MHz	0.03
25°C, 10 MHz	0.025
dielectric strength, volts/mil,	
at 30°C	1000
60°C	780

[a] Reference 47 and 48.

The electrical properties are strongly affected by the ability of poly(vinyl acetate) to absorb water. Whereas dried resin has a dielectric constant, ϵ', of 3.3, and a loss factor, ϵ'', of 0.08 ± 0.02 (at 35°C, 60 Hz), after exposure to 100% humidity the figures are 10 and 2.7.

As with many thermoplastics, strength properties increase with molecular weight; tensile strengths up to 7300 psi may be obtained. The softening point, as determined by the ring and ball method, or by the Kraemer and Sarnow method, also increases with molecular weight, as shown in Table 6.

Table 6. Softening Points and Molecular Weights of Commercial Poly(vinyl acetate)

Grade viscosity[a]	Fikentscher K value	Softening point, °C		Molecular weight, wt av
		K & S	R & B	
1.5	13	65	75	11,000
2.5	19	81	90	18,000
7	32	106	116	45,000
15	42	131	139	90,000
25	48	153	163	140,000
60	58	196		300,000
100	62	230		500,000
800	79			1,500,000

[a] One molar solution in benzene at 20°C, cP.

Poly(vinyl acetate) commercially available in pure, dry form as beads, granules, or lumps is graded according to the viscosity at 20°C of a solution of 86.1 grams, or one mole, of resin dissolved in benzene to make one liter; the viscosity grades correspond to the molecular weights shown in Table 6. In Europe, the Fikentscher "K" value, also derived from viscosity measurements, is used to characterize commercial resins.

When heated above room temperature all viscosity grades become very flexible, and limp at 50°C. They will stand hours of heating at 125°C without change, but at 150°C will gradually darken in color, and at over 225°C will liberate acetic acid forming a brown insoluble resin which in turn will carbonize at a much higher temperature. On cooling below room temperature (10–15°C) poly(vinyl acetates) all become brittle. The brittle point may be lowered by plasticization or copolymerization.

Aging qualities are excellent due to its resistance to oxidation and inertness to the effects of ultraviolet and visible light.

The nmr spectrum of poly(vinyl acetate) in carbon tetrachloride solution at 110°C shows absorptions at 5.14 τ (pentad) of the methine proton; at 8.22 τ (triad) of the methylene group; and at 8.02 τ, 8.04 τ, and 8.06 τ which are the resonances of the acetate methyls in isotactic, heterotactic, and syndiotactic triads respectively. Poly(vinyl acetate) produced by normal free-radical polymerization is completely atactic and noncrystalline. The infrared spectrum is shown in Figure 7.

The spectra of the copolymers of vinyl acetate differ from that of the homopolymer depending on the identity of the comonomer and its proportion.

The chemical properties of poly(vinyl acetate) are those of an aliphatic ester. Thus, acidic or basic hydrolysis produces poly(vinyl alcohol) and acetic acid or the acetate of the basic cation. Industrially, poly(vinyl alcohol) is produced by a base-catalyzed ester interchange with methanol, where, in addition to the polymeric product, methyl

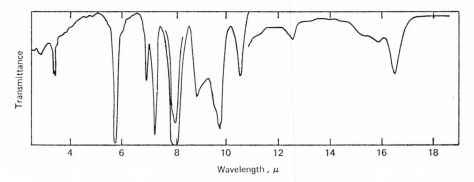

Fig. 7. Infrared spectrum of poly(vinyl acetate).

acetate is formed. The chemical properties of poly(vinyl acetate) may be modified by copolymerization. When a comonomer having a carboxylic acid group or a sulfuric acid group is used, the copolymer becomes soluble in dilute aqueous alkali or ammonia. These copolymers also have better adhesion to metals than homopolymers or neutral copolymers because of the interaction between the acid groups and the metal surface.

Many properties of poly(vinyl acetate) emulsions are determined not by the characteristics of the pure polymer it contains but by the aqueous phase and its contents. The specific gravity at 20°C for all the emulsions is about 1.1. They generally possess a slight odor of residual monomer but if this is removed, by some means such as steam stripping, the emulsions are virtually without odor. The traces of acetic acid which are present can be neutralized with a base such as ammonia, sodium bicarbonate, or triethanolamine, and when the free monomer has been removed the pH of the resulting emulsion may be adjusted to remain constant at 7 or thereabouts. By the judicious selection of monomer concentration, protective colloid, emulsifying and wetting agents, method of polymerization and post-treatment, the properties can be varied to suit the end use. Such properties include average particle size and particle size range, polymer molecular weight, pH, emulsion viscosity, particle charge, adhesion, speed of tack, solvent tolerance, film characteristics and water resistance, and stability to storage, to freezing, to dilution, to mechanical action and to compounding. Various resins, plasticizers, thickening agents, solvents, pigments, extenders, and some dyes may be added, but caution is necessary and it is advisable to adhere closely to the manufacturer's directions. Plasticizers are often added to provide increased flexibility to the dried film, but it is unusual to incorporate these before making the final formulation since their presence may adversely affect stability, particularly in cold weather.

Commercially, emulsions are supplied with viscosities ranging from 200 to 5000 cP, at solids content of about 55%. Particle sizes range from 0.2 to 10 microns. Lower solids–lower viscosity emulsions are also available.

When the emulsion is applied to a surface, water is lost by evaporation, and by absorption if the surface is porous. The particles stick together and eventually coalesce to form a tough, more or less clear, continuous coating. The clarity is often improved by the presence of plasticizer which also enhances the film's resistance to water. Films from most emulsions containing poly(vinyl alcohol) as the protective colloid are liable to reemulsify on contact with water unless they contain relatively large quantities of plasticizer or solvent.

Poly(vinyl acetate) emulsion films have good adhesion to most surfaces, and good binding capacity for pigments and fillers. Plasticized films have good strength and flexibility at ordinary temperatures. The films are unaffected by light, oxygen, chlorine in moderation, and dilute solutions of acids, alkalis, and salts. The films are also inert to oils, fats, waxes, and greases, unless these are largely aromatic in character. Solvents such as acetone, alcohol, ethyl acetate, benzene, and toluene will dissolve or, at any rate, swell the film. An important property of a poly(vinyl acetate) film is its permeability to water vapor. This allows the film to be laid down on a damp surface, the trapped moisture gradually passing through the film without lifting or blistering. The permeability to saturated water vapor at 40°C is 0.021 g/(hr)(100 cm²) for a film of 1 mil thickness.

ECONOMIC ASPECTS

Prices for poly(vinyl acetate) materials are dependent on the form of the polymer, whether it is resin or emulsion, homopolymer or copolymer, as well as which specific product it is. As of 1970 emulsion prices are 24–30¢/dry lb of resin or 12–15¢/wet lb. Copolymer emulsions tend to be higher than homopolymer by 2–3¢/lb. Bead resins and other dry forms are priced at 40–42¢/lb. Specialty copolymers generally obtain a premium price; alkali-soluble resins for example, are about 10¢/lb more than homopolymer. These price ranges are for large shipments. Polymer prices have tended to follow monomer prices; since monomer prices are expected to fall to about 10¢/lb because of the introduction of the ethylene-based processes, there may be a reduction in polymer prices in the near future (54).

Growth in poly(vinyl acetate) consumption is expected to continue. The emulsions dominate the water-based paint market—about 60% of trade sales latex paints are vinyl acetate-based—and are expected to increase their share substantially (49).

The companies listed in Table 7 are among the major suppliers of poly(vinyl acetates) and vinyl acetate copolymers but there are numerous other suppliers. Many other companies produce these polymers in quantities large and small and consume them internally in formulating products.

SPECIFICATIONS AND STANDARDS

The specifications of the commercially available emulsions which are usually listed are tabulated in Table 8, with the ranges of the properties most frequently encountered. There are exceptions to the ranges given. For example, most emulsions have 55–56% solids but some are available at 46–47% with viscosities of 10–50 cP, and others at 59% solids with viscosities in the range listed. Copolymer emulsions are available with 65% solids.

Borax stability is a property of importance in adhesive, paper, and textile applications where borax is a frequently encountered substance. Other emulsion properties tabulated by manufacturers include solvent tolerance (to specific solvents), surface tension, minimum filming temperature, dilution stability, freeze–thaw stability, percent soluble polymer, and molecular weight. Properties of films cast from the emulsions are also sometimes listed; they include clarity, gloss, light stability, water resistance, flexibility, heat-sealing temperature, specific gravity, and bond strength.

Homopolymer resin specifications usually include viscosity grade (one molar solution in benzene at 20°C), 1.5–800 cP ± 10%; volatiles, 1–2%; acidity (as acetic

Table 7. Some Major Poly(vinyl acetate) Producers

Country	Company	Trade Names
USA	Air Reduction	Vinac, Flexbond, Aircoflex
	Borden Chemical	Polyco, Lemac
	Celanese	
	E. I. du Pont de Nemours	Elvacet, Elvace, Elvax
	GAF	Polectron
	W. R. Grace (Dewey and Almy)	Darex, Daratak, Everflex
	Monsanto	Gelva
	National Starch and Chemical	Resyn
	Reichhold Chemical	Wallpol
	Union Carbide Corp.	Bakelite, Ucar, Co-Mer
	U.S.I. Chemicals	Ultrathene
Japan	Kobunshi Chem	
	Nippon Carbide	
	Nippon Gosei	
	Sekisui	
	Showa Denko	
England	ICI	
	Revertex	Emultex
Germany	Farbewerke Hoechst	Mowilith
	Wacker Chemie	Vinnapas
Switzerland	Ebnoether	
	Lonza	Vipolit
Canada	Shawinigan	Gelva
France	Rhône-Poulenc	Rhodopas
Italy	Monte-Edison	Vinavil

Table 8. Poly(vinyl acetate) Emulsion Specifications

Property	Range
solids, %	54–56
viscosity, cP	200–4500
pH	4–6
residual monomer, %	<1.0–0.3
particle size, μ	0.2–3.0
particle charge	neutral or negative
density at 25°C, lb/gal	9.2
borax stability	stable or unstable
mechanical stability	good or excellent

acid), 0.1–0.3%; and softening point (see Table 6). Data are also sometimes given on heat-sealing temperature, tensile strength, elongation and abrasion resistance.

Poly(vinyl acetate) is also available as spray-dried emulsion with average particle size of 2–20 μ, which may be reconstituted to an emulsion by addition of water or added directly to formulations such as concrete. Solutions of resin in methyl and ethyl alcohol at 25–50% solids are also available.

ASTM tests which deal with functions that poly(vinyl acetate) may perform are found as test procedures for the functional class of materials such as paints and lacquers. Poly(vinyl acetate) is nontoxic and approved by the Food and Drug Administration for food-packaging applications.

Emulsions are shipped in 5-gallon pails, 55-gallon drums, tank trucks (3000–4000 gal), and railroad tank cars (6000–8000 gal). Large shipments are insulated against cold. During unloading precautions should be taken to avoid the formation of foam. Latexes corrode ordinary steel and must be stored in tanks having stainless steel, glass, plastic, or coated surfaces. Suitable coatings are baked phenolic, epoxy phenolic, or poly(vinylidene chloride). Storage temperatures should be about room temperature since excessive exposure to high or low temperatures may cause phase separation. Containers should be closed to prevent evaporation of water which will lead to skin formation. Diaphragm pumps and screw pumps are preferred to centrifugal pumps.

Dry resins should be stored at room temperature or below to prevent caking.

USES

Poly(vinyl acetate)-based products are used primarily in adhesive and paint applications. In 1968 the estimated U.S. monomer production was 708 million pounds, and 364 million pounds of poly(vinyl acetate), identified by U.S. Department of Commerce as material to be put into application as poly(vinyl acetate), was produced (30). (The remainder of the monomer was consumed in production of poly(vinyl alcohol), poly(vinyl acetals), copolymers with vinyl chloride, ethylene, acrylonitrile, N-vinyl-pyrrolidone, etc, export, miscellaneous other uses, and production losses.) In recent times, about 250 million pounds per year of poly(vinyl acetate) has been produced in emulsion form and about 110 million lb/yr in resin form. Both figures are on a dry resin basis. These data are summarized in Figure 8, which also contains a partial

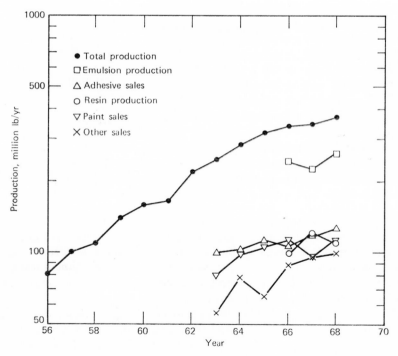

Fig. 8. Poly(vinyl acetate) production and uses. Courtesy U.S. Tariff Commission (30).

breakdown of sales by application area. Poly(vinyl acetate) was sold to the adhesive industry in 1968 at a rate of 125 million lb/yr, and to the paint industry at a rate of 110 million lb/yr. The total of poly(vinyl acetate) sales for uses in applications other than adhesives and paint was 92 million lb/yr in 1968. The actual usage is somewhat greater than the sales in these categories because of internal consumption by some companies U.S. government data for the years 1966–68 show that this "other uses" category includes about 28 million lb/yr in the paper industry and about 10 million lb/yr in the textile industry. Poly(vinyl acetate) exports from the U.S. have been about 2.5 million lb/yr since 1963, except for 1964 when 3.6 million lb were exported.

Adhesives (qv) (50,51). The main areas of poly(vinyl acetate) adhesive use are packaging and wood gluing. The emulsion form of poly(vinyl acetate) is especially suitable for adhesives because of several properties that are peculiar to emulsion systems. The stability of poly(vinyl acetate) emulsions allows them to accept many types of modifying additives without being damaged. For example, solvents, plasticizers, tackifying resins, and fillers can be directly added to homopolymer and copolymer emulsions without having to be pre-emulsified—an impossible feat with the elastomeric adhesive latexes. Homopolymer emulsions containing partially hydrolyzed poly(vinyl alcohol) as a protective colloid can accept greater amounts of these modifying additives than any other type of emulsion, without coagulating.

The stability of the emulsions further permits them to be compounded in simple liquid blending vessels, using simple agitators such as marine-type propellers, paddles, or turbines. Elaborate pre-blending techniques are unnecessary. The adhesives can be adapted to any type of machine application, from spray guns to rollers to extruder-type devices. Different applicators are fairly specific in their viscosity requirements; so are the various substrates receiving the adhesive.

Poly(vinyl acetate) emulsions can be used in high-speed gluing equipment. In contrast to aqueous solutions of natural or synthetic products which lose water slowly in order to set or form an adhesive bond, the emulsions quickly lose a small amount of water, and invert and set rapidly.

Homopolymers stick well to porous or cellulosic surfaces, such as wood, paper, cloth, leather, and ceramics. Homopolymer films tend to creep less than copolymer or terpolymer films. They are especially suitable in adhesives for high-speed packaging operations.

Copolymers wet and adhere well to nonporous surfaces, such as plastics and metals. They form soft, flexible films, in contrast to the tough, horny films formed by homopolymers, and have a greater resistance to water. As the ratio of comonomer to vinyl acetate increases, the variety of plastics to which the copolymer will adhere also increases. Comonomers containing functional groups will often adhere to specific surfaces; for example, carboxyl-containing polymers adhere well to metals.

Crosslinked polymer emulsions accept water-miscible solvents better than straight-chain emulsions. In the film state, they resist water and organic solvents better, and tend to have higher heat-sealing temperatures, which signify a greater resistance to blocking and cold flow. Straight-chain polymers are more sensitive to water and organic solvents than the crosslinked polymers, when they are in the film state. It does not follow, however, that they will have lower heat-sealing temperatures, since they can have high molecular weights, which also contribute to high heat-sealing temperatures. By varying the conditions of polymerization, either straight-chain or crosslinked polymer emulsions can be obtained.

Tack is that property of an adhesive enabling it to form an immediate bond between contacting surfaces when they are brought together. It permits the alignment of an assembly and prevents the adherends from separating before the adhesive sets. Tack is differentiated into two types: (1) wet tack, also called grab or initial tack, is the tack of an adhesive before the liquid carrier (organic solvent or water) has fully evaporated; and (2) dry tack, also called residual tack or pressure sensitivity, is the tack remaining after the liquid carrier has evaporated.

Generally, the forces that work to increase dry tack work also to lower the heat-sealing temperature, and those that decrease dry tack tend to raise the heat-sealing temperature.

Wet tack is often necessary in paper-converting operations. Applicators having little or no pressure on the combining section require emulsion adhesives with enough wet tack to bond strongly at the slightest touch. Emulsions containing poly(vinyl alcohol) as a protective colloid have stronger wet tack than those protected with a cellulose derivative. If still more wet tack is required of the poly(vinyl alcohol)-protected adhesive, both a solvent and a plasticizer can be added, and the solids-content increased.

Dry tack, otherwise known as pressure sensitivity, is needed where two nonporous surfaces are to be bonded. Here, the nature of the adherends does not permit the water in the adhesive to escape by penetration or evaporation, so the adhesive must dry before the adherends are joined, yet still retain enough tack to form a permanent bond. Film-to-film and film-to-foil laminations are good examples of applications requiring dry tack. (Films are of organic materials, see Film materials; foils are thin metal.) Packaging films composed of two or more laminated plies, and decorative packaging materials use pressure-sensitive adhesives in large quantities. Dry tack can be obtained by adding large amounts of both plasticizers and tackifying resins to either homopolymer or copolymer emulsions. It can be increased, but only temporarily, by adding a solvent with a relatively high molecular weight, such as xylene. The copolymer emulsions will have a proportionately greater increase in dry tack than the homopolymer emulsions, since they are a little softer to begin with.

The setting-speed of an adhesive is defined as the time during which the bond formed by the adhesive becomes permanent. Before the setting of an emulsion adhesive can occur at all, the inversion of the emulsion must take place; in other words, it must change from a dispersion of discrete polymer particles in an aqueous continuous phase to a continuous polymer film containing discrete particles of water. The point at which the adhesive has actually set is that point at which an assembly, whether joint or lamination, can no longer be disassembled without damaging one or more of the adherends.

Setting-speed, usually measured in seconds, can be roughly evaluated by drawing a film of the adhesive on a sheet of paper by means of a Bird applicator or an RDS wire-wound rod. A second sheet of paper is immediately placed on the wet adhesive film and a weighted roller is drawn over the assembly. The setting-speed is the time that must pass before the paper fibers are torn when the lamination is pulled apart. Rapid setting-speeds are necessary in many types of packaging adhesives destined for application by high-speed gluing equipment. Here, the rapid inversion possible with poly(vinyl acetate) emulsions, plus their low viscosities, allow them to be compounded into adhesives that not only set rapidly but that also machine easily at high speeds. This machinability permits their application by practically any means, including glue guns, rollers, and spray guns.

On a packaging line, bonding pressure is applied by a compression belt or by a series of rollers. The faster the setting-speed of the adhesive, the shorter the compression section can be. Because floor space is always at a premium, the more efficient package-sealing devices are compact and demand fast-setting adhesives. The terrific speeds demanded by contemporary packaging equipment can only be comprehended by actually watching a modern carton-making operation.

Increases in setting-speed can be made by increasing the solids content; that is, by increasing the amount of water-insoluble substances contained in the emulsions. This "crowds" the aqueous phase of the emulsion, hastening inversion and setting. Adding tackifying resins is one way to crowd the emulsion. Adding plasticizers and solvents is another, but it must be born in mind that these additives mainly increase the setting-speed by softening the polymer particles and hastening their coalescence. Surface-active agents also increase setting-speed by helping the water in an adhesive to penetrate porous surfaces more rapidly. (They can also retard it by stabilizing the emulsion.) The usual way to prepare a high-speed adhesive is to add both a solvent and a plasticizer to the base emulsion.

Poly(vinyl acetate) emulsions are excellent bases for water-resistant paper adhesives destined for use in manufacturing bags, tubes, and cartons. Glue-lap adhesives, which require a moderate-to-high resistance to water, are one example of this type. When routine water-resistance is required, a homopolymer vinyl acetate emulsion containing a cellulosic protective colloid will supply enough for most purposes, unless a water-sensitive substance will also be in the formulation. Next in line are the emulsions containing fully hydrolyzed poly(vinyl alcohol) as a protective colloid, then those containing partially hydrolyzed poly(vinyl alcohol).

When more than routine water-resistance is required, a copolymer vinyl acetate emulsion can be counted on to supply it to a high degree. The plasticizing comonomer in the polymer particles increases their intrinsic coalescing ability; thus they can coalesce more readily than homopolymer particles to a film that will resist water more stubbornly. This resistance to water does not extend to the organic solvents, however, which are better resisted by homopolymer films. The soft copolymers have a lower solubility parameter than homopolymers and are more readily attacked by solvents of low polarity such as hydrocarbons. See Solubility parameters in Supplement Volume.

In spite of their high resistance to water, copolymer emulsions are seldom chosen to bond paper to paper. Cost is a factor here; the cheaper homopolymer emulsions do the job well enough. When copolymers are used in paper adhesives, it is chiefly to join coated or uncoated papers to films, taking advantage of the lower critical surface tension for wetting of the copolymers. Since the majority of packaging adhesives must be able to form strong paper-to-paper bonds, homopolymer vinyl acetate emulsions containing cellulosic protective colloids are always the first choice. The water resistance of any emulsion can be increased by improving its coalescence into a film, either by adding solvents and plasticizers to it, or by adding specific coalescing agents such as diethylene glycol monoethyl ether or diethylene glycol monoethyl ether acetate. Emulsions containing completely hydrolyzed poly(vinyl alcohol) can also incorporate a crosslinking agent, which will further insolubilize them. Crosslinking agents, however, can make the formulation unstable so that it must often be used within a few hours after it has been compounded.

For any adhesive to do its job, it must first thoroughly wet the surface to be bonded; hence, it must be fluid at the time it is applied. Fluidity is no problem with emulsion adhesives. Wetting, however, can be a problem, especially when slick or coated surfaces are to be joined together. A case in point is the manufacture of high-gloss, clay-coated cartons or the adhering of waxed papers.

Copolymer emulsions tend to wet slick surfaces better than homopolymer emulsions because of the extra mobility and softness given the polymer particles by the plasticizing comonomer. Their dried films also tend to conform better with this type of substrate. Homopolymer emulsions containing poly(vinyl alcohol) as a protective colloid wet paper surfaces well, and are the first choice for paper-to-paper packaging adhesives. The setting ability of these emulsions can be increased by incorporating either nonionic wetting agents, or partially hydrolyzed poly(vinyl alcohol), or both. Small amounts of partially hydrolyzed poly(vinyl alcohol) present in an adhesive will also help it to wet lightly waxed papers.

When an adhesive penetrates an adherend, stronger adhesion is possible owing to the greater surface area being exposed to the adhesive. Penetration is regulated according to the type of substrate receiving the adhesive. On a dense, relatively nonporous surface, the penetrating power of the adhesive usually needs to be increased. This can be accomplished by incorporating nonionic wetting agents into it, and by adding water to increase its fluidity and dilute its solids content.

On a soft, porous paper such as chipboard, which is often used for laminating, carton forming and sealing, or tube making, the holdout of the adhesive must be increased—that is, the penetration must be minimized. Excessive penetration of a porous surface can lead to a type of bond failure known as a "starved joint" in which too little adhesive has been left at the adhesive–substrate interface. To prevent such an occurrence, fillers such as China clay, or other large-particle-size clays, or calcium carbonate (whiting), can be added to the emulsion, and the amount of wetting agent reduced.

Solvents that thicken the emulsion will also prevent excessive penetration. High-viscosity fully-hydrolyzed poly(vinyl alcohol) can be added to reduce penetration. Both act by increasing viscosity. The viscosity of an adhesive directly influences its penetration into a substrate: as the viscosity increases, the penetrating power decreases. It also determines the amount of mileage, or spread, that can be obtained. An optimum viscosity exists for each substrate and each set of machine conditions, which must be achieved in order to manufacture an efficient adhesive. Poly(vinyl acetate) emulsions are frequently too low in viscosity to be metered efficiently or to perform well as adhesives by themselves. They must be bodied to working viscosities, and one way to do this is by adding thickeners.

Depending on the type of thickener added, other rheological properties besides viscosity can be controlled. For example, pseudoplastic adhesives (those thinning under high shear) can be prepared by adding polyacrylates, such as sodium or ammonium polyacrylate. Starch and cellulosic thickeners will also confer pseudoplasticity. Some thickeners can prevent adhesives from spitting and throwing during high-speed applications, and permit them to transfer cleanly, breaking short rather than drawing to fibers. Poly(vinyl alcohol) is a good example of this type of thickener.

Blocking and cold flow are unwelcome properties. Blocking refers to dried adhesive surfaces that have become sticky, causing unwanted bonding. Cold flow occurs with a polymer having a low softening point; if the temperature of the adhesive assembly warms to the softening range of the polymer, the bond will slip. The temperature at which a dried adhesive film will form an instantaneous bond between two surfaces when heat is applied, is called its heat-sealing temperature. This property is closely related to the blocking temperature and to the temperature at which cold flow, or "creep," can occur. Whatever increases or decreases the temperature of one will have the same effect on the temperature of the others.

The heat-sealing temperature of an emulsion is related to the thermoplasticity of the poly(vinyl acetate) particles dispersed in it. Thermoplastic polymers soften with the application of heat; those of relatively high molecular weight or those that are crosslinked soften at higher temperatures than those of low molecular weight. In addition, homopolymers soften at higher temperatures than do copolymers of like molecular weight.

Another factor that affects the heat-sealing temperature of an emulsion is the amount of poly(vinyl alcohol) it contains, if any, for poly(vinyl alcohol) is a polymer having a high melting point. High heat-sealing temperatures, which are desirable in an emulsion adhesive since they indicate resistance to blocking and cold flow, are usually attainable with emulsions containing large amounts of poly(vinyl alcohol) as a protective colloid. To further raise the heat-sealing temperature hard extenders such as clay or whiting or substances with high softening points such as ethylcellulose are added. Intermediate heat-sealing temperatures can be obtained by mixing emulsions having high and low heat-sealing temperatures.

Fillers are added to emulsion adhesives to build the total solids content, to reduce penetration into a porous substrate, and to lower costs. Depending on their individual properties, they can also modify the adhesive in other ways: they can add stiffness and strength, or decrease tack and blocking. As a special case, uncooked starch fillers appear to reduce cold-flow in wood adhesives.

Clays and other mineral fillers impart stiffness to adhesive films. Clays and carbonates, in particular, can increase the holdout of adhesives, which is another way of saying they can decrease their penetration into a substrate. A highly plastic clay, bentonite (qv), makes adhesives thixotropic: they become fluid under conditions of high shear, but thicken rapidly once they have been applied.

Plasticizers are added to emulsion adhesives to modify several properties of both the emulsion and the finished adhesive film. By softening the polymer particles dispersed in the emulsion, and increasing their mobility, plasticizers cause them to flow together more easily. This usually increases the viscosity of the emulsion, and tends to destabilize it for faster breaking and setting speeds at the time it is applied. In addition, the increased softness and mobility help the emulsion to wet smooth, nonporous surfaces such as films, foils, and coated papers, thereby increasing its adhesion to them. At the same time, the softened polymer particles coalesce more rapidly and at a lower temperature than is possible with the unplasticized emulsion. This improved coalescence increases the water resistance of the adhesive film. Plasticizers are usually high-boiling esters such as phthalates. Phosphate esters are useful as fire-retardant plasticizers.

Solvents perform several functions in emulsion adhesives: they promote adhesion to solvent-sensitive surfaces; they can increase the viscosity of the emulsion and intensify the tack of the wet adhesive; and they improve the coalescing properties of the film.

Low-boiling solvents impart only wet tack to the adhesive film, whereas high-

boiling solvents confer both dry and wet tack and also lower the heat-sealing temperature. Some high-boiling solvents behave as plasticizers but only temporarily, since
they eventually evaporate from the film. This can be helpful in the case of a wood
adhesive or a packaging adhesive that requires a fast setting-speed. The solvent can
impart the necessary speed to the wet adhesive, but, because it evaporates, it will not
cause the dried bond to creep, which often happens with a plasticized film that ages
under stress.

Solvents promote adhesion to solvent-sensitive adherends by softening and partially dissolving them, thus allowing the adhesive to wet or penetrate the surface, or
both. Some adherends of this type are plastic films, and films of reconstituted cellulose, such as cellophane and cellulose acetate. In the case of coated adherends that
resist adhesion—such as printed, lacquered, or waxed papers—some solvents will
partially dissolve the coating, thus allowing the adhesive to penetrate to the substrate
Chlorinated solvents, in particular, promote adhesion to waxed surfaces.

Tackifiers increase the tackiness and the setting-speed of adhesives. They increase tackiness by softening the poly(vinyl acetate) polymer, both in the wet and in
the dry adhesive film. They increase setting-speed by raising the total solids-content
of the emulsion, thereby crowding the aqueous phase. In many cases, these resins
increase the affinity of adhesives for specific surfaces, such as plastic and foil, and quite
often lower the softening and heat-sealing temperatures of the adhesive film. Tackifiers are usually rosin or its derivatives, or phenolic resins. Other additives frequently
needed for specific application and service conditions are antifoams, biocides, wetting
agents, and humectants.

Poly(vinyl acetate) dry resins and ethylene–vinyl acetate (EVA) copolymers are
used in solvent adhesives which may be applied by typical industrial techniques such
as brushing, knife-coating, roll-coating, spraying, or dipping. Proper allowances must
be made for evaporation of solvent during or before bonding. Poly(vinyl acetate)
resins and EVA copolymers having 21–40% vinyl acetate are widely used in hot-melt
adhesive applications. Homopolymers are compounded with 25–35% plasticizer and
25–30% extender resins. These additives increase fluidity and adhesion as well as
reducing cost. EVA resins are mixed with waxes, rubbers and resin to make the hot-
melt adhesive compound. Hot-melt adhesive application processes may be extremely
rapid both in application and setting speed, cause no problems due to solvent or water
evaporation, and give bonds of high water resistance. The adhesive materials have
indefinite shelf lives. Hot-melts are largely used in packaging, laminating, and bookbinding. For the latter application, alkali-soluble copolymers are frequently used to
facilitate the reclamation of scrap paper, a process in which treatment with hot caustic
is used for de-inking paper (56).

Paints (52) (see Coatings, industrial; Paints). Paints based on water emulsions
of poly(vinyl acetate) solutions were being developed in Canada in 1935, and the true
poly(vinyl acetate) dispersions in England, Canada, and Germany about 1939. During and subsequent to World War II their use for both interior and exterior application was expanded until in 1953 about 15 million pounds of vinyl acetate monomer was
consumed in Europe in this field. In the U.S. and Canada, poly(vinyl acetate) latex
paints (emulsion paints) reached commercial development for interior use about 1946.

It was not until 1950 that commercial production of exterior paints based on poly-
(vinyl acetate) emulsions was started. In 1952, 10 million pounds of emulsion was used
in this application, and paints prepared from them are now well established as primers

or sealers and as surface coatings. As of 1970, about 110 million pounds of poly(vinyl acetate) (dry), in emulsion form was used annually in the paint industry in the U.S. Almost all is used in trade sales of latex paint and is equivalent to about 80 million gallons of finished paint, about two thirds of which is for interior use (55).

Poly(vinyl acetate) emulsion paints form flexible, durable films with good adhesion to clean surfaces including wood, plaster, concrete, stone, brick, cinder blocks, asbestos board, asphalt, and tar paper, and wall boards, aluminum, and galvanized iron. Adherence is also good on surfaces already painted if the surfaces are free from dirt, grease, and rust.

Poly(vinyl acetate) latex paints have become the first choice for interior use (55). In 1967, 55% of the 98 million gallons of interior latex paint sold were products using poly(vinyl acetate) homopolymer or copolymer binders. Their ability to satisfy the two requirements of any paint—to provide protection and decoration—is reinforced by several advantages belonging exclusively to latex paints: they do not contain solvents so that physiological harm and fire hazards are eliminated; they are odorless; easy to apply with spray gun, roller-coater, or brush; and dry rapidly. The paint may be thinned with water and brushes or coaters cleaned with soap and tepid water. The paint is usually dry in 20 min to 2 hr, and two coats may be applied the same day.

They are also widely used in exterior paints. Exterior latex trade sales in 1967 were 61 million gallons, of which vinyl acetate-based paints were 54%. Their durability, particularly their resistance to chalking (see Vol. 14, p. 462), far surpasses that of any conventional oleoresinous paints, which chalk soon after application and chalk heavily. Even though a latex-paint film can also chalk soon after application, it will rarely chalk heavily. Poly(vinyl acetate) films owe their good nonchalking properties to their resistance to degradation by ultraviolet light, which degrades an oleoresinous film by depolymerizing and crosslinking the polymer chain. Latex paints that are correctly formulated from quality poly(vinyl acetate) latexes will develop little or no chalk, thus giving maximum tint retention, and will endure a long time before repainting becomes necessary.

The blister resistance of poly(vinyl acetate) paints is another of their important advantages for exterior use. Latex-paint films are permeable enough to permit water vapor, not liquid, to penetrate them, which prevents blistering and peeling. Their film-formation will not be impaired if they are painted on damp surfaces or applied under very humid conditions.

Several types of plasticizing comonomers can be copolymerized with vinyl acetate to produce latexes suitable for manufacturing paints. The older, and more common, are the alkyl maleates and fumarates such as dibutyl maleate and dibutyl fumarate, and the acrylic esters, such as 2-ethylhexyl acrylate. The most recent development has been use of ethylene as another plasticizing comonomer.

The chief reason for using a comonomer in vinyl acetate polymers is to permanently increase the deformability of the paint film, thus permitting it to expand and contract as the dimensions of the substrate change with changes in temperature. This property of deformability is especially important at the low temperatures found in some parts of the United States, in order to prevent exterior paint films from cracking and flaking.

In addition to lowering the T_g, the plasticizing comonomer also softens the polymer particles, thus increasing their mutual compatibility. When a film is formed from a latex paint, the polymer particles must deform and fuse together to form a continuous film, in order to obtain the maximum in satisfactory paint properties. This coalescing ability is, again, directly traceable to the amount of comonomer present. In both interior and exterior paints, the improvement in coalescence that is obtainable by using high comonomer levels results in a general improvement of all the properties necessary for interior and exterior paints. Polymers with up to 40% dibutyl maleate are in use.

The incorporation of high levels of comonomers such as dibutyl maleate or 2-ethylhexyl acrylate must be done with extreme care. If high comonomer levels are used with low-molecular-weight polymers, the resulting paint film will suffer from excessive dirt pickup at high temperatures because it becomes soft and tacky. This will occur even with high pigment loadings. The only remedy is to use high-molecular-weight polymers, since it is only with these that sufficient comonomer can be incorporated to give the tint retention without dirt pickup that is required by exterior paints.

The strength of a polymer is directly proportional to its molecular weight; consequently, the toughest paint films are formed from latex polymers having the highest molecular weights. For exterior paints, the Federal Government has specified a minimum intrinsic viscosity of 0.45 dl/g. Commercial paint polymers usually range from 0.60 dl/g to 1.0 dl/g and above. Commercial emulsions are usually not characterized by intrinsic viscosity in the product literature.

The critical pigment volume concentration (CPVC) of vinyl acetate paints can range from 48% to over 60%. Generally, the smaller and softer the polymer particles, the higher will be the CPVC and the greater the pigment binding capacity of the latex. The most widely used white pigment is titanium dioxide in the form of rutile or of anatase. Zinc oxide has also been used. The color pigments used in latex paints fall into two categories: the organic pigments, which are usually hydrophobic, and the inorganic pigments, which are usually hydrophilic. The hydrophilic pigments, by their nature, are relatively easy to incorporate into water-base paints. The hydrophobic pigments are more difficult to incorporate, but this can be overcome by choosing the correct blend of surfactants for a particular pigment.

The organic colors commonly used are toluidine red, Hansa yellow, phthalocyanine blue and green, pigment green B, and carbon black, among others. Examples of acceptable inorganic colors are: iron oxide red, brown, yellow and black; and chrome oxide green. Colors that should not be used are those which are reactive, partially water soluble, or sensitive to pH changes.

Extender pigments are used as adjuncts to the prime pigments in order to extend the hiding properties of the paint. They also contribute solids and aid in film build, at a reduced raw-materials cost. They are occasionally called functional pigments as they also contribute certain functional properties to the paint, such as buffering and viscosity control. They include calcium carbonates, talcs, clays and silicas. See also Pigments.

In addition to latex and pigment, paint formulations contain dispersants and wetting agents (both surfactants), defoamers, thickeners and protective colloids, freeze–thaw stabilizers, coalescing agents, and biocides.

Surfactants modify the properties of the surface layer of one phase of a system in contact with another, generally by lowering the interfacial tension. In latex paints, they aid color development, adhesion, and freeze–thaw resistance. The viscosity and storage stability of latex paints are also heavily influenced by the proper choice and amount of surfactant used. Careful laboratory pretesting is necessary to determine just which surfactants are suitable for any given formulation. Surfactants tend to be "temperamental," depending on the ingredients used with them; thus, a surfactant that will perform well in one type of formulation may not perform quite as well in another. In latex paints, their chief functions are as pigment dispersants and wetting agents, and as defoamers.

Anionic surfactants such as potassium tripolyphosphate and calcium polyacrylate are used to disperse the prime and extender pigments throughout the paint. They act by placing negative charges on the pigment particles, which then disperse by repelling each other. Lecithin is often used in conjunction with the anionic dispersants to assist in the dispersion of titanium dioxide.

The negative charges placed on the pigment particles by the anionic surfactants are gradually lost as the formulation ages, and it is then that the nonionic surfactants act to stabilize the dispersion. Along with the protective colloids, the nonionic surfactants maintain an even distribution of the prime and extender pigments throughout the paint, thus preventing flocculation and precipitation. Their judicious use is extremely important, because poor pigment dispersion will result in low hiding power, poor texture and improper coalescence. See also Surfactants.

In most cases, the thickeners used to increase the viscosity of a paint also function as protective colloids by surrounding the pigment particles with a protective envelope that guards against coagulation in the presence of electrolytes or excessive shear. As protective colloids, they assist the surfactants in suspending the pigments, and in stabilizing the viscosity of the paint. They are important aids to freeze–thaw stability, and also improve the leveling, brushability, and open time (the time after application during which the paint can still be brushed). By retarding a too-rapid loss of water, they assist in film formation during hot weather or over porous substrates.

The most widely used thickeners–protective colloids are methylcellulose and hydroxyethylcellulose. Both are commercially available in a wide range of viscosities. Low-viscosity cellulosic thickeners tend to give a paint better leveling properties and a longer open time than high-viscosity thickeners. Because large amounts are required to achieve enough body, paints containing them will have maximum freeze–thaw stability and resistance to shear. Because of their water solubility, paints containing them will also tend to be sensitive to water. High-viscosity thickeners are chosen when lowered costs and maximum thickening power are required. Because lesser amounts can be used to successfully body a paint, they are also chosen when maximum water resistance is required of the finished film.

Freeze–thaw stabilizers are usually ethylene glycol or diethylene glycol. They protect the paint against coagulation caused by freezing.

Coalescing agents are added to assist film formation at normal temperatures, and to prevent poor fusion under adverse conditions, such as at low temperatures or over porous substrates. Coalescing agents, or fluxing agents as they are sometimes called, act by softening the discrete polymer particles so that they can flow together and fuse rapidly at the crucial moment of film formation. They can be solvents, plasticizers, and, to a lesser extent, wetting agents. Active solvents, such as diethylene glycol monomethyl ether (Carbitol, Union Carbide); Carbitol acetate; 2-phenoxyethanol (ethylene glycol monophenyl ether, see Vol. 10, pp. 643–4); and Texanol (Eastman, 2,2,4-trimethylpentan-1,3-diol 1-isobutyrate) are most commonly used.

Besides acting as coalescing aids, these agents contribute to the pigment-binding capacity, color uniformity, and enamel holdout (the property of resisting penetration by an overcoat of enamel) of the paint film, since these properties are particularly dependent upon good film formation. Unfortunately, the agents that work for good film coalescence can also impair the freeze–thaw stability of the paint. For this reason they should be used in minimal amounts—just enough to promote close and rapid fusion of the film, but not enough to soften the polymer particles excessively when they are in suspension.

Biocides are necessary to avoid growth of microorganisms on the cellulosic materials present in the paint. Such growth can degrade the quality of the product to a point where it is unusable. Biocides in use in latex paints include chlorinated phenols, mercurials, organic tin compounds, barium metaborate, and zinc oxide.

Except for the use of water as a medium, poly(vinyl acetate) paints are handled similarly to oil-paint systems, and can be manufactured in conventional paint-manufacturing equipment. Copolymer latexes are particularly easy to handle, because they do not require the addition of plasticizers.

Normally, there are three phases to the manufacturing process: (1) preparing the pigment paste; (2) milling or "grinding" the paste; and (3) the let down to the finished paint.

(1) The pigment paste is started from a mixture of water, dispersants, thickener, freeze–thaw stabilizer, and defoamer. After these materials have been thoroughly mixed, the prime pigments and extenders are added incrementally. In order to develop maximum shear during the ensuing milling operation, the paste must be kept fairly heavy.

(2) The pigment paste can be milled or ground most efficiently by a high-speed mill, such as a Morehouse Mill, Day Hy-R-Speed Mill, or Cowles Dissolver. It is possible to use pebble mills if these are available. Roller mills can be used, but are not recommended because the steel rollers rust easily.

(3) The let down is ordinarily carried out in a thinning tank equipped with a low-speed agitator. At this point, the coalescing agent and the balance of dispersants, thickeners, and defoamers are thoroughly mixed into the pigment paste. The required amount of latex is then added.

This is a critical point in the manufacture of the paint. It is extremely important to avoid vortexing since this will introduce air into the paint, which is very difficult to remove.

After a final viscosity adjustment with thickener solution or water, the paint is ready for packaging. It may be necessary to strain the paint before it can be packaged. In these cases, either a nylon or stainless steel strainer can be used.

Latex paints pose one problem in packaging that is not encountered with oleoresinous paints: the water in the formulation tends to corrode the container. Although antirust additives have been tried, they have met with little success; the best way to prevent corrosion is to pretreat paint containers with a suitable protective coating.

Paper. Poly(vinyl acetate) emulsions and resins are used as the binder in coatings for paper and paperboard. The coatings may be clear, colored, or pigmented and are glossy, odorless, tasteless, greaseproof, nonyellowing, and heat-sealable. Conventional paper-coating equipment is used; formulations normally have 60–65% solids with a pigment to binder ratio of 1:5. Printing quality and ink-pick resistance are excellent.

In paper-making, emulsions applied as wet-end additions to the furnish improve the strength and durability of the final product. Plasticized emulsions may be used to give the product toughness and flexibility.

Textiles. Poly(vinyl acetate) emulsions find wide application as textile finishes because of their low cost and good adhesion to natural and synthetic fibers.

In textile piece-goods finishing, the dispersions diluted with water to 1–3% resin are most often used to obtain a stiff or crisp hand on woven cotton fabrics. Concentrations of 2–20% emulsion are recommended for bodying, stiffening, and bonding. Major applications include the stiffening of felts and the binding of nonwoven fabrics. Finishes to improve the snag-resistance and body or hand of nylon hosiery are based on poly(vinyl acetate) emulsions.

Poly(vinyl acetate) emulsions are used to prime-coat fabrics to improve the adhesion of subsequent coatings, or to make them adhere better to plastic film. Plasticized emulsions are applied, generally by roller-coating, to the backs of finished rugs and carpets to bind the tufts in place and to impart stiffness and handle. For upholstery fabrics woven from colored yarns, poly(vinyl acetate) emulsions may be used to bind the tufts of pile fabrics or to prevent slippage of synthetic yarns.

The emulsion formulations are generally applied to cloth by padding from a bath and squeezing off the excess. Modifying a formulation in the pad box, to increase or decrease firmness, for example, may be easily done by adding emulsion or softener. The alkali-soluble vinyl acetate copolymers previously mentioned may be used as warp sizes during weaving.

Nonwoven fabrics are made with binder formulations based on poly(vinyl acetate) emulsions.

Other Uses. Poly(vinyl acetates) are used as binders for numerous materials such as fibers, leather, asbestos, sawdust, sand, clay, etc, to form compositions which may be shaped with heat and pressure. The compressive and tensile strength of concrete is improved by addition of poly(vinyl acetate) emulsions to the water before mixing. About 10% polymer solids to total solids is the best ratio. It also aids adhesion between new and old concrete when patching or resurfacing. Joint cements, taping compounds, caulks and fillers are other construction-industry uses.

Emulsions having added poly(vinyl alcohol) and bichromate are used to make light-sensitive stencil screens for textile printing and ceramic decoration. The resins are used in printing inks, nitrocellulose lacquers, and special high-gloss coatings. Inks made with poly(vinyl acetate) and metallic pigments approach the appearance of foil since the formulations have a high leafing power, do not induce tarnish, and contribute no unwanted color or aging.

Vinyl acetate homopolymers and copolymers have been used for chewing gum bases.

Bibliography

"Vinyl Acetate" in *ECT* 1st ed., under "Vinyl Compounds," Vol. 14, pp. 686–691 by T. P. G. Shaw; pp. 691–698 by K. G. Blaikie and T. P. G. Shaw; pp. 699–709 by K. G. Blaikie and M. S. W. Small, Shawinigan Chemicals Ltd.

1. Ger. Pat. 271,381 (1912), F. Klatte (to Chem. Fabr. Grieshiem-Electron).
2. C. A. Schildknecht, *Vinyl and Related Polymers*, John Wiley & Sons, Inc., New York, 1952, p. 323.
3. *BDSA Quarterly Industry Report, Chemicals*, U.S. Dept. of Commerce, Washington, D.C., April 1969, p. 26.
4. *Vinyl Acetate, Bulletin No. S-56-3*, Celanese Chemical Co., New York, 1969.

5. *Vinyl Acetate Monomer, F-41519*, Union Carbide Corp., New York, June, 1967.
6. *Vinyl Acetate Monomer, BC-6*, Borden Chemical Co., New York, 1969.
7. *Vinyl Acetate Monomer*, Air Reduction Co., New York, 1969.
8. F. S. Dainton and K. J. Irwin, *Trans. Faraday Soc.* **46**, 331 (1950).
9. R. W. Gallant, *Hydrocarbon Process.* **47** (10), 115–121 (1968).
10. L. H. Horsley, *Azeotropic Data, II, Advances in Chemistry Series, No. 35*, American Chemical Society, Washington, D.C., 1962.
11. W. P. Paist, et al., *J. Org. Chem.* **6**, 280 (1941).
12. M. S. Matheson, et al., *J. Am. Chem. Soc.* **71**, 2610 (1949).
13. M. K. Lindemann, "The Mechanism of Vinyl Acetate Polymerization," in *Vinyl Polymerization*, G. E. Ham, ed., Vol. 1, Part 1, Marcel Dekker, Inc., New York, 1967, Chap. 4.
14. J. Brandrup and E. H. Immergut, eds., *Polymer Handbook*, Interscience Publishers, a division of John Wiley & Sons, Inc., New York, 1966.
15. G. O. Morrison and T. P. G. Shaw, *Trans. Electrochem. Soc.* **63**, 425 (1933).
16. D. Swern and E. F. Jordan, Jr., "Vinyl Laurate and Other Vinyl Esters," in *Organic Syntheses*, Collective Vol. 4, N. Rabjohn, ed., John Wiley & Sons, Inc., New York, 1963, p. 977.
17. H. Hopff and M. A. Osman, *Tetrahedron* **24**, 2205 (1968).
18. C. F. Kohll and R. van Helden, *Rec. Trav. Chim.* **86**, 193 (1967).
19. H. E. Simmons and R. D. Smith, *J. Am. Chem. Soc.* **81**, 4256 (1959).
20. C. Walling, *Free Radicals in Solution*, John Wiley & Sons, Inc., New York, 1957, Chap. 6.
21. *Chemical Week*, Aug. 12, 1967, pp. 73–75, updated.
22. *Polymer Report, No. 128*, 4, Institute of Polymer Industry, Tokyo (1968).
23. W. Schwerdtel, *Hydrocarbon Process.* **47** (11), 189 (1968).
24. *Petrol. Refiner* **38** (11), 304 (1959).
25. R. Remirez, *Chem. Eng.* **75** (17), 94 (1968).
26. *Hydrocarbon Process.* **46** (4), 146 (1967).
27. R. van Helden, et al., *Rec. Trav. Chim.* **87**, 961 (1968).
28. *Hydrocarbon Process.* **44** (11), 287 (1965).
29. H. C. Volger, *Rec. Trav. Chim.* **87**, 501 (1968).
30. *Preliminary Report on U.S. Production and Sales of Plastics and Resin Materials*, U.S. Tariff Commission, Washington, D.C., April 15, 1969 and prior reports.
31. H. Bartl, "Polymerization of Vinyl Acetate and Higher Vinyl Esters," in *Methods of Organic Chemistry* (Houben-Weyl), *Macromolecular Materials*, Part 1, E. Muller, ed., Georg Thieme Verlag, Stuttgart, Germany, 1961, pp. 905–918.
32. M. K. Lindemann, *Paint Manuf.* **38** (9), 30–36 (1968).
33. E. Levine, W. Lindlaw, and J. A. Vona, *J. Paint Technol.* **41**, 531 (1969).
34. *Product List*, Air Reduction Co., New York, 1968.
35. *Elvacet Poly(Vinyl Acetate) Brochure*, E. I. du Pont de Nemours & Co., Inc., Wilmington, Del., 1968.
36. *Thermoplastics Division Product Directory*, Borden Chemical Co., New York, 1968.
37. *Gelva Poly(Vinyl Acetate) Technical Bulletin, Publ. No. 6103*, Monsanto Chemical Co., St. Louis, Mo., 1969.
38. E. Tromsdorff and C. E. Schildknecht, "Polymerizations in Suspension," in *Polymer Processes*, C. E. Schildknecht, ed., Interscience Publishers, Inc., New York, 1956, pp. 105–109.
39. R. D. Dunlop and F. E. Reese, *Ind. Eng. Chem.* **40**, 654 (1948).
40. Reference 20, Chap. 4, p. 151.
41. P. J. Flory and F. S. Leutner, *J. Poly. Sci.* **3**, 880 (1948); **5**, 267 (1950).
42. S. P. Pontis and A. M. Deshpande, *Makromol. Chem.* **125**, 48 (1969).
43. Reference 20, Chap. 3, p. 95.
44. J. W. Breitenbach et al., *Monat. Chem.* **99**, 625 (1968).
45. D. Stein, *Makromol. Chem.* **76**, 170 (1964).
46. M. Matsumoto and Y. Ohyanagi, *J. Poly. Sci.* **46**, 441 (1960).
47. D. J. Mead and R. M. Fuoss, *J. Am. Chem. Soc.* **63**, 2832 (1941).
48. S. O. Morgan and Y. A. Yager, *Ind. Eng. Chem.* **32**, 1519 (1940).
49. A. Errico, *Paint and Varnish Prod.* **57**, 31 (1967).
50. *Adhesives Handbook*, Airco Chemical and Plastics Division of Air Reduction Co., Inc., New York, 1969.

51. R. A. Weidener, "Thermoplastic Adhesives," in *Treatise on Adhesion and Adhesives*, Vol. 2, R. L. Patrick, ed., Marcel Dekker, Inc., New York, 1969, Chap. 10. pp. 432–447, 467–471.

52. *Paint Handbook*, Airco Chemical and Plastics Division of Air Reduction Co., Inc., New York, 1969.

53. W. W. Graessley, W. C. Uy, and A. Gandhi, *Ind. Eng. Chem., Fundamentals* **8**, 697 (1969).

54. *Oil, Paint, and Drug Reporter* Dec. 8, 1969, 13; July 21, 1969, 9 and 29.

55. *Am. Paint J.* **53**, 7, 58 (1968).

56. R. K. Kumler, "Paper Stock," in *Handbook of Pulp and Paper Technology*, K. W. Britt, ed., Reinhold Publishing Corp., New York, 1964, pp. 37–43.

57. Y. Landler, *Compt. Rend.* **230**, 539 (1950); *J. Poly. Sci.* **8**, 63 (1952).

58. S. Imoto, J. Ukida, and T. Kominami, *Kobunshi Kagaku* **14**, 101 (1957).

59. D. Stein and G. V. Schulz, *Makromol. Chem.* **52**, 249 (1962).

DAVID RHUM

Air Reduction Co.

POLY(VINYL ALCOHOL)

Poly(vinyl alcohol) is believed to have been first prepared by a German chemist, Willie Hermann. Its first commercial application was undertaken in the 1920s in Germany and it was commercially introduced into the United States in 1939. Chemically, poly(vinyl alcohol) can be broadly classified as a polyhydric alcohol with secondary hydroxyl groups on alternate carbon atoms. Although poly(vinyl alcohol) can be regarded as a polymer of vinyl alcohol, it must be remembered that the theoretical monomer, $CH_2=CHOH$, does not exist, since it rearranges to acetaldehyde, CH_3CHO. The polymer of vinyl alcohol can only be obtained by hydrolysis of some other vinyl polymer. Commercially, the term poly(vinyl alcohol) includes all resins made by the hydrolysis of poly(vinyl acetate).

Variation of two major factors during processing markedly affects the properties of the final product. The properties vary according to the molecular weight of the parent poly(vinyl acetate) and the extent of hydrolysis. The structure of poly(vinyl alcohol) obtained by complete hydrolysis may be represented by $CH_3CHOH(CH_2-CHOH)_n$—, where n is related to the molecular weight of the parent resin. Although the formula for ordinary poly(vinyl alcohol) is commonly written in this manner it must be remembered that this is an idealized structure, and in practice some head-to-head polymerization occurs resulting in a 1,2-glycol structure, $—CH_2CHOHCH_2CHOH-CHOHCH_2CHOHCH_2$—. On partial hydrolysis, proportional amounts of residual CH_3COO— groups are distributed along the chain in place of OH, and the amount of such acetate groups expressed as a percentage is the acetate content; thus, in a poly(vinyl alcohol) of 70% acetate content, 30% of the acetate groups of the original poly(vinyl acetate) were hydrolyzed to hydroxyl groups and 70% remain as acetate groups. The hydroxyl group content of the poly(vinyl alcohol) is of primary concern for commercial applications and samples where 87–89% of the acetoxy groups present have been converted to hydroxyl groups are called partially acetylated; in the fully hydrolyzed grades, a minimum of 98% of the acetoxy groups is replaced by alcohol groups and, in the super hydrolyzed grade, more than 99.7% of the polymer is hydrolyzed to poly(vinyl alcohol).

The chief uses of poly(vinyl alcohol) in the United States are as paper and textile sizes, solvent and oxygen resistant films, adhesives, emulsifiers, and as an intermediate for the manufacture of poly(vinyl butyral) used as a safety glass laminate. See under

Poly(vinyl acetals). In addition to these applications, in Japan, because of the special nature of the chemical industry, poly(vinyl alcohol) fiber is widely used. Lesser applications include light polarizing films, sand molding binders, electrical insulators for luminescent screens, and cosmetric applications.

Poly(vinyl alcohol) is usually shipped in powder or granular form. A large series of grades differing in acetate content and molecular weight is available from several manufacturers. Stereospecific poly(vinyl acetate) when hydrolyzed yields poly(vinyl alcohol) with a high degree of crystallinity, possessing properties different from ordinary poly(vinyl alcohol).

Properties

Physical Properties. Polyvinyl alcohol is stable in color up to 140°C. On prolonged heating to 160–170°C, it darkens in color and becomes insoluble in water. At over 170°C, ether formation occurs; over 200°C, it softens, and at still higher temperatures, decomposes with carbonization. Under higher pressures at 200–250°C, a moldable resin of slightly modified solubility is attained (1,2). Its heat of combustion is 5902 cal/g (3–5). The densities and refractive indexes of resins of various acetate contents are given in Table 1.

Table 1. Density and Refractive Index of Poly(vinyl alcohol)

Acetate in resin, %	d_{20}^{20}	n_{D}^{20}
0	1.329	1.557
5	1.322	1.553
10	1.316	1.548
20	1.301	1.539
30	1.288	1.530
40	1.274	1.521
50	1.260	1.512
60	1.246	1.503
70	1.232	1.494

The coefficient of linear expansion of poly(vinyl alcohol) is $7-12 \times 10^{-5}$ (6).

Water is really the only practical solvent for poly(vinyl alcohol). Water solutions of poly(vinyl alcohol) tolerate substantial amounts of such monohydric alcohols as methanol, ethanol and 2-propanol, the proportion increasing as the percent hydrolysis of the poly(vinyl alcohol) decreases.

Everything else being equal, the differences in the degree of hydrolysis govern the ease with which the poly(vinyl alcohol) will dissolve in water. On the other hand, the lower the degree of polymerization, the higher the solubility. For any specific molecular weight, the solubility is maximum in the 92–94% hydrolysis range.

Since poly(vinyl alcohol) is water sensitive, gas transmission rates are affected by the water content of the film. Test data should be applied only to applications where humidity and other conditions are the same. Table 2 gives permeability coefficients for oxygen and some other compounds at 50% RH, 25°C, and atmospheric pressure. Poly(vinyl alcohol) will not actually dissolve in cold water, but when solution is obtained in hot water by agitation the resin does not precipitate on cooling. Over a period of time solutions of low acetate content polyvinyl alcohol will increase in viscosity

Table 2. Permeability Coefficients for Gases Through PVA Film (8)

Gas	Permeability coefficient[a]
oxygen	0.72×10^{-10}
carbon dioxide	1.20×10^{-10}
water	$2900-14{,}900 \times 10^{-10}$
hydrogen	2.14×10^{-10}
acetylene	3.56×10^{-10}

[a] $\dfrac{cm(STP) \times cm}{cm^2 \times sec \times cmHg}$

and may even gel; the phenomenon is concentration and molecular weight dependent. The upper part of Figure 1 shows the variation in properties occurring with change in molecular weight when the hydrolysis of the parent product is kept constant, and the lower part shows the variation in properties when the hydrolysis is varied keeping the molecular weight constant. Generally speaking, the variations are much the same for both, with a few exceptions; the viscosity of poly(vinyl alcohol) solutions is principally determined by the molecular weight (as the molecular weight increases, so does the viscosity).

A number of dyes and other organic compounds when added to solutions of completely hydrolyzed grades of poly(vinyl alcohol), form thermally reversible gels (7). Congo Red is one of the most effective gelling agents; 3% based on the dry resin weight, added to a 10% solution of completely hydrolyzed poly(vinyl alcohol) produces a composition which is fluid and easily applied at room temperature. Most gelling agents are not effective with partially hydrolyzed grades of poly(vinyl alcohol). Congo Red, Pontamine Gast Red F, Pontamine Bordeaux B, Pontamine Green 2 GB, Pontamine Brown D 3 GN, and Pontamine Orange R all give colored gels. Resorcinol, catechol, phloroglucinol, gallic acid, salicylanilide and 2,4-dihydroxybenzoic acid give colorless, thermally reversible gels.

Poly(vinyl alcohols) of low acetate content are insoluble in and almost unaffected at ordinary temperatures by gasoline, acetylene, butane, propane, sulfur dioxide, kerosene, benzene, xylene, methylene dichloride, trichloroethylene, carbon tetrachloride,

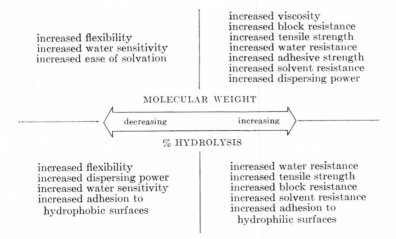

Fig. 1. Properties of poly(vinyl alcohol).

chlorofluorocarbons, monochlorobenzene, methanol (anhydrous), ethyl alcohol (anhydrous), ethylene glycol, acetone, furfural, methyl acetate, ethyl acetoacetate, dioxane, formamide, diacetone alcohol, fenchyl alcohol, cyclohexanol, 1-butanol, butyl Carbitol (diethylene glycol monobutyl ether), diethylene glycol, cresylic acid, monoethanolamine, triethanolamine, morpholine, quinoline, pyridine, nitrobenzene, ether, and aniline (9).

There are only a few types of organic compounds having any appreciable solvent action on the completely hydrolyzed grades of poly(vinyl alcohol). These compounds are often used as plasticizers and in most cases also function as humectants. The presence of one or more hydroxyl groups in the solvent enhances activity. Examples include glycerol, ethylene glycol and a few of the lower polyethylene glycols. Also useful are amines such as the ethanolamines, and ethanolamine salts and amides, including formamide, N-(2-hydroxyethyl)formamide and N-(2-hydroxyethyl)acetamide. Most, if not all, of the solvents require heat to dissolve even small amounts of poly(vinyl alcohol). Compounds with lower solvent power, but still useful as plasticizers, are sorbitol and urea.

Poly(vinyl alcohol) above a 45% acetate content is soluble in cresylic acid and above an 80% acetate content in liquid sulfur dioxide. Poly(vinyl alcohols) of low acetate content, and depending somewhat on the degree of polymerization, are soluble in hot solvents, such as acetic acid, formamide, glycerol, glycol, 2,2'-thiodiethanol, acetamide, urea, amino hydroxy compounds, and phenol; on cooling, these solutions gel. However, if the phenol contains a trace of water, the solution remains liquid on cooling (1).

The osmotic pressure (10) of a poly(vinyl alcohol) solution of 1.2 g/100 ml water (low acetate content, low molecular weight) is 0.008 atm. The viscosities in poises of poly(vinyl alcohol) of various acetate contents in 4, 8, and 12% solutions in water at 20°C are given in Table 3.

Table 3. Viscosity of Poly(vinyl alcohol) Solutions, 20°C

Molecular weight parent PV acetate	Acetate content, %	Viscosity, poises		
		4% sol	8% sol	12% sol
10,000	20	0.0165	0.026	0.045
15,000	20	0.024	0.054	0.125
33,000	5	0.048	0.18	0.65
33,000	20	0.045	0.16	0.52
135,000	5	0.20	5.5	28
135,000	30	0.30		
400,000	1	0.50	16	40
400,000	40	0.70		

The abrasion resistance of poly(vinyl alcohol) is 10 times that of rubber (1,9) and its hardness (6) as measured by the Shore durometer, is 10–100 (plasticized). The tensile strength and elongation (at break) of low-acetate material are given in Table 4 (6). The cold flow (9) (percent deformation of a 0.5 in. cube subjected to 4000 psi for 24 hr at 120°F and then allowed 24 hr to recover) of poly(vinyl alcohol) is 63.4%.

Table 4. Tensile Strength and Elongation (at break)

Temperature, °C	Extruded		Pressed	
	tensile strength, psi	elongation, %	tensile strength, psi	elongation, %
−32	7650	30	7400	40
−18	6770	85	6760	85
0	6760	99	3640	267
25	5200	225	2110	445
46	3530	255	1610	520
66	3090	300	1490	590
80	2110	340	1250	620

The heat sealing temperatures (11) of resins of various acetate contents under pressures of 10 psi for 30 sec are:

acetate, %	25	35	45	55	65	75
heat sealing temperature, °C	195	180	163	148	133	118

The thermal conductivity (9) of poly(vinyl alcohol) in cal/(sec)(cm²)(°C/cm) is 5×10^{-4} for the extruded resin and 18×10^{-4} for the molded resin. The difference in values is believed to be a function of imperfections developed during the processing step and environmental variations. The specific heat for poly(vinyl alcohol) varies from 0.005 cal/(deg)(g) at 10°K to 0.256 cal/(deg) (g) at 245°C (6). At the latter temperature the entropy is 0.2655 cal/(deg)(g) and the enthalpy is 32.45 cal/g.

The dielectric constants (17,9) and power factors at various temperatures are given in Table 5. The electrical resistivity (9) in ohm-cm is $3.1 \times 3.8 \times 10^7$.

Table 5. Dielectric Constant and Power Factor

Temperature, °C	Dielectric constant	Power factor (10^6 Hz)
25	5.9	0.12
50	7.5	0.16
75	10.8	0.18
87.5	13.6	0.176

The infrared absorption spectrum was determined by Blout and Karplus (18). The light transmitted by a 0.001 in. thick poly(vinyl alcohol) film at various wave lengths is given below:

wave length, Å	2536	3130	3650
transmission, %	77.5	72.9	81.1

Chemical. Monomeric vinyl alcohol, $CH_2{=}CHOH$, is the enol of acetaldehyde (qv) and is unknown in the free state, although some of its derivatives such as mercury compounds are fairly stable. Poly(vinyl alcohol) was first mentioned by Herrmann and Haehnel (1,2,12) in 1927; they made it by saponification of poly(vinyl esters).

Poly(vinyl alcohol) undergoes chemical reactions very similar to those of low molecular weight aliphatic alcohols. For example, it reacts with acid chlorides or anhydrides to form esters and with aldehydes or ketones to form acetals (see Poly-(vinyl acetals).) In many instances it is possible to obtain a completely new resin by reacting all of the hydroxyl groups; even with smaller degrees of reaction, the properties of the resin may be considerably altered, while retaining, if desired, the water solubility. Further interesting property variations are achieved by using a partially hydrolyzed poly(vinyl alcohol), thus retaining some of the original acetate groups.

Poly(vinyl alcohol), reacted with sulfur trioxide, yields $-OH + SO_3 \rightarrow -OSO_3H$, poly(vinyl hydrogen sulfate), which is highly water soluble (13). Alkane- or arene-sulfonyl chlorides, when reacted with poly(vinyl alcohol), yield sulfonates $-OH + RSO_3H \rightarrow -OSO_3R$ (14,15). Acid chlorides in the presence of alkalies or alkoxides form the expected esters (16), $-OH + RCOCl \rightarrow -O-COR$. The esters formed in this way from unsaturated acid chlorides, such as cinnamoyl chloride, are organic solvent soluble, film forming, light-sensitive materials (19).

Copolymers of poly(vinyl alcohol) with ethylene prepared by complete hydrolysis of the vinyl acetate copolymer have been reacted with phenyl dineopentyl phosphite $(C_5H_{11}O)_2POC_6H_5$ by displacement of phenol to give flame-retardant phosphite esters (20). Polymeric drying "oils" for use in the surface coatings industry can be prepared by reaction with long chain unsaturated fatty acids (4). Polymeric acids, such as polyacrylic acid, react with poly(vinyl alcohol) to form esters having insoluble gel characteristics whose mechanical properties vary reversibly with pH and temperature (22). Controlled gelation of poly(vinyl alcohol) is obtained by the addition of trivalent titanium compounds and oxidizing agents. Slow conversion of the titanium compound to the tetravalent state gels the mixture (23). Dianhydrides prepared from maleic anhydride and diolefins crosslink poly(vinyl alcohol) copolymers to yield elastomeric compositions (24). Other chemical systems such as glyoxal and melamine resin are often used to crosslink poly(vinyl alcohol). Crosslinked poly(vinyl alcohol) is obtained by copolymerization of vinyl acetate with a di-unsaturated dibasic acid (33) or a dialkyl acetal (34) followed by hydrolysis.

Poly(vinyl alcohol) can be etherified using a number of reagents including alkyl or aryl halides to yield films whose permeability at 50% etherification is only one-tenth that of pure polyvinyl alcohol (25). The cyanoethylation of poly(vinyl alcohol) is accomplished by catalyzing the addition of acrylonitrile with sodium hydroxide or sodium cyanide; the products are rough, rubbery resins, useful in adhesives and coatings (26). Acrylonitrile can be polymerized in an aqueous medium containing poly-(vinyl alcohol) to form a graft copolymer of reduced water sensitivity, capable of being spun into fibers or molded (27). The reaction of poly(vinyl alcohol) and ethylene oxide can be carried out in aqueous solution with an alkaline catalyst to yield the hydroxyethyl ether (28,29). The gelation tendency of polyvinyl aqueous solutions is thereby reduced (30). The reaction between poly(vinyl alcohol) and thiourea forms the thiol derivative. The latter can be reacted with metals such as silver, mercury, and platinum, allowing them to be isolated from solutions (31).

The reaction of poly(vinyl alcohol) with aldehydes yields acetals. The butyral is well known as a safety glass laminate. The formal, because of its high tensile strength, water resistance, adhesion and flexibility, has wide application in such products as wire enamel, sponges, and textile coatings (32). See Poly(vinyl acetals).

It has been found possible to modify poly(vinyl alcohol) by the grafting of other monomers in such a way as to improve the product for film and coating applications. This is usually done in solution using free radical or ionic catalysts. In this way, vinylidene chloride (35,36), ethylenimine (37), acrylate esters (38), vinyl chloride (36,39), acrylonitrile (40), aldehydes (41) and alkali cellulose (42) have been grafted to poly(vinyl alcohol).

Poly(vinyl alcohol) is not attacked by compressed oxygen at temperatures down to −70°C. With 30% hydrogen peroxide, acetic acid and carbon dioxide but no succinic acid are formed (1,44). Potassium permanganate oxidizes poly(vinyl alcohol) only very slightly if at all. Chromic acid yields a brown insoluble mass which contains some of both the starting materials. This is made use of in printing techniques by using the acid as ink and then washing away the untreated area. Concentrated nitric acid forms polyvinyl nitrate in the cold but oxidizes it on heating with the evolution of nitric oxide and formation of oxalic acid (45). Periodic acid does not oxidize normal poly(vinyl alcohol) but the 1,2-glycol structure is broken down (1).

Because of its resistance to oxidation, poly(vinyl alcohol) of low and medium acetate contents is practically immune to deterioration by ageing (43).

Toxicity

Metabolism tests on mice show no poisonous effect (43). The data on intravenous, intramuscular or subcutaneous injections appear to be contradictory. Grigoryan et al. studied the effect of colloidal poly(vinyl alcohol) solutions when used as plasma substitutes; high molecular weight material caused virtually no side effects, but the low caused some temporary changes (29). Danishevskii et al. (3) report poly(vinyl alcohol) to be a compound of relatively low toxicity which mostly affected the eyes and kidneys after continuous injection. Hueper and his associates (46), Kainer (1) and Zaeva et al. (47) report little or no harmful effects from intravenous, intramuscular or subcutaneous injections. Other workers (48–54) noticed various toxic symptoms indicating widespread tissue and organ degeneration.

Certain investigators considered poly(vinyl alcohol) to be unsuitable for use as a blood substitute (6) causing a decrease in the amount of red corpuscles and an increase in the sedimentation rate, to eventual coagulation (1,46) when used in large amounts. On the other hand, other investigators have found that poly(vinyl alcohol) can be used successfully as a plasma substituting material with virtually no toxic effects (55–58).

Humans show no harmful effects when poly(vinyl alcohol) film is used in surgery or as a carrier for medicaments (1). Because of its minimal oral toxicity, it has been allowed as a component of food packaging adhesives (59) and for use in coatings that come into direct contact with food (60). Nevertheless, poly(vinyl alcohol) is not broken down in the body and will remain as a foreign substance (6); hence, care should be exercised not to inhale the dust into the lungs.

External applications of poly(vinyl alcohol) for various purposes have been approved by the Food and Drug Administration Agency in view of its inert character (see Application, cosmetic). The use of poly(vinyl alcohol) in application for internal consumption, such as food and drugs, must be cleared with the Food and Drug Administration.

Manufacture

Poly(vinyl alcohol) is produced by either acid or alkaline hydrolysis of poly(vinyl esters), usually the acetate. The resins made by the two processes differ somewhat, and the acid process product may contain some ether groups. The nature of the acid in the starting vinyl ester is known to give different products. For example, poly(vinyl trifluoroacetate) yields a highly ordered poly(vinyl alcohol) (61–63).

Alkaline Process. A general procedure usually involves the mixing of a methanol solution of the polyester with methanolic sodium methoxide, $-OCOCH_3 + HOCH_3 \rightarrow CH_3OCOCH_3 + -OH$. A rapid reaction takes place at room temperature resulting in a gel. The extent of hydrolysis is controlled by the degree of agitation, the amount of alkaline agent employed, and the temperature. The gel is broken up and the alkali neutralized by a variety of means. Various procedures to overcome the difficulties of attaining rapid and high conversions of ester to alcohol in the gel system are employed, including different mechanical procedures such as the use of belts, special kneaders and agitators. In addition, a variety of alkaline agents may be employed, including sodium and potassium hydroxide, sodium carbonate, Lewis alkalis, etc (64–74).

Acid Process. Hydrolysis by either free acid or acid catalysts has been reported to yield poly(vinyl alcohols) having special properties. For example, less degradation of the molecular weight has been reported by acid catalysis. A typical process involves mixing methanolic polyester solution and either sulfuric or hydrochloric acid, followed by a 3 hr heating period at 57–59°C. Poly(vinyl alcohol) precipitates from the mixture and the extent of hydrolysis is controlled by the time the polymer is allowed to contact the acid before neutralization. The product is filtered, washed with methanol and dried (75–78). Other acidic agents are sometimes employed such as 2-naphthalenesulfonic acid (79) or Lewis acids (73). The acid hydrolysis of poly(vinyl alkyl ethers) has been reported to yield a stereospecific alcohol if the starting polymer was stereospecific (80,81).

The preparation of stereoregular poly(vinyl alcohol) can be effected in a number of other ways. The hydrolysis of special esters such as vinyl formate or vinyl esters of fluorinated acids (82–84) or the use of special catalysts (85–89) affects the stereoregularity (90). Solvent modification varying the dielectric constant of the medium also appears to provide a method for regulating the stereoisomerism of poly(vinyl alcohol) (91). A review of preparation methods for stereoregular poly(vinyl alcohol) and of procedures for tacticity analysis has been prepared (92).

The hydrolysis of cyclopolymerized divinyl formal has been shown to lead to a poly(vinyl alcohol) with a higher 1,2-glycol structure content than that in commercial poly(vinyl alcohol) (93).

Poly(vinyl alcohol) sells in bulk between 32–34 cents a pound. Special grades occasionally sell at premium prices. The 1968 sales of poly(vinyl alcohol) in the United States totaled 60 million pounds which can be broken down as follows:

textiles	17 million lbs
adhesives	29 million lbs
paper	9 million lbs
film	2 million lbs
miscellaneous	3 million lbs

An annual rate of increase of sales of approximately 10% for the next few years is forecast.

The types of poly(vinyl alcohol) commercially available can be divided into two broad categories; completely hydrolyzed and partially hydrolyzed resins. Those containing about 4% residual acetate content or less are referred to as completely hydrolyzed. Partially hydrolyzed grades usually contain 20% or more residual acetate groups. For specialty applications, a number of fine variations within these broad categories are commercially available.

Poly(vinyl alcohol) resins are marketed under various names: Alvyl, Alvylol (Société Nobel française), Darex (Dewey and Almy Chemical Co.), Elvanol (DuPont Co.), Gelvatol (Monsanto Co.), Lemol (Borden Co.), Moviol (Hoechst), Poval (Kurashiki Rayon Co.), Polyviol (Wacker-Chemie GH), Resistoflex (Resistoflex Corp.), Rhodoviol (Société des Usines chimiques Rhône-Poulenc), Solva (Monsanto Corp.), Solvar (Monsanto, Ltd.), Vinarol (Farbwerke Hoechst AG), Vinavilal (Rhodiatoce S.p.A), Vinol (Air Reduction Co.).

Applications

Most of the uses for poly(vinyl alcohol) involve extrusion of the resin or its application as a surface coating. Because of its intractable character, the resin must be plasticized to allow extrusion to take place without decomposition. Poly(vinyl alcohol) hose is made in this way. The largest majority of applications for poly(vinyl alcohol) involves the use of aqueous solutions for coating, casting and dipping procedure. Aqueous film casting is complicated by the limited solubility of poly(vinyl alcohol), even in hot water, 50% concentration being about the maximum useful limit and even this must be worked close to 100°C. Under pressure operations are performed above 100°C.

The solution process depends on the degree of acetylation. For high concentrations of partially acetylated poly(vinyl alcohol), cold water, preferably below 75°F, is vigorously stirred and the powder added slowly. These procedures minimize lumping or balling. In order to shorten solution time, the dispersion is heated to a temperature between 180 and 190°F. At that temperature solution is usually complete within 20–30 min. This procedure can be satisfactorily employed for completely hydrolyzed and for medium acetate grades. However, some medium and high acetate products require the addition of some alcohol or acetone to aid solution.

Poly(vinyl alcohol) films have found applications in the photographic field serving for supporting base or coatings (94–96). Special chemical treatment of poly(vinyl alcohol) to provide hardened coatings has been employed (97). The polymer has served particularly in the formation of oriented light polarizing films (98,99).

The combination of high tensile strength and ease of water solubility have made poly(vinyl alcohol) useful in packaging measured portions of bleach and laundry detergents for household applications (100–102).

Poly(vinyl alcohol) has been used either as a homopolymer or as a copolymer for the preparation of sausage casings. For food wrapping purposes, poly(vinyl alcohol) is sometimes improved by being coated with other resins such as a vinylidene chloride–vinyl chloride copolymer (103) or a vinylidene chloride–acrylonitrile copolymer (104). Other film improvements can be effected by orientation (105), heat treatment (106), or squeezing between rollers (107).

Studies have been made of the permeability of poly(vinyl alcohol) films to moisture and water vapor (108). In addition to being of interest in the food field, such information has assisted in the development of improved membranes for artificial kidney dialysis (109–111).

Film. (See also Film materials.) Poly(vinyl alcohol) has been made into films that are capable of serving as membranes for the desalination of water (112–114). Copolymers of poly(vinyl alcohol) with styrenesulfonic acid yield cationic-selective membranes (115,116). Derivatives with (a) 1,2-epoxy-3-diethylaminopropane (117); (b) thiourea; (c) chloromethyl ether and trimethylamine or pyridine (118); (d) formaldehyde and salicyclic acid (119); (e) polyfunctional acids (120), and (f) carbamoylmethyltrimethyl-ammonium hydrochloride (121) form anionic or cationic membranes.

Adhesives and Binders. Poly(vinyl alcohol) has been used as a superior adhesive in applications where starch and other low cost materials were formerly employed. Paper and cardboard applications usually employ a formulated system which comprises the addition of crosslinking agents, inert diluents and inorganic salts (122–128). Variations on the additives employed in the mixtures allow poly(vinyl alcohol) to be used for joining wall board, leather and cloth (129,130), cork and polystyrene foam (131).

Poly(vinyl alcohol) has found wide usage as an intermediate in the making of poly(vinyl butyral) used as an adhesive laminate for safety glass (119,131,132).

Poly(vinyl alcohol) is used as an additive to make improved mortars where superior adhesion or high tensile strength in thin sections is desired (133,134,135).

Electrical. Poly(vinyl alcohol) has been used for a variety of electrical applications where advantage can be taken of its dielectric properties. Poly(vinyl alcohol) and a number of derivatives form films having a high dielectric constant making them useful as phosphor-containing layers in electroluminescent devices. Particularly useful are cyanoethylated poly(vinyl alcohol) resins (136,137).

The use of poly(vinyl alcohol) in electrolytic plating baths has been reported to provide improved metal coatings, particularly for zinc and cadmium coats (138–143). Superior beryllium and chromium metal coatings have been plated from baths containing poly(vinyl alcohol) (132).

Textiles. Poly(vinyl alcohol) has found application as a replacement for starch and other natural resins as a sizing agent for cotton or mixtures of cotton and synthetic fibers. Despite the higher price of poly(vinyl alcohol), the improved product and the lower concentrations employed have favored its usage (144–148,149).

Wash and wear type textiles need chemical treatment to provide resistance to creasing. Aqueous solutions or dispersions of poly(vinyl alcohol) have been found useful for this purpose (150,151). Heat and a variety of crosslinking agents including formaldehyde (18) are often employed to enhance the resin's utility (152,153). Cellulosic fabrics treated with a mixture of poly(vinyl alcohol) and thermosetting polymers have produced improved antishrink wash-resistant finishes (154,155).

Paper Treatments. Poly(vinyl alcohol) exhibits excellent adhesion and compatibility with cellulosic materials and has found considerable use in adhesive formulations for paper bond, paper bag seam and bottom sealing, book padding, tube winding, case sealing, and carbon manufacturing as well as for laminating solid fiberboard.

As a paper adhesive, poly(vinyl alcohol) when contrasted with natural adhesives such as starches, gums, dextrines, animal glues and casein, has the following advantages: stronger bonds and, in many cases, more water-resistant paper-to-paper bonds, wider ranging compatibility with other adhesive resins, high degree of resistance to bacterial attack and lower cost owing to lesser amounts of material.

Poly(vinyl alcohol) is useful for paper sizing because it is an easily applied, water-soluble resin, providing outstanding resistance to greases, oils and solvents, ease of film formation, high binding strength and high abrasion resistance. The coatings can be mixed with inert extender solids, other resins and/or pigments. Pretreatment of the paper surface with gelling or crosslinking agents reduces the amount of poly(vinyl alcohol) needed and provides superior coatings (35,156–160). Paper materials coated in this manner are useful in the packaging of foods, greasy and oily products and chemicals. Alternatively, poly(vinyl alcohol) is often used in paper applications by intimate mixing with the paper pulp which is then converted into paper either in the normal way or by pressing the mixed extruded fibers (161–163).

Fibers

Fibers prepared from poly(vinyl alcohol) have found application because of their high tensile strength, ability to absorb moisture and attractive feel. The fiber is prepared by extruding a water solution of the polymer into a coagulating bath, usually containing inorganic salts (10,173,174). The fiber is then treated with other organic chemicals including aldehydes and ketones to impart additional properties such as flame proofing (175,176), water insolubility (177–179) and additional chemical resistance (180). Commercial development of poly(vinyl alcohol) fiber in the United States has been undertaken but competitive fibers have proved more attractive; particularly from the price aspect. However, in Japan the fiber is widely used in many textile applications; particularly clothing. Although in recent years polyester fibers have made inroads in some markets, Japan's peculiar chemical economics and lack of natural raw materials explain the volume production of poly(vinyl alcohol) in Japan and in no other country.

It is possible to prepare poly(vinyl alcohol) fibers that are easily water soluble. Advantage is taken of their special properties in weaving special patterns into textiles. When the fabric is washed after weaving has been completed, an open structure or highly porous fabric results. Novel effects are thus obtained that would be almost impossible to prepare in any other way.

Miscellaneous

The absorption of poly(vinyl alcohol) in a variety of soil aggregates has been studied (164). Poly(vinyl alcohol) graft or copolymers with polar monomers which have been converted to either potassium, sodium or calcium derivatives have been reported as useful for the stabilization of agricultural soils (165).

Graft polymers derived from poly(vinyl alcohol) and acrylonitrile or acrylamide are useful as scale-forming inhibitors in condensers employing hard water (131,143).

Aqueous solutions of poly(vinyl alcohol) and sodium borate have been claimed to gel drilling muds in order to increase the sheer thickening properties of solution (166). Detergent, wear-resistant oils are supposed to be inhibited against sludging and oxidation by the addition of small percentage of oil-soluble, water-insoluble nonionic polymeric alcohols. Such alcohols are copolymers of α-olefins having at least ten carbon atoms per molecule with vinyl alcohol (167).

Poly(vinyl alcohol) has been found useful in heat treating processes for the hardening of metals. The quenching capacity and characteristics of 5–40% poly(vinyl alcohol) solutions with respect to construction and tool steels was investigated (168). A patented process has been described particularly adapted for the hardening of crankshafts using a quench bath of poly(vinyl alcohol) and water (169).

Poly(vinyl alcohol) has been recommended to accelerate the sedimentation of particulate matter in paper, sewage, and industrial effluents (170). It has also been found to give improved dispersibility of pigments in water-based paints and in such formulations often serves as a nonionic emulsifying agent (171,172).

Cosmetic skin cleansing creams (181) have been formulated using poly(vinyl alcohol). The resin has also found application as an ingredient in a topically applied composition for protection of the skin against organic solvents (182).

Bibliography

"Polyvinyl alcohol" in *ECT* 1st ed., Vol. 14, pp. 710–715, by T. P. G. Shaw and L. M. Germain, Shawinigan Chemicals, Limited.

1. F. Kainer, *Polyvinylalkohole*, Enke, Stuttgart, 1949.
2. U.S. Pat. 1,672,156 (June 5, 1928), W. O. Herrmann and W. Haehnel (to Consort f. elek. Ind.).
3. S. L. Danishevskii, A. Ya. Broitman, W. E. Gavrilova, and E. G. Robaschevskaya, *Toksikol. Vysokomol. Mater. Khim. Syr'ya Ikh Sin. Gos. Nauch.-Issled. Inst. Polim. Plast. Mass.* 94–105 (1966).
4. C. E. Hall and O. Hall, *Am. J. Pathol.* **41**, 247–257 (1962).
5. C. E. Hall and O. Hall, *Texas Rept. Biol. Med.* **23** (2), 423–434 (1965).
6. *Elvanol*, Du Pont, Wilmington, Del., 1953.
7. U.S. Pat. 2,249,536; 2,249,537; and 2,249,538 (to Eastman Kodak Co.).
8. ASTM Method D1434-66-ASTM Standards, Part 27.
9. E. S. Peierls, *Mod. Plastics* **18**, (6), 53 (1941).
10. R. H. Wagner, *Ind. Eng. Chem., Anal. Ed.* **16**, 520 (1944).
11. U.S. Pat. 2,424,110 (July 15, 1947), G. O. Morrison and T.P.G. Shaw (to Shawinigan Chemicals Ltd.).
12. Ger. Pat. 450, 286 (Oct. 5, 1927), W. Haehnel and W. O. Herrmann (to Consort. f. elek. Ind.).
13. U.S. Pat. 2,623,037, R. V. Jones (to Phillips Petroleum Co.); Japan Pat. 7371 (1951), S. Okamura and T. Motoyama.
14. D. D. Reynolds and W. O. Kenyon, *J. Am. Chem. Soc.* **72**, 1584–1598 (1950).
15. Michel Lagache, *Ann. Chim. (Paris)* **13** (1), 5–52 (1956).
16. U.S. Pat. 3,329,664 (Cl. 260-91.3) (July 4, 1967), Minoru Tsuda (to Japan, Bureau of Industrial Technics).
17. W. Holzmuller, *Kunststoffe* **30**, 177 (1940).
18. E. R. Blout and R. J. Karplus, *J. Am. Chem. Soc.* **70**, 862 (1948).
19. U.S. Pat. 2,725,372, L. M. Minsk, U.S. Pat. 2,747,997, J. G. Smith and E. I. Sundeen, U.S. Pat. 2,728,745, A.C. Smith and C. C. Unruh (all to Eastman Kodak Co.).

20. U.S. Pat. 3,318,855 (May 9, 1967), Lester Friedman (to Union Carbide Corp.).
21. A. J. Seavell, *J. Oil and Colour Chemists Assoc.* **39**, 99 (1956).
22. H. Nogucki, *Bussciron Kenkyu* **69** (1953); *Chem. Abst.* **48**, 5606 (1954).
23. Neth. Appl. 6,503,659 (Sept. 24, 1965), (to E. I. duPont de Nemours & Co.).
24. U.S. 3,326,841 (June 20, 1967), Bruce W. Hotten (to Chevron Research Co.).
25. *Chem. Abst.* **49**, 11585 h (1955); S. N. Ushakov and E. M. Lavent'eva, *J. Appl. Chem., U.S.S.R.* **28**, 388 (1955).
26. U.S. Pat. 2,341,553, R. C. Houtz (to Du Pont); Y. Takamatsu et al., Japan Pat. 6942 (1951); J. F. Wright and L. M. Minsk, *J. Am. Chem. Soc.* **75**, 98 (1953).
27. Brit. Pat. 742,900 (1958), E. Jones, L. Morgan, and S. Todd (to Imperial Chemical Industries, Ltd.).
28. Kitty Ettre and Peter F. Varadi, *Anal. Chem.* **34**, 752–757 (1962).
29. S. S. Grigoryan, S. N. Allaverdyan, and A. E. Akokyan, *Z. Kh. Partev.* (U.S.S.R.) *Sb. Tr., Mauch.-Issled. Inst. Gematol. Pereliv. Krovi, Tiflis* (1967), 10–11, 113–116.
30. S. G. Cohen et al., *J. Polymer Sci.* **11**, 193 (1953); G. Champetier and M. Lagache, *Comp. Rend.* **241**, 1135 (1955).
31. Y. Nakamura, *J. Chem. Soc. Japan, Ind. Chem. Sect.* **58**, 269 (1955); *Chem. Abst.* **49**, 143764 (1955).
32. U.S. Pat. Re 20,430 G. O. Morrison, F. W. Skirrow, and K. G. Blaikie; *Chem. Eng.* **62**, 108 (Jan. 1955).
33. Ger. 1.042,237 (Oct. 1958), Josef Heckmaier and Eduard Bergmeister (to Wacker-Chemie G.m.b.H.).
34. U.S.S.R. Pat. 134, 868 (Jan. 1961), S. N. Usakov.
35. Neth. Pat. 6,505,348 (Dec. 1965), (to E. I. duPont de Nemours & Co.).
36. Belg. 670,858 (Jan. 1967), Dietrich Hardt and Herbert Bartl (to Farbenfabriken Bayer A.-G.).
37. U. S. 2,972,606 (Feb. 1961), Robert J. Hartmand and Edward J. Fujiwara (to Wyandotte Chemicals Corp.).
38. U.S. Pat. 3,300,546 (Jan. 1967), Robert L. Baechtold (to American Cyanamid Co.).
39. Seizo Okamura and Hiroshi Asakura, *Amer. Chem. Soc., Div. Polymer Chem.*, Preprints, (2). 649–654 (1966).
40. Fr. Pat. 1,476,489 (April 1967), (to Kurashiki Rayon Co., Ltd.).
41. Hiroshi Takida and Takatsugu Moriyama, *Kobunshi Kagaku* **24** (261) 57–67 (1967).
42. U.S. Pat. 3,341,483 (Sept. 1967); Albert Zilkha, Bena Felt, and Akiva Bar-Nun (to Yissum Research Development).
43. N. Platzer, *Mod. Plastics* **28** (7), 95 (1951).
44. H. Staudinger and H. Warth, *J. prakt. Chem.* **155**, 261 (1940).
45. C. S. Marvel and C. E. Denoon, Jr., *J. Am. Chem. Soc.* **60**, 1045 (1938).
46. W. C. Hueper, J. W. Landsberg, and L. D. Eskridge, *J. Pharmacol. Exptl. Therap.*, **70**, 201 (1940).
47. G. N. Zaeva, M. D. Babina, V. I. Fedorova, and V. A. Shchirskaya, *Toksikol. Novykh Prom. Khim. Veshchestv* **5**, 136–149 (1963).
48. W. C. Hueper, *Arch. Pathol.* **67**, 589–617 (1959).
49. C. E. Hall and O. Hall, *Experientia* **18** (1), 38–40 (1962).
50. Charles E. Hall and Octavia Hall, *Lab. Invest.* **12** (7), 721–736 (1963).
51. Charles E. Hall and Octavia Hall, *Lab. Invest.* **13** (3), 227–238 (1964).
52. G. N. Zaeva, V. I. Fedorova, M. D. Babina, and V. A. Shchirskaya, *Toksikol. i Gigiena Vysoko-molekul. Soedin. i Khim. Syr'ya, Ispol'z dlya ikh Sinteza.*
53. Kichihei Miyasaki, Yoshinobu Kobayaski, Takuji Yoshikawa, Takao Konda, Michio Inoue, Muneo Takigushi, and Mitsuyoshi Muragushi, *Kobe Ika-Daigaku Kiyo* **13**, 320–322 (1958).
54. C. E. Hall and O. Hall, *Texas Rept. Biol. Med.* **21**, 16–27 (1963).
55. I. M. Khlebrikova and E. A. Senchilo, *Vestn. Khirurg.* **93** (9), 40–46 (1964).
56. L. G. Bogomolova and Z. A. Chaplygina, *Probl. Gematol. i Pereli. Krovi* **5** (5), 24–31 (1960).
57. G. P. Weil, *J. Pharm. Belg.* **18** (45), 445–482 (1963).
58. Z. A. Chaplygina, *Sb. Nauch. Tr. Leningrad. Nauch. Issled. Inst. Pereliv. Krovi* **14**, 584–594 (1963).
59. *Federal Register* **32**, 8523 (June 1967).
60. *Vinol, Polyvinyl Alcohol*, Air Reduction Co.

61. Howard C. Haas, Eugene S. Emerson, and Norman W. Schuler, *J. Polymer Sci.* **22**, 291–302 (1956).
62. U.S. Pat. 3,141,003 (July 1964), Clifford A. Neros and Nelson V. Seeger (to Diamond Alkali Co.).
63. L. Alexandru and M. Opris, *Rev. Chim. (Bucharest)* **13** (5), 279–281 (1962).
64. U.S. Pat. 2,478,431 (Aug. 9, 1949), G. S. Stamatoff (to Du Pont).
65. L. Alexandruand, F. Butaciu, *Rev. Chim. (Buchardst)* **13**, (2), 80–83 (1962).
66. Ger. Pat. 1,206,158 (Dec. 1965), Walton B. Tanner (to Du Pont).
67. Brit. Pat. 924,038 (April 1963), (to Kurashiki Rayon Co., Ltd.).
68. U. S. Pat. 3,300,460 (Jan. 1967), Alphonse J. Vacca (to Cumberland Chemical Corp.).
69. Brit. Pat. 1,048,120 (Nov. 9, 1966), Shiro Masuda and Kozo Konishi (to Electro Chemical Industrial Co., Ltd.).
70. T. Rosner and Z. Joffe, *Szczecin. Tow. Nauk.,Wydz. Nauk Mat. Tech.* **5**, 75–79 (1966).
71. U.S. Pat. 2,950,271 (Aug. 1960), James M. Snyder (to Du Pont).
72. Brit. Pat. 797,127 (June 1958), (to Lonza Elektrizitätswerke und Chemische Fabriken A.-G.).
73. Japan Pat. 20,191 (1965), Naoynki Sakado and Keiji Koike (to Shin-Etsu Chemical Industry Co., Ltd.).
74. T. Rosner, Z. Joffe, K. Pawlaczyk, and H. Zarzycka, *Szczecin Tow. Nauk., Wydz. Nauk Mat. Tech.* **5**, 56–66 (1966).
75. Japan Pat. 11,246 (1961), Ikuo Igarashi, Hitashi Takagi, and Minoru Shimizu (to Shin-Nippon Nitrogen Fertilizer Co.).
76. Japan Pat. 29,353 (1964), Tadaski Maebotoke, Tetso Ikeda, and Eiichi Chizaki (Japan Carbide Industries Co., Inc.).
77. U.S.S.R. Pat. 171, 561 (May 1965), A. G. Azyadyan.
78. Belg. Pat. 657, 268 (June 1965) (to Farbwerke Hoechst A.-G.).
79. U.S.S.R. Pat. 110, 139 (Feb. 1958), G. N. Freidlin, O.Ya Fedotova, and I. P. Losev.
80. Belg. Pat. 636,139 (Feb. 1964), Roland J. Kern (to Monsanto Co.).
81. Shunsuke Murahashi, Shunichi Nozakura, Masao Sumi, Saburo Fuji, and Keiji Matsumura, *Kobunshi Kagaku* **23** (255), 550–558 (1966).
82. Brit. Pat. 882,776 (Nov. 1962), Kurashiki Rayon Co., Ltd.
83. Japan Pat. 9854 (1965), Junji Ukich and Masakazo Matsumoto (to Kurashiki Rayon Co., Ltd.).
84. E. N. Rostovskii, L. D. Budovskaya, A. V. Sidorovich, and E. V. Kuvshinskii, *Vysokomol. Soedin., Ser. B* **9**, (1), 4 (1967).
85. Brit. Pat. 895,153 (May 1962) (to Kureha Chemical Works, Ltd.).
86. U.S. Pat. 3,036,054 (May 1962), Ora L. Wheeler and Harold D. Smyser (to Air Reduction Co., Inc.).
87. U.S. Pat. 3,121,705 (Feb. 1964), Arthur I. Lowell and Orville L. Mageli (to Wallace & Tiernan Inc.).
88. Tadashi Nakata, Takayuki Otsu, and Minoru Imoto, *J. Macromol. Chem.* **1** (3), 553–562 (1966).
89. Masao Ishii and Hidenari Suyama, *Sen 1 Gakkaishi* **18**, 22–27 (1962).
90. Brit. Pat. 819,291 (Sept. 1959), (to Hercules Powder Co.).
91. J. W. L. Fordham, *J. Polymer Sci.* **39**, 321–334 (1959).
92. Henryk Sibinski, *Wiad. Chem.* **20** (11), 677–692 (1966).
93. Yuji Minoura and Motonori Mito, *J. Polymer Sci. Pt. A,* **3** (6), 2149–2163 (1965).
94. U.S.S.R. Pat. 180, 713 (March 1966), M. A. Berchenko and A. S. Kalshnikova.
95. U.S. Pat. 3,314,814 (April 1967), Douglas A. Newman (to Columbia Ribbon & Carbon Manufacturing Co., Inc.).
96. Ger. Pat. 1,203,604 (Oct. 1965), Wolfgang Himmelmann and Alexander Riebel (to Agfa A.-G.).
97. U.S. Pat. 3,330,664 (July 1967), Louis M. Minsk and Hyman L. Cohen (to Eastman Kodak Co.).
98. Brit. Pat. 1,084,820 (Sept. 1967), Alvin M. Marks and Mortimer M. Marks.
99. A. S. Makas, *J. Opt. Soc. Am.* **52**, 43–44 (1962).
100. U.S. Pat. 3,322,674 (May 1967), Jack Friedman.
101. U.S. Pat.3, 257,348 (June 1966), Campbell F. Epes, Leonard R. Corazzi, and Alexander J. Marsh (to Reynolds Metals Co.).

102. U.S. Pat. 3,186,869 (June 1965), Jack Friedman.
103. U.S. Pat. 3,186,657 (June 1964), Daniel S. Dixler and Robert A. Eodice (to Air Reduction Co., Inc.).
104. Fr. Pat. 1,348,116 (Jan. 1964), Charles A. Heiberger and Daniel S. Dixler (to Air Reduction Co., Inc.).
105. U.S. Pat. 3,048,511 (Aug. 1962), Jay F. Strawinski.
106. Naomichi Takahashi and Kenji Onozato, *Kogyo Kagaku Zasshi* **65**, 2062–2064 (1962).
107. U.S. Pat. 2,837,770 (June 1958), Willy O. Herrmann and Heinz Winkler (to Consortium fur electrochemische Industrie G.m.b.H.).
108. Marie Korte-Falinski, *J. Chem. Phys.* **59**, (1), 27–35 (1962).
109. R. A. Markle, R. D. Falb, and R. I. Leininger, *Trans. Am. Soc. Artificial Internal Organs* **10**, 22–25 (1964).
110. R. A. Markle, R. D. Falb, and R. I. Leininger, *Rubber Plastics Age* **45**, 800–801 (1964).
111. M. Odian and E. F. Lonard, *Amer. Soc. Artificial Internal Organs* **14**, 19–23 (1968).
112. G. Thau, R. Block, and O. Kedem, *Desalination* **1** (2), 129–138 (1966).
113. C. E. Reid and H. G. Spencer, *J. Phys. Chem.* **64**, 1587–1588 (1960).
114. Lucius Gilman, *Natl. Acad. Sci.-Natl. Res. Council, Publ. No. 942*, 301–307 (1963).
115. Masato Nishimura, Akemi Aoyama, and Mizuho Sugihara, *Kagaku To Kogyo (Osaka)* **39** (11), 689–698 (1965).
116. Masato Nishmura, Mikio Yabu, and Mizuho Sugihara, *Kogyo Kagaku Zasshi* **70** (3), 393–398 (1967).
117. Eric Selegny, Yves Merle, and Michel Metayer, *Bull. Soc. Chim. France* 2400 (1966).
118. A. S. Tevlina, N. I. Skripchenko, and G. S. Kolesnikov, *Plasticheskie Massy* 66–69 (1966).
119. U.S. Pat. 3,313,779 (April 1967), Le Roy A. White.
120. U. S. Pat. 3,312,665 (April 1967), John R. Caldwell and Edward Hill (to Eastman Kodak Co.).
121. U.S. Pat. 2,805,196 (Sept. 1957), Hermannus G. Roebersen and Cornelis van Bochove (to Nederlandsche Centrale Organisatie voor Toegepast-Natuur Wetenschappelijk Onderzoek ten behoeve van Nyverheid, Handel en Verkerr.).
122. Neth. Appl. 6,610,579 (Feb. 1967) (to Du Pont).
123. Fr. Pat. 1,295,562 (June 1956), Claude Chatelus.
124. U.S. Pat. 2,764,568 (Sept. 1956), Raymond L. Hawkins.
125. U.S. Pat. 3,112,235 (Nov. 1963), Samuel E. Blanchard (to Borden Co.).
126. Japan Pat. 10,023 (1966), Akira Ioka, Akira Tokutaniyama, Atsushi Aochima, and Tetsuro Dosono, (to Asahi Chemical Industry Co.).
127. U. S. Pat. 3,311,581 (March 1967), Andrew E. Pierce.
128. U.S. Pat. 3,135,648 (June 1964), Raymond L. Hawkins (to Air Reduction Co., Inc.).
129. N. K. Baramboim and M. A. Nuriev, *Nauchn. Tr. Mosk. Tekhnol. Inst. Legokoi Prom.* **26**, 77–80 (1962).
130. U.S. Pat. 3,303,147 (Feb. 1967), Howard S. Elden (to National Gypsum Co.).
131. L. G. Tebelev and S. A. Glikman, *Vysokomol. Soedin.*, *Ser. A* **9** (4) 723–729 (1967).
132. U.S. Pat. 3,350,285 (1967), Susumu Nishigaki, Takeshi Kusaka, and Masahori Sato (to NGK Insulators, Ltd.).
133. Japan Pat. 1333 (Feb. 1957), Teiji Shindo.
134. Shin Nakamura and Seihachi Osawa, *Semento Gijutsu Nempo* **17**, 200–204 (1963).
135. U.S. Pat. 3,030,258 (April 1962), Herman B. Wagner (to The Council of America, Inc.).
136. U.S. Pat. 3,312,851 (April 1967), Leonard C. Floers and David Marschik (to Westinghouse Electric Corp.).
137. M. P. Kozlov and M. V. Prokof'eva, *Plasticheskie Massy* **10**, 17–20 (1966).
138. R. Peruzzi, R. Piontelli, and A. Porta, *Electrochim. Metal.* **1** (4), 423–434 (1966).
139. Brit. Pat. 1,059,569 (Feb. 1967), Tony E. Such and Harry F. Davison (to W. Canbing & Co., Ltd.).
140. Ger. Pat. 936,124 (Nov. 1965), Artur Kutzelnigg and Siegfried Gruttner, (to Dr. Hesse & Cie. Spezialfabrik für Galvanotechnik).
141. French Pat. 1,390,051 (Feb. 1965), A. G. Schering.
142. U.S. Pat. 3,329,512 (July 1967), Charles R. Shipley, Jr. and Michael Gulia (to Shipley Co. Inc.).
143. Neth. Pat. 6,605,265 (Oct. 1966), (to Imperial Chemical Industries Ltd.).

144. Edward P. Czerwin, *Mod. Text. Mag.* **47** (12), 29–30; **32,** 34 (1966).
145. Hans Hadert, *Staerke* **13** (12) 396–397 (1966).
146. F. Linke and H. Seidel, *Melliand Textilber.* **44** (10), 1134–1138 (1963).
147. C. R. Blumenstein, *Text. Ind.* **130** (7) 63–69, **97** (1966).
148. H. Jalke, *Textilveredlung* **2** (4) 164–169 (1967).
149. Stanley M. Suchecki and Walter N. Rozelle, *Text. Ind.* **131** (7), 122–123 (1967).
150. Ger. Pat. 1,116,236 (May 1966), Joh. A. Benckiser (to Chemische Fabrik G.m.b.H.).
151. E. A. Osminin, B. N. Mel'nikov, and P. V. Moryganov, *Tekst. Prom. (Moscow)* **26** (11), 58–60 (1966).
152. Austrian Pat. 235,238 (Aug. 1964), Hermann Zorn, Josef Helm, and Rudolf Steidl (to Getzner Mutter & Cie).
153. Ger. Pat. 857,943 (Dec. 1952), Erich Torke and Josef Konig (to Phrix-Werke A.-G.).
154. A. S. Lifentseva and N. P. Solov'ev, *Tektil'n Prom.* **25** (7), 57–58 (1965).
155. U. S. Pat. 3,267,549 (Aug. 1966), Edward R. Band (to DHJ Industries Inc.).
156. B. A. Beardwood and E. P. Czerwin, *Tappi* **43,** 944–952 (1960).
157. Isamu Yoshino, *Sen-i Gakkaishi* **14,** 8–11 (1958).
158. U.S. Pat. 3,298,862 (January 1967), William P. Fairchild and John E. Robison (to Kelco Co.).
159. U.S. Pat. 3,251,709 (May 1966), Francis A. Bonzagni (to Monsanto Co.).
160. G. P. Colgan and J. J. Latimer, *Tappi* **44,** (11) 818–822 (1961).
161. Yukio Nakaba, Sigeo Oyama, and Ryukichi Matsuo, *Kami-Pa Gikyoski* **21** (10), 545–556 (1967).
162. S. M. Ivanov and G. M. Gorskii, *Bumazhn. Prom.* **39** (1) 4–6 (1964).
163. Ger. Pat. 1,146,740 (April 1963), Hiroshi Iwasaki (to Kurashiki Rayon Co., Ltd.).
164. B. G. Williams, D. J. Greenland, and J. P. Quirk, *Adelaide Aust. J. Soil Res.* **4** (2), 131–143 (1966).
165. G. Martin Guzman and F. Arranz, *Anales Real. Soc. Espan. Fis. Quim (Madrid) Ser. B* **59** (6), 471–482 (1963).
166. U. S. Pat. 3,299,952 (Jan. 1967), (C. 166–122), Joseph G. Savins (to Mobil Oil Corp.).
167. Brit. Pat. 853,535 (Nov. 1960) (to N. V. de Bataafsche Petroleum Maatschappij).
168. M. Stanescu, *Metalurgia (Bucharest)* **17** (6), 300–303 (1965).
169. U.S. Pat. 3,305,409 (Feb. 1967), Philip E. Cary (to International Harvester Co.).
170. Witold Kowalski and Ewa Rej, *Zeszyly Nauk Politch. Slask., Chem.* **26,** 3–10 (1965).
171. J. N. Coker, *Ind. Eng. Chem.* **49,** 382–385 (1967).
172. Brit. Pat. 1,077,422 (July 1967), Japan Paint Co., Inc.
173. Pol. Pat. 48,099 (Feb. 1964), Waclaw Sopiela, Marian Sobolewski, Edward Maslowski, Zibigniew Rybicki, and Jerzy Cypryk (to Institut Wlokien Sztucznych i Syntetye enzych).
174. T. Rosner, J. Slonecki, and A. Morawski, *Szcecin. Tow. Nauk. Wdyz. Nauk Mat. Tech.* **5,** 118–135 (1966).
175. N. M. Volgina, A. I. Meos, L. A. Vol'f, and E. E. Nifant'ev, *Zh. Prikl. Khim.* **40** (1) 209–210 (1967).
176. U.S.S.R. Pat. 189,515 (Nov. 1966), N. F. Orlov, L. A. Vol'f, M. V. Androsova, and Yu. K. Kirilenko.
177. U.S. Pat. 3,084,989 (April 1963), Hitoshi Ave and Yasuji Ono ($\frac{1}{4}$ to Air Reduction Co., Inc. and $\frac{3}{4}$ to Kurashiki Rayon Co., Ltd.).
178. Hidenari Suyama and Hajime Haioka, *Sen-i Gakkaishi* **18,** 207–213 (1962).
179. K. E. Perepelkin and L. E. Utevskii, *Khim. Volokna* **6,** 7–10 (1962).
180. Fr. Pat. 1,459,247 (Nov. 1966), Paul Couchoud (to Société Rhodiacetà).
181. Austrian Pat. 202,705 (April 1959), Edith Lauda and Friedl Schreyer.
182. Ger. (East) Pat. 11,595 (May 1956), Hans L. Rohn and Hans E. Schwarzkopf.

MORTON LEEDS
Air Reduction Company, Ltd.

POLY(VINYL CHLORIDE)

The formation of poly(vinyl chloride) was first observed by Liebig in 1835 and reported by Regnault (1). They exposed 1,2-dichloroethane in an alcoholic solution of potassium hydroxide for several days to sunlight and obtained a white precipitate. After repeated careful analysis, they assigned it the formula C_2H_3Cl. Baumann (2) investigated the wave-length dependence of the rate of formation of the precipitate and found the highest rate under the influence of ultraviolet irradiation. He also correctly determined the density of the product as 1.406 g/cm^3. These early investigators did not recognize the polymeric character or the commercial importance of the material. Ostromuislenskii (3) suggested the production of synthetic rubbers by the action of ultraviolet irradiation upon vinyl chloride in organic solvents. He also postulated the radical mechanism of the polymerization process. The application of poly(vinyl chloride) as film, fiber, and lacquer was described in a patent by Klatte (4). The technique of copolymerization and the use of peroxides as polymerization initiators were introduced subsequently (5). Wide areas of nonrigid applications were opened up through the discovery of the plasticizing effect by Semon (6). The important principle of heat stabilization was discovered by Susich and Fikentscher (7). Industrial production of poly(vinyl chloride) was started in Germany in 1931 (Badische Anilin- und Soda Fabrik, I. G. Farbenindustrie, A.G.), in the United States in 1933 (Union Carbide Corp.), in Japan in 1939 (Shin Nippon Chisso Hiryo), in Great Britain and France in 1940 (Imperial Chemical Industries, Ltd., and Péchiney-Saint-Gobain), and in Italy in 1951 (Montecatini, S. G.).

Poly(vinyl chloride) is now one of the three major plastics, between polyethylene and polystyrene. World production in 1968 was estimated at about 10 billion lb (8). The expected annual growth rate during the next five years is 8–12% (9,10).

Several monographs on poly(vinyl chloride) have appeared (11–16b). Additional information may be found in references 17–19. For the properties and manufacture of the monomer, vinyl chloride, see Vol. 5, p. 171.

Polymerization

Bulk Polymerization. In this technique, the monomer is polymerized in the presence of a free-radical initiator only. The resulting polymer is insoluble in the remaining monomer, leading to a heterogeneous reaction medium (20–26). The rate of polymerization increases with increasing conversion (20). This acceleration is caused by the precipitated polymer and can be induced by the addition of preformed polymer. The effect is proportional to the two-thirds power of the polymer concentration present (24) and is explained by the trapping of growing polymer chains in the precipitated polymer particles. This reduces the rate of diffusion of the chains and consequently the probability of two active chains meeting, thus lowering the rate of termination (27). However, so far it was not possible to demonstrate the presence of these radicals by electron-spin resonance measurements (28,29). The following polymerization mechanism is proposed (19,30):

Reactions in the Liquid Phase

initiation $\quad$ I $\xrightarrow{k_i}$ R· $\hfill$ (1)

propagation $\quad$ R· + M $\xrightarrow{k_p}$ R· $\hfill$ (2)

chain transfer to a
$\quad$ monomer $\qquad\qquad$ R· + M → P + M· $\hfill$ (3)

termination $\quad$ R· + R· $\xrightarrow{k_t}$ P $\hfill$ (4)

Reactions in the Polymer Particles

radical trapped in
$\quad$ polymer particle $\qquad$ R· + P → (R·) + P $\hfill$ (5)

propagation $\quad$ (R·) + M → (R·) $\hfill$ (6)

chain transfer to a
$\quad$ monomer $\qquad\qquad$ (R·) + M → P + M· $\hfill$ (7)

termination $\quad$ (R·) + (R·) → P $\hfill$ (8)

$\qquad\qquad\qquad$ (R·) + M· → P $\hfill$ (9)

Reactions of the Monomeric Radicals in the Polymer Particles

propagation $\quad$ M· + M → (R·) $\hfill$ (10)

chain transfer to
$\quad$ polymer $\qquad\qquad$ M· + P → (R·) $\hfill$ (11)

escape into the
$\quad$ liquid phase $\qquad$ M· → R· $\hfill$ (12)

In this scheme, I stands for an initiator molecule, R· for a chain radical in the liquid phase, M for a monomer molecule, P for a polymer molecule, M· for a monomer radical in the liquid phase, and (R·) for a chain radical trapped in a polymer particle. The symbols k_i, k_p, and k_t indicate the reaction-rate coefficients for initiation, propagation, and termination. Absolute values for k_p and k_t at 60°C are 1.23×10^5 and 2.3×10^{10} liter/(mole)(sec), respectively (73). Assuming a steady-state concentration for all free radicals in the system, a homogeneous polymerization medium and making various other simplifying assumptions (27), the following expression for the degree of conversion, c, as a function of time, t, is obtained.

$$c = k_p[\mathrm{M}]/[\mathrm{M}]_0 \left(f \frac{k_i[\mathrm{I}]}{k_t} \right)^{\frac{1}{2}} \left[1 + b k_p[\mathrm{M}] \left(f \frac{k_i[\mathrm{I}]}{k_t} \right)^{\frac{1}{2}} t \right] t \qquad (13)$$

[I] represents the concentration of initiator, [M] the concentration of monomer at time t, $[\mathrm{M}]_0$ the initial monomer concentration, f the efficiency of the initiator, and b a constant. Equation 13 agrees well with experimental time-conversion curves up to conversions of 20%. Assuming a heterogeneous polymerization system, consisting of a dilute, monomer-rich phase and a concentrated, polymer-rich phase, with polymerization occurring in both phases simultaneously the following equation can be derived (23,31,32):

$$c = \frac{1}{q} \exp \left[(qk_i[\text{I}]^{1/2}t) - 1 \right] \tag{14}$$

with $q = -(1 + A - AQ)$, and A representing the monomer–polymer ratio in the concentrated phase and Q the ratio of the polymerization rates in the concentrated and the dilute phase. Equation 14 fits experimental values up to about 60% conversion.

The order of reaction with respect to the initiator is found experimentally to be between 0.46 and 0.58 for a variety of initiators (22,24,27,33–35). An exception is the case of cyclohexanone peroxide, where a first-order dependence was established (33,36). Within rather wide limits (for benzoyl peroxide at 50°C from 10^{-5} to 10^{-3} moles/mole of monomer) the molecular weight of the resulting polymer is independent of the initiator concentration (22,24,36,37). Only at high peroxide concentrations, the molecular weight decreases (37). The average molecular weight is also nearly independent of the degree of conversion (25). With decreasing temperature, the molecular weight of the bulk polymer increases up to a maximum value at about −30°; beyond this point, however, further decrease in polymerization temperature leads to decreasing molecular weights (34,36,38). Bulk polymerization under increasing pressure results in higher reaction rates (39). For example, at 5000 kg/cm² the rate is about 3.3 times that at 3000 kg/cm². From this pressure dependence, the activation volume for the polymerization of vinyl chloride is calculated to be −16.7 cm²/mole. The polymers obtained at increasing pressure show increasing molecular weight and density (39a).

From the above mentioned independence of the molecular weight upon initiator concentration and conversion, it can be concluded that chain transfer plays an important part in the bulk polymerization of vinyl chloride. The chain-transfer constant to the monomer, C_M (defined as the ratio of the rate coefficient of transfer to monomer to that of chain propagation) is 6.25×10^{-4} at 30°C; 1.35×10^{-3} at 50°C; 1.48×10^{-3} at 60°C; and 2.38×10^{-3} at 70°C (26,36,40). This means that, at 30°C, on the average, one in every 1600 reactions between monomer and chain radical leads to termination of the chain and transfer of the radical activity to the monomer, as indicated in equation 3. The temperature dependence of C_M suggests an activation enthalphy of 7300 cal/mole for this reaction (26). The chain-transfer constant to the polymer can be estimated from kinetic experiments on model compounds for the polymer, ie 2,4-dichloropentane for the dimer and 2,4,6-trichloroheptane for the trimer (29,41). At 50°C a value of 5×10^{-4} is found. Measurements on deuterated monomers show that the β-hydrogen atom is more active in the transfer reactions than the α-hydrogen or the chlorine atom (29).

Bulk polymerization techniques had, in the past, little commercial importance. Heterogeneous reactions of this type are difficult to control at higher degrees of conversions, when the mixture is extremely viscous and local overheating is a problem, leading to low molecular weights and broad molecular-weight distributions (42). These difficulties have recently been overcome in a two-stage process (43–45, 47), which is shown in a simplified flow diagram in Figure 1 (43). Sufficient vinyl chloride for a day's operation is transferred from the storage tank into the day tank. A portion, together with approx 0.015 wt % initiator, is pumped into the prepolymerizer, a vertical, stain-

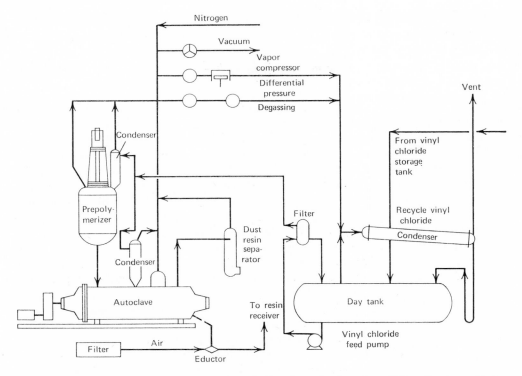

Fig. 1. Simplified flow diagram of a two-step bulk-polymerization plant (43).   Courtesy, *Hydrocarbon Processing.*

less steel-clad vessel, equipped with a flat-blade turbine stirrer and baffles.   Here, the monomer is polymerized to a conversion of 7–10% at a temperature of 40–70°C.   The considerable heat of reaction, amounting to 22,500 cal/mole, is removed by vaporizing excess monomer, which is then recondensed.   The reaction products are polymer beads which serve as seed for continued polymerization in the second step.   The structure of these beads has great influence upon the properties of the final product.   Indications are that the number of particles remains constant throughout the entire reaction (46).   In the second step, the mixture from the prepolymerizer, together with an equal amount of fresh monomer, are transferred into the autoclave, a horizontal reactor holding approx 750 ft³, equipped with slowly rotating agitator blades.   Here, the reaction is carried to a conversion of 90%.   Beyond approximately 20% conversion, the reaction medium is essentially powdery and the heat of reaction is removed by cooling the autoclave jacket and the agitator shaft, and by a reflux condenser.   Reaction time is 5–9 hr.   Unreacted monomer is then removed by a vacuum and recovered by vapor compression and condensation in the recycle condenser.   The resin is transferred to the resin receiver by means of an air eductor.   Fines are collected in a dust separator and removed through a manhole.

It is reported that these bulk resins show several desirable features (43,47).   Due to the absence of any additives during polymerization, the polymer particles have high porosity and clarity.   Shape and size are uniform, with granules of 0.5–1 μ diam.   This leads to better fusion, greater transparency, impact strength, heat and light stability, and improved electrical properties, at production costs claimed to be slightly below conventional homopolymers. Investment costs for a 70-million-lb/yr plant are quoted at approx $6 million, excluding product storage, engineering, and financing costs (1968 conditions). Existing bulk-production facilities have an estimated capacity of 500 million lb.

Emulsion Polymerization. In this technique, vinyl chloride is emulsified in water by means of surface-active agents. Most of the monomer is thus present as emulsion droplets, while a smaller amount is dissolved in a fraction of the laminar soap micelles, ie, colloidal aggregates which are formed above a critical emulsifier concentration. Water-soluble initiators are added, and polymerization begins when a radical enters a monomer-swollen micelle (48,49). Additional monomer is supplied to the growing latex particle by diffusion through the aqueous phase from the monomer emulsion droplets; the stability of the latex is maintained by adsorption of additional detergent from the remaining micelles. Chain termination takes place within each latex particle by the usual radical–radical interaction. One of the principal advantages of this technique is the possibility of obtaining high-molecular-weight polymers at very rapid rates. A quantitative theory for this true emulsion polymerization was proposed by Smith and Ewart (50). Under certain assumptions, the theory postulates that the overall reaction rate is proportional to the number, N, of latex particles per ml of emulsion, while this number in turn depends upon the six-tenths power of the soap concentration, $[S]$, and the four-tenths power of the initiator concentration, $[I]$. The average number of radicals per latex particle, $\bar{n}$, is given as one-half. A sharp maximum in the rate of polymerization is expected at the time when the micelles disappear from the system (51).

The kinetics of the emulsion polymerization of vinyl chloride does not follow this scheme. The reaction rate shows the same autocatalytic behavior as in the case of bulk polymerization (19,52). The number N depends on $[S]$, but the functionality varies greatly with type and concentration of the emulsifier (52–54). However, the rate of polymerization is practically independent of N (52,55); the order with respect to $[I]$ is found to be between 0.5 and 0.7 for various initiators (54,55). Very low values for $\bar{n}$ are obtained and termination seems to occur without participation of a second radical (56). Vinyl chloride at its saturation pressure has a solubility of 0.6 wt % in water at 50°C. This relatively high solubility is used to explain the strong deviations from true emulsion-polymerization kinetics by surmising that a major portion (55) or practically all of the polymerization (57) takes place in the aqueous solution. Arguments against this interpretation are derived from the observation that a reduction in the pressure of vinyl chloride leads to an increase in the reaction rate (52) and the fact that monomers with higher solubility in water than vinyl chloride show only minor deviations from the Smith-Ewart theory (58). It is also argued that the insolubility of poly(vinyl chloride) in its own monomer affects the kinetics of the process.

Emulsion polymerization has considerable technical importance in the production of poly(vinyl chloride), although the expected advantages in rate and molecular weight are not realized. It is preferred in cases where very fine particles are desired (0.5–2 μ). An estimated 15% of the U.S. production is manufactured by this process (47); the percentage in Europe is probably higher. A wide variety of emulsifier and initiator systems are reported in the literature (15,59). Preferred detergents are alkyl sulfates, alkane sulfonates, and fatty acid soaps. Higher molecular weights are sometimes obtained with emulsifiers of increasing chain length. Unsaturation in the detergent leads to a decrease in the rate of polymerization. The introduction of heteroatoms into the paraffin chain has little effect (60). Widely used initiators are hydrogen peroxide, organic peroxides, peroxydisulfates, and redox systems. Oxygen is strictly excluded and the pH of the mixture is maintained between 6 and 8. Conversion is carried to a relatively high percentage, corresponding to a final solids contents of 50–55% in continuous or discontinuous reactors. Emulsion resins in the 0.2–5μ particle-size range, ie, plastisol-type products, are normally spray dried and ground. Emulsion resins of smaller particle size usually reach the market in the form of latex systems. Larger-size emulsion resins (for calendering and extrusion) are mostly recovered by coagulation with salts, followed by washing, drying, and sieving. Contrary to bulk resins, emulsion polymers may contain residual detergent. This has an adverse effect on clarity, compatibility with compounding adjuvants, electrical resistivity, water absorption, and weathering of the final product.

Solution and Precipitation Polymerization. Solution polymerization is carried out in a medium that is a solvent for both the monomer and the polymer. In precipitation polymerization the polymer is insoluble in the system.

Solution polymerization of vinyl chloride is generally possible only at initial monomer concentrations, $[M_0]$, below a characteristic value and only up to certain degrees of conversion. For chlorobenzene this limit is at 7.7 moles/liter (61,62); for tetrahydrofuran at 7.6 moles/liter (30,61); and for 1,2-dichloroethane at 7.9 moles/liter (61,63), all at 50°C. In the homogeneous region, the order of the reaction with respect to initiator is one-half, and with respect to monomer it is one, with the exception of the chlorobenzene system, where the solvent acts as a strong retarder. Here the respective exponents are 0.56 and 1.23 (62). Transfer constants for solvent and monomer in tetrahydrofuran are $C_S = 3 \times 10^{-3}$ and $C_M = 5 \times 10^{-3}$ (30) and in 1,2-dichloroethane $C_S = 4.5 \times 10^{-4}$ and $C_M = 1.1 \times 10^{-3}$, respectively (63). The latter value is similar to that found for heterogeneous bulk polymerization (40); this suggests that the homogeneity of the solution system increases only the magnitude of k_t, while it practically does not affect k_p (63). Addition of small amounts of effective chain-transfer agents, such as carbon tetrachloride or dodecanethiol (lauryl mercaptan) leads to a marked acceleration of the solution polymerization, although not to the same extent as in bulk (64).

Beyond a certain degree of conversion, the above systems become heterogeneous, due to the formation of a concentrated polymer phase. This phase separation occurs at conversions which become progressively lower as the ratio monomer/solvent and the molecular weight of the resulting polymer increase. The phase separation is accompanied by a strong decrease in the rate of polymerization. Reactions that are started above the limiting monomer concentration are heterogeneous from the onset of the process and show the autocatalytic effect discussed under Bulk polymerization. Other agents, such as methanol, lead to precipitation at any monomer concentration (65). As in the case of bulk polymerization, the rate in all of these systems increases upon application of pressure (39).

An estimated 3% of the U.S. production is manufactured by precipitation polymerization (47). There are procedures using methanol (66), benzene (67), acetone (68), and methanol/water (69) at temperatures between 40 and 60°C, and also halogenated media, such as 1,1,2,2-tetrafluorodichloroethane, between 0 and 80°C (70). The resins are largely free of impurities and additives and have lower molecular weights, due to the chain-transfer activity of the solvents.

Suspension Polymerization. In this technique the monomer is finely dispersed in water by vigorous agitation. Polymerization is started in the droplets by means of monomer-soluble initiators. Addition of a suspension stabilizer minimizes coalescence of the growing particles by forming a protective monolayer around them. The stabilizer should also increase the viscosity of the water phase without appreciably changing its surface tension. Consequently, a stabilizer should be soluble in water, but insoluble in vinyl chloride. The kinetics of suspension polymerization is identical with that of bulk polymerization (32,71). It shows autocatalytic behavior and the molecular weight of the resulting polymer is practically independent of the concentration of the initiator and of the degree of conversion. Equation 14 (see p. 371) describes the reaction well up to conversions of approximately 60%. Molecular weights also decrease with increasing reaction temperature. The net activation enthalpy of the process is mainly determined by the activation enthalpy for the creation of the free radicals. In the case of most conventional peroxides, the activation enthalpy is around 20,000 cal/mole (71), while for a monomer-soluble redox catalyst it is as low as 6500 cal/mole (72). The former number is close to the value of 21,800 cal/mole reported for bulk polymerization with benzoyl peroxide (24), and the latter is on the same order of magnitude as the value of 3500 cal/mole found for the photosensitized solution polymerization of vinyl chloride (73). The activation enthalpy increases with increasing pressure (74).

Suspension polymerization has great technical importance; an estimated 82% of the U.S. production is manufactured in this way (47). Figure 2 shows a flow sheet for a typical batch process (75). Most reactors presently in use are water-jacketed and glass-lined to reduce polymer buildup on the walls. Reactor sizes vary from 2000 to 6000 gal. Ratios of water to vinyl chloride range from 1.5:1 to 4:1. Low ratios allow higher monomer charges for a given reactor, while high water contents facilitate better temperature control and thus permit higher conversions. The latter method yields a more porous resin. The desired quantity of monomer is measured in the weigh tank and transferred into the reactor containing the proper amount of water. Initiator, suspending agent, and buffer are charged into the reactor via the charge bomb. Agitation is started and the mixture is brought to the reaction temperature, in general 45–55°C. The reaction is usually carried to 90% conversion and cooling water is used to remove the heat of polymerization. The mixture is then transferred to the dump tank, where it is stripped of unreacted monomer by the application of vacuum. This monomer is recovered and recycled into the day tank. Several runs are combined in the blend tank to assure a more uniform product. The bulk of the water is separated in a centrifuge and discharged and the resin is dried in a stream of hot air in a rotary drier. The product is separated from the wet air stream in the cyclone separator, from which it is screened and sent to storage. The wet air stream containing the fines is passed through a filter.

A large variety of suspending agents are mentioned in the literature (15). The three most popular compounds are poly(vinyl alcohol), gelatin, and methylcellulose. They are used in concentrations of 0.05–0.5 parts per hundred parts of monomer (phm). The minimum amount of poly(vinyl alcohol) applicable is determined by its molecular weight; low-molecular-weight polymer (lmw) at 0.005 phm is already an effective stabilizer, while high-molecular-weight material at the same level is ineffective (76). Increasing concentrations of suspending agents lead to decreasing polymer-particle sizes (72,76,77). The desired particle size is 50–150 μ. In some cases, emulsifying agents are added in minor concentrations (0.01–0.03 phm) to obtain a more por-

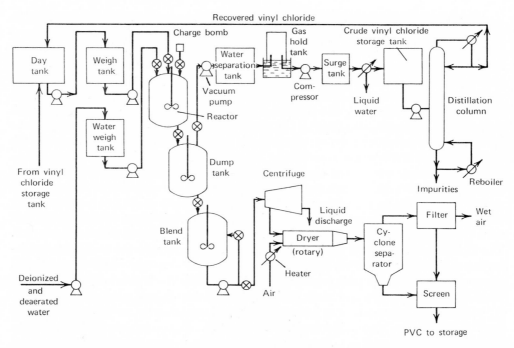

Fig. 2. Simplified flow diagram of a batch-type suspension-polymerization plant (75). Courtesy *Chemical Engineering.*

ous resin (78).   Other procedures are carried out in the absence of any suspending or emulsifying agents and the stability of the systems is maintained by vigorous stirring (79) or by the use of special water-soluble initiators, such as peroxysulfates and silver salts (80).

Initiators are employed in concentrations of 0.03–0.1 phm.   In a patent procedure the initiator is prepared in situ by reacting haloformates (0.005–1 phm) with sufficient amounts of peroxide (81).   Among the most frequently used systems are lauroyl peroxide (LP), azobisisobutyronitrile (AIBN), and isopropyl peroxycarbonate (IPP). The latter is the most active among the three, with a half-life of 2 hr at 50°C, as compared to 48 hr for LP.   Rates of conversions at that temperature range from about 8%/hr for LP to 16%/hr for IPP.

Special Methods of Polymerization.   The original method of photochemical initiation (1–3) can be improved by the use of sensitizers (73,82) or ultraviolet irradiation (83).   A patent procedure utilizes the application of γ-rays from a ^{60}Co source (83).   In the case of emulsion polymerization, the reaction rate is unexpectedly found to be proportional to the 1.22 power of the dose rate and the 0.11 power of the emulsifier concentration, while the polymer molecular weight depends upon the 0.06 power of that concentration (84).   Bulk polymerization of vinyl chloride by γ-rays, however, follows the expected dependence of the 0.5–0.6 power of the dose rate.   Over a temperature range of −78 to 110°C, a maximum degree of polymerization is obtained at 5°C (85).

The use of metal organic compounds as catalysts for the polymerization of vinyl chloride has been known since 1930 (86).   Since then a large number of patents and publications have appeared in this field and many are summarized in references 15 and 19. In principle, two classes of catalysts are employed: complexes between alkyls of metals of groups I–III in the periodic system of elements with halides and other derivatives of transition metals of groups IV–VIII (Ziegler-Natta coordination catalysts), and metal alkyls with certain other activating additives. The mechanism of these polymerizations is not always well understood. For *tert*-butylmagnesium chloride in tetrahydrofuran (87) and for butyllithium and butyllithium–butyl iodide complexes in heptane (88), an anionic mechanism is assumed. In the first case, the following three step reaction mechanism is proposed (87):

Instantaneous initiation

$$R^-, {}^+MgCl + M \rightarrow R\text{—}CHClCH_2{}^-, {}^+MgCl \tag{15}$$

Propagation

$$RCHClCH_2{}^-, {}^+MgCl + M \rightarrow R\text{—}(CHClCH_2)\text{—}CHClCH_2{}^-, {}^+MgCl \tag{16}$$

Termination

$$R(CHClCH_2)_n CHClCH_2{}^-, {}^+MgCl \rightarrow R(CHClCH_2)_n\text{—}CH{=}CH_2 + MgCl_2 \tag{17}$$

where R stands for $—C(CH_3)_3$.

The usual Ziegler-Natta catalysts, such as titanium trichloride–triethylaluminum, decompose vinyl chloride to acetylene and hydrochloric acid (89). Special systems, such as titanium tetrabutoxide–aluminum alkyl chlorides (90,91), trialkylaluminum–chlorine (92), titanium tetrachloride–diethylaluminum ethoxide (93), and diethylaluminum butoxide (94) are believed to polymerize vinyl chloride via a radical mechanism, although in at least one case a partial contribution via an ionic-coordinated mechanism is claimed (91). The best yields are obtained with aluminum alkyls of low Lewis-acidity with addition of nucleophilic compounds, such as amines or ethers. The reaction rate is found to depend upon the 0.53 power of the catalyst concentration and the first power of the monomer concentration (93). The systems tributylboron–oxygen (34), triethylboron–hydrogen peroxide (95), thionyl chloride–*tert*-butyl hydroperoxide (96), and tetraethyllead–ammonium hexanitratocerate, $(NH_4)_2(Ce(NO_3)_6)$ (97), also generate free radicals, in the latter case, for instance, according to the following equation:

$$Pb(C_2H_5)_4 + Ce^{4+} \rightarrow Pb(C_2H_5)_3{}^+ + Ce^{3+} + C_2H_5\cdot \tag{18}$$

Many of the above combinations can be used for the polymerization of vinyl chloride at subzero temperatures.

Copolymerization. Vinyl chloride can be copolymerized with a large variety of monomers (M_2), such as other unsaturated halogenated hydrocarbons, styrene and its halogenated derivatives, vinyl esters and ethers, olefins, dienes, esters and other derivatives of acrylic and methacrylic acids, olefinic dicarboxylic acids and esters, and various heterocompounds. Review articles (15–18,98a,99) present details on the preparation and tables of copolymerization ratios, *r*, and *e* and *Q* values (99–101). See Vol. 16, p. 233. Table 1 gives these quantities for a number of combinations of technical interest.

The most important product is the copolymer of vinyl chloride and vinyl acetate. Commercial resins contain between 3 and 20 wt % vinyl acetate and are generally manufactured by a suspension or solution (precipitation) technique. In the latter case, acetone, butyl acetate, and ethylene dichloride are solvents for the copolymer, while

Table 1. Copolymerization Parameters of Vinyl Chloride

M_2	r_1	r_2	Temp. °C	Reference	e	Q
acrylic acid	0.107	6.8	60	102	0.77	1.15
acrylonitrile	0.04	2.7	60	103	1.20	0.60
butadiene	0.035	8.8	50	104	−1.05	2.39
n-butyl acrylate	0.07	4.4	45	105	1.06	0.50
diethyl fumarate	0.12	0.47	60	106	1.25	0.61
dimethyl itaconate	0.053	5.0	50	107	1.34	1.03
ethylene	3.21	0.21	50	108	−0.20	0.015
isobutylene	2.05	0.08	60	109, 102a	−0.96	0.033
isoprene					−1.22	3.33
maleic anhydride	0.296	0.008	75	110	2.25	0.23
methacrylic acid	0.034	23.8	60	102	0.65	2.34
methacrylonitrile	0.11	2.38	60	102b	0.81	1.12
methyl acrylate	0.12	4.4	50	111	0.60	0.42
methyl methacrylate	0.1	10	68	112	0.40	0.74
octyl acrylate	0.12	4.8	45	105	1.07	0.35
propylene	2.27	0.3		113, 102a	−0.78	0.002
styrene	0.02	17	60	114	−0.80	1.0
vinyl acetate	1.68	0.23	60	115	−0.22	0.026
N-vinylcarbazole	0.17	4.8	50	116	−1.40	0.41
vinyl chloride					0.20	0.044
vinylidene chloride	0.3	3.2	60	117	0.36	0.22
vinyl isobutyl ether	2.0	0.02	50	118		
N-vinylpyrrolidone	0.53	0.38	50	119	−1.14	0.14

benzene, butane, cyclohexane, and toluene are nonsolvents. Due to the higher reactivity of vinyl chloride, any given mixture of the two monomers leads to copolymers with increasing vinyl acetate content. If a homogeneous chemical composition is preferred, a portion of the vinyl chloride must be added continuously during the polymerization, thus keeping the monomer ratio constant. The amount to be proportioned for any desired copolymer is calculated from copolymerization kinetics (120). Molecular weight of the product may be controlled by the addition of chain-transfer agents, such as trichloroethylene. The introduction of a third monomer can lead to the formation of terpolymers with desirable properties; for example, a small percentage (approx 2 wt %) of maleic anhydride facilitates adhesion and curing; monomers containing an oxirane ring also lead to curable compositions. The properties of the products can be varied by partial or complete saponification of the acetate moieties.

Copolymers with olefins are gaining increasing importance. Two wt % of ethylene can be introduced by suspension polymerization of a mixture of vinyl chloride and ethylene in the mole ratio of 1:0.11 in the presence of tributylboron (121). Copolymers with up to 10 wt % of ethylene are obtained by suspension polymerization employing organic peroxides or redox systems (122). Similar procedures are used to prepare copolymers with up to 10 wt % propylene (123). The chain-transfer activity of the olefin is compensated for by the application of more active initiator systems and lower polymerization temperatures, giving products in the conventional range of molecular weights (124). As in the case of homopolymerization (39), increase in pressure accelerates the reaction rate (125).

Copolymers with vinyl alkyl ethers, such as vinyl isobutyl ether (126) or vinyl lauryl and cetyl ether (127), can also be prepared by essentially conventional suspension-polymerization methods.   Other commercial or semicommercial products contain acrylonitrile, octyl acrylate, and vinylidene chloride.

Graft copolymerization is carried out either by reacting various monomers in the presence of poly(vinyl chloride) or by polymerizing vinyl chloride onto other polymeric substrates.   In the first group, comonomers, such as acrylonitrile (128) or ethyl acrylate + acrylonitrile (129), are used.   In the second group, vinyl chloride is, for example, grafted onto noncrystalline polyethylene (130,131), polypropylene (132), copolymers of vinyl acetate and ethylene (133), or chlorinated polyethylene.   These products are also prepared by essentially conventional solution, suspension, or emulsion techniques.

Polymer Properties

Structure.   In principle, addition of the vinyl chloride monomer units during polymerization can occur either in head-to-tail fashion, resulting in 1,3 positions for the chlorine atoms

$$-[CH_2CHClCH_2CHCl]_n \tag{19}$$

or head-to-head, tail-to-tail, placing the chlorine atoms in 1,2 positions, as follows:

$$-[CH_2CHClCHClCH_2]_n \tag{20}$$

Dehalogenation of structure (20) would be expected to be complete, for each chlorine atom can pair off with a chlorine on an adjacent carbon atom.   But with structure 19 it is not so; assuming that dechlorination starts at random positions along the chain, some chlorine is left unable to pair off.   Statistical considerations show that random removal of pairs of chlorine atoms from structure (19) would leave behind 13.5% of the halogen atoms (134).   Treatment of dilute solutions of poly(vinyl chloride) with zinc actually removes up to 87% of the chlorine, proving the essential head-to-tail arrangement of the polymer (135).   A product corresponding to structure (20) can be prepared by addition chlorination of *cis*-1,4-polybutadiene to a chlorine level of 56.6% (theoretical 56.8%).   The infrared spectrum of this product differs greatly from that of conventional poly(vinyl chloride), especially in the 1333 cm^{-1} region (136).

Three categories of end groups are possible in the polymer.   Saturated groups are formed by chain transfer with monomer and polymer and by termination through disproportionation:

$$\text{---}CH_3; \quad \text{---}CH_2Cl; \quad \text{---}CHCl_2 \tag{21}$$

Unsaturated chain ends are due to termination by disproportionation and to chain transfer to monomer:

$$-CH_2CHClCH{=}CHCl; \quad -CCl{=}CH_2; \quad -CH_2CHClCH{=}CH_2 \tag{22}$$

Initiator or solvent (chain-transfer agent) fragments, represented by R, can be incorporated in the terminal group:

$$-CH_2R; \quad -CHClR \tag{23}$$

Due to the high transfer activity of the monomer, about 60% of the polymer molecules are estimated to have unsaturated end groups (137).   For the same reason, the percentage of chain ends containing initiator fragments is low (138); the amount of solvent fragments depends upon its transfer activity.

Long-chain branching can be caused by the incorporation of the terminal double bond of a polymer molecule into a growing chain:

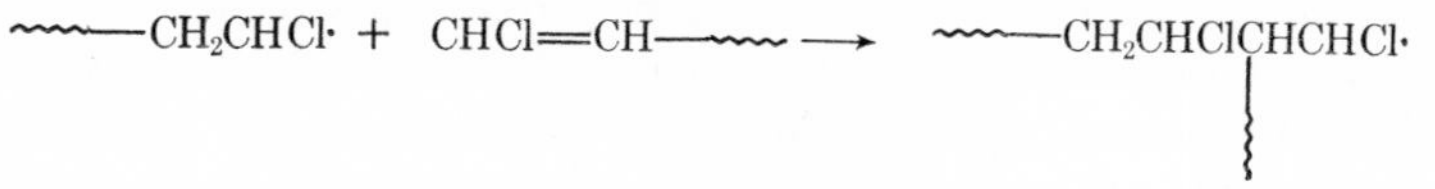

$$\text{\textasciitilde\textasciitilde\textasciitilde—CH}_2\text{CHCl·} + \text{CHCl}\!=\!\text{CH—\textasciitilde\textasciitilde} \longrightarrow \text{\textasciitilde\textasciitilde—CH}_2\text{CHClCHCHCl·} \qquad (24)$$

or by intermolecular chain transfer to polymer:

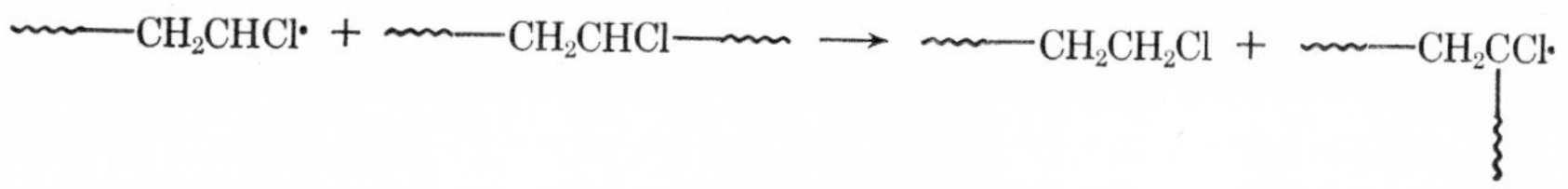

$$\text{\textasciitilde\textasciitilde—CH}_2\text{CHCl·} + \text{\textasciitilde\textasciitilde—CH}_2\text{CHCl—\textasciitilde\textasciitilde} \longrightarrow \text{\textasciitilde\textasciitilde—CH}_2\text{CH}_2\text{Cl} + \text{\textasciitilde\textasciitilde—CH}_2\text{CCl·} \qquad (25)$$

Intramolecular chain transfer (back-biting) leads to the formation of short side chains:

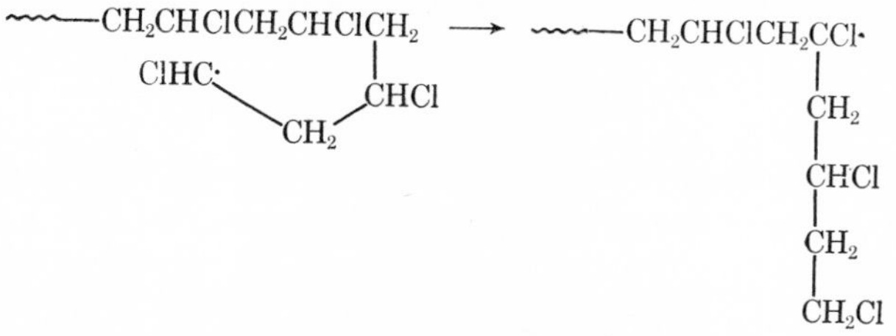

$$(26)$$

For the determination of total branching, the polymer is subjected to reductive hydrogenation, with chlorine removal, by means of lithium aluminium hydride. The ratio of methyl to methylene groups is measured by infrared spectroscopy, using the bands at 1378 and 1350, or 1370 and 1386 cm^{-1}, respectively. For conventional resins, polymerized in bulk or suspension at temperatures of 50–90°C, between 0.2 and 2 branches per 100 carbon atoms are found (138–141). The percentage distribution between long and short chain branching is not established. With decreasing temperature of polymerization, the number of branches decreases, because the enthalpy of activation for propagation is smaller than that for chain transfer. The degree of branching of vinyl chloride–vinyl acetate copolymers is expected to be higher than that of the homopolymer.

The tactical placement of the monomer units with respect to their next neighbors affects the properties of the resulting polymer. The following two possibilities exist:

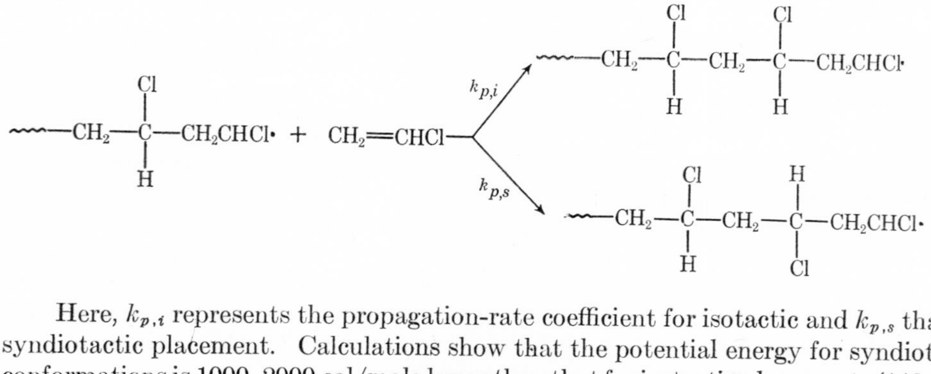

Here, $k_{p,i}$ represents the propagation-rate coefficient for isotactic and $k_{p,s}$ that for syndiotactic placement. Calculations show that the potential energy for syndiotactic conformations is 1000–2000 cal/mole lower than that for isotactic placements (142,143).

Consequently, the free energy of activation for syndiotactic placement, ΔG_s, must be lower than ΔG_i, and from absolute reaction-rate theory it follows that the ratio

$$k_{p,s}/k_{p,i} = \exp\left[(\Delta G_i - \Delta G_s)/\text{RT}\right] \tag{28a}$$

$$k_{p,s}/k_{p,i} = \exp\left[-(\Delta S_i - \Delta S_s)/\text{R}\right]\exp\left[(\Delta H_i - \Delta H_s)/\text{RT}\right] \tag{28b}$$

must increase with decreasing temperature (the symbols ΔS and ΔH stand for the respective entropies and enthalpies of activation). This means that with decreasing temperature of polymerization, the degree of syndiotacticity of poly(vinyl chloride) should increase for purely thermodynamic reasons, without necessarily requiring adsorption or complexing by a catalyst. The extent of syndiotactic placement can be determined by an analysis of the NMR spectrum (144) or by infrared spectroscopy (145). Measurements on several series of polymers prepared in bulk or suspension via radical initiation at temperatures from 120 to $-80°$C show some discrepancies, but demonstrate clearly an increase in syndiotactic placement from approx 50 to 65%. Experimental values for the difference in activation enthalpy ($\Delta H_i - \Delta H_s$), range from 310 to 600 cal/mole, while the difference in activation entropy ($\Delta S_i - \Delta S_s$), is around 0.6 eu (146).

The syndiotactic sequences in poly(vinyl chloride) are capable of crystallization. The degree of crystallinity can be determined from x-ray patterns and is found to correlate with the degree of syndiotacticity. The unit cell is orthorhombic with axes of 10.6, 5.4, and 5.1 Å (147); the dimensions of the crystallites are from 50 to 100 Å (146). It is also possible to prepare single crystals of lmo poly(vinyl chloride) (148). The degree of crystallinity of a sample depends on previous treatments and is up to 5% for conventional and 25–30% for low-temperature resins. Considerably higher degrees of crystallinity are found in samples polymerized in the presence of aliphatic aldehydes (149). This is probably due to the exceptionally low degree of polymerization and branching obtained under these conditions (150).

The degree of crystallinity in percent, X_c, can also be determined from a measurement of the specific volume (reciprocal density) of an unknown sample, $\bar{v}$, if the specific volumes of a completely crystalline, $\bar{v}_c$, and a completely amorphous sample, $\bar{v}_a$, are known. The following relation is used:

$$X_c = 100\,\frac{\bar{v} - \bar{v}_a}{\bar{v}_c - \bar{v}_a} \tag{29}$$

Experimental literature values are 0.694 for $\bar{v}_c$ (147) and 0.722 for $\bar{v}_a$ (140), the latter in good agreement with a theoretical value calculated under the assumption of additivity of group increments for molar volumes (151). Other authors put the two values at 0.671 and 0.709, respectively, using a different annealing technique (152).

Solution Properties. Table 2 presents a series of solvents for poly(vinyl chloride) and most of its copolymers. The Hildebrand solubility parameters, δ (153) and the Flory-Huggins interaction parameters, χ_1 (154), are measures for the quality of a solvent (see Solubility parameter in Supplement Volume). For vinyl chloride monomer, $\delta = 7.8$ and $\chi_1 = 0.88$ at 50°C (159); for poly(vinyl chloride), δ values in the literature range from 9.4 to 10.8, while for a copolymer with 13 wt % vinyl acetate, $\delta = 10.4$ (160). According to the theory, good solvents have small or negative χ_1-values and δ identical with that of the polymer. Other criteria for solvent power are a high electron-donor capacity, low steric hindrance to approach to the polymer molecules and sufficient bulk for keeping the polymer chains well separated. A large number of sol-

Table 2. Solvents for Poly(vinyl chloride)

Solvent	Density,[a] g/ml	Refractive index,[a] n_D	Parameters		Reference
			δ, $(cal/ml)^{1/2}$	χ_1	
anisole	0.9954	1.5179		0.515[53]	155
				0.49[76]	155
benzyl alcohol	1.0419	1.5396	12.1		
bromobenzene	1.5219[0]	1.5598			
1-bromonaphthalene	1.4887[17]	1.6588[19]	10.6	0.40[53]	155
				0.38[76]	155
butyl acetate	0.8813	1.39406	8.5	0.40[53]	155
				0.41[76]	155
γ-butyrolactone	1.1286[16]	1.4343[26]	14.5		
chlorobenzene	1.1064	1.5248	9.5	0.514[25]	156
				0.53[53]	155
				0.50[76]	155
cyclohexanone	0.9462	1.45097	9.9	0.176[25]	157
cyclohexene oxide	0.975[15]		9.3		
cyclopentanone	0.9504[18]	1.4366	10.4	0.253[25]	157
				0.19[25]	158
				0.259[40]	157
diamyl sebacate[b]	0.9510[18]			0.24[53]	155
				0.24[76]	155
1,2-dibromoethane	2.180	1.5389	10.4	0.468[25]	156
di(2-butoxyethanol) phthalate[b]	1.063	1.483[25]	8.0	0.44[53]	155
				0.42[76]	155
dibutyl phthalate[b]	1.049	1.490[25]	8.5	−0.04[53]	155
				−0.01[76]	155
o-dichlorobenzene	1.3059	1.5515	10.0		
1,2-dichloroethane	1.256	1.4448	9.9	0.428[25]	156
				0.46[53]	155
				0.43[76]	155
diethyl phthalate[b]	1.1202[18]	1.503[18]	10.0	0.42[53]	155
				0.40[76]	155
diethyl sebacate[b]	0.9646	1.4359		0.17[53]	155
				0.18[76]	155
dihexyl adipate[b]	0.933[25]	1.439[25]		0.19[53]	155
				0.23[76]	155
dihexyl phthalate[b]	1.007	1.487	8.9	−0.13[53]	155
				−0.09[76]	155
dihexyl sebacate[b]				0.35[53]	155
				0.36[76]	155
diisopropyl ketone	0.8108	1.417	8.0		
dimethylformamide	0.9445[25]	1.4269[25]	12.1		
dimethyl sebacate	0.9882[28]	1.4355[28]		0.34[53]	155
				0.34[76]	155
dimethyl sulfoxide	1.1014		12.0		
di-n-octyl succinate[b]				0.39[53]	155
				0.39[76]	155

vent combinations fulfill the conditions and show even synergistic action, acetone–carbon disulfidide, acetone–perchloroethylene, and mixtures of *N,N*-disubstituted amides with cyclic ethers, aliphatic ketones, alkyl-substituted pyridines, and aromatic hydrocarbons (161).　　Cyclohexanone and tetrahydrofuran are good solvents for physico-chemical measurements, most commonly used.

Table 2 (*continued*)

Solvent	Density,[a] g/ml	Refractive index,[a] n_D	Parameters		Reference
			δ, (cal/ml)$^{1/2}$	χ_1	
di(2-ethylhexyl) phthalate (DOP)[b]	0.986	1.485[25]	7.9	0.01[53]	155
				0.03[76]	155
dioxane	1.03375	1.42241	10.0	0.503[25]	156
				0.55[25]	158
				0.52[27]	154
ethyl acetoacetate	1.0288[17]	1.41937		0.49[53]	155
				0.46[76]	155
ethylene oxide	0.882[10]	1.3597	9.9		
isophorone	0.9229	1.4789[22]	9.1		
mesityl oxide	0.8578	1.4440	9.0	0.428[25]	156
methyl amyl ketone	0.8111	1.41156[15]	8.5	0.18[53]	155
				0.17[76]	155
methyl ethyl ketone	0.80473	1.37850	9.3	0.465[25]	156
methyl isopropyl ketone	0.8031	1.38788[16]	8.5	0.426[25]	156
methyl n-propyl ketone	0.809	1.3895	8.7	0.415[25]	156
morpholine	0.9994	1.4545	10.2	0.24[25]	158
nitrobenzene	1.2037	1.5562	11.0	0.30[25]	158
				0.29[53]	155
				0.29[76]	155
1-nitropropane	1.0081[24]	1.4003[24]	10.3	0.44[53]	155
				0.42[76]	155
β-propiolactone	1.1460	1.4131	13.3		
tetrahydrofuran	0.8898	1.4091	9.1	0.364[25]	156
				0.291[25]	157
				0.26[25]	158
				0.21[25]	158
				0.14[27]	154
tetrahydrofurfuryl alcohol	1.0544	1.4517			
tetrahydropyran			8.6	0.355[25]	158
tributyl phosphate[b]	0.9727[25]	1.4224[25]	7.9	−0.65[53]	155
				−0.53[76]	155
tricresyl phosphate[b]	1.165	1.555[25]	8.7	0.38[53]	155
				0.38[76]	155
triethyl phosphate	1.0725[19]	1.4067[17]		0.13[53]	155
				0.15[76]	155
trimethylene oxide			9.6		
xylene			8.8		

[a] Measured at 20°C, unless otherwise indicated.

[b] Used as plasticizers.

It is practically impossible to obtain a truly molecular-disperse solution of poly-(vinyl chloride). Even in solutions in good solvents, prepared by heating to elevated temperatures for several hours, supermolecular structures (microgels) can be detected by light-scattering or ultracentrifugation. They seem to consist of highly branched or crosslinked particles of approx 2,500 Å diam and molecular weights in the order of 10^8. The amount of microgel increases with time and may be due to oxidative crosslinking (162). Crystalline polymers have lower solubility and show crystalline aggregates in solution (163). In poor solvents, all poly(vinyl chloride) samples form clusters of several molecules which can contain from 15 to 50% of all polymer chains present (162).

Table 3 presents a list of fractionations that have been performed on poly(vinyl chloride), using the following methods: fractional precipitation (FP), fractional solution (FS), chromatographic fractionation (CF), gel-permeation chromatography (GPC), thermal diffusion (TD), and turbidimetric titration (TT). The solution anomalies, discussed above, also affect the results of the fractionations. Often the fraction with the highest molecular weight behaves differently from all others; ie, it requires a smaller volume of nonsolvent, precipitates in the form of a powder and dissolves with difficulty or incompletely. The amount of this fraction increases with decreasing polymerization temperature of the original resin. The inhomogeneity of the fractions of higher molecular weight—expressed as the ratio of the weight-average-molecular weight, M_w, to the number-average-molecular weight, M_n—is often approaching or even exceeding that of the original sample. M_w/M_n for the lower molecular-weight fractions is usually small. In some cases, the third or fourth fraction apparently has the highest molecular weight ("inversion"). All these anomalies can be qualitatively explained by the formation of aggregates (180, 191), as discussed previously. Some specific results of fractionation experiments may also be mentioned. The molecular-weight distribution of a suspension resin seems to be independent of the degree of conversion during the polymerization (167); in the case of an emulsion resin, the distribution broadened with increasing conversion and the maximum shifted toward lower molecular weights (165).

The molecular weight of poly(vinyl chloride) is most frequently determined by measuring the limiting viscosity number, $[\eta]$. This number is related to the viscosity

Table 3. Fractionations of Poly(vinyl chloride)

Method	Solvent/nonsolvent	Temperature, °C	Reference
FP	tetrahydrofuran/water	50	164
FP	tetrahydrofuran/water	40	165, 166
FP	tetrahydrofuran/water	30	167, 168, 210
FP	tetrahydrofuran/water		169, 170
FP	cyclohexanone/ethanol		171
FP	cyclohexanone/n-butanol		172
FP	cyclohexanone/ethylene glycol		173, 209
FP	methyl cyclohexanone/ethanol		174
FP	chlorobenzene/gasoline[a]		175
FP	chlorobenzene + acetone/methanol		175
FS	tetrahydrofuran/water	44	176
FS	methyl isobutyl ketone[b]		177
FS	cyclohexanone/methanol	30	178
FS	tetrahydrofuran + methyl ethyl ketone/chloroform		179
FS	cyclohexanone/ethylene glycol	130 or 70	180
CF	tetrahydrofuran + water/water	60–25	181
CF	cyclohexanone + methanol/methanol	60–25	182
GPC	tetrahydrofuran[c]	25	179, 183–186
GPC	tetrahydrofuran + benzene[d]	25	187
TD	cyclohexanone	565[e]	188, 189
TT	cyclohexanone/heptane + carbon tetrachloride		190

[a] According to chlorine content.
[b] Separation of copolymers.
[c] On crosslinked PST.
[d] On Santocel A.
[e] Per centimeter.

of the solvent, η_0, and that of the dilute solution, η, of concentration c in g/ml, as follows:

$$\frac{\eta - \eta_0}{\eta_0 c} = [\eta] + k'[\eta]^2 c + \ldots \tag{30}$$

where k' is the Huggins slope constant (192). Table 4 gives values of k' for various solvents.

Table 4. Huggins Slope Constants for Poly(vinyl chloride), Measured at 25°C (158)

Solvent	k'	Solvent	k'
cyclopentanone	0.31	nitrobenzene	0.44
cyclohexanone	0.36	tetrahydropyran	0.44
tetrahydrofuran	0.37	chlorobenzene,	
morpholine	0.41	measured at 30°C[a]	0.68

[a] Reference 193.

In dilute solutions, η_0 and η can be approximated by the capillary flow times of the solvent, t_0, and of the solution, t. The limiting viscosity number is related to the molecular weight of the polymer by equations of the type proposed by Mark-Houwink-Sakurada:

$$[\eta] = K \times M^a \tag{31}$$

Values for the constants K and a have to be determined by absolute methods, such as osmometry (OS), light scattering (LS), Archibald ultracentrifugation (AU), or sedimentation-diffusion (SD). The first technique yields M_n, the others M_w. Table 5 gives the values for K and a for a variety of solvents at the indicated temperatures. Benzyl alcohol at 155.4°C represents a pseudo-ideal solvent (θ point). The table also shows the number of whole polymers or fractions and the absolute method used in these studies, together with the applicable molecular- weight range. The wide scatter of the data for any given solvent–temperature system is in part again caused by the solution anomalies of the polymer (208). The osmotic method is especially sensitive to molecular-weight inhomogeneity. In one case, a linear decrease of a with increasing M_w/M_n is found (166). A smoothed equation for the determination of M_w by viscosity measurements in cyclohexanone at 25°C, obtained by averaging the data of several authors, is proposed (213):

$$[\eta] = 2.69 \times 10^{-2} \times M_w^{0.72} \tag{31a}$$

In the older scientific literature and in industrial practice, the molecular weight is often characterized by the Fikentscher k value (214):

$$\log (\eta/\eta_0) = \left[\frac{75\, k^2}{1 + 1.5\, kc_g} + k \right] \times c_g \tag{32}$$

Here, c_g is the polymer concentration in g/100 ml. At $c_g = 1$, the relation is as follows:

$$k = -0.0166 + 0.066[6.25 + 750 \times \log(\eta/\eta_0)]^{1/2} \tag{32a}$$

The relation with $[\eta]$ is given by

$$[\eta] = 0.2303\, k(75k + 1) \tag{32b}$$

Since k is a small number, 1000 k is often used and designated as K-value.

Table 5. Relationship of Molecular Weight to Limiting Viscosity Number for Poly(vinyl chloride)

Solvent	Temperature, °C	$K \times 10^3$, ml/g	a	Number of samples Whole polymers	Number of samples Fractions	Mol wt range $\times 10^{-4}$	Method	Reference
benzyl alcohol	155.4	156	0.50		9	4–35	LS	194
chlorobenzene	15	15.7	0.52		8	2–13	AU	193
	30	69.4	0.61		8	2–13	AU	193
cyclohexanone	20	11.6	0.85		6	3–10	OS	195
		13.7	1.0	5	7	7–13	OS	165
		112.5	0.63	3	5	9–15	OS	165
		8.3	0.84		several	1–12	OS	196
	25	15.6	0.76		several		OS	197
		19.5	0.79		several		OS	198
		24	0.77		13	3–14	OS	199
		204	0.56		32	2–15	OS	200
		208	0.56		6	6–22	OS	170
		1.1	1.0	11		2–14	LS	202
		174	0.55		6	15–52	LS	170
		0.84	0.87		several	36–110	LS	189
		6.0	0.84		several	0.8–20	LS	205
		13.8	0.78		several	1–30	LS	208
		1.13	1.0	1		10	SD	201
	27	1.1	1.0	4		3–10	OS	203
	30	21.6	0.75	11				
		0.54	1.08	9				
		2.15	0.96	9		3–20	OS	166
		30.0	0.73	9				
		1.78	1.0	10				
		14.1	0.82		8	2–13	AU	193
tetrahydrofuran	20	163	0.92		46	2–17	OS	164
	25	53.2	0.67		12	6–38	OS	204
		49.8	0.69		5	4–18	LS	206
		16.3	0.77		7	2–17	LS	169
		15.0	0.77		several	1–30	LS	208
	30	21.9	0.54		16	5–30	LS	168
		63.8	0.65		several	3–32	LS	207
		12.7	0.82		8	2–13	AU	193

Measurements of $[\eta]$ allow the determination of the conformational factor, σ, which is defined as

$$\sigma = [\bar{L}^2]_0^{1/2} / [\bar{L}^2]_{0,f}^{1/2} \tag{33}$$

where $[\bar{L}^2]_0$ represents the unperturbed mean-square displacement length and $[\bar{L}^2]_{0,f}$ the mean-square displacement length for the case of free internal rotation around a fixed valence angle, θ. The latter quantity can be calculated from the following relation:

$$[\bar{L}^2]_{0,f} = nl^2(1 + \cos \theta)(1 - \cos \theta)^{-1} \tag{33a}$$

where n and l are the number and length of the carbon—carbon bonds. For poly(vinyl chloride) it is $l = 1.54$ Å and $\cos \theta = 0.333$. Representative values for σ are: 2.04 in tetrahydrofuran at 30°C; 2.05 in cyclohexanone at 30°C; 2.17 in chlorobenzene at 30°C; 2.26 at 15°C (193); and 2.09 in benzyl alcohol at 155.4°C (194), averaging to approx 2.1. This relatively large value is mainly due to the dipole interactions of the

C—Cl bonds (193) and to a lesser extent to the bulkiness of the Cl substituent. An increase of σ with decreasing polymerization temperature of the sample is claimed in the literature (168).

For the viscosity of concentrated solutions or poly(vinyl chloride) melts the following relation is valid:

$$\eta = K' \times M_w^{3.4} \tag{34}$$

provided that the critical condition for the formation of a transient network structure through chain entanglements is fulfilled (215). For poly(vinyl chloride) this condition is (211,212):

$$\phi_2 M_w = 4000 \tag{35}$$

Here, ϕ_2 represents the volume fraction of polymer. For poly(vinyl chloride) melts (containing 2 phr Thermolite (MnT Chemicals, Inc., American Can Co.) T-31 stabilizer) the viscosity can be correlated experimentally with $[\eta]$ in cyclohexanone at 30°C, as follows:

$$\log \eta = 4.0 \times \log [\eta] + C(T) \tag{36}$$

where $C(T)$ is a function of melt temperature (216).

Thermal Properties. The glass-transition temperature, T_g, of poly(vinyl chloride) depends upon the polymerization temperature of the resin. Table 6 presents T_g values measured by dilatometry (140), differential-scanning calorimetry (146), and dynamic mechanical methods (217,218) on samples polymerized at the indicated temperatures. The data follow an approximately linear decrease with increasing polymerization temperature, t_{pol}, showing the effects of molecular weight and the amount of crystalline material present:

$$T_g = 93.4 - 0.20 \times t_{pol} \tag{37}$$

Table 6. Glass-Transition, T_g, and Melting-Point, T_m, Temperatures of Poly(vinyl chloride)

Polymerization temperature, °C	T_g, °C	T_m, °C	Reference
125	68	155	217
90	75		217
60	85		140
50	82		146,218
40	80	220	217
0	95		146
−10	90	265	217
−15	105	285	140
−20	95		218
−30	100		146
−40	105		146,218
−50	106		146
−60	110		218
−75		310	140
−80	100	>300	217

The glass-transition temperature of pure commercial poly(vinyl chloride) resins is around 81°C (219,220). In the case of copolymers, the glass-transition temperature can be approximated by a weighted average of the transition temperatures $T_{g,1}$ and $T_{g,2}$ of the homopolymer components (221,222):

$$\frac{1}{T_g} = \frac{w_1}{T_{g,1}} + \frac{w_2}{T_{g,2}} \tag{38}$$

where w_1 and w_2 represent the weight fractions. Examples for measured values are: 90°C for a copolymer of 60% vinyl chloride and 40% acrylonitrile (223); 63 and 59°C for copolymers with 5 and 10% vinyl acetate, respectively (224); and 60 and 38°C for copolymers with 10 and 20% ethylene, respectively (225).

The melting-point temperature, T_m, of poly(vinyl chloride) cannot be measured directly because of the thermal instability of the resin (see below). It is usually determined from the melting temperature T_m^*, of solutions of the polymer in plasticizers, according to the relation (226):

$$\frac{1}{T_m} = \frac{1}{T_m^*} - \left(\frac{R}{\Delta H_u}\right)\left(\frac{V_u}{V_1}\right)(v_1 - \chi_1 v_1^2) \tag{39}$$

Here, R is the gas constant, ΔH_u the heat of fusion per mole of repeating units of the polymer, V_u and V_1 are the molar volumes of repeating unit and diluent, v_1 is the volume fraction and χ_1 the interaction parameter (see Table 2) of the diluent. A value for ΔH_u of 785 cal/mole has been reported (140). Table 6 shows that the melting-point temperatures thus obtained again increase linearly with decreasing polymerization temperature, t_{pol}, of the resin, according to the approximate relation

$$T_m = 257.5 - 0.81 \times t_{pol} \tag{40}$$

The linear coefficient of expansion, α, of poly(vinyl chloride) is 6–8 $\times$ 10^{-5} (°C^{-1}) below T_g and 20–22 $\times$ 10^{-5} (°C^{-1}) above T_g. The specific heat values at constant pressure, c_p, in these two ranges are 0.25–0.30 and 0.42 cal/(g)(°C), respectively. Thermal conductivity at 20°C is 3.8 $\times$ 10^{-4} cal/(cm^2)(sec)(°C) (227). For detailed thermodynamic functions, see reference 228.

At temperatures above 100°C poly(vinyl chloride) begins to decompose at a noticeable rate. This process is discussed in detail in several monographs and review articles (229–234). Upon heating poly(vinyl chloride) hydrogen chloride is evolved and the resin becomes discolored, brittle, and finally insoluble. The rate of decomposition depends among other variables, upon the surrounding atmosphere, the temperature, and the molecular weight of the polymer. In oxygen, the rate is higher than in nitrogen, but the resin shows less discoloration at an equal extent of reaction. In an inert atmosphere, the molecular weight of the polymer increases from the beginning of the reaction, while in oxygen it goes through a minimum. At constant oxygen pressure, the rate is proportional to the square root of the partial pressure of the gas (235). The activation enthalpy in this case is approx 24,000 cal/mole, while in a nitrogen atmosphere it is as high as 33,000 cal/mole (235,236). The effect of temperature upon the evolution of HCl and the limiting-viscosity number, starting with a resin of $[\eta] = 110$, after 4 hr exposure at the indicated temperature, is shown in Table 7 (237).

The rate of HCl evolution at a given temperature increases with decreasing number-average molecular weight of the starting resin, according to the following relation (236):

$$\text{micromoles HCl/(g)(hr)} = a + b/M_n \tag{41}$$

Table 7. Effect of Temperature upon Decomposition of Poly(vinyl chloride)

Temperature, °C	Moles HCl/mole resin	$[\eta]$, ml/g
130	0.10	110
150	0.53	110
170	4.77	120
190	27.5	160

In nitrogen, $a = 0$, $b = 7 \times 10^5$; in air, $a = 6$, $b = 6 \times 10^5$; in oxygen, $a = 18$, $b = 3 \times 10^5$. All values are measured at 182°C.

The mechanism of the thermal degradation of poly(vinyl chloride) is not fully understood. It decomposes more readily than any of its model compounds. This is demonstrated in Table 8, where the gas-phase decomposition temperatures, T_D, of a series of models are summarized. T_D is defined as the temperature of initiation of decomposition, as measured by ir spectrometry (238).

Table 8. Decomposition Temperatures of Model Compounds for Poly(vinyl chloride)

Compound	Structure	T_D, °C
3-chloropentene-2	$CH_3CH{=}CClCH_2CH_3$	400
2,4-dichloropentane	$CH_3CHClCHCHClCH_3$	360
2-chloropropane	$CH_3CHClCH_3$	340
4-chloro-1-hexene	$CH_2{=}CHCH_2CHClCH_2CH_3$	325
3-chloro-1-pentene	$CH_2{=}CHCHClCH_2CH_3$	280
2-methyl-2-chloropropane	$CH_3CCl(CH_3)CH_3$	240
4-chlorohexene-2	$CH_3CH{=}CHCHClCH_2CH_3$	150

At 150°C, poly(vinyl chloride) and 4-chloro-2-hexene have the same initial decomposition-rate constant (229). These data indicate that allylic and tertiary chlorines may be the temperature-sensitive groupings in poly(vinyl chloride). They can be formed in the polymer by disproportionation or chain transfer to monomer (cf eq. 22) and by branching reactions (cf eqs. 25 and 26). Once dehydrohalogenation of an allyl chloride end group has started, new allylic groupings are formed and the reaction can proceed in a "zipper" fashion, leading to polyene structures, which are largely responsible for the discoloration (237):

$$—CH_2CHClCH_2CHClCH_2CHClCH{=}CH—$$

$$\xrightarrow{-\,HCl} —CH_2CHClCH_2CHClCH{=}CHCH{=}CH—$$

$$\xrightarrow{-\,HCl} —CH_2CHClCH{=}CHCH{=}CHCH{=}CH— \qquad (42)$$

Network formation can occur by elimination of HCl between chains. In the presence of oxygen, additional reactions take place, such as formation of peroxides and β-chloroketones, as well as chain scission and network formation via radical mechanisms. Besides the latter, unimolecular elimination and ionic mechanisms are claimed in the decomposition reactions of poly(vinyl chloride). The HCl formed during decomposition has an accelerating effect on further degradation (239). Chlorides of certain metals, such as iron, barium, and zinc, are used as catalysts for the dehydrohalogenation.

The frequency distribution of the polyene sequences can be determined by uv spectroscopy. At low degrees of dehydrohalogenation and a reaction temperature of 180°C, average sequence lengths of 5–10 conjugated double bonds are found. The number of longer sequences decreases continually, with maximum lengths of 25–30 (236,240). With increasing temperature and time of degradation, the frequency distribution shifts toward lower sequences (241). At high degrees of conversion, aromatic decomposition products, such as benzene and toluene, are formed (242).

The thermal degradation of poly(vinyl chloride) can be delayed or slowed down by the addition of heat stabilizers. Four principal groups of stabilizers can be distinguished:

1. Metal salts and soaps. Examples are lead hydrogen phosphate, $PbHPO_4$, lead stearate, and barium, cadmium, and zinc laurates and stearates. Part of the stabilizing activity of these materials is based on their ability to bind HCl:

$$(RCOO)_x M + x HCl \rightarrow MCl_x + x RCOOH \qquad (43)$$

The resulting metal chlorides are usually complexed by added chelating agents, such as alkyl and aryl phosphites. The stabilizer also acts by replacing —Cl by —OCOR (243); the resulting ester group is more stable than the original resin and thus blocks dehydrochlorination.

2. Polyols and nitrogen compounds. Examples are pentaerythritol, sorbitol, benzoguanamine, and melamine. The amino bases have the disadvantage of being catalysts for dehydrochlorination. They act, however, as antioxidants and radical inhibitors (244).

3. Metal organic compounds. Examples are dibutyltin maleate, dibutyltin lauryl mercaptide, and di(n-octyl)tin S,S-bis(isooctylmercaptoacetate). These materials can also react with HCl and with active chlorine atoms in the polymer chain.

4. A fourth group of compounds are called secondary stabilizers because they are effective only in combination with a material in the above three groups of primary stabilizers. This effect is called synergism. It is also found with certain combinations of primary stabilizers, which perform better than similar or even larger amounts of the individual components. Examples of secondary stabilizers are epoxidized oils and various phosphites, such as diphenyl decyl phosphite.

In practice, the thermal stability of polymer compounds is measured by an oven test (see p. 401), with a Brabender Plastograph (245), or by thermogravimetric analysis (246). Comparing various products at similar molecular weights and identical formulations, it is found that copolymers with vinyl acetate and with vinyl cetyl ether are thermally less stable than homopolymers, while ethylene copolymers have approximately the same stability as homopolymers, and propylene copolymers are considerably more stable (246,247).

Mechanical Properties. The mechanical properties of poly(vinyl chloride) depend strongly upon the formulation of the compound and the measurement temperature. If the compound contains only approx 3 phr of stabilizers and lubricant (and perhaps a filler), it is called "rigid vinyl." Typical properties in this case, measured at 23°C (Commercial Standard CS 201-55 (see p. 395)), are given under type 1 in Table 9. Type 2 refers to rigid, high-impact compounds, containing approximately an additional 20 phr of an impact modifier.

Poly(vinyl chloride) can be given a soft, rubbery appearance at room temperature ("flexible vinyl") by the addition of plasticizers. These are defined as additives in-

Table 9. Physical Properties of Rigid Vinyls

Property	Type 1	Type 2
tensile strength, psi	7,000	5,000
flexural strength, psi	11,000	8,500
flexural modulus of elasticity, psi	400,000	300,000
Izod impact strength, ft. lb/in.	0.5	3.0
Rockwell hardness, R-scale	110	100

corporated in a plastic to increase the flexibility, workability, and distensibility. A plasticizer (or softener) may reduce the melt viscosity and lower the glass-transition temperature or the elastic modulus of the compound (248). Primary plasticizers are good solvents for the polymer (cf Table 2) and do not exude excessively from a compound after frequent flexing or on long standing. Secondary plasticizers are nonsolvents or, at best, partial solvents for the polymer and must always be blended with a primary softener to give satisfactory performance (249). Suitable plasticizers should have the following additional properties: low volatility and cost, good electrical and insulation properties and resistance to degradation, chemical inertness, nonflammability and nontoxicity, and a small temperature coefficient of viscosity.

Primary plasticizers reduce the effectiveness of the intermolecular (van der Waals) forces in the polymer. The molecules of poly(vinyl chloride) are polar, and the softener must also be polar in order to interact. For this reason most commercial primary plasticizers are high-boiling esters. The addition of smaller amounts of plasticizer (less than 20 phr) leads to surprising results: the impact strength of the compound is lower than that of the original polymer and the tensile strength goes through a maximum. This behavior is explained by an increase in crystallinity of the composition. Completely plasticized poly(vinyl chloride) still contains a three-dimensional gel structure, the crosslinks of which are formed by crystallites. The mechanical properties of a given plasticized compound can be markedly changed by recrystallization under suitable conditions of temperature and loading (251). Most secondary plasticizers are nonpolar and lower the intermolecular forces by their physical bulk. Table 10 compares the effectiveness of various products (252). The properties are measured in cyclohexanone at 25°C on a homopolymer of intrinsic viscosity 1.15. The compounds consist of 3 phr basic lead carbonate stabilizer and sufficient plasticizer to obtain a modulus of elasticity of 1800 psi at 25°C. Three classes of plasticizers are used: phthalates, $C_6H_4(COOR)(COOR')$, sebacates, $ROOCC_8H_{16}COOR'$, and phosphates, $OP(OR)(OR')(OR'')$.

In Table 10 the viscosity of the softener, η, is given in cSt at 30°C; the flex temperature, T_f, is defined as the temperature at which the modulus of elasticity of the compound is 135,000 psi (253); volatility is the percentage of plasticizer lost from a sheet of $3 \times 1 \times 0.02$ in. size at a temperature of 105°C during 24 hr under an air flow of 200 ft/min; extractability is the percentage of plasticizer extracted from a similar sheet by immersion in water at 25°C for 24 hr.

The data in Table 10 allow several generalizations (252). Phthalates, as a group, are more effective than phosphates or sebacates. The greatest efficiency is found with materials whose functional groups represent the largest portion of the total molecule. Aliphatic groups are more effective than aromatic. Low-temperature flexibility is best obtained by long-chain aliphatic molecules. Volatility correlates reasonably well with

Table 10. Effectiveness of Various Plasticizers

Plasticizer	η, cSt	phr	T_f, °C	Volatility, %	Extrability, %
phthalates					
diethyl	8.7	41	−5.5	55.2	0.76
di-*n*-butyl	13.9	41	−12	36.0	0.45
di-*n*-propyl	13.4	42	−5.5	48.0	0.56
di-*n*-amyl	18.1	42	−13	27.0	0.04
diallyl	8.2	42	−13.5	54.3	2.21
butyl benzyl	32.0	45	−2	14.4	0.00
dimethyl	10.0	46	−5.5	71.3	0.89
butyl hexyl	21.3	46	−14	17.1	0.24
di-(2-ethylbutyl)	33.6	48	−12.5	10.0	0.15
di-*n*-octyl	27.2	49	−29	1.9	0.02
di-*n*-hexyl	33.3	50	−14	11.9	0.15
di-(2-ethylhexyl)	48.4	50	−23	4.1	0.02
dicapryl	44.5	54	−24.5	4.6	0.04
dibenzyl		60	+2.5	4.5	0.05
dilauryl		75	−33	2.5	0.34
dicyclohexyl		88	+7	4.6	0.04
sebacates					
dimethyl	4.9	35	−17.5	74.9	2.73
dibutyl	7.8	37	−34.5	36.4	0.33
dibenzyl		42	−13	9.9	0.04
di-(2-butoxyethyl)		44	−34	9.4	0.26
dihexyl	13.0	54	−47	18.4	0.13
di-(2-ethylhexyl)	17.7	54	−57.5	1.0	0.00
phosphates					
tributyl		38	−25.5	61.5	2.25
tri(2-butoxyethyl)	8.9	44	−31	21.4	2.82
triphenyl		52	−2.5	5.8	0.19
tri(2-ethylhexyl)	11.1	56	−58	5.1	0.02
tricresyl	68	57	+1.5	1.5	0.04
dibutyl lauramide	16.8	43	−38.5	22.1	0.10
dicapryl diglycollate		46	−19	15.6	0.18
dioctyl maleate	13.8	47	−39	37.5	0.24
dioctyl thiodiglycollate	12.6	48	−35.5	4.7	0.20
ethylene glycol dipelargonate	8.9	50	−56		0.25

molecular weight for each series. Extractability by water generally depends upon the water solubility of the respective plasticizer.

A large number of polymers are suggested as plasticizers for poly(vinyl chloride). They have the advantage of negligible migration, but are much less efficient than lmw softeners. In order to facilitate incorporation of the materials into the compound (polyblending), the degree of polymerization of the polymeric plasticizer is kept low. Examples are polyesters produced from adipic, azelaic, or sebacic acids with propylene or butylene glycols and copolymers of butadiene and acrylonitrile (15).

Poly(vinyl chloride) can also be plasticized internally by copolymerization with suitable monomers. These comonomers increase the mobility of the copolymer chain, similar to lmw softeners. It is necessary, however, that the respective homopolymer has a lower glass-transition temperature than poly(vinyl chloride) (cf eq. 38). Examples are vinyl acetate (120), ethylene (121), esters of citraconic, itaconic or mesaconic acid (254), acrylic esters of higher alcohols (255), and vinyl esters of higher fatty acids (256).

Other Physical Properties. The electrical properties of poly(vinyl chloride) products depend to a large extent on the formulation. Typical values for a rigid product are the following: the dielectric constant at 60 Hz and 30°C is 3.54, at 1000 Hz and 30°C it is 3.41; power factors at these two conditions are 3.51 and 2.51%, respectively; dielectric strength = 1080 V/mil; specific resistivity = 10^{15} Ω-cm. The addition of up to 15% of a plasticizer, such as tricresyl phosphate (TCP) does not affect the dielectric constant or volume resistivity significantly. Larger quantities of softener increase the former (approx 10 at 60% TCP) and decrease the latter (approx 10^9 at 60% TCP). Temperature of measurement and thermal history of the sample also affect the results.

Permeability of poly(vinyl chloride) films for gases and vapors are again strongly dependent upon formulation. Typical values for rigid vinyls are the following: the water-vapor transmission from 100 to 50% relative humidity at 23°C is 0.03 g/(24 hr) (100 in.²) (torr/mil); permeability for oxygen = 6.2 and for carbon dioxide = 28.5 cc/(24 hr) (100 in.²)(torr/mil). Gas-transmission rates increase strongly with plasticizer content and with temperature.

Chemical Properties. The most important chemical reaction of poly(vinyl chloride) is its postchlorination. The process can be carried out in an organic medium, such as carbon tetrachloride, at moderate temperatures under the influence of uv irradiation. With increasing chlorine content, the product becomes soluble and is later recovered by precipitation with methanol (257). The reaction can also take place in aqueous suspension with the addition of a swelling agent, such as chloroform or carbon tetrachloride, catalyzed by ultraviolet irradiation (258) or an oil-soluble acyl peroxysulfonate (259).

Other patent procedures describe the postchlorination of poly(vinyl chloride) in the dry state (260) or after addition of smaller amounts of chloroform (261). The reaction can readily be carried to a chlorine content of approx 73 wt %, corresponding to the introduction of an additional chlorine atom per monomer unit. Higher halogen contents can only be obtained with difficulty; by carrying out the reaction in thionyl chloride, $SOCl_2$, up to 75.6 wt % chlorine is reported (262). The chlorine content can be calculated from the measured density, d, in g/ml, of a press-polished sheet containing 2 wt % organotin stabilizer (263), as follows:

$$\text{wt \% Cl} = (65.12 \times d) - 33.37 \tag{44}$$

The glass-transition temperature of a given starting resin increases with increasing degree of chlorination; it follows the approximate relation (in °C), as shown (264):

$$T_g = (3.8 \times \text{wt \% Cl}) - 136 \tag{45}$$

The thermal stability of highly chlorinated poly(vinyl chloride) resins, as measured by weight loss upon heating, is considerably greater than that of the original polymer (265). The mechanical strength of the chlorinated products is slightly higher, the impact strength lower, than that of the starting resin (266).

The structure of postchlorinated poly(vinyl chloride) can be elucidated by chemcal means (267), infrared spectroscopy (268), NMR (269), and pyrolysis–gas chromatography (270).   Chlorine can be incorporated into the polymer chains in the following two ways:

$$\text{\textasciitilde\textasciitilde\textasciitilde}-CH_2CHCl-\text{\textasciitilde\textasciitilde\textasciitilde} \; + \; \tfrac{1}{2}Cl_2 \left[\begin{array}{l} \longrightarrow \text{\textasciitilde\textasciitilde\textasciitilde}-CHClCHCl-\text{\textasciitilde\textasciitilde\textasciitilde} \quad (46) \\[2ex] \longrightarrow \text{\textasciitilde\textasciitilde\textasciitilde}-CH_2CCl_2-\text{\textasciitilde\textasciitilde\textasciitilde} \quad (47) \end{array}\right.$$

This leads either to a 1,2-dichloroethylenic (eq. 46) or a 1,1-dichloroethylenic unit (eq. 47).   It is shown that the measured ratio of 1,2- to 1,1-units formed is higher than the ratio of 2:1 statistically expected from the availability of hydrogen in the methylene and methyne groups.   The experimental ratio is only slightly higher than 2:1 at low degrees of chlorination and reaches 5:1 at high chlorine contents.   Smaller amounts of trisubstituted 1,1,2-units can be found at chlorine levels over 67 wt %.   These ratios are identical for all methods of postchlorination.   However, some physical properties vary with the method used: resins postchlorinated in solution have greater solubility and lower glass-transition temperatures than resins treated in suspension in the presence of swelling agents.   The difference is thought to be due to a different distribution of the halogenated units (271).

Hydrogenation and dehydrochlorination of poly(vinyl chloride) are discussed earlier (see pp. 379 and 389).   Treatment with silver nitrate in glacial acetic acid at 65°C leads to the formation of poly(vinyl acetate) (272).

$$-Cl + AgNO_3 + CH_3COOH \rightarrow -OCOCH_3 + AgCl + HNO_3 \qquad (48)$$

The chlorine atoms can also be exchanged for sulfonamide groups by reacting the resin in aqueous suspension with ammonium sulfite and heating the isolated product in the presence of a dehydration catalyst (273).

$$-Cl + (NH_4)_2SO_3 \rightarrow -SO_2NH_2 + NH_4Cl + H_2O \qquad (49)$$

Reaction of poly(vinyl chloride) with aldehydes (274) and high-boiling petroleum and coal-tar fractions (275,276) are described in the literature.   Grignard-type compounds can be obtained by treatment with magnesium in cyclic ethers.   Subsequent reaction with acetone and esters leads to the formation of tertiary hydroxyl groups (277), while dehydrochlorination followed by reaction with organic peroxyacids introduces epoxy groups (278).

Various methods are used to crosslink poly(vinyl chloride).   Chlorosulfonation of ethylene copolymers by the action of sulfur dioxide and chlorine (279) or of homopolymer with thiourea and subsequent chlorination (280) gives products that can be vulcanized.   Patent procedures for direct crosslinking of poly(vinyl chloride) suggest various polyamines, such as the condensation product of butyraldehyde and aniline (281), hexamethylenetetramine (282), ethylene diamine (283,284), or poly(2,2,4-trimethyldihydroquinoline) (283).   Liquid dithiol compounds can be used to cure poly(vinyl chloride) (285).

Economic Aspects

U.S. production of poly(vinyl chloride) and copolymers for the years 1948–1968 are shown in Figure 3.   The price per lb of general-purpose vinyl resin is given in cents normalized to the purchasing power of the dollar in 1968.   Statistics for selected European countries and Japan are shown in Figure 4.   Estimated world production for 1969

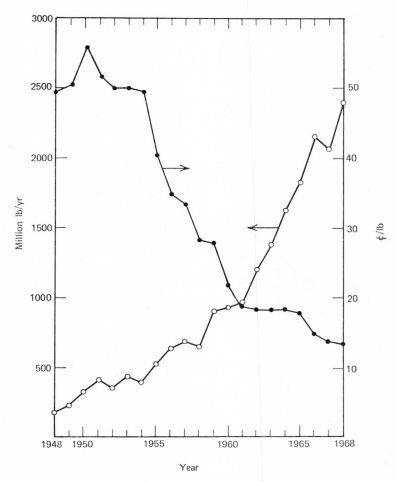

Fig. 3. Total U.S. production of poly(vinyl chloride) and cost of general-purpose resin (calculated to equivalent 1968 dollars).

is 10 billion lb (8). The expected annual growth rate during the next five years is 8–12% (9,10).

Table 11 shows the U.S. consumption for 1966–1968 of poly(vinyl chloride) and copolymers for the various uses in million lb (286,287). Table 12 summarizes the U.S. consumption for 1966–1968 of the principal vinyl compounding additives in million lb (288). Table 13 presents the estimated capacities in million lb for 1968 of the major producers of poly(vinyl chloride) and copolymers (8,288). These figures are, in most cases, considerably higher than the actual production.

Specifications and Standards; Analytical and Test Methods

The Society of the Plastics Industry, with the cooperation of the U.S. Department of Commerce, has developed a series of commercial standards, such as CS 192-53 for general-purpose vinyl plastic film, CS 201-55 for rigid poly(vinyl chloride) sheet, CS 207-57 for dimensions and tolerances for rigid poly(vinyl chloride) pipe, and CS 209-57

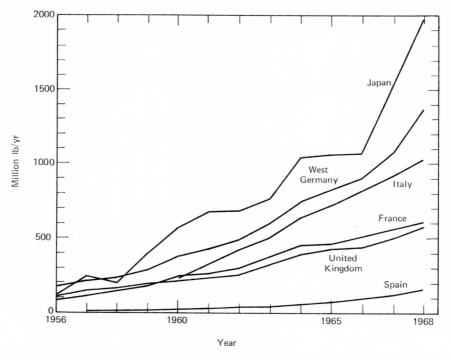

Fig. 4. Poly(vinyl chloride) production of selected countries.

Table 11. U.S. Consumption of Poly(vinyl chloride) and Copolymers, Million lb

Product	1966	1967	1968
film and sheeting			
calendered	438	380	432
extruded	66	103	135
floor coverings			
calendered	300	250	267
coated	70	52	72
cable and wire	226	196	259
other extruded products	366	297	353
paper and textile coatings	117	101	125
phonograph records	114	104	107
plastisols for molding compounds	44	78	103
adhesives and protective coatings	55	72	94
injection and blow molding	64	72	94
other uses	197	302	278
exports	69	71	90
total	2126	2078	2407

Table 12. U.S. Consumption of Compounding Additives, Million lb

	1966	1967	1968
plasticizers	835.0	815.0	900.0
heat stabilizers	58.0	59.0	65.0
light stabilizers	0.3	0.3	0.3
colorants	17.0	21.0	25.0

Table 13. Estimated 1968 Capacities of Major Poly(vinyl chloride) Producers

Country and producer	million lb
Africa	
Morocco	22
South Africa	
African Explosives and Chemical Ind., Ltd.	112
Americas	
Argentina	
Electroclor S.A.I.C.	20
Brazil	
Ind. Quím. Electro-Cloro S.A.	
Canada	70
B. F. Goodrich Canada Ltd.	
Monsanto Canada Ltd.	
Shawinigan Chemicals Ltd.	
Colombia	
Colcarburo	14
Petroquímica Colombiana S.A.	13
Mexico	
Geon de Mexico S.A.	
Monsanto Mexicana S.A.	
Plásticos Omega S.A.	11
Promociones Industriales Mexicanas S.A.	13
Nicaragua	
Nicavinyl	13
Peru	
Sociedad Paramonga Ltda.	17
Puerto Rico	
International Polymer	110
Puerto Rico Chemical Corp.	110
United States	
Airco Chemicals and Plastics Div., Air Reduction Co., Inc.	120
Allied Chemical Corp.	125
American Chemical, Inc.	70
Borden Chemical Co.	240
Diamond Shamrock Corp.	250
Escambia Chemical Corp.	50
Ethyl Corp.	125
Firestone Plastics Co.	210
General Tire & Rubber Co.	75
The B. F. Goodrich Chemical Co.	600
Goodyear Tire and Rubber Co.	90
Great American Plastics	70
Hooker Chemical Corp.	60
Keysor Chemical	50
Monsanto Co.	150
The Pantasote Co.	110
Stauffer Chemical Co.	80
Tenneco Corp.	190
Thompson Plastics	150
Union Carbide Corp.	350
Uniroyal, Inc.	135
Venezuela	
Venezuelan Petrochemical Institute	165
Asia	
Formosa	
China Plastics Corp.	27

(continued)

Table 13 (*continued*)

Country and producer	million lb.
India	
Calico Mills Ltd.	10
Chemicals and Plastics India Ltd.	44
Delhi Cloth and General Mills Co., Ltd.	27
Koyali Refinery	44
National Organic Chemical Industries Ltd.	44
Plastics, Resins and Chemicals Ltd.	31
Iran	
Abadan Petrochemical Co., Ltd.	44
Israel	
Electrochemical Industries Ltd.	17
Japan	
Denki Kagaku Kogyo KK.	145
Japanese Geon Co., Ltd.	190
Kanegafuchi Chemical Industry	175
Kureha Chemical Industry Co., Ltd.	190
Mitsubishi Monsanto Chemical Co.	110
Mitsui Chemical Industry Co., Ltd.	130
Nippon Carbide Industries Co., Inc.	200
Nissin Kagaku Kogyo K.K.	77
Shin-Etsu Chemical Industry Co., Ltd.	115
Shin Nippon Chisso Hiryo K.K.	53
Sumitomo Chemical Co., Ltd.	150
Toagosei Chemical Industry Co., Ltd.	88
Tokuyama Sekisui Kogyo K.K.	26
Korea	
Korean Chemical Co.	33
Philippines	
Mabuhay Vinyl Corp.	13
Australia	
B. F. Goodrich-CSR Chemical Ltd.	
ICI of Australia and New Zealand Ltd.	80
Europe	
Austria	
Halvic Kunststoffwerke G.m.b.H.	44
Belgium	
Badische Phillips Petroleum N.V.	110
Solvic S.A.	200
Bulgaria	66
Czechoslovakia	110
France	
Aquitaine Organico	55
Kuhlmann-Plastugil S.A.	110
Péchiney-St. Gobain	155
Rhône Poulenc S.A.	44
Solvic Sarl	220
Germany	
Badische Anilin- und Soda Fabrik A.G.	260
Chemische Werke Huels A.G.	400
Deutsche Solvay Werke G.m.b.H.	155
Dynamit Nobel A.G.	80
Farbwerke Hoechst A.G.	220
Lonza Werke G.m.b.H.	33
Wacker Chemie G.m.b.H.	350
Greece	
Esso Pappas Industrial Co., A.E.	36

Table 13 (*continued*)

Country and producer	million lb
Hungary	66
Italy	
Manifattura Ceramica Pozzi S.p.A.	110
Montecatini Edison S.p.A.	365
Montesud Petrochímica S.p.A.	125
Polymer Industrie Chimiche S.p.A.	175
Quirinia S.p.A.	66
Società Chimica Ravenna S.p.A.	240
Solvic S.p.A.	88
Netherlands	
Shell Nederland Chemie N.V.	155
Norway	
Norsk Hydro Electrisk	120
Poland	420
Portugal	
CIRES	22
Rumania	180
Spain	
Etino Química S.A.	44
Hispavic Industrial S.A.	66
Reposa	60
Sweden	
Fosfatbolaget	80
Switzerland	
Lonza Ltd.	44
Turkey	
Petkim Petrokimya A.S.	57
United Kingdom	
BXL Plastics Materials Group Ltd.	22
British Geon Ltd.	310
Imperial Chemical Industries Ltd.	400
Staveley Continental Ltd.	55
U.S.S.R.	260
Yugoslavia	33

for vinyl chloride plastic garden hose. Others include vinyl chloride plastic weather strip, vinyl waterstops, vinyl–metal laminates, and flexible foams made from polymers or copolymers of vinyl chloride. Commercial standard CS 201-55 is discussed in some detail in Table 9. The standards and procedures are available from the Superintendent of Documents, U.S. Government Printing Office, Washington, D.C.

The American Society for Testing and Materials, Philadelphia, also publishes specifications and detailed procedures for methods of testing and recommended practices; Table 14 lists the standards for poly(vinyl chloride) and copolymers.

Health and Safety Factors

The use of poly(vinyl chloride) and copolymer compounds in food packaging is regulated by the Food and Drug Administration (FDA) of the U.S. Department of Health, Education, and Welfare. Under the 1958 Food Additives Amendment, the basic resins and all compounding components, such as stabilizers, impact modifiers, lubricants, etc, are classified into one of several groups.

Table 14. ASTM Standards for Poly(vinyl chloride) and Copolymers

specifications	
flexible sheeting	D 1593-61
closed cell sponge	D 1667-64
resins	D 1755-66
rigid compounds	D 1784-65T
plastic pipe	D 1785-67a
rigid sheets	D 1927-67
rigid vinyl acetate copolymer compounds	D 2114-66
vinyl acetate copolymer sheets	D 2123-67
plastic pipe	D 2241-67
flexible molding and extrusion compounds	D 2287-66
threaded pipe fittings	D 2464-67
socket-type pipe fittings	D 2466-67
socket-type pipe fittings	D 2467-67
copolymer resins	D 2474-66T
bell-end pipe	D 2672-68
postchlorinated pipe	D 2846-69T
conditioning procedures	D 618-61
analytical tests	
water-vapor transmission	E 96-66
thickness of solid electrical insulation	D 374-57T
water absorption	D 570-63
coefficient of linear expansion	D 696-44
specific gravity	D 792-66
coefficient of cubical expansion	D 864-52
plasticizers	D 1045-58
specific viscosity	D 1243-66
total chlorine	D 1303-55
density by gradient technique	D 1505-67
particle size	D 1705-61
bulk density	D 1895-65T
infrared spectroscopy of components	D 2124-67
electrical tests	
dielectric strength	D 149-64
dielectric constant and dissipation factor	D 150-65T
electrical resistance	D 257-66
arc resistance	D 495-61
welding performance	D 1789-65
mechanical tests	
impact strength	D 256-56
deformation under load	D 621-64
tensile properties	D 638-68
fatigue	D 671-63T
creep or stress relaxation	D 674-56
compressive properties	D 695-63T
shear strength	D 732-46
mechanical tests	
flexural stiffness	D 747-63
impact strength at nonambient temperature	D 758-48
Rockwell hardness	D 785-65
flexural properties	D 790-66
tensile properties of thin sheets	D 882-67
bearing strength	D 953-54
warpage	D 1181-56
abrasion resistance	D 1242-56
bursting strength of round rigid tubing	D 1599-62T

Table 14 (*continued*)

optical tests	
luminous reflectance, transmittance, and color	E 308-66
haze and luminous transmittance	D 1003-61
permanence tests	
accelerated weathering	E 42-65
chemical resistance	D 543-67
simulated service conditions	D 756-56
permanent effect of heat	D 794-64T
artificial weathering	D 795-65T
volatile loss	D 1203-67
chemical resistance	D 1239-55
outdoor weathering	D 1435-65
environmental stress cracking	D 1693-66
acetone immersion of pipe	D 2152-67
methanol extractability	D 2222-66
thermal tests	
flammability of thin films	D 568-61
flow properties	D 569-59
flammability	D 635-63
heat-deflection temperature	D 648-56
brittleness temperature	D 746-64T
self-extinguishing materials	D 757-65
heat stability	D 793-49
flammability of flexible tubing	D 876-65
melt index	D 1238-65T
Vicat softening point	D 1525-65T
oven heat stability	D 2115-67

Those materials which were in food-contact use before 1958 and which had been accepted for such use by a government agency, such as the FDA or the Meat-Inspection Division of the Department of Agriculture, were classified as "prior sanctioned." These materials can continue to be used, subject to any limitations that might have applied to the prior sanction, unless this status is specifically changed by the FDA.

Another class of materials are those which had been in use for a considerable time before 1958 and which have been declared by experts to be safe for such use. These materials are frequently called "GRAS" (generally recognized as safe). This status also may be changed by the FDA for a reason.

A third class of materials are those which, although used as a component of a food packaging material or other food-contact surface, may not reasonably be expected to become a component of the food, such as a nonmigratory substance. These materials are classified as "nonadditives."

All other materials are subject to the Food-Additives Amendment of 1958, and FDA Regulations defining the material and conditions of safe use are required. These may limit the types of food which are contacted, the temperatures under which the food is held or processed, or the kinds of other materials for which a particular ingredient is being regulated. Thus, under FDA regulation 121.2597 acrylic impact modifiers and MBS impact modifiers may be used with homopolymer, vinyl chloride–propylene copolymers, vinyl chloride–lauryl vinyl ether copolymers, and vinyl chloride–ethylene copolymers; but the octyltin stabilizers (subject to limitations as to quantity and migration) may be used at present only with vinyl chloride homopolymer, and vinyl chloride –propylene copolymer.

The use of poly(vinyl chloride) and copolymers in liquor packaging is regulated by the Alcohol, Tobacco, and Firearms (ATF) Division of the Internal Revenue Service, U.S. Department of the Treasury.

Rigid vinyl is considered self-extinguishing. However, the addition of large amounts (over 25%) of certain plasticizers results in flammable compositions. Introduction of flame retardants, such as phosphate esters, chlorinated hydrocarbons, antimony oxide, or boranes reduces the fire hazard.

Processing and Uses

Poly(vinyl chloride) and its copolymers can be divided into six types: (*1*) general-purpose resins, which are hmw resins used in plasticized form for most applications; (*2*) dispersion resins, which are hmw resins of small particle size suitable for dispersion in plasticizer, and as plastisols and organosol; (*3*) water-dispersed resins, ie latexes; (*4*) solution-type resins, which are lmw resins or copolymers soluble in organic solvents; (*5*) polyblends; (*6*) rigid resins, which are copolymers or hmw resins designed for unplasticized applications.

General Purpose Resins. These are easily modified to meet various requirements. Plasticizers play a most important role. The amount used determines the degree of flexibility, and the type determines the physical and chemical properties. Stabilizers are used to prevent decomposition if the resin is subjected to excessive heat during processing or service life. They also help to prevent degradation by uv irradiation. Small amounts of lubricants are processing aids. They reduce internal friction and minimize sticking of the compound to metal parts during processing. Some are practically insoluble in the resin, but easily dispersible. Most lubricants have polar centers with short carbon chains attached. Examples are monoglycerin stearate, stearyl stearate, and ethylenediamine stearate (289). Fillers, such as clay or whiting, are used to reduce costs. Since the basic resins are transparent, color possibilities are unlimited. Both organic and inorganic color pigments are used. These ingredients are incorporated in the resin by mill, Banbury, or dry-powder mixing (see Mixing).

Extrusion and injection molding and calendering for sheeting, are commonly used fabricating methods.

The basic properties of poly(vinyl chloride), its resistance to weathering, chemical inertness, and electrical resistivity, and the ease of adapting its inherent properties to meet specific requirements, are responsible for its widespread use. Not only its electrical properties, but also its oxidation resistance, nonflammability, and ease of fabrication make it a leading material for wire and cable insulation. Tough cable jackets can be compounded to meet the exacting low-temperature requirements of military field communication wire, as well as the underwater conditions. Other applications are garden hose, gaskets, and welting for shoes, upholstery, and automobiles.

Film and sheeting account for a substantial share of the vinyl resin produced. The attraction of an infinite variety of colors and designs, ease of cleaning, and water resistance have led to uses such as shower curtains, bedspreads, draperies, garment bags, table covers, and place mats. Outdoor uses vary from wading pools to protective covers for outside storage. Poly(vinyl chloride) is adaptable to heat-sealing by high-frequency heating, and most items made from film, such as rainwear and inflatable toys, are fabricated by this method.

Dispersion Resins. *Plastisols* are dispersions of vinyl resins in plasticizer. If the liquid phase contains volatile, nonaqueous materials, the system is called an *organosol*. These dispersion systems depend on the fact that the resin is relatively insoluble in plasticizer at room temperature, yet soluble at elevated temperatures. To obtain optimum physical properties it is necessary to fuse the compound at a temperature of 300–375°F.

Organosol resins have a particle size ranging from 0.2 to 2.0 μ and are usually supplied as agglomerates of individual particles. Ball and pebble mills are effective in reducing these agglomerates, as well as providing a closed system to contain the volatile liquids. The composition of the liquid phase (dispersant/diluent ratio) is critical, and determines the viscosity and stability of the system. The dispersant (plasticizer and volatile solvents such as ketones or esters) tends to wet and swell the resin, while the diluent (aromatic or aliphatic hydrocarbons) controls the swelling and reduces viscosity and cost. All the usual compounding ingredients (fillers, stabilizers, color pigments) may be added with the resin if desired.

Plastisols are mixed in shearing-type rather than high-speed mixers, and at room temperature (preferably below 90°F). Since plasticizer is the sole dispersing medium, its choice determines the flow properties which are vital to success in any application. Stabilizers, fillers, and color pigments are usually predispersed in plasticizer before being added to the plastisol. Because of the relatively high viscosity of plastisols, air entrapped during mixing fails to escape and must be removed for most applications by subjecting the plastisol to a vacuum.

The bulk of vinyl dispersions are used in the coating field on fabric or paper. Coated fabrics comprise the largest portion of the coating field and find their principal use in upholstery. Other applications are in rainwear, table covers, tarpaulins, wall coverings, flooring, protective clothing, rug backing, and conveyor belting. Vinyl coatings on paper are used for table and shelf covers, novelties, furniture decorations, and floor and wall coverings. Coatings on synthetic-rubber-saturated papers are used as shoe innersoles, automobile door linings, and luggage.

Dip-coating is also an effective method of application. Electroplating racks, wire goods such as dish drainers, fabric gloves, automotive seat springs, lamp sockets, and strippable coatings on metal are typical of the products made. With plastisol systems, coatings up to 0.060 in. may be applied in a single dip. In multiple dipping, each coating is usually gelled separately, and final fusion is carried out after all coats have been applied. Heavier coatings may be applied from plastisols using a hot dipping process.

Gelling agents, such as a metallic soap or a high oil-absorption filler are used to prepare *plastigels*, which are plastisols having extremely high viscosities and yield values. Puttylike pastes may be prepared in this way and are applied by grease gun or trowel as sealants. Infrared lamps are used for fusion.

Slush and rotational molding are used to make hollow flexible articles, particularly toys, dolls, doll parts, and balls. A hollow mold is preheated and filled with plastisol. The portion adjacent to the wall is gelled by the heat of the mold. The remaining fluid plastisol is drained out and the mold sent through a fusion oven. After cooling in a water chamber, the article is removed by an air stream, and the operation repeated.

Vinyl sponge and foam are made from plastisols. Flotation equipment and insulation are made from closed-cell sponge which is produced with chemical blowing agents in closed molds under pressure. Open-cell-sponge may be made with a chemical blowing agent without pressure or by dissolving an inert gas in a plastisol under relatively high pressure. Foamed vinyl is used in many applications, such as mattresses, seating, automobile arm rests, and gasketing.

Water-Dispersed Resins. Vinyl latexes are water dispersions of poly(vinyl chloride) or copolymers of vinyl chloride. The individual particles are negatively charged and are about 0.2 μ in diameter. These colloidal dispersions contain about 50% solids and are offered as resin or preplasticized resin latexes, having a pH of 7–9. Latexes may be compounded with plasticizers, fillers, and pigments like other vinyl resins, although the ingredients must be added as alkaline water solutions, emulsions, or dispersions.

After the latex is applied to the paper or fabric, water must be evaporated at about 75–80°F for some latexes, while others must be fused at elevated temperatures of 250–300°F. Embossing, printing, and smoothing are typical of the additional processing steps which may be used.

Vinyl latexes are used as "bridge" coats or tie coats on loosely woven fabric which is subsequently to be coated with an organosol or plastisol. Adhesion is improved and excessive penetration prevented. These coatings are also used for upholstery, luggage, shade cloth, and table covers. The strength, abrasion resistance, and good aging properties of vinyl latexes make them suitable for felt coatings for automobile trunk linings, and as binders and back-sizes for carpets.

Decorative coatings on paper, such as shelf and wallpaper, take advantage of the ease of coloring, clarity, gloss, and light stability of many of these resins. Among other applications are heat-sealable coatings, prime coats on paper-surfaced wallboard, disposable paper sanitary products, rubber-release coatings on paper tapes, and interior and exterior coatings on fiber drums.

In impregnation and saturation operations, vinyl latexes are used as binders and flameproofing agents for nonwoven fabrics, and also for protective clothing and hospital sheeting. Drinking cups, shoe linings, and bookcovers may be made from paper saturated with these latexes.

Another important field of application is that of pigment binding for products such as printing inks, leather finishes, and emulsion paints.

Solution Resins. Coatings made from these resins are used to line beverage containers, closures, food-packaging paper, and beverage dispensers, and on kitchen furniture, medicine cabinets, refrigerator parts, etc. Their toughness and flexibility enable them to withstand drawing, stamping, and forming operations after being coated on flat metal sheets.

Strippable coatings are applied with these resins by dipping or spraying. The films deposited provide excellent protection during storage and shipment. Large, irregular shapes may be covered effectively with a "cocoonlike" coating. The equipment is first covered with a web of pressure-sensitive tape. A special vinyl webbing solution is then sprayed over the tape framework.

Polyblends. Liquid plasticizers, normally used with vinyl resins, have some drawbacks, such as volatility, extractability, and the tendency to migrate to other materials. The polymeric plasticizers (see p. 390) overcome these shortcomings in many cases. They are mixed in a finely divided state with the vinyl resin. The product, called a polyblend, is offered in sheet form. It is normally used as a thermoplastic material, although it can also be vulcanized.

Standard vinyl stabilizers, lubricants, and pigments are added on a cold or hot mill. Polyblends can be extruded, molded, or calendered. If better physical and extraction properties are desired, polyblends can be vulcanized with the usual rubber curing agents. These may be added on a cold mill.

Polyblends are used as film for food packaging, as shoe welt, and in electric cords which come in contact with styrene products, such as clock cases. The most important electrical application is in jackets for coaxial cables. Calendered products are used in the textile and flooring fields. Vulcanized polyblend is used in printing-roll covers, gaskets, and basketball covers.

Polyblends are also available in latex form. As such they are used as a grease-proof coating on paper, a fiber binder, and as a coating on felt for trunk linings.

Rigid Vinyls. These resins are supplied both as powders and as finished compounds in granular form. Those processed without the use of plasticizer are the most important and may be divided into two groups: (a) hmw resins used primarily for corrosion resistance, and (b) lmw resins used for phonograph records. It is necessary to add stabilizer, lubricant, and pigment for processing purposes. Both types may be extruded, calendered, or molded, using somewhat different techniques. Cold forming (drawing and roll forming) shows great potential for future applications (290). Recently developed copolymers (see p. 385) greatly facilitate all forms of processing.

Rigid vinyls are designed to take maximum advantage of the chemical inertness of poly(vinyl chloride). Pipe is made of poly(vinyl chloride) to carry salt water and sour crude oil. Although paraffin sticks and builds up on the inner walls of metal pipe, it does not do so in poly(vinyl chloride) pipe. Poly(vinyl chloride) is also used in chemical plants for pipes, fittings, valves, ducts, hoods, and filter frames. Parts may be joined by the hot gas-welding method, friction welding, or by cementing. They may be sawed, machined, etc, on standard metal or wood-working tools.

In addition to standard grades of hmw resins used in the unplasticized state, a high impact grade is offered. Although the chemical resistance and heat distortion of this type polymer are slightly poorer than the normal impact grade, the increase in impact strength (as measured on the Izod machine in ft lb/in. of notch) from 0.8 to 15.0 makes this resin preferable for most applications. Rigid sheets are used in tank linings, or may be vacuum-formed into a variety of articles, such as advertising and highway signs, aircraft-window frames, luggage, printing plates, chemical containers, refrigerator parts, and housings of all types.

In addition to phonograph records, the lmw copolymers are used in vinyl–asbestos floor tiling, although plasticizer is used with it as part of the binder.

Growth areas for the application of rigid poly(vinyl chloride) and especially some of the more easily processable copolymers are the building and packaging industries. In the former, postchlorinated resins find increasing use for hot-water pipe systems (291). In the latter, clear bottles—food grade and others—show excellent prospects (292,293).

Bibliography

"Vinyl chloride and poly(vinyl chloride)" in *ECT* 1st ed., Vol. 14, pp. 723–735, by C. H. Alexander and G. F. Cohan, The B. F. Goodrich Chemical Co.

1. V. Regnault, *Ann.* **14,** 22 (1835).
2. E. Baumann, *Ann.* **163,** 308 (1872).
3. Brit. Pat. 6,299 (March 13, 1912), I. I. Ostromuislenskii (to Society for Production and Sale of Resin Articles "Bogatyr").
4. Ger. Pat. 281,877 (April 4, 1913), F. Klatte (to Chem. Fabrik Griesheim-Elektron).
5. Austrian Pat. 70,348 (June 1, 1915), (to Chem. Fabrik Griesheim-Elektron).

6. U.S. Pat. 1,929,453 (Oct. 10, 1933), W. L. Semon (to The B. F. Goodrich Co.).
7. Ger. Pat. 659,042 (April 22, 1938), G. v. Susich and H. Fikentscher (to I. G. Farbenindustrie, A.G.).
8. Technical and Marketing Data Cards, *Rubber Plastics Age* **49** (1968).
9. G. P. Parker, *Chem. Eng. News* **47** (36), 64A (1969).
10. *Report on Plastics*, Organisation for Economic Co-operation and Development (OECD), Paris, 1969.
11. W. S. Penn, *PVC Technology*, MacLaren & Sons, Ltd., London, 1962.
12. M. Kaufman, ed., *Advances in PVC Compounding and Processing*, MacLaren & Sons, Ltd., London, 1962.
13. L. R. Whittington, ed., *A Guide to the Literature and Patents Concerning Polyvinyl Chloride Technology*, 2nd ed., The Society of Plastics Engineers, Inc., Stamford, Conn. 1963.
14. L. I. Nass, ed., *Chemistry and Technology of Poly(Vinyl Chloride)*, The Society of Plastics Engineers, Stamford, Conn., in print.
15. H. Kainer, *Polyvinylchlorid und Vinylchlorid-Mischpolymerisate*, Springer-Verlag, Berlin-Heidelberg-New York, 1965.
16. H. A. Sarvetnick, *Polyvinyl Chloride*, Van Nostrand Reinhold Co., New York, 1969.
16a. J. V. Koleski and L. H. Wartman, *Poly(vinyl chloride)*, Gordon & Breach Science Publishers, Inc., New York, 1969.
16b. M. Kaufman, *The History of PVC*, MacLaren & Sons, Ltd., London, 1969.
17. C. E. Schildknecht, *Vinyl and Related Polymers*, John Wiley & Sons, Inc., New York, 1952, Chap. 7.
18. J. T. Barr, Jr., "Vinyl and Vinylidene Chloride Polymers and Copolymers," in W. M. Smith, ed., *Manufacture of Plastics*, Vol. 1, Reinhold Publishing Corp., New York, 1964, Chap. 7.
19. G. Talamini and E. Peggion, "Polymerization of Vinyl Chloride and Vinylidene Chloride" in G. E. Ham, ed., *Vinyl Polymerization*, Vol. 1, Part 1, Marcel Dekker, Inc., New York, 1967, Chap. 5.
20. J. Prat, *Mém. Serv. Chim. État* (*Paris*) **32**, 319 (1945).
21. W. I. Bengough and R. G. W. Norrish, *Nature* **163**, 325 (1949).
22. J. W. Breitenbach and A. Schindler, *Monatsh. Chem.* **80**, 429 (1949).
23. E. Jenckel, H. Eckmans, and B. Rumbach, *Makromol. Chem.* **4**, 15 (1949).
24. W. I. Bengough and R. G. W. Norrish, *Proc. Roy. Soc.* (*London*) *Ser. A* **200**, 301 (1950).
25. E. J. Arlman and W. M. Wagner, *J. Polymer Sci.* **9**, 581 (1951).
26. J. W. Breitenbach, *Makromol. Chem.* **8**, 147 (1952).
27. A. Schindler and J. W. Breitenbach, *Ric. Sci. Suppl.* **25**, 34 (1955).
28. R. J. Abraham, D. W. Ovenall, and D. H. Whiffen, *Arch. Sci.* (*Geneva*) **10**, 84 (1957).
29. S. Enomoto, *J. Polymer Sci. A1* **7**, 1255 (1969).
30. H. S. Mickley, A. S. Michaels, and A. L. Moore, *J. Polymer Sci.* **60**, 121 (1962).
31. G. Talamini, *J. Polymer Sci. A2* **4**, 535 (1966).
32. A. Crosato-Arnaldi, P. Gasparini, and G. Talamini, *Makromol. Chem.* **117**, 140 (1968).
33. F. Danusso, *Ric. Sci. Suppl.* **25**, 46 (1955).
34. G. Talamini and G. Vidotto, *Makromol. Chem.* **50**, 129 (1961).
35. *Ibid.*, **53**, 21 (1962).
36. F. Danusso and D. Sianesi, *Chim. Ind.* (*Milan*) **37**, 278 (1955).
37. F. Danusso and G. Perugini, *Chim. Ind.* (*Milan*) **35**, 881 (1953).
38. G. Talamini and G. Vidotto, *Chim. Ind.* (*Milan*) **46**, 371 (1964).
39. T. Imoto, Y. Ogo, and Y. Hashimoto, *Intern. Chem. Eng.* **8** (3), 564 (1968).
39a. A. Crosato-Arnaldi, G. B. Guarise, and G. Talamini, *Polymer* **10** (5), 385 (1969).
40. F. Danusso, G. Pajaro, and D. Sianesi, *Chim. Ind.* (*Milan*) **41**, 1170 (1959).
41. D. Lim and M. Kolinsky, *J. Polymer Sci.* **53**, 173 (1961).
42. C. E. Schildknecht, "Polymerizations in Bulk," in C. E. Schildknecht, ed., *Polymer Processes*, Interscience Publishers, Inc., New York, 1956, Chap. 2.
43. J.-C. Thomas, *Hydrocarbon Process.* **47** (11), 192 (1968).
44. Fr. Pat. 1,357,736 (April 10, 1964), J.-C. Thomas (to Produits Chimiques Péchiney-Saint Gobain).
45. Fr. Pat. 1,382,072 (Dec. 18, 1964), J.-C. Thomas (to Produits Chimiques Péchiney-Saint Gobain).

46. D. N. Bort, Y. Y. Rylov, N. A. Okladnov, and V. A. Kargin, *Polymer Sci. (USSR) (English Transl.)* **9** (2), 334 (1967).
47. L. T. Camilleri, L. A. Fabrizio, C. L. Warren, and J. A. Wenzel, *Annual Technical Conference of the Society of Plastic Engineers, Detroit, Mich., May 1967*, p. 146.
48. H. Fikentscher and G. Hagen, *Angew. Chem.* **51**, 433 (1938).
49. W. D. Harkins, *J. Am. Chem. Soc.* **69**, 1428 (1947).
50. W. V. Smith and R. H. Ewart, *J. Chem. Phys.* **16**, 592 (1948).
51. J. G. Watterson, A. G. Parts, and D. E. Moore, *Makromol. Chem.* **116**, 1 (1968).
52. J. Ugelstad, P. C. Mork, P. Dahl, and P. Rangnes, *J. Polymer Sci. C* **27**, 49 (1969).
53. J. T. Lazor, *J. Appl. Polymer Sci.* **1**, 11 (1959).
54. E. Peggion, F. Testa, and G. Talamini, *Makromol. Chem.* **71**, 173 (1964).
55. K. Giskehaug, *Society of Chemical Industry Monograph No. 20*, London, 1966, p. 235.
56. H. Gerrens, *Dechema Monograph.* **49**, 53 (1963).
57. B. Jacobi, *Angew. Chem.* **64**, 539 (1952).
58. H. Cherdron, *Kunststoffe* **50** (10), 568 (1960).
59. F. A. Bovey, I. M. Kolthoff, A. I. Medalia, and E. J. Meehan, *Emulsion Polymerization*, Interscience Publishers, a div. of John Wiley & Sons, Inc., New York, 1965.
60. H. Hopff and I. Fakla, *Makromol. Chem.* **88**, 54 (1965).
61. A. Crosato-Arnaldi, G. Talamini, and G. Vidotto, *Makromol. Chem.* **111**, 123 (1968).
62. G. S. Park and D. G. Smith, *Trans. Faraday Soc.* **65**, 1854 (1969).
63. G. Vidotto, A. Crosato-Arnaldi, and G. Talamini, *Trans. Faraday Soc.* **114**, 217 (1968).
64. J. W. Breitenbach, O. F. Olaj, H. Reif, and A. Schindler, *Trans. Faraday Soc.* **122**, 51 (1969).
65. F. Danusso and F. Sabbioni, *Chem. Ind. (Milan)* **37**, 1032 (1955).
66. Brit. Pat. 366,897 (Aug. 6, 1929) (to Imperial Chemical Industries, Ltd.).
67. Ger. Pat. 579,048 (June 20, 1933), A. Voss and E. Dickhäuser (to I.G. Farbenindustrie, A.G.).
68. U.S. Pat. 2,064,565 (Dec. 15, 1936), E. W. Reid (to Union Carbide and Carbon Corp.).
69. Ger. Pat. 676,627 (June 8, 1939), E. Hanschke (to I.G. Farbenindustrie, A.G.).
70. Dutch Pat. 68,18281 (Dec. 19, 1968) (to Wacker-Chemie G.m.b.H.).
71. E. Farber and M. Koral, *Polymer Eng. Sci.* **8** (1), 11 (1968).
72. A. Konishi and K. Nambu, *J. Polymer Sci.* **54**, 209 (1961).
73. G. M. Burnett and W. W. Wright, *Proc. Roy. Soc. (London) Ser. A* **221**, 28 (1954).
74. J. Osugi and K. Hamanoe, *J. Chem. Soc. Japan, Pure Chem. Sect.* **90** (6), 544 (1969).
75. L. F. Albright, *Chem. Eng.* **74** (12), 145 (1967).
76. F. H. Winslow and W. Matreyek, *Ind. Eng. Chem.* **43**, 1108 (1951).
77. S. G. Bankoff and R. N. Shreve, *Ind. Eng. Chem.* **45**, 270 (1953).
78. R. S. Holdsworth, W. M. Smith, and J. T. Barr, *Mod. Plastics* **35** (10), 131 (1958).
79. U.S. Pat. 2,511,593 (June 13, 1950), W. J. Lightfoot (to United States Rubber Co.).
80. U.S. Pat. 2,625,539 (Jan. 13, 1953), V. L. Folt (to The B. F. Goodrich Co.).
81. U.S. Pat. 3,022,281 (Feb. 20, 1962), E. S. Smith (to Goodyear Tire and Rubber Co.).
82. U.S. Pat. 2,661,331 (Dec. 1, 1953), E. G. Howard (to E. I. du Pont de Nemours & Co., Inc.).
83. Brit. Pat. 855,213 (Nov. 30, 1960) (to Imperial Chemical Industries, Ltd.).
84. U. R. Wang, *J. Chinese Chem. Soc. (Taiwan)* **9**, 195 (1962).
85. H. Sobue and H. Kubota, *Makromol. Chem.* **90**, 276 (1966).
86. U.S. Pat. 1,775,882 (Sept. 16, 1930), C. O. Young and S. D. Douglas (to Carbide and Carbon Chemicals Corp.).
87. A. Guyot and P. Q. Tho, in reference 55, p. 69.
88. V. G. Gasan-Zade, V. V. Mazurek, and V. P. Sklizkova, *Vysokomolekul. Soedin.* **A10** (3), 479 (1968).
89. Belg. Pat. 545,968 (Oct. 10, 1956) (to Solvay and Cie.).
90. G. P. Budanova and V. V. Mazurek, *Vysokomolekul. Soedin.* **A9** (11), 2393 (1967).
91. N. Yamazaki and S. Kambara, *J. Polymer Sci. C* **22**, 75 (1968).
92. K. S. Minsker, Y. A. Sangalov, and G. A. Razuwayev, *J. Polymer Sci. C* **16**, 1489 (1967).
93. J. Ulbricht, J. Giesemann, and M. Gebauer, *Angew. Makromol. Chem.* **3**, 69 (1968).
94. Belg. Pat. 629,687 (Oct. 21, 1963) (to Sicedison S.p.A.).
95. J. Furukawa, T. Tsuruta, and S. Shiatani, *J. Polymer Sci.* **40**, 237 (1959).
96. H. Minato, H. Iwai, K. Hashimoto, and T. Yasui, *J. Polymer Sci. C* **23**, 761 (1968).
97. G. Borsini and C. Nicora, *J. Polymer Sci. A1* **6**, 21 (1968).

98. G. E. Ham, ed., *Copolymerization*, Interscience Publishers, a div. of John Wiley & Sons, Inc., New York, 1964.

98a. J. F. Gabett and W. M. Smith, *Copolymerizations Employing Vinyl Chloride or Vinylidene Chloride as Principal Components*, in reference 98, Chap. 10.

99. M. W. Kline and E. N. Skiest, "Polymerization and Manufacture of Vinyl Chloride Co-polymers," in L. I. Nass, ed., *Chemistry and Technology of Poly(Vinyl Chloride) and Related Compositions*, The Society of Plastics Engineers, Stamford, Conn., in press, Chap. 4.

100. H. Mark, B. Immergut, E. H. Immergut, L. J. Young, and K. I. Beynon, "Copolymerization Reactivity Ratios," in reference 98, Appendix A.

101. L. J. Young, "Copolymerization Reactivity Ratios," in J. Brandrup and E. H. Immergut, eds., *Polymer Handbook*, Interscience Publishers, a div. of John Wiley & Sons, Inc., New York, 1966, Chap. II, p. 332.

102. A. G. Kronmann, I. V. Pasmanik, B. I. Fedoseyev, and V. A. Kargin, *Vysokomolekul. Soedin.* **A9** (11), 2503 (1967).

102a. A. R. Cain, *Polymer Preprints* **11** (1), 312 (1970).

102b. *Methacrylonitrile*, Tech. Bull., Standard Oil Co. of Ohio, 1969.

103. N. Ashikari and A. Nishimura, *J. Polymer Sci.* **31**, 250 (1958).

104. G. V. Tkachenko, P. H. Khomikovskii, A. D. Abkin, and S. S. Medvedev, *Zhur. Fiz. Khim.* **31**, 242 (1957); *Chem. Abstr.* **51**, 15229b, 1957.

105. G. V. Tkachenko, L. V. Stupen, L. P. Kofman, and L. Z. Frolova, *Zhur. Fiz. Khim.* **31**, 2676 (1957).

106. F. M. Lewis and F. R. Mayo, *J. Am. Chem. Soc.* **70**, 1533 (1948).

107. S. Nagai, T. Uno, and K. Yoshida, *Chem. High Polymers (Japan)* **15**, 550 (1958).

108. C. E. Wilkes, J. C. Westfahl, and R. H. Backderf, *J. Polymer Sci. A1* **7**, 23 (1969).

109. F. M. Lewis, C. Walling, W. Cummings, E. R. Briggs, and W. J. Wenish, *J. Am. Chem. Soc.* **70**, 1527 (1948).

110. M. C. deWilde and G. Smets, *J. Polymer Sci.* **5**, 253 (1950).

111. C. S. Marvel and R. Schwen, *J. Am. Chem. Soc.* **79**, 6003 (1957).

112. P. Agron, T. Alfrey, Jr., J. Bohrer, H. Haas, and H. Wechsler, *J. Polymer Sci.* **3**, 157 (1948).

113. K. Matsuoka, K. Takemoto, and M. Imoto, *J. Chem. Soc. Japan. Ind. Chem. Sect.* **68**, 1941 (1965).

114. K. W. Doak, *J. Am. Chem. Soc.* **70**, 1525 (1948).

115. F. R. Mayo, C. Walling, F. M. Lewis, and W. F. Hulse, *J. Am. Chem. Soc.* **70**, 1523 (1948).

116. M. Imoto and S. Shimizu, *Chem. High Polymers (Japan)* **18**, 747 (1961).

117. F. R. Mayo, F. M. Lewis, and C. Walling, *J. Am. Chem. Soc.* **70**, 1529 (1948).

118. E. C. Chapin, G. E. Ham, and C. L. Mills, *J. Polymer Sci.* **4**, 597 (1949).

119. J. F. Bork and L. E. Coleman, *J. Polymer Sci.* **43**, 413 (1960).

120. C. M. Thomas and J. R. Hinds, *Brit. Plastics* **31**, 522 (1958).

121. U.S. Pat. 3,051,689 (Aug. 28, 1962), N. L. Zutty (to Union Carbide Corp.).

122. U.S. Pat. 3,468,840 (Sept. 23, 1969), C. A. Heiberger and L. Fishbein (to Air Reduction Co., Inc.).

123. U.S. Pat. 3,468,858 (Sept. 23, 1969), C. A. Heiberger and L. Fishbein (to Air Reduction Co., Inc.).

124. M. J. R. Cantow, C. W. Cline, C. A. Heiberger, D. Th. A. Huibers, and R. Phillips, *Modern Plastics* **46** (6), 126 (1969).

125. J. Osugi and K. Hamanoe, *J. Chem. Soc. Japan, Pure Chem. Sect.* **90** (6), 552 (1969).

126. Ger. Pat. 1,046,882 (Dec. 18, 1958), H. Fikentscher and W. Huebler (to Badische Anilin- und Soda Fabrik, A.G.).

127. U.S. Pat. 3,168,594 (Feb. 2, 1965), Y. Hoshi and M. Onozuka (to Kureha Kasei Co., Ltd.).

128. U.S. Pat. 2,983,675 (May 9, 1961), J. Gabilly and M. Jobard (to Produits Chimiques Péchiney-Saint Gobain).

129. U.S. Pat. 3,254,049 (May 31, 1966), A. J. Cole and R. Reichard (to Firestone Tire and Rubber Co.).

130. U.S. Pat. 2,947,719 (Aug. 2, 1960), F. M. Rugg and J. E. Potts (to Union Carbide Corp.).

131. U.S. Pat. 3,347,956 (Oct. 17, 1967), L. Rademacher (to Monsanto Co.).

132. Brit. Pat. 852,042 (Oct. 19, 1960) (to Montecatini Società Generale per l'Industria Mineraria i Chimica).

133. U.S. Pat. 3,358,054 (Dec. 12, 1967), D. Hardt and H. Bartl (to Farbenfabriken Bayer, A.G.).
134. P. J. Flory, *J. Am. Chem. Soc.* **61**, 1518 (1939).
135. C. S. Marvel, J. H. Sample, and M. F. Roy, *J. Am. Chem. Soc.* **61**, 3241 (1939).
136. F. E. Bailey, Jr., J. P. Henry, R. D. Lundberg, and J. M. Whelan, *J. Polymer Sci. B* **2**, 447 (1964).
137. B. Baum and L. H. Wartman, *J. Polymer Sci.* **28**, 537 (1958).
138. G. Bier and H. Kraemer, *Kunststoffe* **46**, 498 (1956).
139. J. D. Cotman, Jr., *Ann. N.Y. Acad. Sci.* **57**, 417 (1953).
140. A. Nakajima, H. Hamada, and S. Hayashi, *Makromol. Chem.* **95**, 40 (1966).
141. G. Boccato, A. Rigo, G. Talamini, and F. Zilio-Grandi, *Makromol. Chem.* **108**, 218 (1967).
142. J. W. L. Fordham, *J. Polymer Sci.* **39**, 321 (1959).
143. Y. V. Glazkovskii and Y. G. Papulov, *Vysokomolekul. Soedin* **A10**, 492 (1968).
144. U. Johnsen, *J. Polymer Sci.* **54**, S8 (1961).
145. J. W. L. Fordham, P. H. Burleigh, and C. L. Sturm, *J. Polymer Sci.* **41**, 73 (1959).
146. G. Pezzin, *Plastics Polymers* **37**, 295 (1969).
147. G. Natta and P. Corradini, *J. Polymer Sci.* **20**, 251 (1956).
148. R. W. Smith and C. E. Wilkes, *J. Polymer Sci. B* **5**, 433 (1967).
149. P. H. Burleigh, *J. Am. Chem. Soc.* **82**, 749 (1960).
150. G. M. Burnett, F. L. Ross, and J. N. Hay, *J. Polymer Sci. A1* **5**, 1467 (1967).
151. D. W. van Krevelen and P. J. Hoftyzer, *J. Appl. Polymer Sci.* **13**, 871 (1969).
152. A. N. Kostyuchenko, N. A. Okladnov, and V. P. Lebedev, *Vysokomolekul. Soedin.* **A10** (11), 2604 (1968).
153. J. Hildebrand and R. Scott, *The Solubility of Nonelectrolytes*, Reinhold Publishing Corp., New York, 1949.
154. M. L. Huggins, *Physical Chemistry of High Polymers*, John Wiley and Sons, Inc., New York, 1958.
155. P. Doty and H. S. Zable, *J. Polymer Sci.* **1**, 90 (1946).
156. G. M. Bristow and W. F. Watson, *Trans. Faraday Soc.* **54**, 1742 (1958).
157. Z. Mencik, *Collection Czech. Chem. Commun.* **24**, 3291 (1959).
158. W. R. Moore and R. J. Hutchinson, *J. Appl. Polymer Sci.* **8**, 2619 (1964).
159. H. Gerrens, W. Fink, and E. Koehnlein, *J. Polymer Sci. C* **16** (5), 2781 (1967).
160. H. Burrell, *J. Paint Technol.* **40**, 197 (1968).
161. R. L. Adelman and I. M. Klein, *J. Polymer Sci.* **31**, 77 (1958).
162. P. Kratochvil, V. Petrus, P. Munk, M. Bohdanecki, and K. Solc, *J. Polymer Sci. C* **16**, 1257 (1967).
163. A. Crugnola and F. Danusso, *J. Polymer Sci. B* **6**, 535 (1968).
164. H. Batzer and A. Nisch, *Makromol. Chem.* **22**, 131 (1957).
165. G. Bier and H. Kraemer, *Makromol. Chem.* **18/19**, 151 (1955).
166. R. Endo, *Chem. High Polymers (Japan)* **18**, 143 (1961).
167. G. Pezzin, G. Talamini, and G. V. Vidotto, *Makromol. Chem.* **43**, 12 (1961).
168. A. Nakajima and K. Kato, *Makromol. Chem.* **95**, 52 (1966).
169. M. Freeman and P. P. Manning, *J. Polymer Sci. A* **2**, 2017 (1964).
170. W. R. Moore and R. J. Hutchinson, *Nature* **200**, 1097 (1963).
171. L. de Brouckere, E. Bidaine, and A. van der Heyden, *Bull. Soc. Chim. Belges* **58**, 418 (1949).
172. P. Doty, H. L. Wagner, and S. Singer, *J. Phys. Chem.* **51**, 32 (1947).
173. Z. Mencik, *Chem. Zvesti* **9**, 165 (1955).
174. A. Peterlin, *Simposio Int. Chim. Macromol., Milano-Torino, 1954.*
175. B. N. Rutowski and W. W. Tschebotarwski, *Kunststoffe* **41**, 230 (1951).
176. H. Staudinger and M. Haeberle, *Makromol. Chem.* **9**, 48 (1952).
177. W. J. Langford and D. J. Vaughan, *Nature* **184**, 116 (1959).
178. G. Martin Guzman and J. M. Fatou, *Anales Real. Soc. Espan. Fis. Quim. (Madrid) Ser. B* **54**, 263 (1958).
179. H. L. Pedersen, *J. Polymer Sci. B* **5**, 239 (1967).
180. C. Garbuglio, A. Mula, and L. Chinellato, *J. Polymer Sci. C* **16**, 1529 (1967).
181. R. Endo, *Chem. High Polymers (Japan)* **18**, 477 (1961).
182. M. A. Crook and F. S. Walker, *Nature* **198**, 1163 (1963).
183. L. E. Maley, *J. Polymer Sci. C* **8**, 253 (1965).

184. H. L. Pedersen and J. Lyngaae-Jorgensen, *Proc. Fifth Intern. Seminar, Gel Permeation Chromatography, London, 1968.*

185. J. Lyngaae-Jorgensen, *Proc. Sixth Intern. Seminar, Gel Permeation Chromatography, Monte Carlo, 1969.*

186. C. L. Rohn, *J. Polymer Sci. A2* **5**, 547 (1967).

187. D. MacCallum, *Makromol. Chem.* **100**, 117 (1967).

188. G. Langhammer, *Svensk. Kem. Tidskr.* **69**, 328 (1957).

189. G. Martin Guzman and J. M. Fatou, *Anales Real. Soc. Espan. Fis. Quim. (Madrid) Ser. B* **54**, 609 (1958).

190. A. Oth and V. Desreux, *Bull. Soc. Chim. Belges* **63**, 261 (1954).

191. P. Kratochvil, M. Bohdanecky, K. Solc, M. Kolinsky, M. Ryska, and D. Lim, *J. Polymer Sci. C* **23**, 9 (1968).

192. M. L. Huggins, *J. Am. Chem. Soc.* **64**, 2716 (1942).

193. A. Nakazawa, T. Matsuo, and H. Inagaki, *Bull. Inst. Chem. Res. Kyoto Univ.* **44** (4), 354 (1966).

194. M. Sato, Y. Koshiishi, and M. Asahina, *J. Polymer Sci. B* **1**, 233 (1963).

195. J. W. Breitenbach, E. L. Forster, and A. J. Renner, *Kolloid-Z.* **127**, 1 (1952).

196. F. Krasovek, *Rept. J. Stefan Inst.* **3**, 203 (1956).

197. M. Fournier and H. Thiesse, *Compt. Rend.* **222**, 1437 (1946).

198. J. Hengstenberg, *Angew. Chem.* **62**, 26 (1950).

199. F. Danusso, G. Moraglio, and S. Gazzera, *Chim. Ind. (Milan)* **36**, 883 (1954).

200. Z. Mencik, *Collection Czech. Chem. Commun.* **21**, 517 (1956).

201. D. J. Mead and R. M. Fuoss, *J. Am. Chem. Soc.* **64**, 277 (1942).

202. C. Ciampa and H. Schwindt, *Makromol. Chem.* **21**, 169 (1956).

203. R. M. Fuoss, *J. Phys. Chem.* **47**, 59 (1943).

204. R. Endo, R. Goto, and M. Takeda, *Chem. High Polymers (Japan)* **22**, 405 (1965).

205. G. Pezzin, G. Talamini, and N. Gligo, *Chim. Ind. (Milan)* **46**, 648 (1964).

206. A. Takahashi, M. Obara, and I. Kagawa, *J. Chem. Soc. Japan, Ind. Chem. Sect.* **66**, 960 (1963).

207. M. Sato, Y. Koshiishi, and M. Asahina, *11th Ann. Meeting Soc. High Polymers Japan, May 26, 1962.*

208. M. Bohdanecki, K. Solc, P. Kratochvil, M. Kolinsky, M. Ryska, and D. Lim, *J. Polymer Sci. A2* **5**, 343 (1967).

209. G. Garbulio, G. Pezzin, and G. Badoni, *Chim. Ind. (Milan)* **46**, 797 (1964).

210. G. Pezzin, G. Sanmartin, and F. Zilio-Grandi, *J. Appl. Polymer Sci.* **11**, 1539 (1967).

211. G. Pezzin and N. Gligo, *J. Appl. Polymer Sci.* **10**, 1 (1966).

212. G. Pezzin, *J. Appl. Polymer Sci.* **10**, 21 (1966).

213. G. Schroeter, in K. Krekeler and G. Wick, eds., *Polyvinylchlorid*, Vol. 2, Carl Hanser Verlag, Munich, 1963, Chap. 2.

214. H. Fikentscher, *Cellulosechem.* **13**, 58 (1932).

215. R. S. Porter and J. F. Johnson, *Chem. Rev.* **66** (1), 1 (1966).

216. C. L. Sieglaff, *SPE (Soc. Plastics Engrs.) Trans.* **4** (2), 129 (1964).

217. F. P. Reding, E. R. Walter, and F. J. Welch, *J. Polymer Sci.* **56**, 225 (1962).

218. G. Garbuglio, A. Rodella, G. C. Borsini, and E. Gallinella, *Chim. Ind. (Milan)* **46**, 166 (1964).

219. J. J. Keavney and E. C. Eberlin, *J. Appl. Polymer Sci.* **3**, 47 (1960).

220. R. F. Boyer and R. S. Spencer, *J. Polymer Sci.* **2**, 157 (1947).

221. R. F. Boyer, *Rubber Chem. Technol.* **36**, 1303 (1963).

222. J. M. Barton, *Royal Aircraft Establishment, Tech. Rept. RAE-TR-67187* (AD-662536), Farnborough, England, 1967.

223. E. W. Rugeley, T. A. Feild, Jr., and G. H. Fremon, *Ind. Eng. Chem.* **40**, 1724 (1948).

224. V. L. Simril, *J. Polymer Sci.* **2**, 142 (1947).

225. F. P. Reding, J. A. Faucher, and R. D. Whitman, *J. Polymer Sci.* **57**, 483 (1962).

226. P. J. Flory, *Trans. Faraday Soc.* **51**, 848 (1955).

227. K. Eiermann, *Kunststoffe* **51**, 512 (1961).

228. B. V. Lebedev, I. B. Rabinowich, and V. A. Budarina, *Vysokomolekul. Soedin.* **A9** (3), 488 (1967).

229. G. Y. Gordon, *Stabilization of Synthetic High Polymers*, Israel Program for Scientific Translations, Ltd., Jerusalem, 1964, Chap. 5.

230. M. B. Neiman, ed., *Aging and Stabilization of Polymers*, Consultants Bureau, New York, 1965, Chap. 6.
231. N. A. J. Platzer, ed., *Stabilization of Polymers and Stabilizer Processes*, American Chemical Society, Washington, 1968, Chaps. 1–6.
232. W. C. Geddes, *Rubber Chem. Technol.* **40,** 177 (1967).
233. D. Braun, *Chemiker-Ztg.* **92,** 101 (1968).
234. M. Onozuka and M. Asahina, "Reviews in Macromolecular Chemistry," *J. Macromol. Sci.* **C3** (2), 235 (1969).
235. G. Talamini and G. Pezzin, *Makromol. Chem.* **39,** 26 (1960).
236. E. J. Arlman, *J. Polymer Sci.* **12,** 543 (1954).
237. B. Baum, *SPE (Soc. Plastics Engrs.) J.* **17** (1), 71 (1961).
238. M. Asahina and M. Onozuka, *J. Polymer Sci. A* **2,** 3505, 3515 (1964).
239. S. van der Ven and W. F. de Wit, *Angew. Makromol. Chem.* **8,** 143 (1969).
240. D. Braun and M. Thallmaier, *Makromol. Chem.* **99,** 59 (1966).
241. M. Thallmaier and D. Braun, *Makromol. Chem.* **108,** 241 (1967).
242. R. R. Straus, S. Stromberg, and B. G. Achhammer, *J. Polymer Sci.* **35,** 355 (1959).
243. A. H. Frye and R. W. Horst, *J. Polymer Sci.* **40,** 419 (1959).
244. K. U. Ingold, *Chem. Rev.* **61,** 564 (1961).
245. J. B. DeCoste, *SPE (Soc. Plastics Engrs.) J.* **21,** 764 (1965).
246. L. Weintraub, J. Zufall, and C. A. Heiberger, *Polymer Eng. Sci.* **8** (1), 64 (1968).
247. C. A. Heiberger, R. Phillips, and M. J. R. Cantow, *Polymer Eng. Sci.* **9** (6), 445 (1969).
248. M. L. Huggins, *J. Polymer Sci.* **8,** 257 (1952).
249. A. K. Doolittle, *The Technology of Solvents and Plasticizers*, John Wiley & Sons, Inc., New York, 1954, Chap. 15.
250. G. Gruenwald, *Kunststoffe* **50,** 381 (1960).
251. T. Alfrey, Jr., N. Wiederhorn, R. Stein, and A. Tobolsky, *Ind. Eng. Chem.* **41,** 701 (1949).
252. R. R. Lawrence and E. B. McIntyre, *Ind. Eng. Chem.* **41,** 689 (1949).
253. R. F. Clash and R. M. Berg, *Modern Plastics* **21,** 119 (1944).
254. U.S. Pat. 2,419,122 (April 15, 1947), F. W. Cox (to Wingfoot Corp.).
255. U.S. Pat. 2,636,024 (April 21, 1953), R. J. Wolf (to The B. F. Goodrich Co.).
256. W. S. Port, E. F. Jordan, Jr., W. E. Palm, L. P. Witnauer, J. E. Hansen, and D. Swern, *Ind. Eng. Chem.* **47,** 472 (1955).
257. U.S. Pat. 1,982,765 (Dec. 4, 1934), C. Schoenburg (to I.G. Farbenindustrie, A.G.).
258. Brit. Pat. 893,288 (April 4, 1962), (to The B. F. Goodrich Co.).
259. U.S. Pat. 3,328,371 (June 27, 1967), L. A. Beer (to Monsanto Co.).
260. U.S. Pat. 2,590,651 (March 25, 1952), D. S. Rosenberg (to Hooker Electrochemical Co.).
261. Belg. Pat. 712,900 (Jan. 29, 1969), (to Manifatura Ceramica Pozzi S.p.A.).
262. H. Fuchs and D. Louis, *Makromol. Chem.* **22,** 1 (1957).
263. V. Heidingsfeld, V. Kuska, and J. Zelinger, *Angew. Makromol. Chem.* **3,** 141 (1968).
264. K. Fukawa, *Vinyl Polymers (Japan)* **9** (2), 7 (1969).
265. P. Berticat, *J. Chim. Phys.* **64,** 892 (1967).
266. G. Bier, *Kunststoffe* **55,** 694 (1965).
267. H. Kaltwasser and W. Klose, *Plaste Kautschuk* **13,** 515 (1966).
268. H. Germar, *Makromol. Chem.* **86,** 89 (1965).
269. G. Svegliado and F. Zilio Grandi, *J. Appl. Polymer Sci.* **13,** 1113 (1969).
270. S. Tsuge, T. Okumoto, and T. Takeuchi, *Macromolecules* **2,** 277 (1969).
271. W. Trautvetter, *Kunststoffe Plastics* **13** (2), 54 (1966).
272. H. E. Fierz-David and H. Zollinger, *Helv. Chim. Acta* **28,** 455 (1945).
273. U.S. Pat. 2,750,358 (June 12, 1956), H. F. Park (to Monsanto Chemical Co.).
274. Ger. Pat. (East) 8,685, H. Demus.
275. Ger. Pat. 757,293, W. Muehlendyck, W. Weiss and F. Eisenstecken (to Kohle- und Eisenforschung G.m.b.H.).
276. Ger. Pat. 883,498, H. Baehr (to Badische Anilin- und Soda Fabrik A.G.).
277. U.S. Pat. 3,041,323 (June 26, 1962), H. E. Ramsden (to Metal and Thermit Corp.).
278. U.S. Pat. 3,050,507 (Aug. 21, 1962), R. W. Rees (to Shawinigan Chemicals Ltd.).
279. U.S. Pat. 2,586,363 (Feb. 19, 1952), A. McAlevy (to E. I. du Pont de Nemours & Co., Inc.).

280. Belg. Pat. 593,523 (Jan. 30, 1961), W. Wehr and K. Schneider (to Solvay and Cie).
281. U.S. Pat. 2,405,008 (July 30, 1946), K. L. Berry and J. W. Hill (to E. I. du Pont de Nemours & Co., Inc.).
282. U.S. Pat. 2,427,070 (Sept. 9, 1947), L. F. Reuter (to The B. F. Goodrich Co.).
283. U.S. Pat. 2,451,174 (Oct. 12, 1948), L. F. Reuter (to The B. F. Goodrich Co.).
284. U.S. Pat. 2,514,185 (July 4, 1950), K. C. Eberly (to Firestone Tire & Rubber Co.).
285. Y. Nakamura and K. Mori, *J. Polymer Sci. A1* **6,** 3269 (1968).
286. Anonymous, *Mod. Plastics* **45** (5), 86 (1968).
287. *Ibid.,* **46** (1), 25 (1969).
288. Anonymous, *SPE (Soc. Plastics Engrs.) J.* **24** (12), 33 (1968).
289. G. J. van Veersen, *Kunststoffe* **59** (3), 180 (1969).
290. R. J. Fabian, *Mater. Eng.* **70** (2), 42 (1969).
291. Anonymous, *Brit. Plastics* **41** (9), 123 (1968).
292. S. Wood, *Mod. Plastics* **45** (12), 78 (1968).
293. Anonymous, *Mod. Packaging* **42** (5), 78 (1969).

Manfred J. R. Cantow
Air Reduction Co., Inc.

VINYL ETHER MONOMERS AND POLYMERS

Vinyl ether products were first developed commercially by Badische Anilin- und Soda Fabrik A.G. (BASF) and received special attention after World War II as a part of so called "Reppe acetylene chemistry" (1). They have been useful as anesthetics and as organic intermediates, but have greatest interest scientifically and technically in polymerizations. Because of high sensitivity of the polymer structures and properties to conditions of polymerization, satisfactory commercial high polymer products have been rather slow in development. So far the outstanding industrial polymer products have been ethyl vinyl ether polymers of Union Carbide, isobutyl vinyl ether polymers of BASF and General Aniline and Film Corp. (GAF) and vinyl methyl ether-maleic anhydride interpolymers of GAF. The present use of these products in the U.S. amounts (1970) to several million pounds per year at prices of $0.50–1.00 per pound. Superior, stabilized interpolymers for use in pressure sensitive adhesives, coatings and high impact plastics are under development.

Among the monomers available are methyl, ethyl, butyl, isobutyl, 2-ethylhexyl, methoxyethyl, dodecyl, cetyl, octadecyl and chloroethyl vinyl ethers as well as divinyl ether. The lower alkyl vinyl ether monomers used in largest volume sell for about $0.30 per lb. Air Reduction Company (Airco) supplies ethyl vinyl ether and 2,2,2-trifluorethyl vinyl ether as anesthetics. The largest volume chemical intermediates manufactured from vinyl ethers at this time are malonaldehyde tetraacetal and glutaraldehyde. Other alkyl vinyl ethers are used as intermediates for manufacture of pharmaceuticals, flavors, oil modifiers and polishes.

Synthesis and properties of alkyl vinyl ether monomers have been reviewed (2,3). The Reppe vinylation process in liquid phase is suitable for manufacture of alkyl vinyl ethers from primary alcohols, but the pyrolysis of acetals has the advantages of lower pressures and fewer safety precautions:

$$\text{Vinylation process:} \quad \underset{\text{diluted}}{CH{\equiv}CH} + HOR \xrightarrow[130–180°]{KOR} CH_2{=}CHOR$$

$$\text{Acetal process:} \quad CH_3CH(OR)_2 \xrightarrow[\text{Catalyst}]{200–300°} CH_2{=}CHOR + ROH$$

Instead of employing the usual pyrolysis conditions the acetal vapor may be passed into Arochlor 1242 (Monsanto), a chlorinated diphenyl, at 180°C containing a little toluenesulfonic acid (4).

A number of modifications of the vinylation process have been developed. For greater safety the acetylene and nitrogen can be saturated with the alcohol before being compressed (5). The acetylene may be predissolved in the alcohol before passing to the heated reaction zone, eg at 150°C and 10 atm (6). Acetylene and the lower alcohol may be reacted in vapor phase at about 200°C over a fluidized bed of calcium alcoholate (7). Calcium carbide may be added to compensate for catalyst consumed. Alcohols of 1–5 carbon atoms may be reacted with acetylene in a higher alcohol as solvent at 160–200°C and with 5–15% alkali metal hydroxide (8). Yields of vinyl ethers from secondary and tertiary alcohols are not very favorable by vinylations using acetylene.

Vinyl ethers may be made also by alcoholysis of vinyl acetate in presence of

$$CH_2=CHOCOCH_3 + ROH \rightarrow CH_2=CHOR + CH_3COOH$$

sulfuric acid and mercuric salts (9). The preparation of vinyl alkyl ethers is possible using ethylene, acetic acid, alcohol, and oxygen with palladium catalysts, but commercial use has not been attractive. Vinyl ethers of secondary alcohols as well as vinyl ethers of primary alcohols having 6 or more C atoms can be made by vapor phase alcoholysis of methyl or ethyl vinyl ether by the higher alcohol, eg at 325°C over porous clay or nickel catalysts (10).

Vinyl ethers have been prepared by alkaline vinylation with acetylene under pressure from cyclohexanol, decahydro-2-naphthol, hydroabietinol, and terpene alcohols, but they have not shown commercial promise. The 1,2- and 1,3-diols such as ethylene glycol and 1,3-butanediol do not vinylate smoothly with acetylene but may react violently to form cyclic acetals. Vinyl ethers have been prepared from acetylene and unsaturated long chain alcohols derived from linseed and soybean oils and their copolymers have been evaluated (11).

Reaction of acetylene with phenols in presence of zinc or cadmium salts does not give vinyl ethers, but instead resinous condensation products such as the rubber tackifier Koresin (GAF) (12). However, in presence of ZnO or CdO and KOH, acetylene can react with phenol to give vinyl phenyl ether (13). Pressures of 200–500 psi and temperatures of 220–230°C are suitable.

The purity of vinyl ethers can be estimated by iodimetry (14) and gas chromatography. Aldehyde, alcohol, and acetal impurities should be removed from commercial vinyl ethers before their use in polymerizations. Successive washings with cold water at pH 7–8 should be made until the Tollens test for aldehyde is negative. The monomer may then be dried over KOH or lime and distilled from dry alkali or metallic sodium. Divinyl ether can be made by pyrolysis of bis(2-dichloroethyl)ether in water-free gas phase at 600–800°C (15). Divinyl ether and the ethers of glycols may polymerize to crosslinked polymers if stored in light without antioxidant-type stabilizers such as arylamines or phenols.

Properties and Reactions of Vinyl Ethers

With the exception of methyl vinyl ether the vinyl ether monomers are normally colorless liquids or low melting solids. See Table 1. Vinyl octadecyl ether is a white

Table 1. Some Vinyl Ether Monomers, CH_2=CHOR

Vinyl Ether	Boiling point	Melting point, °C	Sp gr (20/4)	Refractive index n_D^{20} (°C)	Solubility in water at 20°C, or other property
methyl	6°C	−122	0.7511	1.3947 (−25)	1.5%
ethyl	36	−115	0.7541	1.3739 (25)	0.9%
n-propyl	65		0.7674	1.3908	
isopropyl	55	−140	0.7534	1.3840	0.6%; surface tension, 18.72 dyne/cm
n-butyl	94	−113	0.7803	1.3997 (25)	0.3%; surface tension, 21.95 dyne/cm
isobutyl	83	−112	0.7692	1.3938 (25)	0.2%; heat capacity, 0.55 cal/(°C)(g)
sec-butyl	81		0.7715	1.3970	
t-butyl	75		0.7691	1.3922	
n-hexyl	143		0.7966	1.4171	surface tension, 24.5 dyne/cm
n-octyl	58 (4)		0.8020	1.4268	
2-ethylhexyl	76 (32)	−100	0.8108	1.4247 (25)	
decyl	101 (10)	−41	0.8123	1.4346	
cetyl	160 (2)	+16	0.822_{15}^{27}	1.4444 (25)	polymer mp, 36°C
octadecyl	182 (3)	+30			polymer mp, about 50°C
divinyl	28°C	−101	0.773	1.3989	
2-chloroethyl	109	−70	1.0493	1.4381	0.6%
phenyl	157			1.5226	
2,2,2-Trifluoroethyl	43		1.13	1.3188	0.4%

crystalline solid. The lower alkyl vinyl ethers are sparingly soluble in water, but are miscible with a wide range of organic solvents.

The vinyl ethers have rather pleasant odors. The more volatile compounds possess anesthetic action, but they are believed to be essentially nontoxic. Divinyl ether and ethyl vinyl ether are used as anesthetics of more rapid action than diethyl ether; 2,2,2-trifluoroethyl vinyl ether (Fluoromar, Airco) is valuable as an anesthetic of low flammability (16).

The double bonds, as well as the ether groups of vinyl ethers are reactive. Vinyl ethers are Lewis basic so that the double bonds very readily add halogens, hydrogen halides, and other Lewis acidic electrophilic agents. However, vinyl ethers do not undergo self addition (homopolymerization) nor do they add to basic substances except in presence of catalysts.

The lower alkyl vinyl ethers hydrolyze slowly in water at room temperature and more rapidly in presence of mineral acids forming acetaldehyde:

$$CH_2=CHOR + H_2O \xrightarrow{H^+} [CH_2=CHOH] \xrightarrow{H^+} CH_3CHO$$

Among vinyl ethers which are relatively resistant to hydrolysis by dilute aqueous acids are 2-ethylhexyl, phenyl, 2,2,2-trifluoroethyl, and 2-chloroethyl vinyl ethers.

The vinyl alkyl ethers that hydrolyze readily can be estimated by the acetaldehyde so formed. Vinyl ether content can be determined also by reacting with an excess of mercuric acetate in methanol to form a mercury addition compound with liberation of acetic acid which can be titrated with alkali (17). In order to avoid hydrolysis vinyl ethers are usually stored with added alkali, eg 0.1% triethanolamine. The basic stabilizer, as well as impurities such as alcohol, acetaldehyde, and acetals, should be removed before use in polymerizations by washing with water or very dilute KOH (pH 8) followed by drying over solid KOH or CaO and distillation from KOH or metallic sodium. In purifying and handling vinyl ethers contact with concentrated mineral acids or strong Lewis acid agents must be avoided to prevent violent exothermic polymerizations.

When heated in vapor phase, eg at 300°C over thoria on alumina, alkyl vinyl ethers may isomerize to aldehydes $CH_2{=}CHOCH_3 \rightarrow CH_3CH_2CHO$ (18). In general the alkyl vinyl ethers do not form peroxides rapidly on storage in air but α-hydroperoxyethers can be prepared by reaction of vinyl ethers with hydrogen peroxide (19). Peroxides can be identified in vinyl ethers by reaction with KI in dioxane (20).

Vinyl ethers add alcohols and phenols in presence of acid catalysts to form acetals (21):

$$ROCH{=}CH_2 + R'OH \xrightarrow{H^+} CH_3CH(OR)(OR')$$

Similarly with carboxylic acids they form acylals (ether esters), $CH_3CH(OR)(OCOR')$, (22). With mercaptans (see Thiols), thioacetals, $CH_3CH(SR)OR'$ are formed (23). In synthetic work reactive OH or SH groups can be protected temporarily by acetal formation using a vinyl ether.

Vinyl methyl ether adds methyl orthoformate to give *malonaldehyde tetramethyl acetal*, an important pharmaceutical intermediate:

$$CH_3OCH{=}CH_2 + CH(OCH_3)_3 \rightarrow (CH_3O)_2CHCH_2CH(OCH_3)_2$$

Boron trifluoride etherate may be used as catalyst at 25 to 40°C (24). The product has been used as a dialdehyde-forming agent in synthesis of sulfadiazine.

Dimethyl chloroacetal can be prepared readily (25):

$$CH_3OCH{=}CH_2 + Cl_2 + CH_3OH \xrightarrow{NaOH} ClCH_2CH(OCH_3)_2 + NaCl + 2H_2O$$

It has been employed as an intermediate with thiourea in the synthesis of sulfathiazole. The chloroacetal is a bifunctional reagent which can react either as an alkyl halide or as a source of aldehyde, or both, depending upon reaction conditions. Chloroacetal reacts with ammonia under pressure at 130°C forming an aminoacetal (26).

The alkyl vinyl ethers react with acetals in presence of acid catalysts to form 1:1 adducts and/or acetal-terminated low polymers depending upon conditions (27):

$$CH_2{=}CHOR + CH_3CH(OCH_3)_2 \xrightarrow{BF_3} \underset{\underset{OCH_3}{|}}{CH_3CH}\underset{\underset{OCH_3}{|}}{CH_2CHOR}$$

Trimethoxybutane so derived can be heated with phosphoric acid to form 1-methoxy-1,3-butadiene (28).

Liquid low polymers of vinyl ethers terminated by acetal groups, such as polymethoxy dimethyl acetals (PMAC), have had industrial interest (29). They can be hydrolyzed to aldehydes and hydrogenated to polyalkoxy alcohols:

$$\underset{\underset{OR}{|}}{CH_3(CHCH_2)_n}CH(OR)_2 \xrightarrow[H_2O]{H^+} \underset{\underset{OR}{|}}{CH_3(CHCH_2)_n}CHO \xrightarrow{H_2} \underset{\underset{OR}{|}}{CH_3(CHCH_2)_n}CH_2OH$$

The polymethoxy acetals and related alcohols have been evaluated in hydraulic and lubricating fluids and as plasticizers.

Alkyl vinyl ethers add aromatic amides and aromatic sulfonamides (30).   In

$$CH_2{=}CHOC_4H_9 + C_6H_5CONH_2 \xrightarrow{100°C} CH_3CH(NHCOC_6H_5)_2 + C_4H_9OH$$

presence of mercuric salts alkyl vinyl ethers add to 2-aminoethanol to form oxazolidine and to ethylene diamine to form imidazolidine (31).

Methyl vinyl ether undergoes Diels-Alder addition with acrolein to form dihydro-pyrans from which glutaraldehyde, $OHC(CH_2)_3CHO$, can be obtained (32).   Union Carbide manufactures this crosslinking agent for gelatin and leather.

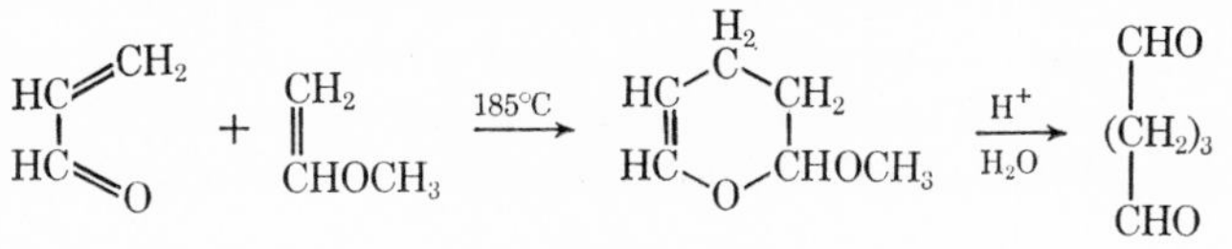

By this route 1,5-pentanediol can also be manufactured.

Alkyl vinyl ethers can undergo cyclic additions to form isoxazolines (33).   Ethyl vinyl ether adds to diphenylketene $(C_6H_5)_2C{=}C{=}O$ (34) and to sulfonyl isocyanates, RSO_2NCO (35) to form cyclic compounds.

In presence of benzoyl peroxide, CCl_4 adds to alkyl vinyl ethers to give CCl_3CH_2-$CHCl(OR)$ which on heating at 130°C loses HCl to form $CCl_2{=}CHCHCl(OR)$ (36). The latter hydrolyzes to β,β-dichloroacrolein and by heating with dilute NaOH both compounds yield chloroacetylene (spontaneously flammable).   By portionwise addition of n-butyl vinyl ether to CCl_4 containing a little pyridine at 77°C the 1:1 molar addition compound was obtained in high yield with minimum formation of polymers (37).   When ethyl vinyl ether and CCl_4 in 4:1 molar ratio were heated at 65°C up to 63.5% yield of the addition compound was recovered along with a polymer of molecular weight 900 (38).   At lower temperatures and under free radical conditions with UV radiation or benzoyl peroxide the alkyl vinyl ethers can "copolymerize" readily with CCl_4 or $CHCl_3$ to form polymers of moderate molecular weight containing more chlorine than can be accounted for by normal telomerization (C. E. Schildknecht unpublished work).   For such processes in which a telomer acts as a telogen the name "intertelomerization" is suggested (39).

Shostakovskii and coworkers in Russia published many papers on reactions of vinyl ethers but the industrial applications in Russia apparently have not been outstanding (40).

Vinyl ethers may react as sources of acetaldehyde.   For example methyl vinyl ether has been reacted with ammonia to give 2-methyl-5-ethylpyridine in presence of catalysts which slowly liberate HF (41).   Methyl vinyl ether acts as a source of acetaldehyde in reaction with benzaldehyde to give cinnamaldehyde dimethyl acetal (42). The acetal was heated with 5% p-toluenesulfonic acid to liberate free cinnamaldehyde.

Homopolymerizations of Vinyl Ethers

Vinyl alkyl ethers do not yield homopolymers of high molecular weight by free radical initiation such as by heating with peroxides or irradiation by ultraviolet light. Methods of homopolymerization of vinyl ethers by use of Lewis acid catalysts developed by BASF are disclosed in patents and reports of Allied investigators at the end of World War II (43).   Viscous liquid polymers of low molecular weight were made from methyl, ethyl, propyl, and butyl vinyl ethers using catalysts such as BF_3 etherate at 0°–80°C. Rubberlike amorphous polymers from isobutyl vinyl ether (Oppanol C) were prepared at low temperatures from liquid propane solution by rapid polymerization with addition of BF_3 gas (44). The hardest polymers were obtained from polymerizations completed within a few seconds from vinyl ethers having highly branched alkyl groups, eg $(CH_3)_3CCH(CH_3)OCH=CH_2$.

The kinetics of cationic polymerizations of vinyl ethers at higher temperatures forming viscous liquid polymers have been studied (46). Cyclohexyl vinyl ether polymerized very rapidly, but 2-chloroethyl vinyl ether polymerized slowly at 25°C with iodine as catalyst. Studies of the adhesive properties of vinyl isobutyl ether polymers in relation to polymerization conditions in laboratories of GAF led to the first examples of stereopolymerizations (47). Rapid polymerizations at −70 to −40°C using BF_3 gave substantially amorphous rubberlike high polymers from vinyl isobutyl ether. In contrast, slow growth polymerizations of the same monomer during ½ to several hours with BF_3 etherate catalyst formed harder, less rubbery stereoregular polymers which were normally crystalline by x-ray diffraction and showed cold drawing. Liquid propane and granular dry ice could be used as solvent and refrigerant in both types of polymerizations. Polyvinyl methyl ethers showing crystallinity were prepared slowly with BF_3 etherate at low temperatures from solvents containing chloroalkanes such as $CHCl_3$ (48). The polymers were different in properties from the amorphous methyl vinyl ether polymers prepared by BASF by homogeneous solution polymerization with Lewis acid catalysts. Formation of isotactic segments of alkyl vinyl ether polymer may be represented:

$$CH_2=CH \rightarrow +CH_2-C-CH_2-C-CH_2-C+$$
$$\quad | \qquad\qquad H\backslash \quad H\backslash \quad H\backslash$$
$$\quad OR \qquad\qquad OR \quad OR \quad OR$$

Stereoisomerism and factors influencing stereoregulation (49) or so called stereospecific polymerizations, have had growing interest because of subsequent development of crystalline polymers from 1-olefins, styrene, acrylics and other monomers from which only amorphous polymers had been known. The amorphous polymers having irregular steric configurations were called *atactic* while the more stereoregular polymers were called *isotactic* (50), (see Vol. 16, pp. 244,245). Catalysts related to Ziegler catalysts, eg diethyl aluminum chloride or Grignard reagents could be used. The latter catalysts were found to require oxygen as a cocatalyst to give isotactic isobutyl vinyl ether polymer; water as cocatalyst gave amorphous polymer (51). Slow polymerizations of vinyl isobutyl ether to crystalline polymers were obtained from $TiCl_4$–AlR_3 catalysts of the Ziegler type (52).

Japanese polymerizations of isobutyl and methyl vinyl ether in hexane–chloroform and hexane–toluene mixtures at −74 to −78°C with BF_3—etherate catalyst gave moderately isotactic polymers, apparently under homogeneous conditions (53). Polymerizations from hydrocarbon diluent may be homogeneous at first but often pass into a gel phase at higher conversion (54). Confirmation that the proliferous or growth type heterogeneous cationic polymerization with weak Lewis acids enhances stereoregularity has been reported (55). Other alkyl vinyl ethers have been polymerized at low temperatures under similar conditions to those described for isobutyl vinyl ether. Many of these have shown less tendency to form stereoregular polymers. Vinyl *n*-butyl ether polymerized at −100°C or above, and vinyl isopropyl ether at −110°C or above (56), but normally crystalline polymers are not obtained from these monomers under conditions which give crystalline vinyl isobutyl ether polymers. Brittle solid polymers of vinyl *t*-butyl ether showing crystallinity were prepared from homogeneous and heterogeneous systems using BF₃ etherate catalyst (57). These polymers generally have rather low molecular weights, eg intrinsic viscosity below 0.3 (58). Crystallization of vinyl *t*-butyl ether polymers may be enhanced by annealing at 120°C.

Polymerization of benzyl vinyl ether in toluene at −78°C by BF₃ etherate gave highly isotactic polymers (59). The stereoregularity of the polymers was unaffected by catalyst concentration but was higher from lower initial monomer concentration, and decreased only slightly with higher reaction temperatures. Additions of a polar solvent such as nitroethane decreased tacticity and greatly reduced the polymer molecular weight. Small amounts of water added to the polymerizations decreased the rates, but had little effect upon tacticity.

Homopolymerizations at Higher Temperatures

Only viscous liquid low polymers were obtained from lower alkyl vinyl ethers by I. G. Farben Industrie and earlier workers at 0°C and above. Polymers of fairly high specific viscosity were obtained by slow controlled polymerization of isopropyl vinyl ether in solution at 0 to 50°C using gallium chloride, GaCl₃, as a weak Lewis acid catalyst (60). Moseley was able to prepare solid high polymers from vinyl ethyl ether in hydrocarbon solvent at 0°C and above using complex catalysts containing sulfuric acid and an aluminum salt. This led to the commercial PVEE products of Union Carbide.

Substantially amorphous methyl vinyl ether polymers have attractive adhesive and combustion properties for use as binders in solid propellants. Such polymers may be made using a catalyst of ZnCl₂ with *t*-butyl chloride (61). For methyl and ethyl vinyl ether polymers in 2-butanone, molecular weights may be calculated from intrinsic viscosities: $[\eta] = 1.37 \times 10^{-3}M^{0.54}$ (62). In general the amorphous polymers of lower alkyl vinyl ethers of intrinsic viscosities below 0.5 and weight average molecular weights below 40,000 are balsam-like or viscous liquids.

Using as catalysts reaction products of aluminum alkoxides and sulfuric acid, methyl vinyl ether polymers of crystallinity up to 48% were obtained from methylene chloride solution after 5–30 hr at 0°C (63). High polymers of even higher crystallinity and insoluble both in methanol and in cold water were made from polymerization in methylene chloride with titanium tetrafluoride and titanium tetraisopropoxide at 0°C for 2 hr followed by 25°C for 16 hr (64). Vinyl methyl ether isotactic polymers of T_m as high as 150°C were prepared using catalysts of low solubility made from ball milling triisobutylaluminum and vanadium trichloride in heptane (65). Isotactic polymers were similarly prepared from other vinyl ethers, including vinyl *t*-butyl ether which gave polymers of T_m above 238°C. Complexes of sulfuric acid were used at 0°C to obtain crystalline high polymers of vinyl 2-methoxyethyl ether which were insoluble in water above 75°C (66).

Certain Grignard reagents are suitable catalysts for polymerizing vinyl ethers at relatively high temperatures. They tend to give rubber-like polymers of moderate molecular weight and relatively low tacticity. Vinyl isobutyl ether and cyclohexane was added dropwise to *n*-butylmagnesium bromide at 80°C until a highly viscous solution of polymer was obtained (67). Oxygen was found to be a necessary cocatalyst for the Grignard reagent (68). Vinyl ethers can be homopolymerized by ionic mechanism at room temperature under the action of high energy radiation such as gamma rays (69).

Commercial Polyvinyl Ethers

Homopolymers of vinyl alkyl ethers range from oils to viscous sticky liquids to brittle solids depending upon structure and molecular weight (more commonly expressed as specific viscosities or *Fikentscher K* values of solutions). Methyl, ethyl, and isobutyl vinyl ether homopolymers show more or less crystallinity and variations in properties depending upon the conditions of polymerization, but most of the commercial polymers are substantially amorphous (atactic). When the side chain is bulky as in polyvinyl *t*-butyl ether even lower molecular weight polymers are brittle solids, and when the side chain is very long and unbranched, as in octadecyl vinyl ether polymers, the polymer is a crystalline waxy solid even at low degrees of polymerization. Polyvinyl ethers have an exceptionally wide range of solubility in different types of organic solvents. They are not degraded by aqueous acid or base but tend to be degraded to lower molecular weights, especially in light or with prolonged heating, unless well stabilized by anti-oxidants or by UV absorbers. Most valuable for technical applications are the extreme stickiness or tack of the liquid low polymers and the pressure sensitive adhesion of some of the amorphous elastomeric high polymers.

Table 2. Some Commercial Vinyl Ether Homopolymers

Vinyl ether polymer	Polymer viscosity[a]	Tradename	Fields of use
Methyl			
viscous liquid (balsam-like)	$K = 50$[b]	Lutonal M (BASF)	plasticizer for coatings; aqueous adhesive tackifier
viscous liquid (balsam-like)	$\eta_{inh} = 0.3–0.5$	Gantrez M (GAF)	
Ethyl			
viscous liquid	$K = 60$ and lower grades	Lutonal A (BASF)	plasticizing nitrocellulose and natural resin lacquers
elastomeric solid[c] high polymer	(solid and solutions supplied)	PVEE (Union Carbide)	pressure-sensitive adhesive base
Isobutyl			
viscous liquid	$K = 60$	Lutonal I (BASF)	tackifier for adhesives, etc
viscous liquids	$\eta_{sp} = 0.1–0.5$ K 25–50	Gantrez B (GAF)	ditto
elastomeric solid	K 70–130 $\eta_{sp} = 2–6$	Oppanol C (BASF)	pressure-sensitive adhesive base
Octadecyl			
waxy solid	low DP mp $= 50°C$	V-Wax (BASF)	polishing and waxing agents

[a] $\eta_{sp} = (\eta - \eta_0)/\eta_0 c$; $\eta_{inh} = [\ln (\eta/\eta_0)]/c$ where η_0 and η are the viscosities of the solvent and of a solution of concentration c grams per deciliter.

[b] K values are those of Fikentscher (3); the η_{inh} values are from 1.0 g/100 solvent. The viscous liquid polymers are also supplied in high solid solutions, eg 70% in toluene.

[c] See Table 4.

Table 2 shows some commercial polymer products with tradenames. Most of these homopolymer types were introduced in the 1940s. Experience with their limitations and increased knowledge of vinyl polymerization should permit improved, more stable products especially new copolymers.

The following viscosity data for two methyl vinyl ether polymers in solutions were given in a Gantrez M technical booklet of 1967:

Table 3. Viscosity Data for Two Methyl Vinyl Ether Polymers

Properties				
specific viscosity[a]	0.47	0.77	0.47	0.77
K-value (Fikentscher)	40	50	40	50
solvent	water	water	toluene	toluene
solids in solutions, %	50	50	70	50
approximate viscosity of solutions, P	400	2000	300	150

[a] Note that the specific viscosities were measured using 1.0 g/100 ml benzene. Intrinsic viscosities would be lower. The GAF booklet gave 58 references to properties and suggested applications of PVM. Solubility, compatibility, and applications of Lutonal methyl and ethyl vinyl ether low polymers were discussed in a data sheet of BASF of June 1967.

By employing specially prepared weak Lewis acid catalysts and very carefully controlled reaction conditions Moseley showed the feasibility of preparing alkyl vinyl ether polymers of high molecular weight at 0°C to above room temperatures (70). Polymerizations of this type in hydrocarbon solvent are used by Union Carbide for manufacture of vinyl ethyl ether high polymers (PVEE), the properties and applications of which are described in (71). The grades of PVEE shown in Table 4 are widely used as bases for pressure sensitive adhesive tapes of outstanding tack and stability. However, for many applications their softness and thermoplasticity are limitations. For easier handling, high-solids solutions in hexane are supplied in addition to the solid homopolymers. The PVEE grades are supplied containing 0.5% mono-t-butyl-hydroquinone stabilizer. The polymers are compatible with inexpensive resins useful in modifying adhesive properties such as rosin esters, terpene, and petroleum resins. Up to about 50 parts of such resins may be used with up to 60 parts plasticizer and 50–70 parts PVEE of high molecular weight for pressure sensitive tapes from paper, cellophane, plastic films or textiles. Smaller amounts of the more sticky low viscosity ethyl vinyl ether polymer (EHBC resin) are added to improve the quick tack of the high polymer. Among recent developments are applications to wall coverings and to temporary bonds between mating metal parts during machine shop operations.

Table 4. Properties of Commercial Poly(vinyl ethyl ethers) and Solutions

Polymer type	Low viscosity		High viscosity		Extra-high viscosity	
	EHBC	EDBC	EHBM	EDBM	EHBN	EDBN
solids in hexane, %	80	98	28	98	25	98
approximate reduced viscosity, 20°C	0.3	0.3	4.0	4.0	5.0	5.0
specific gravity, n_{20}^{20}	0.908	0.973	0.747	0.968	0.725	0.968

Vinyl ether polymer formulated solutions of outstanding stability and utility have been supplied by Pittsburgh Plate Glass Corp. Blends with phenol–formaldehyde and acrylic polymers permit more diversified pressure sensitive adhesives (72).

Copolymerizations of Vinyl Ethers

The action of Lewis acid catalysts upon mixtures of monomers does not always result in true copolymers or interpolymers. In many cases mixtures of two different vinyl ethers or a vinyl ether with another type of monomer will form a homopolymer first, the other monomer only reacting subsequently when a higher temperature or higher concentration of catalyst has been reached (39).

Copolymers of vinyl isobutyl ether and allyl vinyl ether made by Lewis acid catalysts hardened slowly in films by further oxidative polymerizations through allyl groups (73). Yuki and coworkers studied copolymerization of *n*-butyl vinyl ether with other vinyl ethers (74). Reactivity was promoted by electron repelling branched alkyl ether groups. Copolymers of unsaturated long chain alkyl vinyl ethers have been evaluated in coatings (75).

Ethylene–vinyl ether block copolymers were prepared by use of Ziegler catalyst (AlR$_3$ + TiCl$_4$) (76). Copolymers of propylene and vinyl ether formed in presence of catalyst from TiCl$_3$ and diethylaluminum chloride (77). The products had improved transparency and dyeability compared to propylene homopolymers. Copolymers of vinyl ethers by ionic initiation have not become very important industrially.

Lewis basic monomers such as vinyl ethers, alpha-methyl styrene, isobutylene, and 1-olefins do not homopolymerize readily by free radical methods, nor do they copolymerize very readily with each other. Vinyl ethers do not copolymerize readily by radical methods with butadiene, isoprene, and styrene monomers. However, vinyl ethers readily form copolymers of high molecular weight by radical initiation with Lewis acidic monomers in which electron-attracting groups are substituted into ethylene. With these acrylic, vinyl ester, vinyl halide, anhydride, and unsaturated acid monomers 1:1 molar copolymerization is favored and more than 50 mole% of vinyl ether units is never obtained in the copolymer products. Internally plasticized interpolymers of vinyl and acrylic monomers with *n*-butyl to *n*-dodecyl vinyl ethers have excellent industrial prospects.

Vinyl chloride–vinyl isobutyl ether copolymers containing 25–30% vinyl ether units under the tradenames Gantrez VC (GAF) and Vinoflex MP400 (BASF) are used in marine paints and corrosion resistant finishes for metals. The copolymers have advantages in solubility in aromatic hydrocarbon, ester and ketone solvents, and have outstanding resistance to alkali. Because of the latter they have found application in traffic paints and finishes for masonry. With alkyds the copolymers may be used for printing and coating polyolefins and other plastics.

Interpolymers of uniform composition can be prepared by copolymerization of vinyl chloride with isobutyl vinyl ether in a neutral aqueous emulsion of persulfate (peroxydisulfate, see Vol. 14, p. 755). In one example an 80:20 monomer mixture was stirred with a solution of 2 parts K$_2$S$_2$O$_8$, 2 parts NaOH and 6 parts sodium hydroxyoctadecanesulfonate in 1700 parts water at 55°C under 6 atm of nitrogen (78). Additional monomer and catalyst could be added continuously during 24 hr of reaction. The interpolymer was precipitated from the latex as a powder. Copolymerization in buffered aqueous suspension with azo and persulfate catalysts and polyvinylpyrrolidone as suspending agent also was disclosed (79). Vinyl chloride–vinyl ethyl ether copolymers have shown favorable properties in plastisols.

Copolymers of vinyl chloride with 2–4% of long alkyl vinyl ethers have been developed in Japan (80), U.S., and Europe (81) for tough, rigid, easy processing plastic especially for use in blown bottles and in vacuum-formed sheeting. Fabrication temperatures are 50°F or more below those of vinyl chloride homopolymers. C_{12} to C_{16} linear alkyl vinyl ethers have had most application. Such a product is Plaskon CG 014 a copolymer with vinyl dodecyl ether of inherent viscosity 0.66 and specific gravity 1.39 made by activated emulsion copolymerization.

Reactivity ratios (see Vol. 16, p. 233) for free radical copolymerization of common monomers with dodecyl and octadecyl vinyl ethers gave r_2 values near zero (82). The r_1 values are given in the table below beginning with the most favorable.

Table 5. Free Radical Copolymerization of Common Monomers

M_1	r_1 with dodecyl vinyl ether	r_1 with octadecyl ether
dibutyl maleate	0.1	—
acrylonitrile	0.9	0.9
vinylidene chloride	1.3	1.5
vinyl chloride	1.9	2.1
vinyl acetate	3.7	—
methyl acrylate	3.8	4.7
methyl methacrylate	10.7	10.8
styrene	55	—

Alkyl vinyl ethers interpolymerize readily under mild conditions with maleic anhydride to form 1:1 molar interpolymers. The methyl vinyl ether–maleic anhydride polymers are conveniently prepared from benzene solution using radical initiators at 50 to 80°C (83) (84). In one example 100 g maleic anhydride was dissolved in 100 ml benzene, the solution was filtered and 2 g benzoyl peroxide was added (85). The solution was stirred at 80°C while methyl vinyl ether was added at 15 g/hr. The interpolymer began to precipitate after 10 min and external cooling was used to maintain the temperature during 4 hr of reaction. The copolymer was washed with fresh benzene and dried in vacuum at 60°C.

Methyl vinyl ether–maleic anhydride interpolymers (MVE–MA) under the tradename Gantrez AN are supplied by GAF in grades of specific viscosity 0.1–3.0 (1.0 g/100 ml methyl ethyl ketone at 25°C) (86). The powder is soluble in ketones and esters, and slowly soluble in water with hydrolysis of the maleic anhydride groups to maleic acid groups. The viscosities of aqueous solutions decrease on storage, especially on exposure to light. Ultraviolet absorbers and water soluble antioxidants such as thiourea may be added to retard degradation. Gradual addition of alkali metal or ammonium hydroxide to aqueous solutions leads to two viscosity peaks. The interpolymers are precipitated by Ca^{2+} and by other heavy metal polyvalent cations. MVE–MA copolymers have found use in hair sprays, textile sizes, thickeners for water paints, and cosmetics.

The maleic anhydride interpolymers can be half-esterified; monoesterification can be carried out by refluxing 4 hr with ethanol, or 18 hr with isopropyl alcohol. Prolonged heating with excess alcohol and an acid catalyst is necessary for full esterification. The half esters are soluble in dilute aqueous alkali and in alkanolamines. GAF supplies ethyl, isopropyl, and n-butyl half ester copolymers as 50% solutions in ethanol or isopropyl alcohol. They are used in hair preparations and other cosmetics.

Passing NH_3 into a slurry of vinyl methyl ether–maleic anhydride interpolymer in benzene gives a half amide–half ammonium salt which has been available:

$$-CH_2CH-CH\underline{\quad}CH- \quad \xrightarrow{\;NH_3\;} \quad -CH_2CH-CH-CH-$$

The maleic acid copolymers and their derivatives can be crosslinked by heating with poly(vinyl alcohols), urea, polyamines, or by polyvalent cations.

Copolymers of maleic anhydride with octadecyl vinyl ether have been supplied in toluene solution (Gantrez AN) as slip agents (antiblocking agents) for inks and coatings.

Fikentscher and coworkers developed methods for free radical copolymerization of vinyl ethers with acrylic monomers in neutral, buffered aqueous systems, thus avoiding hydrolysis of the vinyl ether monomer (87). The monomers could be added slowly during the course of the copolymerization reactions (88). BASF has supplied copolymers of acrylate esters with vinyl ethers in latex dispersions. For example Acronal 400D, a copolymer of butyl acrylate with smaller proportion of vinyl isobutyl ether, is useful in sizing paper and textiles. Copolymers of ethyl acrylate with 2-chloroethyl vinyl ether made in peroxysulfate emulsion have been used as specialty rubbers (89). Terpolymers of acrylate esters, vinyl ethers, and maleic anhydride have been studied (90). Copolymers of ethyl acrylate with monovinyl ethers of glycols are vulcanizable elastomers (91).

Acrylonitrile copolymerizes readily with lower alkyl vinyl ethers using radical initiators in bulk, solution or buffered emulsion (83) (92). Alkyl vinyl ethers are sufficiently Lewis basic to initiate rapid polymerization with vinylidene cyanide (93) and with methyl cyanoacrylate (39).

Minor proportions of lower alkyl vinyl ethers give promising copolymers with ethylene having favorable permeability, clarity and gloss, as well as good extrusion and strength properties. In one patent example, 2.1 mole $\%$ vinyl ethyl ether, was reacted with ethylene containing 0.015% oxygen continuously in a tubular reactor 18 m long at 180°C and 2109 kg/cm² pressure (94). The gaseous monomers were passed at 1135 kg/hr. In another process ethylene was copolymerized with 1.7 mole $\%$ methyl vinyl ether in presence of 7.0 mole $\%$ propane as telogen in an autoclave at 187°C (95). The initiator, t-butyl peroxyisobutyrate, was injected portionwise into the pressure reactor. Attempts to copolymerize ethylene with vinyl isobutyl ether were reported to fail because of the higher chain transfer activity of this monomer.

Lower alkyl vinyl ethers copolymerize fairly readily with vinyl acetate using free radical catalyst in bulk or buffered aqueous dispersions (96). Gantrez AC is such a latex. Rubber-like, vulcanizable copolymers have been prepared from monovinyl ethers of glycols and vinyl esters (97). Terpolymer latexes of vinyl acetate-maleate ester-isobutyl vinyl ether by peroxysulfate emulsion are suitable for water paints and adhesives (98).

Fluorinated vinyl ethers are also of industrial interest. Copolymers of tetrafluoroethylene (TFE) with 35–45 mol $\%$ perfluoromethyl perfluorovinyl ether, $CF_2=$ $CFOCF_3$ have been blended with TFE polymers (99). Latexes obtained by emulsion copolymerizations of TFE and $CF_2=CFOCF_3$ are mobile, colorless, and transparent; the copolymer particles of about 0.2 micron may have refractive index closely matching that of the aqueous phase (100).   2,2,2-Trifluoroethyl vinyl ether, $CF_3CH_2OCH{=}CH_2$, was homopolymerized cationically (101) and was copolymerized with alkyl acrylates, methacrylates, vinyl chloride and vinyl esters (102).   Alkyl vinyl ethers have been copolymerized with fluoroolefins (103).

Bibliography

"Reppe Chemistry" in *ECT* 1st ed., Vol. 11, pp. 651, by J. M. Wilkinson, Jr., Jesse Werner, H. B. Haas, and Hans Beller, GAF.

1. W. Reppe, *Acetylene Chemistry*, U.S. Dept. of Commerce PB 18852-S.   Translation, Charles A. Meyer & Co., New York, 1949.
2. C. E. Schildknecht, A. O. Zoss, and C. McKinley, *Ind. Eng. Chem.* **39,** 180 (1947).
3. C. E. Schildknecht, *Vinyl and Related Polymers*, John Wiley & Sons, Inc., New York, 1952; E. R. Blout et al. eds., *Monomers* II, Interscience Publishers, a division of John Wiley & Sons Inc., New York, 1951.
4. U.S. Pat. 2,667,517 (1954), R. I. Longley (to Monsanto).
5. U.S. Pat. 2,472,084 (1949), H. Beller, R. E. Christ, and F. Wuerth (to GAF).
6. U.S. Pat. 2,617,829 (1952), D. Maragliana (to Montecatini);   U.S. Pat. 2,969,395 (1961), J. J. Nedwick and J. R. Snyder (to Rohm and Haas).
7. U.S. Pat. 3,358,041 (1967), H. O. Mottern and M. W. Leeds (to Airco);   (cf) U.S. Pat. 3,341,- 606 (1967), H. O. Mottern (to Airco).
8. U.S. Pat. 3,370,085 (1968), J. F. Vitcha (to Airco).
9. U.S. Pat. 2,579,411, U.S. Pat. 2,579,412 (1951), R. L. Adelman (to du Pont);   U.S. Pat. 2,984,688 (1961), J. Sixt (to Wacker);   W. H. Watanabe and L. E. Conlon, *J. Am. Chem. Soc.* **79,** 2828 (1957).
10. U.S. Pat. 2,566,415 (1951), R. I. Hoaglin and D. H. Hirsh (to Union Carbide).
11. H. M. Teeter et al., *Ind. Eng. Chem.* **50,** 1703 (1958).
12. A. O. Zoss, W. E. Hanford, and C. E. Schildknecht, *Ind. Eng. Chem.* **41,** 73 (1949).
13. U. S. Pat. 2,615,050 (1952), T. H. Insinger (to Koppers).
14. S. Siggia and R. L. Edsberg, *Anal. Chem.* **20,** 762 (1948).
15. U.S. Pat. 2,832,807 (1958), R. Mittag and J. Smidt (to Consortium fuer Elektrochem. Indust.); (cf) U.S. Pat. 3,256,344 (1966), L. W. McTeer (to Union Carbide).
16. J. Krantz, C. J. Carr, G. Lu, and F. K. Bell, *J. Pharmacol. Exptl. Therap.* **108,** 488 (1953).
17. J. B. Johnson and J. P. Fletcher, *Anal. Chem.* **31,** 1563 (1950).
18. U.S. Pat. 2,642,460 (1953), D. C. Hull et al. (to Eastman Kodak).
19. N. A. Milas, R. L. Peeler, and O. L. Mageli, *J. Am. Chem. Soc.* **76,** 2322 (1954).
20. R. K. Summerbell and D. K. A. Hyde, *J. Org. Chem.* **25,** 1809 (1960).
21. U.S. Pat. 2,543,312 (1950), J. W. Copenhaver (to GAF);   Brit. Pat. 654,166 (1951), R. I. Hoaglin and S. F. Clark (to Union Carbide);   M. G. Voronkov, *Doklady Akad. Nauk. SSSR* **63,** 539 (1948);   *Chem. Abstr.* **43,** 5365 (1949).
22. M. G. Voronkov, *Zhur. Obshchei Kim.* **19,** 293 (1949);   W. Reppe et al., *Ann.* **601,** 108 (1956).
23. Brit. Pat. 642,253 (1950), J. W. Copenhaver (to GAF);   U.S. Pat. 2,551,421 (1951), J. W. Copenhaver (to GAF).
24. U.S. Pat. 2,527,533 (1950), J. W. Copenhaver (to GAF);   R. W. Price and A. Moos, *J. Am. Chem. Soc.* **67,** 207 (1945).
25. U.S. Pat. 2,550,637 (1951), J. W. Copenhaver (to GAF).
26. U.S. Pat. 2,628,254 (1953), J. W. Copenhaver (to GAF).
27. U.S. Pat. 2,165,962 (1939), M. Müller-Cunradi and K. Pieroh (to IG Farbenindustrie); U.S. Pat. 2,487,525 (1949), J. W. Copenhaver (to GAF);   R. I. Hoaglin and D. H. Hirsh; *J. Am. Chem. Soc.* **71,** 3468 (1949).
28. U.S. Pat. 2,905,722 (1959), A. E. Montagna, and D. H. Hirsh (to Union Carbide).
29. U.S. Pat. 2,527,533 (1950), J. W. Copenhaver (to GAF);   U.S. Pat. 2,564,760, U.S. Pat. 2,564,761 (1951), R. I. Hoaglin and S. F. Clark (to Union Carbide);   U.S. Pat. 2,618,663 (1952), S. A. Glickman (to GAF).
30. J. Furukawa et al., *Kogyo Kagaku Zasshi* **60,** 170 (1957);   *Chem. Abstr.* **53,** 6202 (1959).
31. W. H. Watanabe, *J. Am. Chem. Soc.* **79,** 2833 (1957).

32. U.S. Pat. 2,514,168 (1950), C. W. Smith, D. G. Norton, and S. A. Ballard (to Shell Oil); U.S. Pat. 2,546,019 (1951), C. W. Smith, D. G. Norton, and S. A. Ballard (to Shell Oil).
33. R. Paul and S. Tchelitscheff, *Bull. Soc. Chim. France*, 2215 (1962); R. Huisgen, *Angew. Chem. Intern. Ed.* **2**, 565 (1963).
34. C. D. Hurd and R. D. Kimbrough, *J. Am. Chem. Soc.* **82**, 1373 (1960).
35. F. Effenberg and R. Gleiter, *Ber.* **97**, 1576 (1964).
36. E. Levas and E. Levas, *Compt. Rend.* **230**, 1669 (1950); **232**, 521 (1951); *Chem. Abstr.* **45**, 7004 (1951).
37. U.S. Pat. 2,560,219 (1951), S. A. Glickman (to GAF).
38. M. F. Shostakovskii et al., *Chem. Abstr.* **51**, 5730 (1957); (cf) T. S. Nikatina et al., *Chem. Abstr.* **52**, 95 (1958).
39. C. E. Schildknecht, unpublished work.
40. M. F. Shostakovskii, et al., *Chem. Abstr.* **37**, (1943) to **69** (1968).
41. U.S. Pat. 2,706,730 (1955), J. E. Mahan (to Phillips).
42. U.S. Pat. 2,543,312 (1951), J. W. Copenhaver (to GAF).
43. C. E. Schildknecht, *Vinyl and Related Polymers*, John Wiley & Sons, Inc. (1952).
44. U.S. Pat. 2,311,567 (1943), M. Otto, et al. (to Jasco); PB 67,694 or FIAT 944 (post World War II reports).
45. Ger. Pat. 745,030 (1943), M. Müller-Cunradi and K. Pieroh; (cf) U.S. Pat. 2,061,934 (1937), M. Muller-Cunradi and K. Pieroh (to IG Farben Industrie).
46. D. D. Eley et al., *J. Chem. Soc.* 4167 (1952); 3700 (1957).
47. C. E. Schildknecht, S. T. Gross, H. R. Davidson, J. M. Lambert, and A. O. Zoss, *Ind. Eng. Chem.* **40**, 2104 (1948); **50**, 107 (1958); C. E. Schildknecht et al. in J. Elliott, ed., *Macromolecular Syntheses II*, John Wiley & Sons Inc., New York, 1966.
48. C. E. Schildknecht, S. T. Gross, and A. O. Zoss, *Ind. Eng. Chem.* **41**, 1998 (1949).
49. C. E. Schildknecht, *Polymer Eng. and Sci.* **6**, 240 (1966); (cf) U.S. Pat. 2,882,263 (1959), G. Natta (to Montecatini); U.S. Pat. 3,112,300, U.S. Pat. 3,112,301 (1963), G. Natta (to Montecatini); U.S. Pat. 3,419,538 (1968), G. Natta (to Montecatini).
50. G. Natta, et al., *Makromol. Chem.* **18–19**, 455 (1955); *Angew. Chem.* **71**, 205 (1959).
51. H. Hirata and H. Tani, *Polymer* **9**, 59 (1968).
52. J. Lal, *J. Polymer Sci.* **31**, 179 (1968).
53. S. Okamura et al., *J. Polymer Sci.* **33**, 510 (1958); **39**, 507 (1959).
54. G. J. Blake and A. M. Carlson, *J. Polymer Sci.* **A1 4**, 1813 (1966).
55. C. E. Schildknecht and P. H. Dunn, *J. Polymer Sci.* **20**, 597 (1956); A. D. Ketley, *J. Polymer Sci.* **62**, S8 (1962).
56. U.S. Pat. 2,609,364 (1952), U.S. Pat. 2,799,669 (1957), A. O. Zoss (to GAF); (cf) G. Dall'Asta and I. W. Bassi, *Chim e Ind.* (Milan) **43**, 999 (1961).
57. S. Okamura et al. *Makromol. Chem.* **53**, 180 (1962); T. Higashimura et al., *Makromol. Chem.* **86**, 259 (1965); G. Natta et al., *Makromol. Chem.* **89**, 81 (1965).
58. U.S. Pat. 3,278,507 (1966), G. Natta, and G. Dall'Asta (to Montecatini).
59. H. Yuki, S. Murahashi et al., *J. Polymer Sci.* **A1, 7**, 1517 (1969).
60. U.S. Pat. 2,457,661 (1948), F. Grosser (to GAF); *Ind. Eng. Chem.* **41**, 2891 (1949).
61. U.S. Pat. 3,017,260 (1962), G. J. Arquette et al. (to Airco); U.S. Pat. 3,022,280 (1962), J. G. Shukys (to Airco); (cf) U.S. Pat. 3,035,950 (1962), C. G. Long (to Phillips).
62. J. A. Manson and G. J. Arquette, *Makromol. Chem.* **37**, 187 (1960).
63. U.S. Pat. 3,025,282, U.S. Pat. 3,025,283 (1962), D. L. Christman and E. J. Vandenberg (to Hercules); (cf) *J. Polymer Sci.* **C1**, 207 (1963).
64. U.S. Pat. 3,157,626 (1964), R. F. Heck (to Hercules); (cf) U.S. Pat. 3,159,613 (1964), R. F. Heck (to Hercules).
65. Brit. Pat. 820,469 (1966), E. J. Vandenberg; (cf) U.S. Pat. 3,284,426 (1966), E. J. Vandenberg (to Hercules).
66. U.S. Pat. 3,014,013 (1961), R. F. Heck (to Hercules).
67. R. J. Kray, *J. Polymer Sci.* **44**, 264 (1960).
68. M. Bruce and D. W. Farrow, *Polymer* **4**, 407 (1963); G. J. Blake and A. M. Carlson, *J. Polymer Sci.* **4**, 1813 (1966).
69. S. H. Pinner and R. Worrall, *J. Polymer Sci.* **2**, 122 (1959); M. A. Bonin et al., *J. Polymer Sci.* **B2**, 143 (1964); U.S. Pat. 3,264,203 (1966), R. M. Narlock (to Dow Chemical).

70. U.S. Pat. 2,549,921 (1951), S. A. Mosley; (cf) U.S. Pat. 3,295,923 (1967), C. R. Dickey (to Union Carbide); U.S. Pat. 2,984,656 (1961), J. Lal; U.S. Pat. 3,062,789 (1962) (to Goodyear Rubber Co.).

71. *Polyvinyl Ethyl Ether Resins*, Pamphlet, Union Carbide.

72. U.S. Pat. 3,280,217 (1966), W. Lader and D. H. Zang (to Pittsburgh Plate Glass).

73. U.S. Pat. 2,825,719 (1958), K. Herrle et al. (to Badische Anilin- u. Soda-Fabrik A.G.); *Chem. Abstr.* **52**, 9644 (1958).

74. H. Yuki et al., *J. Polymer Sci.* **A1, 7**, 667 (1969).

75. L. E. Gast et al., *J. Am. Oil Chemists' Soc.* **35**, 347 (1958); *Chem. Abstr.* **52**, 15095; U.S. Pat. 3,179,717 (1960), E. J. Dufek et al. (to U.S.A.); G. C. Mustakas et al., *J. Am. Oil Chemists' Soc.* **37**, 68 (1960); *Chem. Abstr.* **54**, 7221 (1960).

76. U.S. Pat. 3,026,290 (1962), E. W. Glueselkamp (to Monsanto).

77. Brit. Pat. 1,063,040 (1967) (to Toyo Rayon).

78. Ger. Pat. 870,035 (1953), H. Rolker (to BASF); *Chem. Abstr.* **52**, 16797 (1958).

79. Ger. Pat. 1,046,882 (1958), H. Fikentscher and W. Huebler (to BASF); *Chem. Abstr.* **55**, 8945 (1961).

80. U.S. Pat. 3,168,594 (1965), Y. Hoshi and M. Onozuka (to Kureha Kasei).

81. French Pat. 1,483,019 (1967), (to Péchiney).

82. G. Akazome et al. (Japan), in *Long Chain Vinyl Ethers*, Data Booklet, GAF, 1967.

83. C. E. Schildknecht, "Vinyl Ethers" Chapter, in E. R. Blout et al., eds., *Monomers* II, Interscience Publishers, a division of John Wiley & Sons Inc., New York, 1951.

84. U.S. Pat. 2,047,398 (1936), A. Voss and E. Dickauser (to IG Farbenindustrie); R. B. Seymour et al., *Ind. Eng. Chem.* **41**, 1509 (1949).

85. Brit. Pat. 712,220 (1954), (to GAF); *Chem. Abstr.* **49**, 4333 (1955); (cf) U.S. Pat. 2,782,182 (1957), R. M. Verburg (to GAF); U.S. Pat. 2,694,697, U.S. Pat. 2,694,698 (1954), F. Grosser (to GAF).

86. Gantrez AN, a booklet, GAF, with 128 references to applications.

87. Ger. Pat. 634,408 (1936), H. Fikentscher; U.S. Pat. 2,016,490 (1935), H. Fikentscher (to, IG Farben Industrie).

88. Ger. Pat. 745,424 (1943), H. Fikentscher and R. Gaeth (to I.G. Farbenindustrie).

89. W. C. Mast and C. H. Fisher, *Ind. Eng. Chem.* **41**, 703, 790 (1949).

90. U.S. Pat. 3,436,378 (1969), J. L. Azorlosa et al. (to GAF).

91. U.S. Pat. 3,146,215 (1964), G. B. Sterling and R. L. Zimmerman (to Dow Chemical).

92. U.S. Pat. 2,436,926 (1948), R. A. Jacobson (to Du Pont); U.S. Pat. 2,436,204 (1948) (Prophylactic), G. F. D'Alelio; (cf) U.S. Pat. 3,227,673 (1966), (to Sohio).

93. H. Gilbert et al., *J. Am. Chem. Soc.* **78**, 1669 (1956).

94. Brit. Pat. 906,249 (1962), W. G. White and R. A. Walther; *Chem. Abstr.* **57**, 16891 (1962); (cf) U.S. Pat. 3,226,374 (1965), (to Union Carbide).

95. U.S. Pat. 3,033,840 (1962), H. W. Strauss (to Du Pont).

96. U.S. Pat. 2,662,016 (1953), P. L. Merz et al. (to Beech-Nut Co.); U.S. Pat. 3,477,980 (1969), W. E. Daniels (to GAF); Brit. Pat. 1,096,569 (1967), K. Julian (to Imperial Chemical Industries); *Chem. Abstr.* **68**, 40317 (1969).

97. U.S. Pat. 3,131,162 (1964), G. B. Sterling and R. L. Zimmerman (to Dow Chemical).

98. Brit. Pat. 828,957 (1960), R. A. W. Pateman and S. A. Miller (to British Oxygen); *Chem. Abstr.* **54**, 12663 (1960).

99. U.S. Pat. 3,484,503 (1969), L. M. Magner and J. O. Punderson (to Du Pont).

100. A. L. Barney, *A.C.S. Polymer Preprints* **10**, 1483 (1969).

101. U.S. Pat. 2,820,025 (1958), C. E. Schildknecht (to Airco); U.S. Pat. 3,365,433 (1968), J. A. Manson and H. Sorkin (to Airco).

102. U.S. Pat. 2,851,449 (1958), U.S. Pat. 2,991,277, U.S. Pat. 2,991,278 (1961), C. E. Schildknecht (to Airco).

103. R. M. Adams and F. A. Bovey, *J. Polymer Sci.* **9**, 481 (1952); U.S. Pat. 2,834,766 (1958), J. M. Hoyt (to Minnesota Mining and Manufacturing Co.); Brit. Pat. 948,998 (1964), (to Du Pont).

C. E. SCHILDKNECHT
Gettysburg College

POLYVINYLPYRROLIDONE

Polyvinylpyrrolidone is a water soluble polymer characterized by its unusual complexing and colloidal properties and its physiological inertness. It is manufactured and sold in the U.S. as PVP. Special pharmaceutical grades of PVP are available under the trademarks Plasdone and Plasdone C (GAF Corporation) manufactured in the U.S. and Kollidon 25 and 17 (Badische Anilin- und Soda-Fabrik) manufactured in Germany. Special grades of PVP for the beverage industry are sold under the trademarks Polyclar L and Polyclar AT (GAF Corporation). Peregal ST leveling and stripping agents (GAF Corporation) and Albigen A (Badische Anilin- und Soda-Fabrik) are aqueous solutions of PVP offered to the textile industry for special applications.

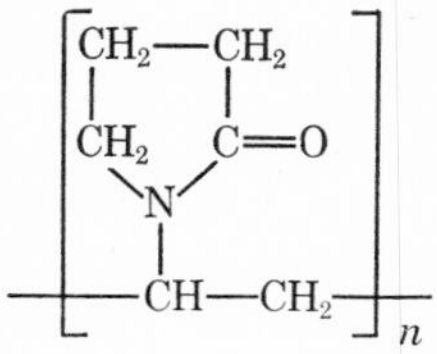

PVP is available as a white free-flowing powder and also in the form of aqueous solutions. It is offered in four viscosity grades in the U.S.

The commercial uses of PVP are related to its outstanding properties. Its film-forming and adhesive qualities are utilized in aerosol hair sprays, adhesives, and lithographic solutions. As a protective colloid, it is used in drug, and detergent formulations, cosmetic preparations, polymerization recipes, and in pigment or dyestuff dispersions. The textile industry makes use of its dye-complexing ability to improve the dyeability of synthetic fibers and as a dye-stripping assistant. Since it complexes with tannin-like compounds, PVP is used as a clarifying agent for vegetable beverages, particularly beer.

PVP was first developed in Germany by Dr. W. Reppe of the I. G. Farben during the 1930s, and was widely used by that country as a blood plasma extender during World War II. In the U.S. PVP was under active commercial development by GAF Corporation beginning in 1950. In early 1956 commercial production was started at GAF's Calvert City, Kentucky, plant. Production of PVP at GAF's second location in Texas City, Texas, was begun in 1969.

Properties

Polyvinylpyrrolidone is manufactured in the U.S. in four viscosity grades, PVP K-15, K-30, K-60, and K-90. The number average molecular weights are about 10,000, 40,000, 160,000, and 360,000, respectively. The relation of limiting viscosity number, $[\eta]$, and weight average molecular weight can be represented by (1)

$$[\eta] = 1.4 \times 10^{-2}\bar{M}w^{0.70}$$

In 1932, Fikentscher (*Cellulose-Chem.* **13**, 58 [1932]) in an attempt to relate the relative viscosity of any polymer solution to concentration, proposed the following formula:

$$\ln n_{\text{rel}} = \left[\frac{k + 75k^2}{1 + 1.5\,kc}\right] c$$

The constant k in this equation, known as Fikentscher's constant, is a function of molecular weight. In the early development of PVP, the polymers were characterized by measuring the relative viscosity and calculating a constant, K, using a slightly modified "Fikentscher's formula" (G. M. Kline, *Mod. Plastics* **137,** Nov. 1945).

$$\frac{\log \eta_{\mathrm{rel}}}{c} = \frac{75K_o{}^2}{1 + 1.5K_oc} + K_o$$

where

$K = 1000K_o$
c = concentration in g/100 ml solution
η_{rel} = viscosity of solution compared with solvent

Since the measurements are made only with $c = 1.00$, the formula reduces to:

$$\log n_{\mathrm{rel}} = \frac{75K_o{}^2}{1 + 1.5K_o} + K_o$$

The use of K value has become well established and is retained today as a means of expressing relative molecular weight.

Aqueous solutions of PVP K-15 have relatively low viscosities, eg, a 50% solution has a viscosity of 350 cSt at 25°C. PVP K-30 in concentrations of 10% or less has little effect on the viscosity of aqueous solutions. The higher molecular weight grades of PVP show thickening action even in concentrations below 10%. The effect of concentration on kinematic viscosity is shown in Figure 1 (2).

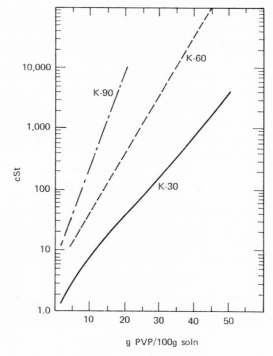

Figure 1.

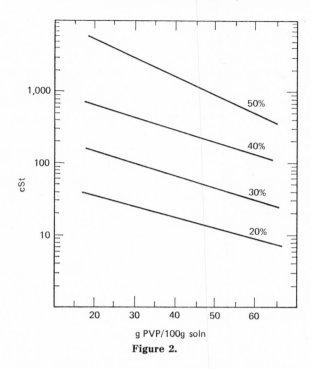

cSt

1,000

100

10

50%

40%

30%

20%

20   30   40   50   60

g PVP/100g soln

Figure 2.

Viscosity does not change appreciably over a pH range from 1–10. However, it increases in concentrated hydrochloric acid and in concentrated nitric acid forms a stable gel (3). Strong caustic precipitates the polymer. Increasing the temperature of aqueous solutions of PVP has considerable effect in lowering the viscosity. For PVP K-90, the equation log viscosity at temperature t_2 = log viscosity at t_1 − 0.01142 (t_2 − t_1) represents the relation of temperature to viscosity. For PVP K-30, see Figure 2 (4). PVP has a heat of solution of −1.15 Kcal/mole (5). Although the structural formula indicates that PVP should be neutral, an aqueous solution is slightly acid with a pH of 4–5, but without buffering action. Electrophoretic experiments show that PVP K-30 slowly migrates to the anode. Electrical conductance measurements (6) in $CHCl_3/CF_3COOH$ show that protonation of PVP increases with acid content. Spectral studies of aqueous PVP solutions (7) in the 210–222 mμ region indicated that below pH 1, PVP has a positive charge and above pH 12, a negative charge indicating enolization.

One of the unusual properties of PVP is its solubility in a wide variety of different solvents (4). The solubility of PVP in water is limited only by the viscosity of the resulting solution. PVP K-30 containing up to 5% water is soluble in alcohols, aliphatic acids, chlorinated solvents containing hydrogen, nitroparaffins, and amines. Desiccated material containing less than 0.5% water is moderately soluble in cyclic ethers such as dioxane or tetrahydrofuran but essentially insoluble in aliphatic ethers. Dry PVP is also moderately soluble in aliphatic esters, ketones, fully chlorinated hydrocarbons such as carbon tetrachloride, and in the aromatic hydrocarbons. It is insoluble in aliphatic hydrocarbons, but solutions can be prepared in such solvents as kerosene, heptane, and Stoddard solvent by using butyl alcohol as a cosolvent. Solutions in the chlorofluoroalkane propellants can be made using 20–30% ethanol.

Films of PVP which are clear, transparent, glossy, and hard can be cast from a number of different solvent systems such as water, methyl alcohol, or chloroform (4). Moisture retained or taken up from the air by PVP acts as a plasticizer for the film. In addition, the hardness of the film may be altered by adding compatible plasticizers without affecting the clarity or luster of the films. PVP films become tacky at 70% rh, and at 50% rh they contain 18% moisture. In the use of PVP films as hair fixatives, the incorporation of plasticizers and humidity control agents is necessary to obtain the best properties.

PVP is compatible with many natural and synthetic resins as well as with many other chemicals and most inorganic salt solutions (4). For example, PVP can be combined in solution or in a film with ethyl cellulose, polyethylene, poly(vinyl chloride), and poly(vinyl alcohol). It is compatible with glycerides such as olive oil and lanolin and with poly(ethylene oxide). Aqueous PVP solutions have good tolerance for inorganic salts such as ammonium chloride, copper sulfate, ferric chloride, and sodium pyrophosphate.

PVP is stable to storage under ordinary conditions (4). It was found to be essentially stable to heating in the air for 16 hr at 100°C. Some darkening in color and a lowering of water solubility are obtained on heating in air at 150°C. However, PVP appears to be quite stable if heated repeatedly at 110–130°C for relatively short intervals. Heating curves on a differential scanning calorimeter (DSC) under air or nitrogen from 35–225°C show no endotherm or exotherm. Aqueous PVP is quite stable if protected from bacteria or molds. Steam sterilization or the use of preservatives such as sorbic acid or phosphoric acid is effective. The stability of aqueous solutions to boiling is an important factor in the sterilization of the product for medical use. The PVP K-30 powder can be stored under ordinary conditions indefinitely, but excessive moisture pick-up must be prevented since the powder is hygroscopic. The equilibrium water content of PVP K-30 varies with relative humidity as shown below.

Relative humidity, %	Grams water absorbed/100 g PVP K-30
28.0	8.3
49.0	16.6
69.0	27.8
89.3	57.2

Small amounts of PVP can effectively stabilize aqueous emulsions, dispersions, and suspensions. PVP is apparently adsorbed as a thin layer on the surface of individual colloidal particles, preventing contact and stabilizing the colloid. Thus, PVP K-90 is effective in controlling the particle size in the suspension polymerization of styrene (8) and vinyl chloride (9). The stabilization of carbon black dispersions by PVP is used to demonstrate the effectiveness of PVP in preventing soil redeposition on cleaned fabrics (10). PVP prevents agglomeration of carbon black in varnish and printing ink formulas (11), and stabilizes colloidal silver against flocculation by sodium bromide (12). PVP solutions peptize and diffuse coarsely dispersed dyes such as Congo Red (13). The presence of 0.1–10% of PVP stabilizes polyacetals against oxygen and heat degradation (14–16). The above are examples of the apparent adsorption of PVP on a variety of substances. In some cases the adsorption becomes sufficiently strong to form complexes.

In aqueous solution, PVP forms complexes with many types of compounds. Insoluble complexes are formed with polyacids such as poly(acrylic acid) or tannic acid (17–20). A 1:1 addition compound of PVP with poly(acrylic acid) was isolated in 70% yield (21). These complexes are insoluble in water, alcohol, and acetone, but are dissolved by dilute alkali which apparently destroys the bond by neutralizing the polyacid. Dyes are strongly held by PVP, and this action accounts for the successful use of PVP as a dye stripping agent. Conversely, the introduction of PVP into other polymers through grafting or merely mixing greatly improves the dyeability of the polymers. PVP forms complexes with many toxins, viruses, drugs, and toxic chemicals, thereby reducing toxicity and irritation (22–29). Chemicals such as potassium cyanate, formamide, nicotine, phenols, chlorbutanol are complexed by PVP as are mercuric chloride, silver oxide, cobaltous oxide, ferric chloride, and iodine (30–34).

The complex with iodine is being marketed because of its excellent germicidal properties combined with the greatly reduced toxicity and absence of the staining action of the iodine (35,36). The iodine is held so tightly in this complex that it is not removed by extraction with chloroform and no appreciable vapor pressure of iodine is apparent above the complex.

Polyclar AT clarifier, a crosslinked form of PVP, forms a complex with hydrogen peroxide sufficiently stable such that it can be dried without destroying the hydrogen peroxide (37).

PVP is permanently insolubilized by heating with strong alkali at 100°C (4). Aqueous PVP will gel when treated with ammonium peroxysulfate at 90°C (18), with hydrazine and hydrogen peroxide (38), or with gamma rays from a Co^{60} source. (39, 40). These gels are apparently formed from permanently crosslinked PVP, since they are substantially insoluble in large amounts of water. The alkaline sodium phosphates will do the same thing. When dried under mild conditions, these gels retain their uniform structure and capacity to swell again by absorbing water. Crosslinked PVP containing glycerine is used as a dialysis membrane (41). The volume of a number of (biological) protein solutions was reduced 5–15 times by dialysis (reverse osmosis) against a concentrated (25–30%) solution of PVP using a cellophane membrane. PVP is preferred because the molecular size is sufficiently large so as not to pass through the membrane, leaving the concentrated protein solution free of PVP (42–44). PVP is also rendered insoluble by the action of oxidizing agents such as bichromate and diazo compounds under the influence of light (45–47). Resorcinol and pyrogallol also precipitate PVP from aqueous solution, but the complex redissolves in additional water and does not precipitate from alcoholic solutions (17).

Toxicity. The acute oral toxicity (48) is $LD_{50} > 100$ g/kg of body weight (white rats and guinea pigs). Acute intravenous toxicity (48): $LD_{50} = 12$–15 g/kg of body weight (white rats and guinea pigs). Chronic oral toxicity (49) was investigated by feeding albino rats 1–10% K-30 by weight in diet for up to twenty-four months. No toxic effect or significant pathology attributable to PVP was observed. Topical skin (49) studies showed that PVP was neither a primary irritant nor a sensitizer. Studies on eye irritation (50) showed that PVP caused no irritation on repeated instillation into rabbit eyes. Inhalation toxicity, as it relates to hair spray, has been extensively studied. PVP has been found to produce no harmful effects (51,52).

Analytical. Percent water is determined by Karl Fischer reagent with powders and using the Cenco moisture balance with aqueous solutions. The Cenco balance is an apparatus in which the sample is exposed to an infrared source of heat while being weighed. The scale readout is adjusted to read directly in percent weight loss. Residual monomer (N-vinylpyrrolidone) is determined by iodimetric titration; ash by ignition; heavy metals by spectrographic emission; arsenic by standard U.S.P. method; nitrogen by either Kjeldahl or Dumas methods; and acetaldehyde by hydroxylamine method. The K value distribution (Plasdone C pharmaceutical grade PVP) is determined by fractional precipitation using methanol as the solvent and diethyl ether as the precipitant.

Determination of PVP in Whole Blood, Plasma, and Urine (53). This colorimetric method is based on the binding of the dye, Brilliant Vital Red, to PVP in aqueous solution and removal of the unbound dye with *n*-butyl alcohol. Other methods are based on complex formation with iodine and spectrophotometric determination at 432 nm (54), and precipitation with trichloroacetic acid (55).

Manufacture

N-Vinyl-2-pyrrolidone, the monomer which is used to make PVP, is a colorless liquid with a freezing point of 13.6°C and a refractive index n_D^{25} of 1.511. It has a boiling point of 96°C at 114 mm and 123°C at 50 mm. Vinylpyrrolidone is completely miscible with water and most organic solvents (56). It is manufactured by the vinylation of 2-pyrrolidone with acetylene in the presence of an alkali metal salt of pyrrolidone (57). See Vol. 16, pp. 853–854.

N-Vinyl-2-pyrrolidone is readily polymerized in bulk solution or suspension to polyvinylpyrrolidone. The polymerization can be catalyzed ionically with boron trifluoride (58), potassium amide (59), or with free radical catalysts such as hydrogen peroxide, benzoyl peroxide, or azobisisobutyronitrile. Highly purified vinylpyrrolidone combines with atmospheric oxygen to give peroxide type compounds which act as polymerization catalyst (60). K_p and k_t, the absolute rate constants for propagation and termination, were determined for the azobisisobutyronitrile catalyzed polymerization at 40–70°C to be $k_p = 1.87 \times 10^3 \times e^{-7100/RT}$ and $k_t = 1.00 \times 10^9 \times e^{-1600/RT}$ (61).

Bulk polymerization was conducted in Germany with hydrogen peroxide as the catalyst. In one example, 35 kg purified vinylpyrrolidone containing 150 cc of 30% hydrogen peroxide was heated to 110°C. The temperature rose rapidly to 190°C because of the exothermic reaction. The molten polymer was poured from the kettle, cooled, and ground. Although this process led to a discolored product which was difficult to purify, it was used to manufacture the grade of polyvinylpyrrolidone needed as a blood plasma expander because the low viscosity obtained was valuable for that application.

Later an improved process was developed at Ludwigshafen by Fikentscher and co-workers (57). In this process a 30% solution of purified vinylpyrrolidone in water was prepared and half of it added to a kettle at 80°C. The remainder was added portionwise during the reaction. The polymerization of this solution was catalyzed by the addition of up to 0.2% hydrogen peroxide and half as much ammonia. The polymerization reaction was completed in 2–3 hours. The ammonia served as a buffer to prevent the solution from becoming acidic enough to cause the splitting off of acetaldehyde from the monomer. In addition, the ammonia has a strong activating effect on the polymerization. The same sort of action can be obtained with amines and their salts. Polyvinylpyrrolidone for Periston, the German blood plasma substitute, was further purified by drying and extracting the dried product with diethyl ether. The redried powder was then dissolved in water to 20% concentration, filtered, and sterilized at 120°C (57).

The rate of the aqueous hydrogen peroxide catalyzed polymerization is greatly affected by the concentration of ammonia, peroxide, and monomer as well as by the pH and temperature (62). The interrelation of these factors has been expressed by Woodhams (63).

$$\text{Rate of polymerization} = K[\text{HOOH}]^{1/2}[\text{NH}_3]^{1/4}[\text{VP}]^{3/2}$$

The rate of polymerization is directly proportional to the monomer concentration up to about 50%. Beyond 60% the rate falls off extremely rapidly. The rate of polymerization is insensitive to pH in the range 7–12. Above 12 the rate slows and above 13 is completely inhibited (63). An induction period is observed in the hydrogen peroxide catalyzed polymerization which can be shortened by excluding oxygen or by increasing the catalyst concentration (62). The use of ammonia or an amine greatly reduces this induction period.

The molecular weight of PVP in the ammonia–hydrogen peroxide system increases slightly with increasing ammonia concentration and is directly proportional to monomer concentration up to about 30%. Above 30% concentration, the molecular weight is dependent almost exclusively upon catalyst and is inversely proportional to the catalyst concentration.

Vinylpyrrolidone has been copolymerized with a wide variety of comonomers in solution and in emulsion systems. Table 1 lists a number of those monomers with their reactivity ratios.

In addition, vinylpyrrolidone has been successfully copolymerized with the following comonomers:

ethylene glycol monovinyl ether	(66)	divinyl carbonate	
ethylene	(67)	cinnamic acid	
laurylacrylamide	(68)	crotonaldehyde	(65)
C_{12}–C_{18} methacrylate			

Table 1. Reactivity Ratios for Free Radical Copolymerization of Vinylpyrrolidone M_1

Comonomer (M_2)	r_1	r_2
acrylonitrile	0.06 ± 0.07	0.18 ± 0.07
allyl alcohol	1.0	0.0
allyl acetate	1.6	0.17
allylidene diacetate (64)	0.92	0.94
crotonic acid (65)	0.85	0.02
maleic anhydride	0.16 ± 0.03	0.08 ± 0.03
methyl methacrylate	0.005 ± 0.05	4.7 ± 0.05
	0.02 ± 0.02	5.0
styrene	0.045 ± 0.05	15.7 ± 0.5
trichloroethylene	0.54 ± 0.04	<0.01
tris(trimethylsiloxy)vinylsilane	4.0	0.1
vinyl acetal	3.30 ± 0.15	0.205 ± 0.015
	2.0	0.24
vinyl chloride	0.38	0.53
vinyl cyclohexyl ether	3.84	0.0
vinyl phenyl ether	4.43	0.22
vinylene carbonate	0.4	0.7

A typical copolymerization recipe employs solvents such as alcohol (67) or benzene, a reaction temperature of 50–75°C and 0.1–1.0% catalyst. Benzoyl peroxide-lauroyl peroxide and azobisisobutyronitrile have been successfully employed. Vinyl acetate copolymers of varying proportions are marketed by Badische Anilin- und Soda-Fabrik and by GAF Corporation. See Table 2.

Table 2. Vinyl Acetate Copolymers

BASF Name	GAF Name	Vinylpyrro-lidone, %	Vinyl acetate, %	Solvent
	PVP/VA E-335	30	70	ethanol
	PVP/VA E-535	50	50	ethanol
	PVP/VA E-635	60	40	ethanol
	PVP/VA E-735	70	30	ethanol
Luviskol VA-37	PVP/VA I-335	30	70	2-propanol
	PVP/VA I-535	50	50	2-propanol
	PVP/VA I-735	70	30	2-propanol
Luviskol VA-64 Pulver	PVP/VA S-630	60	40	powder

PVP is highly susceptible to grafting and has been used to form graft copolymers of styrene and acrylic esters, employing emulsion polymerization techniques (69–71). Improved nylon-6,6 with a high degree of wet-crease recovery has been made by grafting PVP. The PVP is activated with hydrogen peroxide, dried, and then added to molten hexamethylenediammonium adipate at 210°C. The mixture is heated 2 $\frac{1}{2}$ hr to 285°C in a steam atmosphere to give a melt spinnable polymer (72).

Poly(dimethylsiloxane) sheets were irradiated in the presence of vinylpyrrolidone with high energy electrons to form surface and homogeneous graft polymers (73). Cellulose was treated with ferrous salt at a pH of 4–5, washed, immersed in an aqueous solution of vinylpyrrolidone, and treated with hydrogen peroxide. On heating to 70°C, grafting occurred.

PVP can be alkylated with, for example, 1-eicosene ($CH_2 = CH(CH_2)_{17}CH_3$) with the use of a free radical catalyst such as di-*tert*-butyl peroxide to produce polymers with increased oil solubility (74).

Economics. Since there is only one producer in the U.S., PVP production figures are not available. The price structure as of 1969 is shown in Table 3.

Table 3. The Price Structure of PVP

Grade	Bulk price, $ per lb
nonpharmaceutical	
PVP K-15	1.65
PVP K-30	1.36
PVP K-60	1.38[a]
PVP K-90	1.80[a]
PVP K-90 powder	1.85
Polycar AT brand	2.50
pharmaceutical	
Plasdone brands	3.05
Plasdone C (plasma grade) brand	3.63

[a] Based on solids.

Table 4. Nonpharmaceutical Grades

Viscosity grade	Form	K range	Water, %	Ash (dry basis), %	Residual monomer,[a] %
PVP K-15	powder	12–18	5 max	0.02 max	1.0 max
PVP K-30	powder	26–35	5 max	0.02 max	1.0 max
PVP K-60	aqueous solution	50–62	55 max	0.02 max	1.0 max
PVP K-90	aqueous solution	80–100	80 max	0.02 max	1.0 max
PVP K-90	powder	80–100	5 max	0.02 max	1.0 max
Polyclar AT brand	powder	crosslinked	5 max	...	...

[a] Calculated as vinylpyrrolidone, % based on dry solids.

Specifications. The specifications for nonpharmaceutical and pharmaceutical grades of PVP are shown in Tables 4, 5, and 6.

Pharmaceutical grades of PVP (all powders) are marketed under the Plasdone and Plasdone C trademarks. Three viscosity types for nonplasma use are: K-26–28, K-29–32, and K-33–36, having the specifications shown in Table 5.

Table 5. Pharmaceutical Grades for Nonplasma Use

moisture content	less than 5%
unsaturation (calculated as monomer)	less than 1%
ash	less than 0.02%
heavy metals content	less than 20 ppm
arsenic content	less than 2 ppm
nitrogen content	12.6% ± 0.4%

Plasdone C pharmaceutical grade PVP, specifically for plasma use, has the specifications shown in Table 6.

Table 6. Pharmaceutical Grade for Plasma Use

K-value	30 ± 2
K-value distribution	not more than 15% greater than K-41
	not more than 25% lower than K-16
moisture content	less than 5%
unsaturation (calculated as monomer)	less than 1%
ash	less than 0.02%
heavy metals content	less than 20 ppm
arsenic content	less than 2 ppm
nitrogen content	12.6% ± 0.4%
acetaldehyde content	less than 0.5%

Shipping Containers. PVP K-15, K-30, and K-90, powder, are shipped in polyethylene bags inside fiber drums (100 lb net). PVP K-60 and K-90, aqueous solutions, are shipped in aluminum tank cars, and 55 gal polyethylene-lined fiber drums.

Uses

Cosmetics (qv) **and Toiletries.** PVP (75) is used in hand creams, pomades, hair lotions, shaving creams, shampoos (17,31) hair tints (48), pre-electric and after-shave lotions (76,77) etc, because of its emulsifying, thickening, emollient, and dye-solubilizing properties. Its film-forming ability and excellent adhesion to hair has led to its widespread use in aerosol hair sprays (78–82). Dentrifrices containing PVP have enhanced ability to remove dental stains (83, 84).

Textiles. Incorporation of PVP into hydrophobic fibers, such as polyacrylonitrile (85–87), polyesters (88,89), nylon (90,91), viscose (92), natural rubber (93,94) and polypropylene (95–97) greatly increases their dyeability with most classes of dyestuff. PVP K-30 is a stripping agent for removing vat, sulfur, and direct colors from dyed fabrics (98). As a suspending agent, it is used in the scouring of nylon to remove graphite (used in lace manufacture) (77) and in print washes to scavenge loose color. PVP is a suspending agent for titanium dioxide in delustering nylon-6 (77). It is used in fugitive tints (99,100) and as an auxiliary retarding agent in pastel dyeing. PVP is used in sizing glass fibers (101,102). Graft copolymers of PVP with nylon-6,6 exhibit improved wet-crease recovery and moisture regain (103,104).

Detergents. PVP is compatible in clear, liquid, heavy-duty detergent formulations (105). It prevents soil redeposition, particularly on synthetic fibers and resin treated fabrics (10,106,107). It also acts as a loose color scavenger during laundering because it has dye-binding properties.

Beverages. The ability of PVP to complex with certain tannins has been applied to the clarification and chillproofing of various vegetable beverages (108). The addition of 0.01–0.02% to the brew kettle results in improved taste and reduced chill haze while allowing a reduction in the hops used (109,110). PVP is used similarly in wines (111), whiskey, vinegar, etc (112). A special high-molecular-weight crosslinked grade of PVP, Polyclar AT clarifier, insoluble in water, organic solvents, and strong mineral acid and alkali, is particularly effective in clarifying beer (113) and other vegetable beverages. It also controls browning in wines (114).

Pharmaceuticals. A special grade, Plasdone C pharmaceutical grade PVP is marketed for this purpose. Its suspending and drug-retardant properties are utilized in injectible preparations of antibiotics, hormones, analgesics, etc (115,116). PVP is used as tablet binder and tablet coating (116–118). It is used in aqueous or organic-solvent systems, including effervescent products prepared in an anhydrous medium (119). PVP is used in layered tablets and timed-release capsules (120). PVP is effective in stabilizing vitamins (121) and aspirin tablets (122). It acts as a protective colloid and reduces drug irritation in ophthalmic (50) and topical preparations. The ability of PVP to form a stable complex with iodine has led to the development of PVP-Iodine. This product retains the germicidal properties of iodine while drastically reducing the iodine's toxicity toward mammals (36,123). Aqueous solutions of PVP-Iodine form fibers that protect open wounds but are water-washable and do not permanently stain skin, natural fibers, or hard surfaces. PVP-Iodine has been effective as surgical scrub (124) and in treatment of burns (125). Gargles, tinctures, and ointments containing PVP-Iodine are being marketed under the trademark Isodine (Purdue Fredericks Co.).

Plasma Extenders. During World War II, more than a million persons of all ages received emergency administration of PVP solutions to combat shock due to blood loss. In the period 1948–50, thousands of units of PVP were evaluated in clinical tests of human volunteers in the USA. Together with cooperative research by the National Research Council on synthetic plasma expanders, these studies resulted in approval for the emergency stockpiling of PVP by local communities. PVP has been used with outstanding success in the treatment of shock due to severe burns, accidents, or surgical procedures. Moreover, it is nonantigenic, requires no crossmatching and avoids the dangers of infectious hepatitis which are inherent in natural blood. However, plasma or blood of the proper type is still preferred by American medical authorities and PVP solutions are for emergency use when the former are unavailable.

Dyestuff and Pigment Dispersions. In inks based on dyes, PVP improves solubility and gives greater color value per pound of dye. In pigmented inks, PVP is used in the milling operation to give higher tinctorial strength, dispersion stability, and better gloss. It has found use in the preparation of color dispersions for latex paints, paper coatings, and plastics.

Automotive Products. Copolymers of PVP and alkyl acrylates are used in formulations for high-detergency motor oil (126,127), automatic transmission fluids (129,130), and grease (131,132).

Plastics. In two-phase polymerization systems for styrene, vinyl chloride, vinyl esters, acrylics, and other monomers, PVP acts as a particle size regulator, suspending agent, and viscosity modifier (133–135). PVP is also used as a post polymerization additive to improve dyeability and stability of various latexes (93,136). The addition of PVP to polyoxymethylenes (see Acetal resins) improves heat stability of the dried polymer (16).

Miscellaneous. In wax and polish formulations, PVP functions as a protective colloid and film former to give better cleansing action and gloss. In the paper industry, PVP is used in the control of deposition of pitch (137), in the deinking of waste paper (138), in decolorizing rag stock (98), and as a viscosity modifier for paper coatings (139). Its use in lithographic plates (140) and in the preparation of photographic emulsions has been cited (141–143). Uses in water-remoistenable and pressure-sensitive adhesives have been mentioned (17,144,145). PVP functions as a protective colloid in cement mixtures, retarding the separation of water (146), in preventing caking of fertilizer mixes (147), and preparation of conductive coatings for television or cathode-ray tubes (148).

Bibliography

"Polyvinylpyrrolidone" in ECT 1st ed., Vol. 10, pp. 759–764.

1. H. B. Frank and G. B. Levy, *J. Polymer Sci.* **10**, 371 (1953).
2. G. G. Stoner and J. L. Azorlosa, *Paper Am. Chem. Soc. Meeting, Div. Polymer Chem., Dallas, Texas, April 1956.*
3. U.S. Pat. 3,336,981 (Aug. 22, 1967), H. Barron, C. Crowe, and B. Atkins (to Dow Chem. Co.).
4. Technical Bulletin 7543-113, GAF Corp., Dyestuff & Chem. Div.
5. G. Silva, *An. Fac. Quim. Farm. Univ. of Chile* **18**, 126–132 (1966).
6. M. A. Stake and I. Klotz, *Biochemistry* **5** (5), 1726–1729 (1966).
7. G. T. Walker, *Seifen, Oele, Fette, Wachse* **89** (19), 589–590 (1963).
8. Belg. Pat. 668,325 (Dec. 1, 1965), K. Buchholz (to Badische Anilin- und Soda-Fabrik).
9. U.S. Pat. 2,840,549 (June 24, 1958), D. G. McNulty and R. I. Leininger (to Diamond Alkali).
10. F. V. Nevolin and T. G. Tipisova, *Maslob. Zhur. Prom.* **28** (2), 18–20 (1962).
11. E. R. Honak, *Plaste Kautschuk* **11** (6), 372–377 (1964).
12. E. Jirgensons, *Makromol. Chem.* **6**, 30 (1951).
13. H. Bennhold and T. Schubert, *Z. Ges. Exptl. Med.* **113**, 722 (1943).
14. Belg. Pat. 606,878 (Feb. 1962) (to Farbwerke Hoechst).
15. U.S. Pat. 3,204,014 (Aug. 1965), R. Green (to Tenneco Chem. Co.).
16. U.S. Pat. 3,219,727 (Nov. 23, 1965), R. J. Kray and T. J. Dolce (to Celanese).
17. E. Benk, *Chemiker Ztg.* **78**, 41 (1954).
18. U.S. Pat. 2,658,045 (Nov. 3, 1953), C. E. Schildknecht (to GAF Corp.).

19. H. Willenegger, *Schweiz. Med. Wochschr.* **77**, 614 (1947).
20. B. Schille and J. Neel, *J. Chem. Phys.* **60** (4), 475–491 (1963).
21. F. Boyer-Kawenoki, *Compt. Rend.* **263** (3), 203–205 (1966).
22. J. Dieckhoff, *Arch. Kinderheilk.* **139**, 52 (1950).
23. J. Dieckhoff, *Z. Kinderheilk.* **70**, 177 (1951).
24. J. Dieckhoff and H. J. Ludwig, *Arch. Kinderheilk.* **137**, 160 (1949).
25. H. Muhlbauer, *Munch. med. Wochschr.* **94**, 2486 (1952).
26. R. Schubert, *Arzneimittel-Forsch.* **4**, 42 (1949).
27. R. Schubert, *Deut. med. Wochschr.* **76**, 1487 (1951).
28. R. Schubert, *Arzneimittel-Forsch.* **3**, 425 (1949).
29. J. Stroder and J. Hockerts, *Deut. med. Wochschr.* **74**, 282 (1949).
30. J. Weidener, *Deut. med. Wochschr.* **78**, 1192 (1953).
31. J. M. Wilkinson, G. G. Stoner, E. P. Hay, and D. B. Witwer, *Papers CSMA Proc. 40th Mid-Year Meeting, Cincinnati,* 1954.
32. B. N. Kabadi and E. R. Harnarlund, *J. Pharm. Sci.* **55** (10), 1069–1074 (1966).
33. C. K. Bahal and H. B. Kostenbauder, *J. Pharm. Sci.* **53** (9), 1027–1029 (1964).
34. U.S. Pat. 3,082,193 (March 1963), M. Mendelsohn (to Yardney Int. Corp.).
35. Technical Bulletin 7543-004, GAF Corp., Dyestuff & Chem. Div.
36. U.S. Pat. 2,739,922 (March 27, 1956), H. A. Shelanski (to GAF Corp.).
37. U.S. Pat. 3,376,110 (April 2, 1968), D. Shiraeff (to GAF Corp.).
38. Brit. Pat. 1,022,945 (March 16, 1966), E. V. Hort and F. Grosser (to GAF Corp.).
39. A. J. Henglein, *J. Phys. Chem.* **63**, 1852–1858 (1959).
40. A. Charlesky and P. Alexander, *J. Chim. Phys.* **52**, 699–709 (1955).
41. U.S. Pat. 3,276,996 (Oct. 4, 1966), Leon Lazara.
42. R. Plan and M. T. Fayet, *Ann. Biol. Clin.* **19**, 169–170 (1961).
43. D. Leka, R. Moret, R. Sohier, and J. Tigaud, *Am. Inst. Pasteur.* **90** (6), 770–778 (1956).
44. R. Moret, A. Boucherie, and M. Lambert, *Compt. Rend.* **149** (7, 8), 719–721 (1955).
45. U.S. Dept. Comm. Office Tech. Serv. PB Report 4116, 327 (1945).
46. U.S. Dept. Comm. Office Tech. Serv. PB Report 1308, 223 (1945).
47. U.S. Dept. Comm. Office Tech. Serv. PB Report 78256, 616 (1947).
48. H. A. Shelanski, M. V. Shelanski, and A. Cantor, *J. Soc. Cosmetic Chemists* **5**, 129 (1954).
49. M. V. Shelanski, private communication.
50. T. M. Schieffelin & Co., New York, Polygyl Bulletin.
51. P. Alexander, *Mfg. Chemists* **35** (8), 61–65 (Aug. 1964).
52. J. E. Epsom, *Med. News,* 10 (June 19, 1964).
53. F. P. Chinard, *J. Lab. Clin. Med.* **39** (4), 666 (1952).
54. R. L. Larkin and R. E. Kupel, *Am. Ind. Hyg. Assoc. J.* **26** (6), 558–561 (1965).
55. A. Meiger, *Clin. Chem. Acta* **6**, 713–714 (Sept. 1961).
56. Technical Bulletin 7543-037, GAF Corp., Dyestuff & Chem. Div.
57. C. E. Schildknecht, *Vinyl and Related Polymers,* John Wiley & Sons, Inc., New York, 1952, pp. 662–678.
58. C. E. Schildknecht, A. O. Zoss, and F. Grosser, *Ind. Eng. Chem.* **41**, 2891 (1949).
59. *Chem. Abst.* **62**, 640a (1965).
60. F. Cech, *Polymerization and Copolyermization of Vinyl Pyrrolidone,* Univ. Vienna, April 24, 1957.
61. V. A. Agasandyan, E. A. Trosman, Kh. S. Bagdasaryan, A. D. Litmanovich, and V. Ya Shtern, *Vysokomolekul. Soedin.* **8** (9), 1580–1585 (1966).
62. J. W. Copenhaver and M. H. Bigelow, *Acetylene and Carbon Monoxide Chemistry,* Reinhold Publishing Corp., New York, 1949, p. 67–74.
63. R. Thomas Woodhams, *The Kinetics of Polymerization of Vinyl Pyrrolidone,* Polytechnic Institute of Brooklyn, April 1954.
64. T. Minakova, *Izv. Akad. Nauk Kaz SSR, Ser. Khim. Nauk* **10**, 1880 (1965).
65. S. N. Ushakov et al., *Vysokomolekul. Soedin.* **A 9** (8), 1807 (1967).
66. U.S. Pat. 3,149,091 (Sept. 15, 1964), G. Sterling and R. Zimmerman (to Dow Chem. Co.).
67. Fr. Pat. 1,392,354 R. Resz and H. Barth (to Farbenfabriken Baeyer).
68. Fr. Pat. 1,391,571 (March 5, 1965), B. Malone and J. Coon (to Shell Int.).
69. U.S. Pat. 3,244,657 (April 1966), F. Grosser and M. R. Leibowitz (to GAF Corp.).

70. U.S. Pat. 3,244,658 (April 1966), F. Grosser and M. R. Leibowitz (to GAF Corp.).
71. Ger. Pat. 1,156,238 (Oct. 24, 1963), F. Grosser and M. R. Leibowitz (to GAF Corp.).
72. U.S. Pat. 3,287,441 (Nov. 22, 1966), E. Magat (to E. I. du Pont de Nemours).
73. H. Yasuda and M. F. Refojo, *J. Polymer Sci. Part A-2* (12), 5093–5098 (1964).
74. Brit. Pat. 1,076,543, A. Merijan, F. Grosser, and E. V. Hort (to GAF Corp.).
75. Technical Bulletin 7543-066, GAF Corp., Dyestuff & Chem. Div.
76. Ger. Pat. 873,891, H. Kroper (to Badische Anilin- und Soda-Fabrik).
77. R. J. Holmes and D. B. Witwer, Proc. Am. Assoc. Textile Chemists Colorists, *Am. Dyestuff Reptr.* **44**, 702 (1955).
78. *Chem. Eng. News* **34**, 500 (1956).
79. *Drug Trade News* **28**, 43 (1953).
80. *Drug Trade News* **29**, 53 (1954).
81. P. Lenthen, *Am. Perfume Cos.* **81**, 53–57 (1966).
82. R. Micchelli and A. Korte, *Drug Trade News* **26**, 30 (1963).
83. Union S. Africa Pat. 17,493, M. F. Nelson (to Colgate-Palmolive Co.).
84. Belg. Pat. 659,791 (June 16, 1965), J. M. Marcel (to Bleirot).
85. Brit. Pat. 758,937 (to Dow Chem. Co.); Australian Pat. 167,575.
86. U.S. Pat. 3,296,741 (Jan. 13, 1967), H. Brian and P. Hurley (to Dow Chem. Co.).
87. *Chem. Week* **57**, July 8, 1961.
88. Swiss Pat. 382,338 (Nov. 30, 1964), H. Kirner (to Rohner A.G. Pratteln).
89. U.S. Pat. 2,882,255 (April 14, 1959), J. R. Caldwell (to Eastman Kodak).
90. U.S. Pat. 3,105,732 (Oct. 1, 1963), H. Artheil (to Burlington Industries).
91. Fr. Pat. 1,448,353 (Aug. 5, 1966), E. Magat (to E. I. du Pont).
92. U.S. Pat. 3,377,412 (April 9, 1968), N. Franks (to Am. Enka Corp.).
93. Ger. Pat. 1,258,822 (Jan. 18, 1968), J. Swigett (to Uniroyal, Inc.).
94. U.S. Pat. 3,138,431 (Sept. 23, 1964), J. Swigett (to U.S. Rubber Co.).
95. Brit. Pat. 995,802 (June 23, 1965), G. Bryant and N. Zutty (to Union Carbide Corp.).
96. Japan. Pat. 27,636 (Dec. 2, 1964), K. Miyamichi (to Nitto Boseki Co. Ltd.).
97. Brit. Pat. 987,421 (March 31, 1965), J. J. Press.
98. E. C. Hansen, C. A. Bergman, and D. B. Witwer, Proc. Am. Assoc. Textile Chemists Colorists, *Am. Dyestuff Reptr.* **43**, 72 (1954).
99. H. C. Olpin and A. J. Wesson, *J. Soc. Dyers Colourists* **69**, 357 (1953).
100. U.S. Pat. 2,959,461 (Nov. 8, 1960), E. M. Murry.
101. U.S. Pat. 3,207,623 (Sept. 21, 1965), A. Marzocchi and N. S. Janetos (to Owens-Corning).
102. Neth. Pat. 6,400,435 (Sept. 22, 1964) to Pittsburgh Plate Glass Co.
103. U.S. Pat. 3,287,441 (Nov. 22, 1966), E. E. Magat (to E. I. du Pont de Nemours & Co.).
104. U.S. Pat. 3,278,639 (Oct. 11, 1966), O. J. Matray (to E. I. du Pont de Nemours & Co.).
105. B. M. Milwillsky, *Soap Chem. Specialities* **39** (4), 53–56 (1963).
106. L. Eichelmannova, *Prumsyl Potravin* **14**, 660–661 (1963).
107. U.S. Pat. 3,318,816 (May 9, 1967), J. Trowbridge (to Colgate-Palmolive Co.).
108. U S. Pat. 2,688,550 (Sept. 7, 1954), W. D. McFarlane (to Canadian Breweries, Ltd.).
109. U.S. Pat. 2,939,791 (June 7, 1960), W. D. McFarlane (to Canadian Breweries, Ltd.); Brit. Pat. 854,455.
110. W. D. McFarlane, E. Wye, and H. L. Grant, *Papers Europ. Brewing Conven. Cong.*, Elsevier Pub., New York, pp. 298–310.
111. R. A. Clemens and A. J. Martinelli, *Wines and Vines* **39**, 55 (1958).
112. U.S. Pat. 3,222,180 (Dec. 7, 1965), R. Sucietto (to National Distiller & Chem. Co.).
113. Technical Bulletin 7543-085, GAF Corp., Dyestuff & Chem. Div.
114. A. Caputi and R. G. Peterson, *Am. J. Enol. and Viticult.* **16** (1), 9–13 (1965).
115. Technical Bulletin AP-123, GAF Corp., Dyestuff & Chem. Div.
116. M. A. Lesser, *Drug & Cosmetic Ind.* **75**, 1 (1954).
117. U.S. Pat. 2,820,741 (Jan. 21, 1958), C. J. Endicott, T. A. Prickett, and A. A. Dallanis (to Abbott Laboratories).
118. F. J. Prescott, *Drug & Cosmetic Ind.* **97** (4), 497, 621, 622 (1965).
119. U.S. Pat. 3,136,692 (June 9, 1964), F. J. Bandellin (to Strong-Cobb-Arner).
120. U.S. Pat. 3,102,845 (Sept. 3, 1963), M. Fennell (to McNeil Lab.).
121. Brit. Pat. 966,685 (Aug. 12, 1964) (to Dagra Feed Prod.).

122. St. Nikolov, L. Nedeleva, and E. Velrkova, *Farmatsiya* **17** (3), 60–63 (1967).

123. R. C. Bogash, *Bull. Am. Soc. Hosp. Pharm.* **13**, 226 (1956).

124. J. F. Connel and L. M. Rousselot, *Am. J. Surg.* **108**, 849–854 (1964).

125. D. Wynn-Williams and A. Monballin, *Brit. J. Plastic Surg.* **18** (2), 146–150 (1965).

126. Brit. Pat. 1,035,011 (July 6, 1966) (to Shell Inter. Research).

127. U.S. Pat. 3,153,640 (Oct. 20, 1964), E. Barnum and L. Lorenson (to Shell Oil Co.).

128. Neth. Pat. 6,605,998 (Dec. 5, 1966) (to Rohm & Haas Co.).

129. U.S. Pat. 3,175,976 (March 30, 1965), E. G. Foehr (to Calif. Research Corp.).

130. U.S. Pat. 3,311,560 (March 28, 1967), B. Hotten (to Chevron Research Co.).

131. U.S. Pat. 3,159,575 (Dec. 1, 1964), Dean Criddle (to Calif. Research Corp.).

132. Brit. Pat. 1,039,166 (Aug. 17, 1966) (to Chevron Research Co.).

133. Technical Bulletin AP-142 (1963), GAF Corp., Dyestuff & Chem. Div.

134. H. Gerrens, H. Ohlinger, and R. Fricker, *Hundert Jahre BASF, Aus Forsch.*, 443–449 (1965).

135. G. Schulz, *Faber Lack* **66** (11), 621–624 (1960).

136. S. Susskind, *J. Appl. Polymer Sci.* **9** (7), 2451–2458 (1965).

137. U.S. Pat. 3,081,219 (March 12, 1963), T. J. Drennen and L. E. Kelley (to Rohm & Haas Co.).

138. Can. Pat. 694,471 (Sept. 15, 1964), S. U. Hossain and A. A. Pataki (to Abitiki Power and Paper).

139. M. R. Leibowitz, E. M. Wolfe, N. Landberg, and B. Webster, *Tappi* **48**, 73A–76A (Jan. 1968).

140. G. W. Jorgensen, *Proc. Tech. Assoc. Graphic Arts, 4th Annual Meeting, 1952*, p. 94.

141. U.S. Pat. 3,167,429 (Jan. 26, 1965), Marilyn Levy (to Sec. of the Army, U.S.A.).

142. Fr. Pat. 1,468,094 (Feb. 3, 1967), K. Kubodera (to Fuji Photo Film Co.).

143. Fr. Pat. 1,364,027, 1,364,028 (R. P. J. G. Thielaut and J. P. C. G. Dubosc) (to Kodak Pathe).

144. P.B. Report 95355, I. G. Farbenindustrie, Misc. Repts. relating to Kollidons; Bibliog. Tech. Reports, U.S. Dept. Commerce 11, 373 (1949).

145. U.S. Pat. 3,096,202 (July 2, 1963), E. de G. von Arx (to Johnson & Johnson).

146. U.S. Pat. 3,359,225 (Dec. 19, 1967), to Charles P. Weisend.

147. Belg. Pat. 661,598 (July 16, 1965), to Kao Soap Co., Ltd.

148. Neth. Pat. 6,608,315 (Dec. 6, 1966), to Acheson Industries, Inc.

A. S. WOOD
GAF Corporation

2-VINYLPYRIDINE, $CH_2{=}CHC_5H_4N$. See Vol. 16, p. 803.

VINYLTOLUENE, $CH_2{=}CHC_6H_4CH_3$. See Vol. 19, p. 77.

VINYON AND RELATED FIBERS

The Textile Fiber Products Identification Act of 1960 defines "vinyon" as any fiber containing at least 85% by wt of vinyl chloride units. Under this definition all homopolymer PVC fibers as well as the vinyl chloride–vinyl acetate copolymer (which is 86% vinyl chloride) are legally designated as vinyon. In the United States, however, the vinyl acetate copolymer known as Vinyon HH is the only vinyl chloride fiber manufactured in reasonably large quantities. It should be emphasized, however, that the generic term "vinyon fiber" may be applied to homopolymeric PVC or other copolymers containing 85% or more vinyl chloride.

Historically, poly(vinyl chloride) enjoys the distinction of being the first synthetic polymer to be fabricated into man-made fibers. A patent issued in 1913 (1) speaks of making artificial threads by coagulating a solution of poly(vinyl chloride) in chlorobenzene. This work, however, was somewhat premature and serious development of PVC fibers did not begin until the early thirties in both Germany and the United States.

The initial obstacle to commercial development was the insolubility of the material in the solvents commercially available at that time. This problem was overcome in Europe in two ways: The Germans developed a technique of "post-chlorination" of PVC which rendered it soluble in acetone (2), while the French discovered that PVC could be dissolved in mixed solvents, especially in acetone–carbon disulfide mixtures (3). The German fibers were designated PeCe fibers and the French Rhovyl. A subsequent German discovery was the solubility of PVC in tetrahydrofuran, which led to the development of PCU fibers which were not post-chlorinated.

In the United States the solubility problem was solved by the Carbide and Carbon Chemicals Company, by copolymerizing vinyl chloride with vinyl acetate and other monomers. In particular, the copolymer with vinyl acetate was found to be soluble in acetone and was developed into a commercial fiber by this company in cooperation with the American Viscose Corporation. Another copolymer developed by Carbide and Carbon was that using acrylonitrile as comonomer. See Dynel under Acrylic and modacrylic fibers.

In Japan a fiber is being produced, under the name Cordela, consisting of vinyl chloride and vinyl alcohol. The Italian firm Applicazioni Chimica S.p.A. has commercialized a fiber based on the discovery that polymerizing PVC at temperatures substantially below $-20°C$ results in increased tacticity, giving increased crystallinity, higher softening temperatures, and higher glass transition temperatures (8,9). This fiber, Leavin, shows superior heat and solvent resistance (11).

Preparation of Polymers

Post-chlorinated PVC. These fibers, called PeCe, were first produced commercially (12) in 1934 and are still being produced in limited quantity in East Germany. In the process believed to be used, a solution of 7–8% PVC in 1,1,2,2-tetrachloroethane is saturated with chlorine gas at 70°C. The heat of reaction is continuously removed and the temperature kept below 120°C. After 24–40 hr the solution is cooled, excess chlorine and hydrochloric acid are removed under vacuum, and the chlorinated polymer is precipitated with methanol, filtered, washed with methanol, and dried. The chlorine content of the resulting polymer is raised from 56% to 62–65% by this process, and the resulting polymer is soluble in acetone and thus ready for dry or wet spinning.

Vinyl Chloride–Vinyl Acetate Copolymer. This material, Vinyon HH, is a copolymer containing 86% vinyl chloride and 14% vinyl acetate; it is soluble in acetone and hence easily dry-spun into fibers. The polymer is made by free-radical copolymerization of vinyl chloride and vinyl acetate in solution, using a catalyst such as benzoyl peroxide (17,26), at temperatures around 30°C. The molecular weight should be carefully controlled in order to obtain uniformity of fiber properties and optimum solubility in acetone.

The resulting material consists of a random copolymer of vinyl chloride and vinyl acetate, linked in a head-to-tail fashion. Work by Douglas and Stoops (18) showed that the chlorine content of the copolymer, and hence the mole ratio of vinyl chloride to vinyl acetate, did not vary greatly in the molecular weight range from 5,800 to 15,800. Careful control of time and temperature are required to give a reproducible molecular weight and a narrow molecular weight distribution. Control of color is also important for fiber applications, and excellent colorless polymers are said to be produced by this process.

The polymer used in the production of Cordela fibers, according to the patent literature (27), is a mixture of poly(vinyl alcohol) and a copolymer of vinyl chloride and vinyl alcohol. An emulsion of vinyl chloride and poly(vinyl alcohol) is made (about 10:1, VC:PVA) containing a catalyst, eg, potassium peroxysulfate. Polymerization takes place in an autoclave at 45°C, during which the PVC grafts onto the PVA. After polymerization additional PVA is added to bring the mole ratio of PVC to PVA to approximately unity. This resulting mixture then forms the spinning dope from which the fibers are produced.

Spinning

There are three general methods of forming fiber from polymers: wet spinning, dry spinning, and melt spinning. The first two have been used for vinyl fibers, but melt spinning cannot be used because thermal degradation of the polymer takes place at the melting point (or flow point) of PVC (about 175°C). The molten polymer is also degraded by high shear rates commonly encountered in melt extruders. These two factors lead to off-color fibers, and black particles which plug jet holes. In extruding the polymer for plastic articles, film, and sheet, stabilizers are added in large quantities. These also act as plasticizers and would seriously degrade the physical properties if used to form fibers.

In wet and dry spinning of PVC the polymer is first dissolved in a suitable solvent before extrusion through the spinneret at moderate temperatures. In wet spinning, the emerging filaments are contacted with a coagulating bath, while in dry spinning the solvent is evaporated by means of hot air.

As mentioned previously, one of the initial problems of spinning PVC was the selection of a suitable solvent. In the past two or three decades, however, a large number of solvents have become commercially available for PVC. A patent survey of the years 1947–1964 has revealed the following solvents claimed: cyclohexanone, tetrahydrofuran, acetone–CS_2 mixtures, dimethylformamide, thionyl chloride, trichloroethylene–nitromethane mixture, diacetone alcohol mixed with various ketones, N-methyltetrahydrofurfurylamine, mixtures of methyl ethyl ketone and CS_2, tetrahydrofurfuryl alcohol, tetrahydrofuran plus lactones or lactams, and tetrahydrofuran with sulfones or sulfoxides.

In spite of all the solvents patented, there are only two or three solvents or solvent systems currently in commercial use. In dry spinning, a 50:50 or 60:40 solution of benzene and acetone is used in Japan and a 50:50 mixture of CS_2 and acetone in Europe. In wet spinning regular PVC a solution in tetrahydrofuran or tetrahydrofuran–acetone may be used, while for the syndiotactic crystallizable PVC a solution in cyclohexanone is used. For chlorinated PVC and Vinyon HH the solvent is acetone. Cordela is wet-spun from an aqueous emulsion.

In all cases, the spinning dope is made as concentrated as possible (within limits consistent with practical handling of highly viscous solutions) for economic reasons, ie, to minimize the amount of solvent recovery necessary per pound of fiber. Concentrations on the order of 20–25% by wt are normally used in dry spinning, with lower concentrations (10–12%) employed in wet spinning. Various techniques may be used to dissolve the polymer in certain solvents, for example, contacting the polymer with solvent at a low temperature to achieve good dispersion under minimum swelling conditions, and then raising the temperature to obtain dissolution (28). Another technique is to wet the polymer with one component of a two-component solvent before adding the second component. This also gives good dispersion before the polymer mass is allowed to dissolve. Corbiere, Terra, and Paris (29) have published a detailed study of the solvent power of mixed solvents for PVC.

The concentrated solution is very carefully filtered prior to spinning in order to remove gel particles and dirt which would plug holes in the spinnerets and thus shorten jet life. Since the polymer is stable at room temperature, the solution may be stored indefinitely prior to spinning.

Dry spinning is accomplished in a manner and in apparatus similar to that used in spinning cellulose acetate. The polymer solution is heated under pressure, filtered a second time, and extruded through the holes in the spinneret into a long tube or chamber containing heated air which evaporates the solvent. The solvent-laden air may move either cocurrent or, more commonly, countercurrent to the filaments. The air emerging from the tube goes to the solvent-recovery system, which may be either a carbon adsorption bed, water, or a condenser. The emerging filaments are oiled to provide lubrication and antistatic protection and wound on a spool or bobbin in a conventional manner, or else formed into a tow for further processing into staple.

Société Rhovyl, the largest European producer of PVC fiber, uses a variation on the conventional dry-spinning system—the spinning tube forms part of a system which is closed except for the entrance of the filaments through the spinneret and their exit at a small orifice in the bottom of the tube (30). The solvent-laden air passes cocurrent with the filaments to a section below their exit where it is condensed by a cooling section of the tube. The condensed solvent is continuously drawn off, while the air, with considerably less solvent vapor, is reheated and again transferred to the area of the spinning head.

Spinning speeds in dry spinning range from about 200 to 500 m/min. The total denier of the spun yarn, however, is limited by the rate of diffusion of solvent out of the yarn, which in turn is a function of the size of the tube, the velocity of the air, and the nature of the solvent. (The denier of a filament or yarn is the weight, in grams, of 9000 m). In general, only fine-denier yarns for continuous filament may be dry-spun. However, in the case of staple the tow may be made from a number of dry-spun filaments gathered together, or it may be wet-spun. Spinning speeds for wet spinning may be only one-tenth of those used in dry spinning, but the ability to spin many thousands of filaments simultaneously may be more economical in the long run, ie, the total spun denier per minute may be larger even though the individual filaments travel at a slower speed.

In wet spinning the polymer solution emerging from the jet is contacted with a coagulating bath. This bath must be a nonsolvent for the polymer, but miscible with the liquid used to dissolve the polymer in the spinning dope. The spin bath solidifies the stream of polymer into a filament and provides a medium for the diffusion of the solvent out of the thread. These two functions must be nicely balanced. Thus, too rapid a diffusion of solvent will lead to weak, porous fibers containing large voids, while too rapid a coagulation may make removal of the solvent extremely difficult.

In the wet spinning of the syndiotactic fiber Leavin (31) the polymer is dissolved in hot cyclohexanone at 137°C. A hot, 18% solution is extruded through jet holes of 0.1-mm diam into a cold (60°C) bath consisting of 50% water, 24% cyclohexanone, and 26% ethyl alcohol. The drop in temperature, plus the water, coagulates the polymer, while the ethyl alcohol solubilizes the cyclohexanone solvent in the spin bath. On emerging from the spin bath the filaments are washed with a solution of ethyl alcohol to remove all traces of cyclohexanone.

In the case of Cordela, which is a mixture and copolymer of vinyl chloride and vinyl alcohol, the emulsion prepared as previously described, which contains about 60% solid material, is spun into a 40% (w/v) aqueous solution of sodium sulfate. The high salt content presumably breaks the emulsion, permitting the solid particles to coalesce into a continuous fiber which is then dried in air without tension.

Aftertreatment. PVC fibers immediately after spinning are weak and highly extensible. As is the case with most synthetic fibers, it is necessary to stretch the fiber several-fold in order to orient the molecules in the direction of the fiber axis and thus impart strength to the material, usually at a sacrifice of extensibility. Stretching is done in air or water at temperatures substantially above the glass transition temperature of 75°C, and to 200–700% of the unstretched length. Although this improves the strength at room temperature, the stresses which are built into the molecular structure are relieved when the fiber is subsequently heated above the glass transition temperature. The result is a shrinkage of the fiber, which is one of the major defects of PVC as a textile material, ie, lack of dimensional stability toward hot-water laundering.

Types of Products. Regular atactic PVC is available in a wide range of deniers, in both continuous filament and staple. Typical deniers for staple are 1.8, 3.0, 3.5, 5.0, 6.0, 9.0, and 15 den/fil. Typical filament yarn constructions are 75/24, 120/38, 200/64, 800/256 and 3200/1024 den/fil count, which is about 3.1 den/fil. The fibers may also be characterized by their shrinkage at boiling-water temperatures, since this may be controlled by a relaxed shrinkage at high temperatures after the stretching operation. Thus, Rhovyl 55 will shrink 55%, Clevyl T will shrink 25%, and Rhovyl T will shrink 0% at 100°C (32).

The syndiotactic fiber Leavin is available only in staple form, probably because the wet-spinning process would not be economical for fine-denier continuous filament. A full range of deniers is available from 2.5 to 35.0, and shrinkage of these fibers is essentially zero at boiling-water temperatures.

Vinyon HH is available only as staple because the major end use, ie, as a fibrous binder in nonwoven fabrics, demands it in that form even though the fiber is dry-spun. Fibers are offered in 1.5, 3.0, and 5.5 den/fil and a variety of cut lengths from as short as ¼ to 4 in. Cordela, produced by a form of wet spinning, is available as staple or tow in deniers from 1.5 to 60.

Physical Properties

Mechanical Properties. The mechanical properties of a textile fiber are determined by the degree of orientation of the molecules along the fiber axis and by the crystallinity of the fiber, as well as by the inherent properties of the macromolecules, eg, the flexibility of the molecular backbone. In general, the tenacity and modulus are determined by the orientation, while the elongation to break, stiffness, and stability to heat are a function of the crystallinity. In ordinary atactic PVC the crystallinity is of a very low order, hence the material acts like a highly viscous liquid or like a glass, depending on the temperature.

The demarcation temperature between the glassy and viscous liquid states is the glass transition temperature. For ordinary PVC this temperature is 75°C (32). Below this temperature the individual atoms and segments of the molecule can engage only in vibratory or limited rotational movement. Above the transition temperature large segments of the molecules can move and rotate in concert. Hence at temperatures approaching or above 75°C various strains which are built into the fiber may relax. The most important effect is that the orientation of the molecules becomes more random, the fiber shrinks, and the surface becomes soft and tacky. This effect is simultaneously the main drawback and one of the major advantages of ordinary PVC fibers. It is a disadvantage in that it means that articles made of these fibers may not be laundered in hot water or ironed without serious shrinkage and softening. It is an advantage in the construction of various novelty yarns and fabrics, eg, bulked yarns, cloque fabrics, "sculptured" pile fabrics, etc, where the shrinkage is deliberately used to create special effects (32).

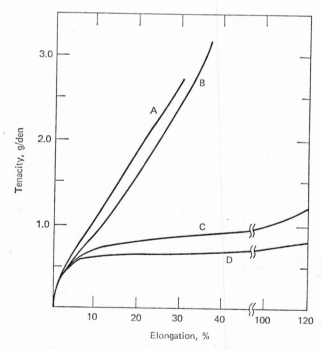

Fig. 1. Stress-strain curves of PVC fibers. A, C, regular PVC; B, Leavin; D, Vinyon HH.

Ordinary PVC yarns and fibers may have tenacities ranging from 1.2 to 3.0 g/den and extensibilities from 12 to 180%. High tenacity is associated with low extensibility and vice versa. Obviously, this range of properties spans the spectrum from wool-like to cottonlike properties. Figure 1 shows the stress-strain curves of regular PVC fibers at the two ends of the scale as well as Leavin and Vinyon HH. Note that in all cases there is a yield point or "knee" in the curve at a few percent elongation. This is the point beyond which the deformation is not recoverable.

The syndiotactic fiber Leavin is distinguished from ordinary PVC by the fact that it is crystallizable. This crystallinity, which develops during the hot stretching or later heat setting of the fiber or fabric, tends to stabilize the molecular configuration of the fiber to the point where much higher temperatures are required to relax the strains produced by stretching. Thus, whereas regular PVC begins to shrink at about 65°C when heated in silicone oil, Leavin does not start to shrink until 120°C is reached (33). This higher crystallinity also makes it possible to develop a higher tenacity and elongation to break, such as 3.6 g/den tenacity at 40% extensibility for 2.5 den Leavin staple. A further advantage is a decreased swelling in organic solvents which makes syndiotactic PVC more resistant to drycleaning agents than regular PVC.

The copolymer fiber Vinyon HH shows the same lack of crystallinity as homopolymer atactic PVC. In addition the presence of vinyl acetate makes the fiber liable to thermal shrinkage at even lower temperatures (50°C) than the homopolymer. This characteristic has been commercially exploited in that the major end use of Vinyon HH staple is as a fibrous binder for nonwoven fabrics. The Vinyon is dispersed and intimately mixed with the other fibers making up the nonwoven, and the application of heat enables the fabric to be sealed to itself. A simple example is in tea bags. Vinyon fibers as currently marketed have poor physical properties: tenacity about 0.6 g/den and extensibility of 125%. These can be improved by stretching, however, if the end use warrants it.

Cordela fiber is said to have tenacities in the range of 2.8–3.3 g/den with extensibilities of 20–24% in the dry state. Wet strengths drop from 2.3 to 2.0 g/den at equivalent extensibilities (19). This sensitivity to water is undoubtedly due to the poly(vinyl alcohol) present in the fiber.

Nonflammability. PVC fibers, in all the forms discussed in this article, are completely nonflammable in the pure state, ie, unmixed with other fibers; in fact, PVC is the only organic synthetic fiber which has this property. When exposed to open flame PVC chars, but it does not propagate the flame, nor does it melt and drip the way many other synthetic fibers do. PVC homopolymer fibers, moreover, will impart nonflammability to blends with other fibers, provided the mixture contains 60% or more PVC.

Moisture Regain. Moisture regain (percent of moisture when the fiber is in equilibrium with air at 70°F and 65% rh) is virtually nil for homopolymer PVC fibers, and is about 0.1% for Vinyon HH and 3.2% for Cordela.

Chemical Resistance. Homopolymer PVC fibers are inert to all chemicals, with the exception of chlorinated organic solvents and certain other organic chemicals, eg, dimethylformamide, cyclohexanone, and tetrahydrofuran, which tend to swell or dissolve the fiber. A greater range of organic chemicals will dissolve Vinyon HH and post-chlorinated PVC, including most low-molecular-weight ketones and aldehydes, and the same general conclusion is probably true regarding Cordela, although no information is yet available on this fiber.

Biological Resistance. All PVC fibers show a high resistance to biological deterioration, ie, they do not mildew or rot.

Weathering. In a study of the effect of atmospheric conditions on the properties of synthetic fibers, PVC fibers showed outstanding ability to withstand weathering (34).

Electrical Properties. The dielectric constant of Vinyon HH is 3.45 at 1 Hz, with a loss factor of 0.016. This low dielectric constant and high resistivity are responsible for the high static electrical charges built up on PVC fibers which are not protected by an antistatic finish. This high charge in turn is said to be responsible for the analgesic and antirheumatic properties of PVC fibers (36).

Dyeing of PVC Fibers

PVC fibers are generally difficult to dye in deep or bright shades because of their hydrophobicity and lack of specific polar groups on the molecule. For this reason the practice of "dope" or "solution" dyeing is often used in PVC fiber production. In this system, a colored pigment is added in finely dispersed form to the spinning solution and the resulting fiber retains the pigment. Such colored fibers are very fast and resistant to sunlight, laundering, etc. On the other hand, this method of achieving colored fibers is not very economical from a manufacturing standpoint unless large quantities of a given shade are made. Otherwise, the cost of cleaning equipment for each new color, matching colors, and maintaining large inventories of various shades becomes prohibitive.

PVC fibers can be dyed by using the same type of dyestuff originally developed for acetate fibers, ie, the dispersed dyes. In this system the dye, which is water-insoluble, is dispersed in water and brought into contact with the fabric or fibers. Because of its greater solubility in the organic fiber phase, the dye dissolves in the fiber to some extent, usually dependent on the temperature. Unfortunately, most regular and post-chlorinated PVC fibers are temperature-sensitive, as mentioned above, and cannot be heated above 60°C in the dyebath without shrinking and losing their strength and deforming the fabric. Some PVC fibers, eg, Rhovyl T, Clevyl T, and Movil, have been allowed to relax in boiling water during manufacture. These fibers may be dyed at the boil to achieve deeper shades. Fibers which have not had this heat treatment can only be dyed to pale shades at low temperatures unless a dyeing assistant or "carrier" is used. The carrier is an organic solvent, eg, an aromatic halide, which is dispersed with the dye and whose function is to swell the fiber and thus increase the penetration of the dye. Of course, such swelling also impairs the mechanical properties of the fibers, but not to the same degree as heat.

Syndiotactic PVC is much more resistant to heat than the atactic form and can be satisfactorily dyed at the boil without the aid of a carrier. In fact, it is claimed that this fiber may be dyed under pressure at still higher temperatures than normal boiling and thus achieve even greater penetration of the dyestuffs (33).

Vinyon HH is subject to the same problem in dyeing as atactic PVC, with the added disadvantage of a slightly lower softening temperature (about 50°C). Since this fiber is generally more soluble than homopolymer PVC, smaller quantities of carrier at lower temperatures are effective in dyeing. The 50% poly(vinyl alcohol) present in Cordela improves the dyeability markedly by providing active dye-sites on the molecule (27). This fiber is said to be dyeable with sulfide, azoic, vat, disperse, and complex salt dyes (19).

Health and Safety Factors

PVC is said to be completely nondermatitic and nonallergenic. Its nonflammable characteristic has encouraged its use in hospitals, sanatoria, and the like. In this connection, the antirheumatic properties of PVC are of interest. It is claimed in both Japan and Europe that the wearing of undergarments or use of bed sheets made of PVC gives significant relief to patients suffering from rheumatism, arthritis, and related painful conditions. This was initially thought to be due to the high negative electric charge generated by PVC in contact with the skin, which is not found with any other fiber (36). However, the emphasis now seems to be on the superior warmth of the garments due to the low thermal conductivity of PVC (40).

Analysis and Testing

A quick screening test for PVC fibers can be made by attempting to burn the sample. If the material burns readily, it is obviously not PVC. A nonflammable fiber, however, may be PVC or another material which has been rendered flame-retardant by means of an additive or special treatment.

Table 1. Major Producers of Poly(vinyl chloride) Fibers (38)

Company	Country	Trade name	Type of fiber	Estimated capacity, millions of lb
Société Rhovyl	France	Rhovyl, Thermovyl, Fibrovyl, Retractyl Clevyl	homopolymer atactic	40
Teijin, Ltd.	Japan	Teviron	homopolymer	15
Polymer Ind.	Italy	Movil	atactic	8
Fiberfabriken Wolfe	East Germany	PeCe	post-chlorinated	6
Acetilen Fiber	Yugoslavia		homopolymer atactic	2
	U.S.S.R.	Khlorin	homopolymer atactic	
A.C.S.A.[a]	Italy	Leavin	syndiotactic homopolymer	7
FMC Corp.	U.S.A.	Vinyon HH	vinyl acetate copolymer	4
Kokoku Rayon Company	Japan	Cordela[b]	vinyl alcohol copolymer	

[a] Subsidiary of Chatillon S.p.A.

[b] Also called Polychlal and Super Enbilon.

An organic chlorine determination by Schöniger combustion or similar method will distinguish between homopolymer post-chlorinated PVC and copolymers. Homopolymer PVC contains about 56% chlorine, post-chlorinated PVC about 62–65% chlorine, and copolymers less than 56%. Vinyon HH contains 47% chlorine and

Cordela contains about 25% chlorine.   Post-chlorinated PVC and Vinyon HH are soluble in acetone, while homopolymer and Cordela are not.

Economic Aspects

The cheap raw material, poly(vinyl chloride), has been one of the decisive factors in the continuing interest in PVC fibers.   Unfortunately, the wet and dry spinning processes required for fiber production are more expensive than the melt spinning used for fibers such as nylon and polyester, enabling the latter fibers to approach PVC in price. Nonetheless, it is probably true that PVC fibers will always be somewhat less expensive than these fibers and will thus always have a certain advantage.   This may not be true of the syndiotactic fibers where the more costly low-temperature polymerization is required and a complex wet-spinning technology is involved.   Of course, the nonflammable characteristic of PVC fibers also gives them a competitive advantage, although a great deal of research and development is currently being devoted to methods of conferring nonflammability on other fibers.

Due to their relatively small percentage of the total fibers market, production figures on PVC fibers are difficult to obtain, usually being reported in the category "other noncellulosics" along with glass, Saran (National Plastic Products Co.), etc. In 1964, however, the two major producers in the world were Société Rhovyl of France, and Teijin, Ltd. of Japan (38).   Table 1 lists the major producers of PVC fibers, as well as estimated capacities and brand names associated with each producer.

Applications

The history of PVC fibers is an interesting study of the effect of economic, political, and technological factors on the production and utilization of textile material. These fibers were initially produced in Germany in the years prior to World War II and were undoubtedly exploited as part of the effort of the German government at this time to become self-sufficient and independent of imports of natural fibers, eg, cotton and wool.   PVC could not compete with these fibers and the newer synthetics in the open market which existed at the end of the war, however, and consequently it fell into disrepute.   PVC is still one of the cheapest organic polymers in the world, though, and this economic fact kept interest in these fibers from lagging and led to research and development into the best ways to utilize this material in the modern world.

The place of PVC fibers in today's textile market is dictated strictly by the economic and technical advantages and disadvantages of the fiber.   PVC is an inexpensive, strong, nonflammable, chemical- and weather-resistant fiber with an unfortunate sensitivity to temperatures below the boiling point of water.   This latter fact has kept PVC out of the really big textile markets for woven materials which can be laundered in hot water, ironed, heat-set, or resin-treated.   Thus, the fiber has sought markets where high temperature resistance is not required and where its virtues give it a definite advantage.

The one unique quality which PVC provides is its nonflammability.   This has led to a broad spectrum of end uses including: curtains, furnishings, upholstering fabrics, women's and children's nightwear, blankets, mattresses, rugs, and industrial fire-protective clothing. Blends with wool or nylon are also used in various clothing applications—it is said that at least 75% PVC is required to impart sufficient nonflammability to blends (37). Automotive and aircraft upholstery fabrics are a growing end use for PVC. It is certain that the nonflammable property of PVC will be of increasing importance in the future as the governments and peoples of the world become more conscious of the flammability of textile materials and their role in causing death and injuries in fires.

The resistance to weather and rot of PVC has led to applications in the field of tents, tarpaulins, nets, and other exposed, outdoor textile articles such as flags. The water-repellent property of PVC in conjunction with its excellent thermal insulation finds application in ski-wear, mountain-climbing and hunting clothing, sleeping bags, and winter underwear. The chemical-resistant quality of PVC is also an advantage in protective clothing for workers in the chemical industry.

The French firm Société Rhovyl has been most aggressive in developing markets for PVC fibers. These have included applications where the heat sensitivity of the fibers has been turned into an advantage. For example, spun yarns made with blends containing PVC staple fibers, when heated above the glass transition temperature of PVC, become quite bulky due to the retraction of the PVC fibers. By heating a fabric woven from PVC the shrinkage of the warp and filling yarns results in a very tight weave. The resulting fabric can be used as a map substrate or for protective clothing for chemical workers. When PVC fibers form only part of the fabric, shrinkage produces novel cloque effects. PVC fibers in a pile fabric can be used to produce a sculptured or relief effect when heat is applied. Woven fabrics of PVC can also be molded to intricate forms under the application of heat and pressure (32).

PVC fibers are also finding increasing markets in the rapidly expanding field of nonwoven fabrics. Nonwovens made of pure PVC may be calendered using heat and pressure to give a smooth surface or various designs may be impressed by using embossing rollers. Needle-punched batts made with PVC fibers in blends with other fibers may be given greater strength by the thermal retraction of the PVC after forming. PVC fibers may also be used as fibrous binders in nonwoven blends with other fibers. Mention has been made of the use of Vinyon HH in the preparation of tea-bag nonwovens where the Vinyon also permits the bags to be heat-sealed, thus dispensing with a costly sewing operation.

The syndiotactic fiber Leavin has been commercially available in large quantities only since late 1967 and hence has not had the opportunity to develop large markets. However, its superior thermal and drycleaning resistance promise more extended applications than the regular PVC. It is noteworthy that Société Rhovyl has plans to commercialize a similar fiber, called Stavinyl. Teijin Ltd. of Japan has also investigated this fiber and has several patents in this field. Similarly the copolymer fiber Cordela, or Polychlal as it is sometimes called, has not been on the market long enough to assess its potential. In this case, the superior moisture absorbance and dyeability are expected to take it into applications beyond those available to homopolymer atactic PVC. The market potential of both these fibers will depend to a certain extent on their price, since the manufacturing process for both fibers is more complex and hence probably more costly than dry-spun regular PVC. In this respect they will be competing against the more extensively used fibers, eg, nylon, polyester, and, above all, acrylics which they resemble.

Bibliography

"Dynel and Vinyon" treated in *ECT* 1st ed., under "Textile Fibers (Acrylic and Vinyl)," Vol. 13, pp. 831-836, by H. L. Carolan, Carbide and Carbon Chemicals Company, a division of Union Carbide and Carbon Corporation.

1. Ger. Pat. 281,877 (1913), F. Klatte (to Chemische Fabrik Griesheim-Elektron).
2. E. Hubert, *Vierjahresplan* **4,** 222 (1940): *Kunststoffe* **30,** 244 (1940).
3. J. Corbieri, *Atomes (Paris)*, 201 (1947).
4. C. S. Marvel et al., *J. Am Chem. Soc.* **61,** 3241 (1939); **64,** 2356 (1942).
5. P. J. Flory, *J. Am. Chem. Soc.* **61,** 1518 (1939).
6. C. S. Fuller, *Chem. Rev.* **26,** 162 (1940).
7. J. W. L. Fordham, *J. Polymer Sci.* **39,** 321 (1959).
8. J. W. L. Fordham, P. A. Burleigh, and C. L. Sturm, *J. Polymer Sci.* **41,** 73 (1959).
9. F. P. Reding, E. R. Walter, and F. J. Welch, *J. Polymer Sci.* **56,** 225 (1962).
10. P. H. Burleigh, *J. Am. Chem. Soc.* **82,** 749 (1960).
11. Anon., *Man-Made Textiles* **43,** 26 (1966).
12. Ger. Pat. 596,911 (1934), C. Schoenburg (to F. G. Farbenindustrie).
13. V. F. Krska, J. Stamburg, and Z. Pelzbauer, *Angew. Makromol. Chem.* **3,** 149 (1968).
14. H. Germar, *Makromol. Chem.* **86,** 89 (1965).
15. J. Petersen and B. Rånby, *Makromol. Chem.* **102,** 83 (1967).
16. W. Pungs, *Kunststoffe* **57,** 317 (1967).
17. U.S. Pat. 2,075,575 (1937), S. D. Douglas (to Carbide and Carbon Chemical Co.).
18. S. D. Douglas and W. H. Stoops, *Ind. Eng. Chem.* **28,** 1152 (1936).
19. N. C. Heimbold, *Textile World* 117 (June 1967).
20. C. F. Ruebensaal, *Chem Eng.* **57,** 102 (Dec. 1950).
21. Brit. Pat. 958,969 (May 27, 1964), (to Monsanto Chemical Co.).
22. Belg. Pat. 661,932 (April 2, 1964), Società Edison.
23. Belg. Pat. 609,856 (Nov. 3, 1961), Società Edison.
24. A. Nakajima, H. Hamada, and S. Hayashi, *Makromol. Chem.* **95,** 40 (1966).
25. G. Borsini and M. Cipolla, *J. Polymer Sci. B* **2,** 291 (1964).
26. U.S. Pat. 2,075,429 (1937), S. D. Douglas (to Carbide and Carbon Chemical Co.).
27. U.S. Pat. 3,111,371 (Nov. 19, 1967), S. Okamura et al. (to Toyo Kogaku Co.).
28. Japan. Pat. 260,309 (Nov. 12, 1958), Teijin Ltd.
29. J. Corbiere, P. Terra, and R. Paris, *J. Polymer Sci.* **8,** 101 (1952).
30. U.S. Pat. 2,472,842 (June 14, 1949), A. Mouchiroud and J. Trillat (to Société Rhodiaceta).
31. Belg. Pat. 630,561 (April 3, 1963), (to Applicazioni Chimica S.p.A.).
32. L. Gord, "PVC Fibers," in H. Mark, S. Atlas, and E. Cernia, eds., *Man-Made Fibers*, Vol. 3, Interscience Publishers, a division of John Wiley & Sons, Inc., New York, 1968.
33. C. Mazzolini, *Fibre Colori* **17,** 214 (1967).
34. J. Lunenschloss and H. Stegherr, *Textil-Praxis* **15,** 939 (Sept. 1960).
35. C. E. Schildknecht, *Vinyl and Related Polymers*, John Wiley & Sons, Inc., New York, 1952, p. 421.
36. C. Frager, *Textile Res. J.* **32,** 168 (1962).
37. Anon., *Skinner's Silk Rayon Record* **38,** 907 (1964).
38. *Man-Made Fiber Producers*, Noyes Development Corp., New York, 1964.
39. Anon., *Chem. Eng. News* 36 (March 13, 1967).
40. Anon., *Textile Industries* 78 (Aug. 1968).

J. P. Dux
American Viscose Division
FMC Corporation

VIRAL INFECTIONS, CHEMOTHERAPY

In recent years many hundreds of compounds with antiviral activity in laboratory test systems have been reported. This has not been reflected in a corresponding number with useful preventive or curative activity against virus diseases of man and animals, however, and compared with the dramatic progress in the control of bacterial diseases, the chemotherapy of virus infections is still in its infancy.

Virus diseases such as rabies, smallpox, influenza, the common cold, and measles, among others, are familiar to most people, but the actual causative agents, the viruses, are less well known to the general public. Luria has described viruses as submicroscopic entities capable of being introduced into specific living cells and reproducing inside such cells only. For the purpose of this article, the cells referred to above are of human or animal origin. Lwoff has further described viruses as being constructed of a nucleic acid core enclosed in a protein coat, the core consisting of either ribonucleic acid (RNA) or deoxyribonucleic acid (DNA), but not both. Thus an animal virus is an inert package of one type of nucleic acid surrounded by a protein coat and capable of reproducing itself only in living cells of animal origin. While possessing these common qualities, viruses vary widely in shape, size, and characteristics. *Vaccinia*, a DNA virus, is a particle 210 by 260 nm, whereas the RNA virus of foot-and-mouth disease is a sphere with a diameter of 10 nm. In addition, some viruses have only a limited number of hosts, such as human rhinoviruses (causing the "common cold") which are generally restricted to man; other viruses, such as members of the arboviruses, infect man, other animals, birds, and anthropods. The goal of viral chemotherapy is to prevent the multiplication of these diverse agents without harm to the host.

Areas of Application

The major effort in virus chemotherapy has been made only since 1950. This recent origin is partly due to the lack of adequate test methods and partly due to the success of vaccines for the control of certain virus diseases. Smallpox, yellow fever, and the more recent polio vaccines have been highly successful in virtually eliminating these diseases whenever used on a broad scale. For these diseases the vaccine approach has been the most practical and effective method. In other situations, however, vaccines have not had as great a success, and control by chemotherapy may be the method of choice.

All viruses contain specific antigens which, when introduced into a host animal, will induce the host over a period of time to make antibodies which can react with the specific virus antigen and render the virus noninfectious. This occurs with natural infections or can be induced by vaccines which consist of inactivated or attenuated viruses still capable of inducing antibody production. Once the antibody has been induced, either naturally or by vaccines, the host will remain resistant to infection by the inducing virus unless the virus undergoes a change in antigenic makeup (antigenic drift). Influenza viruses show a continuous antigenic drift, and for this reason, although influenza vaccines have been available for over twenty years, the disease is far from being controlled. Major shifts occur periodically, as in 1957 with the advent of the A2 Asian-type influenza, and in 1968 with the A2 Hong Kong-type influenza; when such a shift occurs the immunity produced by the older vaccines is of little or no value.

Even between these periods a slow change occurs, and vaccine is seldom available for use against the virus type most sensitive to its action. Studies to date indicate that the efficacy of antiviral compounds is not affected by these antigenic shifts of influenza viruses, and thus chemotherapy offers the best method of coping with such diseases while a vaccine effective against the new variant is developed.

A second area of use is for chemicals effective against the rhinoviruses. These viruses are antigenetically stable; over 100 different strains are known, however, and due to the great specificity of their antigenic makeup, each would require its own vaccine should this approach be taken.

Laboratory studies indicate that certain compounds can show broad-spectrum activity against all rhinovirus strains. In this case the use of such a compound with clinical efficacy would be the method of choice, since treatment could be effective regardless of the strain causing the infection.

A third area for chemotherapy is against a virus such as *Herpes simplex* which causes disease even in the presence of circulating antibody and is thus not amenable to control with vaccines.

In addition, chemicals act rapidly whereas vaccines take several weeks to induce immunity; therefore a chemical could be used in an epidemic along with vaccination to protect the individual until immunity has developed.

Methods

The reproductive cycle of a virus consists of the following steps: (*1*) attachment of the virus to a susceptible cell; (*2*) entrance of the virus into the cell; (*3*) uncoating of the virus to release the infectious core nucleic acid inside the host (this may also occur at the cell surface); (*4*) induction of synthesis of new virus components inside the cell; (*5*) assembly of components into a complete virus; (*6*) release of the virus, usually resulting in destruction of the cell. Illness or death of the host occurs once a sufficient number of cells are affected.

One of the early problems in chemotherapy was to discover test systems other than the natural host for virus screening. In 1949 Enders et al. (1) showed that an intact host was not necessary for virus growth and that viruses could reproduce in cells grown in vitro. In 1952 Dulbecco (2) introduced a quantitative plaque assay for viruses using cultured chick embryo cells, and this was followed by adaptation of a large number of viruses to a variety of cell cultures, leading to widespread screening for antiviral activity. The major tools for screening at present are the use of tissue culture, the chick embryo, and animals, often the mouse. Multiplication of viruses is measured in a variety of ways, including death of the cell or animal, measurement of hemaglutinin produced by certain viruses, and the use of fluorescent antibodies. In animals, weight loss, lung consolidation, antibody rise, and lesion formation are also often used.

The disease process in animals is much more complex than that in tissue culture, and for practical purposes a compound should not be considered as a chemotherapeutic agent until efficacy in an animal has been demonstrated. Most animal studies and preliminary studies in man are carried out with challenge infections. In these studies a virus inoculum with desired characteristics is introduced into the test animal or human volunteer and the effect of the drug on the infection is compared with similarly infected but non-drug-treated control groups. Once efficacy is established by such tests, the drug is then evaluated for the prevention or treatment of naturally occurring infections.

Link, in 1968 (3), has published an excellent report on virus test methods which clearly spells out the care and standardization of methods necessary to obtain reliable results. In laboratory tests hundreds of compounds have shown some antiviral activity, especially in tissue culture or in the chick embryo, with examples of activity against most of the phases of the virus replicative cycle. The majority of these fail to be effective in the animal, however, and of those with animal activity, only a limited number have been tested and shown to be active in man. This article discusses only the few compounds extensively tested and reported efficacious in man. Many review articles are available for consultation on the state of virus chemotherapy at the laboratory stage and a number of these articles are listed in the bibliography.

Antiviral Agents Effective in Man

One of the first reviews on antiviral chemotherapy was published in 1956 by Hurst and Hull (4). At this time the *psittacosis–lymphogranuloma* group was included under viruses and a number of antibiotics were reported to be effective against these organisms. For the true viruses, described as "the smaller viruses," the authors stated that no substance of practical value against man or animal was available. Since that review several chemicals have reached the stage of testing and efficacy in man.

Thiosemicarbazones. The first report of antiviral activity for this series was in 1950 when Hamre and colleagues (5) found that the antituberculosis compound *p*-aminobenzaldehyde thiosemicarbazone provided some protection to chick embryos and mice infected with *vaccinia* virus. A program of analog synthesis followed, and in 1955 Bauer (6) reported that indole-2,3-dione-3-(thiosemicarbazone) (isatin-3-semicarbazone (**1**) provided almost complete protection from mortality of mice infected

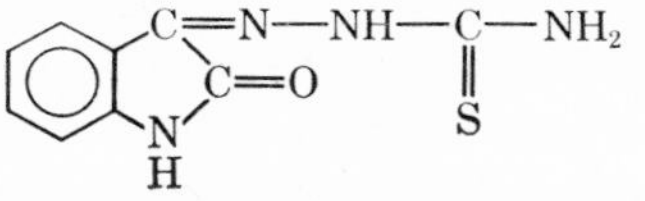

$$\text{C}=\text{N}-\text{NH}-\underset{\underset{\text{S}}{\|}}{\text{C}}-\text{NH}_2$$

(1)

with high doses of *vaccinia* virus. Mode-of-action studies indicated that the compound prevented the formation of certain of the virus components intracellularly, so that even though many of the virus components were synthesized, complete, mature virus was not formed. The interest in this compound stimulated continued studies which resulted in the synthesis of 1-methylisatin-3-thiosemicarbazone (methisazone, cpd 33T57) (**2**).

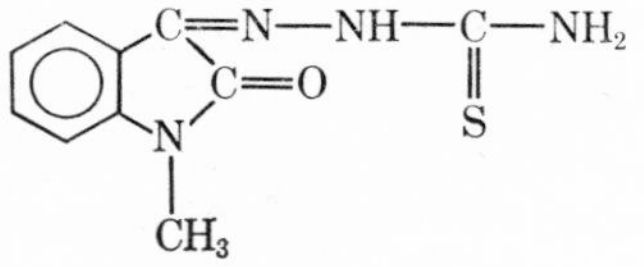

$$\text{C}=\text{N}-\text{NH}-\underset{\underset{\text{S}}{\|}}{\text{C}}-\text{NH}_2$$

(2)

Methisazone showed excellent prophylactic prevention of *vaccinia* infections of mice. In 1963 methisazone was tested prophylactically in a large-scale clinical trial during a smallpox epidemic in Madras (7) and provided a marked reduction in smallpox infections compared with nontreated controls. Subsequent field trials have confirmed these results and the drug also appears to give an additive effect with vaccination greater than that seen with either alone. Other clinical studies suggest that the compound is active against *eczema vaccinatum* and also may have some effect on the lethal complications of *vaccinia gangrenosa*. Curative efficacy against smallpox infections has not been demonstrated. The drug is not licensed for use in the United States, but is available as "Marboran" (Burroughs Wellcome & Co.) for experimental use.

Another compound of the series, 4-bromo-3-methylisothiazole-5-carboxaldehyde thiosemicarbazone (isathiazole thiosemicarbazone, M & B 7714), provided good protection against *vaccinia* and *variola major* (smallpox) infections of mice, with evidence for clinical prophylactic activity against smallpox in man, but with no evidence of curative effect. This material appears to have no advantages over methisazone.

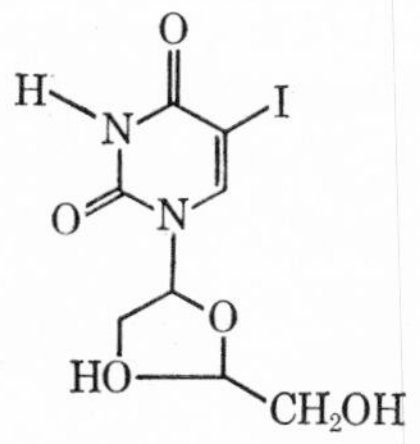

(3)

Nucleosides. 5-Iodo-2′-deoxyuridine (idoxuridine, IDU, IUDR) (3) was first synthesized in 1958 by Prusoff as an experimental drug for the inhibition of tumors. In 1961 Herrmann (8) reported that IDU inhibited plaque formation due to *herpes simplex* virus in tissue culture and in 1962 Kaufman (9) reported the successful treatment with IDU of *herpes simplex* corneal infections (herpetic keratitis) in man. It has now been used in thousands of patients against acute herpetic keratitis with favorable results, but with less significant results in chronic and recurrent herpetic keratitis. Treatment of dermal lesions (cold sores) with topical IDU has yielded conflicting results with approximately half of the studies showing a favorable result and half showing a negative result. IDU appears to block synthesis of viral DNA as well as to be incorporated into an abnormal, presumably nonfunctional, virus DNA. IDU is licensed for the topical treatment of herpetic keratitis and available as an ophthalmic solution as Stoxil (Smith, Kline & French Laboratories).

1-β-D-Arabinofuranosylcytosine (cytosine arabinoside, cytarabine (CA)) (4) has

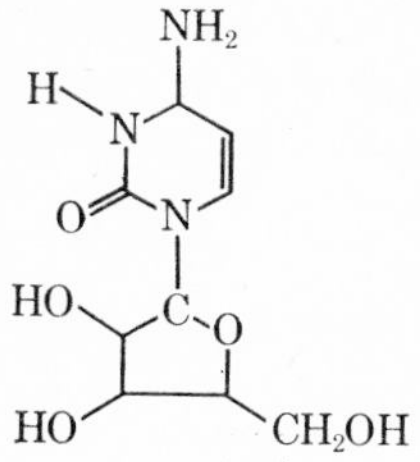

(4)

been effective in laboratory tests against *herpes simplex* and has also shown efficacy by topical application against herpetic keratitis infections in man. It is not licensed and is available only as an experimental drug.

Adamantanes (qv in Supplement Volume). In 1963 Jackson et al. (10) reported that 1-aminoadamantane hydrochloride (amantadine hydrochloride; EXP 105-1) (5)

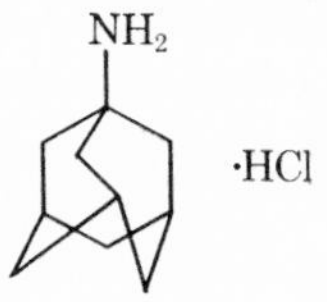

NH₂

·HCl

(5)

influenced the course of challenge influenza A2 infections in man.   In 1964 Davies et al. (11) showed that amantadine hydrochloride was effective against infections with influenza A, A1, and A2, in tissue culture, in ovo, and in mice.   Although there is evidence for activity against a number of other viruses such as *rubella, parainfluenza,* and *vaccinia,* at well-tolerated concentrations the major effect is against the A group of influenza.   In addition to its protective effect against A2 influenza infections of man, amantadine hydrochloride has also shown an antiviral effect against equine influenza in horses and avian influenza A strains in quail and turkeys.   The compound acts at an early stage in the replication cycle by blocking or slowing the penetration of infectious virus into the host cell.

In 1966 amantadine hydrochloride or Symmetrel (E. I. du Pont de Nemours & Co.) was cleared in the United States for use in the prevention of respiratory illness due to susceptible influenza A2 viruses by prophylactic treatment of contacts of patients and when index cases appear in the area (12,13).   This generated a great deal of controversy in the medical field and a report by Sabin in 1967 (14) was highly critical of the practical significance of the test results although further challenge and field studies have added support for the safety and efficacy of amantadine hydrochloride in preventing influenza A2 infections in man (15–17).

In 1968 the Asian-type influenza A2, prevalent in the human population since 1957, underwent an antigenic shift to the Hong Kong influenza A2 type which was sufficiently great to render previous immunity and influenza vaccines ineffective. Although isolates of the Hong Kong A2 virus were found to be sensitive to amantadine hydrochloride in tissue culture, in ovo, and in mice, the lack of clinical data for this type prevented claims for use of the drug against the A2 Hong Kong influenza during the 1968–1969 epidemic caused by this variant.   This is one of the problems found in a new biological area and guide lines remain to be established as to the significance of laboratory tests in relation to efficacy in clinical practice.

Recent reports of clinical and laboratory animal studies indicate that therapeutic treatment of influenza A2 infections exerts a beneficial effect on the course of the disease (18,19).   These studies are in the experimental stage and amantadine hydrochloride has not been cleared for therapeutic use.

α-Methyl-1-adamantanemethylamine hydrochloride (rimantadine hydrochloride

NH₂
|
H—C—CH₃

·HCl

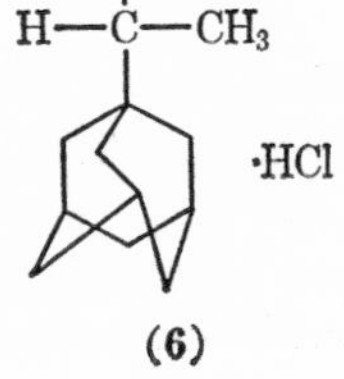

(6)

EXP 126) (**6**), another experimental drug, has been shown to exert a protective effect similar to that of amantadine hydrochloride against challenge infections with influenza A2 in man and in mice (20,21).

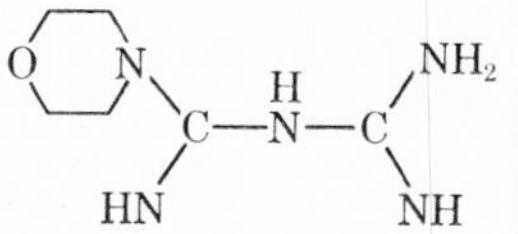

(**7**)

Biguanide. In 1960 Melander (22) reported that ABOB—*N′,N′*-anhydrobis(β-hydroxyethyl)biguanide [moroxydine, abitilguanide, "Virustat" (Labs. Delagrange, Paris, France)] (**7**) had prophylactic activity against influenza A infections of mice as well as having a low order of mouse toxicity. Sjoberg, 1960 (23) reported that prophylactic use of this material suppressed illness in human volunteers during an influenza A2 epidemic. ABOB has since been studied extensively in human volunteers in the United States, Europe, and Japan, with conflicting results. The United States studies failed to show efficacy of ABOB against challenge and natural influenza A infections of man (24), whereas studies in other countries have continued to report weak but significant activity (25). The mode of action of this material is unknown.

ABOB is not licensed in the United States but is available in other countries as a constituent of Flumidin (A. B. Kabi, Stockholm, Sweden), Spenitol (Winthrop, G.m.b.H., Frankfurt am Main, Germany), Vironil (Winthrop Labs., Div. Sterling Pharmaceuticals Pty., Ltd., Sidney, Austrialia), and Virugon (Winthrop, Kingston-on-Thames, England).

Isoquinolines. In 1967 Larin and colleagues (26) reported that two isoquinolines prepared by Chemical Research (Pfizer, Sandwich, Kent, England), exhibited antiviral activity against a number of viruses in tissue culture. These compounds [1-(*p*-chlorophenoxymethyl)- (U.K. 2054) and 1-(*p*-methoxyphenoxymethyl)- (U.K. 2371)

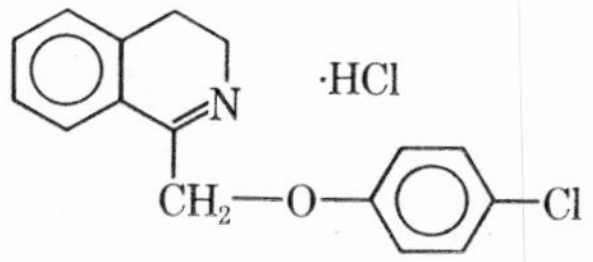

U.K. 2054

3,4-dihydroisoquinoline hydrochloride] were reported to inactivate members of the myxo-virus group but inhibited members of the picornavirus group by blocking an intracellular step in virus replication.

U.K. 2371 has been reported to show a limited effect on clinical illness due to challenge infections of man with influenza B/England/101/62 (27). U.K. 2054 has

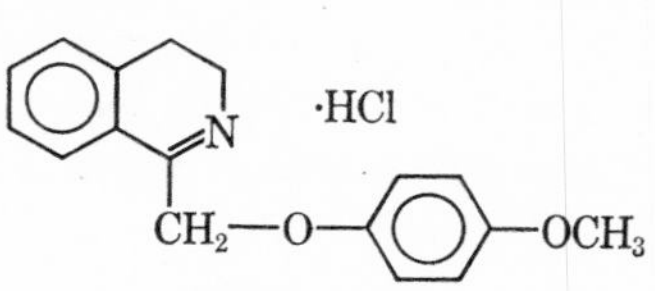

U.K. 2371

also been reported to exert a protective effect in volunteers challenged with influenza B/England/13/65 as well as against natural infections of man with influenza A2/Hong

Kong/1968 (28). The responses to drug treatment were weak and further investigations are necessary for evaluating the significance of the findings.

Interferon

In 1957 Isaacs and Lindenmann (26) published the results of studies of a host-produced antiviral agent they called interferon. As it is now understood, interferons are cell-derived proteins which have certain basic physical and chemical properties and which possess the common characteristic of inducing antiviral resistance in susceptible cells. Interferon is further characterized by having a broad spectrum of antiviral activity, but it is also largely specific for the species in which it is produced. Induction of interferon in cells was originally by means of viruses, but it has since been discovered that a variety of materials, ranging from bacteria and their products to synthetic polymers, will also cause interferon induction.

Although no reports of clinical efficacy of interferon or interferon inducers have been published, their potential as broad-spectrum antiviral agents, as contrasted to the narrower spectrum of the antiviral chemicals discovered to date, has aroused a great deal of interest and resulted in a large amount of experimental work. Two approaches are possible: (1) the use of inducers to cause the host to make its own interferon, and (2) the direct application of interferon derived from cells of human origin. In the United States the major approach to interferon control of virus diseases is by use of interferon inducers which are of two major classes. One class includes double-stranded RNA polymers either from natural sources or synthetic materials such as polyriboinosinic acid–polyribocytidylic acid complexes (Poly I–Poly C) (27). Poly I–Poly C has been effective in herpetic keratitis of rabbit eyes when administered either before or after infection (28). The second class includes synthetic polymers with the following characteristics: a large molecule with a high density of free anionic groups not readily degradable nor readily eliminated by the host (29). Agents with these characteristics include polymers of maleic anhydride, polyacrylates, and carboxylates. Both polyacrylic acid (PAA) and polymethacrylic acid (PMAA) have induced interferon in tissue culture and protected mice from infection with a number of viruses.

Pyran copolymer (a maleic acid/divinyl ether copolymer designated as NSC-46015-C) has been reported by Merigan and Regelson (30) to induce interferon in cancer patients, but only at relatively toxic concentrations. This is the only synthetic interferon inducer reported as tested in man, although many of this type of agents have been reported as effective against a variety of virus challenge infections in laboratory animal systems.

It is not clear whether the synthetic polymers induce the formation of new interferon, cause the release of preformed interferon, or act by both methods. Toxicity problems associated with the inducers have not yet been solved, however, and European workers have pursued the use of interferon directly. It now appears technically feasible to produce large amounts of interferon in cells of human origin, and this approach is still under active consideration. With the rapid progress in this area, it should not be too far in the future before the practical significance of these agents is known.

Future of Antiviral Chemotherapy

After some two decades of study, the limited number of antiviral agents available for the treatment of human diseases appears discouraging. However, the fact that these exist permits a more rational approach to the problem and provides guidance in the selection of test methods in the laboratory that will correlate with efficacy in clinical practice. The antivirals available in the clinic have a narrow spectrum of activity, but in the laboratory there are compounds under study with a wider antiviral spectrum. With he more advanced test methods and accumulating knowledge, there is reason to hope that virus chemotherapy will take its place as one of the means for the control of viral diseases of man and animals.

Bibliography

1. J. F. Enders, T. H. Weller, and F. C. Robbins, *Science* **109**, 85–87 (1949).
2. R. Dulbecco, *Proc. Natl. Acad. Sci. U.S.* **38**, 747–752 (1952).
3. F. Link in K. Maramorosch and H. Koprowski, eds., *Methods in Virology*, Vol. 3, Academic Press, New York, 1968, Chap. 5, pp. 211–278.
4. E. W. Hurst and R. Hull, *Pharmacol. Rev.* **8**, 199–263 (1956).
5. D. Hamre, J. Bernstein, and R. Donovick, *Proc. Soc. Exptl. Biol. Med.* **73**, 275–278 (1950).
6. D. J. Bauer, *Brit. J. Exptl. Pathol.* **36**, 105–114 (1955).
7. D. J. Bauer, L. St. Vincent, C. H. Kempe, and A. W. Downie, *Lancet* **2**, 494–496 (1963).
8. E. C. Herrmann, Jr., *Proc. Soc. Exptl. Biol. Med.* **107**, 142–145 (1961).
9. H. E. Kaufman, *Proc. Soc. Exptl. Biol. Med.* **109**, 251–252 (1962).
10. G. G. Jackson, R. L. Muldoon, and L. W. Akers, *Antimicrobial Agents and Chemotherapy* **1963**, pp. 703–707.
11. W. L. Davies, R. R. Grunert, R. F. Haff, J. W. McGahen, E. M. Neumayer, M. Paulshock, J. C. Watts, T. R. Wood, E. C. Hermann, and C. E. Hoffmann, *Science* **144**, 862–863 (1964).
12. "Council on Drugs," *J.A.M.A.* **201**, 374–375 (1967).
13. Symmetiel, A New "Antiviral" Agent, *The Medical Letter* **9** (2), 5 (1967).
14. A. B. Sabin, *J.A.M.A.* **200** (11), 943–950 (1967).
15. Y. Togo, R. B. Hornick, and A. T. Dawkins, Jr., *J.A.M.A.* **203** (13), 1089–1094 (1968).
16. E. Callmander and L. Hellgren, *J. Clin. Pharm.* **8**, 186–189 (1968).
17. J. F. Finklea, A. V. Hennessy, and F. M. Davenport, *Amer. J. Epidemiol.* **85**, 403–412 (1967).
18. J. L. Schulman, *Proc. Soc. Exptl. Biol. Med.* **128**, 1173–1178 (1968).
19. A. T. Dawkins, L. R. Gallagher, Y. Togo, R. B. Hornick, and B. A. Harris, *J.A.M.A.* **203**, 1095–1099 (1968).
20. J. W. McGahen and C. E. Hoffmann, *Proc. Soc. Exptl. Biol. Med.* **129**, 678–681 (1968).
21. W. L. Wingfield, D. Pollock, and R. R. Grunert, *New Engl. J. Med.* **281** (11), 579–584 (1969).
22. B. Melander, *Antibiot. Chemotherapy* **10**, 34–45 (1960).
23. B. Sjoberg, *Antibiot. Med. Clin. Therapy* **7**, 97–102 (1960).
24. G. G. Jackson, R. L. Muldoon, L. W. Akers, O. Liu, G. C. Johnson, and C. Engel, *Antimicrobial Agents and Chemotherapy* **1961**, pp. 883–891; G. W. Parker, R. B. Stonehill, and A. C. DeGraff, *Antibiot. Chemotherapy* **12**, 155–158 (1962); G. Meiklejohn, W. P. McCann, and S. Shkolnik, *Military Med.* **128**, 890–894 (1963).
25. B. Zetterberg, L. Heller, S. Gustafson, and O. Ringertz, offprint, *Conference on ABOB and Flumidin, Stockholm 1960;* J. Haglind, *Ind. Med. Surg.* **31**, 351–353 (Aug. 1962); E. Dahlgren, T. Ericsson, A. Ivert, O. Karlberg, Y. Rigens, and N. Tologen, *Läkartidningen* **63**, 4903 (1966); O. Kitamoto, M. Kaji, and H. Yoshida, *5th Intern. Congr. Chemotherapy Abstr.* **2**, 1245 (*1967*).
25a. N. M. Larin, A. S. Beare, M. P. Copping, C. R. McDonald, J. K. McDougall, B. Roberts, and J. B. Smith, *Antimicrobial Agents and Chemotherapy* **1967**, pp. 646–653.
25b. A. S. Beare, M. L. Bynoe, and D. A. J. Tyrrell, *Lancet* **1**, 843–845 (1968).
25c. P. N. Meenan and I. B. Hillary, *Lancet* **2**, 614–615 and 641–642 (1969).
26. A. Isaacs and J. Lindemann, *Proc. Royal Soc. London, Ser. B* **147**, 258–267 (1957).
27. G. P. Lanpson, A. A. Tytell, A. K. Field, M. M. Nemes, and M. R. Hilleman, *Natl. Acad. Sci. Proc.* **58**, 782–789 (1967).
28. J. H. Park and S. H. Baron, *Science* **162**, 811–813 (1968).
29. T. C. Merigan and M. S. Finkelstein, *Virology* **35**, 363–374 (1968).
30. T. C. Merigan and W. Regelson, *New Engl. J. Med.* **277**, 1283–1287 (1967).

Review Articles

P. W. Sadler, "Chemotherapy of Viral Diseases," *Pharmacol. Rev.* **15**, 407–447 (1963).

D. G. O'Sullivan, "Viruses and the Chemotherapy of Viral Diseases," *Roy. Inst. Chem. (London), Lectures. Monographs, Repts. 1965* (2).

M. R. Hilleman, "Immunologic, Chemotherapeutic and Interferon Approaches to Control of Virus Diseases," *Amer. J. Med.* **38**, 751–766 (1965).

L. S. Kucera and E. C. Herrmann, Jr., "Antiviral Agents," *Ann. Repts. Med. Chem. 1965*, Chap. 13, pp. 129–135.

E. C. Herrmann, Jr., "Antiviral Agents," *Ann. Repts. Med. Chem. 1966*, Chap. 13, pp. 122–130.

C. E. Hoffmann, "Antiviral Agents," *Ann. Repts. Med. Chem. 1967*, Chap. 12, pp. 116–125.

W. H. Prusoff, "Recent Advances in Chemotherapy of Viral Diseases," *Pharmacol. Rev.* **19**, 209–250 (1967).

C. E. Hoffmann, "Antiviral Agents," *Ann. Repts. Med. Chem. 1968*, Chap. 11, pp. 117–125.

General References

S. Baron, "Mechanism of Recovery from Viral Infections," *Advan. Virus Res.* **10**, 39–64 (1963).

F. M. Burnet and W. M. Stanley, ed., "The Viruses," *Animal Viruses*, Vol. 3, Academic Press, New York, 1959.

P. D. Cooper, "Plaque Assay of Animal Viruses," *Methods in Virology*, Vol. 3, Academic Press, New York, 1967, Chap. 6, pp. 244–311.

N. B. Finter, ed., *Interferons*, W. B. Saunders Co., Philadelphia, Penna., 1966.

F. H. Horsfall, Jr., and I. Tamm, eds., *Viral and Rickettsial Infections of Man*, 4th ed., J. B. Lippincott Company, Philadelphia, Penna., 1965.

D. M. Horstman, *Virology and Epidemiology*, Archon Books, The Shoe String Press, Inc., Hamden, Conn., 1962.

J. Vilček, *Interferon*, Virology Monographs, Springer-Verlag, Inc., New York, 1969.

H. E. Whipple, ed., "Antiviral Substances," *Ann. N.Y. Acad. Sci.* **130**, 1–482 (1965).

B. I. Wilner, *A Classification of the Major Groups of Human and Other Animal Viruses*, 4th ed., Burgess Publishing Co., Minneapolis, Minn., 1969.

Conrad E. Hoffmann
E. I. du Pont de Nemours & Co., Inc.

VISCO. See Surfactants.

VISCOMETRY

Viscometry is the measurement of the irreversible flow behavior of gases, liquids, and solids. The rate of flow, ie, the rate of continuous deformation, is a function of the applied stress resulting from the action of gravity or an externally applied force. The ratio of applied shearing stress to the rate of shear is called the coefficient of viscosity, or simply viscosity, and is a constant, independent of rate of shear for ideal or Newtonian fluids. Viscometry is a branch of the science of rheology (qv). Rheology is concerned with both the reversible and irreversible changes accompanying deformation. The theory of rheology, and consequently viscosity, is based on idealized models and on the assumption that the constants of the derived theoretical and empirical equations do not change with changes in the variables. However, deviations from the ideal are encountered frequently (see Rheology).

The measurement of viscosity is essential to many industrial processes, mechanical engineering calculations, and the more fundamental sciences. Viscosity is the parameter which affects the flow of fluids and gases in pipes and the mixing of materials in stirred reactors. Viscosity is a force to be overcome in molding and extrusion of plastics and elastomers and in applying surface coatings, printing inks, paints, etc. Viscosity is a parameter in determining when flow will be streamlined or turbulent. Lubrication of moving surfaces to eliminate friction and prevent wear is a function of the viscous properties of the fluid. Viscosity is an important quantity in both the biological and physical sciences since the viscosity coefficient is significantly influenced by the size, shape, arrangement, and force between molecules. This article will be concerned primarily with the measurement of viscosity under steady-state conditions for Newtonian fluids, although some of the instruments can be used to measure nonviscous elements such as plasticity and elasticity. For more detailed information the reader is referred to Barr (1); Merrington (2); Dinsdale and Moore (3); Van Wazer, Lyons, Kim, and Colwell (4); and Eirich (5). Methods of measuring the viscosity of materials having industrial importance and for which procedures and specifications have been developed can be found in the appropriate American Society for Testing and Materials (ASTM) test method.

Principles

Newton's Law. Newton postulated in his *Principia* the first hypothesis concerning the magnitude of force required to overcome resistance to flow. It is upon Newton's fundamental assumptions that the theory and practice of viscometry are based. Consider two close parallel plates of infinite extent, each of area A, separated by a layer of liquid of thickness h. Consider one of the plates stationary while the other is moved with a constant velocity, v, in its own plane. In order to keep the velocity constant, a force, f, must be applied to the moving plate. Newton assumed this force to be proportional to the area of the plates and to the velocity gradient of the fluid between the plates. Thus:

$$f \ \alpha A v / h = \eta \ A v / h \tag{1}$$

$$\eta = \delta s / \delta \sigma = f h / A v \tag{2}$$

where s is the shear force and σ is the rate of shear. The constant of proportionality, η, is the viscosity coefficient or viscosity of the fluid. The dimensions of η are g/(cm sec) or dyne sec/cm². This is known as the poise (P) in honor of Poiseuille, one of the pioneers of viscometry. The viscosity of many fluids is reported in centipoise (cP) since the viscosity of water is approximately 0.01 poise at normal temperature and pressure. The reciprocal of viscosity, $1/\eta$, is called the fluidity and is generally denoted by the Greek letter ϕ. Some viscometers give a measurement of the ratio of η/d where d is the density. This ratio is the kinematic viscosity, often designated η_{KV} or simply KV and in the cgs system is called the stoke with the dimension of cm²/sec. The most commonly used kinematic unit is the centistoke (cSt). In the range from 0.001 to 0.1 P it is difficult for the eye to distinguish differences in viscosity. In the range from 1–100 P, it is not difficult to distinguish changes in viscosity of relatively small magnitude. For materials having viscosities greater than 1000 P, the casual observer will consider them solid since they will appear not to flow.

In engineering calculations it may be desirable to express viscosity in units other than cgs; conversion factors are presented in Table 1. In the International System (SI) of units, the dyne having units of g cm/sec² is replaced by the newton with units of kg m/sec². To convert from dynes to newtons, multiply by 1.00×10^{-5}. The SI unit of viscosity replacing the poise is newton sec/in². To convert from poise to the SI unit of viscosity, multiply by 1×10^{-1}.

Effect of Shear Stress. For Newtonian fluids the shear rate is directly proportional to the shear stress, line *A* of Figure 1. Gases, simple organic and inorganic liquids, true solutions, dilute suspensions, and dilute emulsions are generally Newtonian in character. There are, however, many fluids which deviate from direct proportionality between the shear rate and the shear stress, ie, they are non-Newtonian. Concentrated colloidal and other particulate systems, large molecules, solutions of polymers, and other macromolecules in dispersed systems exhibit non-Newtonian flow because of particle interaction, interaction with the continuous phase, particle deformation including disentanglement, and reformation of molecular entanglement. Figure 1 presents idealized curves for a variety of types of non-Newtonian behavior.

Table 1. Conversion Factors for Dynamic Viscosity (6)

Multiply by / To Obtain	Poise $\left[\dfrac{\text{gm mass}}{\text{cm-sec}}\right]$ or $\left[\dfrac{\text{dyne-sec}}{\text{cm}^2}\right]$	Centipoise $\left[\dfrac{\text{gm mass}}{\text{cm-sec}}\right]$ or $\left[\dfrac{\text{dyne-sec}}{\text{cm}^2}\right]$	$\left[\dfrac{\text{slugs}}{\text{ft-sec}}\right]$ or $\left[\dfrac{\text{lb force-sec}}{\text{ft}^2}\right]$	Reyns $\left[\dfrac{\text{slugs}}{\text{in.-sec}}\right]$ or $\left[\dfrac{\text{lb force-sec}}{\text{in.}^2}\right]$	$\left[\dfrac{\text{lb mass}}{\text{ft-sec}}\right]$ or $\left[\dfrac{\text{pdl-sec}}{\text{ft}^2}\right]$
Poise $\left[\dfrac{\text{gm mass}}{\text{cm-sec}}\right]$ or $\left[\dfrac{\text{dyne-sec}}{\text{cm}^2}\right]$	1	0.01	478.8	6.895×10^4	14.88
Centipoise $\left[\dfrac{\text{gm mass}}{\text{cm-sec}}\right]$ or $\left[\dfrac{\text{dyne-sec}}{\text{cm}^2}\right]$	100	1	47,880	6.895×10^6	1,488
$\left[\dfrac{\text{slugs}}{\text{ft-sec}}\right]$ or $\left[\dfrac{\text{lb force-sec}}{\text{ft}^2}\right]$	2.088×10^{-3}	2.088×10^{-5}	1	144	0.03108
Reyns $\left[\dfrac{\text{slugs}}{\text{in.-sec}}\right]$ or $\left[\dfrac{\text{lb force-sec}}{\text{in.}^2}\right]$	1.45×10^{-5}	1.45×10^{-7}	6.95×10^{-3}	1	4.48
$\left[\dfrac{\text{lb mass}}{\text{ft-sec}}\right]$ or $\left[\dfrac{\text{pdl-sec}}{\text{ft}^2}\right]$	0.0672	6.72×10^{-4}	32.17	0.223	1

For example, given Reyns, multiply by 6.895×10^4 to obtain poises.

When the rate of shear increases more than in proportion to the shearing stress, line *B* of Figure 1, the fluid appears to become less viscous as the shear stress becomes larger. Such a material is a pseudoplastic or a shear-thinning liquid. A fluid in which the rate of shear decreases with an increase in the shear stress is called a dilatant or shear-thickening fluid, line *C* of Figure 1.

Some materials are anomalous and will not flow until a critical shear stress is exceeded. This critical shear stress is called yield value, designated as ψ in Figure 1. Once the yield value has been exceeded, direct proportionality between the rate of shear and the shear stress may be observed. Such a material is called a plastic substance or a Bingham body, curve *D* in Figure 1. Materials exhibiting a yield value can also be pseudoplastic and dilatant, curves *E* and *F*, respectively, of Figure 1.

Time of Flow. The phenomenological description of flow behavior discussed above is not complete without the inclusion of a time-dependent effect. Materials which become more fluid with increasing time of flow are said to be thixotropic. If the material exhibits the opposite effect and becomes less fluid with increasing time of flow at a constant shear rate, the effect is called rheopexy. These effects may be spontaneous, reversible, or irreversible. In many cases the apparent viscosity of a thixotropic fluid decreases logarithmically with time to a final limiting value.

Theories of Flow Behavior. The viscosity of a gas, to the extent that it approximates ideality, can be calculated with some accuracy from kinetic theory (see Rheology). Experimental investigations have shown that the viscosity of gases, to the extent that they approach ideality, is nearly independent of pressure from several hundredths of an atmosphere to many atmospheres. The viscosity of gases which are not ideal increases with pressure dependent on the degree of deviation from ideality.

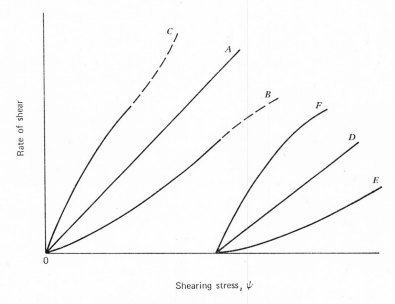

Fig. 1. Idealized flow behavior of fluids.

For gases the viscosity increases with temperature, a result of the increased number of collisions between gas molecules and the increase in the average energy of the collisions.

For liquids the problems associated with the statistical mechanical approaches to basic kinetic and molecular theories of condensed phases are so formidable that the theoretically derived equations are cumbersome and difficult to use in the practical application of rheology. For liquids the viscosity is a direct function of the pressure and an inverse function of the temperature. As a rough approximation, increasing the pressure by 10,000 psi is equivalent to a drop in temperature of 25°C. Many semiempirical and empirical approaches have been taken to formulate equations which will express the viscosity of liquids as a function of temperature and pressure. The familiar Eyring equation expresses viscosity as a function of temperature

$$\eta = (Nh/V)e^{\Delta F^{\ddagger}/RT} \simeq Ae^{\Delta H^{\ddagger}/RT} \tag{3}$$

Table 2. Viscosity–Temperature Relationships

	Subtangent $\dfrac{d\eta}{\eta dt}$	Reference
$\eta = Re^{-\alpha t}$	$-\alpha$	11
$\eta = \dfrac{S}{(a+t)^m}$	$-\dfrac{m}{(a+t)}$	12
$\eta = K \exp\left(\dfrac{b}{t+\theta}\right)$	$-\dfrac{b}{(t+\theta)^2}$	13
$\eta = C \exp\left(A + \dfrac{B}{T}\right)^2$	$-\dfrac{2B}{T}\left(A + \dfrac{B}{T}\right)$	14
$\eta + \gamma = \dfrac{\beta Dl}{T^c}$	$-\dfrac{c \log D}{T^{c+1}}\left(1 + \dfrac{\gamma}{\eta}\right)$	15
$\text{arc sinh}\left(\log_e \dfrac{\eta}{\eta_0}\right) = d - n \log_e T$	$-\dfrac{n}{T}\cosh\left(d - n \log_e T\right)$	16
$\eta = K' \exp \dfrac{b'}{(t+\theta')^2}$	$-\dfrac{2b'}{(t+\theta)^3}$	17

where V is the molar volume, h is Planck's constant, N is Avogadro's number, $\Delta F^{\ddagger}$ is the standard free energy of activation, and $\Delta H^{\ddagger}$ the enthalpy of activation (7). The theoretical derivation of Eyring's equation considers a liquid as an imperfect lattice, and flow as a rate process based upon the motion of molecules into available holes in this liquid lattice. Empirically, $\Delta H^{\ddagger}$ is related to the molar heat of vaporization of the

Table 3. Methods for Estimating the Viscosity of Liquids at Elevated Pressures

	Reference
$\log \dfrac{\eta_p}{\eta_0} = 16 \log\left[1 + \dfrac{0.062\,(10^{+a})\,P\,10^{-3}}{\eta_0^{0.062}}\right]$	18
$\quad$ where $-a = \dfrac{°F}{400} + 0.4$	
$\log \dfrac{\eta_p}{\eta_0} = \alpha P(10^{-4})$	19
$\quad$ where $\alpha = (0.183 + 0.295 \log \eta_0)$	
$\log \dfrac{\eta_p}{\eta_0} = \left[\dfrac{P}{7350}\right]^y (0.228 + A \log \eta_0)$	20
$\quad$ where y and A are functions of the percent carbon in aromatic and naphthenic ring structures, C_A and C_N, respectively: $\log(y - 0.890) = 0.00955\,(C_A + 1.5\,C_N) - 1.930$ $A = 0.002\,C_A + 0.003\,C_N + 0.055$	
$\log \dfrac{\eta_p}{\eta_0} = P\,(10^{-4})\,(0.239 + 0.168\,\eta_0^{0.278})$	21

NOTE: In all of above equations η_p = absolute viscosity at pressure P, η_0 = absolute viscosity at atmospheric pressure, P = pressure in psig.

liquid. For liquids composed of spherical molecules, $\Delta H^{\ddagger}$ is approximately equal to $\frac{1}{3}$ the heat of vaporization (7). In liquids which exhibit secondary bonding or are asymmetrical, $\Delta H^{\ddagger}$ is approximately $\frac{1}{4}$ the heat of vaporization (7). Equation 3 is often limited to relatively narrow temperature ranges since $\Delta H^{\ddagger}$ is temperature dependent, especially in the low temperature ranges where the viscosity is high and is increasing at a large rate as the temperature is lowered.

If holes in the imperfect liquid lattice represent free volume, then the Doolittle equation expresses viscosity as a function of V_f, the fractional free volume in the liquid:

$$\eta = A \, \exp \, (B/V_f) \tag{4}$$

where both A and B are constants (8). Fractional free volume can be defined as

$$V_f = (V_T - V_0)/V_T \tag{5}$$

where V_0 is the extrapolated volume of the liquid at $0°K$ in the absence of any phase change. If the density of a liquid is known as a function of temperature and/or pressure, then equation 4 allows the calculation of viscosity at conditions other than those of the measurement. The Tammann-Hessee equation

$$\eta = A \, \exp \, B/(T - T_0) \tag{6}$$

is similar to equation 4 but expresses V_f in terms of $T - T_0$ where T_0 is the "thermodynamic" glass transition temperature (the temperature at which the configurational entropy becomes zero) or an empirically chosen constant (9).

The Eyring equation and the Doolittle equation, when combined, yield an equation of the form

$$\eta = A \, \exp \, (E_v^*/RT + B^*/V_f) \tag{7}$$

which is reported to fit the variation of viscosity both with temperature and with pressure for a wide variety of fluids over an extended range (10). Other empirical equations applicable principally to hydrocarbon lubricants which express viscosity as a function of temperature and pressure are tabulated in Tables 2 and 3.

Measurement of Viscosity

Generally speaking, there are three fundamental types of viscometers in use. These are: capillary viscometers, rotational viscometers, and falling-sphere viscometers. There are also a large number of specialized instruments designed to fulfill a particular requirement of the experimenter. This discussion will confine itself to the most generally used and available viscometers. A more complete description of modern apparatus can be found in the reference books cited earlier.

CAPILLARY TUBE VISCOMETERS

Capillary tube viscometers are based on Poiseuille's Law which states that the volume of liquid V, per unit time, t, passing through a capillary of radius, r, at moderate velocity and exhibiting laminar flow is inversely proportional to the length of the tube, l, and directly proportional to the radius of the capillary to the fourth power. Poiseuille's Law can be stated as follows:

$$\eta = \pi r^4 pt/8Vl \tag{8}$$

where p is the pressure drop through the capillary. In practice there are several corrections which must be made in order to measure correctly the viscosity of liquids from the time of flow through the capillary.

Capillary tube viscometers are particularly convenient for the measurement of the flow properties of liquids that do not exhibit yield values or time-dependent changes. They are normally employed for precise measurement of Newtonian liquids in the range from 0.01 to 100 P. When instruments are used in which it is possible to vary the pressure drop through the capillary over a wide range, the flow characteristics of non-Newtonian liquids can be deduced from capillary measurements.

A viscometer is considered to be an absolute instrument if its calibration constants can be obtained solely from its dimension and its operating characteristics. While it is theoretically possible to use a capillary viscometer as an absolute instrument, in practice it is extremely difficult to obtain accurate results. The determination of the viscosity of water by Swindells et al. (22), serves as an excellent example of the apparatus and technique required to obtain measurement of the highest accuracy using capillary viscometers. In general, absolute measurements are made only for special standardizing or research requirements. For moderately accurate work in which errors of approximately 0.5% can be tolerated, the Poiseuille equation is an adequate expression for flow even when the capillary length is rather short. For more refined work one must consider the kinetic energy losses, end effects, turbulence, drainage, surface tension effects and, with high viscosity fluids, heat effects. When the materials being measured are viscoelastic or are multiphase, the effect of elastic energy and wall effects, respectively, must be considered.

The largest numerical correction is that due to the kinetic energy required to accelerate the liquid from its velocity in the inlet reservoir to its average velocity in the capillary. The second numerical correction, of a smaller magnitude, is an allowance for viscous resistance incurred by the liquid as a result of velocity gradients in the converging stream entering the capillary. It is normally expressed as a hypothetical increase in the capillary length which is proportional to capillary radius. The equation for fluid flow through a capillary incorporating the two correction terms considered above is:

$$\eta = \pi r^4 p / 8Q(l + nr) - mdQ/8\pi(l + nr) \tag{9}$$

$$KV = \eta/d = \pi r^4 h\, gt / 8V(l + nr) - mV/8\pi(l + nr)t \tag{10}$$

$$KV = Ct - B/t \tag{11}$$

where η = liquid viscosity in poise, r = capillary radius in cm, p = pressure drop through the capillary tube in dynes/cm^2, Q = rate of flow of liquid through the tube in cm^3/sec, l = length of a capillary tube in cm, d = liquid density in g/cm^3, m = kinetic energy coefficient, h = mean head during measurement, V = volume of liquid delivered from the upper reservoir bulb, t = time for the liquid to pass the fiducial marks, g = acceleration due to gravity, n = the so-called "Couette" constant such that nr is the hypothetical increase in the capillary length needed to account for convergence friction, and C and B = instrument constants obtained from measurements on known material.

While it is extremely difficult to obtain accurate capillary viscometer constants from the dimensions and operating conditions of the apparatus, it is a comparatively easy matter to calibrate capillary viscometers by standardizing with liquids of known viscosity.   Comparatively simple apparatus can be used for such relative measurements.   The primary reference liquid is water at 20°C because it is easily prepared in a pure form and its viscosity is known to a high degree of accuracy (ASTM D445).   A number of secondary standards are available from the ASTM.   These liquids provide

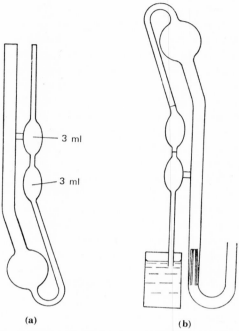

(a)   (b)

Fig. 2.   Cannon-Fenske viscometer for transparent liquids (**a**) and the method of introducing samples to the viscometer (**b**).

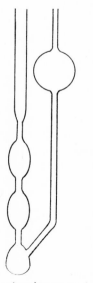

Fig. 3.   Cannon-Fenske viscometer for opaque liquids.

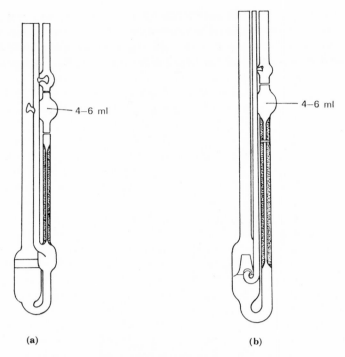

4–6 ml

4–6 ml

(a)   (b)

Fig. 4. Ubbelohde viscometer (**a**) and SIL (Standard Inspection Laboratory) viscometer (**b**) for transparent liquids.

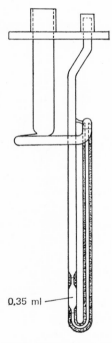

0.35 ml

Fig. 5. Zeitfuchs cross-arm viscometer.

a wide range of viscosities and a means of calibrating the viscometer with a fluid having surface tension characteristics similar to those of the fluid being measured.

Gravity-Driven Viscometers. Capillary viscometers fall into two classes depending upon whether the hydrostatic pressure itself induces flow through the capillary or an external pressure is applied to force the liquid through the tube. The first type of instrument is more commonly used. Gravity-driven viscosity tubes of several different types are shown in Figures 2–5. The design of these viscometers is described in ASTM D445.

The Cannon-Fenske style viscometer (Fig. 2a and b) have been shown to be excellent instruments for general use. A long capillary and a small upper reservoir bulb result in a small kinetic energy correction while the large diameter of the lower reservoir minimizes head errors due to volume changes from temperature effect or improper adjustment of initial volume. The greater length of the capillary with the consequently larger head of driving liquid also gives rise to smaller percentage corrections than would be the case for shorter tubes. The Cannon-Fenske viscometers are constructed so that the lower reservoir is directly under the upper reservoir. This condition has been shown to give the minimum change in head resulting from error in the vertical alignment of the viscometer. A Cannon-Fenske viscometer for opaque liquids is shown in Figure 3. The liquid flows from bulb A up through bulbs B and C.

The Ubbelohde design (Fig. 4a) was first proposed as a means of eliminating the surface tension corrections. However, it has been pointed out that this cannot be done (23). Nonetheless, the instrument has one advantage. In using the Ubbelohde viscometer a roughly measured amount of liquid can be introduced and allowed to reach thermal equilibrium. The capillary, upper reservoir, and the side tube are then filled by applying pressure to the lower reservoir until the liquid is above the upper fiducial mark. When the pressure is released, the liquid flows rapidly out of the large side tube and the reservoir under the capillary is emptied. The time for the liquid to flow out of the upper bulb through the capillary is then measured. By this means the lower liquid volume is made unimportant and the volume initially added to the instrument does not have to be precisely measured.

The SIL viscometer (Fig. 4b) has the advantage that the flow arm permits very precise volume determinations. The Zeitfuchs cross-arm viscometer (Fig. 5) is useful in measuring the viscosity of very dark-colored oils, the meniscus of which cannot be observed through the oil film as the liquid drains. The fiducial reservoir is at the bottom of the tube and the time of rise of the meniscus into the previously unwet bulb is measured.

Recently several commercial viscometers have been made available which allow for some degree of automation in the operation of capillary instruments. The time of flow is measured by interruption of a light beam and the efflux time automatically displayed. Another version automatically fills the viscometer, can be programmed to obtain repetitive readings, and provides a wash cycle between fillings as well as printing out the efflux times.

For convenience in use and in calibration it is advisable to employ a set of viscometers with the dimensions adjusted so that the constant C in equation 11 for successive instruments differs by a factor of 3. For many instruments it will be found that the correction B/t is very small relative to the first term in equation 11 and can be neglected except for measurement on materials having viscosities less than 2 cSt. For these materials this correction can be neglected if the efflux time is greater than 300 sec. Table 4 summarizes the dimensions and viscosity ranges for various glass capillary viscometers.

In utilizing capillary viscometers care should be taken to see that the capillary tubes are free of dirt, dust, particles, and are free-draining. It is advisable to filter all liquids and acids used for cleaning before they are introduced into the tube. The primary method of calibration requires that all viscometers be standardized against water at 20°C. This is possible only for the lowest instruments in the series. However, after these have been calibrated they can be directly compared with the next higher viscometer using a suitable liquid of higher viscosity. This can be continued with increasingly more viscous oils up to tubes having a constant of 10 cSt/sec.

Table 4. Dimensions and Viscosity Ranges for Glass Capillary Viscometers

Viscosity range, cSt	Approx constant, C	Respective capillary diameters, mm			
		Cannon-Fenske	Ubbelohde	SIL	Cross-Arm
0.6–3.0	0.003	0.31 ± 0.02		0.41 ± 0.02	0.28
1.0–5.0	0.005	0.42 ± 0.02			
2.0–10	0.01	0.63 ± 0.02	0.58 ± 0.02	0.61 ± 0.02	0.38
6–30	0.03	0.78 ± 0.02	0.77 ± 0.02	0.73 ± 0.02	0.50
10–50	0.05		0.87 ± 0.02	0.91 ± 0.02	
20–100	0.1	1.02 ± 0.02	1.10 ± 0.03	1.14 ± 0.03	0.67
60–300	0.3	1.26 ± 0.02	1.43 ± 0.03	1.50 ± 0.03	0.88
100–500	0.5	1.48 ± 0.02	1.64 ± 0.03	1.71 ± 0.03	
200–1,000	1.0	1.88 ± 0.02	1.95 ± 0.03	2.03 ± 0.03	1.20
600–3,000	3.0	2.20 ± 0.05	2.67 ± 0.04	2.80 ± 0.03	1.42
1,000–5,000	5.0	3.10 ± 0.05	3.06 ± 0.04	3.06 ± 0.03	
2,000–10,000	10.0	4.0 ± 0.04	3.62 ± 0.04	3.79 ± 0.04	1.93
6×10^3–3×10^4	30.0				2.52
2×10^4–1×10^5	100.0				3.06
		Volume of bulb, mm			
		3.15 ± 0.15	4–6	4–6	0.35
		Length of capillary, mm			
		73 ± 3	90 ± 5	145 ± 1	210
				127 ± 3	165

A number of corrections must be made in order to relate flow times in capillary viscometers to viscosity through equation 11. These have been discussed in detail (24). The liquid is generally introduced into the U-tube viscometer by pipeting at room temperature which is usually different from the test temperature. The volume change of heating or cooling the pipet to the required test temperature must therefore be taken into account. For lubricating oil, pipeting at 70°F with an instrument used at 210°F, the correction can amount to 0.1–0.4% of the observed flow time.

Calibration relative to water or a standardized oil is made at a single temperature. When the viscometer is used at a different temperature, a correction must be made for changes in dimensions of the tube due to expansion and contraction. The order of magnitude of this correction is very small and is usually combined with the correction for the filling volume.

In the U-tube viscometer the driving head of liquid is opposed by the weight of the air column in the opposite arm. Since the weight of this air changes with tempera-

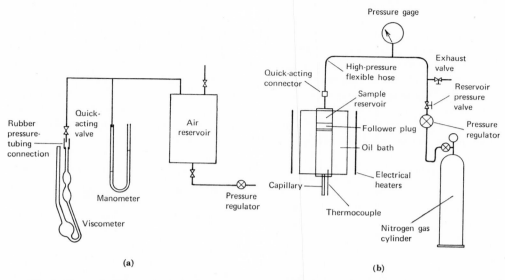

Fig. 6. Pressure-driven viscometers: (**a**) an arrangement for applying external pressure to glass capillary viscometers; (**b**) schematic diagram showing the principle of the Burrell-Severs extrusion rheometer.

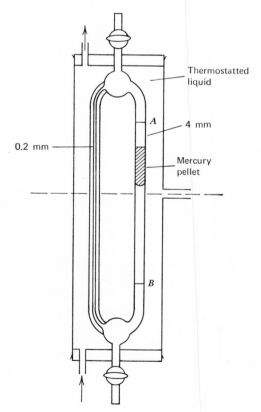

Fig 7. Rankine gas viscometer.

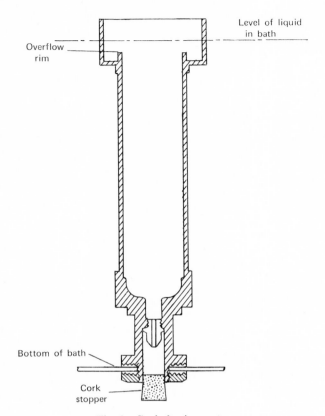

Fig. 8. Saybolt viscometer.

ture and humidity, a correction must be made. This correction is on the order of magnitude of 0.05% when a material having a density of 0.8 is run at 100°F.

The constants C and B of equation 14 are constant only if the volume does not change. Appreciable changes in liquid head result from the viscometer being used in a nonvertical position.

Pressure-Driven Viscometers. Fundamentally, pressure-driven capillary viscometers are not different from the gravity-driven type. Rather than being operated at atmospheric pressure, the liquid is forced through the capillary under pressure. The advantages of a pressure-driven capillary viscometer are: (1) the absolute viscosity can be determined with accuracy without the necessity of obtaining accurate density data, (2) fewer tube assemblies are needed to cover a wide viscosity range since the time intervals can be adjusted by altering the applied pressure, and (3) surface tension effects and other head corrections are minimized when the applied pressure is substantial. The disadvantages of pressure-driven capillary viscometers are: (1) they are inconvenient to use compared with the simple gravity-driven instruments, (2) it is difficult to maintain constant pressure on the liquid, (3) drainage corrections needed for variations in the rate of emptying the reservoir bulbs are large and uncertain, and (4) dissolved gases have considerable effect on liquid viscosity.

Figure 6a is a schematic diagram for applying pressure to a capillary viscometer. Pressure-driven capillary viscometers (extrusion rheometers) are used to measure the

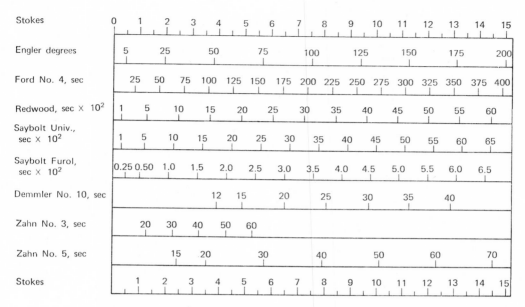

Fig. 9. Viscosity conversion chart for orifice viscometers.

bulk viscosity of polymers and other viscous materials. The simple melt indexer, a mechanically loaded piston in a heated cylinder which forces the melt through the capillary orifice, is an example of apparatus used industrially for this purpose. Figure 6b is a schematic diagram showing the principles of an extrusion rheometer when the cylinder is loaded by means of gas pressure.

Gas Viscometers. Capillary viscometers can also be used to measure the viscosity of gases. Figure 7 shows such an instrument. The apparatus is filled with a gas at the desired pressure and the time required for the mercury pellet to fall between marks on the larger tube is measured. The details of operation and various factors influencing precision and accuracy have been discussed in the literature (1).

Capillary Viscometers for Process Control. The capillary viscometers are excellent for precise work. However, for control of industrial processes and products where it is not necessary to have a precise measure of viscosity, these instruments are rather cumbersome to use. Therefore, a wide variety of instruments adapted to the needs of various process industries such as petroleum, paint and varnish, printing inks, etc, were designed. The design of these viscometers consists of short capillaries or orifices and the viscosity is generally recorded as the efflux time of a fixed volume of liquid. Historically the most commonly used viscometer of this type in the United States was the Saybolt Universal Viscometer. Viscosities of petroleum oils are often given in terms of Saybolt Universal Seconds (SUS), even though the viscosity is measured in a standard capillary viscometer. Conversion between kinematic and Saybolt units may be made by means of tables (ASTM D446). Figure 8 illustrates the construction of a Saybolt oil tube. In addition to the Saybolt viscometers, other specialized industrial capillary viscometers are listed in Table 5 together with some of their characteristics and areas of application. The chart of Figure 9 gives a rough approximation of the conversion of the units of these viscometers to stokes. A more accurate conversion can be made to kinematic viscosity via equation 12 which in form is

Table 5. Industrial Capillary-type Viscometers

Type	Major areas of application	Viscosity ranges, cSt	Viscosity units	Thermostat	Approximate constants	
					k	K
Barbey (France)	petroleum products	1–4,000	ml/hr	no		
Demmler	wire enamel	4–20,000	sec			
Engler (German)	tar, petroleum	1–1,500	sec/200 ml or Engler degree	yes	0.073	0.0631
Ford Cup	paint, varnish	1–1,500	sec	yes or no	0.013–0.037	4–8
Gardner (one shot)	paint, varnish		sec	no	0.04	8
Narsh Funnel	drilling mud		sec/946 ml			
Redwood (England) No. 1	petroleum products	1–500	sec/60 ml	yes	0.00264 0.00247	1.9 for 40 < t < 85 sec 0.65 for 85 < t < 200 sec
No. 2	petroleum products	500–5,000	sec/60 ml	yes	0.026	0.40
Saybolt Universal	petroleum products	1–400	sec/60 ml	yes	0.00226	1.95 for t < 100 sec
Saybolt Furol	petroleum products	400–4,000	sec/60 ml	yes	0.00220	1.30 for t > 100 sec

similar to equation 11. However, the coefficients in this case are not true constants and it is sometimes necessary to calibrate the instrument for various ranges of viscosities. The approximate constants, k and K, are tabulated in Table 5.

$$KV = \eta/P = kt - K/t \tag{12}$$

ROTATIONAL VISCOMETERS

Rotational viscometers are comprised of two members, separated by the material under test, which are able to rotate relative to one another upon a common axis of symmetry. One of the members, when rotated, produces shearing action on the liquid which is transmitted to the other members. The torque required to produce a given angular velocity or the angular velocity resulting from a given torque is a measurement of the viscosity. The equation relating angular velocity and torque to viscosity for the case of concentric cylinders and Newtonian fluids is:

$$\eta = (M/\Omega \, 4\pi h) \, (1/R_b{}^2 - 1/R_c{}^2) = kM/\Omega \tag{13}$$

where M is the torque exerted on the inner cylinder, h is the length of the inner cylinder, Ω is the relative angular velocity of the cylinder in radians per second, R_b is the radius of the inner cylinder wall, R_c is the radius of the outer cylinder wall, and k is an instrument constant.

Rotational viscometers are, by virtue of their mechanical design, especially suited for the study of the nonideal flow properties of liquids and, in particular, deviations from Newtonian flow. Viscosity by definition is the slope of the plot of stress, s, against rate of shear, σ (eq. 2). For liquids exhibiting a yield value ψ the rate of shear at the inner cylinder is

$$\sigma = (s_b - \psi)\Omega/(Mk_1 - \psi \, k_2) \tag{14}$$

where $k_1 = l/(4\pi h)(1/R_b{}^2 - 1/R_c{}^2)$ and $k_2 = \ln (R_c/R_b)$ and s_b is the shear stress at the inner cylinder. To obtain η a plot is made of the quantity σ, equation 14, against the shear stress at the inner cylinder given by

$$s_b = M/2 \, \pi \, R_b{}^2 \, h \tag{15}$$

Account must be taken of the drag on the inner cylinder due to end effects which appear as a correction to h. The best method to arrive at the correction, h_0, is to measure the angular velocity Ω and torque M at several values of h. A plot is then made of M/Ω against h and extrapolated to a value of h_0 at $M/\Omega = 0$. Equation 13 then becomes

$$\eta = [M/\Omega \, 4\pi(h + h_0)](1/R_b{}^2 - 1/R_c{}^2) \tag{16}$$

and h in equation 15 becomes $(h + h_0)$. The literature should be consulted for other non-Newtonian flow equations and correction factors (4,5).

Rotational viscometers are mechanically more complicated than capillary viscometers. They are normally employed for materials in the range above 50 P although they can be used for gases. For work of the highest precision, their design and construction becomes very critical. However, for routine applications where precision is less essential, rotational viscometers have many features which make them convenient to use and more versatile than capillary instruments. Rotational viscometers are capable of making measurements under steady-state conditions and are thus used to study the flow properties of non-Newtonian materials. The construction of rotational viscometers makes it possible to make measurements at different shear rates and as a function of time.   The literature contains many different designs for rotational viscometers (4).   A number of these are available commercially.

Although the principle of rotational viscometers is simple, it is difficult to construct a coaxial cylinder viscometer for precise and accurate measurement in spite of

Fig. 10. Brookfield Synchro-Lectric viscometer.

Fig. 11. Stormer viscometer.

Fig. 12. MacMichael viscometer.

the fact that equation 13 leads one to assume that the absolute viscosity can be determined directly from the dimensions of the instrument. The greatest problem lies in the elimination of end effects. One of the principal differences in construction of rotational viscometers is a result of the attempt to design around these end effects. Because of the difficulties inherent in all the mechanical methods, the most common approach is simply to make a cylinder long with respect to its diameter.

A number of commercial viscometers are available that are based upon the cylinder principle but modified for routine industrial use. In general these instruments are intended for relative measurement only and are not designed to eliminate end effects or to produce results of high accuracy and precision. Many of the more modern instruments are equipped to record automatically the torque as a function of time or shear rate.

A commonly used rotational viscometer is the Brookfield Synchro-Lectric shown in Figure 10. In the Brookfield viscometer the rates of revolution may be varied from $\frac{1}{2}$ rpm up to 100 rpm. Spindles come in different lengths and different diameters so that viscosities from 0.5–64×10^4 cP may be measured depending upon the model. Other rotational viscometers having general utility are the Stormer, illustrated in Figure 11, and the MacMichael, illustrated in Figure 12. The Stormer viscometer applies a constant torque to the inner cylinder and the rate of rotation is measured.

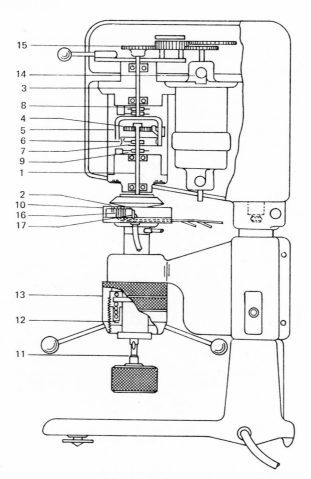

Fig. 13. Ferranti-Shirley cone-and-plate viscometer

1. cone spindle
2. cone
3. driving spindle
4. torque spring
5. bridge housing
6. potentiometer, for
   torque dynamometer
7. wiper for potentiometer
8. slip rings

9. slip rings
10. plate
11. micrometer
12. nut
13. screw for raising plate
14. driving motor
15. gearing
16. thermocouple
17. waterjacket

The shear stress is applied by attaching weights to a string and permitting free-fall through a vertical distance.

More sophisticated rotational viscometers are needed for research work. One example is the Ferranti-Shirley cone-and-plate viscometer, illustrated schematically in Figure 13. The fluid is sheared in the space formed when the apex of the upper or conical section just touches the stationary, flat, lower plate. The fluid is held in the gap by its own surface tension. The cone and plate principle produces a uniform rate of shear at all points and thus this instrument is well suited for measurement of non-

Newtonian fluids since the rate of shear is directly proportional to the angular velocity of the cone. The Ferranti-Shirley viscometer can be used with a programmer and an X–Y recorder. The rotational speed can be programmed and the output recorded as torque against speed in rpm as illustrated in Figure 14.

A special kind of rotational viscometer which provides many advantages in precision and sensitivity is the oscillating viscometer. In this type the rotating element is mounted in such a way that it will oscillate around its axis as a torsional pendulum. The amount of dampening caused by viscous drag then measures the viscosity of the specimen.

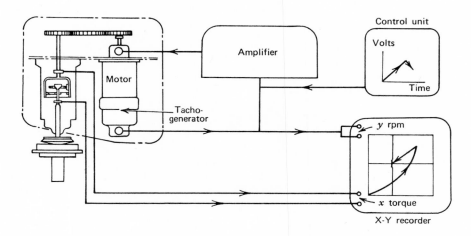

Fig. 14. Schematic diagram of Ferranti-Shirley recording system.

Mooney Viscosity. The Mooney viscometer is used as a standard instrument industrially to measure the flow characteristics of rubber and other elastomers (ASTM D1646). The Mooney instrument is a rotational shearing disc viscometer. The measurement of viscosity is given in terms of arbitrary Mooney units. The torque required to rotate the disc is indicated on a linear scale. One Mooney unit is a displacement of 0.001 inch. Mooney readings are dependent upon rotor size, temperature, and time at which the reading is made after the test is started.

FALLING-BODY VISCOMETERS

Falling-sphere viscometers are based on the application of Stokes' Law. If a body is moved through a fluid, the fluid contiguous with the body can be considered as having zero velocity with respect to the solid. Adjacent layers are set in motion by the viscous drag. In 1845 Stokes derived the relation pertaining to the motion of a sphere in a liquid

$$f = 6\ \pi r s \eta \tag{17}$$

where f is the force causing the sphere to move through the fluid at constant velocity, s, and r is the radius of the sphere. If the sphere is allowed to fall freely through the fluid, it will accelerate until the viscous force exactly balances the force due to gravity,

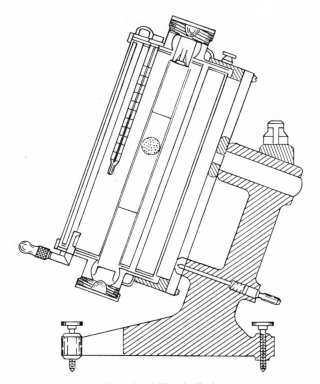

Fig. 15. Höppler falling-ball viscometer.

g. Stokes' equation relating viscosity to the fall of a solid body through a liquid can be written:

$$\eta = 2r^2(d_s - d_l)g/9s \tag{18}$$

where d_s is the density of the sphere and d_l is the density of the liquid.

Viscometers of the falling-body type are useful over an extremely wide viscosity range. They will satisfactorily measure viscosities of gases as well as of heavy asphalt and tars. In general they are not capable of as high precision and accuracy as the capillary and rotational viscometers, but are excellent instruments for routine work, particularly for viscous liquids. Since the driving force causing the body to fall or rise depends upon the difference in densities between the body and the liquid, it is necessary to determine these densities with some precision in order to translate the rate of rise or fall into units of viscosity. Since falling-body viscometers do not allow a straightforward analysis of shear rates, they are usually used only with Newtonian fluids. Although it is theoretically possible for falling-sphere viscometers to make absolute measurements, it is difficult to obtain in practice the precision in the sphere and the uniformity of tube walls that is required.

The viscosity of a fluid can be determined from the rate of fall of a sphere by the following formula:

$$KV = \eta/d_l = Kt(d_s - d_l) \tag{19}$$

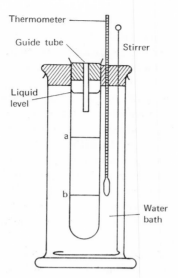

Thermometer

Guide tube

Stirrer

Liquid level

a

b

Water bath

Fig. 16. Falling-sphere viscometer.

where K is a proportionality constant, d_s is the density of the sphere, d_l is the density of the liquid, and t is the time of fall for a given distance. A falling-ball viscometer is calibrated with liquids of known viscosity by the step-up procedure. An example of the Höppler falling-ball viscometer is given in Figure 15. A series of balls of different diameter may be used so that a viscosity in the range of 0.01–25,000 P may be covered.

A simple falling-ball viscometer may be constructed from a test tube with a short guide tube placed through a stopper in the mouth of the test tube as illustrated in Figure 16.

It is possible to use the rate of rise of a bubble in the liquid to measure viscosity, and this principle is employed for routine viscometric control although the theory surrounding this technique is complicated and not well understood. In its simplest form a rising-bubble viscometer can be made by capping a test tube, leaving a small airspace between the top of the liquid level and the cap. When inverted, the air bubble will rise through the liquid and by timing the rate of rise the viscosity may be estimated. A commercial form of the rising-bubble viscometer consists of a row of equal-sized bubble tubes forming a graduated series of standards for comparison with an unknown. The tubes are filled with liquids having a known kinematic viscosity at 25°C. In operation one simply inverts the tubes until one finds which tube exhibits the same rate of bubble rise as the unknown. There are five sets of Gardner bubble viscometers and they cover the range of 5×10^{-3} to 10^3 St.

There is an ASTM procedure (D1545) which describes the use of rising-bubble tubes for the measurement of viscosity in the range of 40–1000 St. Listed in ASTM D1545 is a conversion table which relates the now historical Gardner-Holdt letter-standards to stokes.

Temperature Control. Temperature coefficients of viscosity are very high and generally tend to increase as the viscosity of the material increases. In working with most materials a temperature control within ±0.01°C is required for measurements intended to be consistently accurate to within 0.1% of their true value. In measurements for which 0.5% accuracy is sufficient, a variation of ±0.05°C is tolerable. The predominant role of temperature has been recognized by the principal standardizing bodies of the world, and rigid specifications for temperature control have been included in the standard methods. To assure temperature reproducibility among laboratories, elaborate specifications for thermometers to be used in viscosity measurements have been set up (ASTM E-1).

Careful attention must be paid to the design of constant temperature baths, particularly in regard to the occurrence of temperature gradients in the bath liquid. It is not generally appreciated that a bath in which stirring is vigorous and in which the temperature at one point is controlled to within $0.01°C$ can have stable temperature differences in other parts of the bath amounting to several hundredths of a degree. This condition is usually brought about by large heat losses through the bath walls. It can be eliminated by improving the insulation and by introducing the greater part of the compensating heat uniformly over the walls of the bath.

Another factor which is often overlooked is the necessity for the sample to equilibrate completely with the constant temperature bath. The time required for thermal equilibration is surprisingly long. A viscometer already in a bath may have its temperature equilibrium upset by immersing in the bath another viscometer which is either hotter or colder than the bath temperature.

Viscosity of Mixtures and Suspensions

The viscosity of mixtures of liquids can be approximated by the expression

$$\log KV_{1,2} = N_1 \log KV_1 + N_2 \log KV_2 \tag{20}$$

provided there are no strong interactions between the components; where N is the mole fraction of components 1 and 2, etc. For liquid mixtures of approximately the same molecular weight, the mole fraction may be replaced by the weight or volume fraction (25). The standard ASTM viscosity temperature chart may also be used to estimate the viscosity of blends at a given temperature, provided the viscosities of the component liquids are known (ASTM D341).

For high-molecular-weight polymers of narrow molecular-weight distribution, the equation for the viscosity is of the form

$$\eta = KM^a \tag{21}$$

where K is a control, M is the molecular weight, and a has the value of unity for low molecular weights and a value of 3.5 for higher molecular weight materials to account for the molecular entanglement. When the molecular-weight distribution is broad, M in equation 21 must be replaced by a weight average or average molecular weight of a higher order (26).

For solutions of polymers the viscosity relative to that of a solvent is defined by

$$\eta_r = \eta/\eta_0 = 1 + [\eta]c + k'[\eta]^2c^2 + k''[\eta]^3c^3 \tag{22}$$

where c is the concentration of polymer in grams/deciliter, $[\eta]$ is the intrinsic viscosity, and k', k'', etc, are interaction coefficients (27). The intrinsic viscosity can be determined by plotting $(\eta_r - 1)/c$ against c and extrapolating to a polymer concentration of 0. In some cases a plot of $(\eta_r - 1)/c$ may show significant deviation from linearity as c approaches 0. This may be due to absorption of the polymer on the walls of the capillary. A correction can be made by measuring the viscosity using capillaries of varying diameter.

For suspension of solids in liquids the classical Einstein equation (28) relates the relative viscosity η_r to the concentration of the solid phase

$$\eta_r = \eta/\eta_0 = 1 + k_1 v_2 + k_2 v_2{}^2 + k_3 v_2{}^3 \cdots \tag{23}$$

where η_0 is the viscosity of the liquid phase, v is the volumetric concentration of the solid phase, and k_1, k_2 ... are interaction constants. Einstein ascribed a value of 2.5 to k_1 and a value of 14.1 has been ascribed to k_2 by Eirich and Simha (28). Equation 23 assumes a Newtonian liquid phase; smooth, rigid, uniform spheres; that the volume of the solid phase is small compared to that of the liquid phase; and that there is no interaction between particles. In systems where there is considerable interaction between the solid and liquid phases or in concentrated liquid suspensions, equation 23 must be modified to compensate for entanglement and "absorbed liquids" carried along with the solid phase, thus effectively descreasing the liquid volume. An equation applicable to this case is

$$\eta/\eta_0 = 1 - [kv_2/(1 - Sv_2)] \tag{24}$$

where k is a constant and S is the so-called sediment volume (28). The volume of free liquid is $(1 - Sv_2)$.

Bibliography

"Viscometry" in ECT 1st ed., Vol. 14, pp. 756–776 by F. H. Stross and P. E. Porter, Shell Development Corporation.

1. G. Barr, *A Monograph of Viscometry*, Oxford University Press, New York, 1931.
2. A. C. Merrington, *Viscometry*, Edward Arnold, London, 1959.
3. A. Dinsdale and R. Moore, *Viscometry and Its Measurement* (The Institute of Physics and the Physical Society), Reinhold Publishing Corp., New York, 1963.
4. J. R. Van Wazer, J. W. Lyons, K. Y. Kim, and R. E. Colwell, *Viscosity and Flow Measurement— A Laboratory Handbook of Rheology*, Interscience Publishers, a division of John Wiley & Sons, Inc., New York, 1963.
5. F. R. Eirich, ed., *Rheology: Theory and Application*, Academic Press, Inc., New York, Vol. 1 (1956); Vol. 2 (1958); Vol. 3 (1960); Vol. 4 (1968).
6. P. O. Teichman, *Prod. Eng.* **18** (8), 146 (1947).
7. S. Glasstone, K. J. Laidler, and H. Eyring, *Theory of Rate Processes*, 1st ed., McGraw-Hill Book Co., Inc., New York, 1951, pp. 477–551.
8. A. K. Doolittle, *J. Appl. Phys.* **22**, 1471 (1951); **23**, 236 (1952).
9. A. A. Miller, *J. Polymer Sci.*, Part A-2, **6**, 1161 (1968); **6**, 249 (1968).
10. P. B. Macedo and T. A. Litovitz, *J. Chem. Phys.* **42** (1), 245 (1965).
11. O. Reynolds, *Phil. Trans. Roy. Soc. London, Ser. B* **177**, 157 (1886).
12. K. F. Slotte, *Wied. Ann.* **14**, 13 (1881).
13. H. Vogel, *Physik Z.* **22**, 645 (1921).
14. K. S. Ranaiye, *All-Union Conf. on Friction and Wear in Machines*, 2nd Conf., Moscow (Inst. Mach. Design) **2**, 545 (1948).
15. C. Walther, *Maschinenbau Tech.* **10**, 671 (1931).
16. H. Umstatter, *Arch. Tech. Messen*, **178**, 9122 (1950).
17. R. F. Crouch and A. Cameron, *J. Inst. Petrol.* **47** (453), 307 (1961).
18. P. S. Y. Chu and A. Cameron, "Pressure Viscosity Characteristics of Lubricating Oils," *J. Inst. Petrol.* **48** (461), 147 (1962).
19. R. C. Worster, discussion of paper by A. E. Bingham, "Some Problems of Fluids for Hydraulic Power Transmissions," *Proc. Inst. Mech. Engrs.* **165**, 269 (1951).

20. C. J. A. Roelands, J. C. Vlugter, and H. I. Waterman, *Trans. A.S.M.E., J. Basic Eng.* **85** (4), 601 (1963).
21. B. Kouzel, *Hydrocarbon Process. Petrol. Refiner* **44** (3), 120 (1965).
22. J. R. Swindells, J. F. Coe, Jr., and T. B. Godfrey, *J. Res. Natl. Bur. Std.* **48**, 1 (1952).
23. G. Barr, *J. Phys. Soc. (London)* **58**, 575 (1946).
24. J. F. Swindells, R. C. Hardy, and R. L. Cottington, *J. Res. Natl. Bur. Std.* **52**, 105 (1954).
25. A. Bondi, *Physical Chemistry of Lubricating Oils*, Reinhold Publishing Corp., New York, 1951, p. 46.
26. F. Bueche, *Physical Properties of Polymers*, Interscience Publishers, a division of John Wiley & Sons, Inc., New York, 1962, p. 78.
27. H. van Oene in "Characterization of Macromolecular Structure," *Publication 1573*, Natl. Acad. Sci., Washington, D.C., 1968, pp. 353–367.
28. S. G. Ward, *J. Oil & Colour Chemists' Assoc.* **38**, 9 (1955).

RICHARD S. STEARNS
Sun Oil Company

VISCOSE. See Rayon.

VISCOSITY. See Rheology; Viscometry.

VISCOSITY BREAKING, VISBREAKING. See Vol. 15, p. 22.

VITAMINS

SURVEY

Vitamins are chemical substances essential to the physiologic processes of animals, including man, effective in small amounts, catalytic in activity and they can, as a rule, not be synthesized by the organism. They must be provided by exogenous sources which may be natural foods, synthetic preparations, or produced by bacterial fermentation or by treatment such as ultraviolet radiation of natural vitamin precursors called provitamins.

Since the vitamins do not constitute a homogeneous chemical group of compounds the term is essentially a functional designation.

The name was coined by Funk (1) in 1911 who believed the antiberiberi factor of Eijkman to be a chemical *amine* essential for the preservation of life (Latin *vita*).

The presence of certain "accessory factors" had also been postulated by Hopkins (2) who observed that a synthetic diet composed of carbohydrates, proteins, fat and water was deficient in sustaining normal life. The existence of more than one vitamin was also indicated by the studies of McCollum and Davis (3) who discovered a fat-

Table 1. Vitamins, Vitamers and Their Characteristics (4)

| Vitamin | | Principal vitamers | | |
Common name	Principal synonyms	Natural	Synthetic	Characteristics
vitamin A	axerophthol antiophthalmic factor vitamin A_1 retinol	α-, β-, and γ-carotene cryptoxanthine vitamin A esters vitamin A_2 neo-vitamin A	vitamin A acid vitamin A ketone vitamin A esters	prevention of visual defects; affects skin and mucosal integrity
thiamine	vitamin B_1 aneurin	thiamine pyrophosphate (cocarboxylase) thiamine orthophosphate	thiamine disulfide acylated thiamine analog with methyl group in position 6 instead of 2 in pyrimidine ring	prevention of beriberi; antineuritic, enters into carbohydrate metabolism (cocarboxylase)
riboflavin	vitamin B_2 vitamin G lacto-, ovo-, verdo-, or hepatoflavin	riboflavin mononucleotide riboflavin dinucleotide	7-methyl-9-(D-1'-ribityl)-isoxalloxazine 6-methyl-9-(D-1'-ribityl)-isoxalloxazine 6-ethyl-7-methyl-9-(D-1'-ribityl)-isoalloxazine	required for cellular oxidation and reduction mechanisms
niacin	nicotinic acid P-P factor antipellagra factor anti-black-tongue factor	niacinamide NAD(CoI) NADP(CoII) N^1-methylnicotinamide	coramine esters of niacin	prevention of pellagra; vasodilator effects; affects carbohydrate metabolism (codehydrase)
vitamin B_6	pyridoxine anti-acrodynia factor adermin	pyridoxal pyridoxamine pyridoxal phosphate		prevents dermatitis, infantile convulsions; component of amino acid metabolism (decarboxylase)
pantothenic acid	chick antidermatitis factor filtrate factor		esters	component of acetyl transferase system
biotin	vitamin H anti-egg-white injury factor bios II coenzyme R	dethiobiotin	sulfoxide of biotin esters	required for growth of many microorganisms
pteroylglutamic acid	folic acid folacin anti-anemia factor vitamin B_c	pteroyltri(tetra, etc)-glutamic acid fermentation *L. casei* factor liver *L. casei* factor	pteroic acid	anti-anemia requirement
p-aminobenzoic acid	PABA chromotrichia factor vitamin B_x anti-gray-hair factor			

(continued)

Table 1 (*continued*)

| Vitamin | | Principal vitamers | | |
Common name	Principal synonyms	Natural	Synthetic	Characteristics
vitamin B_{12}	cobalamin cyanocobalamin	hydroxycobalamin (B_{12a}) aquocobalamin ($B_{12a,b,d}$)	nitritocobalamin	anti-pernicious anemia "extrinsic factor"
choline	sinkalin bilineurine fagin amanitin	methionine + ethanolamine betaine + ethanolamine	analog containing phosphorus instead of nitrogen arsenocholine methyl-diethyl homolog triethyl homologs	participates in acetylcholine and lecithin anabolism
inositol	bios I mouse antialopecia factor rat antispectacled eye factor	phytin soybean cephalin	methyl inositol inositol hexaacetate	affects lipid metabolism
ascorbic acid	vitamin C antiscorbutic vitamin	dehydroascorbic acid	6-deoxyascorbic acid isoascorbic acid L-fucoascorbic acid	anti-scurvy; activation of cell metabolism
vitamin D	antirachitic vitamin	vitamin D_2 (calciferol, ergocalciferol) vitamin D_3 (cholecalciferol)	irradiated ergosterol (D_2) irradiated 7-dehydrocholesterol (D_3) irradiated 22-dihydroergosterol (D_4) irradiated 7-dehydrositosterol (D_5)	antirachitic factor, regulation of calcium and phosphorus metabolism
α-tocopherol	vitamin E antisterility vitamin	β- and γ-tocopherols tocopherol esters	esters racemic α-tocopherol analogs with NH_2 in place of OH, or ethyl substituents in place of methyl	antisterility; cell respiration; antioxidant
vitamin K	vitamin K_1 phylloquinone antihemorrhagic vitamin coagulation vitamin	vitamin K_2	menadione menadione sodium bisulfite esters of the hydroquinone forms	blood clotting

soluble factor A and a water-soluble factor B, later renamed vitamin A and vitamin B, respectively.

A basic classification of the vitamins by solubility characteristics has been retained as useful and at present (1970) vitamins A, D, E and K are those of the fat-soluble group, while the water-soluble group is comprised of vitamin C and the B vitamins.

Continuing chemical knowledge of the constitution of the vitamins has led to the development of natural and synthetic derivatives and a distinction is made between vitamin activity and vitamin content. The major vitamins, structurally related compounds possessing similar activity, called vitamers, and their characteristics are listed in Table 1.

Table 2. Recommended Daily Dietary Allowances (6)

Age, yr	Weight, kg	Height, cm	Fat-soluble vitamins			Water-soluble vitamins						
			Vitamin A activ, IU	Vitamin D, IU	Vitamin E activ, IU	Ascorbic acid, mg	Folacin, mg	Niacin, mg equiv	Riboflavin, mg	Thiamin, mg	Vitamin B6, mg	Vitamin B12, µg
infants												
0–⅙	4	55	1500	400	5	35	0.05	5	0.4	0.2	0.2	1.0
⅙–½	7	63	1500	400	5	35	0.05	7	0.5	0.4	0.3	1.5
½–1	9	72	1500	400	5	35	0.1	8	0.6	0.5	0.4	2.0
children												
1–2	12	81	2000	400	10	40	0.1	8	0.6	0.6	0.5	2.0
2–3	14	91	2000	400	10	40	0.2	8	0.7	0.6	0.6	2.5
4–6	19	110	2500	400	10	40	0.2	11	0.9	0.8	0.9	4
8–10	28	131	3500	400	15	40	0.3	15	1.2	1.1	1.2	5
males												
10–12	35	140	4500	400	20	40	0.4	17	1.3	1.3	1.4	5
14–18	59	170	5000	400	25	55	0.4	20	1.5	1.5	1.8	5
22–35	70	175	5000		30	60	0.4	18	1.7	1.4	2.0	5
35–55	70	173	5000		30	60	0.4	17	1.7	1.3	2.0	5
55–75+	70	171	5000		30	60	0.4	14	1.7	1.2	2.0	6
females												
10–12	35	142	4500	400	20	40	0.4	15	1.3	1.1	1.4	5
14–16	52	157	5000	400	25	50	0.4	16	1.4	1.2	1.8	5
22–35	58	163	5000		25	55	0.4	13	1.5	1.0	2.0	5
35–55	58	160	5000		25	55	0.4	13	1.5	1.0	2.0	5
55–75+	58	157	5000		25	55	0.4	13	1.5	1.0	2.0	6

The essential role of the vitamins becomes evident from the deficiency symptoms produced by their lack in the animal organism. Several of the vitamins exhibit a species specificity as exemplified by the fact that man, monkeys and guinea pigs require vitamin C in their diet, whereas dogs and rats are capable of endogenous synthesis of this substance. In recent years the physiologic importance of the vitamins has been intensely studied with the result that the vitamins have been found to participate in a variety of enzymic processes. Thus thiamine acts as a precursor of cocarboxylase, niacin takes part in the formation of nicotinamide-adenine dinucleotide (NAD) and nicotinamide-adenine dinucleotide phosphate (NADP), vitamin K is an essential component in the mechanism of blood coagulation, and vitamin A is a natural substrate of rhodopsin.

In view of these findings Folkers (5) suggested that the definition of vitamins might be modified as follows: "A vitamin is an organic substance of nutritional nature which in low concentrations, as an intrinsic part of enzyme systems, catalyzes reactions required by the organism, and the organism may or may not have a capacity for the biosynthesis of the substance."

Table 3.   Toxicity of Vitamins

Vitamin	LD$_{50}$	Route	Animal
thiamine	125 mg/kg	intravenous	mice
	250 mg/kg	intravenous	rats
	300 mg/kg	intravenous	rabbits
	350 mg/kg	intravenous	dogs
nicotinic acid	4–5 g/kg	subcutaneous	mice, rats
	5–7 g/kg	oral	mice, rats
riboflavin	560 mg/kg	intraperitoneal	rats
pyridoxine HCl	3.7 g/kg	subcutaneous	rats
	5.5 g/kg	oral	rats
pantothenic acid	2.7 g/kg	subcutaneous	mice
	3.4 g/kg	subcutaneous	rats
	2 g/kg	intravenous	rabbits
pteroylglutamic acid	600 mg/kg	intravenous	mice
	500 mg/kg	intravenous	rats
	410 mg/kg	intravenous	rabbits
	120 mg/kg	intravenous	guinea pigs
choline	320 mg/kg	intraperitoneal	mice
	6.7 g/kg	oral	rats
menadione	0.5 g/kg	oral	mice

The requirements for the vitamin intake of man have been of particular interest and a tabulation of recommended dietary allowances is periodically published by the Food and Nutrition Board of the National Academy of Sciences-National Research Council. A selection is shown in Table 2 (6).

Vitamin deficiencies or total lack of vitamins (avitaminosis) may lead to pathologic conditions in the organism. Excesses of vitamins beyond the body's ability to utilize them are generally excreted, but some untoward effects have been observed in states of hypervitaminosis. However these are primarily confined to the fat-soluble vitamins which are capable of being stored in the organism.

As a class, vitamins are of a low order of toxicity as evidenced by a representative sampling of LD$_{50}$ values in various animal species (Table 3).

For nearly all natural vitamins the structure has been elucidated, and their synthesis has also been accomplished. The process by which any given vitamin is commercially produced is determined by the state of technology, availability of materials, and the economics. The major source of most vitamins commercially available is synthesis. However, riboflavin is more economically produced by fermentation, vitamin B_{12} continues to be derived by microbial fermentation, inositol is obtained by isoation from natural sources, and vitamin A and K are produced by both synthesis and isolation.

Although vitamins are adequately available in natural food stuffs to satisfy normal nutritional requirements, supplemental use of vitamin preparations to compensate for inadequate diets of man and domestic animals, medicinal use in stress and disease conditions, and enrichment of certain foods, such as flour, bread and milk, whose natural vitamin content becomes diminished by the mechanics of processing, have led to the continually increasing manufacture of vitamins. Table 4 summarizes vitamin production in the United States since 1945.

Table 4. Production and Sales of Vitamins (7)

Year	Production, lb	Sales, lb	Value, $
1945	2,945,000	2,516,000	48,933,000
1950	3,455,000	2,981,000	58,714,000
1955	6,139,000	5,131,000	82,811,000
1960	11,063,000	7,995,000	68,684,000
1963	14,874,000	10,519,000	64,474,000
1964	14,105,000	10,629,000	60,578,000
1965	16,297,000	12,028,000	65,366,000
1966	17,582,000	12,042,000	70,752,000
1967	17,568,000	11,108,000	65,847,000

In addition to the establishment of recommended daily allowances for vitamins in man as shown in Table 2, the National Research Council has also established therapeutic vitamin requirements for conditions of stress and the Committee on Animal Nutrition has determined vitamin needs for livestock and domestic animals.

In order to make meaningful quantitative statements regarding vitamin requirements, proper assay methods had to be devised and quantitative standards needed to be adopted. The tools of analytical chemistry, colorimetric and fluorimetric methods, microbiological and animal assay are in current use. The biological activity of specific weights of pure vitamin preparations has become the unit of activity, called an International Unit (IU). The International Units or expressions based on actual weight are utilized as U.S. Pharmacopeia (U.S.P.) units.

Antagonists of specific vitamin activity are called antivitamins. Many such compounds are known to occur naturally, others have been synthesized. However, only a limited number of antivitamins has been found to be effective in the human or animal organism. The action of the antivitamins may be characterized either as that of an antimetabolite or of a competitive inhibitor. Important examples are: avidin, a protein of egg white, capable of inactivating biotin by combining with the vitamin and rendering it unabsorbable; pyrithiamine and oxythiamine analogs and antivitamins of thiamine; dicumarol, a clinically important antivitamin K; and sulfanilamide, an antagonist of para-aminobenzoic acid. Certain folic acid antagonists such as aminopterin and methotrelate have taken on particular significance because of their chemotherapeutic utility as antineoplastic agents.

Bibliography

"Vitamins, Survey" in *ECT* 1st ed., Vol. 14, pp. 777–791, by H. R. Rosenberg and Mollie G. Weller, E. I. du Pont de Nemours & Co., Inc.

1. C. Funk, *J. Physiol.* **43,** 395 (1911).
2. F. G. Hopkins, *J. Physiol.* **44,** 425 (1912).
3. E. V. McCollum and M. Davis, *J. Biol. Chem.* **23,** 181 (1915).
4. Adapted from *Hawk's Physiological Chemistry*, B. L. Oser, ed., 14th ed, Blakiston Division, McGraw-Hill Book Co., Inc., New York, 1965.
5. K. Folkers, *Intern. Z. Vitaminforsch.* **37,** 526 (1967).
6. "Recommended Dietary Allowances," *National Research Council Publ. No. 1694*, Washington, D. C., rev. 1968.
7. *Statistical Abstract of the United States*, 90th ed., U.S. Dept. of Commerce, 1969.

Peter W. Fryth
Hoffmann-LaRoche, Inc.

VITAMIN A

Vitamin A is the isoprenoid polyene alcohol all-*trans*-3,7-dimethyl-9-(2′6′6′-trimethyl-1′-cyclohexen-1′-yl) 2,4,6,8-nonatetraene-1-ol. The most recent nomenclature changes recommended by the International Union of Pure and Applied Chemistry (1) and adopted by *Chemical Abstracts* prescribes the base name *retinol* for vitamin A and the abandonment of the Geneva numbering system above, for that shown in (**1**). The latter is intended to conform with the system officially used for the carotenoids to show the relationship between the two.

(**1**)

All derivatives and isomers of vitamin A have been renamed on this basis; eg retinene or vitamin A aldehyde becomes *retinal*, vitamin A acid becomes retinoic acid, vitamin A₂ or 3-dehydrovitamin A becomes 3-*dehydroretinol*, etc. The stereoisomers of vitamin A are designated by using the first carbon number of the double bond involved; eg 13-*cis*, 9-*cis*, 9,13-di-*cis*, etc.

Vitamin A is characterized physiologically by its ability to produce normal growth and cure eye disease in rats deficient in vitamin A in the standard rat-growth assay (2). A number of synthetic products and derivatives of vitamin A can partially fulfill the requirements of this assay and have been loosely referred to as vitamins A. But none have qualitatively and quantitatively duplicated the effect of vitamin A in this test.

Apart from its fundamental role in vision, the metabolic function of vitamin A is not well understood. It is conceded to play an indispensable role in general metabolism, most facets of which remain to be discovered.

Vitamin A occurs in nature in animal organism only, accompanied by minor amounts of the various cis isomers. It is usually found esterified as the palmitate and occurs most abundantly in fish-liver oils.

Accompanying vitamin A in fish-liver oils is 3-dehydroretinol (**2**), or vitamin A$_2$, as it is commonly called. It is more abundant in fresh-water fish than salt-water animals. Physiologically it is just as effective as vitamin A.

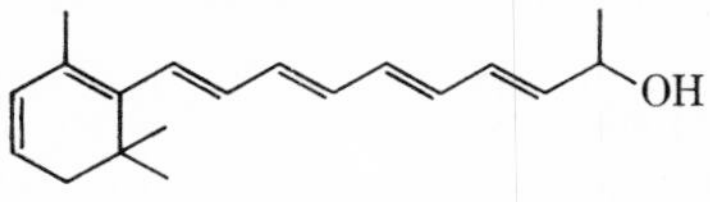

(**2**)

Structurally vitamin A is very closely related to the carotenoids and is an oxidation product of those carotenoids having at least one unsubstituted 1,1,3-trimethylcyclo-hex-2-en-2-yl group, such as α-, β-, and γ-carotene (**3**).

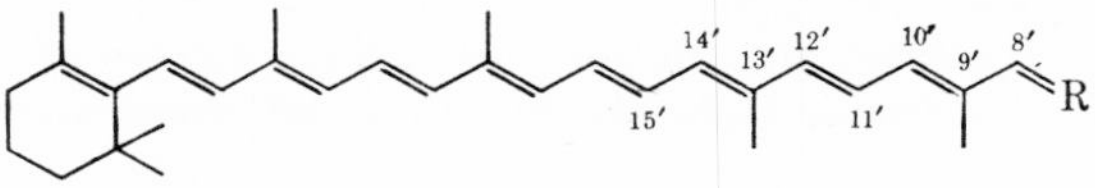

carotene (**3**)

where R =

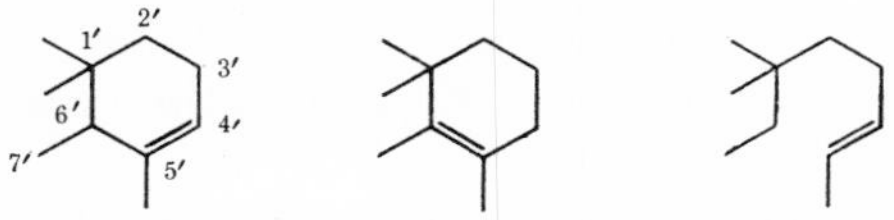

α β γ

Thus, when β-carotene is fed to animals it is metabolically degraded to vitamin A. For this reason carotenoids which produce vitamin A in the animal body are called pro-vitamins A. They are widely distributed in plants.

Vitamin A and the carotenoids structurally obey the "isoprene rule," and can formally be considered as terpenoids. However, because of their many special features they are normally treated as a class by themselves.

History. In 1913, E. V. McCollum and M. Davis, and, almost simultaneously, T. B. Osborn and L. B. Mendel, published the results of experiments which showed that rats could not develop normally on a purified diet of proteins, fats, carbohydrates, and minerals, unless butter, egg-yolk extract, or cod-liver oil was added. They concluded, therefore, that these substances contained a hitherto unrecognized factor (or factors) indispensable for normal growth and development. Similar experiments and observations had been made earlier (1906–1907) by F. G. Hopkins in England, but unfortunately these results were not published until much later. The new factor was called "fat-soluble A" to distinguish it from another factor, "water-soluble B" found shortly afterward to be likewise indispensable for normal nutrition.

Another fat-soluble factor, distributed in much the same manner as fat-soluble A (in the nonsaponifiable fraction of butter and cod-liver oil) was found in 1919 by E. J. Mellanby to be responsible for the prevention of rickets. This new factor was eventually distinguished from fat-soluble A by E. V. McCollum, who showed that the growth-promoting activity of cod-liver oil was completely destroyed by aeration at 100°C without affecting the antirachitic activity present. (See Vitamin D.) In 1937 the existence of a second growth-promoting substance was recognized by R. A. Morton and called vitamin A$_2$.

During the 1920s, the importance of the biological role of vitamin A became well established, although numerous attempts at isolation and chemical identification failed. Positive correlations between its growth-promoting activity, spectral extinction at

325–328 nm and the intensity of the blue color formed with antimony trichloride, permitted a reasonably accurate estimation of the vitamin without resorting to the laborious bioassay. Much richer sources than cod liver were found in other fish, especially halibut, from which concentrates containing at least 50% of the vitamin could be made.

In 1920 it was first demonstrated by H. Steenbock that carotene, a normal constituent of plants, can adequately replace vitamin A in the diet. Subsequently, many varieties of vegetables were found to be a potent source of vitamin A activity. In 1930 T. Moore unequivocally demonstrated the conversion of β-carotene to vitamin A in rats, thereby establishing the role of certain carotenoids as provitamins A. Chemical degradation studies of β-carotene and of vitamin A, going on in other laboratories at the same time, led to the recognition of the intimate chemical relationship between the two, and in 1931 the correct structural formulas for the two compounds were proposed by P. Karrer. That of vitamin A was shortly afterward confirmed by Karrer by identifying a specimen of fully hydrogenated vitamin A with perhydrovitamin A obtained by synthesis. The formula of β-carotene was later confirmed by other workers.

First direct evidence for the basic role of vitamin A in vision was reported by G. Wald in 1935 and the participation of retinal in this process by R. A. Morton and T. W. Goodwin in 1944.

A crystalline form of vitamin A was first isolated by H. N. Holmes and R. E. Corbet in 1937 as a methanolate. The pure unsolvated form and several of its pure crystalline esters were eventually obtained by J. G. Baxter and C. D. Robeson in 1942, who also isolated crystalline 13-*cis*-vitamin A in 1946.

O. Isler and his co-workers were the first to synthesize the pure crystalline vitamin in 1947. Previous syntheses by R. Kuhn in Germany in 1937, and by N. Milas in the United States in 1940 yielded impure, though biologically active, products. In 1950 the synthesis of β-carotene was reported by Karrer in Switzerland and H. Inhoffen in Germany and in 1952 that of vitamin A_2 by E. R. H. Jones in England.

Today practically all the vitamin A and β-carotene on the market is of synthetic origin.

Stereochemical Aspects

The four double bonds in the vitamin A or vitamin A_2 side chain theoretically permit sixteen possible geometric isomers, and the nine double bonds in the side chain of β-carotene allow for 272. For many years it was assumed that a cis configuration of type I (see Fig. 1) in an isoprenoid polyene would be too unstable, because of the steric hindrance involved, to be capable of existence. Therefore only cis configurations of type II (see Fig. 1), as well as that involving the central double bond in the carotenoids, were considered possible. This limited the number of expected stereoisomers of vitamin A and β-carotene to 4 and 20, respectively (3).

However, it was subsequently shown by synthesis that these "forbidden" cis configurations in isoprenoid polyenes are not only possible but are considerably more stable than had been presumed (4,5). This finding reestablished 16 and 272 as the maximum number of stereoisomers that may theoretically be expected for the vitamins A and β-carotene, respectively.

A cis configuration around the 7-double bond would theoretically lead to abnormally large interference between the *gem*-dimethyl group of the ring and the 9-methyl type I type II

Fig. 1. The types of cis configuration possible in isoprenoid polyenes.

group of the side-chain. For this reason it is presently considered that such cis isomers, if produced, would be at most capable of only transient existence.

To date all three possible "unhindered" cis isomers of vitamin A (type-I) are known, the 9-*cis* (**4**), the 13-*cis* (**5**), and the 9,13-di-*cis* (**6**). Only two "hindered" isomers (type-II) are known, the 11-*cis* (**7**) and 11,13-di-*cis* (**8**) (see Scheme 1). With the exception of the last named, all have been found to occur naturally.

Scheme 1.

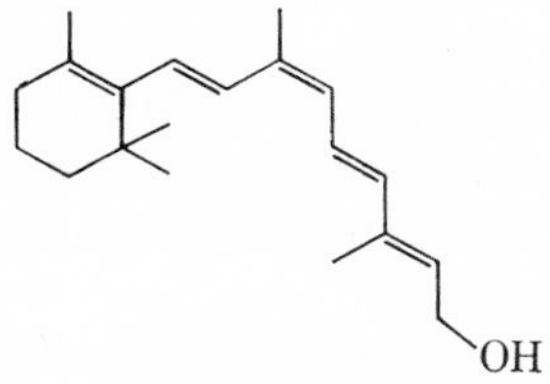

9-*cis* (**4**)

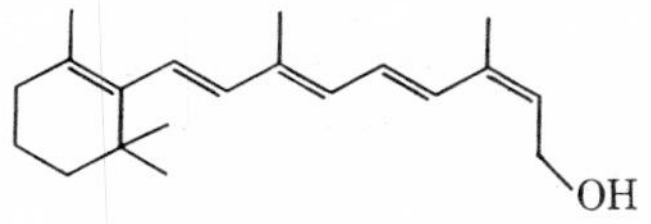

13-*cis* (**5**)

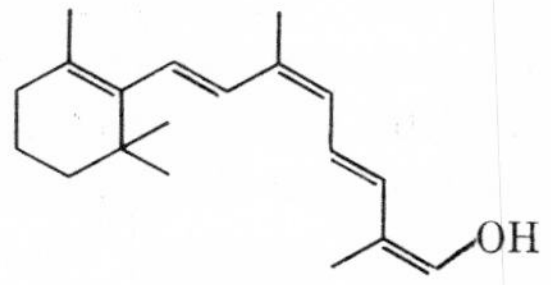

9,13-di-*cis* (**6**)

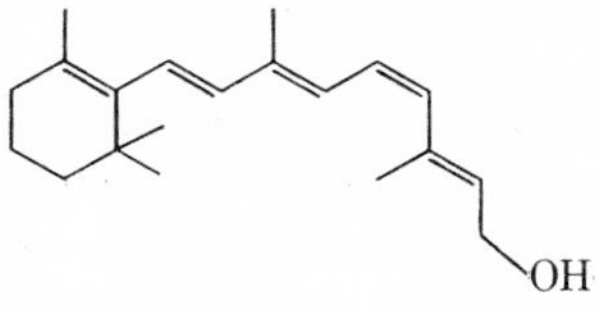

11-*cis* (**7**)

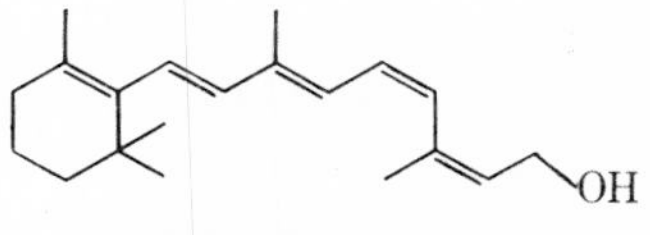

11,13-di-*cis* (**8**)

The analogous cis isomers of vitamin A_2 are also known. All are of synthetic origin. There is evidence for the existence of 13-*cis*-vitamin A_2 in nature (6,6a).

The carotenoids exist in nature predominantly in the all-trans form (**3**). Because of the extraordinary ease of cis-trans isomerization in the carotenoids, the existence of cis-isomers in this group is most difficult to establish. A few have been identified with certainty. Most arise as artifacts of the isolation procedure.

Cis isomers of the vitamins A and A_2 and of the carotenoids can be obtained by stereoselective synthesis and cis-trans isomerizations. The syntheses will be outlined in subsequent sections. The isomerizations are generally carried out by exposing the compound to heat, light, catalysis by iodine or acids, and, in some instances, simply by solution in an inert solvent.

In the vitamins A, the 13-double bond readily equilibrates to a mixture of cis and trans forms on iodine catalysis but the 9-double bond is surprisingly inert toward any means of isomerization. The corresponding 9-double bond in the carotenoids, on the other hand, isomerizes readily.

Isomerization of the hindered cis bonds to the trans form proceeds very readily. The reverse shift, ie trans to hindered cis, has been observed only in the case of retinal, which yields 11-cis retinal upon irradiation with sunlight.

Separation of stereoisomers from mixtures in both the vitamins and provitamins A is best accomplished by chromatography on neutral or basic adsorbents, such as alumina, secondary calcium phosphate, or magnesia.

Vitamin A

Properties. Vitamin A is a pale yellow crystalline solid (mp, 63–64°C), which can be purified by crystallization from ethyl formate or petroleum ether at low temperature (7). It crystallizes from methanol as the methanolate (mp, 8°C). Vitamin A is insoluble in water but dissolves readily in the usual fat solvents. The partition coefficient between petroleum ether and 83% ethanol is 0.82. Vitamin A can be isolated from mixtures by chromatography, because it is readily adsorbed on alumina, magnesia, and other neutral or alkaline adsorbents; it is destroyed by acidic adsorbents. It fluoresces in ultraviolet radiation.

Vitamin A shows a very characteristic uv absorption spectrum shown in Figure 2, which consists of a single-peaked absorption band (λ_{max}^{EtOH} 325 nm, $E_{1cm}^{1\%}$ 1,830, ϵ 52,800 (8)) rather than the usual triple-peaked band of conjugated polyene alcohols and hydrocarbons. Its absorption maximum is also considerably displaced toward the shorter wave lengths from that usually shown by equally substituted conjugated pentaenes (4). These unique spectral features of vitamin A are due to steric hindrance in the molecule between the *gem*-dimethyl group of the ring and the 8-hydrogen. Such hindrance is present in all β-ionylidene compounds (4). *Retro*vitamin A (**10**) on the other hand, in which such hindrance is absent, shows the normal three-peaked curve of a conjugated pentaene, as shown in Figure 2 (4).

As a conjugated polyene, vitamin A shows the usual reactions of conjugated double bonds. It is oxidized readily by atmospheric oxygen, especially in the presence of light and heat. In an inert atmosphere it is destroyed by ultraviolet radiation but is quite stable to heat, and can be distilled in high vacuum (b_{0-005} 120–125°C) (9). Oil solutions can be stabilized toward oxidation by means of antioxidants such as α-tocopherol (see Vitamin E). The last two double bonds of the side chain readily participate in a Diels-Alder reaction with maleic anhydride (8). Catalytic hydrogenation yields the biologically inactive perhydro compound.

As a primary alcohol, vitamin A forms the following esters and ethers, which can be used for characterization: acetate (mp, 57–58°C); palmitate (mp, 27–28°C); *p*-phenylazobenzoate (mp, 79–80°C); anthraquinone-2-carboxylate (mp, 122–123°C) (10); of the known ethers, only the methyl (mp, 34–35°C) and the phenyl (mp, 90–92°C), are solids (11). The esters and ethers are more stable than the free alcohol. The acetate is the basis of definition of the International and USP units of biological activity.

Vitamin A is not affected by bases, but its allylic hydroxyl group and conjugated polyene character make it very sensitive to acids. Dilute acids at room temperature cause dehydration with the formation of anhydrovitamin A (**9**) ($\lambda\lambda_{max}^{EtOH}$ 350, 368, and 390 nm) (12), shown in Scheme 2. This compound is encountered very often in

Scheme 2.

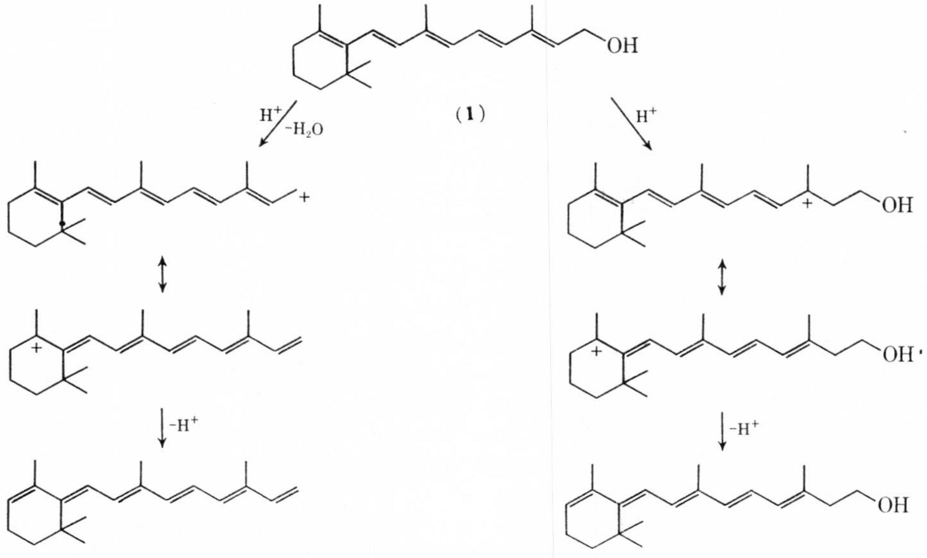

anhydrovitamin A (**9**)   retrovitamin A (**10**)

chemical manipulations with vitamin A and is frequently present as an artifact in column chromatography of vitamin A. It has no significant vitamin A activity. The esters and ethers of vitamin A undergo the same reaction.

In the presence of concentrated hydrobromic acid at low temperature, vitamin A is readily isomerized to *retro*vitamin A (**10**), a golden yellow oil ($\lambda\lambda_{max}^{EtOH}$ 332, 348, and 366 nm) (13), as shown in Scheme 2.

Oxidation with specially prepared manganese dioxide (14) yields the corresponding aldehyde, retinal (15,16). Oppenauer oxidation via aluminum *tert*-butoxide and acetone also yields retinal which, in this case, reacts with the acetone present to give retinylideneacetone (17).

With antimony trichloride in chloroform solution, vitamin A gives a characteristic blue color ($\lambda_{max}^{CHCl_3}$ 620) (Carr-Price test), which is the basis of an analytical procedure (see p. 505). Other strong Lewis acids, like arsenic trichloride, stannic chloride, and boron trifluoride, and certain acid earths, also give intense colors with vitamin A.

On treatment with catalytic quantities of iodine, vitamin A is isomerized to 13-*cis*-vitamin A to the extent of 30% (10).

Synthesis. Since 1950, numerous methods have been developed for the synthesis of vitamin A. Two excellent reviews on this subject have been written by Isler (6a,18). Several of these processes are commercially feasible and the one which dominates the market now was developed by Isler (18) and is outlined in Scheme 3. This scheme was originally proposed by Heilbron (19) and in a somewhat different form by Milas (20).

Scheme 3.

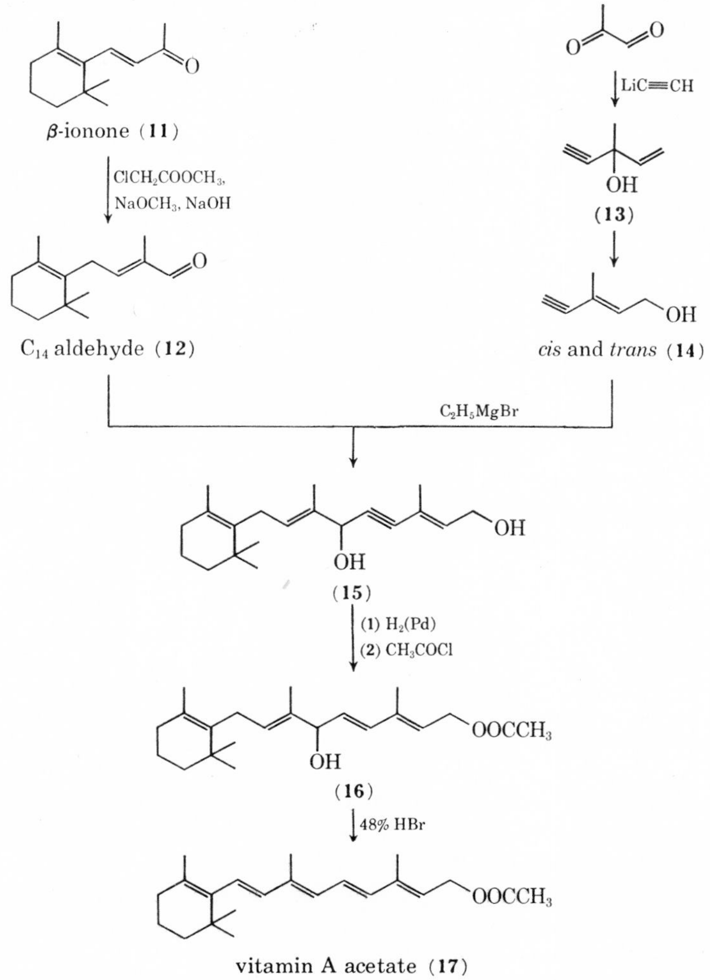

β-ionone (**11**)

C$_{14}$ aldehyde (**12**)

(**13**)

cis and *trans* (**14**)

C$_2$H$_5$MgBr

(**15**)

(1) H$_2$(Pd)
(2) CH$_3$COCl

(**16**)

48% HBr

vitamin A acetate (**17**)

The starting material β-ionone (**11**), is a well-known commercial product employed for many years in the perfume industry (see Vol. 19, p. 829). This is condensed in a Darzens reaction with methyl or ethyl chloroacetate. The resulting glycidic ester (not shown in Scheme 3) is simultaneously hydrolyzed, decarboxylated, and rearranged to the C$_{14}$ aldehyde (**12**) by the unique method of simply stirring the crude Darzens reaction mixture with methanolic alkali at 5°C. The high yield of aldehyde obtained in this way is one of the features of this synthesis that helped make it commercially practical. Next, methyl vinyl ketone is condensed in liquid ammonia with lithium acetylide to yield 3-methylpent-1-en-4-yn-3-ol (**13**). This is allylically rearranged in dilute acid to 3-methylpent-2-en-4-yn-1-ol (**14**) in cis and trans forms. Conversion of *cis* (**14**) to the Grignard reagent by means of ethylmagnesium bromide, followed by condensation with the C$_{14}$ aldehyde gives the carbon skeleton of vitamin A (**15**). The acetylenic bond is then selectively semihydrogenated by means of a specially poisoned palladium catalyst and the primary hydroxyl group of the glycol is acetylated to give

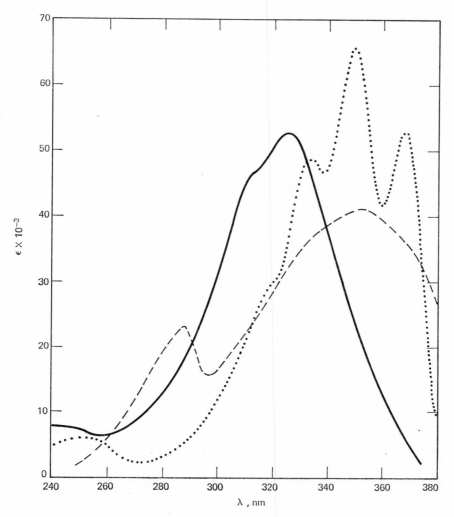

Fig. 2. Ultraviolet absorption curves in 95% ethanol of vitamin A (solid curve), vitamin A₂ (dashed curve), and retrovitamin A (dotted curve).

(**16**). The glycol monoacetate is shaken in chloroform or methylene dichloride solution at −25°C with concentrated hydrobromic acid, thus forming the fifth double bond. The viscous syrup obtained is dissolved in anhydrous ethyl formate or ethyl alcohol, from which vitamin A acetate (**17**) crystallizes on cooling. The free alcohol is then obtained by alkaline hydrolysis of the acetate. The palmitate, an important commercial form of vitamin A, is obtained by transesterification of the acetate with methyl palmitate (6a). The yields in each step, commencing with the condensation of the C_{14} aldehyde, are reported to be better than 80% (11, 21–25).

The Cis Isomers of Vitamin A

Properties. 13-*cis-Vitamin A* (**5**) (8,10). This isomer, formerly called neovitamin A, shows about 75% of the biological potency of vitamin A in the rat-growth test.

It is a pale-yellow crystalline solid (mp, 58–59°C); p-phenylazobenzoate (mp, 94–96°C); anthraquinone-2-carboxylate (mp, 134–136°C). Its uv absorption curve (λ_{max}^{EtOH} 328 nm, $E_{1cm}^{1\%}$ 1,690, ϵ 48,300) is very similar to that of vitamin A. The shift of its absorption maximum to a higher wave length is characteristic of polyenes having a terminal cis configuration (5). Its physical and chemical properties are similar to those of vitamin A, but some of its reactions, such as the formation of anhydrovitamin A (9) and the condensation with maleic anhydride, are much slower as a consequence of the terminal cis configuration. The latter reaction is the basis for its determination in mixtures with vitamin A. Oxidation with manganese dioxide yields 13-*cis*-retinal. In the Carr-Price test it gives the same blue color observed with vitamin A.

13-*cis*-Vitamin A can be separated from vitamin A by chromatography, being less adsorbed than vitamin A on alumina, magnesia, or sodium aluminum silicate. It can be isomerized to vitamin A by catalytic quantities of iodine, which produces the same equilibrium mixture of the two isomers as that obtained from vitamin A (see above).

9-cis-Vitamin A (**4**) (8). This isomer shows only about one quarter the biological potency of vitamin A (26). It is a pale-yellow crystalline solid (mp, 81.5–82.5°C); p-phenylazobenzoate (mp, 79–80°C). Its uv absorption curve shows a single-peaked absorption band (λ_{max}^{EtOH} 323 nm, ϵ 42,300) and a small cis peak around 260 nm. Since neither of the last two double bonds of the side chain is cis, reaction with maleic anhydride is fast, as compared to 13-*cis*-vitamin A. Oxidation with manganese dioxide yields 9-*cis*-retinal. The Carr-Price color is the same as observed with the other isomers. Iodine does not affect the 9-cis configuration.

9,13-di-cis-Vitamin A (**6**) (8). This isomer has the same biological potency as the 9-cis isomer (26). It is a pale-yellow crystalline solid (mp, 58–59°C); p-phenylazobenzoate (mp, 91.5–92.5°C). Its uv absorption curve (λ_{max}^{EtOH} 324 nm, ϵ 39,500, cis peak about 260 nm) is very similar to that of the 9-cis isomer. Reaction with maleic anhydride is relatively slow because of the terminal cis configuration. Oxidation with manganese dioxide yields the corresponding di-*cis*-retinal. The Carr-Price color is the same as observed with the other isomers. Catalysis with iodine in the dark produces an equilibrium mixture with the 9-cis isomer.

11-cis-Vitamin A (**7**) (27). This is a sterically hindered cis isomer, showing about one quarter the biological potency of the all-trans isomer (26). It has not yet been obtained crystalline. The chromatographically purified oil shows λ_{max}^{EtOH} 321 nm, ϵ 32,500, and a cis peak at 233 nm; the p-phenylazobenzoate melts at 67°C. Oxidation with manganese dioxide gives 11-*cis*-retinal, the key component in vision (see p. 506). Iodine catalysis completely inverts the 11-cis configuration. 11-*cis*-Vitamin A reacts slowly with maleic anhydride. The Carr-Price color is the same as with the other isomers.

11,13-di-cis-Vitamin A (**8**) (27). This is also a sterically hindered cis isomer, showing about 15% the biological potency of vitamin A (26). It is a crystalline solid (mp, 86–88°C; λ_{max}^{EtOH} 311 nm, ϵ 26,000); p-phenylazobenzoate (mp, 99–100°C); antraquinone-2-carboxylate (mp, 139–140°C). Oxidation with manganese dioxide yields the unstable corresponding aldehyde. Iodine produces the all-trans isomer. Reaction with maleic anhydride is only about one sixth as fast as that with the 11-cis isomer. The Carr-Price color is the same as observed with the other isomers.

Synthesis. 9-*cis*-, 13-*cis*-, *and* 9,13-*di-cis-Vitamin A*. While these steric forms of vitamin A are of no commercial importance, their synthesis and identification by Robeson (8) was a milestone in elaborating the stereochemistry of vitamin A. The stereoselective reaction sequence leading to these isomers is shown in Scheme 4.

Scheme 4.

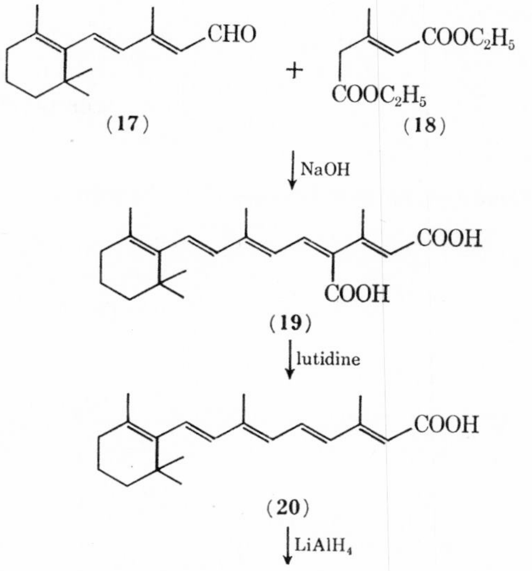
CHO
+
$COOC_2H_5$
$COOC_2H_5$
(17)
(18)
NaOH
COOH
COOH
(19)
lutidine
COOH
(20)
$LiAlH_4$

13-*cis*-vitamin A (**5**)

The stereoselectivity is based on the fact that β-methylglutaconic ester (**18**) produces a 13-cis configuration in the final product irrespective of which steric form of (**18**) is used. By starting with 9-*cis* or 9-*trans*-β-ionylideneacetaldehyde (**17**) the stereochemistry of the 9-double bond was controlled. The 9-mono-cis isomer was eventually obtained by isomerization of the 9,13-di-cis isomer catalyzed with iodine.

11-cis- and 11,13-di-cis-Vitamin A. These two forms represent the only known sterically hindered cis isomers of vitamin A. They are of no commercial significance but their importance lays in establishing the structure of the steric form of vitamin A responsible for vision (see under Biological action). They were synthesized by Oroshnik (27) by a modification of the reaction sequence in Scheme 3, as outlined in scheme 5.

Scheme 5.

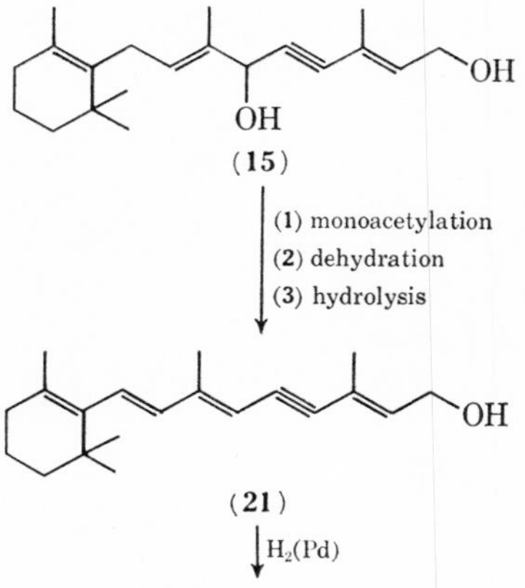
OH
OH
(15)
(1) monoacetylation
(2) dehydration
(3) hydrolysis
OH
(21)
$H_2(Pd)$

11-*cis*-vitamin A (**7**)

The acetylenic glycol (**15**) was produced in two steric forms, 13-cis and 13-trans, by condensing the C_{14} aldehyde (**12**) with *cis*- or *trans*-(**14**), respectively, as shown in Scheme 3. Each form was then carried through the same sequence of reactions to give 13-*cis*- and 13-*trans*-11-dehydrovitamin A (**21**), respectively. By catalytic semihydrogenation of the 11-acetylenic bond a cis ethylenic bond is stereospecifically produced in each case to yield 11,13-di-*cis*- and 11-mono-*cis*-vitamin A, respectively.

Derivatives and Isomers of Vitamin A

The Retinals. All of the stereoisomers of retinal corresponding to the known steric forms of vitamin A are known. They are obtained in each case by the oxidation of the corresponding vitamin A with specially activated manganese dioxide (28,29). (See Scheme 6).

The 11,13-di-cis isomer is very unstable and isomerizes too rapidly to permit determination of physical constants (29). The others are crystalline solids showing sharp melting points, typical ultraviolet absorption curves, and yield conventional aldehyde derivatives (28).

The all-trans and 11-cis isomers are the key components of the visual cycle (see under Biological action) and are interconvertible in vivo. In the rat-growth assay the all-trans and 13-cis isomer show over 90% of the potency of vitamin A; the 9-cis and 9,13-di-cis less than 20%; the 11-cis about 50% and the 11,13-di-cis about 30% (26).

The Retinoic Acids. All of the retinoic acids corresponding to the known stereoisomers of vitamin A are known (6a). All have been obtained by synthesis and are crystalline solids. The all-trans isomer is equivalent to vitamin A as a growth promoter in the standard rat assay but fails to cure eye disease. Wald has shown that, unlike the corresponding alcohol and aldehyde, all-*trans*-retinoic acid cannot function in vision (30). Nothing is known of the physiological effects of the cis isomers. None of the retinoic acids have been found to occur in nature.

Vitamin A Hydrocarbon. This derivative is known only through synthesis and is claimed to be biologically active (31,32). It is a crystalline solid (mp, 75–76°C) and shows a uv absorption spectrum (λ_{max}^{MeOH} 325 nm, ϵ 50,000) which is similar to that of vitamin A.

Vitamin A Epoxides (33,34). The 5,6-epoxide is obtained by the action of monoperoxyphthalic acid on vitamin A acetate, followed by saponification of the acetate group. The corresponding aldehyde, obtained by manganese dioxide oxidation of the alcohol, was found biologically equivalent to all-*trans*-retinal. The 5,6-epoxides rearrange to the 5,8-isomers on treatment with ethanolic hydrochloric acid.

α-**Vitamin A** (**8**). This isomer of vitamin A, in which the ring double bond is shifted to the 4,5-position, out of conjugation with the side chain, is known only through synthesis. No information is available on its physiological activity. The methyl ether (**35**) shows only 1–3% of the activity of vitamin A.

*Retro*vitamin A (**11**) (13). This isomer, so called because the entire conjugated pentaene system is shifted one carbon atom back toward the ring, occurs as an artifact in the synthesis of vitamin A (see Scheme 2). It sometimes contaminates commercial preparations of synthetic vitamin A and is 12–21% as potent physiologically as vitamin A (36).

Acetylenic Analogs of Vitamin A. Both possible analogs the 7,8- and 11,12- are known through synthesis (13,27). They show 40 and 15%, respectively, of the biological potency of vitamin A.

Vitamin A_2

Vitamin A_2, all-*trans*-3-dehydroretinol, is a crystalline solid (mp, 63–65°C) showing a characteristic uv absorption curve ($\lambda\lambda_{max}$ 350 nm) with subsidiary peaks at 286 and 276 nm (6a) (see Fig. 2, dashed curve). Like vitamin A it is sterically hindered at the ring–side-chain junction which accounts for its spectral differences from normal conjugated hexaenes (4).

Its reactions are similar to those of vitamin A. It is readily esterified, is oxidized to the aldehyde with manganese dioxide, reacts with maleic anhydride, and gives an intense color in the Carr-Price test ($\lambda_{max}^{CHCl_3}$ 693 nm).

Because of the additional double bond vitamin A_2 is more sensitive to atmospheric oxidation than vitamin A. When treated with acid in ethanol it rearranges to 3-ethoxyanhydrovitamin A (6a). For historical reasons this is called "anhydrovitamin A_2," although it is not a hydrocarbon.

All of the stereoisomers of vitamin A_2 corresponding to those known for vitamin A have been synthesized, as well as some of their corresponding aldehydes and acids (6a). The methods are analogous to those utilized for the synthesis of the various vitamin A isomers, employing the 3-dehydroanalogs of the starting materials. The 9-cis bond in the vitamin A_2 series appears just as resistant to isomerization catalyzed by iodine as that in the vitamin A series.

Economically vitamin A_2 is of no importance since it plays only an incidental role in mammalian metabolism.

The Provitamins A

No official definition exists for a provitamin A. Usage has confined the term to those naturally occurring carotenoids which are transformed in vivo to vitamin A and, as a consequence, possess vitamin A activity. The Council on Pharmacy and Chemistry of the American Medical Association regards the term "provitamin A" as a synonym for α-, β-, or γ-carotene (**3**) and for cryptoxanthin, 3-hydroxy-β-carotene (**37**), but these are not the only carotenoids that show provitamin A activity, as shown in Table 1. In the broad sense, any compound that can be converted in the animal body to

Table 1. Some Provitamin A Carotenoids

Compound	Empirical formula	Source	Vitamin A activity, %/mole
α-carotene	$C_{40}H_{56}$	most plants	54
β-carotene	$C_{40}H_{56}$	most plants	100
γ-carotene	$C_{40}H_{56}$	carrots (rare)	42
cryptoxanthin	$C_{40}H_{56}O$	yellow corn, butter	57
aphanin	$C_{40}H_{54}O$	blue algae	50
aphanicin	$C_{40}H_{56}O$	*Aphanizomenon flosaquae*	26
myxoxanthin	$C_{40}H_{54}O$	blue algae	active
echinenone	$C_{40}H_{56}O$	sea urchin	active
torularhodin	$C_{37}H_{48}O_2$	red yeast	active
mutatochrome (citroxanthin)	$C_{40}H_{56}O$	orange peels	active
α-carotene epoxide	$C_{40}H_{56}O$	*Ranunculus acer*	active

vitamin A should be called a provitamin A. This would include such products as β-apo-2-carotenal and retinal, which have definitely been proven to yield vitamin A in vivo. Many other biologically active degradation products of β-ionylidene carotenoids are undoubedly also provitamins A. A number of active synthetic analogs, homologs, and derivatives of vitamin A and of β-carotene have been referred to as provitamins A in some of the literature, but such classification should be deferred until it is shown whether these compounds are converted in vivo to vitamin A or are active per se.

The carotenoids are yellow to red polyene pigments occurring in all green tissues of plants, as well as in many species of fungi, bacteria, and algae. They also occur in animals, which acquire them by ingestion of vegetable products. Carotenoids are not synthesized in the animal organism. A very comprehensive review of the chemistry of the carotenoids has been written by Karrer (38) and one on their biochemistry by Goodwin (39) (see also reference 6a).

Structurally, the carotenoids are characterized by a long polyene chain composed of recurring 2-methyl-2-butene units ("isoprene units") linearly joined to each other in a head-to-tail fashion by double bonds. The center of the molecule always consists of two such units joined "tail-to-tail." α-, β-, and γ-carotene (3) are typical examples of C_{40} carotenoids. In other examples one or both of the rings may contain hydroxyl or keto groups or both. The lower carotenoids, such as bixin, C_{24}, and crocetin, C_{20}, are dicarboxylic acids and may be regarded as symmetrical degradation products of C_{40} precursors.

Some biologically active carotenoids and their sources are listed in Table 1. The prime essentials for provitamin A activity are an unsubstituted β-ionone ring and an unsaturated side chain equivalent at least to that of vitamin A. The only exceptions to this rule are the carotenoid epoxides. The activity of these derivatives is presumed to be due to the ability of the animal organism to remove the epoxide group (38).

The fact that β-carotene, the most active provitamin A, consists of two vitamin A moieties symmetrically linked at their terminal side-chain carbons led to the postulation that the provitamin A carotenoids owe their activity to in vivo cleavage at the 15,15'-double bond in the center of the molecule. Although β-carotene should, on this basis, give rise to two molecules of vitamin A, the preponderance of bioassay data shows only one molecule of vitamin A to be metabolically available, thus lending support to the theory that metabolic degradation to vitamin A starts at one terminus of the polyene chain. International activity standards show a 1:2 relationship between vitamin A and β-carotene on a weight basis and approximately a 1:1 relationship on a molar basis. The fact that those carotenoids having only one unsubstituted β-ionone ring show at most only half the activity of β-carotene lends support to the theory of central fission in vivo. It has been pointed out, however, that all such carotenoids should show an activity equivalent to that of α-carotene, whereas in fact there are significant differences. The question of terminal versus central fission is still unsettled (40).

Economically, the most important provitamin A carotenoid is β-carotene. It is the most widely distributed carotenoid in nature, and has been found in the leaves of all plant species so far examined. The all-trans isomer is a dark-red crystalline solid (mp, 181–182°C), showing a characteristic uv absorption spectrum ($\lambda\lambda_{max}^{CS_2}$ 450, 485, and 520 nm). It gives the standard blue color ($\lambda_{max}^{CHCl_3}$ 590 nm) in the Carr-Price reaction and shows the typical reactions of conjugated double bonds, ie sensitivity to oxygen, light, heat, etc. It is soluble in the usual fat solvents.

β-carotene is presently synthesized industrially by a modification of the Inhoffen route (6a), starting with the C_{14} aldehyde (**12**). The sequence is a rather long one, and a much simpler approach has been developed by Sarnecki (41) based on vitamin A as a starting material, as shown in Scheme 6.

Scheme 6.

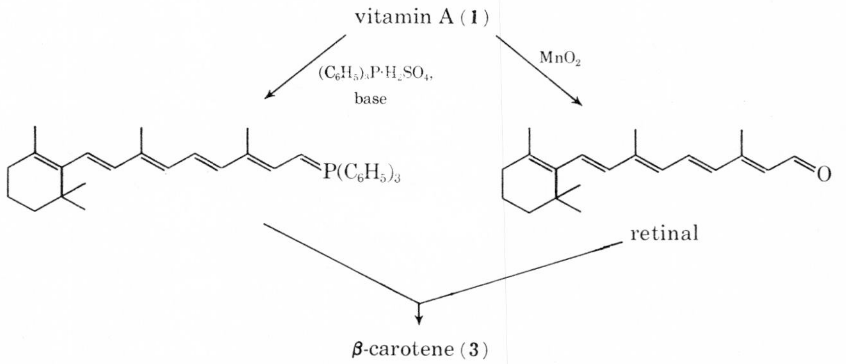

The product is obtained as a mixture of 15,15′-cis and trans isomers which is then fully transified catalytically with iodine or by heat.

A comprehensive review of cis-trans isomerism in the carotenoids has been written by Zechmeister (42).

Occurrence

Although the vitamins and provitamins A are widely distributed in nature, no natural source can compete with synthetic vitamin A or β-carotene. Prior to the advent of synthetic vitamin A, liver oils from the shark, halibut, tunny, and whale constituted a vital source of vitamin A concentrate and cod-liver oil an important source of vitamin A containing oil. Today there is only a very limited market for fish-liver oils and the supply is entirely from the Japanese industry.

The market for natural carotene is likewise very limited. Alfalfa, carrots, and palm oils have been prime sources of provitamins A. Endeavors to develop a low-cost microbiological source of carotenoids (43) have not fulfilled expectations.

The importance of the natural occurrence of vitamin A and the carotenoids lies in their distribution in foods and the value thus imparted to food items. Table 2 lists some typical foods of both animal and vegetable origin and their vitamin A or provitamin A content.

Standards

The Committee on Biological Standardization of the World Health Organization defined the International Unit (IU) of vitamin A as the biological activity of 0.344γ of vitamin A acetate (45). This is equivalent to 0.300γ of the free alcohol. Thus, one gram of vitamin A contains 3.33×10^6 IU.

The *U.S. Pharmacopeia* standard (46) is equivalent. Authentic specimens of a standard solution are available from U.S. Pharmacopeia Reference Standards and contain 34.4 mg of vitamin A acetate per gram of oil solution. The IV of provitamin A activity is defined as 0.600γ of pure all-*trans*-β-carotene (47).

Table 2. Vitamin A and Provitamin A Content of Certain Foods, 10 IU/100g Food (44)

Food	Vitamin A	Provitamin A
butter	3,300	
cheeses	1,100–1,500	
cream	480–1,540	
milk (whole)	140–160	
eggs (whole, raw)	1,180	
margarine	3,300	
calf's liver	22,500	
white fish (raw)	2,260	
halibut (raw)	440	
eel (raw)	1,610	
herring (raw)	105	
oyster (raw)	310	
apples (raw)		50
apricots (raw)		2,500
bananas (raw)		333
oranges (raw)		84
prunes (dried)		1,666
asparagus (raw)		833
broccoli tops (raw)		4,166
cabbage		500
carrots (raw, mature)		20,000
cauliflower (raw)		50
lettuce (raw)		1,666
parsley (raw)		13,333
pumpkin (raw)		2,500
spinach (raw)		10,000
watercress (raw)		5,000

Physicochemical constants for pure vitamin A, vitamin A acetate, and β-carotene are now well established and can be used for the accurate determination of potency.

Assay

Vitamin A. The assay of vitamin A can be accomplished by any one of three accepted methods: biological, colorimetric, or spectrophotometric.

Biological Methods (2). The original assay procedure, still widely used, is based on the biological response of vitamin A-deficient rats. This method lacks specificity in that it measures not only the vitamin A present but the provitamins as well. Rats are fed on a vitamin A-free diet until they show a cessation of growth. One group is then separated and its diet supplemented by a predetermined dose of standard vitamin A. For a more accurate assay two or three groups on different levels of the standard may be used. The others are divided further into two or more groups and fed different levels of the sample being tested. The rate of growth (ie increase in body weight) is the criterion of response officially used. At the end of four weeks, the assay is terminated, and the increase in weight of the animals fed the unknown sample is compared with that of the standard group. From these data the vitamin A content of the sample can be calculated. If provitamins are present in the sample, these must be determined separately by other methods and the value obtained subtracted from the bioassay value.

The greatest source of error in this assay is the individual differences in the response of the test animals. (See Bioassay.) Most reliable bioassays are, for these reasons, best carried out in laboratories routinely performing them.

A liver-storage test for vitamin A has been proposed (2), which offers a partial solution of the bioassay problem. The basis of the procedure is the well-recognized fact that the amount of vitamin A stored in the liver varies directly with the quantity fed. Weanling rats are placed on a standard vitamin A-free diet for a period of six days. The animals are then separated into groups and on two successive days treated with the vitamin A standard or test sample. Forty-eight hours after the last dose, the livers are removed and analyzed by the Carr-Price procedure.

Compared to the rat-growth assay, this method saves labor, time, and cost. Furthermore, there is excellent agreement between the values obtained by the two procedures (2). The chief limitation of the liver-storage test lies in the shortcomings of the Carr-Price assay.

Colorimetric Method (Carr-Price) (48). The colorimetric assay for vitamin A is based on the Carr-Price reaction. The intensity of the blue color (λ_{max} 620 nm; "blue value") formed when vitamin A and antimony trichloride are mixed in chloroform solution is measured and compared to that given by a standard solution of vitamin A. This assay is subject to a number of serious sources of error, chief among which is the speed with which the color develops and fades, making time a vital factor in the procedure. In addition, small amounts of moisture or unsaturated substances hamper the formation of the blue color or produce interfering colors. Nevertheless, with experience and adequate provisions for these factors, accurate and reproducible assays of vitamin A can be made. One of its advantages that gives it wider applicability than the bioassay and spectrophotometric assays is its specificity for Vitamin A. The USP relies upon it as an identification test for vitamin A to be carried out in conjunction with the spectrophotometric assay.

Spectrophotometric Method (48). The characteristic uv absorption spectrum of vitamin A affords the most accurate and reproducible means of vitamin A determination. The extinction at 325 nm, compared to that of a standard sample, gives the vitamin A content directly. This method is applicable to solutions of the pure (or nearly pure) vitamin. Errors due to other materials absorbing in the region 300–350 nm can be corrected by a procedure developed by Morton and Stubbs, based on absorption measurements at 310 and 334 nm, as well as at 325 nm, and thereby estimating the difference of the curve obtained from that of pure vitamin A. When the correction is too large, however, separation of the vitamin A from the extraneous absorbing material is necessary prior to assay.

Provitamin A Carotenoids. These carotenoids can be assayed biologically, by the solvent-partition method, or by chromatography (49). The Carr-Price assay is not applicable to the provitamin A carotenoids, since they all give different colors with this reagent. The laborious nature of the bioassay and the nonspecificity of the solvent-partition method have made chromatography the method of choice. With a 1:3 mixture of magnesia and diatomaceous earth, Celite or Super-Cel (Johns-Manville Corp.), as absorbent and a 2–5% solution of acetone in petroleum ether as eluent, the separation of the carotenes from other carotenoids and extraneous materials can be rapidly and accurately accomplished. The eluted carotenes are then determined spectrophotometrically.

Vitamin A and Carotene. The determination of vitamin A and carotene in the presence of each other is frequently necessary in products, such as butter, milk, and eggs, and, clinically, in blood serum. A Carr-Price determination is first made for the vitamin A on an ether extract of the material. The carotene is then chromatographically separated from another aliquot of the ether extract and determined spectrometrically.

Biological Action

Although it is generally conceded that vitamin A plays a vital role in general metabolism, the only biological function where its action is clearly understood is that of vision. This has been due largely to the work of Wald (50) and Morton (51). The role of vitamin A in vision is outlined in basic form in Scheme 7.

Scheme 7.

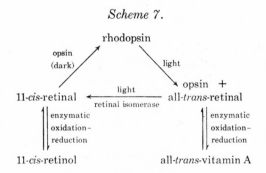

Vitamin A is transported by the bloodstream to the retina. There it is enzymically oxidized to all-*trans*-retinal. This is then isomerized, also enzymically, to 11-*cis*-retinal. Only this cis form of retinal can combine with the protein opsin (ever present in the eye) to form rhodopsin, a purple pigment (visual purple). This is the photoreceptor in the rod cells of the retina and is sensitive to light of low intensity. In the cone cells of the retina the photoreceptors, which are sensitive to light of high intensity and different colors, are composed of 11-*cis*-retinal and different proteins. The pigment in this case is called iodopsin.

Both rhodopsin and iodopsin, when exposed to light, are bleached; ie they are decomposed to retinal and the original protein. The retinal is isomerized in this process to the all-trans form. The energy released in this chemical break-up is imparted to the optic nerve, resulting in vision. The all-*trans*-retinal can then be reisomerized in the eye to the 11-cis form and the cycle repeated.

In fresh-water fish vitamin A_2 functions in an analogous manner, forming the corresponding 3-dehydro analogs of rhodopsin and iodopsin.

Other compounds related to vitamin A, which are not convertible in vivo to retinal, cannot function in vision although still remaining effective in promoting growth in the standard rat-growth assay. Notable among these has been retinoic acid. Rats maintained on this compound in place of vitamin A grow normally but go blind (30). This discovery has spurred considerable research aimed at finding the role of vitamin A in specific metabolic reactions. Current interest tends to emphasize its role in effecting membrane action of the individual cell and subcellular particles (52).

Aside from night blindness vitamin A deficiency in humans is not associated with any clear cut general disease syndrome as in the case of other vitamin deficiencies, eg beriberi (vitamin B_1), pellagra (nicotinic acid), or scurvy (nicotinic acid). Some clinical

Table 3. Recommended Daily Dietary Requirements of Vitamin A (55)

Children, yr	IU	Adults	IU
1	1500	men (70 kg)	5000
1–3	2000	sedentary	5000
4–6	2500	physically active	5000
7–9	3500	heavy work	5000
10–12	4500		
13–15		women (56 kg)	
girls	5000	sedentary	5000
boys	5000	moderately active	5000
16–20		very active	5000
girls	5000	pregnancy,	
boys	6000	latter half	6000
		lactation	8000

manifestations that have been associated with vitamin A deficiency are hyperkeratosis of the skin, and conjunctival xerosis. In man severe cases of vitamin A deficiency are very rare because the adult human has enormous capacity to store and utilize vitamin A previously acquired. Recommended daily requirements of vitamin A are listed in Table 3.

In animals lack of vitamin A causes a distinct cessation of nonskeletal growth, the appearance of xerophthalmia and cornification of many epithelia, such as those of the respiratory, digestive, urinary, and vaginal tracts. Additional manifestations may be susceptibility to infections, nephrolithiasis, and degenerative changes in the spinal and cranial nerves (53). For a review of the pharmacology and toxicology of vitamin A see reference 54.

Market Forms

The synthetic acetate and palmitate of vitamin A are the derivatives most generally used. Formulations of these depend on the use, ie pharmaceuticals, food fortification, or animal feeds.

In pharmaceuticals, oral or parenteral preparations are usually water-dispersed oilfree vitamin, utilizing a nonionic dispersing agent. Tablets and capsules for oral use employ finely dispersed vitamin A esters in a matrix of gelatin and sugar. Such formulations of vitamin A are quite stable toward oxidation and are, in addition, water dispersible.

Stable gelatinized powder is also used in the food industry for the fortification of vitamin A. Fatty foods, such as margarine, cooking oils, and milk, utilize oil concentrates of vitamin A.

Animal feeds, the largest consumer of vitamin A, utilize both the oil concentrate and the stable gelatinized form. Water dispersions are used in medicated feeds.

β-Carotene is used in food fortification when a yellow color is required or desirable as in the case of margarine, butter, cheese, salad dressing, etc. Water dispersions of β-carotene have been used in fruit juices and carbonated beverages.

Price Trends

In 1949, when the market was dominated by natural vitamin A, the price was 50¢/million IU. Synthetic vitamin A fulfilled the world requirements so effectively

Table 4. U.S. Production of Vitamin A[a]

Year	lb	Year	lb
1950	18,000	1962	290,000
1956	132,000	1969[b]	420,000

[a] Based on data published by the U. S. Tariff Commission.

[b] Estimated.

that the price dropped as low as 4¢/million IU (oil concentrate) in 1969, while consumption increased more than twentyfold, as shown in Table 4. The price remained stable in the late 1960s.

The market for β-carotene is much more limited and is dominated by the margarine industry. The 1969 price was approx 12¢/million IU.

Bibliography

"Vitamins (vitamin A)" in *ECT* 1st ed., Vol. 14, pp. 791–813, by W. Oroshnik, Ortho Research Foundation.

1. IUPAC, "Commission on the Nomenclature of Biological Chemistry," *J. Am. Chem. Soc.* **82** 5581–5583 (1960).
2. P. L. Harris, *Vitamins Hormones* **18**, 341 (1960).
3. L. Zechmeister, *Chem. Rev.* **34**, 267 (1944).
4. W. Oroshnik, G. Karmas, and A. D. Mebane, *J. Am. Chem. Soc.* **74**, 295 (1952).
5. W. Oroshnik and A. D. Mebane, *J. Am. Chem. Soc.* **76**, 5719 (1954).
6. W. H. Sebrell and Robert S. Harris, eds., *The Vitamins*, Vol. I, 2nd ed., Academic Press, Inc., New York, 1967.
6a. U. Schwieter and O. Isler, in reference 6, Chap. I(2); O. Isler, H. Kläui, and U. Solmes, in reference 6, Chap. I(3).
7. J. G. Baxter and C. D. Robeson, *J. Am. Chem. Soc.* **64**, 2407, 2411 (1942).
8. C. D. Robeson, J. D. Cawley, L. Weisler, M. H. Stern, C. C. Eddinger, and A. J. Chechak, *J. Am. Chem. Soc.* **77**, 4111 (1955).
9. K. C. D. Hickman, *Ind. Eng. Chem.* **29**, 968, 1107 (1937).
10. C. D. Robeson and J. G. Baxter, *J. Am. Chem. Soc.* **69**, 136 (1947).
11. O. Isler, A. Ronco, W. Guex, N. C. Hindley, W. Huber, K. Dialer, and M. Koffler, *Helv. Chim. Acta* **32**, 489 (1949).
12. E. M. Schatz, J. C. Cawley, and N. D. Embree, *J. Am. Chem. Soc.* **65**, 901 (1943).
13. R. H. Beutel, D. F. Hinkley, and P. I. Pollack, *J. Am. Chem. Soc.* **77**, 5166 (1955).
14. J. Attenburrow, A. F. B. Cameron, J. H. Chapman, R. M. Evans, B. A. Hems, A. B. A. Jansen, and T. Walker, *J. Chem. Soc.* **1952**, 1094.
15. S. Ball, T. W. Goodwin, and R. A. Morton, *Biochem. J.* **42**, 516 (1948).
16. C. D. Robeson, W. P. Blum, J. M. Dieterle, J. D. Cawley, and J. G. Baxter, *J. Am. Chem. Soc.* **77**, 4120–4125 (1955).
17. J. W. Batty, A. Buraway, S. H. Harper, I. M. Heilbron, and W. E. Jones, *J. Chem. Soc.* **1938**, 175.
18. O. Isler and P. Schudel, "Synthetic Methods in the Carotenoid and Vitamin A Fields," in R. A. Raphael, E. C. Taylor, and H. Wynberg, eds., *Advances in Organic Chemistry: Methods and Results*, Vol. IV, Interscience Publishers, a div. of John Wiley & Sons, Inc., New York, 1963, p. 115.
19. I. M. Heilbron, A. W. Johnson, E. R. H. Jones, and A. Spinks, *J. Chem. Soc.* **1942**, 727.
20. N. A. Milas, *Vitamins Hormones* **5**, 1 (1947).
21. O. Isler, W. Huber, A Ronco, and M. Koffler, *Helv. Chim. Acta* **30**, 1911 (1947).
22. U.S. Pat. 2,541,739 (Oct. 19, 1948), O. Isler (to Hoffmann-La Roche, Inc.).
23. U.S. Pat, 2,451,740 (Oct. 19, 1948), H. Lindlar (to Hoffmann-La Roche, Inc.).
24. U.S. Pat. 2,610,207 (Sept. 9, 1952), H. Lindlar and R. Rügg (to Hoffmann-La Roche, Inc.).

25. U.S. Pat. 2,610,208 (Sept. 9, 1952), J. D. Surmatis (to Hoffmann-La Roche, Inc.).

26. S. R. Ames, *Ann. Rev. Biochem.* **27**, 375 (1958).

27. W. Oroshnik, *J. Am. Chem. Soc.* **78**, 2651 (1956).

28. C. D. Robeson, W. P. Blum, J. M. Dieterle, J. D. Cawley, and J. G. Baxter, *J. Am. Chem. Soc.* **77**, 4120 (1955).

29. G. Wald, P. K. Brown, R. Hubbard, and W. Oroshnik, *Proc. Natl. Acad. Sci.* **41**, 438 (1955).

30. J. E. Dowling and G. Wald, *Vitamins Hormones* **18**, 387 (1960).

31. Ger. Pat. 1,022,583 (1958), H. Pommer and W. Sarnecki (to BASF).

32. U. S. Pat. 2,835,713 (1958), C. D. Robeson (to Eastman Kodak Co.).

33. F. B. Jungawala and H. R. Cama, *Biochem. J.* **95**, 17 (1965).

34. M. R. Lakshmanan, F. B. Jungawala, and H. R. Cama, *Biochem. J.* **95**, 27 (1965).

35. W. Oroshnik, *J. Am. Chem. Soc.* **76**, 5499 (1954).

36. T. N. R. Varma, P. E. Erdody, and T. K. Murray, *J. Pharm. Pharmacol.* **17**, 474 (1965).

37. *New and Nonofficial Remedies*, American Medical Association, Chicago, Ill., 1954, p. 546.

38. P. Karrer and E. Jucker, *Carotenoids* (translated and revised by E. Braude), Elsevier Publishing Co., Amsterdam, 1950.

39. T. W. Goodwin, *The Comparative Biochemistry of the Carotenoids*, Chapman and Hall, Ltd., London, 1952.

40. J. Ganguly and S. K. Murthy, in reference 6, Chap. I(4).

41. Ger. Pat. 1,158,505 (1964), W. Sarnecki, A. Nürrenbach, and W. Reif (to BASF).

42. L. Zechmeister, *Cis-trans Isomeric Carotenoids, Vitamin A, and Arylpolyenes*, Springer Verlag, Vienna-New York, 1962.

43. *Chem. Eng. News* **44**, 44 (1966).

44. B. K. Watt and A. L. Merrill, *U.S. Dept. Agr. Agr. Handbook* **198**, 8 (1966).

45. Expert Committee on Biological Standarization, *World Health Organization Tech. Rept. Ser.* **187**, 10 (1960).

46. *U.S. Pharmacopeia XVI*, Mack Publishing Co., Easton, Pa., 1960, pp. 837, 938.

47. Expert Committee on Biological Standardization, *World Health Organization Tech. Rept. Ser.* **222**, 10 (1961).

48. M. Koffler and S. H. Rubin, *Vitamins Hormones* **18**, 315 (1960).

49. M. Freed, *Methods of Vitamin Assay*, 3rd ed., John Wiley & Sons, Inc., New York, 1966.

50. G. Wald, *Vitamins Hormones* **18**, 417 (1960).

51. R. A. Morton and G. A. Pitt, *Fortschr. Chem. Org. Naturstoffe* **14**, 244 (1957).

52. O. A. Roels, in reference 6, Chap. I(5).

53. T. Moore, in reference 6, Chap. 1(9).

54. *Ibid.*, Chap. I(10).

55. *Natl. Acad. Sci. Natl. Res. Council Publ.* **589** (1958).

WILLIAM OROSHNIK
Chemo Dynamics

NICOTINIC ACID

Nicotinic acid (acidum nicotinicum, 3-pyridinecarboxylic acid, pyridine-β-carboxylic acid, PP (pellagra preventing) factor, vitamin PP, antipellagra vitamin) is also known by various commercial names (1), ie Akotin, Niconacid, Nicotinipca, Nicyl, etc. This compound is commonly known as niacin, especially with reference to its role as a vitamin, as the name nicotinic acid (still used in Europe) is too easily confused with the similarly named compound nicotine. Niacin and niacinamide are considered part of the vitamin-B complex.

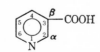

nicotinic acid (niacin);
3-pyridinecarboxylic acid

Nicotinic acid exists in minute amounts in all living systems and, in its amide form, participates in biological electron transfer as the functional portion of the nucleotides NAD (nicotinamide adenine dinucleotide) and NADP (nicotinamide adenine dinucleotide phosphate), which act as coenzymes for many enzymatic oxidation-reduction reactions.

The name nicotinic acid derived from the fact that this chemical was first obtained in 1867 in Germany by C. Huber from the oxidation of nicotine with potassium dichromate:

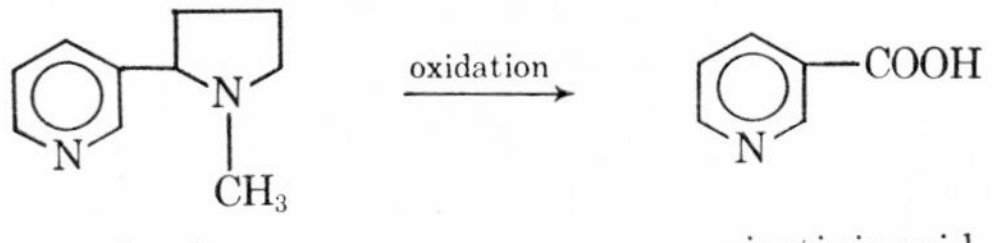

Both the acid and its amide are used as additives in flour or bread, as the human daily requirement of this vitamin is 12–25 mg.

It is used primarily as a therapeutic agent against pellagra in humans, and black-tongue in dogs and other animals; as a vasodilator and to correct blood pressure changes; as a powerful agent to decrease cholesterol and free fatty acid levels in the blood; and as a stimulant in tooth eruption, improving the odontogenic process.

Industrially it has been employed as an anticorrosion agent for aluminum–zinc alloys in contact with acid solutions; it has been recommended as a brightener in zinc electroplating. Nicotinic acid is made commercially from easily available materials and its commercial price is \$1.24/lb (1969).

History

The lack of niacin in the diet causes pellagra in man and blacktongue in dogs. Pellagra has been known to exist in Spain since 1735 (2), but only in the twentieth

Table 1. Chronology of Niacin and Its Deficiency

Year	Development
1735	pellagra was recognized by the Royal Spanish Physician Don Gaspar Casal
1830	5% of the people of Northern Italy were suffering from pellagra
1850	endemic disease in all of France
1863	pellagra reported in the United States by Gray in New York and Tyler in Massachusetts
1867	Huber prepared nicotinic acid by the oxidation of nicotine
1870	Huber established a formula of nicotinic acid, $C_6H_5O_2N$
1912	already a toll of 30,000 cases of pellagra with a 40% fatality rate in the United States
1911–1913	Funk isolated niacin from yeast
1912	Suzuki et al. isolated niacin from rice bran
1915	Goldberger proved that pellagra could be prevented by the proper diet and produced pellagra experimentally in prison volunteers
1925	Goldberger established that pellagra is caused by a vitamin deficiency
1935	Warburg and Christian found nicotinamide in red blood cells coenzyme
1937	Elvehjem and co-workers established the relationships between nicotinamide, active liver extracts and the diseases due to vitamin deficiency (canine blacktongue) (5)

century was it reported in the southeastern United States, where 170,000 cases occurred annually from 1910 to 1935.

Corn contains niacin but in a bound form which makes the vitamin unavailable. The pre-Columbian civilizations from Mexico (Aztec, Toltec, Mayan) devised a procedure to cook the corn in the presence of lime water, which improved the plastic properties of the dough in the preparation of "tortillas," and, without their knowledge, created an excellent process to bring out the niacin in free form; thus, pellagra was prevented among ethnic groups whose nutritional pattern is based on corn. These chemical transformations have been confirmed by Kodicek et al. (3); a modernized version of the general procedure is described in detail in a patent (4).

A brief chronology of niacin and niacin deficiency is given in Table 1.

Physical Properties

Nicotinic acid occurs as white needle-shaped crystals or crystalline powder, odorless or with a slight odor, and with a tart taste. It is soluble in water (1 g in 60 ml at room temperature) but the solubility varies according to temperature, the nicotinic acid being freely soluble in boiling water and ethanol. Slobodin and Gol'dman (6) give the solubility data shown in Table 2.

Table 2. Solubility of Nicotinic Acid, g/100 ml

Temperature, °C	Distilled water	Ethanol
0	0.86	0.57
15	1.3	0.92
38	2.47	2.1
61	4.06	4.2
78	6.0	7.6
100	9.76	

Nicotinic acid is also soluble in glycerol, propylene glycol, alkaline solutions, and dilute acids, but insoluble in ether. It melts at 235–237°C, subliming without decomposition. Nicotinic acid has a pK of 4.76 and its saturated solution shows a pH value of 2.7. At 22°C it shows two acid dissociation constants: 1.12–1.23×10^{-5} for the carboxyl proton and 3.55×10^{-11} for the basic quaternary N proton (7). Nicotinic acid shows a uv absorption maximum at 263 nm.

Chemical Properties

Nicotinic acid is nonhygroscopic and very stable in air, resisting autoclaving at 120°C for 20 min. However, in an alkaline medium and at high temperature it undergoes decarboxylation to form pyridine. Niacin is stable against oxidation in the presence of the usual oxidizing agents, even at elevated temperatures. Ultraviolet irradiation at 235.7 nm produces striking spectral changes and biologically inactive products (8).

The amount of nicotinic acid in different edible products is more stable if the pH is maintained near 4.5–5.0. In honey kept at 28–30°C for one year, an 8% loss was observed (9).

Since it has a carboxyl group, nicotinic acid forms salts with several metals, such as sodium, calcium, copper, aluminum, etc. Heavy metals, such as copper, are usually employed for its analysis, purification, or recovery from dilute solutions; aluminum salts may be administered therapeutically (10). As a pyridine derivative, nicotinic acid also forms pyridinium salts, ie hydrochlorides or salts obtained from strong acids other than hydrochloric. Nevertheless, nicotinic acid in solution is not in the zwitterion form.

Occurrence

Nicotinic acid has been found in small amounts in all kinds of bacteria, plants, and animals. Appreciable quantities can be found in animal organs, such as liver, kidney, brain, blood, and adrenal glands. Yeast and fungi are also rich sources, as well as alfalfa, several legumes, whole cereals, and corn. Elvehjem (11) reported the data shown in Table 3.

Table 3. Nicotinic Acid Content of Some Edible Products, mg/100 g

brewer's dry yeast	44.7	skim milk	
bottom dry yeast	35.6	powder	10.5
baker's dry yeast	25.7	salmon (canned)	6.0
kidney (fresh)	19.4	whole wheat	5.3
beef liver (fresh)	17.8	soybean meal	4.8
raw peanut meal	13.0	egg yolk (dry)	4.0
sheep liver (fresh)	12.5	whole corn	1.3, 1.18[a]
pig liver (fresh)	11.8	potatoes	1.15[a]

[a] See reference 12.

It has been found by Cravioto et al. (13) that corn shows variations in nicotinic acid content and an average determination of several samples of Mexican corn indicates 2 mg/100 g. As mentioned before, the vitamin is free only after acid or alkaline hydrolysis, which liberates the niacin from a peptide combination, making it unavailable even through digestion.

Animals submitted to a maize diet show signs of nicotinic acid deficiency, but these symptoms are cured by feeding them "tortilla'" in which the niacin was liberated by treatment of the corn with alkali (3).

An interesting discovery made by Cravioto et al. (14) is that raw coffee beans containing small amounts of nicotinic acid show a large increase in vitamin content after having been roasted in the usual manner. Thus, 100 g of roasted beans may contain up to 36.8 mg of niacin. The reason for this increase seems to be the conversion of trigonelline (see p. 535) into nicotinic acid by heat.

Nicotinic acid and nicotinamide are normally found combined in the form of larger molecules linked to a pentose sugar (D-ribose), phosphoric acid, and sometimes the purine nucleotide adenylic acid, forming the compounds shown in Table 4, from which they are split off by the isolation procedures.

The last two compounds in Table 4 are important coenzymes which are very active in biological dehydrogenations and have received different names in the chemical literature. Nicotinamide adenine dinucleotide (NAD) has also been called diphosphopyridine nucleotide (DPN), Harden and Young's coenzyme, cozymase, cozymase I, codehydrase I, and coenzyme I (Co I). Nicotinamide adenine dinucleo-

Table 4. Biologically Active Natural Forms of Combined Nicotinic Acid and Nicotinamide

Name of compound	Composition
nicotinamide nucleoside	nicotinamide—ribose
nicotinamide mononucleotide (NMN)	nicotinamide—ribose—phosphate
nicotinamide adenine dinucleotide (NAD)	nicotinamide—ribose—phosphate | adenine—ribose—phosphate
nicotinamide adenine dinucleotide phosphate (NADP)	phosphate | nicotinamide—ribose—phosphate | adenine—ribose—phosphate

tide phosphate (NADP) has also been called triphosphopyridine nucleotide (TPN), Warburg and Christian's coenzyme, phospho-cozymase, cozymase II, codehydrogenase II, and coenzyme II (Co II). Only the names nicotinamide adenine dinucleotide (NAD) and nicotinamide adenine dinucleotide phosphate (NADP) are officially recommended by the Commission on Enzymes of the International Union of Biochemistry (15).

Preparation and Manufacture

Nicotinic acid can be made by using pyridines or nonpyridines as starting materials. In general, procedures for preparing nicotinic acid can also be employed to make isonicotinic acid (4-pyridinecarboxylic acid), provided the starting material is substituted in the 4 (or γ)- rather than in the 3 (or β)-position. Some methods produce a mixture of both the 4- and 3-isomers, which can then be separated by any of several patented techniques. There are basically four known syntheses for the preparation of nicotinic acid:

1. The oxidation of substituted pyridines. This includes the classical Huber method for the oxidation of nicotine, which led to the discovery of nicotinic acid. A large variety of raw materials may be used, including (a) alkyl-substituted pyridines (eg, 3-picoline), several lutidines (dimethylpyridines), and the widely used 2-methyl-5-ethylpyridine (MEP); (b) aromatic substituted pyridines (eg, quinoline); and (c) heterocyclic substituted pyridines, eg nicotine and anabasine (2-(3-pyridyl)piperidine).

2. The carboxylation of pyridine. This method consists of three steps: sulfonation of pyridine; replacement of the sulfonic group by cyanide to form a nitrile; and hydrolysis. Nicotinamide is an intermediate product and thus could be obtained by stepwise controlled hydrolysis.

3. Condensation of aldehydes and ammonia, followed by oxidation. This method has two variations: the use of acetaldehyde (paraldehyde) and ammonia, forming a substituted pyridine which is then oxidized, and the Hantzsch synthesis employing a β-keto ester, an aldehyde, and ammonia with further oxidation.

4. A complex synthesis using p-xylene, ammonium sulfate, and ammonium sulfide. After autoclaving and cooling, a mixture of compounds (naphthoic acid, succinic acid, p-tert-butylbenzoic acid, and nicotinic acid) is obtained.

There are many other synthetic pathways but some of them could be classified as simple curiosities without practical applications. However, niacin required for nutritional purposes, such as feeding cattle, can be obtained from yeast extracts. A promising biological procedure calls for the use of a bacterial fermentation of crude oils and paraffin wax to produce an edible product rich in niacin and also containing thiamine, riboflavine, pyridoxine, pantothenic acid, and vitamin B_{12} (16). In general, biological syntheses are useful only when a vitamin complex is desired and no further purification is contemplated.

The first method, the **oxidation of substituted pyridines,** with many modifications and employing a great variety of conditions, is still the one most widely used for industrial production of nicotinic acid. High temperatures and pressures have been employed with various oxidative agents and catalysts, including nitric acid, air, sulfuric acid, permanganates, dichromates, chlorine, ozone, sodium hypochlorite, selenium dioxide, sulfuric anhydride, vanadium pentoxide, molybdenum tetroxide, chromium trioxide, nitrogen dioxide, manganese dioxide, copper and mercury salts, and the electrochemical oxidation in an acid medium.

Another important commercial process is the **oxidation of 3-picoline,** illustrated by the following equation:

This reaction is generally carried out by the use of crude 3-picoline, potassium permanganate, chlorine, and air or electrochemical oxidation. When potassium permanganate is used, the 3-picoline fraction (with 3- and 4-picolines, and 2,6-lutidine) can be used, producing a 70% yield of pyridinecarboxylic acids. The 2,6-lutidine could interfere but it can be easily eliminated as the urea adduct (17). During this procedure strong alkaline solutions are obtained and, on lowering the pH with strong mineral acids, the solubility of nicotinic acid increases and 10–20% more can be recovered (18).

In the chlorine oxidation the gas is passed into the concentrated aqueous solution of 3-picoline hydrochloride in the presence of light at 100–110°C. Nicotinic acid hydrochloride saturates the solution and separates out.

The use of air and ammonia has produced good yields of nicotinonitriles (19), and a mixture of the 3- and 4-picolines (20) treated with a fivefold excess of ammonia and a tenfold excess of air at 300–450°C in the presence of vanadium pentoxide on a kaolin carrier produced 85% of the corresponding nitriles (21).

The electrochemical oxidation calls for technical-grade 3-picoline, a ceramic diaphragm, sulfuric acid, and an electric flow of 5 A/dm^2 at 40–45°C, keeping the anodic solution acid, the middle solution neutral, and the cathodic liquid alkaline (22,23).

Oxidation of the 2,4-dimethylpyridine (2,4-lutidine) with boiling nitric acid and treatment with potassium permanganate, produces isocinchomeronic acid. This acid, heated for 8 hr above its melting point, gives nicotinic acid (24).

The **oxidation of 2-methyl-5-ethylpyridine** (MEP or aldehydine) is probably the procedure most commonly employed to manufacture nicotinic acid, since the price of the starting material is low. MEP is commercially available and is easily made by the condensation of paraldehyde and ammonia (25). From this viewpoint, this synthesis could be thought of as belonging to the third group of procedures (nonpyridine starting materials). The oxidation is carried out with nitric acid at temperatures ranging from 185 to 253°C and at pressures from 250 to 750 psi. The nitric acid can then be regenerated by oxidation with air. The yields vary from 60 to 70 % and the following reactions take place:

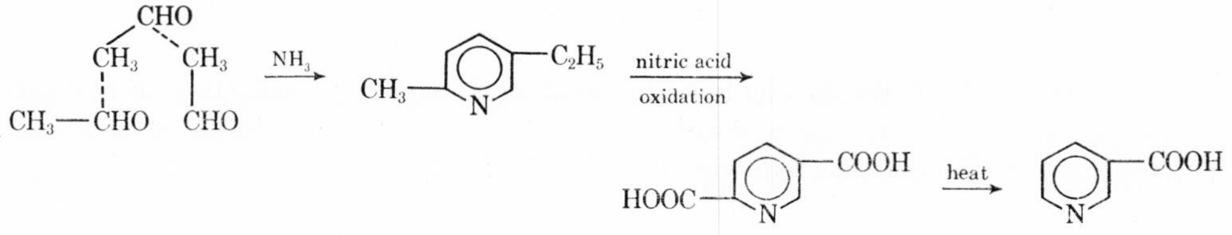

A continuous process uses 6 moles of nitric acid per mole of MEP to be oxidized (26). A variation of this method is the use of 5-acetyl-2-methylpyridine, employing potassium permanganate and sulfur dioxide with silver as catalyst (27), or autoclaving the intermediate, 2,5-pyridinedicarboxylic acid, at 175–180°C for 1.5 hr (28).

If sulfuric acid with selenium dioxide is used instead of nitric acid, 17 lb of boiling concentrated sulfuric acid oxidizes 1 lb of MEP to crude nicotinic acid with a yield of 80% (29–31). Copper salts (32) and lead dioxide (33) have also been used.

In the **oxidation of quinoline,** the general method can be indicated as follows:

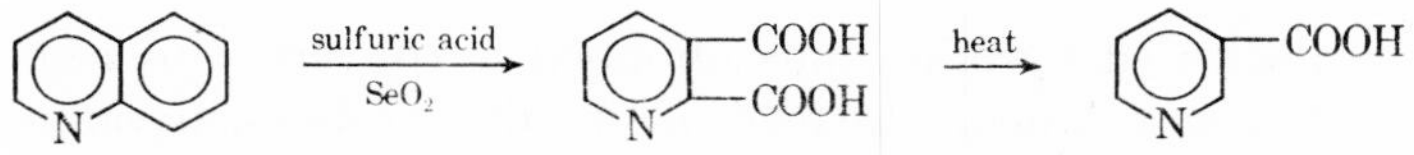

The most common oxidant is sulfuric acid with selenium dioxide because the oxidation is more selective and a 75% yield can be obtained (34). The oxidation is carried out in a continuous reactor using a countercurrent of sulfuric anhydride at 200°C for 9 hr (35). A mixture of nitric–sulfuric acids with molybdenum tetroxide (36) or vanadium pentoxide (37) as catalysts has also been employed.

Quinoline plus 65% oleum (fuming sulfuric acid) produces quinoline-8-sulfonic acid, which is an intermediate in the synthesis of nicotinic acid or 8-hydroxyquinoline, depending on whether it is treated with sulfuric acid or with sodium hydroxide, respectively (38).

In a good procedure quinoline or quinaldine (2-methylquinoline) is heated with nitric acid at 40°C and a mixture of air with 8% ozone is bubbled in until 8.9 parts of the latter is absorbed. Quinolinic acid is obtained in 90% yield and converted to nicotinic acid by sublimation at 220–230°C (39).

The **oxidation of nicotine,** the oldest method known, is still in use in some European countries but is no longer employed in the United States. Nitric acid is the oxidizing agent and yields of 68.4% have been obtained (40,41).

The **carboxylation of pyridine** is a good method, but one that is not very attractive industrially because of the high cost of the starting material. Pyridine is sulfonated with sulfuric acid at 300°C in the presence of vanadium sulfate (42). The sulfonic acid is fused with potassium cyanide and potassium hydroxide to yield 3-cyanopyridine. Acid hydrolysis of this intermediate yields nicotinic acid; however, by controlled alkaline hydrolysis it is possible to form nicotinamide. This process thus yields two important products.

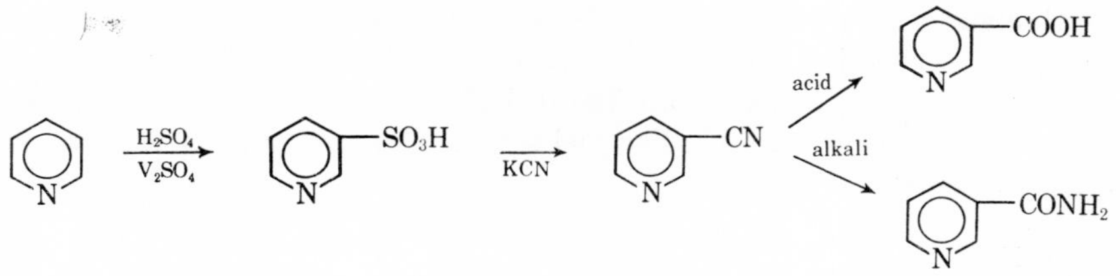

The **Hantzsch synthesis** involves the condensation of two molecules of a β-keto ester or a β-diketone with one molecule of an aldehyde and one molecule of ammonia. The ring closure occurs in the following manner (43):

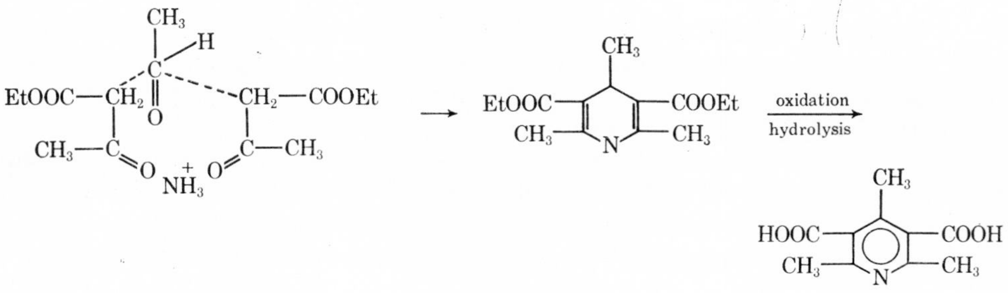

The polysubstituted pyridine thus obtained is converted to nicotinic acid by oxidation and decarboxylation. Instead of dihydropyridines, pyridines are formed directly if hydroxylamine is used instead of ammonia (44).

In the fourth procedure listed—not an important one commercially—crude *p*-xylene, ammonium sulfate, ammonium sulfide, and water are heated in an autoclave at 315°C for 1 hr, cooled, and steam distilled. The precipitated sulfur is removed, and the mixture acidified with dilute hydrochloric acid. A mixture of acids, including nicotinic, naphthoic, succinic, and *p-tert*-butylbenzoic, is obtained (45).

Nicotinic acid made by any of the above-mentioned processes may be purified by azeotropic distillation with pure solvents or with mixtures of tetralin (1,2,3,4-tetrahydronaphthalene) or diphenyl (46), or by sublimation at 210–220°C (47).

Economic Aspects

The production of niacin and niacinamide has been increasing since 1940 at a rapid rate; however, after 1955, the total production in the United States has remained more or less constant. In 1940 about 4000 lb of nicotinamide was produced, and three years later the annual production was 160,000 lb. In 1961 the production of niacinamide increased to 724,000 lb, but three years later no more than 802,000 lb was manufactured.

With cheaper raw materials and better processes, the unit value has dropped considerably. In 1941 one lb of niacin cost $6.94; twenty years later the price had gone down to $1.40.

Tables 5–7 indicate the production, sales, sales value, and unit value for niacin, niacinamide, and their hydrochlorides, as given by the U.S. Tariff Commission.

Niacin and niacinamide are sold as two grades, USP and technical. The USP grade is available in drums; nicotinic acid must carry a white label for shipping.

Table 5. U.S. Production and Sales of Niacin, 1957–1964[a]

		Sales		
Year	Production, 1000 lb	Quantity, 1000 lb	Value, 1000 $	Unit value, $/lb
1957	2204	1708	4128	2.42
1958	2242	1321	2776	2.10
1959	2548	1490	2730	1.83
1960	2435	1244	1742	1.40
1961	2307	1979	2766	1.40
1962	1893	1679	2087	1.24
1963	2542	1629	1884	1.16
1964	1549	1218	1505	1.24
1965	1828	1461	1812	1.24

[a] Source: U.S. Tariff Commission, *Synthetic Organic Chemicals, U.S. Production and Sales, 1957–1964.*

Table 6. U.S. Production and Sales of Nicotinamide, 1957–1964[a]

		Sales		
Year	Production, 1000 lb	Quantity, 1000 lb	Value, 1000 $	Unit value, $/lb
1957	825	544	2003	3.68
1958	715	613	2205	3.60
1959	1023	671	2143	3.19
1960	580	501	1462	2.92
1961	724	649	1655	2.55
1962	706	548	1346	2.46
1963	783	773	1456	1.88
1964	802	757	1369	1.81
1965	889	858	1754	1.92

[a] Source: U.S. Tariff Commission, *Synthetic Organic Chemicals, U.S. Production and Sales, 1957–1964.*

Table 7. U.S. Production and Sales of Niacin, Niacinamide, and Niacinamide Hydrochloride, 1952–1956[a]

		Sales		
Year	Production, 1000 lb	Quantity, 1000 lb	Value, 1000 $	Unit value, $/lb
1952	2122	1567	5675	3.62
1953	1893	1660	6037	3.64
1954	2206	1632	5812	3.56
1955	2318	1961	6344	3.24
1956	2721	2098	6711	3.20

[a] Source: U.S. Tariff Commission, *Synthetic Organic Chemicals, U.S. Production and Sales, 1952–1956.*

Analytical and Test Methods

The identification of niacin and niacinamide can be easily accomplished by the following spot test (48): place a drop of solution on filter paper, spray the zone with a 1.5% alcoholic solution of *o*-tolidine, and place the paper in an atmosphere of cyanogen bromide, generally a large-mouthed flask containing crystals of this reagent (poison!), keeping it tightly closed all the time in the cold. As little as 1 γ can be detected when a red spot appears. Niacinamide and other pyridines also give a positive test, but isonicotinic acid gives a purple color (151). This reaction has been used to distinguish the human tubercle bacillus from the bovine type (48), as the latter does not synthesize appreciable amounts of niacin.

When large samples are available, the *U.S. Dispensatory* (49) suggests melting 10 mg of the sample plus 20 mg of 2,4-dinitrochlorobenzene by gentle heating. The melt is cooled and 3 ml of alcoholic potassium hydroxide are added. A deep red color is a positive test. In an alternative procedure copper sulfate is added to a neutral solution of the acid. A positive reaction is shown if a blue precipitate forms gradually.

For quantitative determinations various procedures are recommended, many of them using the classical König reaction which condenses the niacin (or other pyridine derivatives) with cyanogen bromide and some aromatic amine, such as *p*-aminopropiophenone (50), 1-napthylamine (51), aniline, and many others. Polarographic determinations (52) are also recommended.

However, the most reliable method, because it supplies not only chemical but also biological data, is the microbiological assay, using *Lactobacillus arabinosus* or a similar organism (53–55). The growth may be estimated by turbidity reading or by titration of the lactic acid formed.

Biochemical Aspects

A nutritional deficiency of niacin results in a complex of symptoms called pellagra in humans and canine blacktongue in dogs. Pellagra is characterized by dementia, diarrhea, and dermatitis and has been correlated with warm climates and maize diets. The association of pellagra with a maize diet puzzled investigators until a relationship between tryptophan and nicotinic acid was established (56,57). It was found that a low growth rate induced in rats by a corn diet deficient in nicotinic acid could be returned to normal by the addition of 1 mg nicotinic acid or 50 mg L-tryptophan to each 100 g of diet. Pellagra appears, therefore, when the diet is deficient in both nicotinic acid and tryptophan.

Nicotinic acid is unusual among the vitamins in that its biochemical significance was known before its nutritional importance was assessed. Harden and Young established in 1904 that a dialyzable cofactor was required for yeast extracts to convert glucose to ethanol (58). This cofactor has since been shown to be NAD.

Warburg and Christian in a study of the oxidation of glucose-6-phosphate by mammalian erythrocytes established the requirement for a similar dialyzable cofactor which could replace the Harden and Young factor (59). This coenzyme was composed of 1 mole of nicotinamide, 1 mole of adenine, 2 moles of pentose, and 3 moles of phosphate. Enzymes have been isolated which catalyze the interconversion of NAD to NADP in the presence of adenosine triphosphate (ATP) (60).

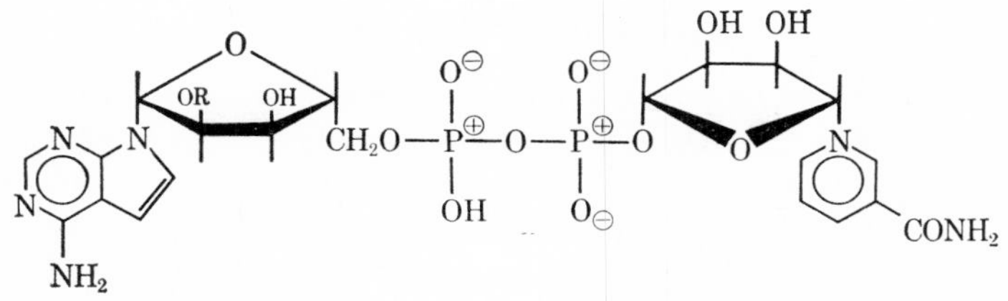

Fig. 1. Structure of nicotinamide adenine dinucleotide, NAD (where R = H) and nicotinamide adenine dinucleotide phosphate, NADP (where R = PO₃H₂).

Approximately half of all oxidation-reduction reactions are catalyzed by dehydrogenases utilizing either NAD or NADP (see Fig. 1) as coenzymes. These pyridine–nucleotide-dependent enzymes are typically highly specific for either NAD or NADP.

Oxidized NAD has an absorption maximum near 260 nm, which is the absorption of the adenine and nicotinamide portions of the molecule. However, upon reduction, the absorbance at 260 decreases and a new band appears with a maximum at 340 nm. This band is characteristic of dihydronicotinamide and indicates that the nicotinamide moiety is the site at which reduction takes place. This reduction was demonstrated to occur at the 4-position of the nicotinamide moiety (61). On reduction to NADH, a new asymmetric center is produced and the stereospecificity of the introduction of the additional hydrogen has been shown to be specific for a variety of enzymes. Thus, NAD- and NADP-linked dehydrogenases can be classed as either A or B forms with respect to their requirement for either the α or β stereoisomer of reduced nicotinamide adenine dinucleotide. Using deuterium-labeled reduced NAD⁺, Vennesland and Westheimer were able to show that alcohol dehydrogenase and lactic dehydrogenase both require the α-form of NADH (62). There is only one known instance of an enzyme which can exchange hydrogen on both sides of the nicotinamide moiety. This enzyme is lipoyldehydrogenase. Stereospecificity is exhibited not only with respect to the coenzyme; but, at least in the case of alcohol dehydrogenase, it is also exhibited with respect to the hydrogen at the 1-position of ethanol. Thus, at least in this one case, stereospecificity with respect to substrate and coenzyme is present. In a formal sense, reactions dependent on NAD and NADP involve the transfer of a hydride ion, that is, a hydrogen nucleus possessing two electrons between the substrate and the coenzyme, as shown below.

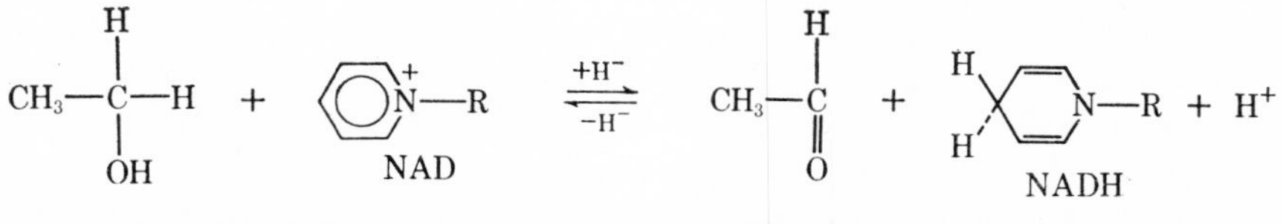

where R = portion of NAD molecule that does not participtate in the reaction

Whether in fact a transfer of the hydride ion occurs or, the alternative, transfer of a hydrogen ion succeeded by transfer of an electron in a second step, is not known. However, Westheimer has studied several model reactions which strongly suggest that the reduction occurs by the transfer of a hydride ion, rather than by a free radical route.

Biosynthesis of NAD and the Pyridine Nucleotide Cycle. Scheme 1 summarizes the synthesis, breakdown, and resynthesis of NAD. Early workers proposed the biosynthesis of NAD from nicotinamide, ribose-1-phosphate, and ATP (63). However, a

Scheme 1.

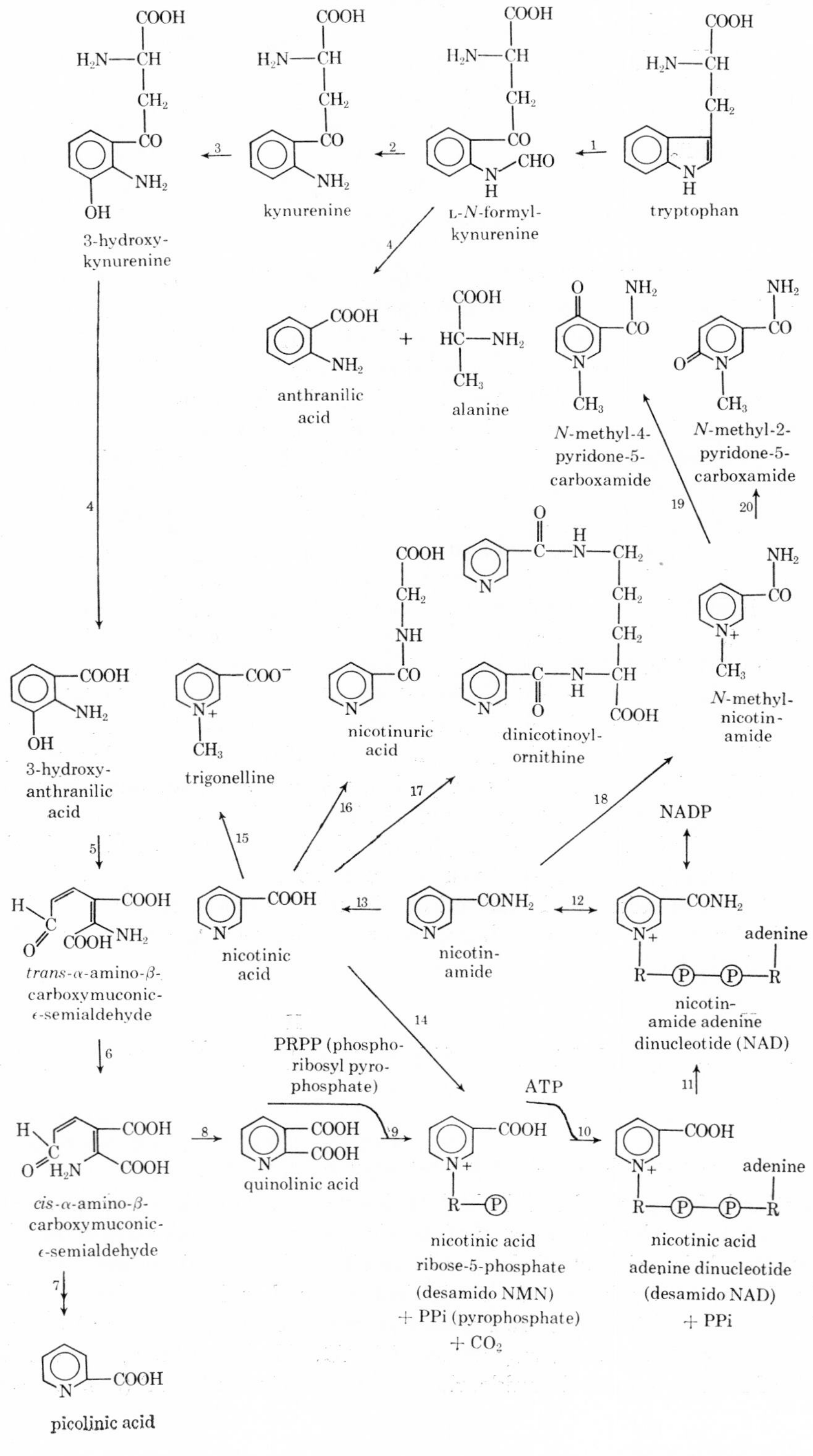

3-hydroxy-kynurenine

kynurenine

L-*N*-formyl-kynurenine

tryptophan

anthranilic acid

alanine

N-methyl-4-pyridone-5-carboxamide

N-methyl-2-pyridone-5-carboxamide

3-hydroxy-anthranilic acid

trigonelline

nicotinuric acid

dinicotinoyl-ornithine

N-methyl-nicotin-amide

NADP

trans-α-amino-β-carboxymuconic-ε-semialdehyde

nicotinic acid

nicotin-amide

nicotin-amide adenine dinucleotide (NAD)

PRPP (phospho-ribosyl pyro-phosphate)

ATP

adenine

cis-α-amino-β-carboxymuconic-ε-semialdehyde

quinolinic acid

nicotinic acid ribose-5-phosphate (desamido NMN) + PPi (pyrophosphate) + CO₂

nicotinic acid adenine dinucleotide (desamido NAD) + PPi

adenine

picolinic acid

much more physiologically significant synthetic mechanism was discovered in human erythrocytes where nicotinamide and phosphoribosylpyrophosphate condensed to yield nicotinamide mononucleotide and pyrophosphate (64). However, this reaction was discovered to require a high concentration of nicotinamide and the high apparent K_m, therefore, made it unlikely that this reaction could be significant in the synthesis of NAD. Subsequently, it was discovered that nicotinic acid could be converted to nicotinate mononucleotide (NMN) and that the enzyme had a very strong affinity for the nicotinic acid. These observations led to the conclusion that nicotinate was the precursor of choice for the synthesis of NAD (Preiss-Handler pathway) (65). However, it has recently become clear that this, in fact, represents a salvage pathway, permitting nicotinate to be recycled to form NAD. The de novo biosynthesis appears to result from the production of quinolinic acid from either tryptophan (66) or aspartate and

Table 8. Key to Reactions in Scheme 1

Reaction	Enzyme	Organism
1	tryptophan pyrrolase	mammals, *Neurospora*, *Pseudomonas*
2	formamidase	mammals, *Neurospora*, *Pseudomonas*
3	kynurenine-3-hydroxylase	mammals, *Neurospora*, *Pseudomonas*
4	kynureninase	mammals, *Neurospora*, *Pseudomonas*
5	3-hydroxyanthranilic oxygenase	mammals, *Neurospora*, *Pseudomonas*
6	cis-trans isomerase[a]	mammals, *Neurospora*, *Pseudomonas*
7	nonenzymatic two-step condensation	cat livers
8	probably no enzyme involved, fast reaction	
9	quinolate-phosphoribosyl pyrophosphate transferase	
10	NAD-pyrophosphorylase	red blood cells
11	NAD-synthetase	mammals
12	NAD-ase	mammals
13	nicotinamide desamidase	mammals, birds, insects, bacteria
14	nicotinate-phosphoribosyl transferase	red blood cells, yeast
15	transmethylase	wheat, barley, corn, *Torula*
16	nicotinuric acid synthetase	mammals
17	not known	birds
18	nicotinamide methyl transferase	mammals
19	xanthine oxidase	mammalian liver
20	aldehyde oxidase	mammalian liver

[a] Name not commonly accepted.

glycerol, with the condensation of phosphoribosylpyrophosphate to yield nicotinate mononucleotide. The enzyme responsible, quinolinate-phosphoribosyl pyrophosphate transferase, was first observed in liver homogenate by Nishizuka (67). The discovery of a hepatic nicotinamide deamidase permits closure of the cycle from nicotinamide to nicotinate (68). There remains, however, some question as to whether NAD biosynthesis occurs via nicotinic acid under physiological conditions, since most studies have occurred under nicotinamide challenge conditions, ie high concentrations. However, when high concentrations of nicotinate were used in perfused rat-liver experiments it was concluded that nicotinic acid is converted to NAD solely through the Preiss-Handler pathway and that little or no deamidation of nicotinamide occurs in perfused liver (69). The nicotinamide was methylated under these conditions. In addition, there are kinetic parameters of nicotinamide deamidase which apparently make it difficult to accept it as operating in the biosynthesis of NAD at the physiological levels of nicotinamide (70).

It seems, however, that the question of whether nicotinate or nicotinamide is the more adequate precursor of NAD is made less important by the fact that the de novo synthesis appears not to involve either free nicotinate or free nicotinamide but rather occurs from tryptophan through quinolate to nicotinate mononucleotide. These observations (71) explain the early nutritional studies which indicated that dietary tryptophan in large enough quantity could abolish the nicotinic acid requirement in animals. Tryptophan has been shown to be the precursor of quinolinic acid in mammals, *Neurospora*, and aerobically-grown yeast. Quinolinic acid appears to be formed from aspartate and glycerol in higher plants and many microorganisms. The reactions responsible for the formation of quinolinic acid from aspartate and glycerol have not yet been elucidated. Scheme 1 summarizes those known from tryptophan. The oxidation of tryptophan is initiated by the inducible enzyme tryptophan pyrrolase. The levels of this enzyme are elevated in mammals by the administration of α-methyltryptophan or glucocorticoids. The product L-N-formylkynurenine can form kynurenine or anthranilic acid. In mammals, anthranilic acid is excreted as an end product in the urine. Kynurenine forms 3-hydroxykynurenine which then can form 3-hydroxyanthranilic acid. Some important metabolic relationships of nicotinic acid and nicotinamide to tryptophan and quinolinic acid are summarized in Scheme 1. Table 8 gives the key to these reactions. The first evidence of nicotinic acid synthesized by 3-carbon plus a 4-carbon precursor was demonstrated in *E. coli* (72) and a similar mechanism was reported for *M. tuberculosis* (72a). A hypothesis (73) for the biosynthesis in higher plants of nicotinic acid from a 3-carbon plus a 4-carbon precursor is summarized in Scheme 2 (73a).

Scheme 2.

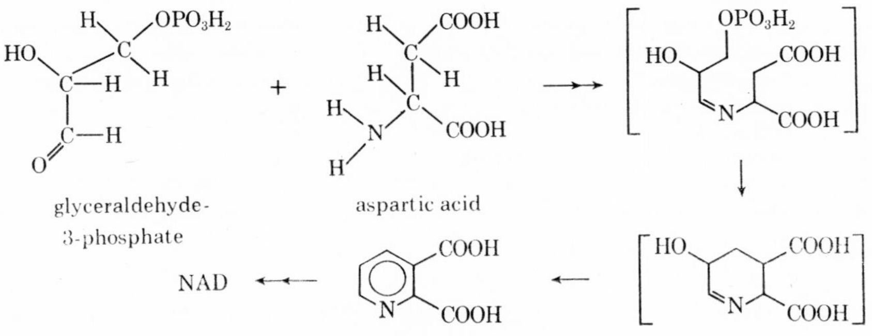

glyceraldehyde-3-phosphate

aspartic acid

NAD

quinolinic acid

The primary degradation product of NAD in several bacterial strains has been shown to be nicotinamide (74). In mammals, a major metabolite of nicotinic acid has been shown to be N-methyl-4-pyridone-5-carboxamide (75). It has been proposed that 4-pyridone-5-carboxamide arises in vivo from N-methyl-nicotinamide, which is formed from nicotinamide on methylation by S-adenosyl-L-methionine. A subsequent oxidation introduces oxygen derived solely from water. Some species appear to form 2-pyridone-5-carboxamide in preference to the 4-derivative as, for example, in the rabbit, and these species differences are being investigated (76). The synthesis of 2-pyridone-5-carboxamide appears to be catalyzed by aldehyde oxidase (77).

6-Hydroxynicotinic acid and 6-hydroxynicotinamide have been reported as minor excreted metabolites of nicotinate or nicotinamide. Their physiological significance has not been determined (78). These metabolites are excreted in the urine by rats. There are several aspects of nicotinamide metabolism that are currently under investigation. Gholson has noted the widespread occurrence of the rapid breakdown and resynthesis of NAD from nicotinamide. This recycling phenomenon has an unknown function (79). It is suggested that there may be some relationship between the production of nicotinamide and the nicotinamide–adenine–dinucleotide transglycosidases from mammalian nuclei (80). These enzymes catalyze a direct polymerization of the ADP–ribose portion of NAD with the simultaneous release of nicotinamide (80). These studies have also indicated that an enzymic adenosine–diphosphate (ADP) ribosylation of histone can occur (81).

In addition to the main pathway for the biosynthesis of NAD in erythrocytes, liver, and some microorganisms, there is evidence that, for instance, in Ehrlich ascites tumor cells and rat mammary tissue, this path may not be used, and that the incorporation of nicotinamide into NAD occurs via the formation of nicotinamide mononucleotide (NMN) through the action of NMN-pyrophosphorylase, with the subsequent NMN-adenyl transferase adding to the NMN. These studies also indicate that the tryptophan pathway, ie the quinolinic acid pathway, is inactive in rat mammary glands (82).

Toxicity

Large amounts of niacin are toxic and, with susceptible persons, elevated niacin doses may cause flushing, burning, stinging, and a sensation of heat on the skin within half an hour. One hundred times the therapeutic dose of niacin is fatal. Apparently niacinamide does not produce such effects. When diluted solutions of niacin or niacinamide are injected, they cause a quickening of respiration and, thereby, a reduced air intake, which in turn causes restlessness. These symptoms pass after 1–3 hr (83).

The LD_{50} of sodium nicotinate for rats is 7 g/kg orally and 5 g/kg subcutaneously. Minimal lethal doses, in g per kg, were 2.2–3.5 for piglets; 2.0–3.2 for mice; 1.6–2.5 for rabbits; and 3.4–5.0 for lambs.

Lower values were obtained for niacinamide. Generally speaking, toxic quantities of niacin correspond to 70–100 times the therapeutic doses, and 200–400 times the nutritional dose for cattle feeding.

Abnormal carbohydrate metabolism was reported in patients receiving a large dose of niacin for prolonged periods as a hypocholesterolemic agent. Apparently it aggravates diabetes and causes abnormal liver function (84).

Uses

The obvious use for niacin (the antipellagra vitamin) is the cure or prevention of pellagra. This disease has been disappearing quickly with the use of enriched diets. "Enriched" products contain added nutritional factors like niacin (or niacinamide), thiamine, riboflavin, or iron. The amounts to be added are governed by Federal standards (85) (see Table 9), as well as state laws.

The average requirement for humans is 5 mg/1000 cal and the usual prophylactic dose is 25 mg of the acid or its amide. In acute pellagra up to 1000 mg of the amide has been given daily in doses. The normal therapeutic dose is about half of that, and aluminum nicotinate is one of the preferred forms of administration (86). Sometimes niacinamide is used, as it does not possess the vasodilatory effect.

Besides its use against pellagra, niacin has proved to be an excellent cure against high cholesterol levels in the blood and organs. In rabbits niacin has shown a selective action against cholesterol deposits in the heart, liver, and blood vessels (87); in dogs it acted as a powerful anti-atherogenic drug, producing 35% less average cholesterol than the controls (88); and in humans, the total cholesterol decreased about 25% when aluminum and magnesium nicotinates were administered (89). Presumably the effect of niacin is to enhance the oxidation of this steroid by liver mitochondria (90).

Many drugs have been prepared to administer niacin in a better and more active form, in order to combat hypercholesterolemia. Among them we have triphenylethyl nicotinate (91) or the 2,5,6-trimethyl-3-phytyl-1,4-hydroquinone dinicotinate, which acts like niacin and has vitamin E functions (92). Studies with rabbits indicate that the effect of niacinamide is greater than that of the niacin alone (93).

Table 9. Niacin (or Niacinamide) Requirements for Enriched Cereal Products (85)

Enriched product	Required content, mg/lb	
	Min	Max
bread and rolls or buns	10.0	15.0
flour	16.0	20.0
farina	16.0	20.0
macaroni products	27.0	34.0
noodle products	27.0	34.0
corn meals	16.0	24.0
corn grits	16.0	24.0
rice	16.0	32.0

The pharmacological action of niacin includes the healing of broken blood vessels and activity as spasmolytic agent and vasodilator (94,95), or as a drug for decreasing the concentration of free fatty acids in the plasma (96). The use of niacin has also been recommended for its effect in improving the odontogenic process, producing more rapid tooth eruption, greater strength, increased calcification, and fewer caries (97).

As a simple but biologically very active anion, niacin has been utilized to administer pantothenic acid as nicotinate (98) or chloramphenicol salt, with excellent results against chloramphenicol-resistant *Shigella dysenteriae* (99). By the same token, using its vitaminic properties, it can be converted into a chemotherapeutical agent by the addition of fluorine; thus, it forms 5-fluoronicotinic acid and 5-fluoronicotinamide, which are effective against Gram-positive and Gram-negative bacteria (100).

A very considerable use of niacin is in animal feeds. The supplement of nicotinic acid greatly improves the health and growth of the animals and permits high ratios of hybrid corn in the diet.

Industrially, nicotinic acid is a very useful anticorrosion agent for metals, mainly zinc and aluminum, in contact with acids (101,102). The light metal bobbins for spinning rayon are in contact with sulfuric acid; the addition of 5% nicotinic acid diminished corrosion by 75% (103).

Finally, excellent results are obtained in zinc electroplating when nicotinic acid or related compounds are employed (104).

Nicotinic Acid Derivatives

NICOTINAMIDE

Nicotinamide (3-pyridinecarboxamide, pyridinecarboxamide, vitamin B_3, vitamin PP); also known commercially (1) as Aminicotin, Nicamindon, Nicofort, Nicotamide, etc, was isolated by two groups of investigators (105,106). It is converted to nicotinic acid on losing the amino group, occurs as widely distributed as niacin, and is usually found in combined form.

nicotinamide, niacinamide

Nicotinamide and nicotinic acid can be distinguished easily by their physical and chemical properties; they have some different physiological effects.

Physical and Chemical Properties. Niacinamide is a white crystalline powder, odorless or nearly so, with a bitter taste. It is very soluble in water (1 g in 1 ml) and ethanol (1 g in 1.5 ml), soluble in glycerol (1 g in 10 ml), butanol, amyl alcohol, and ethylene glycol, less soluble in ethyl acetate, acetone, and chloroform, and very slightly soluble in benzene and ethyl ether. It has a melting point of 131–132°C and distills at 150–160°C at 5×10^{-4} mm Hg. It shows an absorption maximum at 261.5 nm like nicotinic acid, and two pK values of 10.6 and 13.5 for the pyridyl$^+$CONH$_2$ and pyridyl$^+$CONH$_3^+$ forms, respectively. Aqueous solutions are neutral to litmus, being very stable even under autoclaving conditions; they hydrolyze easily in acid or alkaline media, forming nicotinic acid. Niacinamide in dry form is stable below 50°C.

Its occurrence in natural forms is discussed above under Nicotinic acid. The compounds can be separated from each other by extraction of the amide with organic solvents from an aqueous solution or by paper chromatography (107).

Preparation and Manufacture. *Direct Amidation of Nicotinic Acid* can be accomplished by formation of the acid chloride and exchange of the chlorine by ammonia. Thionyl chloride is the halogenating agent.

This method, however, requires the complete absence of water or moisture.

A better procedure is the amidation with molten urea at 230°C (108). There is a patent which is based on the use of equal parts of niacin and urea heated together at 180–220°C for 30 min, followed by crystallization of the crude product (109). A variation of this process is the reaction of dry ammonia gas with nicotinic acid at elevated temperature, preferably in the vicinity of the melting point (110).

Ammonolysis of nicotinic esters follows the usual method, ie commercially available esters (eg methyl nicotinate), are autoclaved with gaseous ammonia for 24 hr. The crude nicotinamide is washed with methanol to give an 81% yield (111). If the temperature is raised to 200–270°C in a sealed reactor (800 psi) up to 96–99% yield can be obtained (112).

These methods are all batchwise, and they can be made continuous by using proportionating pumps to feed an excess of liquid ammonia and the esters dissolved in methanol (113).   A good choice for nicotinic ester is the tetrahydrofurfuryl nicotinate, which reacts at room temperature (114).

A modification of these processes is to first hydrolyze the esters with sulfuric acid and methanol at 130°C just before the addition of ammonia (115).

Partial hydrolysis of 3-cyanopyridine (3-nicotinonitrile) is another method (see p. 513).   The suggestion has been made that some ion-exchange resins be used as catalysts for this hydration (116).

For all these procedures, the separation of the nonreacted nicotinic acid is the major problem.   Some purification methods use mixtures of solvents, but ammonia and cationic resins have also been employed (117).

Substituted nicotinamides can be made by heating nicotinic acid and amines for 1–3 hr at 160–180°C with polyphosphoric acid.   By this method combined molecules with mixed therapeutic properties have been made, such as the pyrazolone–nicotinamide derivatives (118).

Analysis.   The best analytical method is identification of the pyridine ring by condensation with some aromatic amine and cyanogen bromide (see also p. 534). This produces a typical color and is the well-known König reaction, the basis for the USP method (see below).   The acid and the amide can be distinguished by their melting point, which are 100°C apart.

The USP method (119) suggests dissolving 2 mg of nicotinamide in 2 ml of water, adding 1 drop of ammonia solution plus 5 ml of cyanogen bromide solution and 2 ml of sulfanilic acid solutions.   A golden yellow color indicates a positive test.

An alternative method uses 20 mg of the amide plus 5 ml of sodium hydroxide, heating the mixture, and testing for the odor of ammonia.   In another procedure, which permits identification and quantitative determination of the niacinamide content of drugs, the Dragendorff reagent is used (120).

Small amounts of the amide can be determined by measuring the absorbancy at 426 nm (121) or by microbiological methods using *Lactobacillus* (122,123).

Biological Aspects and Uses.   Biologically, niacinamide has exactly the same function as the acid (see also Table 4) and, from the medical point of view, the amide is preferred for the treatment of pellagra as this substance has no vasodilatory effect. Nicotinamide is an important precursor for NAD formation and its addition to tissue cultures produced a 150% increase in NAD biosynthesis (124) and a sevenfold rise in the NAD content of livers from treated cats (125).

Niacinamide has been recommended for the treatment of tuberculosis; mixed with pantothenic acid and resorcinol it has been used to treat some scalp infections (126). The derivative 5-fluoronicotinamide has been tested against some types of leukemia, and nicotinamide is used to neutralize some side effects caused by the fluoro derivative in human brain (127).

3-ACETYLPYRIDINE

β-Acetylpyridine (3-acetylpyridine or methyl-3-pyridyl ketone) is a liquid with a peculiar odor, melting at 13–14°C, and boiling at 220°C at 760 mm Hg or 90–92°C at 5 mm Hg.   3-Acetylpyridine is soluble in water, ethanol, and acids.

3-acetylpyridine

It may be prepared by dry distillation of a mixture of calcium nicotinate and calcium acetate, recovering the ketone as the phenylhydrazone. An alternative method uses the condensation of ethyl nicotinate and ethyl acetate in the presence of sodium ethoxide, followed by hydrolysis.

Biologically, 3-acetylpyridine is an antagonist for niacin and produces vitamin deficiency in dogs (128). When this substance is administered to mammals, an exchange reaction takes place in the molecule of NAD, forming the analog compound APAD (acetylpyridine adenine dinucleotide), as shown below:

$$\begin{matrix} \text{nicotinamide} \\ | \\ \text{adenine} \\ \text{dinucleotide} \end{matrix} + \text{3-acetylpyridine} \rightarrow \text{nicotinamide} + \begin{matrix} \text{3-acetylpyridine} \\ | \\ \text{adenine} \\ \text{dinucleotide} \end{matrix}$$

This analog produces bad effects in the central nervous system of rats. In vitro the analog and its phosphate react in a slower fashion than the corresponding coenzyme (129–132). The APAD has some important uses in enzymology, being a useful tool in discriminating between multiple forms of dehydrogenases (133–135). It is employed also for the determination of substrates (136). This "false" coenzyme can be detected in small amounts by a fluorometric method (137).

The elimination of 3-acetylpyridine by the body proceeds through a series of reactions, giving 1-(3-pyridyl)ethanolglucuronide and some pyridones, the main products of excretion (138).

PYRIDINE-3-SULFONIC ACID

This compound is an intermediate in the preparation of niacin by the sulfonation of pyridine. It is a white crystalline substance, readily soluble in alkaline solutions, and melts at 357°C. The easiest way to prepare this acid is by sulfonation of pyridine with sulfuric acid containing 20% sulfuric anhydride (SO_3) and mercuric sulfate at 220–240°C (139).

Some studies indicate that pyridine-3-sulfonic acid is also an antagonist for niacin, but in other cases, because of difficulties in permeability, it is considered an inert substance. It is also an antibacterial drug, having produced good results in inhibiting the development of dental caries without alteration in the diet of rats (140), and having shown synergistic activity with 6-mercaptopurine and indole-3-propionic acid against *Mycobacterium tuberculosis* (141).

ISONICOTINIC ACID

Isonicotinic acid (acidum isonicotinicum, γ-pyridinecarboxylic acid, 4-pyridinecarboxylic acid, γ-picolinic acid) is a position isomer of nicotinic acid.

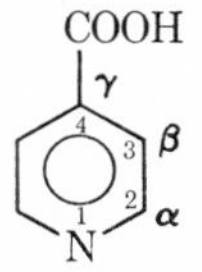

isonicotinic acid,
(4-pyridinecarboxylic acid)

It forms white crystalline platelets with an unusually high melting point of 319°C and sublimes at 260°C under 15 mm Hg pressure. Isonicotinic acid is scarcely soluble in water (0.52 g/100 ml at 20°C) and insoluble in benzene, ethyl ether, or boiling alcohol.

The pH of its saturated solution is 3.6. Isonicotinic acid has an absorption maximum at 261 nm, and a minimum at 234 nm.

The general methods of preparation resemble those of nicotinic acid, and often a mixture of both isomers is obtained which can be separated easily (142). The simplest procedure is the oxidation of 4-picoline with potassium permanganate in alkaline solution. However, boiling in aqueous sulfuric acid at 45–60°C with no more than 10% excess of potassium permanganate for 1 hr gives an 85–90% yield (143,144).

Another method uses copper nitrate as the oxidizing agent (145). Lutidine (2,6-dimethylpyridine) can be oxidized with nitric acid at 200°C in the presence of sulfuric acid forming a mixture of compounds with isonicotinic acid as the main product (146). The high-boiling fraction from the light pyridine bases can be employed, using sulfuric acid and selenium with a 25% yield of isonicotinic acid (147). If γ-picoline, p-formaldehyde, and sodium hydroxide are refluxed for 48 hr and dilute nitric acid is added dropwise, isonicotinic acid is formed (148).

As indicated before (see p. 516), substituted pyridines can be oxidized by electrochemical procedures to produce the corresponding carboxylic acids. In this case the substance of choice is the methylol derivative of 4-picoline (149).

Mixtures of nicotinic and isonicotinic acid can be separated by difference in solubilities of the free acids and their calcium and sodium salts (150).

Isonicotinic acid is commercially available in 1-, 5-, 50-, and 100-lb containers.

Isonicotinic acid is the closest relative to isonicotinhydrazide which makes it an important intermediate. Its ability to form analogous coenzymes which function in the cell has been demonstrated. In *Lactobacillus arabinosus* isonicotinic acid can replace the nicotinamide moiety of nicotinamide adenine dinucleotide; the new compound is biochemically active (151).

Cancer research has shown that a pyridine ring with a substituent in the 4-position could be a potential inhibitor of glycolysis, the main mechanism by which cancer cells derive their energy (152).

ISONICOTINIC ACID HYDRAZIDE

Isonicotinic acid hydrazide (INH, isoniazid, isonicotinylhydrazine, isonicotinoylhydrazine, 4-pyridinecarboxylic acid hydrazide, pyridine-γ-carboxylic acid hydrazide) (153), is the most commonly used antitubercular drug, being about 100 times as active in vitro as streptomycin. Its bacteriostatic action against *Mycobacterium tuberculosis* (the tubercle bacillus) is noticeable in a 1:60,000,000 dilution (154). It is usually employed in combination or simultaneously with other chemotherapeutic agents.

isonicotinic acid hydrazide,

isoniazid (INH)

This drug is water-soluble and easily absorbed through the digestive tract and could be administered orally, parenterally, or even intrapleurally. Large amounts may have toxic effects. It is easily made from ethyl isonicotinate (or 4-cyanopyridine) and hydrazine (155) (see below).

History. Isoniazid has been known since 1912, when Meyer and Mally first prepared it by heating isonicotinic acid and hydrazine together above 300°C (156). For forty years there was no interest for this compound. In 1952, isoniazid was first reported as a powerful tuberculostatic agent; this discovery was made simultaneously by three pharmaceutical companies: Farbenfabriken-Bayer, A. G., Hoffmann-La Roche, Inc., and E. R. Squibb & Sons, Inc. (157). All of them were engaged in the study of thiosemicarbazones. Interest was centered in amithiozone, and, in order to make the pyridine conjugates, it was necessary to make isonicotinic acid hydrazide which was automatically tested in vitro with surprising results.

The hydrazines, mainly phenylhydrazine, were known to be active in vitro, but their toxicity limits their use. The role of nicotinamide seemed interesting in vivo, but the good results were interpreted as vitaminic effect in spite of the fact that good results were also found in vitro (158).

Once the excellence of isoniazid as a chemotherapeutic agent was demonstrated, hundreds of studies were undertaken by different investigators from industry, hospitals, and universities to better determine its mode of action and its effects on enzymes, organs, and whole organisms; there is also a search for derivatives with less toxicity, mixed effects, or with still greater antitubercular activity (159–162).

Physical Properties. Isoniazid is a white, odorless, crystalline powder, with a melting point of 171–172°C. It is quite soluble in water (14 g in 100 ml at 25°C, 26 g at 40°C) and in ethanol (2 g in 100 ml at 25°C, 10 g in boiling ethanol), and slightly soluble in chloroform (0.1 g in 100 ml), but completely insoluble in ethyl ether or benzene. It has an absorption maximum at 266 nm and a minimum at 234 nm. Its extinction coefficient, $E_{cm}^{1\%}$, is 378 (aqueous solution).

Chemical Properties. Isoniazid is stable in the presence of atmospheric oxygen and can be autoclaved at 120°C for 30 min. Light and air may affect it slowly. However, no hydrolysis is detected chromatographically after 2 days at 37°C at pH 7.2; it retains its full activity after 15 days at 37°C (163). In water solutions it acts as a weak acid. The pH of a 1% solution is about 6. Isoniazid is not stable in liquid media in the presence of ferrous or manganous ions, especially under alkaline conditions.

When isoniazid is heated in the presence of anhydrous sodium carbonate, it liberates pyridine. It can be oxidized with various reagents, such as iodine, perchloric acid, phosphomolybdic acid, and potassium ferricyanide. Oxidation is employed for its quantitative determination.

Preparation and Manufacture. The basic method of preparation is the condensation of hydrazine with a 4- or γ-substituted pyridine. Practically all the known methods are only minor variations of this. The most frequently employed or better known are shown below.

Condensation of Isonicotinic Acid and Hydrazine. This is obviously the simplest method. The main problem is the removal of water, which is accomplished by the use of azeotropic mixtures and distillation (164–166).

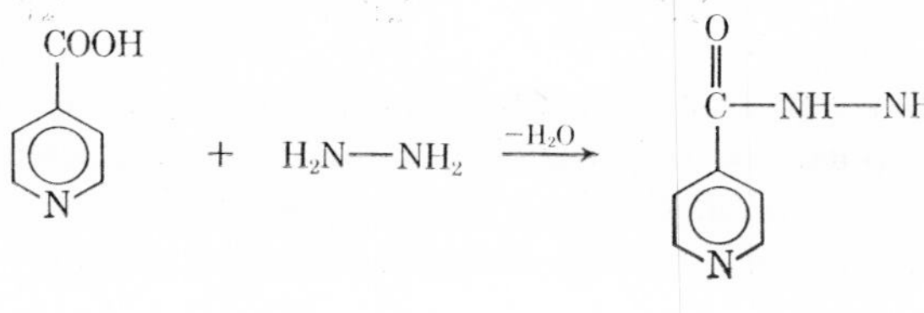

Hydrolysis of Isonicotinoyl Esters (or Nitriles), followed by Condensation with Hydrazine (155).

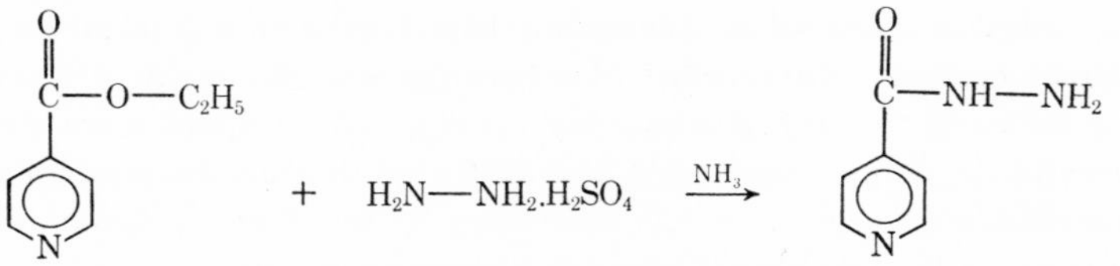

This reaction can be carried out in the presence of ammonia at room temperature for 7 days with a 76% yield (167).

Oxidation of 4-Alkylpyridines, Carboxylic Chlorination, and Condensation with Hydrazine.

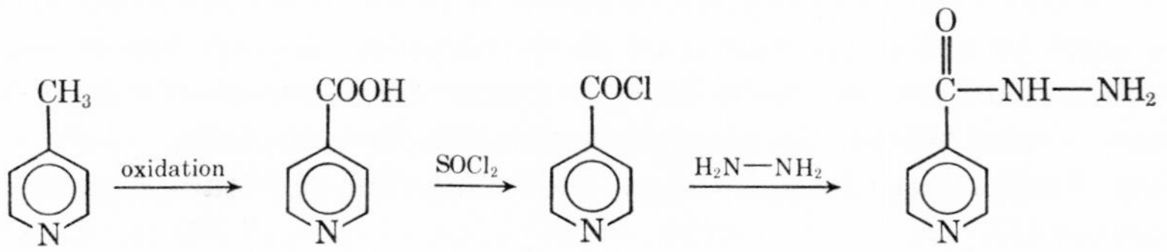

γ-Picoline and 70% sulfuric acid are heated at 60°C. Manganous dioxide is added in portions and heating is continued at the same temperature with agitation for 5 hr. The reaction mixture is diluted with water, alkalized with sodium carbonate, and the isonicotinic acid is precipitated with hydrochloric acid. The dry isonicotinic acid is mixed with thionyl chloride and refluxed for 4 hr; the excess thionyl chloride is removed by distillation. The residue is treated with hydrazine in anhydrous benzene, refluxed for 6 hr, distilled to dryness, and recrystallized from ethanol (168).

There are minor variations of this procedure, such as treating the isonicotinoyl chloride with ethanol and sodium hydroxide to first produce ethyl isonicotinate, which reacts with hydrazine in ethanol at 100°C to form isoniazid (169).

Oxidation of (Polyalkyl)pyridines and Ammonification. In this process, 2,4-dimethylpyridine is oxidized with sulfuric acid and selenium by heating for 1 hr at 270–310°C, cooling, and adding methanol. The mixture is heated at 160°C for 6 hr, poured on ice, and neutralized with ammonia. Three compounds are extracted from the oily product with ethyl ether: methyl isonicotinate, isonicotinamide, and isonicotinic hydrazide (170). The reactions are as follows:

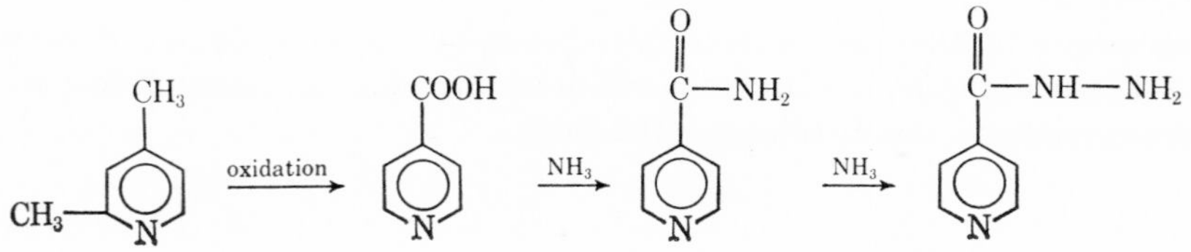

In any of the above-mentioned processes, slow cooling and careful washing techniques are of great importance for producing an odorless and colorless compound (171).

Analysis. *Qualitative Identification.* Besides the physical and chemical properties already given, *USP XVII* (172) recommends the following tests:

(a) Mix 50 mg of the sample with 500 mg of anhydrous sodium carbonate, heating the mixture in a test tube over a small flame. The presence of pyridine odor corresponds to a positive test.

(b) To a solution containing 100 mg of the compound in 5 ml of water add a hot solution of 100 mg of vanillin in 10 ml of water, rubbing the inside of the vessel with a glass rod and allowing the solution to stand for 1 hr. A positive test is the formation of a yellow precipitate which melts at 228–231°C after filtration and recrystallization from 5 ml hot 70% ethanol.

Also, it is possible to recognize isonicotinic acid or its hydrazide by placing a drop of its solution on a piece of filter paper, spraying the spot with 1% o-tolidine in water with 1% acetic acid, and introducing the paper into a flask with some crystals of cyanogen bromide. A violet-red (magenta) color is a positive test (173).

Quantitative Determination. The methods for the quantitative estimation of isoniazid follow several lines:

(a) When sufficient amounts of more or less pure substance are present, the best procedure is the quantitative oxidation of the sample. A large variety of oxidizing agents have been employed, including potassium ferricyanide after acid hydrolysis (174); phosphomolybdic acid (175); iodine in alkaline medium, where each ml of $0.1N$ iodine corresponds to 3.429 mg of isoniazid (176); perchloric acid (177,178); and sodium nitrite (179).

(b) Spectrophotometric methods, based on specific absorbancy at 298 nm when in alkaline medium, are used. The limit for determination is $0.5 \, \gamma$/ml (180).

(c) Polarographic determinations are carried out at pH 6.52 and give good results at concentrations from 10^{-3} to $10^{-4}M$ (181).

(d) Special procedures designed for biological fluids are available which include column fractionation, the use of enzymes, etc, before color development. This gives good results for urine analysis (182). According to Kelly and Poet (183), isoniazid or its derivatives are extracted from alkaline plasma with a mixture of isoamyl alcohol, ammonium sulfate, and ether, and after further purification, the isoniazid is determined by spectrophotometry.

(e) Microbiological determinations (184) use a susceptible organism like *Mycobacterium tuberculosis* H₃₇Rv, *Mycobacterium phlei*, or *Mycobacterium butyricum* in a diffusion-plate assay (185). Biological fluids may be analyzed. The limit for determination is 5 µg/ml.

Biological Aspects. Isoniazid is an excellent therapeutic agent against the tubercle bacillus (*Mycobacterium tuberculosis*); it shows a tuberculostatic effect in vitro from 0.015 to 0.25 µg/ml. The effect is selective for acidfast bacteria and can be demonstrated with the electron microscope (186).

Some strains of *Mycobacterium* develop resistance to isoniazid both in vivo and in vitro. This can be prevented by the use of streptomycin, this action being reciprocal. However, the concentrations employed never should be subeffective. *p*-Aminosalicylic acid (PAS) can also be used as the companion for INH instead of streptomycin.

From 150 to 300 mg of isoniazid daily plus 1 g of streptomycin twice a week is normal therapeutic practice; this produces better results, as interpreted by x-ray studies, clinical observation, and bacteriological analysis, than isoniazid alone. The most recommended combination is 200 mg of isoniazid plus 10–12 mg of *p*-aminosalicylic acid daily. The use of the three drugs simultaneously is not advisable because of the remote possibility of creating triple-resistant organisms. However, in desperate cases the three substances may produce some effective action against the disease. Therapy should last from 12 to 18 months in order to produce radical cures.

Isoniazid is absorbed rapidly and 1–6 hr after its administration it can be demonstrated in blood or urine, also passing the placental barrier.

Apparently isoniazid is acetylated in the body, the acetyl group being attached to the hydrazine end. This intermediate is hydrolyzed, forming isonicotinic acid and acetylhydrazine, which is finally converted into 1,2-diacetylhydrazine. The intermediary compounds and isoniazid as such can be found in human urine. The second acetylation step is missing in the dog (187). Other reports indicate that isonicotinoylglycine or diisonicotinic hydrazide are also catabolic products derived from isoniazid and are eliminated through the urine (188).

Mechanism of Action. In hundreds of laboratories a tremendous effort is made to elucidate the mode of action of isoniazid. Although there are literally thousands of research projects with the same aim, it can safely be said that the mode of action has not yet been completely established.

Radioactive isoniazid has been shown to combine only with susceptible bacteria not reacting with resistant organisms (189). Small amounts of isoniazid are bacteriostatic (2 μg/ml in 2 hr) and larger quantities are required for bactericidal action (190).

Pyridoxal or pyridoxamine seem to be the most effective agents antagonizing the action of isoniazid in vitro (191), but on a smaller scale ketones and ketoacids exert some antagonistic action.

There are many reports of activities against a great variety of enzymes; this could simply be a nonspecific adsorption of the substance by the bacterial cell, which then suffers morphological and physiological changes. This may result from a reaction between the drug and pyridoxal in vivo, with the formation of a Schiff base (192).

It has been completely established (193–195) that there is a close relationship among the following three facts: (a) catalase deficiency, (b) resistance to isoniazid, and (c) less virulence. Genetic studies have demonstrated that there is no correlation between resistance to isoniazid and to streptomycin (196).

It has been postulated that the hydrazine part of the molecule forms a complex (chelate) with iron or copper which is related to an enzyme system containing pyridoxal. This statement is supported by the findings that some bacterial enzymes (not from *Mycobacterium*), which are isoniazid-resistant, become susceptible to the isoniazid–copper complex (197). Isoniazid could be adsorbed on the catalase before penetrating and attacking other enzyme systems of the bacterial cell. There could be some incidental action against the catalase of the host, as its activity was seen to decrease in tuberculous patients receiving isoniazid (198). Therefore, adsorption and penetration of isoniazid should be more difficult if there is little or no catalase in the bacterial cell.

There are many changes discovered in the bacteria related to its lipid and amino acid contents, its nucleic acids complexes, etc. Some obvious changes observed in *Mycobacterium tuberculosis* due to the action of isoniazid include the following:

Increase in:

synthesis of leucine, isoleucine, and phenylalanine (199),
incorporation of ^{35}S (200),
NADase content (201), and
^{14}C incorporation into leucine (202).

Decrease in:

> synthesis of arginine, glutamic acid, lysine and threonine (199),
> incorporation of ^{32}P (200),
> nucleic acid synthesis (203),
> acidfastness (203),
> NAD content (201),
> lipid synthesis from acetate and total lipid content (204,205),
> incorporation of ^{35}S and ^{14}C into protein (201), and
> dehydrogenase activity (206).

From the viewpoint of the tuberculous patient, there are some bad effects due to side reactions derived from the relationship between pyridoxine and isoniazid. Some transaminations could be impaired (207–209), epileptiform convulsions could develop if the diet is pyridoxin deficient (210), and special treatment should be given to tuberculous diabetics, as the isoniazid acts apparently at the hexokinase level (211).

Toxicity. The toxicity of isoniazid in rats is LD_{50} 340–650 mg/kg orally (1,154) and 300 mg/kg subcutaneously (154).

The most noticeable effect of toxic action is the presence of neurotoxic symptoms including peripheral neuritis, muscular paresis of arms and legs, etc. This occurs in about 10% of patients receiving 10 mg/kg of body weight. Special caution must be taken in case of renal damage. There are some reports indicating immunological reactions during the use of isoniazid, but they can be considered as minor effects when compared with those of other drugs.

The neurotoxic action has been related to a pyridoxine deficiency, and it has been shown in vitro that pyridoxine neutralizes the INH effect against *M. tuberculosis* (191).

Ingestion of 5 g of this drug produced vomiting, convulsions, coma, oliguria, and death after 70 hr (212). α-Ketoglutarate or pyruvate, or mixtures of these, are the best antidotes for isoniazid poisoning (213).

Some studies were made in order to investigate a possible oncogenic effect under continuous use. Mice injected with 9, 30, or 90 mg for 3 months did not show any reaction. However, investigations after two years showed more tumors were present than in the controls (214).

Uses. Isoniazid is one of the universal drugs used to fight all forms of tuberculosis. It should always be used in combination with another drug, preferably streptomycin or *p*-aminosalicylic acid (PAS).

Many combinations have been prepared in order to increase the potency of isoniazid and one of them, Marsilid (1-isonicotinyl-2-isopropylhydrazine or iproniazid), was highly effective but too toxic.

An increasing number of isoniazid derivatives are being studied and even marketed finding clinical application as psychotherapeutic drugs rather than as antitubercular preparations.

NIKETHAMIDE

N,N-Diethylnicotinamide (niacin diethylamide, 3-pyridinecarboxylic acid diethylamide, pyridine-β-carboxylic acid diethylamide, nicotinic acid diethylamide) is a synthetic drug which finds application as a stimulant of the central nervous system and has useful analeptic effects, helping to restore the activity of the heart and the respiratory system.   The commercial product is known as Coramine.

nikethamide; niacin diethylamide

Physical and Chemical Properties.   Nikethamide is a slightly viscous, clear, colorless or pale yellowish liquid, which looks like a crystalline solid when standing in the cold.   It has a characteristic aromatic odor and a peculiar delicate bitter taste, followed by a mild sensation of warmth.   The physical constants are given in Table 10.

Table 10. Physical Constants of Nikethamide

Constant	Value
melting point, °C	24–26
freezing point, °C	22–24
boiling point, °C	
at 760 mm Hg	296–300 (dec)
at 23–25 mm Hg	170–175
at 10 mm Hg	158–159
at 3 mm Hg	128–129
density, g/ml	
at 15°C	1.064–1.067
at 25°C	1.058–1.066
refractive index, n_D	
at 20°C	1.525–1.526
at 25°C	1.522–1.524

Nikethamide is miscible with water, ether, chloroform, ethanol, and acetone. A 25% aqueous solution (the usual therapeutic concentration) has a pH of 6.0–6.5 and the nikethamide can be precipitated by sodium carbonate solutions.

Like most pyridine derivatives, it reacts with cyanogen bromide and an aromatic amine to form a colored adduct.   The usual method is to heat the mixture for 2 min at 80°C, allowing the sample to cool and stand in the dark for 15 min and then read a sample on the spectrophotometer.   The peaks for the most common related pyridines are: pyridine, 378.3 nm; niacin, 352.4 nm; niacinamide, 392.5 nm; nicotine, 378.0 nm; and nikethamide, 393.2 nm.

Preparation and Manufacture.   Most methods employ the condensation of diethylamine and a suitable derivative of nicotinic acid.   Nicotinic acid itself may be used with a dehydrating agent such as phosphorus pentoxide (215).   A Swedish patent claims a 90–94% yield of nikethamide (216).   The reaction proceeds as follows:

This condensation is facilitated by using N,N-diethylbenzenesulfonamide as catalyst (217).

The most common procedure is based on the preparation of nicotinic acid chloride followed by condensation with diethylamine or its hydrochloride (218), as shown below.

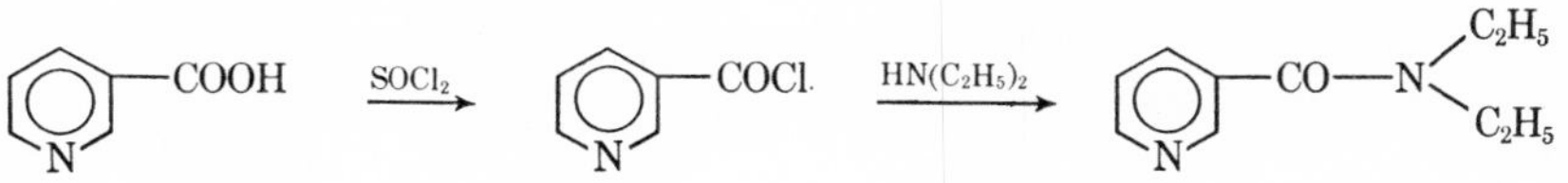

Other methods employ ethyl nicotinate or quinolinic anhydride, which loses a carboxyl group before condensing with diethylamine.

Nikethamide is sold as USP or technical grades.

Analysis. Determination of nikethamide may be accomplished by paper chromatography using butyl alcohol and hydrochloric acid as developer and making the spots visible with Dragendorff's reagent (219). The determined R_f value is 0.32.

Biological Aspects. Besides its antipellagra activity, derived from the fact that nikethamide is a diethyl derivative of the antipellagra vitamin, this drug stimulates the circulatory system and respiration. Nikethamide has a specific action against nervous-system and heart failure. It increases the vasomotor tone and blood pressure, counteracts cardiac insufficiency, and reinforces digitalis action, permitting a reduction in dose (220). Some studies indicate, however, that even though it increases coronary blood flow, it does so at the expense of mechanical efficiency, producing undesirable effects like convulsions and delayed death (221). Similar pharmacological results were obtained with isolated frog hearts (222).

Nikethamide acts in the lower nervous centers, especially the medulla and spinal cord. It produces a potent stimulating action, elevating blood sugar and diminishing intestinal contractions (223). Nikethamide has been shown to be effective as a respiratory stimulant at low but not at high altitudes (224).

Toxicity. The oral LD in rabbits is 650 mg/kg. Large doses and overdoses (ten times the normal dose) cause convulsions, respiratory failure, and death. Apnea is caused by large doses administered intravenously (225). A vesicant effect can be shown with 0.5 ml of a 25% aqueous solution injected subcutaneously in a rabbit ear. At the end of 1 hr there is a pale spot, after 3–4 hr blistering (226).

Uses. Nikethamide is a good respiratory stimulant, especially in excess depression of the central nervous system as occurs in accidents, poisoning, large doses of anesthetics, or collapse. It is not quite as effective as other drugs, such as pentylenetetrazole, bemegride (β-ethyl-β-methylglutarimide), or picrotoxin (227). The use of nikethamide is beneficial against an overdose of barbiturates or morphine and in the shock therapy of schizophrenia (228). The usual dose is 1–5 ml of a 25% aqueous solution, administered orally, intramuscularly, or intravenously.

TRIGONELLINE

This is the *N*-methylbetaine of nicotinic acid, and is also known as coffearine, caffearine, gynesine, trigonelline, etc.

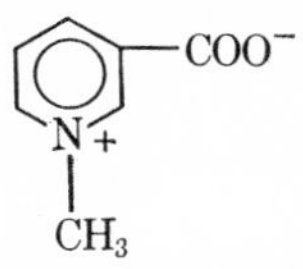

trigonelline

This compound was discovered in the seed of *Trigonella* (a leguminous plant). It is the most widely distributed of the natural pyridine derivatives, found especially in the coffee bean (see p. 512). It is pharmacologically inert and is excreted intact if given to man or dog. Trigonelline can be prepared by treating nicotinic acid with methyliodide and removing the halogen with silver.

N-METHYLNICOTINAMIDE

This compound is the amide of trigonelline and is the major urinary metabolite from niacin or niacinamide in man. It is produced in the liver, methionine being the methyl donor compound. It is oxidized in the liver to the 2-pyridone derivative (see Scheme 1). Trigonelline and *N*-methylnicotinamide do not have anti-blacktongue activity.

NICOTINURIC ACID

Nicotinuric acid (see Scheme 1), also known as nicotinoylglycine, is the detoxification product and urinary metabolite of niacin in the dog. It resembles hippuric acid (benzoylglycine) as it is a false peptide made in the liver by condensation of glycine and a cyclic acid. This product is formed in several mammals (dog, rabbit, rat, mouse, hamster), reptiles (turtle), and amphibians (frog). Birds eliminate dinicotinoylornithine (see Scheme 1), a compound very similar to ornithuric acid.

Bibliography

"Nicotinic Acid and Nicotinamide" in *ECT* 1st ed., Vol. 9, pp. 305–313, by A. P. Sachs, Consultant, and J. F. Couch, U.S. Department of Agriculture.

1. *The Merck Index of Chemical Compounds and Drugs*, 7th ed., Merck and Co., Inc., Rahway, N.J., 1960, pp. 719–720.
2. A. White, P. Handler, and E. L. Smith, *Principles of Biochemistry*, McGraw-Hill Book Co., New York, 1959, pp. 954–957.
3. E. Kodicek, R. Braude, S. K. Kon, and K. G. Mitchell, *Brit. J. Nutr.* **13**, 363–384 (1959).
4. U.S. Pat 2,584,893 (Feb. 5, 1952), W. R. Lloyd and R. Millares (to Armour Research Foundation).
5. C. A. Elvehjem, R. J. Madden, F. M. Strong, and D. W. Woolley, *J. Am. Chem. Soc.* **59**, 1767 (1937); *J. Biol. Chem.* **123**, 137 (1938).
6. Y. M. Slobodin and M. M. Gol'dman, *Zh. Prikl. Khim.* **21**, 859 (1948); *Chem. Abstr.* **43**, 6207 (1949).
7. G. Arroyave and R. Bressani, in M. Florkin and E. H. Stotz, eds., *Comprehensive Biochemistry*, Vol. 11, Elsevier Publishing Co., Amsterdam, 1963, p. 38.
8. D. Abelson, E. Parthe, K. W. Lee, and A. Boyle, *Biochem. J.* **96**, 840–852 (1965).
9. M. Y. Kalmi and K. Sohonie, *J. Nutr. Dietet. (India)* **2**, 9–11 (1965).
10. J. P. Miale, *Current Therap. Res. Clin. Exptl.* **7**, 392–400 (1965).
11. C. A. Elvehjem, *Physiol. Rev.* **20**, 249–271 (1940).
12. A. Birzu, H. Aizicovici, and V. Morosanu, *Igiena (Bucharest)* **14**, 81–84 (1965); *Chem. Abstr.* **63**, 12232 (1965).
13. R. Cravioto, G. Massieu H., J. Guzman G., and J. Calvo T., *Ciencia (Mex.)* **11**, 129–155 (1951).
14. R. Cravioto, J. Guzman G., and L. M. Suarez, *Ciencia (Mex.)* **15**, 24–26 (1955).
15. M. Florkin and E. H. Stotz, eds., *Comprehensive Biochemistry*, Vol. 13, Elsevier Publishing Co., Amsterdam, 1964, pp. 14–20.
16. A. Champagnat, *Rev. Inst. Franc. Petrole Ann. Combust. Liquides* **17**, 1372–1377 (1962); *Chem. Abstr.* **58**, 9591 (1963).
17. Indian Pat. 67,468 (Feb. 8, 1961), A. Lahiri, H. S. Rao, A. Panicker, R. Panicker, and G. N. Kulsrestha (to Council for Scientific Industrial Research); *Chem. Abstr.* **55**, 14883 (1961).

18. Polish Pat. 44,164 (Jan. 18, 1961), A. Kotarski (to Instytut Chemii Ogolnej); *Chem. Abstr.* **57**, 13779 (1962).

19. U.S. Pat. 2,510,605 (June 6, 1950), J. Berg and V. L. King (to American Cyanamid Co.).

20. B. Lipka et al., *Chim. Ind. (Paris)* **93**, 683–686 (1965); *Chem. Abstr.* **65**, 681 (1966).

21. Polish Pat. 42,867 (March 14, 1960), E. Treszczanowicz, B. Lipka, and B. Buszynska (to Instytut Chemii Ogolnej); *Chem. Abstr.* **55**, 6502 (1961).

22. S. S. Kruglikov and V. G. Khomyakov, *Tr. Mosk. Khim. Tekhnol. Inst.* **32**, 194–200 (1961); *Chem. Abstr.* **57**, 16542 (1962).

23. U.S.S.R. Pat. 122,147 (Aug. 25, 1959), V. G. Khomyakov and S. S. Kruglikov; *Chem. Abstr.* **54**, 5701 (1960).

24. R. Lukes, J. Jizba, and V. Galik, *Collection Czech. Chem. Commun.* **26**, 3044–3050 (1961); *Chem. Abstr.* **56**, 11564 (1962).

25. F. Brody and P. R. Ruby, "Synthetic and Natural Sources of the Pyridine Ring," in E. Klingsberg, ed., *Pyridine and Derivatives*, Part I, Chap. II., Interscience Publishers, Inc., New York, 1960, pp. 475, 484.

26. U.S. Pat. 2,905,688 (Sept. 22, 1959), G. M. Illich (to Abbott Laboratories).

27. F. Binns and G. A. Swan, *J. Chem. Soc.* **1962**, 2831–2832.

28. Japan. Pat. 20,555 (Sept. 13, 1965), K. Uda, A. Sakurai, and K. Sakabibara (to Dainnippon Celluloid Co. Ltd.); *Chem. Abstr.* **64**, 2069 (1966).

29. T. E. Jordan, *Ind. Eng. Chem.* **44**, 332–333 (1952).

30. U.S. Pat. 2,436,660 (Feb. 24, 1948), M. B. Mueller (to Allied Chemical & Dye Corp.).

31. U.S. Pat. 2,476,004 (July 12, 1949), W. O. Teeters (to Allied Chemical & Dye Corp.).

32. B. F. Ustavshchikov, M. I. Farberov, A. M. Kutlin, and G. S. Levskaya, *Uch. Zap. Yaroslavsk. Teknol. Inst.* **5**, 71–75 (1960); *Chem. Abstr.* **57**, 11153 (1962).

33. Ger. Pat. 1,071,085 (Dec. 17, 1959), W. Schwarze.

34. R. C. Elderfield, ed., *Heterocyclic Compounds*, Vol. 4, John Wiley & Sons, Inc., New York, 1952, p. 259.

35. U.S. Pat. 3,261,843 (July 19, 1966), W. Swietoslawski, J. Bialek, A. Bylicki, and A. Kotarski; *Chem. Abstr.* **65**, 12177 (1966).

36. U.S.S.R. Pat. 148,411 (July 13, 1962), E. S. Zhdanovich, I. B. Kaplina, and N. A. Preobrazenskii; *Chem. Abstr.* **58**, 9031 (1963).

37. U.S.S.R. Pat. 132,225 (Oct. 5, 1960), A. A. Kondrashova, L. O. Shnaĭdman, S. V. Balatsenko, and E. S. Zhdanovich; *Chem. Abstr.* **55**, 10477 (1961).

38. U.S. Pat. 2,999,094 (Oct. 19, 1959), J. O'Brochta (to Koppers Co., Inc.).

39. U.S. Pat. 2,964,529 (Dec. 13, 1960), M. G. Sturrock, E. L. Cline, K. R. Robinson, and K. A. Zercher (to Koppers Co., Inc.).

40. Y. I. Chumakov, A. I. Mednikov, and R. I. Virnik, *Zh. Prikl. Khim.* **35**, 602–605 (1962); *Chem. Abstr.* **57**, 3403 (1962).

41. Y. I. Chumakov, L. A. Rusakova, A. I. Mednikov, and R. I. Virnik, *Metody Polucheniya Khim. Reaktivov i Preparatov, Gos. Kom. Sov. Min. S.S.S.R. Po Khim.* **7**, 79–82 (1963); *Chem. Abstr.* **61**, 5604 (1964).

42. D. J. Cram and G. S. Hammond, *Organic Chemistry*, McGraw-Hill Book Co., New York, 1964, p. 592.

43. R. C. Fuson, *Advanced Organic Chemistry*, John Wiley & Sons, Inc., New York, 1950, p. 546.

44. A. R. Katritzky and J. M. Lagowski, *Heterocyclic Chemistry*, John Wiley & Sons, Inc., New York, 1960, p. 42.

45. Ger. Pat. 1,109,683 (Feb. 4, 1954), W. G. Toland (to California Research Corp.); *Chem. Abstr.* **56**, 8536 (1962).

46. Brit. Pat. 714,664 (Sept. 1, 1954), Roche Products Ltd.; *Chem. Abstr.* **49**, 12544 (1955).

47. U.S. Pat. 2,916,494 (Dec. 8, 1959), J. O'Brochta (to Koppers Co., Inc.).

48. J. M. Gutierrez-Vazquez, *Rev. Latinoam. Microbiol. (Mexico)* **3**, 41–45 (1960).

49. A. Osol and G. E. Farrar, eds., *U.S. Dispensatory*, 24th ed., J. B. Lippincott Co., Philadelphia, Pa., 1947.

50. T. C. Appanna, *Indian J. Med. Res.* **52**, 1188–1191 (1964); *Chem. Abstr.* **62**, 8311 (1965).

51. R. C. R. Barreto and V. Lorian, *Presse Med.* **71**, 873 (1963).

52. G. P. Tikhomirova, S. L. Belen'kaya, R. G. Madievskaya, and O. A. Kurochkina, *Vopr. Pitaniya* **24**, 32–37 (1965); *Chem. Abstr.* **62**, 12978 (1965).

53. E. E. Snell and L. D. Wright, *J. Biol. Chem.* **139**, 675 (1941).

54. B. C. Johnson, *J. Biol. Chem.* **159,** 227 (1945).

55. W. L. Williams, *J. Biol. Chem.* **166,** 397 (1946).

56. W. A. Krehl, P. S. Sarma, L. J. Teply, and C. A. Elvehjem, *Science* **101,** 489 (1945).

57. W. A. Krehl, P. S. Sarma, L. J. Teply, and C. A. Elvehjem, *J. Nutr.* **31,** 85 (1946).

58. A. Harden and W. Young, *Proc. Roy. Soc. (London) Ser. B* **81,** 528 (1909).

59. O. Warburg and W. Christian, *Biochem. Z.* **242,** 206 (1931).

60. A. Kornberg, *J. Biol. Chem.* **182,** 805 (1950). T. P. Wang and N. O. Kaplan, *J. Biol. Chem.* **206,** 311 (1954).

61. F. H. Westheimer, H. F. Fisher, E. E. Conn, and B. Vennesland, *J. Am. Chem. Soc.* **73,** 2403 (1956). H. F. Fisher, E. E. Conn, B. Vennesland, and F. H. Westheimer, *J. Biol. Chem.* **202,** 687 (1953). F. A. Loewus, P. Ofner, H. F. Fisher, F. H. Westheimer, and B. Vennesland, *J. Biol. Chem.* **202,** 699 (1953). M. E. Pullman, A. San Pietro, and S. P. Colowick, *J. Biol. Chem.* **206,** 129 (1954).

62. B. Vennesland and F. H. Westheimer, "Hydrogen Transport and Steric Specificity in Reactions Catalyzed by Pyridine Nucleotide Dehydrogenases," in W. D. McElroy and B. Glass, eds., *Mechanisms of Enzyme Action*, John Hopkins University Press, Baltimore, Md., 1954.

63. J. W. Rowen and A. Kornberg, *J. Biol. Chem.* **193,** 497 (1951).

64. J. Preiss and P. Handler, *J. Biol. Chem.* **225,** 759 (1957).

65. *Ibid.*, **233,** 488, 493 (1958).

66. F. Lingen and P. Vollprecht, *Z. Physiol. Chem.* **339,** 64–74 (1964). E. F. Müller, *Chemiker-Ztg.* **85,** 145–147 (1961). H. Suomalainen, T. Nurminen, K. Vihervaara, and E. Oura, *J. Inst. Brewing* **71,** 227–231 (1965).

67. Y. Nishizuka and O. Hayaishi, *J. Biol. Chem.* **238,** 483 (1963).

68. B. Petrack, P. Greengard, A. Craston, and H. J. Kalinsky, *Biochem. Biophys. Res. Commun.* **13,** 472 (1963).

69. Y. Hagino, S. J. Lan, C. Y. Ng, and L. M. Henderson, *J. Biol. Chem.* **243,** 49 (1968).

70. L. S. Dietrich, O. Muniz, and M. Powanda, *J. Vitaminol. (Kyoto)* **14,** 123 (1968).

71. R. K. Gholson, *Nature* **212,** 933 (1966).

72. M. Ortega and G. Brown, *J. Biol. Chem.* **235,** 2939–2945 (1960).

72a. C. del Río-E. and H. Patino, *J. Bact.* **84,** 871–873 (1962). C. del Río-E. and H. Patino, *Bact. Proc.* **1964,** 108.

73. E. B. T. Leete, *Advan. Enzymol.* **32,** 395 (1969).

73a. J. Fleeker and R. U. Byerrum, *J. Biol. Chem.* **240,** 499 (1965); *Ibid.*, **242,** 3042 (1967). T. M. Jackanicz and R. U. Byerrum, *J. Biol. Chem.* **241,** 1296 (1966). D. Gross, A. Seige, R. Stecher, A. Zureck, and H. R. Schütte, *Z. Naturforsch.* **20b,** 1116 (1965).

74. N. Iizuka and D. Mizuno, *Biochem. Biophys. Acta* **148,** 320 (1967).

75. M. L. W. Chang and B. C. Johnson, *J. Biol. Chem.* **234,** 1817 (1969). G. P. Quinn and P. Greengard, *Arch. Biochem. Biophys.* **115,** 146 (1956).

76. S. Chaykin, *Biochem. Biophys. Acta* **82,** 633 (1964).

77. K. V. Rajagopalan, I. Fridovich, and P. Handler, *J. Biol. Chem.* **237,** 922 (1962).

78. Y. C. Lee, R. K. Gholson, and N. Raica, *J. Biol. Chem.* **244,** 3277 (1969).

79. R. K. Gholson, *Nature* **212,** 933 (1966).

80. Y. Nishizuka, K. Ueda, K. Nakazawa, R. H. Reeder, T. Honjo, and O. Hayashi, *J. Biol. Chem.* **243,** 3765 (1968).

81. Y. Nishizuka, K. Ueda, T. Honjo, and O. Hayaishi, *J. Biol. Chem.* **243,** 3765 (1968).

82. A. L. Greenbaum and S. Pinder, *Biochem. J.* **107,** 55 (1968).

83. I. E. Mozgov and V. V. Sobolev, *Materialy IX-Oi Nauchn. Konf. Farmakol. Moscow* **1964,** 38–39; *Chem. Abstr.* **63,** 12213 (1965).

84. R. Ishigami, S. Shiotani, T. Kawata, H. Natori, Y. Akiyama, J. Omoto, H. Okada, I. Kadota, and Y. Nishimura, *Bitamin* **32,** 232–239 (1965); *Chem. Abstr.* **61,** 3535 (1964); **62,** 16861 (1965); **63,** 18910 (1965).

85. Title 21, Code of Federal Regulations (21 CFR), Definitions and Standards under the Federal Food, Drug, and Cosmetic Act, U.S. Dept. of Health, Education, and Welfare, Food and Drug Administration, Washington, D.C., Part 15, *Cereal Flours and Related Products* (Oct. 1964), pp. 4, 8, 12, 13. *Ibid.*, Part 16, *Macaroni and Noodle Products* (Oct. 1964), pp. 7, 10. *Ibid.*, Part 17, *Bakery Products* (reissued Sept. 1963), p. 4.

86. *Federal Register* **28,** 10908–10909 (Oct. 11, 1963).

87. P. Lipton and J. G. Michels, *Geriatrics* **20,** 379–387 (1965).

88. J. Vasko, R. Younger, and S. E. Stephenson, *Surg. Forum* **14**, 308–309 (1963).
89. I. I. Lieber, *Prensa Med. Arg.* **49**, 91–97 (1962).
90. D. Kritchevsky and S. A. Tepper, *Arch. Intern. Pharmacodyn.* **147**, 564–568 (1964).
91. Japan. Pat. 17,023 (Aug. 3, 1965), K. Saruto and H. Hagiwara (to Yoshitomi Pharmaceutical Industries, Ltd.); *Chem. Abstr.* **63**, 18038 (1965).
92. Japan. Pat. 17,022 (Aug. 3, 1965), H. Nakano, A. Morimoto, and H. Yoshimitsu (to Fujisawa Pharmaceutical Co. Ltd.); *Chem. Abstr.* **63**, 18038 (1965).
93. R. Fontenot, H. Redetzki, and R. Deupree, *Proc. Soc. Exptl. Biol. Med.* **119**, 1053–1055 (1965).
94. U.S. Pat. 3,193,461 (July 6, 1965), M. E. Eisen (to Sophie E. Gordon).
95. Fr. Pat. M3319 (June 21, 1965), R. Vacher-Collomb; *Chem. Abstr.* **63**, 11515 (1965).
96. L. A. Carlson and J. Ostman, *Acta Med. Scand.* **178**, 7179 (1965).
97. E. Bernardini Jaramillo, *Arch. Inst. Farmae Exptl.* (*Madrid*) **15**, 191–208 (1963).
98. Japan. Pat. 18,109 (Aug. 16, 1965), T. Irikura and S. Sato (to Kyorin Pharmaceutical Co. Ltd.); *Chem. Abstr.* **63**, 18037 (1965).
99. M. Suzuki, Y. Murase, Y. Baba, and S. Hattori, *Takamine Kenkyusho Nempo* **10**, 19–23 (1958); *Chem. Abstr.* **55**, 1630 (1961).
100. F. Streightoff, *J. Bact.* **85**, 42–48 (1963).
101. K. G. Sheth, J. Sundararajan, and T. L. Rama Char, *Jamshedpur* (*India*) **1961**, 252–256; *Chem. Abstr.* **56**, 13932 (1962).
102. K. G. Sheth and T. L. Rama Char, *Corrosion* **18**, 218t–223t (1962); *Chem. Abstr.* **56**, 3249. (1962).
103. J. Nemcova, *Chem. Prumysl* **15**, 497–498 (1965); *Chem. Abstr.* **64**, 856 (1966).
104. Ger. Pat. 1,109,479 (Jan. 3, 1959), M. Koeppel (to Riedel and Co.); *Chem. Abstr.* **56**, 4525 (1962).
105. H. von Euler, F. Schlenck, L. Melzer, and B. Högberg, *Z. Physiol. Chem.* **258**, 212 (1939).
106. P. Karrer and H. Keller, *Helv. Chim. Acta* **22**, 1292 (1939).
107. C. Yandfsky and D. M. Bonner, *J. Biol. Chem.* **190**, 211 (1951).
108. E. Cherbullez and F. Landolt, *Helv. Chim. Acta* **29**, 1438 (1946).
109. U.S. Pat. 2,419,831 (April 29, 1947), P. W. Garbo.
110. U.S. Pat. 2,412,749 (Dec. 17, 1946), E. F. Pike and R. S. Shane (to Gelatin Products Corp.).
111. Brit. Pat. 822,579 (Oct. 28, 1959), Aries Associates, Inc.; *Chem. Abstr.* **54**, 4632 (1960).
112. U.S. Pat. 2,617,805 (Nov. 11, 1952), L. J. Wissow.
113. U.S. Pat. 2,923,715 (Feb. 2, 1960), M. Freifelder, G. M. Illich, and R. M. Robinson (to Abbott Laboratories).
114. U.S. Pat. 2,932,648 (April 12, 1960), M. Freifelder and I. W. Sangrelet (to Abbott Laboratories).
115. I. B. Chekmareva, E. S. Zhdanovich, and N. M. Zhenskii, *Med. Prom. SSSR* **19**, 11–15 (1965); *Chem. Abstr.* **63**, 11488 (1965).
116. U.S.S.R. Pat. 164,601 (Aug. 19, 1964), I. B. Chekmareva, E. S. Zhdanovich, G. I. Sazonova, and N. A. Preobrazhenskii; *Chem. Abstr.* **62**, 9115 (1965).
117. U.S.S.R. Pat. 170,509 (April 23, 1965), I. B. Chekmareva, E. S. Zhdanovich, T. A. Pan'shina, and N. A. Preobrazhenskii; *Chem. Abstr.* **63**, 9923 (1965).
118. Fr. Pat. 1,364,605 (June 26, 1964), L. Vossen; *Chem. Abstr.* **63**, 1793 (1965).
119. *The U.S. Pharmacopeia XVI*, Mack Printing Co., Easton, Pa., 1960.
120. L. T. Ikramow, *Dokl. Akad. Nauk Uz. SSR* **21**, 40–42 (1964); *Chem. Abstr.* **58**, 10040 (1963).
121. A. Sebesta and L. Dalma, *Sci. Aliment.* **10**, 77–78 (1964); *Chem. Abstr.* **62**, 3324 (1965).
122. I. Yamane, Y. Matsuya, and K. Miyakawa, *Kosankinbyo Kenkyu Zasshi* **14**, 164–169 (1960); *Chem. Abstr.* **55**, 15613 (1961).
123. *Ibid.*, 170–178 (1960); *Chem. Abstr.* **55**, 15614 (1961).
124. S. Seki, *Acta Med. Okayama* **17**, 153–173 (1963); *Chem. Abstr.* **61**, 7537 (1964).
125. P. Feigelson, J. N. Williams, Jr., and C. A. Elvehjem, *Proc. Soc. Exptl. Biol. Med.* **78**, 34 (1951). N. O. Kaplan, A. Goldin, S. R. Humphreys, N. M. Ciotti, and F. E. Stolzenbach, *J. Biol. Chem.* **219**, 287 (1956). M. V. Narurkan, U. S. Kumpta, and M. B. Sahasrabudke, *Brit. J. Cancer* **11**, 482 (1957).
126. Fr. Pat. M3203 (April 26, 1965), C. J. Kleinwicks; *Chem. Abstr.* **63**, 12980 (1965).
127. J. H. Burchenal, *Proc. Intern. Congr. Hemtol., 8th, Tokyo, 1960* **1**, 457–461; *Chem. Abstr.* **58**, 13007 (1963).
128. S. Y. Mikhlin and M. F. Nesterin, *Patol. Fiziol. Eksperim. Terapiya* **7**, 61–64 (1963); *Chem. Abstr.* **61**, 1148 (1964).

129. A. Brunnemann, H. Coper, and H. Herken, *Arch. Exptl. Pathol. Pharmakol.* **244**, 223–236 (1962).

130. A. Brunnemann, H. Coper, and H. Herken, *Arch. Exptl. Pathol. Pharmakol.* **245**, 541–550 (1963).

131. H. Coper and D. Neubert, *J. Neurochem.* **10**, 513–522 (1963).

132. F. Willing, V. Neuhoff, and H. Herken, *Arch. Exptl. Pathol. Pharmakol.* **247**, 254–266 (1964).

133. E. Goldberg, *Ann. N.Y. Acad. Sci.* **121**, 560–570 (1964).

134. G. di Sabato and M. Ottesen, *Biochem.* **4**, 422–428 (1965).

135. A. Yoshida, *J. Biol. Chem.* **240**, 1118–1124 (1965).

136. H. Holzer and H. D. Soeling, *Biochem. Z.* **336**, 201–214 (1962).

137. V. Neuhoff and H. Herken, *Naturwiss.* **49**, 519 (1962).

138. V. Neuhoff and F. Koehler, *Naturwiss.* **52**, 475–476 (1965).

139. M. van Ammers and H. J. den Hertog, *Rec. Trav. Chim.* **78**, 586 (1959).

140. S. Dreizen, C. N. Spirakis, and R. E. Sloan, *Intern. Z. Vitaminforsch.* **33**, 321–326 (1963).

141. R. M. Bechtle and G. H. Scherr, *Antibiot. Chemotherapy* **9**, 715–721 (1959).

142. Austrian Pat. 242,699 (Sept. 27, 1965), Loba Chemie; *Chem. Abstr.* **63**, 18037 (1965).

143. Sh. Levi and E. Mutafchieva, *Farmatsiya (Sofia)* **15**, 89–92 (1965); *Chem. Abstr.* **63**, 11483 (1965).

144. H. Gilman and H. S. Broadbent, *J. Am. Chem. Soc.* **70**, 2757 (1948).

145. Japan. Pat. 23,792 (Oct. 19, 1965), K. Uda, A. Sakurai, and K. Sakakibara (to Dainippon Celluloid Co. Ltd.); *Chem. Abstr.* **64**, 3502 (1966).

146. Polish Pat. 47,460 (Sept. 20, 1963), W. Swietolawski, J. Bialek, and A. Bylicki; *Chem. Abstr.* **61**, 5614 (1964).

147. E. M. Gepshtein, *Nauchn. Tr. Vost. Nauchn. Issled. Uglekhim. Inst.* **16**, 30–35 (1963); *Chem. Abstr.* **61**, 8091 (1964).

148. Indian Pat. 66,383 (Oct. 12, 1960), K. R. Doraswamy and B. Srinivasan (to Karamchand Premchand Ltd.); *Chem. Abstr.* **55**, 10478 (1961).

149. S. S. Kruglikov, V. G. Khomyakov, and L. I. Kazakova, *Tr. Mosk. Khim. Tekhnol. Inst.* **32**, 201–206 (1961); *Chem. Abstr.* **57**, 16543 (1962).

150. U.S. Pat. 3,147,269 (Sept. 1, 1964), E. Katscher and W. Moroz (to International Chemical Corp.).

151. C. del Río-E. and J. Sanchez, *Abstr. Comm., Intern. Congr. Biochem., Moscow, 1961*, p. 295.

152. W. C. J. Ross, *J. Chem. Soc.* **1966** C, 1816–1821.

153. J. R. A. Pollock and R. Stevens, *Dictionary of Organic Compounds*, Eyre & Spottiswoode Publishing Co. Ltd., London, 1965.

154. G. Ceriotti, *Farm. Sci. Tec. (Pavia)* **7**, 146–148 (1952).

155. U.S. Pat. 2,830,994 (1958), to Distillers Corp. Ltd.

156. H. Meyer and J. Mally, *Monatsh. Chem.* **33**, 393 (1912); *Chem. Abstr.* **6**, 2073 (1912).

157. E. S. Long, *The Chemistry and Chemotherapy of Tuberculosis*, 3rd ed., The Williams & Wilkins Co., Baltimore, 1958.

158. S. Kushner, H. Dalalian, R. T. Cassell, J. L. Sanjurjo, D. Mackenzie, and Y. Subbarow, *J. Org. Chem.* **13**, 834 (1948).

159. Ger. Pat. 1,191,817 (April 29, 1965), O. Zima and V. Wereder (to E. Merck, A.G.); *Chem. Abstr.* **63**, 8325 (1965).

160. P. Gheorghiu, V. Stroescu, C. Demetrescu, H. Benes, and R. Ciobaru, *Compt. Rend. Soc. Biol.* **159**, 366–370 (1965); *Chem. Abstr.* **63**, 10531 (1965).

161. S. Kakimoto and I. Tone, *J. Med. Chem.* **8**, 868 (1965); *Chem. Abstr.* **54**, 21079 (1960).

162. W. H. Fox, *Science* **116**, 129 (1952); **118**, 497 (1953).

163. G. Siefert, *Arb. Paul-Ehrlich-Inst. Georg-Speyer-Hause Ferdinand-Blum Inst. Frankfurt a.M.* **57**, 140–145 (1962); *Chem. Abstr.* **61**, 11050 (1964).

164. Dan. Pat. 87,228 (July 27, 1959), Holger B. Thomassen (to A. Dumex (Dumex, Ltd.)); *Chem. Abstr.* **54**, 7742 (1960).

165. U.S.S.R. Pat. 106,220 (July 25, 1957), L. N. Pavlov, P. A. Gangrskii, and V. M. Polyachenko; *Chem. Abstr.* **51**, 16563 (1957).

166. Ger. Pat. 1,116,667 (Nov. 19, 1961), H. Pasedach and M. Seefelder (to Badische Anilin und Soda Fabrik, A.G.); *Chem. Abstr.* **56**, 14247 (1962).

167. Japan. Pat. 7472 (Nov. 15, 1954), S. Takizawa; *Chem. Abstr.* **50**, 9447 (1956).

168. Span. Pat. 202,450 (May 28, 1952), to D. Magrane, S.A.; *Chem. Abstr.* **50**, 17323 (1956).

169. G. Lock, *Pharm. Ind.* **14**, 366–368 (1952); *Chem. Abstr.* **47**, 10531 (1953).

170. Japan. Pat. 2629 (April 21 1955), M. Ishikawa (to Shionogi Drug Manufacturing Co.); *Chem. Abstr.* **51**, 14832 (1957).

171. T. Urbanski et al., *Roczniki Chem.* **27**, 161–166 (1953).

172. *U.S. Pharmacopeia XVII*, Mack Publishing Co., Easton, Pa., 1965, p. 327.

173. C. del Río E., unpublished results (1961).

174. J. A. C. van Pinxteren and M. E. Verloop, *Pharm. Weekblad* **99**, 1125–1133 (1964); *Chem. Abstr.* **63**, 1659 (1965).

175. H. Ozawa and A. Kiyomoto, *J. Pharm. Soc. Japan* **72**, 1059–1060 (1952); *Chem. Abstr.* **46**, 11047 (1952).

176. T. Canbäck, *J. Pharm. Pharmacol.* **4**, 407 (1952); *Chem. Abstr.* **46**, 8325 (1952).

177. J. F. Alicino, *J. Am. Pharm. Assoc.* **41**, 401–402 (1952).

178. E. Kühni, M. Jacob, and H. Grossglauser, *Pharm. Acta Helv.* **29**, 233–250 (1954); *Chem. Abstr.* **49**, 3469 (1955).

179. M. Businelli and B. Rocchi, *Farm. Sci. Tec. (Pavia)* **7**, 153–160 (1952); *Chem. Abstr.* **46**, 9019 (1952).

180. D. S. Goldman, *Science* **120**, 315–316 (1954); *Chem. Abstr.* **49**, 1280 (1955).

181. I. Tachi and Y. Nagata, *Kagaku No Ryoiki* **6**, 490–491 (1952); *Chem. Abstr.* **46**, 11045 (1952).

182. J. H. Peters, K. S. Miller, and P. Brown, *Anal. Biochem.* **12**, 379–394 (1965).

183. J. M. Kelly and R. B. Poet, *Am. Rev. Tuberc.* **65**, 484 (1952).

184. R. Bönicke, *Arch. Exptl. Path. Pharmakol.* **216**, 490 (1952).

185. B. Tabenkin, B. Dolan, and M. G. Johnson, *Proc. Soc. Exptl. Biol. Med.* **80**, 613–615 (1952).

186. K. Fujiwara, *Kekkaku* **26**, 144 (1959); *Chem. Abstr.* **61**, 6230 (1964).

187. A. S. Yard and H. McKennis, *J. Med. Pharm. Chem.* **5**, 196–203 (1962).

188. B. P. Lisboa and R. Boenicke, *Tuberk. Forschungsinst. Borstel, Jahresber.* **5**, 401–407 (1961); *Chem. Abstr.* **60**, 842 (1964).

189. W. R. Barclay, R. H. Ebert, and D. Koch-Weser, *Am. Rev. Tuberc.* **67**, 490 (1953).

190. A. R. Armstrong, *Am. Rev. Respirat. Diseases* **81**, 498–503 (1960).

191. H. Pope, *Am. Rev. Tuberc.* **68**, 938 (1953).

192. H. Eiichi, *Yakugaku Zasshi* **79**, 1343 (1959); *Chem. Abstr.* **54**, 6864 (1960).

193. G. Middlebrook and M. L. Cohn, *Science* **118**, 297 (1953).

194. K. Suszko, *Poznan. Towarz. Przyjaciol Nauk. Wydzial Lekar., Prace Komisji Med. Doswiadczalnej.* **25**, 243–266 (1963); *Chem. Abstr.* **60**, 9644 (1964).

195. F. Pinamonti, *Lotta Contro Tuberc.* **31**, 135–139 (1961); *Chem. Abstr.* **58**, 14455 (1963).

196. M. Tsukamura, M. Hasimoto, and V. Noda, *Am. Rev. Respirat. Diseases* **81**, 403–406 (1960).

197. J. Mauron and E. Bujarde, *Helv. Chim. Acta* **46**, 1895–1906 (1963).

198. E. Gajewska et al., *Polish Med. Sci. Hist. Bull.* **6**, 24, 26–29 (1963); *Chem. Abstr.* **59**, 2014 (1963).

199. H. P. Willet, *Am. Rev. Respirat. Diseases* **80**, 404–409 (1959).

200. M. Tsukamura and M. Mizuno, *Kekkaku* **37**, 29–34 (1962); *Chem. Abstr.* **57**, 2670 (1962).

201. A. Bekierkunst, *Science* **152**, 525–526 (1966).

202. M. Tsukamura, *Igaku To Seibutsugaku* **63**, 24–28 (1962); *Chem. Abstr.* **60**, 14837 (1964).

203. P. R. J. Gangadharam, F. M. Harold, and W. B. Schaefer, *Nature* **198**, 712–714 (1963).

204. M. Tsukamura and S. Tsukamura, *Japan J. Tuberc.* **10**, 81–85 (1962); *Chem. Abstr.* **60**, 8373 (1964).

205. K. Munakata, *Sci. Rept. Res. Inst. Tohoku Univ.* **11**, 93–105 (1962); *Chem. Abstr.* **59**, 5537 (1963).

206. E. Panescu, *Studii Cercetari Biochim.* **4**, 543–550 (1961); *Chem. Abstr.* **63**, 6044 (1965).

207. M. R. Alioto and M. Ayala, *Boll. Soc. Ital. Biol. Sper.* **36**, 327–329 (1960); *Chem. Abstr.* **56**, 12241 (1962).

208. M. R. Alioto, *Ital. J. Biochem.* **8**, 361–376 (1959).

209. M. R. Alioto, *Biochim. Appl.* **9**, 333–343 (1962); *Chem. Abstr.* **60**, 7315 (1964).

210. F. Rosen, *Proc. Soc. Exptl. Biol. Med.* **88**, 243–246 (1955).

211. W. Friedel, *Beitr. Klin. Tuberk.* **128**, 146–154 (1964); *Chem. Abstr.* **61**, 9925 (1964).

212. Th. Kelso, H. W. Toll, D. C. Pinkerton, and L. C. Kier, *J. Forensic Sci.* **10**, 313–317 (1965); *Chem. Abstr.* **63**, 13916 (1965).

213. V. Holecek, J. Suva, and J. Safanda, *Rozhledy Tuberk.* **25**, 479–482 (1965); *Chem. Abstr.* **64**, 4136 (1966).

214. H. C. Engbaek, M. Weis Bentzon, H. Heegaard, and O. Christensen, *Acta Pathol. Microbiol. Scand.* **65**, 69–83 (1965); *Chem. Abstr.* **64**, 8797 (1966).
215. E. H. Rodd, *Chemistry of Carbon Compounds. Heterocyclic Compounds*, Vol. 4, Elsevier Publishing Co., Amsterdam, 1957, p. 560.
216. Swed. Pat. 133,964 (Nov. 27, 1951), R. Nylander.
217. Brit. Pat. 587,810 (May 6, 1947), W. F. Short and P. Oxley.
218. F. Giral and C. A. Rojahn, *Prod. Quim. Farm. Ed. Atlante Mex.* **1956**, 1839.
219. H. Kaiser and H. Jori, *Pharm. Ztg. Nachr.* **89**, 331 (1953); *Chem. Abstr.* **47**, 12492 (1953).
220. A. Gatti, *Minerva Med.* **39**, 304–305; *Chem. Abstr.* **42**, 4270 (1948).
221. J. E. Eckenhoff and J. H. Hafkenschiel, *J. Pharmacol. Exptl. Therap.* **91**, 362–369 (1947).
222. F. Guerra, *Arch. Inst. Cardiol. (Mexico)* **15**, 366–374 (1945).
223. F. Hahn, *Arch. Exptl. Pathol. Pharmakol.* **205**, 552–562 (1948); *Chem. Abstr.* **43**, 7137 (1949).
224. F. R. Blood and F. E. D'Amour, *Am. J. Physiol.* **156**, 52–61 (1949).
225. N. Nagasaki, *Japan. J. Pharmacol.* **2**, 97–101 (1953); *Chem. Abstr.* **47**, 10127 (1953).
226. A. Carlsson and F. Serin, *Kgl. Fysiograf. Sallskap. Lund, Forh.* **19**, 233–235 (1949); *Chem. Abstr.* **44**, 9572 (1950).
227. *Remington's Pharmaceutical Sciences*, 13th ed., Mack Publishing Co., Easton, Pa., 1965, p. 1217.
228. R. C. Elderfield, *Heterocyclic Compounds*, John Wiley & Sons, Inc., New York, 1950, p. 565.

CARLOS DEL RÍO-ESTRADA
Universidad Nacional de Mexico
HARRY W. DOUGHERTY
(Biological aspects of nicotinic acid)
Merck Sharp & Dohme

VITAMIN B$_{12}$

Vitamin B$_{12}$ (cyanocobalamin) is a red crystalline cyano cobalt complex. It is required for normal blood formation, for certain other fundamental metabolic processes, for neural function, and for human and animal growth and maintenance. Human requirements are extremely small—less than a thousandth as great as those for most of the other B vitamins.

With the discovery in 1926 that pernicious anemia could be controlled by the administration of whole liver, efforts were begun to obtain the active principle in pure form. An extensive review of the purification work before 1945 has been published (1). The discovery of a microbiological assay with *Lactobacillus lactis* for the antipernicious anemia factor gave a criterion for determination of the potency of concentrates and allowed more rapid evaluation of advances made toward isolation. In 1948 Folkers and co-workers announced the isolation of pure crystalline vitamin B$_{12}$ from liver (2), and also the discovery that it could be produced by fermentations (3,4), thus affording a practical large-scale source of the vitamin both as active concentrates and as the pure compound.

It was further demonstrated that vitamin B$_{12}$ is present in a wide variety of animal tissues, and it soon became evident that many other research efforts directed toward isolation of new animal growth factors, etc, were actually concerned with vitamin B$_{12}$. However, assumption of too great a specificity of the *L. lactis* assay led to erroneous conclusions, and it is now known that some of the reported vitamin B$_{12}$ activity of natural materials is due to other substances.

Structure. The structure of vitamin B$_{12}$ was first announced during the summer of 1955 at a British Chemical Society Symposium and at the Third International Biochemical Congress in Brussels. Teams under the leadership of Alexander Todd (Cambridge) and Lester Smith (Glasgow) (5) and Dorothy Hodgkin (Oxford) (6) were responsible for this achievement. The structure is shown in diagrammatic form below.

(Cyanocobalamin, $C_{63}H_{88}CoN_{14}O_{14}P$)

Physical and Chemical Properties (7). Vitamin B$_{12}$ forms red crystals which do not melt below 300°C, but darken to black at about 210–220°C; refractive indexes: α, 1.616; β, 1.652; and γ, 1.664. The crystals normally contain about 12% moisture and are soluble in water to the extent of 1.25% at 25°C. Vitamin B$_{12}$ is soluble in alcohols and phenols, but insoluble in acetone, chloroform, and ether.

Aqueous solutions show absorption maxima at 278 ($E_{1\,cm}^{1\%}$ 115), 361 ($E_{1\,cm}^{1\%}$ 204), and 550 ($E_{1\,cm}^{1\%}$ 63) nm, which are practically independent of pH. The optical activity in aqueous solution is $[\alpha]_{6563}^{23} = -59° \pm 9°$ and $[\alpha]_{6438}^{20} = -110° \pm 10°$.

Vitamin B$_{12}$ is unstable toward both oxidizing and reducing agents. Optimum stability in aqueous solutions is at pH 4.5–5.0. No significant loss of activity occurs at room temperature for periods of 2 years or longer. However, at elevated temperatures or a higher or lower pH decomposition is more rapid.

Occurrence

Vitamin B$_{12}$ occurs in practically all animal products, probably in the form of a protein complex, in larger amounts in those that are rich in protein. Liver, kidney, and some seafoods are the richest sources. Fermented materials such as rumen contents and feces are particularly rich sources, and it is known to be produced by many actinomycetes and other bacteria, but in general not by molds (8). It is possible that it is entirely a product of fermentation and that that found in animal tissues is either ingested or produced by intestinal fermentation.

In general, plant material contains little or no vitamin B$_{12}$. An exception is algae, and it has been suggested that these plants may be the source of the relatively large amounts of the vitamin found in some shellfish and fish.

Literature reports of the vitamin B$_{12}$ contents of natural materials should in many cases be qualified because of the presence of other substances which could give a response in the particular assay used. This is particularly important in the case of fermentation broths, and in some cases it has been demonstrated that the major portion of the activity measured microbiologically is due to something other than vitamin B$_{12}$ (9).

Table 1 lists the contents of vitamin B$_{12}$-active substance of a variety of natural materials (10).

Production of vitamin B$_{12}$ and other B$_{12}$-active substance has been shown for microorganisms belonging to the Schizomycetes, *Torula* and *Eremothecium*, including bacteria of the genera *Streptomyces*, *Alcaligenes*, *Bacillus*, *Pseudomonas*, *Mycobacterium*, and *Escherichia* (8, 11).

Table 1.   Content of Vitamin B$_{12}$-Active Substances of Some Natural Materials

Material	Vitamin B$_{12}$ activity γ/kg (dry)	Material	Vitamin B$_{12}$ activity γ/kg (dry)
milk	3–5 (liter fresh)	clams	2500
whole milk powder	10–26	fish meal	170–1500
cheese	28	salmon kidney	18,000
egg yolk	28	beef	
green leafy vegetables	0 or traces	round	55–79
(celery, pepper, kale,		liver	500
tomato, cabbage,		kidney	500
kohlrabi, broccoli,		heart	250
leek)		mutton	88
algae	60–280	horsemeat	70–75
filet of sole	13	chicken meat	52
oysters	150–2800		

An extensive review of the sources of vitamin B$_{12}$ has been published (12).

Commercial Production

Vitamin B$_{12}$ is produced almost exclusive by fermentation, and its production has become one of the important fermentation industries in the United States. It is generally produced by primary fermentation, but in some instances is obtained from broths used primarily for production of antibiotics. This is largely a matter of economics, since maximum production of the antibiotic and maximum production of cobalamin generally do not occur under the same fermentation conditions.

Another source is sewage, which is known to contain cobalamin-like compounds in addition to vitamin B$_{12}$. With one exception (13), the chemical and biological properties of these cobalamin-like compounds are not known. A process for extracting the vitamin B$_{12}$ from sewage has been patented (14), and concentrates containing as much as 40 mg/lb have been obtained.

Fermentation.   Microorganisms that have been reported to be used for commercial production (15) are *Streptomyces griseus*, *S. olivaceus*, and *S. aureofaciens*, and the bacteria *Bacillus megatherium*, *Propionobacterium freudenreichii*, and a mixed fermentation of a *Proteus* sp. and a *Pseudomonas* sp. Discussions of the formation of vitamin B$_{12}$-active compounds by a number of other organisms have been published.

Fermentations to produce vitamin B$_{12}$ are generally carried out by submerged culture in an aerated and stirred medium. The fermentation time is usually 3–5 days.

The following description is based on a process for the production of vitamin B$_{12}$ by *S. olivaceus* (16) which was developed by the Fermentation Division of the Northern Regional Research Laboratories. Pure cultures must be used because contamination invariably results in very low yields. All equipment must be sterile and transfers carried out aseptically.

Media for the growth of these organisms consist of carbohydrate, proteinaceous material, and a source of cobalt and other salts. A typical medium is:

Distillers' solubles	4.0%
Dextrose	0.5–1%
CaCO$_3$	0.5%
CoCl$_2$·6H$_2$O	1.5–10 ppm
(pH adjusted to about 7 with NaOH)	

Most forms of distillers' solubles, soybean meal, yeast, casein, etc, are satisfactory under the proper conditions. Cobalt supplementation is essential for maximum yields of cobalamin, although not necessarily for growth of the organism. This has been shown to be true for organisms tested (17) and is the subject of a patent (18).

In some instances cyanide is added to the medium to aid in converting other cobalamins to vitamin B$_{12}$ (19).

The medium may be made up and sterilized batchwise directly in the fermentor (for example, 1 hr at 250°F) or sterilized continuously as the fermentor is charged (for example, 13 min at 330°F) by mixing directly with live steam. During sterilization steam is blown into openings, and all transfer lines are kept filled with live steam when not in use to ensure sterility.

Pure cultures of the organisms to be used are maintained on agar slants and are grown first in 100–250 ml of medium in shake flasks and then in two or three successively larger batches to give the required amount of inoculum culture. Inoculation of the fermentors with about 5% of the volume of medium is satisfactory.

During fermentation the temperature is maintained at about 80°F. Within limits, the rate of growth of the culture, but not the final yield of cobalamin, depends on the rate of agitation and aeration. Too rapid aeration causes excessive foaming, and the optimum rate is about 0.5 volume air/volume medium per min. The air is sterilized by passing through columns of activated charcoal. Foaming is a serious problem, particularly at the beginning and end of the fermentation. It is controlled by addition of sterile defoaming agents such as soybean oil, corn oil, lard oil, or silicones as required. In some cases a defoamer is added when the medium is prepared.

During the first 24 hr of the fermentation the pH falls a few tenths of a unit, the sugar concentration decreases rapidly, and the mixture becomes very thick as a result of mycelium growth and probably also of the formation of viscous materials such as polysaccharides. The concentration of cobalamin in the beer rises very slowly for about 20 hr and then much more rapidly. After 2–4 days when the available nutrients are consumed, lysis of the mycelium begins, the pH rises to about 8, and the mash becomes quite thin. Little or no additional cobalamin is formed after lysis begins. The mash is then stabilized by reducing the pH to about 5 with sulfuric acid and adding small quantities of a reducing agent such as sodium sulfite.

Yields of cobalamin are usually in the range of 1–2 mg/l in the final broth.

During the major part of the fermentation period, most of the cobalamin is present in the mycelium, but at the end a considerable portion is in solution. Heating the mixture to boiling at pH 5 or below frees the cobalamin quantitatively from the mycelium.

Essentials of the production of cobalamins by other bacterial fermentations are similar to those for *Streptomyces* although there are differences in media, conditions, time, etc, for optimum yields (20).

Work-up of the final broth containing cobalamins is determined by the type of product to be produced.

Production of Vitamin B$_{12}$ Feed Supplements. In order to prepare low-potency concentrates of vitamin B$_{12}$ for use as feed supplements, usually all that is necessary is to evaporate the final fermentation broth to dryness. These final broths contain about 3% solids and are first evaporated *in vacuo* to a solids content of 15–20% and the resulting syrups are drum-dried or spray-dried. Differences in conditions during this processing do not seriously affect the potency of the product, but evaporating the broth at temperatures not above 180°F, drum-drying such that the exit gases are not above 200°F, and cooling the product as rapidly as possible result in about 10% greater activity (16). The concentrates obtained in this way by cobalt-supplemented fermentation are reported to contain 10–30 mg/lb of cobalamin (15).

In addition to cobalamins these products may also contain other vitamins, minerals, and antibiotics, depending on the source and method of production (15, 21). For use as a feed supplement the composition of the major portion of the solids is important for determining its nutritive value. The product prepared by evaporating whole broth from *Bacillus megatherium* has much the same composition as yeast (qv) with respect to amino acids and B vitamins (20).

Production of Crystalline Vitamin B$_{12}$ from Fermentation Broths. Processes used commercially for the production of vitamin B$_{12}$ have not been described in detail, and most of the published information is found in the patent literature.

One of the first steps is to treat the filtered broth with cyanide, which converts all the cobalamins into cyanocobalamin (22), although this step may also be carried out after some concentration has been achieved. The cyanocobalamin is next adsorbed from solution. Several substances have been reported for this purpose: charcoal (23), fuller's earth, bentonite (24), and ion-exchange resins (25). Elution from the adsorbent is usually accomplished by the use of an aqueous solution of materials ranging from organic bases to hydrochloric acid. Water, water–acetone, and solutions of sodium cyanide or sodium thiocyanate have been used.

Some type of extraction is usually carried out next, countercurrent distribution between cresol (26), amylphenol (27), or benzyl alcohol (28) and water or single extraction into an organic solvent such as a phenol (29). Precipitation as a copper or zinc cyanide-cyanocobalamin complex has been reported (30).

Another purification step applicable to aqueous concentrates consists in dissolving a zinc salt in the slightly acid solution and then raising the pH to precipitate zinc hydroxide, which removes many impurities (31).

Chromatography on alumina and final crystallization from methanol–acetone, ethanol–acetone, or acetone–water usually complete the process (4,32).

The only industrial yield figures available are again those reported in patents (32,33). For example, a yield of 65% from broth to a 5% concentrate is reported in one case, and an overall yield of about 4% of crystalline material in another.

There are product patents on crystalline vitamin B$_{12}$ (4), and vitamin B$_{12}$ concentrates (11).

Assay

Physical, chemical, biological, microbiological, and radioactive tracer methods have been developed for the determination of vitamin B$_{12}$. The radioactive tracer method is generally the most reliable, unless the absence of all interfering substances is definite. It determines only vitamin B$_{12}$ but is applicable to the determination of total cobalamins by first converting them to cyanocobalamin by treatment with sodium nitrite and sodium cyanide. A known amount of pure radioactive cyanocobalamin is then added to the sample and equilibrated with cyanocobalamin present. Addition of zinc acetate to an acid solution of the sample and then precipitation of zinc hydroxide by raising the pH to 8.5 removes many impurities. The material is next extracted into a cresol-carbon tetrachloride mixture, effecting considerable purification. After addition of 1-butanol and more carbon tetrachloride, the sample is extracted back into water and the aqueous solution passed through a column of ion-exchange resin (IRP-400 and IR-120). Only those fractions showing a red color are retained. The concentration and purity of the vitamin B$_{12}$ are determined by magnitude and ratios of absorbancies at 548, 430, and 361 nm. If desired, a benzyl alcohol–water distribution may be carried out as an additional criterion of purity or to effect further purification, particularly when compounds with absorption spectra similar to B$_{12}$ may be present. From measurement of the radioactivity of the solution of purified cyanocobalamin, the amount present in the original sample is calculated. Samples containing as little as 100 γ in concentrations as low as 0.1 γ/ml can be assayed.

Biochemical Function and Uses

Vitamin B$_{12}$ is required for normal blood formation, growth, certain other fundamental metabolic processes, and neural function and its maintenance. A tissue deficiency of this vitamin causes pernicious anemia, which in its completely developed stages is characterized by macrocytosis, megaloblastic bone marrow, leukopenia, neurological and mental changes, endocrine dysfunction, disturbances in fat metabolism, atrophic gastritis, mucous membrane disorders, and occasionally other disturbances.

A deficiency may result either from a dietary insufficiency of vitamin B$_{12}$ or from inability to use that ingested because of a lack of intrinsic factor, a protein which assists absorption of the small amounts of vitamin B$_{12}$ normally present in the intestine. Parenteral administration of small doses, oral administration of small doses with intrinsic factor, or massive oral doses without intrinsic factor produce rapid remissions of all symptoms of pernicious anemia with few exceptions.

Part of the action of vitamin B$_{12}$ is closely related to folic acid which in sufficient amounts produces hematopoietic responses in pernicious anemia equivalent to vitamin B$_{12}$ but does not cause remission of the other manifestations, particularly the neurological complications, which are benefited only by vitamin B$_{12}$. It has been suggested that administration of folic acid in cases of vitamin B$_{12}$ deficiency may result in rapid depletion of the last small reserve of the vitamin B$_{12}$ in the body, thus causing severe neurological symptoms.

Treatment with vitamin B$_{12}$ is of value in nutritional macrocytic anemia and megaloblastic anemia of infancy, trigeminal neuralgia and diabetic neuritis, sprue and herpes zoster. It has been reported useful in certain skin and liver diseases, and a variety of other diseases in which results are not well defined.

Estimates of the human daily requirement of vitamin B$_{12}$ range from less than 1γ to 5γ per day. Children because of their higher metabolic rate may require several times this amount. It is one of the least toxic of therapeutic agents, and doses of 100 mg/kg (intraperitoneal) in rats and 10 mg orally in humans have shown no toxic effects.

Vitamin B$_{12}$ appears to be essential for the growth of all animals and poultry and is responsible for most, if not all, of the activity of the "animal protein factor," a factor or factors which permit(s) growth of animals or chicks on rations containing no animal protein (see Vol. 8, p. 869). Ruminant animals do not have a dietary requirement for the vitamin because they are capable of synthesizing it in the rumen.

There is evidence to indicate that vitamin B$_{12}$ is involved in carbohydrate and fat metabolism, reduction of sulfides to sulfhydryl (mercapto) compounds, nucleic acid metabolism, protein metabolism, transmethylation, and other one- or two-carbon atom transfers, but no specific enzyme or coenzyme function is known as is the case with most of the other B vitamins.

Since the daily requirement and organ contents of vitamin B$_{12}$ are thousands of times lower than those of most of the other B vitamins, it seems logical to reason that its mechanism of action must be quite different.

Extensive reviews of the biochemical function and clinical and nutritional uses of vitamin B$_{12}$ have been published (34, 10).

Bibliography

"Vitamin B$_{12}$" in ECT 1st ed., Vol. 14, pp. 813–828, by F. M. Robinson, Merck & Co., Inc.

1. Y. Subba-Row, A. B. Hastings, and M. Elkin, *Vitamins Hormones* **3**, 237 (1948).
2. E. L. Rickes, N. G. Brink, F. R. Koniuszy, T. R. Wood, and K. Folkers, *Science* **107**, 396 (1948).
3. *Ibid.*, **108**, 634 (1948).
4. U.S. Pat. 2,563,794 (Aug. 7, 1951) E. L. Rickes and T. R. Wood (to Merck & Co.).
5. R. Bonnett, J. R. Cannon, A. W. Johnson, I. Sutherland, A. R. Todd, and E. L. Smith, *Nature* **176**, 328 (1955).
6. D. C. Hodgkin, J. Pickworth, J. H. Robertson, K. N. Trueblood, R. J. Prosen, and J. G. White, *Nature*, **176**, 325 (1955).
7. K. Folkers and D. E. Wolf, *Vitamins Hormones* **12**, 1 (1954).
8. H. H. Hall, J. C. Benjamen, H. M. Bricker, R. J. Gill, W. C.Haynes, and H. M. Tsuchiya, *Bacteriological Proc., 50th Meeting, Soc. Am. Bacteriologists*, Baltimore, 1950.
9. L. Chaiet, T. Miller and A. E. Boley, *J. Agr. Food Chem.*, **2**, 784 (1954).
10. W. H. Sebrell, Jr. and R. S. Harris, *The Vitamins*, Academic Press, Inc., New York, 1954, pp. 396–523.
11. U.S. Pat. 2,703,302 (March 1, 1955) E. L. Rickes and T. R. Wood (to Merck & Co.).
12. M. A. Darken, *Botan. Rev.* **19**, 99 (1953).
13. W. Friederich and K. Bernhauer, *Z. Naturforsch.* **96**, 686 (1954).
14. U.S. Pat. 2,646,386 (July 21, 1953) C. S. Miner and B. Wolnak (to Sewerage Commission of Milwaukee).
15. A. S. Hester and G. E. Ward, *Ind. Eng. Chem.* **46**, 238 (1954).
16. V. F. Pfeiffer, C. Vojnovich, and E. H. Heger, *Ind. Eng. Chem.* **46**, 843 (1954).
17. D. H. Hendlin and M. Ruger, *Science* **111**, 542 (1950).
18. U.S. Pat. 2,595,499 (May 6, 1952) T. R. Wood and D. Hendlin (to Merck & Co.).
19. U.S. Pat. 2,650,896 (Sept. 1, 1953) L. E. McDaniel and H. B. Woodruff (to Merck & Co.).

20. J. A. Garibaldi, K. Ijechi, N. S. Snell, and J. C. Lewis, *Ind. Eng. Chem.* **45,** 838 (1953).
21. U.S. Pat. 2,619,420 (Nov. 25, 1952), T. H. Jukes (to American Cyanamid Co.).
22. U.S. Pat. 2,530,416 (Nov. 21, 1950), F. J. Wolf (to Merck & Co.).
23. U.S. Pat. 2,505,053 (April 25, 1950), F. A. Kuehl and L. Chaiet (to Merck & Co.).
24. U.S. Pat. 2,626,888 (Jan. 27, 1953), S. Kutosh, G. B. Hughey, and R. Malcolmson (to Merck & Co.).
25. U.S. Pat. 2,628,186 (Feb. 10, 1953), W. Shive (to Research Corp.).
26. U. S. Pat. 2,609,325 (Sept. 2, 1952), N. G. Brink and T. R. Wood (to Merck & Co.).
27. U.S. Pat. 2,594,314 (April 29, 1952), F. R. Koniuszy, N. G. Brink, and K. Folkers (to Merck & Co.).
28. U.S. Pat. 2,607,717 (Aug. 19, 1952), N. G. Brink and F. J. Wolf (to Merck & Co.).
29. U.S. Pat. 2,678,000 (May 18, 1954), R. G. Denkewalter (to Merck & Co.).
30. U.S. Pat. 2,621,114 (Dec. 9, 1952), A. J. Holland (to Merck & Co.).
31. U.S. Pat. 2,653,000 (Sept. 29, 1953), A. J. Holland (to Merck & Co.).
32. U.S. Pat. 2,582,589 (Jan. 15, 1952), H. H. Fricke (to Abbott Laboratories).
33. J. E. Ford, E. S. Holdsworth, and S. H. Kon, *Biochem. J.* **59,** 87 (1955).
34. T. H. Jukes and E. L. R. Stokstad, *Vitamins Hormones* **9,** 1 (1951).

VITAMIN D

The term "Vitamin D" is used to define vitaminlike substances with antirachitic activity. Since there are several related structures with this type of activity, it is common to speak of a D-vitamin when referring to a single member of the group and to the vitamins D when referring to the group as a whole. At present, D-vitamin activity is restricted to select *seco*-steroids, two of which are of major importance. The first, vitamin D_2, ergocalciferol, is produced by the irradiation of the provitamin ergosterol and is active in man, cattle, swine, and dogs, but not in poultry. The other, vitamin D_3, cholecalciferol, is produced by the irradiation of 7-dehydrocholesterol, and is active in all species; it is the major D-vitamin constituent of cod-liver oil. The term "D-activated plant sterols" indicates a supplement providing vitamin D_2-type activity while "D-activated animal sterols" indicates vitamin D_3-type of activity. Although the term calciferol was formerly reserved for vitamin D_2, the more precise term *ergocalciferol* is now preferred and calciferol is used generically in naming various D vitamins and their isomers (9). The D-vitamins belong to the class of fat-soluble vitamins. See also Steroids, particularly the section on naturally occurring sterols (pp. 835–838).

History

Like many other physiological factors, vitamin D started out by masquerading under the guise of a related material. In 1807 the English physician Bardsley, noted the salutory effects of cod-liver oil on osteomalacia, an adult form of rickets (10). In 1919 Mellanby demonstrated a cause-and-effect relationship between a nutritional substance and rickets by producing the disease in dogs and then either preventing it or curing it through the use of cod-liver oil (11). Mellanby ascribed the therapeutic activity to vitamin A. It then remained for McCollum in 1922 to demonstrate the presence of two factors in the oil; one labile to the combination of heat and aeration, and one relatively stable under these conditions. The less stable factor was shown to be vitamin A, while the other was proposed as a new factor, "vitamin D" (12). The well-known biochemists F. Hopkins (1906) and C. Funk (1914) were also contributors to the early concept of a dietetic factor in rickets.

By the end of the 19th century Palm had associated the incidence of rickets with the geographical factor of sunlight (13). Somewhat later, Huldschinsky demonstrated that rickets could be cured either with sunlight or with the ultraviolet rays of a mercury-vapor lamp (14). Hess and his associates arrived at the conclusion that the curative effects of sunlight and cod-liver oil were part of the same phenomenon. Steenbock at the University of Wisconsin (15) and Hess at Columbia University (16) independently discovered the possibility of enhancing the antirachitic factor by the irradiation of certain foods.

Hess later found that sterol moieties, such as plant sterols and cholesterol, could provide the D-vitamin activity upon irradiation. It was subsequently shown by other workers that pure cholesterol alone would not provide the vitamin factor after irradiation, and that an impurity in cholesterol had the same absorption maximum (282 nm) as did ergosterol. Windaus and his new collaborator Hess then demonstrated that ergosterol could be irradiated to the active product. They considered ergosterol the precursor or antirachitic provitamin. Windaus, with the aid of many graduate students, eventually elucidated the structure of ergosterol, a C_{28}-steroid with conjugated double bonds at positions 5 and 7 and an additional double bond in the side chain. The structure of the vitamin derived from it was eventually established through the efforts of Windaus, Heilbron, and Spring, and this was designated as vitamin D_2. It was isolated in the pure form at about the same time in England by Askew (17) and in Germany by Windaus (18). The active product obtained by the irradiation of impure cholesterol, however, was not identical with vitamin D_2, despite the similarity of the ultraviolet spectra (19). Windaus found that 7-dehydrocholesterol could be irradiated to a material of similar potency to vitamin D_2 (in the rat). Brockmann then isolated the active factor from fish-liver oil and showed this to be identical with the irradiated 7-dehydrocholesterol (19a). The new factor was designated as vitamin D_3. The differentiation was now clear: vitamin D_2, the product obtained by the irradiation of plant sterols, had a carbon skeleton derived from ergosterol, while vitamin D_3, the product obtained by the irradiation of 7-dehydrocholesterol or by isolation from fish-liver oil, had a carbon skeleton derived from cholesterol.

Another facet in the history of the vitamin may have been opened by the discovery in 1968 (by DeLuca and associates) of a physiologically active metabolite of vitamin D_3, 25-hydroxycholecalciferol (37). It appears that this is the compound actually carrying out the physiological function of the vitamin in vivo.

Occurrence

Vitamin D occurs in small amounts in most animals. Significant quantities are found mostly in the liver and viscera. Milk and eggs are also natural sources of the vitamin although most milk sold in the United States is now fortified because of its low natural content.

Commercially useful quantities are found in nature only in fish. The concentration in fish-liver oil varies considerably with the species and season and, to a certain extent, with other factors, such as age, climate, and food supply. Halibut livers, for example, give a high yield of low-potency oil in the summer months, and less oil with more vitamin D in the winter months. Fish heavy with body oil are the richest natural source of vitamin D.

The liver-oil content of different species may vary from 100 IU/g in the Atlantic cod to 45,000 IU/g in high-potency species of the order *Percomorphi*, eg, the oriental tuna. Sturgeon and gray-sole oils are practically devoid of activity. A list giving activity by species can be found in reference 6. The vitamin D fraction of fish-liver oil usually contains a good portion in the form of various esters.

While the fat obtained from fish contains relatively large amounts of vitamin D, the fat of other animals contains little or none. Exceptions to this rule are found in the fat of species feeding on fish, for example, certain birds.

Vitamin D occurs only rarely in plants and consequently significant amounts are not found in ordinary fresh vegetables. Mushrooms do contain small amounts even when grown in the dark, and this can be increased by growth in sunlight (6). The antirachitic properties of hay, dried alfalfa, and cacao shells are due to contamination with yeasts or molds capable of synthesizing relatively large amounts of provitamin D. This is partly converted into vitamin D during the process of sun-curing.

The vitamin D obtained from fish is predominantly D_3 but it is most often a mixture. Cod-liver oil with a chick-rat efficacy (see p. 567) of 100% (the same as pure vitamin D_3) was shown to contain at least four active fractions by molecular distillation, vitamin D_2 being present to the extent of about 10%. Although the origin of vitamin D in fish is still not settled, it would appear that little or no vitamin can be synthesized in other species without the action of sunlight. The vitamin D_3 of birds is probably obtained by the action of sunlight on the legs and less probably by the action of sunshine on the oils of the sebaceous glands, the vitamin being orally ingested during preening (5,10a). Among invertebrates the particular vitamin grouping varies with the species, but again vitamin D_3 appears to predominate. In humans, the vitamin arises from ingested sources or from irradiation of the provitamin present in the skin.

Structure and Properties of the D-Vitamins

Scheme 1 shows various structural formulations of vitamin D. The s-cis form (2) is apparently much less stable than the preferred s-trans form (1), but shows the relationship of vitamin D to the sterols. The side chain is frequently drawn in the horizontal manner, as shown in (2), but elongated forms, such as (3), are probably more representative.

Scheme 1.

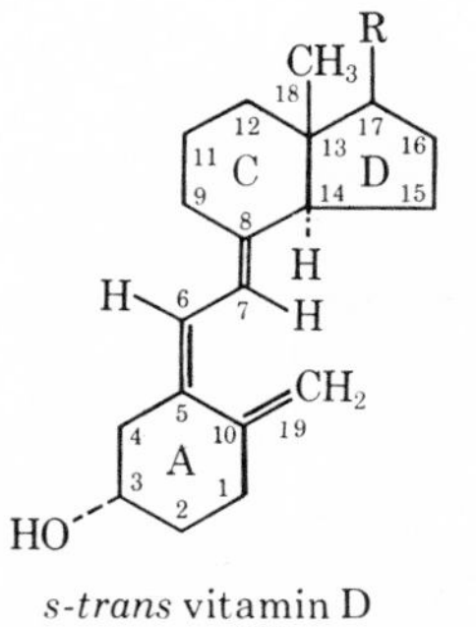

s-trans vitamin D prototype (1)

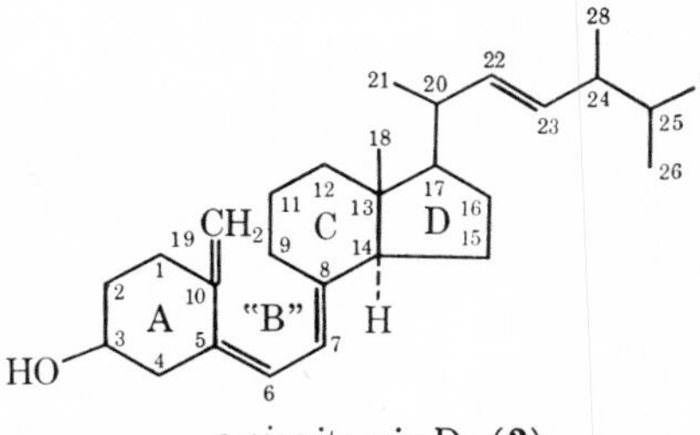

s-cis vitamin D_2 (2)

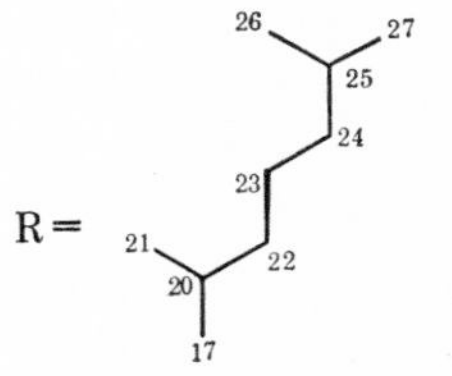

vitamin D_3 side chain (3)

Vitamin D_2 (ergocalciferol, calciferol) (1) (2), 3β-Hydroxy-9,10-seco-6-(cis)-6,7-s-trans-ergosta-5,7,10 (19), 22-tetraene, $C_{28}H_{44}O$, contains three rings, numbered and lettered in the manner of the parent steroid (2,9). There are four exocyclic double bonds. Ring A is six-membered and is connected by a dienyl bridge to the 6-membered ring of a hydrindane. A methylene on ring A completes a trienyl chromophore while the fourth double bond is isolated in the side chain. Rings C and D, the hydrindane, are joined in the trans manner as they are in ergostane. An angular methyl at C-13 projects above the plane of the ring (β) and lies opposite to an α-hydrogen at C-14. The side chain is also known to lie above the plane of the ring (2,9).

The term "seco" is employed to indicate that the compound is derived from a steroid by the opening of a ring, in the present instance opening of ergostane at the 9,10 bond. The molecule has been shown to exist in the "stretched out" form, ie the 5,6- and 7,8-double bonds lie trans with respect to the 6,7-single bond, as in (1). The trans conformation is denoted by the 6,7-*s-trans* term in the systematic name. Evidence for this conformation comes from x-ray crystallography of the iodonitrobenzoate (20), and from the high extinction value of the ultraviolet absorption maximum. Ring A is so oriented that the 4,5 bond and the C-6 hydrogen are cis. The related trans compound is described later. Although the OH group is considered β, it actually projects behind the ring in the s-trans form as compared to the angular methyl group at C-13. If the A ring is rotated with the 6,7 bond as an axis, however, the atoms of the molecule are in the physical relationship found in the steroid from which the vitamin is theoretically derived and named. In this conformation, ie (2), the OH and C-13 methyl both project above the plane, hence the "β" nomenclature.

Table 1. Physical Properties of Vitamin D_2 and D_3

Property	Vitamin D_2	Vitamin D_3	Reference[a]
melting point, °C	115–118	84–85	
color and form	colorless prisms, from acetone	fine colorless needles, from dil acetone	
optical rotation, $[\alpha]_D^{20}$			
acetone	+82.6°	+83.3°	
alcohol	+103°		
chloroform	+52°	+51.9°	(21)
ether	+91.2°		
petroleum ether	+33.3°		
temp coeff of rotation per °C, in alcohol	0.515		(22)
uv max, nm	264.5	264.5	
specific absorption, $E_{1cm}^{1\%}$	458.9 ± 7.5	473.2 ± 7.8	
potency,[b] IU/g	40×10^6	40×10^6	
biological activity	mammals	mammals, birds	
chick-rat efficacy, %	1–2[c]	100	
solubility, g/100 ml			(18,23)
acetone at 7°C	7		
acetone at 26°C	25		
abs alcohol at 26°C	28		
ethyl acetate at 26°C	31		
water	insol		

[a] Adapted from reference 4, except where indicated otherwise.

[b] The international standard for vitamin D is an oil solution of activated 7-dehydrocholesterol. The IU is defined as the biological activity of 0.025 μg of pure cholecalciferol.

[c] Recent studies have claimed an efficacy as high as 10% (110).

Vitamin D_3 (cholecalciferol (**3**), 3β-Hydroxy-9,10-*seco*-6-(cis)-6,7-*s-trans*-cholesta-5,7,10 (19)-triene $C_{27}H_{44}O$, differs from its D_2 analog in the absence of both a double bond and a methyl group in the side chain. The physical properties, listed in Table 1, are quite similar, but the D_2 form lacks the strong activity in birds found in the D_3 form.

Other D-Vitamins. *Vitamin D_1* is an early misnomer, the compound (mp, 124°C) being actually a molecular compound of vitamin D_2 and lumisterol$_2$, as proved by Windaus' elegant melting-point curve (24).

Vitamin D_4, $C_{28}H_{46}O$,

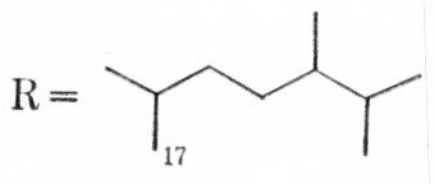

R =

has somewhat less activity than vitamin D_3 (about 30×10^6 IU/g) (mp, 96–98°C; $[\alpha]_D^{18} +89.3°$ in acetone). It was prepared by the irradiation of 22,23-dihydroergosterol by Windaus and Trautmann (25).

Vitamin D_5, $C_{29}H_{48}O$,

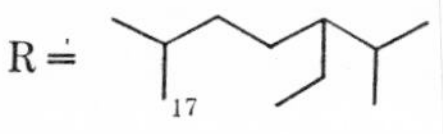

R =

has very little activity and was prepared by the irradiation of 7-dehydro-β-sitosterol by Wunderlich (26).

Vitamin D_6, $C_{29}H_{46}O$ or $C_{29}H_{46}O_2$,

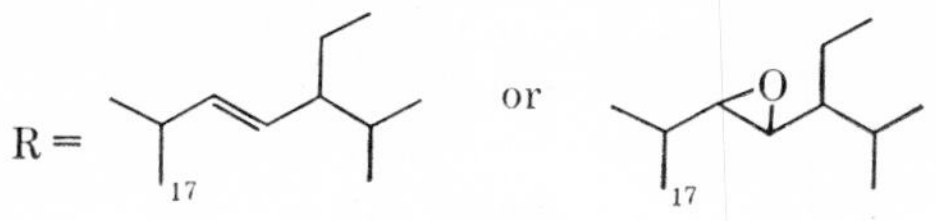

R =   or

is the irradiation product of 7-dehydrostigmasterol (rye-germ oil) (27) and its 22,23-epoxide (28). It has little activity, if any.

Vitamin D_7, $C_{28}H_{46}O$,

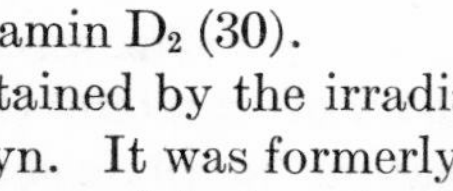

R =

has about $\frac{1}{10}$ the activity of vitamin D_2 and was prepared by irradiation of 7-dehydrocampesterol (29). The term D_7 was also applied by Raoul and associates to a cod-liver-oil fraction obtained by ether extraction of a potassium ethylate solution of the oil, and to certain derivatives of cholesterol prepared by the same group via the Salkowski reaction (a sulfuric acid treatment of a solution of the sterol in chloroform). The activity was about $\frac{1}{80}$ that of vitamin D_2 (30).

Vitamin D_m is a product obtained by the irradiation of sterols derived from the mussel, *Modiolus demissus* Dillwyn. It was formerly an important source of D_3-type activity for poultry feed. The mussel is found off the coast of New Jersey. The vitamin was isolated through the efforts of Petering and Waddell and was differentiated from vitamin D_3 despite the presence of substantial quantities of the latter in the irradiation product (31). The activity is about 30×10^6 IU/g. The melting point of the 3,5-dinitrobenzoate is 128–128.5°C and $[\alpha]_D^{25} + 90.8°$ (chloroform).

Ketone 250, $C_{27}H_{46}O_3$, is a hydroxy ketone; mp, 73°C; $[\alpha]_D$ +30° (alc); uv max, 250 nm (ϵ = 22,000) (32,33). It is reported to have about 1/10 the activity of vitamin D_3 in both rats and chicks; its calcium enolate has an activity equal to that of vitamin D_3 (see reference 3, pp. 163–168). It must be stated, however, that other investigators have had difficulty in reproducing this work (34).

Nor-compounds, obtained by irradiating 5,7-norcholestadiene-3β-ol (35) and Δ-5,7,22-cholestatrien-3β-ol (28-norergosterol) (36), show vitamin D activity.

25-Hydroxycholecalciferol, $C_{27}H_{44}O_2$,

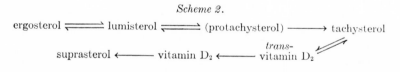

is reported to have 1.4 times the activity of vitamin D_3 (37). It was isolated by fractionation of hog plasma extracts by DeLuca and Schnoes and appears to be the active metabolite in vivo. It was characterized by its mass spectrum (mol wt = 400.3340) and its NMR curve. It was synthesized by DeLuca's group by several routes including the application of the Hunziker-Müllner technique to 25-hydroxy-cholesteryl diacetate (see under 7-dehydrocholesterol). This gave, after removal of the acetate, cholesta-5,7-dien-3β,25-diol. Irradiation of the latter gave the active product (38). Other syntheses have been reported (39,40). The corresponding vitamin D_2 analog has also been isolated (41).

Photochemical and Thermal Conversions

Windaus originally postulated that the irradiation of ergosterol proceeded along the pathway shown in Scheme 2 which illustrates the older concept of the irradiation sequence. The formation of the various products is a function of wavelength and duration of irradiation (42).

Scheme 2.

ergosterol $\rightleftharpoons$ lumisterol $\rightleftharpoons$ (protachysterol) $\longrightarrow$ tachysterol

suprasterol $\longleftarrow$ vitamin D_2 $\longleftarrow$ *trans-* vitamin D_2

About twenty years later, the isolation of a key intermediate on the pathway from ergosterol to vitamin D_2 was announced by Velluz. The results of intensive studies by Velluz (42a) and by Havinga (43) using tracer techniques, kinetic measurements, and quantum-efficiency determinations, finally led to the conclusion that many of the products thought to be on the direct pathway to the vitamin were merely reaction side-products of the new intermediate, preergocalciferol. The term provitamin is now reserved for the substance to be irradiated, and the term previtamin is used for the intermediate.

The discovery of preergocalciferol by Velluz was a result of his studies of the effect of temperature on the photochemistry of ergosterol (44). By maintaining the solution at 20°C during and after the irradiation, he obtained an intermediate which would give the vitamin upon mild heating and without the necessity of additional irradiation.

Preergocalciferol (**5**) was isolated as an oil, $[\alpha]_D$ +43° (chloroform), uv max 262 nm (ϵ = 9,000); the 3,5-dinitrobenzoate melts at 104°C, $[\alpha]_D$ +45° (chloroform). The compound has four double bonds but does not give formaldehyde on oxidation and lacks a characteristic band near 11.1 μ, indicating the absence of a CH_2= group. The low ultraviolet extinction coefficient (less than half that of tachysterol) is indicative of a cis orientation of the 6,7-double bond. Iodine, a typical cis-trans isomerization catalyst, instantly converts the compound into tachysterol₂, the trans isomer.

Precholecalciferol, the related intermediate of the vitamin D₃ series, was isolated as the 3,5-dinitrobenzoate (mp 110–111°C, $[\alpha]_D^{18}$ +52° (chloroform)) by Velluz in 1949 (44,45).

The current concept of the irradiation sequence of ergosterol is shown in Scheme 3. The measured quantum yields are included. In this sequence ergosterol is seen to undergo an initial photoactivated ring opening to preergocalciferol which is then converted to the vitamin by a thermally induced isomerization and double-bond rearrangement (43). The conversion of 7-dehydrocalciferol to cholecalciferol is assumed to occur in an analogous manner.

Scheme 3.

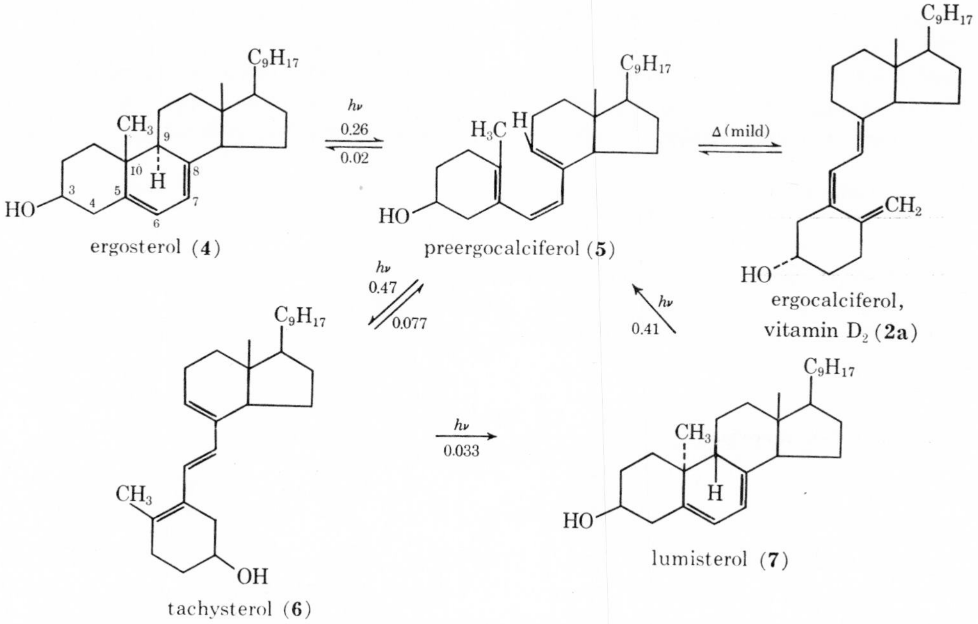

ergosterol (**4**)

preergocalciferol (**5**)

ergocalciferol, vitamin D₂ (**2a**)

tachysterol (**6**)

lumisterol (**7**)

The effect of syn-anti isomerism on the ease of ring opening and closing in structures such as these was studied by Dauben and Fonken (46). Following a suggestion by Oosterhoff cited by Havinga and co-workers (43), the characteristics of the highest occupied π molecular orbital were utilized by Woodward and Hoffman, and by Zimmerman and others in interpreting these reactions (47,48). Some of the properties of the relevant systems are summarized in Table 2.

The *p* orbitals contributing to the molecular orbital have positive and negative lobes indicated by the shaded and unshaded portions. In relating the orbital state to the ring opening or closing, the pattern of the arrangement of the end lobes is noted. In ring closing, the σ orbital is considered to arise from the coalescing of two lobes of the same sign, a bonding interaction. The coalescence of two lobes of opposite sign is considered antibonding and would give rise to a bond of much higher energy. The bond forming may be thought of as arising from a coalescence in either a conrotary

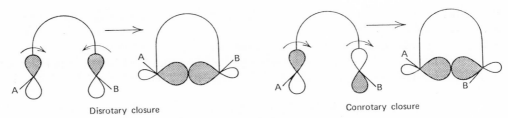

Fig. 1. Ring-closing modes of orbital lobes.

mode in which the lobes rotate in the same clock direction or in a disrotary mode in which the lobes rotate in opposite clock directions, as illustrated in Figure 1.

In structures, such as the unsaturated steroids or *seco*-steroids of the vitamin D series, where the ends of the conjugated diene system are fused onto other ring systems, conrotary closure of the new ring is associated with anti-positioning of the two attached groups, and disrotary closure is associated with syn-positioning. The same considerations apply to ring openings. This can be related to the photoexcited state of pre-ergocalciferol which has oppositely arranged lobes in ψ_4, the highest occupied molecular orbital. Ring closure of this excited state preferentially occurs in the conrotary mode giving rise to antisubstituents, such as those found in ergosterol (9α-H, 10β-CH$_3$).

Table 2. Orbital-Lobe Characteristics and Ring Closures of Relevant Polyenes

System	Triene	Diene
reaction		
end lobes, highest occupied π orbital		
excited state	ψ_4	χ_3
ground state	ψ_3	χ_2
photochemical reaction preference	conrotary	disrotary
thermal reaction preference	disrotary	conrotary

Similarly, photoexcited ergosterol with its combination of antisubstituents and oppositely disposed end lobes in the key orbital (ψ_4) preferentially opens in the conrotary manner.

Mild heating of the previtamin (preergocalciferol) affords ergocalciferol, vitamin D_2 (**2**). Irradiating the previtamin instead of heating it, on the other hand, gives tachysterol$_2$, its 6,7-trans isomer and a by-product in the manufacture of vitamin D_2. Further irradiation of tachysterol$_2$ affords lumisterol$_2$, the product of another trienyl conrotary photoactivated ring-closure to anti substituents. In lumisterol, the C-9 H and C-10 CH_3 groups are β-α compared to the α-β relationship found in ergosterol. Irradiation of lumisterol$_2$ leads irreversibly to preergocalciferol in good quantum yield. The lack of conversion of preergocalciferol to lumisterol has not been satisfactorily explained.

Heating ergocalciferol (**2**) affords pyro- and isopyrocalciferol (8a) and (8b), respectively, as shown in Scheme 4. These compounds are syn isomers of ergosterol having both the C-9 H and C-10 CH_3 in either the α- or β-position. The transformation is presumed to pass through preergocalciferol (43). It is to be noted from Table 2 that the highest occupied π orbital for these thermal reactions is ψ_3, and the preferred ring closing is now disrotary with the attending preference for syn derivatives.

Scheme 4.

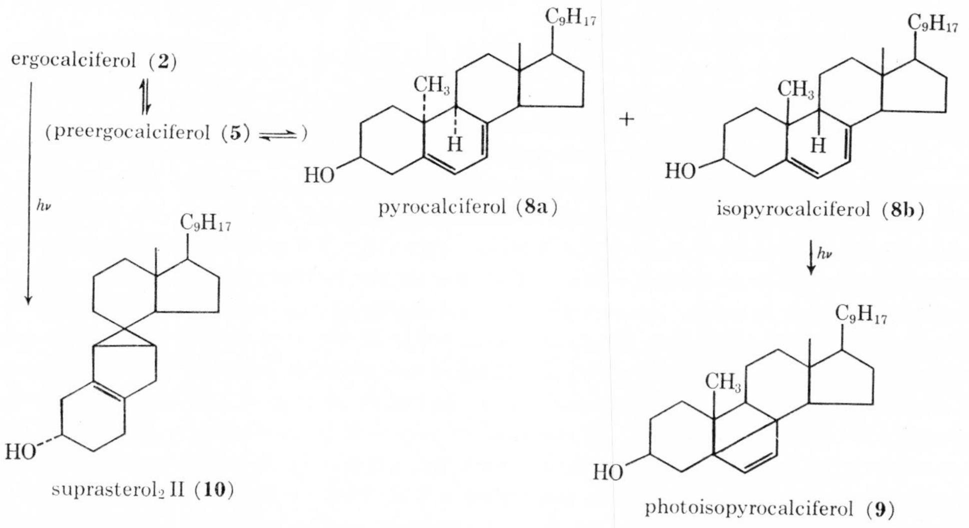

ergocalciferol (**2**)

(preergocalciferol (**5**) ⇌)

pyrocalciferol (**8a**)

isopyrocalciferol (**8b**)

suprasterol$_2$ II (**10**)

photoisopyrocalciferol (**9**)

Photoexcitation of the isopyro compound with its syn substituents does not lead to ring opening as does the irradiation of ergosterol. Here again the ψ_4 orbital is involved in the excited state, but the preferred combination of antisubstituents and a conrotary opening is absent due to the syn orientation of the substituents. The butadienyl system present from C-5 to C-8, on the other hand, has end lobes with similar orientation in the highest occupied π orbital of the excited state (X_3); disrotary closure and syn orientation are now preferred. On irradiation, the butadiene portion of the B ring closes to form the bicyclohexene ring of photoisopyrocalciferol (46). Dauben and Fonken postulate a transition state which could collapse either to the bicyclohexene derivative or to the previtamin open-ring compound depending on the syn-anti disposition of the substituents (46).

Irradiation transforms ergocalciferol into an inert series made up of toxisterol$_2$, suprasterol$_2$ I, and suprasterol$_2$ II. The last has been assigned a structure (**10**) containing two five-membered rings, two six-membered rings, and one three-membered ring (46), as shown in Scheme 4.

Provitamins D

The provitamins D are compounds that can be activated to a vitamin D structure. Chemically they are 3-hydroxysteroids with double bonds at C-5 (6) and C-7. Unlike the D-vitamins, themselves, ergosterol, 7-dehydrocholesterol, and most of the other provitamins are precipitated by digitonin. The provitamins D are not known to have any physiological function of their own. They can, however, be synthesized in the animal kingdom, as well as by plant life. In many animals the appropriate provitamin can be transported within the body and deposited in the skin where sunlight can reach it and convert it into vitamin D. Certain foods and feeds contain small amounts of provitamins D, and it is assumed that the provitamins are also absorbed via the intestinal tract. There seem to be a considerable number of naturally occurring provitamins D, but the exact number and the constitution of many of them are unknown. A number of provitamins have been synthesized from available sterols. Ergosterol (4) and 7-dehydrocholesterol (17) are described in further detail below. Others that have been sufficiently characterized may be identified with reference to the vitamin D which they produce on irradiation (see p. 553).

Isolation. Procedures for isolation are commercially important since ergosterol is at present produced exclusively from biological sources. The commercial product is about 90–100% pure and often contains up to 5% of 5,6-dihydroergosterol.

In general the isolation of provitamins D from natural sources involves two different steps: (a) the isolation of the total sterols and (b) the separation of the provitamins from other sterols. The isolation of the total sterols is usually a relatively simple process, and consists in extracting the total fat, saponifying it and then reextracting the unsaponifiable portion with an ether. The sterols are in the unsaponifiable portion. An alternate route involves the saponification of the total material followed by the isolation of the nonsaponifiable fraction. Separation of the sterols from the unsaponifiable fraction is carried out by crystallization from a suitable solvent, such as acetone or alcohol. Ethylene dichloride, alone or mixed with methanol has been used commercially for recrystallization. In the case of yeasts it is particularly difficult to remove the ergosterol by simple extraction since this gives only about one-quarter of the ergosterol. In industry, therefore, the ergosterol is obtained by preliminary digestion with hot alkalies or with amines (50–55). Variations in the isolation procedure have been developed. For example, after saponification, the fatty acids may be precipitated as calcium salts which tend to adsorb the sterols. The latter are then recovered from the dried precipitate by solvent extraction.

Separation of the provitamins from other sterols is usually a difficult problem and success depends largely upon the type and amount of provitamin D present. The purification of ergosterol is not too difficult since the crude sterols from which it is obtained contain about 90% of the provitamin, and it is usually accomplished by crystallization from benzene–alcohol. Where very little provitamin is present, on the other hand, isolation is often difficult or impossible. Such is the case with the spinal cord of cattle, a major source of cholesterol. Here the sterol content is essentially all cholesterol, and the 7-dehydrocholesterol is present only to the extent of about 0.1%. Isolation is usually more successful where the provitamin D is present in a concentration of at least 4–5%. The usual method is fractional adsorption of the sterols or of

their esters.   In some cases, alumina permits an almost quantitative separation (56).
Of scientific interest is the method of forming a difficultly soluble condensation product
of the provitamin with a dienophile, such as maleic or citraconic anhydride.   After
separation from the much more soluble sterols, the adduct is split by thermal decomposition into the provitamin D and the acid anhydride.

Occurrence.   The unsaponifiable fractions of the fat of nearly all plant and animal
tissue exhibit the provitamin absorption spectrum.   Table 3 gives the concentration of
provitamin in a variety of plant and animal tissues (6).   In general the provitamin of

Table 3.   Occurrence of the Provitamins D in Selected Plants and Animals,
Parts per Thousand of Total Sterol

cottonseed oil	28	*Modiolvs demissus,*	
rye grass	15	ribbed mussel	370
wheat germ oil	10	*Ostrea virginica,* oyster	80
carrot	1.7	*Ostrea edulis,* oyster	34
cabbage	0.5	*Asterias rubens,* starfish	3.8
skin, pig	46	common sponges	20
skin, chicken feet	25	common coral	10
skin, rat	19	*Aspergillus niger,* mold	~1000
skin, mouse	9	*Cortinellus shiitake,*	
skin, calf	7	mushroom	~1000
skin, human adult	4.2	*Claviceps purpurea,* ergot	900
skin, human infant	1.5	*Saccharomyces cerevisiae,* yeast	800
liver, Japanese tuna	11	*Pencillium puberulum,* mold	280
liver, Atlantic cod	4.4	*Fucus vesiculosus,* alga,	
liver, shark	1.0	seaweed	0.8
liver, halibut	0.60	*Tubifex* (Sp.), waterworm	210
liver, tuna	1.0	*Lumbricus terrestris,*	
eggs, Chinese duck	60	earthworm	170
eggs, cod (roe)	5.5	*Tenebrio molitor,* mealworm	120
eggs, hen	1.6	*Gyronomus* (Sp.), gnat	61
wool fat, sheep	3.9	*Cancer pagurus,* common crab	15
milk, cow	2.3	*Daphnia* (Sp.), water flea	7.5
pancreas, beef	1.8	*Musca domestica,* housefly	7.0
blood serum, cow	1.5	*Crangon vulgaris,* shrimp	3.8
spinal cord, beef	1.2	*Homarus vulgaris,* lobster	2.5
herring oil	0.50	*Helix pomatia,* edible snail	97
heart, calf	0.32	*Arion empiricorum,* slug,	
gallstones, man	0.25	red road snail	220
Mytilus edulis,		*Littorina littorea,* periwinkle	170
sea mussel	100	*Sepia* (Sp.), cuttlefish	12

the vegetable kingdom is ergosterol, although 22-dihydroergosterol has also been isolated.   In some cases the presence of provitamin of plant origin in the animal species
can be directly related to absorption of ergosterol from feeding material.   Such is the
case with hen's eggs which contain ergosterol.   It has been shown specifically that
chickens can absorb ingested ergosterol, a distinction from other animal species studied.

It is assumed that the provitamin of vertebrates is essentially 7-dehydrocholesterol
which is absorbed as such in higher animals or can be formed in vivo in the gut wall and
stored in the liver (49).

Properties and Manufacture.   *Ergosterol,* $C_{28}H_{44}O$ (mp 165°C; $[\alpha]_D^{20}$ −130°
in $CHCl_3$),

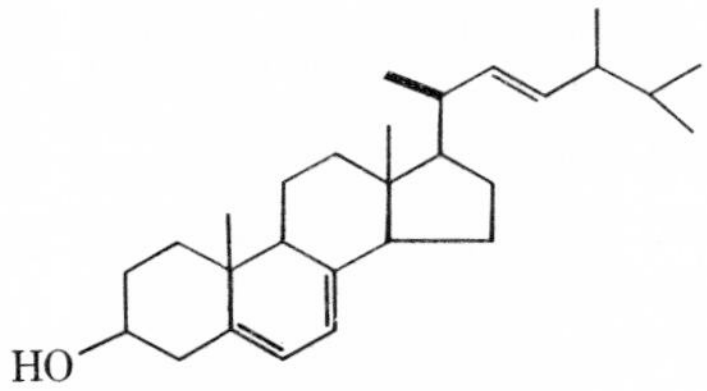

ergosterol (**4**)

crystallizes from alcohols and acetone in small colorless plates with water of crystallization, and from anhydrous solvents, such as ethyl acetate or ether, in the form of fine needles (see also Vol. 20, p. 836). The melting point varies according to the degree of hydration. The best crystallized preparation contains 1.5 moles of water and melts at 168°C (57). Complete dehydration is very difficult to achieve and results in a product with a melting range of 166–183°C. Ergosterol distills in high vacuum at 250°C without decomposition. The uv absorption spectrum has maxima (in alcohol) at 271, 281, and 293 nm. Hogness (92) reported an additional maximum at 262 nm, but Huber (92a) found only an inflection. The molar extinction coefficient at 281 nm was found to be 10,600 by Hogness and 11,500 by Huber.

Among the best natural sources of ergosterol are yeasts and molds. Brewer's and baker's yeast (*Saccharomyces cerevisiae*) are typical raw materials employed for industrial production. See also Yeast. Other sources of commercial interest are *Aspergillus niger* from citric acid fermentation and *Penicillium mycelia* from the fermentation of antibiotics (58). The yields from these molds are low, however, and the yeasts remain the important source. Extensive studies have shown that the yield can be influenced to an appreciable extent by the conditions under which the yeast is grown and by the particular strain employed. The production of 1 lb of ergosterol requires about 310 lb of compressed yeast or 95 lb of yeast solids.

7-Dehydrocholesterol, $C_{27}H_{44}O$ (mp 150–151°C, $[\alpha]_D^{20}$ −113.6° in $CHCl_3$),

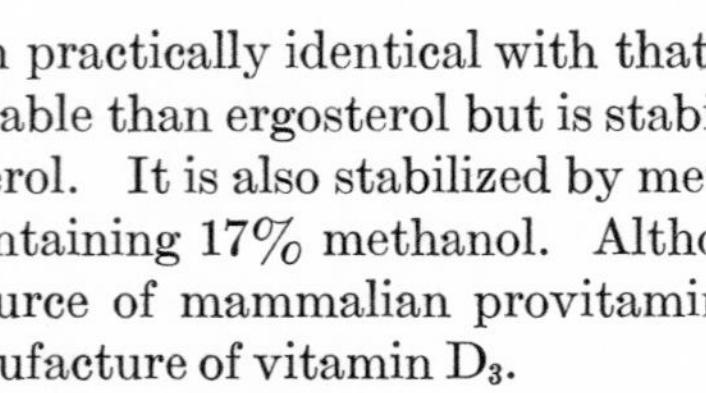

7-dehydrocholesterol (**17**)

has an absorption spectrum practically identical with that of ergosterol. The D_3 provitamin is noticeably less stable than ergosterol but is stabilized to a remarkable extent by the presence of cholesterol. It is also stabilized by methanol, and the product has been supplied as a paste containing 17% methanol. Although 7-dehydrocholesterol is undoubtedly the major source of mammalian provitamin, the synthetic material is invariably used for the manufacture of vitamin D_3.

There are two principal commercial routes. The Windaus procedure (59), shown in Scheme 5, originally gave only a 4% yield, but was improved considerably by subsequent modifications. A cholesteryl ester, usually the acetate, is the starting material which is oxidized to a 7-keto derivative with a chromium compound. Chromium trioxide, the original reagent, gives a higher yield when used with glacial rather than 90% acetic acid (60). Higher yields have also been claimed for *t*-butyl chromate (61). A patent has issued claiming a good yield with sodium chromate in a mixture of acetic acid and acetic anhydride (62). Reduction of the ketone with aluminum isopropoxide gives a mixture of the α and β 7-hydroxy compounds containing 80–90% of the α-form. The acetoxy group is concomitantly converted to the 3-hydroxy group by interchange of the acid residue with the isopropyl alcohol. Lithium aluminum hydride has also been mentioned as a reducing agent (63). The 7-hydroxycholesterol is dibenzoylated and the 7-benzoyloxy group is converted to a 7,8-double bond by splitting out the elements of benzoic acid thermally, or by treatment with dimethylaniline (64). Hydrolysis then gives the desired product. Isodehydrocholesterol (18) is a by-product.

Scheme 5.

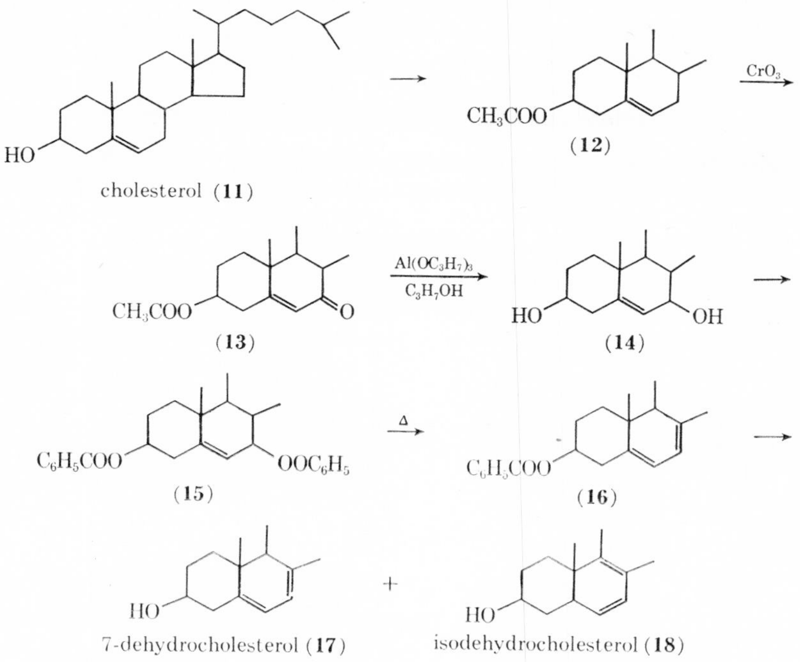

cholesterol (11) → (12) → (13) → (14) →

(15) → (16) → 7-dehydrocholesterol (17) + isodehydrocholesterol (18)

The second route gives somewhat higher yields and is much shorter. It involves the Ziegler allylic bromination of a cholesteryl ester followed by a dehydrohalogenation step (see Scheme 6) (65). The benzoylated ester is commonly used and treated with *N*-bromosuccinimide in petroleum ether (66), bromine in carbon disulfide (photocatalyzed) (67), or, more recently, with *N,N*-dibromo-5,5-dimethylhydantoin in an alkane solvent (72). Hindered nitrogenous bases, such as *s*-collidine, dimethylaniline, or quinaldine, are used to achieve dehydrohalogenation (68–71). Trimethyl phosphite was also used for the dehydrohalogenation and the combination of the dibromohydantoin and trimethyl phosphite gave a crude product containing 51% of the provitamin ester (21) (72).

Scheme 6.

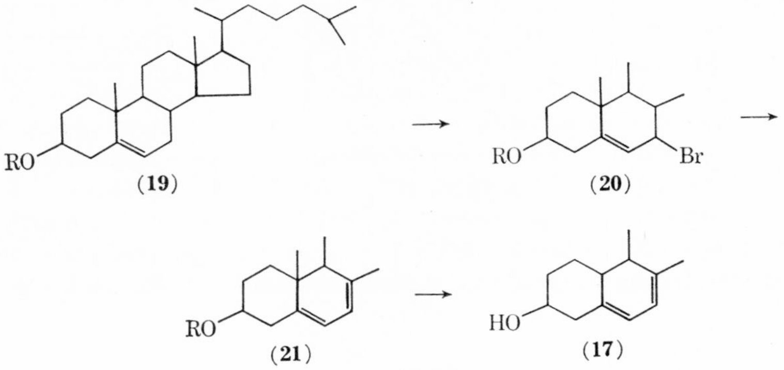

(19)   →   (20)   →

RO   RO   Br

(21)   →   (17)

RO   HO

The method of Wildi involves passing oxygen into a solution of cholesteryl acetate in cumene at 130°C in the presence of a small initial amount of cumene hydroperoxide (73). The 7-ketocholesteryl acetate (13) is isolated easily in pure form in 27% yield.

Synthesis of Ergocalciferol and Cholecalciferol

Ergocalciferol. (See section on Structure and properties for numbering and special usage of cis, β, etc.) The partial synthesis of ergocalciferol is shown in Scheme 7.

Scheme 7.

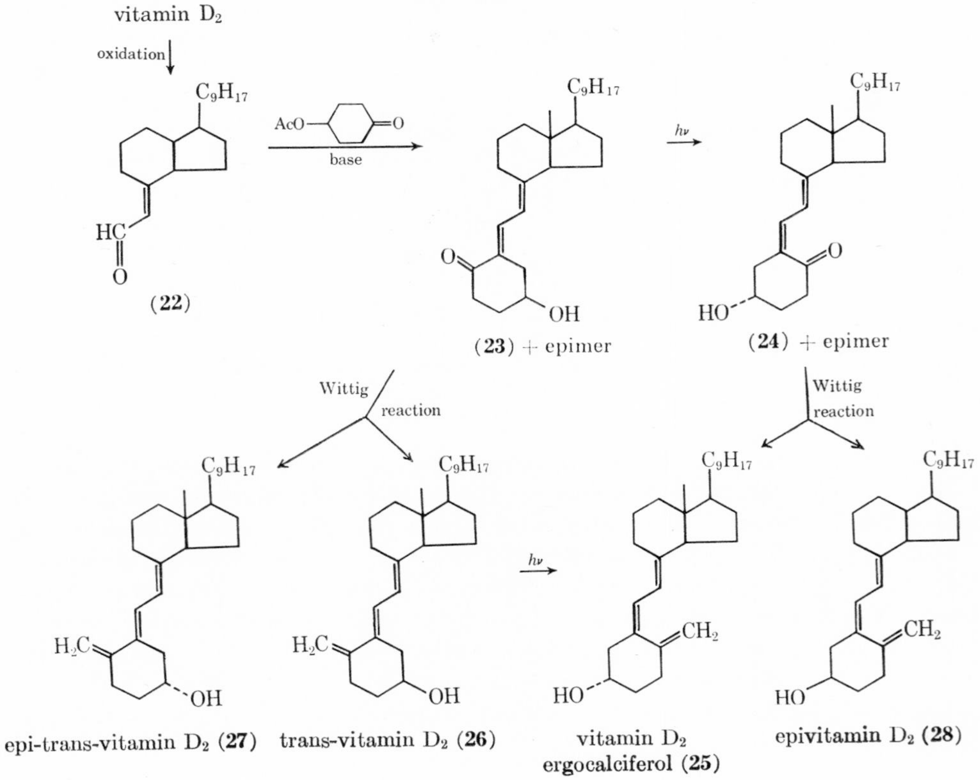

vitamin D$_2$

oxidation ↓

(22)   (23) + epimer   (24) + epimer

epi-trans-vitamin D$_2$ (27)   trans-vitamin D$_2$ (26)   vitamin D$_2$ ergocalciferol (25)   epivitamin D$_2$ (28)

The starting aldehyde (22) was a known degradation product of vitamin D$_2$ obtained by Heilbron via oxidation with permanganate or chromic acid (73a). In the partial synthesis of vitamin D$_2$ reported by Harrison and Lythgoe (74) the *trans*-hydroxyketone (23) (actually, an α and β OH-epimer pair) was first photoisomerized to

Scheme 8.

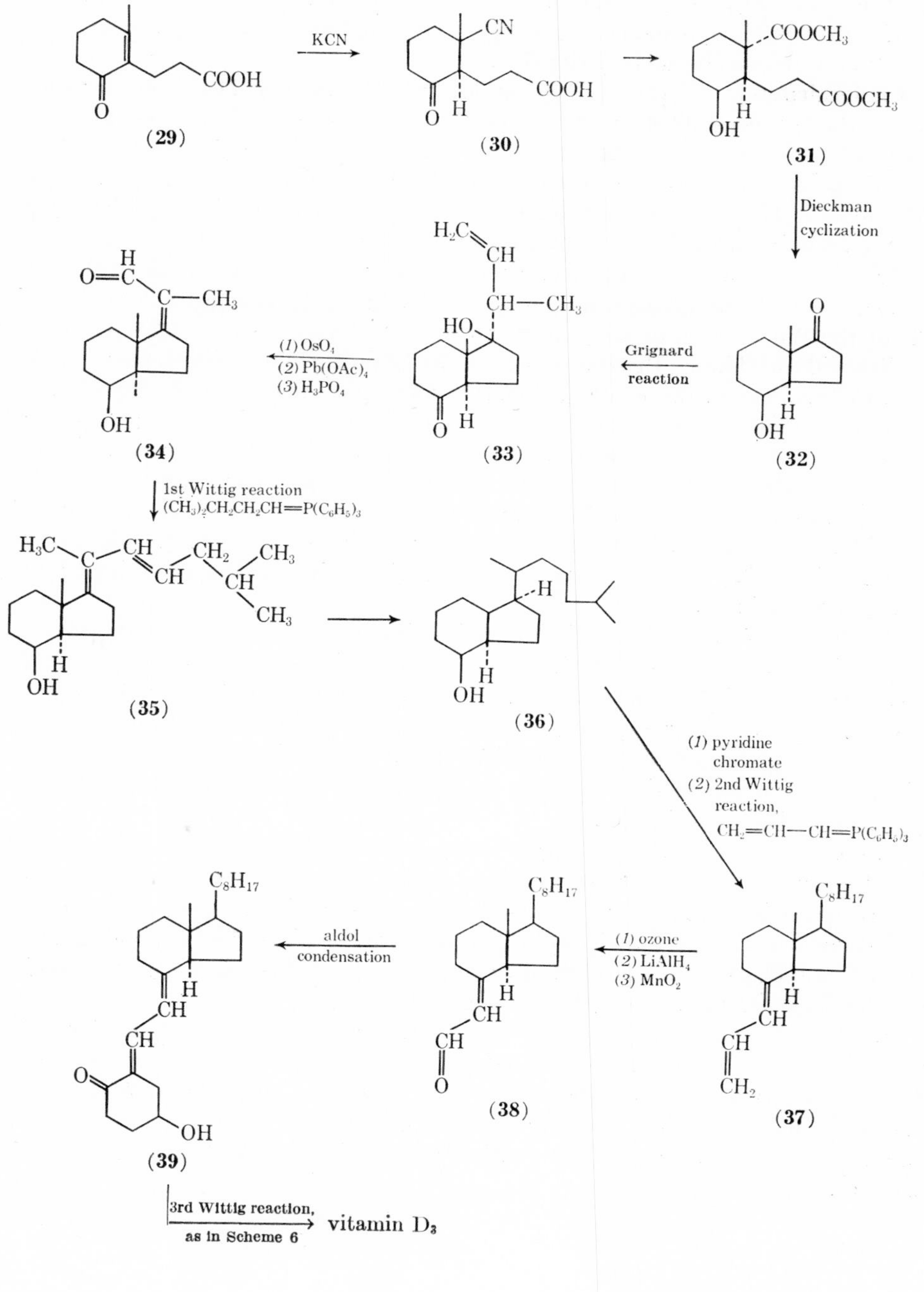

(29)

KCN

(30)

(31)

Dieckman cyclization

(34)

(1) OsO$_4$
(2) Pb(OAc)$_4$
(3) H$_3$PO$_4$

(33)

Grignard reaction

(32)

1st Wittig reaction
(CH$_3$)$_2$CH$_2$CH$_2$CH=P(C$_6$H$_5$)$_3$

(35)

(36)

(1) pyridine chromate
(2) 2nd Wittig reaction,
CH$_2$=CH—CH=P(C$_6$H$_5$)$_3$

(39)

aldol condensation

(38)

(1) ozone
(2) LiAlH$_4$
(3) MnO$_2$

(37)

3rd Wittig reaction,
as in Scheme 6 $\longrightarrow$ vitamin D$_3$

the corresponding *cis*-ketone pair, the synthetically required epimer being shown as (**24**). The ketone pair was then treated with a Wittig reagent (methylenetriphenylphosphorane) to obtain a mixture of vitamin D_2 (3β-OH) (**25**) and its epimer (3α-OH) (**28**). The vitamin was isolated from its epimer by chromatographic separation of the esters. In his partial synthesis, Inhoffen first isolated the β-epimer of the *trans*-hydroxyketone (**23**) and then converted it to the corresponding methylene derivative (**26**), the trans isomer of vitamin D_2, and isomerization of epi-trans-vitamin D_2 (**27**) gave epi-vitamin D_2 (**28**). Isomerization of the trans product gave ergocalciferol, vitamin D_2. He also isolated epi- and epi-*trans*-ergocalciferols (**28**) and (**27**) (75). The sequence of reactions is parallel in the D_3 series (75).

Cholecalciferol. Vitamin D_3 (cholecalciferol) was the first D vitamin for which a total synthesis was established. Since the vitamin had been prepared previously from natural cholesterol via the sequence 7-bromocholesterol, 7-dehydrocholesterol, and irradiation, the total syntheses of cholesterol by Woodward and by Robinson (1952–1953) (see Vol. 18, p. 856) constituted the first such pathway. In 1959 a direct synthesis of the vitamin skeleton was announced by Inhoffen, as shown in Scheme 8.

This overall synthesis is particularly noteworthy in its extensive use of the Wittig reaction, a new synthetic tool discovered in the 1950s. Inhoffen stated at the time that without the Wittig reaction no complete synthesis of vitamin D could have been accomplished (76). The synthesis consisted of the following four major stages:

1. Construction of the nucleus, 8-methyl-*trans*-hydrindanol-one-1 (**32**) (77). The 1,4-addition of KCN across the double bond of 3-methyl-2-(2-carboxyethyl)-2-cyclohexenone (Nientzescu acid) (**29**) yielded the nitrile (**30**) which, after hydrolysis, yielded a product which was isomerized to the pure *trans*-acid racemate. The *trans*-acid was resolved and the desired antipode esterified and reduced to give the hydroxydimethyl ester (**31**). Cyclization, hydrolysis and decarboxylation afforded the dextrorotatory hydrindanolone (**32**) which was identical with the hydrindanolone obtained by degradation of vitamin D_3.

2. Attachment of the isooctyl side-chain with its asymmetric carbon atom (78). This was initiated by an anomalous Grignard alkylation with crotyl magnesium bromide. Hydroxylation, oxidation, and dehydration afforded the nor-aldehyde (**34**) (3-carbon chain). The first Wittig reaction converted the aldehydic oxygen into an isoamylidene residue yielding the 1,4-dione (**35**). Catalytic reduction then gave the desired saturated side chain possessing the proper (β) orientation with respect to the angular methyl group as in (**36**).

3. Formation of the vinyl aldehyde group with bridging carbon atoms at 6 and 7. The saturated C_{18}-alcohol (**36**) was oxidized to the corresponding keto compound, and use of a second Wittig reaction replaced the oxygen and added a three-carbon diene to the six-membered ring (**37**). Selective ozonolysis of the distal double bond and reduction with $LiAlH_4$ gave the nor-alcohol which was then oxidized to a C_{20}-vinylaldehyde (**38**), identical to the vitamin D_3 degradation product.

4. Conversion of the aldehyde to the vitamin in its natural configuration (75) by previously described methods. Aldol condensation of (**38**) with 4-hydroxycyclohexanone gave the *trans*-hydroxyketone (**39**) which, upon treatment with a third Wittig reagent and photoisomerization in a manner analogous to that used for the partial synthesis of vitamin D_2 (see Scheme 6) gave the synthetic cholecalciferol identical with the natural substance.

Manufacture of Vitamin D

Most of the vitamin D manufactured in the United States is now produced by ultraviolet irradiation of solutions of relatively pure provitamin. The D_2- and D_3-vitamins, the only D-vitamins of commercial importance, are thus produced either from ergosterol or 7-dehydrocholesterol, respectively.

In a typical process of irradiation, recovered provitamin and sufficient new material to make up the desired charge are dissolved in peroxide-free ether and the solution is passed through a clarifying filter. The clear liquid is pumped past a bank of high pressure uv-lamps called "burners" by means of an inert gas and recycled until the desired degree of irradiation is attained. The burner output is a mixture of previtamin, vitamin, unreacted provitamin, and irradiation by-products. It is concentrated and then charged into a crystallizing vessel. The remaining ether is removed and the residual mass is slurried with methanol and chilled. The unreacted provitamin, which is largely insoluble in the methanol, is filtered off and returned to the process. The methanol is evaporated and the viscous mass poured into pans where it is allowed to "set-up" or "cure." The product, now called the "resin" or "gum," is a pale yellow to dark-amber mass which fractures easily when cold and has a marked resemblance to wood rosin. The bulk of the resin produced in this manner is used directly in formulations and a smaller quantity is crystallized.

Harris estimated that 7.5×10^{13} quanta of light are required for the production of one USP unit of vitamin D_2 from ergosterol (80). This was later reestimated at 9.3×10^{13} quanta because of problems with the older standards (see reference 6, p. 176). A variety of light sources or burners have been employed in the past, including carbon arcs, metal-cored carbon rods, magnesium arcs, and mercury-vapor lamps. The high-pressure mercury-vapor lamp is widely employed and appears to be the most satisfactory source when used in conjunction with a cell such as the one illustrated in Figure 2.

This setup utilizes a liquid-cooled quartz chamber equipped for continuous flow of the solution to be irradiated. Since light of wavelength 275-300 nm apparently

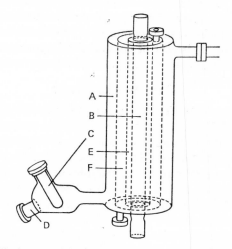

Fig. 2. Quartz irradiation vessel for the conversion of provitamin D to vitamin D.

produces the best yields of vitamin D and a small amount of by-products, filters have been used to limit the passage of undesired wavelengths. Light below 275 nm can be filtered out by aromatic compounds or by a 5% lead acetate solution; other means are available for removing light above 300 nm. Glass filters are also useful for this purpose. See also Photochemical technology.

Although dry irradiation is possible, poor yields are obtained because of the high degree of surface irradiation compared to the light reaching the inner part of the crystals and therefore a solvent is preferred. Ethyl ether is most commonly employed, permitting more rapid activation than is possible with alcohol. Since the temperature coefficient of activation is very small, the low boiling point of the ether is not a drawback although the flammability is definitely a safety problem. Solvents, including cyclohexane and dioxane, have been recommended as alternatives. Photosensitizers, such as eosin, erythrosin and dibromodinitrofluorescein, have been suggested. Various stabilizers, including BHA and BHT (see Vol. 2, p. 593) may be added. For best yields, the concentration of provitamin is kept low, ie 1% or less, and the activation is conducted so as to convert 30–60% of the provitamin. Yields are in the 30–60% range.

Although the provitamins can be irradiated in the oil base used as the carrier of a "vitamin in oil" preparation, the occurrence of various side reactions and the lower optical transparency of the oil combine to render the normal irradiation in ether the cleaner, more efficient process. The vitamin in oil is alternatively prepared by adding to the oil the alcoholic solution remaining after removal of the unreacted provitamin and then evaporating off the alcohol under vacuum or simply by dissolving solidified resin in the oil. To prepare the crystalline USP product, the crude resin is converted to an ester which is then purified by crystallization. After hydrolysis of the ester, the recovered vitamin is crystallized from a suitable solvent. Acetonitrile has been found to be an excellent solvent for this step (81).

The irradiation of various food products for vitamin D fortification has been replaced to a large extent by addition of manufactured vitamin. Thus, the irradiation of milk, which was of considerable importance in the past, has now been superseded by the addition of special formulations containing the concentrated vitamin in a milk or propylene glycol base. The desired unitage is simply added to the raw milk prior to pasteurization. This system overcomes the effect of irradiation on the taste and odor of milk and is both more efficient and economical. The irradiation of yeast to produce a material rich in vitamin D_2 is still of commercial importance, however, and this is accomplished by irradiating layers of an aqueous slurry and spray-drying the product.

Standards and Assay

The Vitamins. Pure vitamin D_3 was adopted in 1949–1951 as the new international (82) as well as the USP standard. By definition, one unit of vitamin D activity (one IU) is represented by 0.025 μg of vitamin D_3. One gram of vitamin D_3 equals 40 million IU or 40 million USP units. The international chick unit (ICU) of vitamin D is also identical with the USP unit (83). Samples of USP reference standard may be purchased from USP Reference Standards, New York. (Care must be taken to distinguish the above units from "standard units," a unit of commerce. One standard unit is equal to one million USP units.)

Prior to the adoption of crystalline vitamin D_3, the D vitamin standard was an olive oil solution of ergosterol irradiated under standard conditions. The change from vitamin D_2 to D_3 was made because of the efficacy of the latter in both the rat and the chick. A useful table of conversion factors for a number of the older units can be found in reference 5, p. 147 and in reference 6, p. 222.

There are three principal methods of assay for the D vitamins:

The *physical method* involves determination of the characteristic uv absorption spectrum at 265 nm. It is fairly accurate in the absence of other substances that absorb in the same region. Such interfering substances can be removed from concentrates of vitamin D derived from fish oils or from irradiation procedures provided they contain at least 50,000 IU/g (84).

The *chemical method* is based on the development of a salmon-pink color after removal of interfering substances. It involves the following steps (see reference 4, p. 358): (a) Preparation of the standard (USP cholecalciferol in isooctane); (b) saponification with aqueous alcoholic KOH; (c) extraction with petroleum ether; (d) removal of the solvent; (e) chromatographic removal of interferences (via three different columns); (f) development of the color with antimony trichloride; and (g) colorimetric determination of intensity (photoelectric colorimeter or spectrophotometer).

The chemical method is the official one in the *USP* since revision *XVI* (84a) but it is still not the official method of the AOAC (Association of Official Analytical Chemists). With suitable modification, it has been found satisfactory for assay of relatively high-potency materials, such as fortified milks, cereals, feeds, feed supplements, and the usual vitamin preparations, including drops, tablets, etc.

Various other chemical methods have been summarized with the following addendum: "Although considerable progress has been made in the development of chemical methods of analysis for vitamin D, those available still lack the sensitivity and specificity of biological assays, and none is yet available which specifically assays for vitamin D_3. Until these disadvantages have been overcome, a vital need will remain for the rat biological assay which measures total vitamin D activity, and the chick biological assay which measures vitamin D_3 activity" (reference 4, p. 350).

Other general methods of assay are also under development, including paper chromatography (85), thin-layer chromatography (86), vapor-phase chromatography (87), infared analysis (88,89) and NMR spectroscopy (90). The last two can be used to distinguish between vitamins D_2 and D_3.

The *biological method* also distinguishes between D_2 and D_3 and it is sufficiently accurate for determination of low concentrations. It is the official method of the AOAC (91). It must be noted, however, that it is costly, time-consuming and accurate only within the limits of biological assay. Rats and chicks are used as test animals. Weanling rats are placed on a rachitogenic diet and different levels of the unknown are fed to some animals while known amounts of reference standard are fed to others. The criterion of healing is the development after 7 days of a line of new calcification through the rachitic metaphysis made visible by staining with silver nitrate. This gives the total vitamin D concentration. Chick assays are performed when D_3 potency is required. According to the AOAC protocol, the potency is determined on the basis of the percentage of ash in the tibia after a 21-day period. Statistical treatment of the data and multiple levels are recommended.

The Provitamins. The most accurate method for the determination of provitamins D is *spectroscopic analysis*. The characteristic absorption spectrum of the provitamins in the ultraviolet region is the same for all provitamins on a molecular basis. This method, while quite accurate for the determination of provitamins D in sterol mixtures isolated from natural sources, cannot distinguish between individual provitamins (92).

There are a number of *chemical methods* that have long been associated with the qualitative or semiquantitative determination of provitamins D. They are listed below and are all based on chemical reactions given by sterols with conjugated double bonds in or about the same positions of the double bonds as in the provitamin D molecule.

1. The reversed Salkowski reaction. The provitamin D is dissolved in chloroform and concentrated sulfuric acid is added. The acid becomes deep red while the chloroform layer remains colorless. Sterols lacking the conjugated double bonds, such as cholesterol and sitosterol, give the opposite color reactions, ie the chloroform becomes red, while the acid exhibits a green fluorescence.

2. The Liebermann-Burchard reaction. The provitamin D is dissolved in chloroform, and acetic anhydride and concentrated sulfuric acid are added dropwise. A red color develops which progressively changes from blue-violet to green. This test has been developed for quantitative determinations (93). Cholesterol gives a similar change of color but the initial red color remains unchanged for a longer period of time.

3. The Tortelli-Jaffé reaction. A solution of provitamin D in acetic acid is mixed with a 2% solution of bromine in chloroform. A green color develops. This reaction is given by steroids with a ditertiary double bond, by the D-vitamins and by substances which react or isomerize to give such bonds (see reference 3, p. 114).

4. The Rosenheim reaction. To a solution of provitamin D in chloroform a solution of trichloroacetic acid in water is added. A red color develops which changes slowly into a light blue. This color reaction becomes considerably more sensitive when, before the addition of the trichloroacetic acid, lead tetraacetate in glacial acetic acid is added to the chloroform solution of the provitamin. An intense green fluorescence occurs which is, however, not given by provitamin D esters. Thus, provitamins D can be differentiated from their esters by this method. This modified color reaction can be employed for the detection of provitamins D in an amount on the order of 0.1 μg. This test has been developed for quantitative determinations (94).

5. Chloral hydrate reaction. Crystals of provitamin D, when heated slowly with crystals of chloral hydrate, melt about 50°C, and the mixture becomes first red, then green, and finally deep blue. Sterols, such as cholesterol, do not give any color reaction with chloral hydrate.

6. The antimony trichloride reaction. Provitamin D dissolved in chloroform and mixed with a solution of antimony trichloride in chloroform yields a red color.

7. The Chugaev reaction. A solution of provitamin D in glacial acetic acid with an excess of acetyl chloride and zinc chloride yields, upon heating to the boiling point, an eosin-red color with a greenish yellow fluorescence. The sensitivity is reported to be 1:80,000.

Biological Functions and Vitamin D Deficiency

Although the effect of vitamin D on general growth has been emphasized repeatedly (95) the most particular and striking effect of the vitamin is on the development of bones. The clinical picture of vitamin D deficiency is known as rickets and is characterized by a failure of the mineralization process to keep pace with the formation of the organic matrix of new bone. Deposition may be retarded or stopped resulting in the cessation of bone growth and the eventual withdrawal of previously deposited material. This can result in considerable softening. Normal blood serum is supersaturated with respect to bone mineral. Rachitic serum is subsaturated and has a low $(Ca^{2+})(HPO_4^{2-})$ product (96). Very small amounts of vitamin D (or its metabolite) together with parathyroid hormone regulate the mobilization of mineral from bone. Vitamin D (or, according to DeLuca (97), more probably its metabolite 25-hydroxy-cholecalciferol) increases the absorption of calcium and, to a lesser extent, the absorption of phosphate through the small intestine, thus maintaining the $(Ca^{2+})(HPO_4^{2-})$ product from ingested calcium. The difference in activity between vitamin D_2 and vitamin D_3 in the chick has been clarified to some extent. Using ^{14}C vitamin D_2 and also tritiated vitamin D_3, it has been shown that the lower activity of D_2 is not caused by any lack of absorption but rather by a much more rapid metabolism and excretion into the feces via the bile (98). Subsequent examination of the bile revealed a conjugated metabolite (99).

Vitamin D deficiency results in typical skeletal changes which are noticeable in the ribs, forearms, wrists, and legs. In babies, retardation of the calcification of the fontanelles is especially characteristic. Instead of calcification of the bones, excessive cartilage formation sets in, especially at the ends of the long bones, and this results in enlargement of the joints. Curvature of the affected bones occurs in later stages of the disease. Deficiency of vitamin D is also a contributing factor to the occurrence of dental caries. Deficiency of vitamin D in children may be involved in *infantile tetany*, and occurs when the faulty phosphorus metabolism is relieved while the calcium metabolism is still impaired. Rickets of the adult is known as *osteomalacia* and is characterized by a decalcification of the bones which leads to brittleness.

An improper supply of vitamin D is hazardous to all domestic animals, and the greater part of commercial production is utilized for preventing vitamin D deficiencies in livestock. Chicks and turkey poults are especially sensitive and, when deficient, soon become inactive and apathetic, displaying ruffled feathers and a warped gait. Deformities of the breastbone are soon observed, and the long bones bend or break with little resistance. Eventually the bones of the legs and feet fail to support the body and the animal loses its ability to stand. The symptoms in young pigs and calves are similar, if somewhat less severe. Eggs produced by hens that are raised on suboptimal doses have thin shells and do not hatch normally. Production itself declines. Vitamin D is also essential for the maintenance, reproduction, and lactation of dairy stock.

Requirements. Because of the tremendous importance of vitamin D for growth and maintenance of good health, the requirements for man and for farm animals have been studied extensively. The National Research Council has issued certain recommendations for dietary allowances for the maintenance of good nutrition in healthy persons in the United States. According to the 1968 revision of its report, the vitamin D requirement for optimum health is estimated to be about 400 IU per day, regardless of age. This amount of vitamin D gives ample protection from rickets, provided a sufficient amount of the other essential nutrients, including calcium and phosphorus, is supplied (99).

The nutritional requirements of domestic animals have been reviewed in detail by various subcommittees of the Council. Requirements for poultry, swine, dairy cattle, beef cattle, sheep, horses, foxes and minks, and dogs, rabbits and laboratory animals, are available in specific reports. Some representative values per kg of feed are as follows: starting and growing chicks, 200 IU; laying and breeding hens, 500 IU; turkeys, 400 IU; ducks, 220 IU; and swine, 125–200 IU. Calves required about 660 IU per 100 kg of body weight (100–102).

Toxicity. Excessive amounts of vitamin D—on the order of 1,000 to 3,000 IU/ (kg)(day)—are potentially dangerous to children and adults and may lead to hypercalcemia with attendant complications. Hypersensitivity may be an additional factor (99). These are large doses, however, and are usually associated with the treatment of certain diseases.

Commercial Utilization

Vitamin D is commercially available in a wide variety of forms. Cod-liver oil, a former favorite among the low-potency products, and percomorph-liver oil, a relatively high-potency product, have had a marked decline in position in recent years. NF cod-liver oil sold for $1.50–1.70/gal in 1970.

Most of the vitamin D now sold is synthetic. Vitamin D_2, either in microcrystalline form or as a concentrate, is still widely used in pharmaceutical preparations, although vitamin D_3 is now preferred by many authorities. Vitamin D_2 in the form of irradiated yeast is manufactured as a feed supplement for cattle, swine, and dogs but its usage has shown serious decline in recent years. Vitamin D_3 has rapidly captured a good share of the total market because it is as active as the D_2 form in man and is useful for all animals.

Crystalline vitamin D is used to a certain extent for medicinal preparations and concentrates in the pharmaceutical industry and for the fortification of fresh, evaporated, and nonfat dry milks. Preparations based on use of the resin have come into increasing favor, however, because of lower costs. Both the D_2 and D_3 resins are available commercially. Essentially all of the pasteurized milk sold in the United States is fortified with vitamin D. Other important fortified products include cereals, margarine, and animal feed.

Aside from the standard supplement preparations, such as tonics, drops, capsules, and tablets, oil-based injectables are available for human and animal use. Combinations are widely employed. An A and D combination is available for human use, for example, and an A, D, and E combination for animals. Preemulsified products are utilized for increased bioavailability. In recent years, emulsifiers have been added directly to the oil type of preparation as this also improves the biological availability. Water miscible formulations have been developed (103).

For animal feeds, use is often made of simple solutions in oil or of preparations of the vitamin in oil on dry carriers, such as corn or flour. These suffer from lack of long-range stability especially in high mineral feeds. A number of stable forms which overcome this objection have been patented and are sold commercially. These include beadlets of dry suspensions in gelatin, in carbohydrates, in wax, and in cellulosic derivatives (104–108).

The U.S. Tariff Commission lists the 1968 U.S. production of vitamin D_2 at 21.6 × 10^{12} USP units or 1900 lb (109). This is a small fraction of the total vitamin D market in the United States which is estimated at ten times this amount. A substantial portion of the D_3 is imported, and it is estimated that with all uses included, over 80% of the sales are of vitamin D_3. The Commission Report lists a price of vitamin D_2 at 0.9¢ per 10^6 USP units. Trade prices vary somewhat. Vitamin D_3 usually sells for a slightly higher price than D_2.

Bibliography

"Vitamins (vitamin D)" in *ECT* 1st ed., Vol. 14, pp. 828–849, by H. R. Rosenberg, E. I. du Pont de Nemours & Co., Inc.

NOTE: References 1–9 are general references.

1. H. F. DeLuca, "Mechanism of Action and Metabolic Fate of Vitamin D," in R. S. Harris, I. G. Wool, and J. A. Loraine, eds., *Vitamins and Hormones*, Vol. 25, Academic Press, Inc., New York, 1967, pp. 353–367.
2. S. F. Dyke, *The Chemistry of the Vitamins*, Interscience Publishers, a div. of John Wiley & Sons, Inc., New York, 1965, Chap. 14.
3. L. F. Fieser and M. Fieser, *Steroids*, Reinhold Publishing Co., New York, 1959.
4. M. Freed, *Methods of Vitamin Assay*, 3rd ed., Interscience Publishers, a div. of John Wiley & Sons, Inc., New York, 1966.
5. H. R. Rosenberg, *Chemistry and Physiology of the Vitamins* (revised reprint), Interscience Publishers, Inc., New York, 1945.
6. W. H. Sebrell, Jr., and R. S. Harris, *The Vitamins*, Vol. II, Academic Press Inc., New York, 1954, Chap. 6.
7. C. W. Shoppee, *Chemistry of the Steroids*, Butterworths and Co. (Publishers), Washington, 1964.
8. A. F. Wagner and F. Folkers, *Vitamins and Coenzymes*, Interscience Publishers, a div. of John Wiley & Sons, Inc., New York, 1964, Chap. XVII.
9. International Union of Pure and Applied Chemistry, "Definitive Rules for Nomenclature of Steroids," *J. Am. Chem. Soc.* **82**, 5577 (1960); "Definitive Rules for the Nomenclature of Vitamins," *ibid.*, 5581.
10. J. H. Bennett, *Treatise on the Oleum Jecoris Aselli or Cod Liver Oil*, Edinburgh, 1848.
10a. H. R. Rosenberg, *Arch. Biochem. Biophys.* **42**, 7 (1953).
11. E. Mellanby, *J. Physiol.* (*London*) **52**, LIII (1919); *Lancet* **196**, 407 (1919).
12. E. V. McCollum, N. Simmonds, J. E. Becker, and P. G. Shipley, *J. Biol. Chem.* **53**, 253 (1922); *ibid.*, **65**, 97 (1925); *Bull. Johns Hopkins Hosp.* **33**, 229 (1922).
13. T. A. Palm, *The Practioner* **45**, 270 (1890).
14. K. Huldschinsky, *Deut. Med. Wochschr.* **45**, 712 (1919).
15. H. Steenbock, *Science* **60**, 224 (1924).
16. A. F. Hess and M. Weinstock, *J. Biol. Chem.* **62**, 301 (1924); *ibid.*, **63**, 297 (1925).
17. F. A. Askew, R. B. Bourdillon, H. M. Bruce, R. K. Callow, J. St. L. Philpot, and T. A. Webster, *Proc. Roy. Soc.* (*London*) **B109**, 488 (1932).
18. A. Windaus, O. Linsert, A. Luttringhaus, and G. Weidlich, *Ann.* **492**, 226 (1932).
19. J. Waddell, *J. Biol. Chem.* **105**, 711 (1934).
19a. H. Brockmann, *Z. Physiol. Chem.* **241**, 104 (1936).
20. D. Crawfoot and J. D. Dunitz, *Nature* **162**, 608 (1948); D. C. Hodgkin, M. S. Webster, and J. D. Dunitz, *Chem. Ind.* (*London*) **1957**, 1149.
21. *The Merck Index*, 8th ed., Merck and Co., Rahway, N.J., 1968, p. 1113.
22. F. W. Anderson, A. L. Bacharach, and E. L. Smith, *Analyst* **62**, 430 (1937).
23. H. Penau and G. Hageman, *Helv. Chim. Acta* **29**, 1366 (1946).
24. A. Windaus, K. Dithmar, and E. Fernholz, *Ann.* **493**, 259 (1932).
25. A. Windaus, and G. Trautmann, *Z. Physiol. Chem.* **247**, 185 (1937).
26. W. Wunderlich, *Z. Physiol. Chem.* **241**, 116 (1936).
27. Q. Linsert, *Z. Physiol. Chem.* **241**, 125 (1936); G. A. D. Haslewood, *Biochem. J.* **33**, 454 (1939).
28. K. Dimroth and J. Paland, *Chem. Ber.* **72**, 187 (1939).
29. W. L. Ruigh, *J. Am. Chem. Soc.* **64**, 1900 (1942).
30. Y. Raoul, N. Le Boulch, J. Chopin, P. Meunier, and A. Guerillot-Vinet, *Compt. Rend.* **235**, 1439, 1704 (1952).
31. H. G. Petering and J. Waddell, *J. Biol. Chem.* **191**, 765 (1951); H. G. Petering, *J. Org. Chem.* **22**, 808 (1957).
32. Y. Raoul, N. Le Boulch, C. Baron, R. Bazier, and A. Guerillot-Vinet, *Bull. Soc. Chim. Biol.* **38**, 495, 885 (1956).
33. Y. Raoul, N. Le Boulch, C. Baron, J. Chopin, and A. Guerillot-Vinet, *Bull. Soc. Chim. Biol.* **36**, 1265 (1954).

34. H. F. DeLuca, private communication, 1970.

35. C. G. Alberti, B. Camerino, and L. Mamoli, *Helv. Chim. Acta* **32**, 2038 (1949).

36. J. van der Vliet, *Rec. Trav. Chim.* **67**, 246 (1948); *ibid.*, **67**, 665 (1948).

37. J. W. Blunt, Y. Tanaka, and H. F. DeLuca, *Proc. Natl. Acad. Sci. U.S.* **61**, 1503 (1968); J. W. Blunt, H. F. DeLuca, and H. K. Schnoes, *Chem. Commun.* **1968**, 801.

38. J. W. Blunt and H. F. DeLuca, *Biochemistry* **8**, 671 (1969).

39. J. A. Campbell, D. M. Squires, and J. C. Babcock, *Steroids* **13**, 567 (1969).

40. S. J. Halkes and N. P. Van Vliet, *Rec. Trav. Chim.* **88**, 1080 (1969).

41. T. Suda, H. F. DeLuca, H. K. Schnoes, and J. W. Blunt, *Biochem. Biophys. Res. Commun.* **35**, 182 (1969).

42. A. Windaus, F. Von Werder, A. Lüttringhaus, and E. Fernholz, *Ann.* **499**, 188 (1932).

42a. L. Velluz, G. Amiard, and B. Goffinet, *Bull. Soc. Chim. France* **22**, 1341 (1955).

43. E. Havinga, R. J. DeKock, and M. P. Rappoldt, *Tetrahedron* **11**, 276, (1960); E. Havinga and J. L. M. A. Schlatmann, *Tetrahedron* **16**, 146 (1961).

44. L. Velluz, G. Amiard, and A. Petit, *Bull. Soc. Chim. France* **16**, 501 (1949); L. Velluz and G. Amiard, *Bull. Soc. Chim. France* **22**, 205 (1955).

45. U.S. Pat. 2,693,475 (Nov. 2, 1954), L. Velluz and G. Amiard (to Usines Chimiques des Laboratoires Français).

46. W. G. Dauben, I. Bell, T. W. Hutton, G. F. Laws, A Rheiner, Jr., and H. Urschler, *J. Am. Chem. Soc.* **80**, 4116 (1958); W. G. Dauben and G. J. Fonken, *J. Am. Chem. Soc.* **81**, 4060 (1959).

47. R. Hoffman and R. B. Woodward, *Accounts Chem. Res.* **1**, 17 (1968); R. B. Woodward and R. Hoffman, *J. Am. Chem. Soc.* **87**, 395, 2511 (1965).

48. H. E. Zimmerman, *J. Am. Chem. Soc.* **88**, 1564, 1566 (1966); *Science* **153**, 837 (1966).

49. M. Scott, J. Glover, and R. A. Morton, *Nature* **163**, 530 (1949).

50. U.S. Pat. 2,395,115 (Feb. 19, 1946), K. J. Goering (to Anheuser-Busch Inc.).

51. U.S. Pat. 2,874,171 (Feb. 17, 1959), H. A. Nelson (to Upjohn Co.).

52. U.S. Pat. 3,006,932 (Oct. 31, 1961), J. Green, S. A. Price, and E. E. Edwin (to Vitamins Ltd.).

53. U.S. Pat. 2,794,035 (May 28, 1957), O. Hummel (to Zellstoff-Fabrik Waldhof).

54. U.S. Pat. 2,865,934 (Dec. 23, 1958), R. A. Fisher (to Bioferm. Corp.).

55. Ger. Pat. 1,252,674 (Oct. 26, 1967), K. Petzoldt, K. Kieslich, and H. J. Koch (to Schering A.G.).

56. U.S. Pat. 2,266,674 (Dec. 16, 1941), A. G. Boer, J. van Niekerk, E. H. Reerink, and A. van Wijk (to Hartford National Bank and Trust Co.).

57. C. E. Bills and E. M. Honeywell, *J. Biol. Chem.* **80**, 15 (1928).

58. U.S. Pat. 2,730,536 (Jan. 10, 1956), R. J. Feeney (to Charles Pfizer and Co., Inc.).

59. U.S. Pat. 2,098,984 (Nov. 16, 1937), A. Windaus and F. Schenck (to Winthrop Chemical Co., Inc.).

60. L. F. Fieser, *J. Am. Chem. Soc.* **75**, 4394 (1953).

61. K. Bloch, *Helv. Chim. Acta* **36**, 1611 (1953).

62. U.S. Pat. 2,505,646 (April 25, 1950), W. C. Meuly (to E. I. du Pont de Nemours & Co., Inc.).

63. Brit. Pat. 656,388 (Aug. 22, 1951), (to Løvens Kemiske Fabrik).

64. J. L. J. van de Vliervoet, P. Westerhof, J. A. Keverling Buisman, and E. Havinga, *Rec. Trav. Chim.* **75**, 1179 (1956).

65. U.S. Pat. 2,441,091 (May 4, 1968), J. van der Vliet and W. Stevens (to Hartford National Bank and Trust Co.).

66. S. Bernstein, L. J. Binovi, L. Dorfman, K. J. Sax, and Y. Subbarow, *J. Org. Chem.* **14**, 433 (1949); U.S. Pat. 2,498,390 (Feb. 21, 1950), S. Bernstein and K. J. Sax (to American Cyanamid Co.).

67. U.S. Pat. 2,633,451 (Mar. 31, 1953), H. Schaltegger (to Dr. A. Wander A. G.).

68. U.S. Pat. 2,546,787 (Mar. 27, 1951), W. L. Ruigh and D. H. Gould (to Nopco Chemical Co.).

69. U.S. Pat. 2,546,788 (March 27, 1951), K. Schaaf (to Nopco Chemical Co.).

70. U.S. Pat. 3,032,562 (May 1, 1962), H. Klein and R. Kapp (to Nopco Chemical Co.).

71. U.S. Pat. 3,037,996 (June 5, 1962), H. Klein and R. Zapp (to Nopco Chemical Co.).

72. F. Hunziker and F. X. Müllner, *Helv. Chim. Acta* **41**, 70 (1958).

73. U.S. Pat. 2,698,853 (Jan. 4, 1955), B. S. Wildi (to Monsanto Chemical Co.).

73a. I. M. Heilbron, K. M. Samant, and F. S. Spring, *Nature* **135**, 1072 (1935).

74. I. T. Harrison and B. Lythgoe, *J. Chem. Soc.* **1958**, 837, 843.

75. H. H. Inhoffen, K. Irmscher, H. Hirschfeld, U. Stache, and A. Kreutzer, *Chem. Ber.* **91**, 2309 (1958); *J. Chem. Soc.* **1959**, 385.

76. H. H. Inhoffen, *Angew. Chem.* **72**, 875 (1960).

77. H. H. Inhoffen, S. Schütz, P. Rossberg, O. Berges, K-H. Nordsiek, H. Plenio, and E. Höroldt, *Chem. Ber.* **91**, 2626 (1958).

78. H. H. Inhoffen, H. Burkhardt, and G. Quinkert, *Chem. Ber.* **92**, 1564 (1959).

79. H. H. Inhoffen, G. Quinkert, and S. Schütz, *Chem. Ber.* **90**, 1283 (1957).

80. R. S. Harris, J. W. M. Bunker, and L. M. Mosher, *J. Am. Chem. Soc.* **60**, 2579 (1938).

81. U.S. Pat. 3,334,118 (Aug. 1967), K. H. Schaaf, S. Schmukler, and H. Klein (to Nopco Chemical Co.).

82. World Health Organ. Expert Comm. on Biological Standardization, "Report of the Subcommittee on Fat-Soluble Vitamins, 1949," WHO/BS/65 Rev. 1; *Chronique WHO* **3**, 147 (1965).

83. E. M. Nelson, *J. Assoc. Offic. Agr. Chemists* **32**, 801 (1949).

84. D. T. Ewing, M. J. Powell, R. A. Brown, and A. D. Emmett, *Anal. Chem.* **20**, 317 (1948).

84a. *United States Pharmacopeia*, Mack Publishing Co., Easton, Pa., 1960.

85. K. Kodicek and D. R. Ashly, *Biochem. J.* **57**, xii–xiii (1954).

86. L. T. Heaysman and E. R. Sawyer, *Analyst* **89**, 529 (1964).

87. A. J. Sheppard, D. E. LaCroix, and A. R. Prosser, *J. Assoc. Offic. Anal. Chemists* **51**, 834 (1968).

88. J. Carol, *J. Pharm. Sci*, **50**, 451 (1961).

89. W. W. Morris, Jr., J. B. Wilkie, S. W. Jones, and L. Friedman, *Anal. Chem.* **34**, 381 (1962).

90. R. Strohecker and H. M. Henning, *Vitamin Assay-Tested Methods*, CRC Press, Cleveland, O., 1966, p. 281.

91. *Official Methods of Analysis*, Association of Official Agricultural Chemists, 10th ed., 1965, Washington, p. 779, 784.

92. T. R. Hogness, A. E. Sidwell, Jr., and F. P. Zscheile, Jr., *J. Biol. Chem.* **120**, 239 (1937). W. Huber, G. W. Ewing, and J. Kriger, *J. Am. Chem. Soc.* **67**, 609 (1945).

93. R. K. Callow, *Biochem. J.* **25**, 87 (1931).

94. A. van Christiani and V. Anger, *Chem. Ber.* **72**, 1124 (1939).

95. H. Steenbock and D. C. Herting, *J. Nutr.* **57**, 449 (1955).

96. J. Howland and B. Kramer, *Trans. Am. Pediat. Soc.* **34**, 204 (1922).

97. H. F. DeLuca, *Federation Proc.* **28**, 1678 (1969).

98. M. H. Imrie, P. F. Neville, A. Snellgrove, and H. F. DeLuca, *Arch. Biochem. Biophys.* **120**, 525 (1967).

99. "Recommended Dietary Allowances," 7th ed., *Natl. Acad. Sci. Natl. Res. Council* **1694** (1968).

100. "Nutritional Requirements of Poultry," 5th ed., *Natl. Acad. Sci. Natl. Res. Council* **1345** (1966).

101. "Nutritional Requirements of Swine," 6th ed., *Natl. Acad. Sci. Natl. Res. Council* **1599** (1968).

102. "Nutritional Requirements of Dairy Cattle," 3rd ed., *Natl. Acad. Sci. Natl. Res. Council* **1349** (1966).

103. U.S. Pat. 2,417,299 (March 11, 1947) L. Freedman and E. Green (to U.S. Vitamin Corp.).

104. U.S. Pat. 2,702,262 (Feb. 15, 1955) A. Bavley and A. E. Timreck (to Charles Pfizer and Co., Inc.).

105. U.S. Pats. 2,777,797 and 2,777,798 (Jan. 15, 1957) M. Hochberg and M. J. MacMillan (to Nopco Chemical Co.).

106. U.S. Pat. 2,827,452 (March 18, 1958) H. Schlenk, D. M. Sand, and J. A. Tillotson (to University of Minnesota).

107. U.S. Pat. 3,067,104 (Dec. 4, 1962) M. Hochberg and C. Ely (to Nopco Chemical Co.).

108. U.S. Pat. 3,143,475 (Aug. 4, 1964) A. Koff and P. F. Widmer (to Hoffmann-LaRoche, Inc.).

109. U.S. Tariff Commission, *U.S. Sales and Production of Medicinal Chemicals*, Washington, D.C., 1968.

110. P. S. Chen, Jr., and H. B. Bosmann, *J. Nutr.* **83**, 133 (1964).

SHELDON B. GREENBAUM
Diamond Shamrock Chemical Co.

VITAMIN E

The term "vitamin E" was used originally to denote a partially characterized material in vegetable oils that was essential for the rat to maintain fertility. Discovery of this fat-soluble vitamin in 1922 and early pioneer investigations of its natural distribution, isolation, and identification were due primarily to Evans, Burr, and Emerson, of the University of California, and to Mattill and Olcott, of the University of Iowa. More than one naturally occurring substance and several synthetic organic compounds that acted like or had a sparing effect on vitamin E in the body were found subsequently. Also, the symptoms of vitamin E deprivation were discovered to vary according to the species of animal. Consequently, "vitamin E activity" was used for many years to designate, in a general manner, a type of physiological activity that could cure or prevent any vitamin E-deficiency symptom in animals.

Four naturally occurring compounds possessing vitamin E activity were finally isolated and identified. These substances were designated α-, β-, γ-, and δ-tocopherols ("tocopherol" is derived from Greek words meaning "to bring forth offspring" and is pronounced tō-kŏf-ēr-ōl). All are methyl derivatives of tocol [2-methyl-2-(4',8',12'-trimethyltridecyl)-6-chromanol] (structures (1)–(4); *denotes point of asymmetry). Within the last decade, four additional compounds analogous to the tocopherols have been characterized (1). These compounds are methyl derivatives of tocotrienol [2-methyl-2-(4',8',12'-trimethyltrideca-3',7',11'-trienyl)-6-chromanol] and contain three unsaturated bonds in the side chain. α-Tocotrienol is shown as (5).

Although the tocopherols and tocotrienols are closely related chemically, they have quite different biological properties. α-Tocopherol, with a completely methylated aromatic ring and a saturated side chain, has the highest biological activity. Because of its high potency and because it is the predominant tocopherol in animal tissues,

(1) α-Tocopherol (5,7,8-trimethyltocol)

(2) β-Tocopherol
(5,8-dimethyltocol)

(3) γ-Tocopherol
(7,8-dimethyltocol)

(4) δ-Tocopherol
(8-methyltocol)

(5) α-Tocotrienol (5,7,8-trimethyltocotrienol)

the term "α-tocopherol" is now widely used instead of "vitamin E." The latter term is reserved as a generic descriptor for tocol and tocotrienol derivatives having qualitatively the biological activity of α-tocopherol.

Both α-tocopherol and α-tocopheryl acetate, the forms most used commercially, are clear, odorless, quite viscous, slightly yellow oils. The naturally occurring d form is the most active isomer physiologically; the racemic synthetic dl-α-tocopherol and its esters are less potent on a weight basis than the corresponding d forms. α-Tocopheryl acetate is the principal commercial form of vitamin E in medicine and for domestic animals as a source of vitamin E activity. The tocopherols are used in food technology as antioxidants to retard the development of rancidity in fatty materials.

Properties

A characteristic physical property of α-tocopherol, $C_{29}H_{50}O_2$, formula weight 430.72, is its fat solubility. Both α-tocopherol and α-tocopheryl acetate are freely soluble in glyceride oils, ethyl alcohol, chloroform, and acetone. They are insoluble in water. The optical activity of natural d-α-tocopherol is very slight. The specific rotation of d-α-tocopherol in 2,2,4-trimethylpentane is only $+0.16°$. A much greater specific rotation, $[\alpha]_D{}^{25} = +31.5°$, is shown by a compound obtained from d-α-tocopherol by oxidation with alkaline potassium hexacyanoferrate(III) (2). Because the specific rotation is $0°$ for the corresponding oxidation product from synthetic dl-α-tocopherol, a method is provided for distinguishing between the d and dl forms.

The ultraviolet absorption spectrum for α-tocopherol in ethyl alcohol shows λ_{max} at 292 nm ($A_{1\ cm}{}^{1\%} = 75.8$) and λ_{min} at 256 nm. Acetylation of the phenolic hydroxyl group shifts λ_{max} to 284 nm ($A_{1\ cm}{}^{1\%} = 43.6$) and λ_{min} to 255 nm. In the infrared region, a band at about 8 μm is characteristic of the chromane structure of

Table 1. Some Physical Properties of Commercial Vitamin E Products

Vitamin E Product	Sp gr $_{25°C}{}^{25}$	$n_D{}^{20}$	A (1%, 1 cm) in ethyl alcohol
d-α-Tocopheryl acetate[a]; melts at about 25°C	0.950–0.964	1.4940–1.4985	40–44 at 284 nm
d-α-Tocopheryl acetate concentrate; contains at least 25% d-α-tocopheryl acetate liquid	0.932–0.955	varies[b]	varies[b]
mixed tocopherols concentrate; contains at least 34% total tocopherol; at least half of tocopherols is d-α-tocopherol, liquid	0.932–0.955	varies[b]	varies[b]
dl-α-Tocopherol,[a] liquid	0.947–0.958	1.5030–1.5070	71–76 at 292 nm
dl-α-Tocopheryl acetate, liquid	0.950–0.964	1.4940–1.4985	40–44 at 284 nm

[a] According to Tentative Rules for Nomenclature recommended by committees of the American Institute of Nutrition and International Union of Nutritional Sciences, the compound isolated from natural sources, also known as d-α-tocopherol, 2-D,4′D,8′D-α-tocopherol or 2-R,4′R,8′R-α-tocopherol, should be designated simply α-tocopherol. The mixture of the eight possible stereoisomers commonly synthesized from racemic isophytol and also known as dl-α-tocopherol, 2-DL,4′DL,8′DL-α-tocopherol, or 2-RS,4′RS,8′RS-α-tocopherol, should be designated as all-$rac.$-α-tocopherol (4).

[b] With potency of concentrate.

tocopherols. The magnetic susceptibility of α-tocopherol has been determined as $\chi = -0.773 \times 10^{-6}$.

Allowable ranges for several physical properties of commercial vitamin E products are shown in Table 1; these values are taken from the 12th edition of the National Formulary (3a). Another important product, especially for tablets and dry-fill capsules, is d-α-tocopheryl acid succinate, a white, crystalline solid, mp 73–78°C, $A_{1\ cm}^{1\%}$ = 35–40 at 284 nm in ethyl alcohol.

Oxidation of α-tocopherol is catalyzed by light and is also accelerated by unsaturated fatty acids, metal salts, or alkalies. Oxidation proceeds according to reactions shown in Scheme 1 (5). Further oxidation (chromic oxide) yields dimethylmaleic anhydride, 2,3-butanedione, acetone, a C_{16} acid, a C_{18} ketone, and a C_{21} lactone. From these products, Fernholz (6) in 1938 correctly postulated the structure of α-tocopherol. Other reaction products of the tocopherols have been described by Nilsson (7) and Skinner et al. (8).

Scheme 1

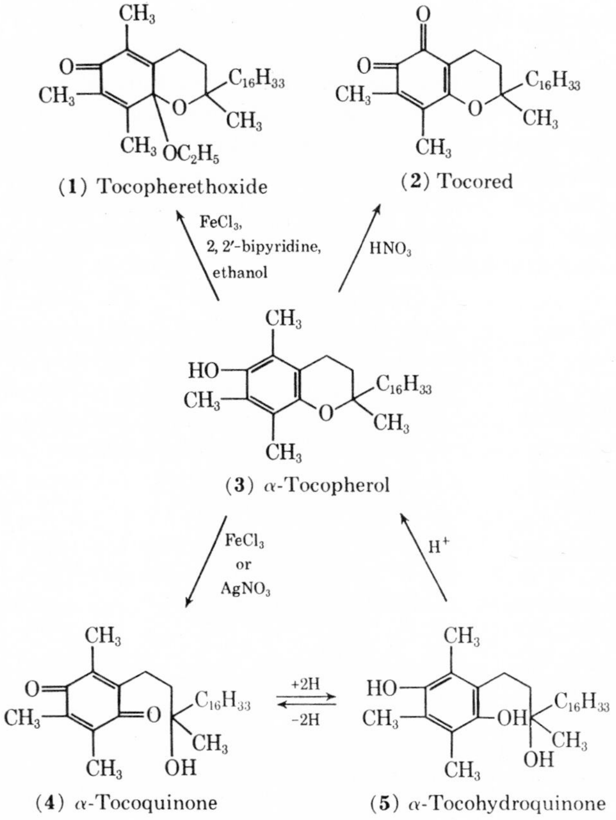

(1) Tocopherethoxide   (2) Tocored

(3) α-Tocopherol

(4) α-Tocoquinone   (5) α-Tocohydroquinone

Evidence for the appearance of the products in Scheme 1 during the metabolism of α-tocopherol is sound only for α-tocoquinone (4). Two other products formed *in vivo* are α-tocopheronic acid and its lactone (9) (see structures 6 and 7). In addition, a dimer and a trimer have been isolated from liver following administration of α-

tocopherol. Structure (**8**) is that proposed by Nelan and Robeson (2) for the dimer obtained by oxidation of α-tocopherol with alkaline potassium hexacyanoferrate(III).

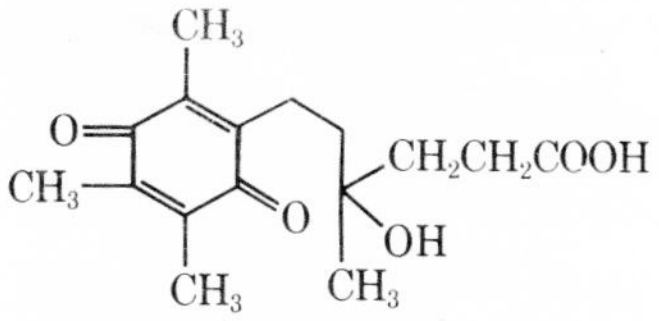

(**6**) α-Tocopheronic acid

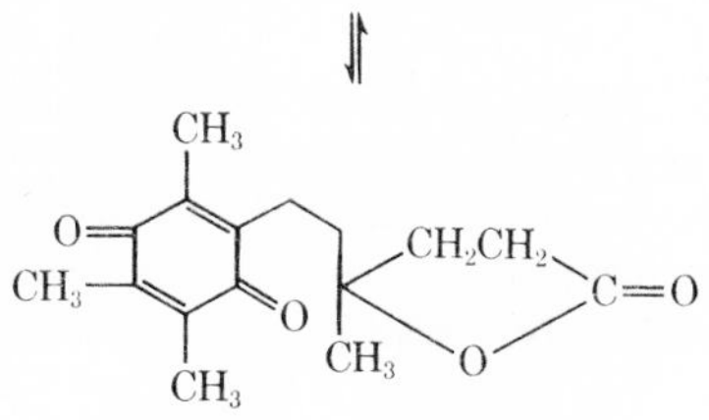

(**7**) α-Tocopheronolactone

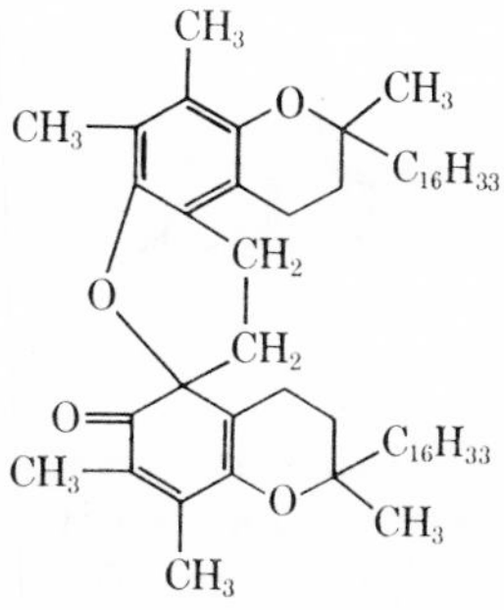

(**8**) α-Tocopherol dimer

The ready reversibility of α-tocopherol with certain of its oxidation products suggests that vitamin E could act in the body as an integral part of some oxidation-reduction system or enzyme. Michaelis and Wollman demonstrated the existence of a free semiquinone radical of α-tocopherol by ultraviolet irradiation at low temperatures (10). Structure (**9**) shows free radical formation from α-tocopherol as postulated by Skinner (11). These free radicals might well be the critical factors in one or more oxidation-reduction reactions *in vivo*.

α-Tocohydroquinone (α-tocopherylhydroquinone) and α-tocopherethoxide show activity in preventing and curing muscle dystrophy in hamsters, while α-tocoquinone (α-tocopherylquinone), as well as α-tocohydroquinone, is effective in the rabbit. The rat can utilize all three compounds to a slight degree for preserving reproductive function, but other oxidized products, including the lactone and dimer, are ineffective. Esterification of α-tocopherol with aliphatic acids ordinarily does not markedly affect biological activity, but etherification almost completely eliminates biopotency. The tocopheramines and N-methyltocopheramines show varying degrees of activity; some are as active as α-tocopherol (12). α-Tocopherothiol shows activity in the erythrocyte hemolysis test. Other derivatives of α-tocopherol in which ring methyl groups are substituted or in which the phytyl side chain is modified are less potent than α-tocopherol itself.

Occurrence

The tocopherols are widely distributed in foods in an unesterified form (13) and occur in the highest concentration in the cereal grain oils (14). For example, crude corn and wheat oils may contain 200 mg of tocopherol/100 g. On the other hand, certain vegetable oils, such as coconut and olive, are practically devoid of tocopherol. The proportion of the most active form, α-tocopherol, also varies widely. Whereas 90% of the tocopherol in safflower oil is α-tocopherol, only 20% of corn and soybean oil tocopherols is α-tocopherol. The γ form predominates in corn oil, whereas the γ and δ forms predominate in soybean oil. Wheat oil tocopherols contain about 65%

of the β form. The major portion of the estimated per capita daily intake of α-tocopherol (15 mg) in the U.S., based on total purchases of food, is derived from margarine, salad oils, and shortening. Other foods supply relatively little of the α-tocopherol ingested, as shown in Table 2 (15). Based on assays of foods as eaten, Bunnell et al.

(9)

(16) calculated that an average American diet may supply only 7 mg of α-tocopherol per day. Farm animals receive most of their α-tocopherol from fresh grains and alfalfa leaf meal and, of course, fresh grasses if the animals have pasture available. The amount in eggs, milk, or meat depends on the amount of α-tocopherol received by the animal. Thus dairy products and meats are highly variable in their content of this vitamin.

Table 2. Fat and d-α-Tocopherol in Various Food Groups Available for Consumption in the United States in 1960

	Food consumption		d-α-Tocopherol	
	Total, g/day	fat, g/day	Content, mg/g fat	Amount, mg/day
Visible fats				
butter	9	7.19	0.016	0.115
lard	10	9.55	0.023	0.220
margarine	12	9.42	0.102	0.961
shortening	16	15.62	0.100	1.562
other fats and oils	14	14.14	0.500	7.063
subtotal	61	55.92		9.921
Other food fats				
dairy products	453	23.31	0.016	0.378
eggs	37	5.33	0.107	0.572
meat, poultry, fish	257	52.08	0.017	0.893
beans, peas, nuts, etc	21	5.33	0.093	0.496
fruits and vegetables	544	1.74	0.917	1.597
grain products	198	2.23	0.489	1.092
subtotal	1510	90.02		5.028
total	1571	145.94		14.949

Appreciable losses of vitamin E may occur during the processing of foods and during cooking (17). The amount of tocopherol left in refined salad oil depends on the severity of the refining process. Losses during the processing of cereal grains depend partially on the amount of germ that is retained (18). The α-tocopherol content of bleached wheat flour is very low. Proper cooking of vegetables does not cause much loss of α-tocopherol, but the portions of the plant with the highest content of α-tocopherol are frequently discarded as inedible (19). Frozen and fresh vegetables have similar vitamin E levels, but canning lowers the level substantially. Losses of α-tocopherol during dehydration, irradiation, or canning of meats may range up to 50% (20).

Isolation and Synthesis

The naturally occurring tocopherols are obtained from vegetable oil sources. An important step, technically, is the concentration by molecular distillation (21). For example, alkali-refined soybean oil containing 0.19% of a mixture of α-, γ-, and δ-tocopherols is distilled in a centrifugal-type molecular still and the tocopherol fraction, which distils below 240°C under 0.004 mm pressure, is collected. After as much as possible of the sterols and other substances in the fraction have been removed by crystallization from acetone at -10°C and the glycerides have been removed by saponification, the tocopherols in the unsaponifiable matter are further concentrated by a second molecular distillation. Thus, a fraction containing at least 60% mixed tocopherols is obtained (22).

Concentrates of mixed tocopherols can be obtained from vegetable oil sources by one or more of the following treatments (23): esterification, saponification, fractional extraction, ion exchange, and precipitation of sterols with hydrogen halides. The non-α-tocopherols can be converted to the more biologically active α-tocopherol by introducing methyl substituents into the aromatic ring (24) by means such as chloromethylation, formylation, hydroxyalkylation, or the Mannich reaction, each followed by reduction.

The four tocopherols that occur naturally have been synthesized in their *dl* forms; one, *dl*-α-tocopheryl acetate, is produced commercially. Although the commercial methods have not been described in detail, the processes probably resemble the procedures reported almost simultaneously in 1938 by Karrer, Fritzsche, Ringier, and Salomon, of Zürich (25); Bergel, Jacob, Todd, and Work, of Manchester (26); and Smith, Ungnade, and Prichard, of Minnesota (27). These authors condensed trimethylhydroquinone with phytol (see Chlorophyll) (26), phytyl halide (25,27), or phytadiene (27). Later, synthetic isophytol was used, and this compound now appears to be the intermediate used most commonly (Scheme 2). The condensation can be carried out in acetic acid or in an inert solvent, such as benzene, with an acidic catalyst, such as zinc chloride, formic acid, or boron trifluoride ethyl etherate.

Scheme 2

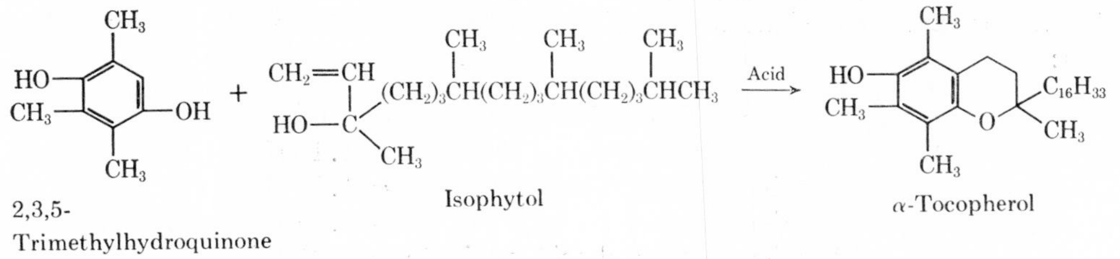

2,3,5-Trimethylhydroquinone   Isophytol α-Tocopherol

Synthetic α-tocopherols may differ in their stereoisomer content. Racemic α-tocopherol synthesized from isophytol is a mixture of the eight possible stereoisomers (see **1**). The form synthesized from natural phytol is a mixture of only two of the eight possible stereoisomers and can be termed 2RS,4′R,8′R-α-tocopherol, where *d*-α-tocopherol is 2R,4′R,8′R-α-tocopherol (28). The configuration of the 2 carbon is of prime importance for biological activity, the *l* epimer being 21% as active as the *d* epimer (29). The 2*d* and 2*l* epimers can be separated by fractional crystallization of mixtures (23). The *dl* forms of β-, γ-, and δ-tocopherols can be synthesized by the procedures used for the synthesis of *dl*-α-tocopherol. Instead of trimethylhydroquinone, the appropriate dimethylhydroquinone or monomethylhydroquinone is used. Thus, 2,5-dimethylhydroquinone yields *dl*-β-tocopherol, 2,3-dimethylhydroquinone yields *dl*-γ-tocopherol, and methylhydroquinone yields *dl*-δ-tocopherol and other monomethyl tocols. The yields of these compounds are usually less than that obtained in the synthesis of α-tocopherol because the hydroquinones can condense with more than one molecule of phytol. Monoesters, such as the monobenzoates, of the hydroquinones give improved yields. A number of other procedures have been used for the synthesis of tocol homologs and analogs (24). Esters of the tocopherols are readily prepared. Typical laboratory procedures for the acetate (30) and the succinate (31) have been described.

Economics

Statistics on the domestic production of both natural and synthetic preparations of vitamin E have only recently become available (32). Both natural and synthetic α-tocopherols are being manufactured in Europe and Asia. The price of vitamin E has steadily decreased, while usage has increased. Table 3 shows some prices ((33a,b) for pharmaceutical-grade α-tocopheryl acetate from 1948 to 1970. The prices for the *dl* form have been lower than those for the *d* forms since 1954, reflecting the higher potency of the *d* forms established when National Formulary defined the International Unit (3b).

Table 3. Price of α-Tocopheryl Acetate in Bulk ($/kg)

	1948	1951	1953	1954	1960	1970
d-α-Tocopheryl acetate	250	250	225	185	122	68
dl-α-Tocopheryl acetate	750	350	225	136	90	50

Specifications and standards for various tocopherol preparations intended for pharmaceutical use are listed in the National Formulary (3a) (see also Table 1). They may be labeled in terms of mg of *d*- or *dl*-α-tocopherol or α-tocopheryl acetate, whichever is appropriate, or in International Units. Label claims for tocopherol preparations, when expressed in International Units, IU, are based on the following equivalents (3b):

$$
\begin{aligned}
1 \text{ mg } \textit{dl}\text{-}\alpha\text{-tocopheryl acetate} &= 1.0 \text{ IU}\\
1 \text{ mg } \textit{dl}\text{-}\alpha\text{-tocopherol} &= 1.1 \text{ IU}\\
1 \text{ mg } \textit{d}\text{-}\alpha\text{-tocopheryl acetate} &= 1.36 \text{ IU}\\
1 \text{ mg } \textit{d}\text{-}\alpha\text{-tocopherol} &= 1.49 \text{ IU}\\
1 \text{ mg } \textit{d}\text{-}\alpha\text{-tocopheryl acid succinate} &= 1.21 \text{ IU}
\end{aligned}
$$

Some revision of these equivalents may be necessary in the light of recent biological studies.

Assay

Chemical and Physical Methods. The methods of analysis for pure or highly concentrated free α-tocopherol preparations are rather simple. Reaction with the Emmerie-Engel reagent [2,2′-bipyridine(α,α′-dipyridyl) and ferric chloride] gives a red color from combination of the bipyridine with the ferrous ions resulting from reduction of the ferric ions; this color is directly proportional to the amount of tocopherol present. Substitution of 4,7-diphenyl-1,10-phenanthroline (bathophenanthroline) for the 2,2′-bipyridine provides greater sensitivity. Titration with ceric sulfate is a good alternative procedure for pure forms. α-Tocopherol in 0.5 N alcoholic sulfuric acid is titrated with 0.01 N ceric sulfate to a blue end point with diphenylamine as indicator. Spectrophotometric measurement is sometimes used, based on the measurement of the specific absorbancy, which is proportional to the α-tocopherol concentration. Analysis of lower potency concentrates, original vegetable oils, or foods and feeds is considerably more difficult. The Emmerie-Engel color reaction is nonspecific and occurs to varying degrees with each of the tocopherols and tocotrienols. Many nontocopherol reducing materials obtained along with the tocopherols in the original extract from foods also react. Hence, the method of assay must provide for isolation of the α-tocopherol from other interfering compounds. Determination of the other tocopherols and tocotrienols is often a secondary goal.

Considerable research effort by many investigators has resulted in a variety of approaches to vitamin E assay, so that the analyst can choose the technique to fit a particular analytical problem (34,35). Available methods accomplish the separation of the tocopherols and tocotrienols by column chromatography on magnesium hydrogen phosphate (36), paper chromatography (37), thin-layer chromatography (35) and gas–liquid chromatography (38). By the first three methods, the isolated compounds can be measured colorimetrically. Analysis of the α-tocopherol esters commonly used in vitamin preparations and enriched foods and feeds requires an initial hydrolysis. This step is extremely critical because tocopherols are sensitive to oxygen under alkaline conditions. However, if oxygen is excluded from the system and an effective antioxidant is added (37), the esters may be saponified without loss of tocopherol. Hydrolysis of the ester prior to any other treatment permits analysis for total tocopherol content. Hydrolysis subsequent to oxidative destruction of the unesterified tocopherols permits analysis for only the added α-tocopherol ester (39).

Bioassay Methods. The most specific procedure for determining α-tocopherol activity is a modification of the Evans resorption gestation method (40). Female rats are raised on a vitamin E-free diet and mated with normal males. Conception and the first half of pregnancy are normal. However, unless a vitamin E supplement is administered within the first 10–12 days of pregnancy, the embryos die and are resorbed. The mother rat is apparently unharmed and may be used repeatedly. If the dose of α-tocopherol administered as a test supplement is above a certain critical amount, pregnancy proceeds and terminates normally. The critical dose of α-tocopherol varies somewhat between assays (0.3–1.0 mg), but the response is an all-or-none type. The young are either alive or dead, a criterion which is easily determined. Consequently, the bioassay properly controlled and carefully conducted is specific and reasonably accurate.

The other procedure best suited for biological determination of vitamin E activity is the hemolytic test in rats (41). This test is based on the fact that adult rats receiving adequate vitamin E in their diet have red blood cells that are resistant to hemolysis induced by chemical action of such compounds as dialuric acid and hydrogen peroxide. Changing the animals to a vitamin E-deficient diet for several days causes the red blood cells to become susceptible to hemolysis. A satisfactory bioassay consists of the administration of various levels of α-tocopherol (both standard and test substance), either prophylactically or curatively, to the rats during the depletion period and measurement of the change in degree of red cell hemolysis that results. The drawback of this test is that it is not specific for α-tocopherol and may respond to compounds that do not prevent other physiological manifestations of vitamin E deficiency.

Biochemical Function and Clinical Evaluation

The exact mechanism whereby α-tocopherol functions in the body as vitamin E is unknown. It probably acts as an antioxidant controlling redox reactions in a variety of tissues and organs, particularly in preventing peroxidation of unsaturated lipids. That this is the exclusive role is being challenged (42), however, and the high isomeric specificity of the α-tocopherol structure in physiological reactions suggests other roles as an integral part of the biological machinery. Experiments with vitamin E–deficient animals have revealed perhaps the largest variety of disorders associated with the nutritional lack of any single vitamin. Vitamin E deficiency affects the reproductive system (testicular degeneration, defective development of the embryo); the muscular system (skeletal muscular dystrophy, cardiac necrosis and fibrosis, ceroid pigment in smooth muscle); the circulatory system (exudative diathesis, erythrocyte hemolysis, anemias); the skeletal system (incisor depigmentation); and the nervous system (encephalomalacia, deposition of lipofuscin pigment). In addition, vitamin E deficiency may be manifested by a number of other conditions, such as discolored adipose tissue, liver necrosis, lung hemorrhage, kidney nephrosis and postmortem autolysis, and creatinuria. The gross and microscopic pathology which results in the tissues and organs of vitamin E-deficient animals of various species has been described by Mason (43,44).

Vitamin E has also been found to ameliorate the adverse effects of various toxic agents (eg, silver nitrate and chlorinated solvents) and a number of suboptimal dietary (eg, low protein) and environmental (eg, hyperoxia) conditions. In recent years, an intriguing but poorly understood interplay between α-tocopherol and selenium has been uncovered (45). Some synthetic antioxidants and hexahydrocoenzyme Q_4 have also been reported to show vitamin E-like activity, perhaps at least partially because of a sparing effect on low levels of α-tocopherol.

Signs of vitamin E deficiency in man were not positively identified until surprisingly late in the studies on this vitamin, beginning with the studies of Gordon and deMetry (46) on children and confirmed on adults by Horwitt and coworkers (47). The incidence of such signs in man appears relatively low, but they are sufficiently numerous and varied to establish a positive correlation between vitamin E deficiency in man and that in other animals (48). These studies also enabled the establishment of recommended daily allowances for vitamin E, ranging from 5 IU for infants to 30 IU for adult men and pregnant or lactating women (49).

Well-founded evidence of the need for therapeutic amounts of vitamin E in humans exists for several conditions (50), namely, intermitten claudication, fat malabsorption syndromes, excessive intakes of polyunsaturated fats, and certain anemias in the young. Tocopherol administration also deserves serious consideration in stasis ulcers and habitual abortion. Subnormal reproductive functions in both males and females were among the clinical problems first treated with α-tocopherol, but results were conflicting. The broad group of diseases of collagen tissue, the so-called collagenoses, have been treated empirically with vitamin E; and such treatment may have value in such conditions as lupus erythematosus, primary fibrositis, Dupuytren's contracture, and Peyronie's disease. Infants, particularly prematures, deserve special attention. They are poorly supplied with vitamin E because the placenta acts as a barrier preventing the passage of α-tocopherol from the mother to the fetus. Conversely, colostrum, and early milk are very rich in tocopherol and apparently are designed to supply infants relatively large amounts of vitamin E during the first few days of life. Artificial feeding with formulas based on cow's milk supplies relatively little α-tocopherol. As a result, many infant feeding preparations are now supplemented with vitamin E. The importance of adequate vitamin E nourishment in the young has been emphasized by the observation of tocopherol-responsive hemolytic anemia in premature infants and macrocytic anemia in malnourished children.

α-Tocopherol administered orally in massive doses is well tolerated; and no symptoms of toxicity have been reported in a number of species, including rats, mice, rabbits, dogs, cats, and monkeys (51). Adult humans have been given levels as high as 500–600 mg/kg body weight daily for five months. Parenteral administration of α-tocopherol as an oil concentrate is not recommended. Local reactions at the site of injection are common, and utilization of the tocopherol is very poor. On the other hand, injectable preparations of vitamin E in water-dispersible emulsifiers are well utilized.

Uses

Both α-tocopherol and esters of α-tocopherol are used in multivitamin and therapeutic-type capsules. The crystalline α-tocopheryl acid succinate or the oily forms embedded in materials such as acacia are particularly useful in the formulation of multivitamin tablets and dry-fill capsules. α-Tocopherol or its esters dispersed or solubilized in hydrophilic media are advantageous for malabsorption conditions and as injectable preparations. Also, specialty items, such as vitamin E-containing ointments, salves, and suppositories, provide other outlets for α-tocopherol in the pharmaceutical field. Doses for human use range from a few milligrams for dietary supplementation to 500–1000 mg for certain therapeutic applications. Many infant formulas are also supplemented with α-tocopheryl acetate to furnish 5–10 IU per reconstituted quart.

Approximately two-thirds of the vitamin E produced commercially is utilized in animal feeds, primarily by the poultry industry. Increasing amounts are being used, however, in supplements and concentrates for calves, lambs, swine, and beef and dairy cattle as well as for fortifying rations of fur-bearing animals, such as mink, and pets, particularly cats. The vitamin E naturally present in feedstuffs is relatively unstable in stored, mixed feeds. To replace tocopherol that might have been destroyed and to ensure that the animal's requirements are met, stable α-tocopheryl acetate is added at levels ranging from 2.5 to 20 IU/lb of feed. The form of addition usually is either a dry, granular, free-flowing, nondusting material containing from 20,000 to 125,000 IU per pound or a high-potency oil concentrate.

Unesterified tocopherols have been used as antioxidants to some extent by food technologists and by pharmaceutical formulators. They are particularly effective in essential oils, rendered animal fats, and mineral oil. In pharmaceutical preparations, tocopherols stabilize other substances, eg, vitamin A and unsaturated lipids during shelf life and also in the gastrointestinal tract. Although not the best antioxidant *per se*, α-tocopherol is uniquely suited for this purpose because it is absorbed and deposited in body tissues, whereas other antioxidants are destroyed or excreted.

Bibliography

"Vitamin E" in *ECT* 1st ed., Vol. 14, pp. 849–858, by P. L. Harris and N. D. Embree, Eastman, Kodak Co.

1. J. F. Pennock, F. W. Hemming, and J. D. Kerr, *Biochem. Biophys. Res. Commun.* **17,** 542–548 (1964).
2. D. R. Nelan and C. D. Robeson, *J. Am. Chem. Soc.* **84,** 2963–2965 (1962).
3a. *National Formulary XII*, American Pharmaceutical Association, Washington, D. C., 1965, pp. 404–409.
3b. *Ibid.*, p. 506.
4. *J. Nutr.* **99,** 245–248 (1969).
5. R. A. Morton, *Wiss. Veroeff. Deut. Ges. Ernaehr.* **16,** 1–13 (1967).
6. E. Fernholz, *J. Am. Chem. Soc.* **60,** 700–705 (1938).
7. J. L. G. Nilsson, *Acta Pharm. Suecica* **6,** 1–24 (1969).
8. W. A. Skinner, R. M. Parkhurst, J. Scholler, and K. Schwarz, *J. Med. Chem.* **12,** 64–66 (1969).
9. E. J. Simon, A. Eisengart, L. Sundheim, and A. T. Milhorat, *J. Biol. Chem.* **221,** 807–817 (1956).
10. L. Michaelis and S. H. Wollman, *Biochim. Biophys. Acta* **4,** 156–159 (1950).
11. W. A. Skinner, *Biochem. Biophys. Res. Commun.* **15,** 469–472 (1964).
12. J. G. Bieri and E. L. Prival, *Biochem.* **6,** 2153–2158 (1967).
13. M. W. Dicks, *Vitamin E Content of Foods and Feeds for Human and Animal Consumption, Bulletin 435*, Agricultural Experiment Station, University of Wyoming, Laramie, 1965.
14. D. C. Herting and E. E. Drury, *J. Nutr.* **81,** 335–342 (1963).
15. P. L. Harris and N. D. Embree, *Am. J. Clin. Nutr.* **13,** 385–392 (1963).
16. R. H. Bunnell, J. Keating, A. Quaresimo, and G. K. Parman, *Am. J. Clin. Nutr.* **17,** 1–10 (1965).
17. R. S. Harris, "Influences of Storage and Processing on the Retention of Vitamin E in Foods" in R. S. Harris and I. G. Wool, eds., *Vitamins and Hormones*, Vol. 20, Academic Press, New York, 1962, pp. 603–619.
18. D. C. Herting and E. E. Drury, *J. Agr. Food Chem.* **17,** 785–790 (1969).
19. V. H. Booth and M. P. Bradford, *Intern. Z. Vitaminforsch.* **33,** 276–278 (1963).
20. M. H. Thomas and D. H. Calloway, *J. Am. Dietet. Assoc.* **39,** 105–116 (1961).
21. N. D. Embree, *Chem. Rev.* **29,** 317–332 (1941).
22. M. H. Stern, C. D. Robeson, L. Weisler, and J. G. Baxter, *J. Am. Chem. Soc.* **69,** 869–874 (1947).
23. T. Rubel, "Vitamin E Manufacture," *Chemical Process Review No. 39*, Noyes Development Corp., Park Ridge, N. J., 1969.
24. O. Isler, P. Schudel, H. Mayer, J. Würsch, and R. Rüegg, "Chemistry of Vitamin E" in R. S. Harris and I. G. Wool, eds., *Vitamins and Hormones*, Vol. 20, Academic Press, New York, 1962, pp. 389–405.
25. P. Karrer, H. Fritzsche, B. H. Ringier, and H. Salomon, *Helv. Chim. Acta* **21,** 520–525 (1938).
26. F. Bergel, A. Jacob, A. R. Todd, and T. S. Work, *Nature* **142,** 36 (1938).
27. L. I. Smith, H. E. Ungnade, and W. W. Prichard, *Sci.* **88,** 37–38 (1938).
28. H. Mayer, P. Schudel, R. Rüegg, and O. Isler, *Helv. Chim. Acta* **46,** 650–671 (1963).
29. S. R. Ames, M. I. Ludwig, D. R. Nelan, and C. D. Robeson, *Biochem.* **2,** 188–190 (1963).
30. C. D. Robeson, *J. Am. Chem. Soc.* **64,** 1487 (1942).
31. L. I. Smith, W. B. Renfrow, and J. W. Opie, *J. Am. Chem. Soc.* **64,** 1084–1086 (1942).
32. U. S. Tariff Commission, "United States Production and Sales of Synthetic Organic Chemicals," U. S. Government Printing Office, Washington, D. C., TC Publication 248, 1968; Preliminary Publication, 1969.
33a. *Hi-Lo Chemical Price Issue. Sect. 2, Oil Paint Drug Report*, Schnell Publishing Co., New York, 1962, pp. 306, 308.
33b. *Oil Paint Drug Rep.* **197** (2), 37 (1970).
34. R. H. Bunnell, "Vitamin E Assay by Chemical Methods" in P. György and W. N. Pearson, eds., *The Vitamins*, Vol. 6, 2nd ed., Academic Press, New York, 1967, Chap. 6.
35. J. G. Bieri, "Chromatography of Tocopherols" in G. V. Marinetti, ed., *Lipid Chromatographic Analysis*, Vol. 2, Marcel Dekker, Inc., New York, 1969, Chap. 8.

36. F. Bro-Rasmussen and W. Hjarde, *Acta Chem. Scand.* **11**, 34–43, 44–52 (1957).

37. Vitamin E Panel, Analytical Methods Committee, *Anal.* **84**, 356–372 (1959).

38. H. T. Slover, J. Lehmann, and R. J. Valis, *J. Am. Oil Chemists Soc.* **46**, 417–420 (1969).

39. S. R. Ames and F. H. Tinkler, *J. Assoc. Offic. Agr. Chemists* **45**, 425–433 (1962).

40. K. E. Mason and P. L. Harris, *Biol. Symp.* **12**, 459–483 (1947).

41. C. I. Bliss and P. György, "Bioassays of Vitamin E" in P. György and W. N. Pearson, eds., *The Vitamins*, Vol. 6, 2nd ed., Academic Press, New York, 1967, Chap. 6.

42. J. Green and J. Bunyan, *Nutr. Abstr. Rev.* **39**, 321–345 (1969).

43. K. E. Mason, "Physiological Action of Vitamin E and Its Homologues" in R. S. Harris and K. V. Thimann, eds., *Vitamins and Hormones*, Academic Press, Inc., New York, 1944, pp. 107–153.

44. K. E. Mason, P. L. Harris, R. S. Harris, and H. A. Mattill, "The Tocopherols" in W. H. Sebrell, Jr., and R. S. Harris, eds., *The Vitamins*, Vol. 3, Academic Press, New York, 1954, Chap. 17.

45. M. L. Scott, *Ann. N. Y. Acad. Science* **138**, Art. 1, 82–89 (1966).

46. H. H. Gordon and J. P. deMetry, *Proc. Soc. Exp. Biol. Med.* **79**, 446–450 (1952).

47. M. K. Horwitt, "Interrelations between Vitamin E and Polyunsaturated Fatty Acids in Adult Men" in R. S. Harris and I. G. Wool, eds., *Vitamins and Hormones*, Vol. 20, Academic Press, New York, 1962, pp. 541–558.

48. D. C. Herting, *Am. J. Clin. Nutr.* **19**, 210–218 (1966).

49. Food and Nutrition Board, National Research Council, *Recommended Dietary Allowances*, 7th ed., Publication 1694, National Academy of Sciences, Washington, D. C., 1968.

50. J. Marks, "Critical Appraisal of the Therapeutic Value of α-Tocopherol" in R. S. Harris and I. G. Wool, eds., *Vitamins and Hormones*, Vol. 20, Academic Press, New York, 1962, pp. 573–598.

51. V. Demole, *Intern. Z. Vitaminforsch.* **8**, 338–341 (1939).

DAVID C. HERTING
Tennessee Eastman Co.

VITAMIN K

Nutritionally-induced hemorrhage was discovered by Dam in 1929 and confirmed by McFarlane and his associates in 1931. The hemorrhagic syndrome was induced by feeding chicks a lipid-deficient diet and was characterized by subcutaneous and intramuscular hemorrhage, anemia, prolonged blood clotting time, and a particularly high mortality rate. By 1935, it was shown that a fat-soluble factor in green leaves and certain vegetables protected the chick from nutritionally-induced hemorrhage, whereas the then known fat-soluble vitamins A, D, and E did not. On the basis of these observations, Dam proposed the existence of a lipid-soluble *Koagulations-vitamin* that eventually became known as vitamin K. At first there appeared to be only two K-vitamins. One, designated vitamin K_1, was isolated from the chloroplasts of green plants; the second, designated vitamin K_2, was of microbial origin and isolated from putrefying plant and animal materials. Eventually it was learned that a large number of structurally related 1,4-naphthoquinones were effective in protecting the chick against nutritionally-induced hemorrhage. Accordingly, each was described as a K-vitamin with a distinguishing subscript. For example, 2-methyl-1,4-naphthoquinone became known as vitamin K_3, and 4-amino-3-methyl-1-naphthol hydrochloride was referred to as vitamin K_7. Today, however, it is customary to reserve the vitamin K designation for products of nature and to designate the others by chemical or generic nomenclature.

Shortly after vitamins K_1 and K_2 were isolated, it was established that each is a 3-substituted 2-methyl-1,4-naphthoquinone; the sole difference between the two compounds is the length and degree of unsaturation of a polyisoprenoid substituent at the 3-position. In vitamin K_1 only the first isoprenoid unit is unsaturated, whereas in vitamin K_2 each isoprenoid unit has one double bond. These designations were adequate until the last decade when it was found that vitamin K_1 was only one member of a homologous series of biologically active 3-substituted 2-methyl-1,4-naphthoquinones and vitamin K_2 was simply one member of another biologically active homologous series. Accordingly, it became necessary to add a parenthetical subscript to the original names to designate the number of carbon atoms in the 3-substituent. By this system, the original vitamins are named vitamin $K_{1(20)}$ and vitamin $K_{2(35)}$. In the biochemical literature, however, it is popular to use the name phylloquinone or the simple abbreviation K for vitamin $K_{1(20)}$ and the name menaquinone-35 (or menaquinone-7) or the abbreviation MK-35 (or MK-7) for vitamin $K_{2(35)}$.

Among the commercially important compounds possessing vitamin K activity are vitamin $K_{1(20)}$ [phytonadione (USP XVII)] in the form of oil, tablet, or stabilized sterile emulsion; menadione (2-methyl-1,4-naphthoquinone) in powder or tablet form; and the two water-soluble salts, menadione sodium bisulfite and tetrasodium dihydromenadione diphosphate.

Vitamin $K_{1(20)}$ (Phylloquinone)

ISOLATION

Extraction and preliminary purification of the vitamin from dried alfalfa meal was reported by Almquist (1), Dam (2), and Doisy (3,4) and their collaborators during the period 1936–1938. Isolation of vitamin $K_{1(20)}$ in what proved to be a pure form, however, was not announced until a year later by Dam and Karrer (5,7) and by Doisy (6) and their collaborators. The essential features of these isolation procedures included: (a) extraction of the factor by ether, hexane, or preferably petroleum ether, (b) removal of chlorophyll by stirring the extract with powdered zinc carbonate or by passing it through a cation exchange gel, (c) four or five chromatographic resolutions on columns of magnesium sulfate, zinc carbonate, activated carbon, or cation and anion exchange complexes, and (d) molecular distillation.

The isolation procedure was simplified significantly by Fieser (8) after it had been established that the vitamin is a derivative of 1,4-naphthoquinone. The vitamin in the crude extract is converted to the hydroquinone form by reduction. The latter is less soluble in petroleum ether than the substances that ordinarily accompany it in natural concentrates and is isolated as a colorless waxy solid. Alternatively, the dihydrovitamin, being a weakly acidic compound, can be selectively extracted into Claisen's methanolic alkali solution (potassium hydroxide). After the alkaline extract is diluted with water, the dihydrovitamin is extracted into ether and then oxidized to the quinone form.

CHEMICAL AND PHYSICAL PROPERTIES

Vitamin $K_{1(20)}$ is a mobile yellow oil that can be crystallized from acetone or ethanol to yield yellow rosettes melting at about $-20°C$. Its molecular formula is $C_{31}H_{46}O_2$; degradation and synthesis established the structure as 2-methyl-3-phytyl-1,4-naphthoquinone (1).

(1)

The double bond at the 2'-3'-position is in the trans configuration (9), and the asymmetric centers at the 7'- and 11'-positions are in the same R-configuration as the corresponding atoms in natural phytol (10). The vitamin is insoluble in water and sparingly soluble in methanol or ethanol. It dissolves readily in petroleum ether, benzene, ether, dioxane, chloroform, acetone, vegetable oils, and other lipid solvents. Since the vitamin is stable to air and moisture, an ether solution of the quinone can be concentrated to dryness at atmospheric pressure without decomposition, and the vitamin can be distilled at pressures of the order of one micron without significant decomposition. Unfortunately, vitamin $K_{1(20)}$ is very sensitive to ultraviolet radiation and even to diffuse daylight; light radiation in the region 400–800 nm, however, has no appreciable effect (11).

The vitamin in benzene solution is weakly levorotatory ($[\alpha]_D^{20} \approx -0.4°$). In an 80% alcoholic solution that is $0.02N$ in acetic acid and $0.02N$ in sodium acetate at 25°C, the oxidation-reduction potential of the vitamin is 0.005 V (12); at 20°C in 95% alcoholic solution that is $0.2N$ in hydrochloric acid and $0.02N$ in lithium chloride, it is 0.363 V (13). The half-wave potential for reduction of vitamin $K_{1(20)}$ at the dropping mercury electrode in isopropyl alcohol and $0.1N$ aqueous potassium chloride is −0.58 V (14). In solvents such as hexane or "isooctane" (2,2,4-trimethylpentane) vitamin $K_{1(20)}$ has an ultraviolet absorption spectrum characterized by maxima at 239 nm, 243–244 nm (log ϵ 4.25–4.27), 248–249 nm (log ϵ 4.28), 261–263 (log ϵ 4.24), 270 nm (log ϵ 4.24) and 325–328 nm (log ϵ 3.49–3.50) (15,16). The infrared absorption spectrum of vitamin $K_{1(20)}$ includes maxima at 1660 cm^{-1} (quinone carbonyl stretch), 1618 cm^{-1} (quinone double bond), and 1595 cm^{-1} and 1590 cm^{-1} (aromatic vibrations) (17). The nuclear magnetic resonance spectrum (18) of a 5% solution of vitamin $K_{1(20)}$ is characterized by bands at 2.00 τ (C-5 and C-8 protons), 2.39 τ (C-6 and C-7 protons), 5.00 τ (C-2' proton), 6.68 τ (C-1' protons), 7.86 τ (2-methyl group), 8.23 τ (3'-methyl group) and 9.14 τ (side chain methyl groups other than 3').

Vitamin $K_{1(20)}$ is stable in dilute acid but is readily decomposed by alkali. An alcoholic alkali solution of the vitamin slowly develops a violet-blue coloration that gradually changes into a dull red color (8). This reaction, in which a small amount of phthiocol (3-hydroxy-2-methyl-1,4-naphthoquinone) forms, is the basis of the Dam-Karrer color test.

Reduction of the quinone moiety by catalytic hydrogenation over a deactivated palladium catalyst (16,19) or by sodium hydrosulfite ($Na_2S_2O_4$) (8) yields dihydrovitamin $K_{1(20)}$ (2).

(2)

This waxy, petroleum–ether insoluble product can be reoxidized smoothly to the quinone form by shaking an ether solution of the reduction product with air, or more conveniently by treatment of the solution with silver oxide in the presence of anhydrous magnesium sulfate. Reduction under acetylating conditions converts the vitamin to crystalline dihydrovitamin $K_{1(20)}$ diacetate, mp 62–63°C (20). Hydrogenation of a methanolic solution of vitamin $K_{1(20)}$ over Raney nickel results in saturation of the phytyl side chain and reduction of the quinone to a hydroquinone moiety, whereas reduction of the compound in acetic solution in the presence of platinum results in its reduction to 2-methyl-3-β,γ-dihydrophytyl-5,6,7,8-tetrahydro-1,4-naphthohydroquinone (12). The vitamin may also be reductively cyclized to the naphthotocopherol derivative (**3**) by refluxing an acetic acid solution of the quinone with stannous chloride and concentrated hydrochloric acid or zinc and sulfuric acid (21). Treatment of dihydrovitamin $K_{1(20)}$ with sulfuric acid in acetic acid solution also leads to the naphthotocopherol analog (**3**).

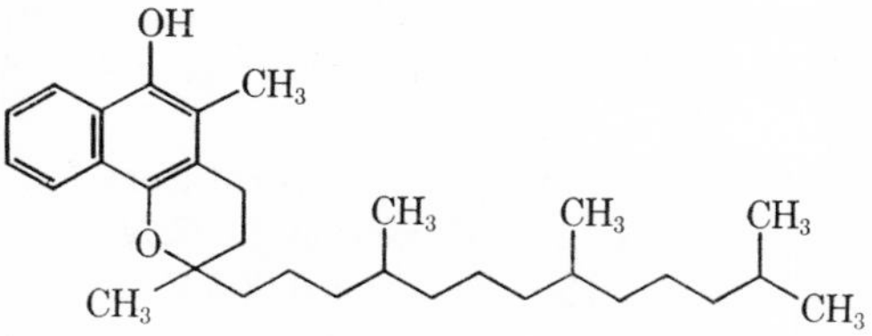

(**3**)

Oxidative degradation of vitamin $K_{1(20)}$ contributed largely to the elucidation of its structure (22,23). Mild chromic acid oxidation of the vitamin yields 2-methyl-1,4-naphthoquinone-3-acetic acid, whereas the use of excess chromic acid gives phthalic acid. Ozonolysis of dihydrovitamin $K_{1(20)}$ diacetate yields 6,10,14-trimethyl-2-pentadecanone and 1,4-diacetoxy-2-methylnaphthalene-3-acetaldehyde. Treatment of vitamin $K_{1(20)}$ with hydrogen peroxide in an alkaline medium results in epoxidation at the 2,3-position of the quinone moiety (see vitamin $K_{1(20)}$ oxide, below).

SYNTHESIS

Phytylation of 2-methyl-1,4-naphthoquinone or 2-methyl-1,4-naphthohydroquinone in the primary step for the synthesis of vitamin $K_{1(20)}$ was described in five independent publications during 1939. Of these, a synthesis described by Fieser (8) was developed into a commercial process and proved to be most useful for about fifteen years. In the first step, 2-methyl-1,4-naphthohydroquinone (**4**) is condensed with phytol (**5**) in the presence of oxalic acid in dioxane solution at 75°C.

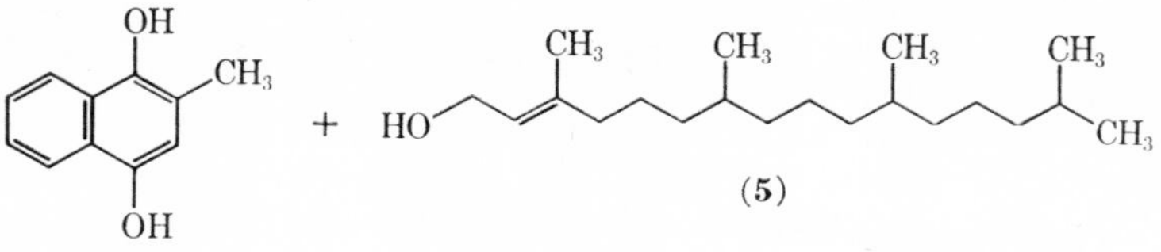

(**4**)   (**5**)

After a 36-hr-reaction period, unchanged 2-methyl-1,4-naphthohydroquinone is removed by extraction with dilute alkali, and the product, dihydrovitamin $K_{1(20)}$ (**2**), is purified by precipitation from petroleum ether or extraction into Claisen's alkali. In the final step, the dihydrovitamin (**2**) is oxidized to vitamin $K_{1(20)}$ by silver oxide. Based on phytol, the overall yield of the process is 29%. Major by-products of the process include phytadiene and 2-methyl-2-phytyl-2,3-dihydro-1,4-naphthoquinone. The latter has, however, been converted to vitamin $K_{1(20)}$ (21).

By 1954, boron trifluoride etherate and potassium acid sulfate were shown to be more efficient catalysts for the condensation than oxalic acid (15,16). Their superiority rests largely on the fact that oxalic acid is prone to esterify phytol and in effect remove it from the condensation reaction. In fact, this ester is the source of the phytadiene by-product that forms during distillation of the crude condensation product. In addition, it was learned that ethers and certain esters of phytol (15,16,24), as well as isophytol (16) are satisfactory phytylating agents. The formation of 2-methyl-2-phytyl-2,3-dihydro-1,4-naphthoquinone was eliminated by substituting 1-O-acetyl-2-methyl-1,4-naphthohydroquinone for 2-methyl-1,4-naphthohydroquinone in the condensation. An additional advantage of this modification became evident when it was learned that 1-O-acetyldihydrovitamin $K_{1(20)}$ is a better intermediate than dihydro-vitamin $K_{1(20)}$ since the former is far less prone to oxidation during the condensation reaction. Consequently, a reduction step prior to purification of the condensation product was no longer necessary, and monoacetyldihydrovitamin $K_{1(20)}$ could be selectively extracted from the crude product by Claisen's alkali. In this medium, the 1-acetyl group is hydrolyzed readily, and the product is sufficiently pure for facile oxidation to vitamin $K_{1(20)}$. By incorporating these features in the original process, the overall yield of vitamin $K_{1(20)}$ from phytol was raised to 66% (15).

ASSAY

The physicochemical and biological methods for the determination of vitamin $K_{1(20)}$ have been reviewed and evaluated by Dam and Sondergaard (25). The method of choice will usually be dictated by several factors. For example, certain procedures are particularly adaptable for the routine evaluation of a large number of samples. Some require partial purification of the sample before assay, whereas others require precise knowledge of the particular K-vitamin present in the sample.

Volumetric Analysis. Vitamin $K_{1(20)}$ and related quinones may be measured quantitatively by hydrogenation of the vitamin over Raney nickel in the presence of phenosafranine as indicator, followed by reoxidation with a standardized solution of 2,6-dichlorophenol-indophenol, which also serves as indicator for the titration (26).

Ultraviolet Spectroscopy. Measurement of the absorbance of a hexane solution of vitamin $K_{1(20)}$ at 249 nm ranks among the more accurate assays. The sample must be free of impurities that absorb in this region of the spectrum, and the hexane must be a special spectroscopic grade. Characteristic isosbestic points (intersections) are also found in a combined plot of the ultraviolet absorption spectra of vitamin $K_{1(20)}$ and dihydrovitamin $K_{1(20)}$. These points at 253 nm and 280 nm have been used to distinguish between the K-vitamins and the ubiquinones and plastoquinones.

Colorimetry. The Irreverre-Sullivan color reaction involves treatment of vitamin $K_{1(20)}$ with sodium diethyldithiocarbonate and sodium ethylate in 95% ethanol (27). The intensity of the deep blue color that develops is measured at 575 nm. Since the highest intensity is attained after 5–6 min and the color begins to fade after 8 min, it is customary to measure the absorbance every minute for a 10 min period and to use the highest reading. Unfortunately, the ubiquinones and plastoquinones also give a color (565 nm) in this reaction. It is customary, therefore, to determine the ubiquinone and plastoquinone content of a given sample by the Craven's color test (ethyl cyanoacetate and ammonium hydroxide), and to make the appropriate correction. Vitamin $K_{1(20)}$ does not react in the latter test.

Polarography. In aqueous isopropyl alcohol containing potassium chloride, vitamin $K_{1(20)}$ shows a sharply defined wave at -0.58 V (14). This method is particularly suitable for pharmaceutical preparations.

Nuclear Magnetic Resonance. This method is particularly useful for the K-vitamins since the degree of unsaturation and number of isoprenoid units of the side chain can be established with high accuracy.

Biological Assays. These methods are generally more sensitive than the chemical and physical methods but are not as accurate. They are specific but cannot distinguish between the various K-vitamins. The chick is considered to be the best animal for biological assay of the vitamin because coprophagy is more readily prevented, and the requirement for vitamin K is roughly five times greater than that of other laboratory mammals. The assay can be based on a curative or prophylactic principle, and the actual clotting time may be determined in whole blood samples or in plasma samples derived therefrom. When the coagulation time is measured in plasma, the assay is usually referred to as "the prothrombin-time assay."

In practice, three-week old chicks are put on a vitamin K deficient diet and begin to show moderate subcutaneous hemorrhage after two weeks. At the same time, another group of chicks is placed on the same diet supplemented with vitamin $K_{1(20)}$. Blood samples are collected from the carotid artery. The blood sample is warmed to 37°C and a preheated solution of thromboplastin-calcium chloride is added. The mixture is agitated gently and the interval from the time of addition of thromboplastin to time of clot formation is measured. The clotting time for each chick on the deficient diet is recorded as a percentage of the clotting time for the normal chicks. Next, a measured amount of the sample to be assayed is introduced through the esophagus and into the crop of each deficient chick. After about 20 hr, a second blood sample is taken from the other carotid artery, and the clotting time is measured. The resulting clotting time percentages are converted to vitamin $K_{1(20)}$ units using a predetermined curve relating the amount of vitamin $K_{1(20)}$ ingested per gram of body weight and the percent of normal clotting time.

There is also an assay to establish the ability of compounds to overcome the anticoagulant activity of vitamin K antagonists. In this case, the chick is reared on a normal diet and dosed with a coumarin analog. Coagulation time of a blood specimen is determined about two days later and the sample is administered orally or intravenously. Six blood samples are taken at hourly intervals and a final blood specimen is taken after 24 hr for measurement of coagulation time.

Vitamin $K_{1(20)}$ Oxide

Vitamin $K_{1(20)}$ oxide (**6**) has been prepared from vitamin $K_{1(20)}$ in 82% yield by treating the vitamin with hydrogen peroxide and aqueous sodium carbonate at 75°C (28).

The oxide has the same high order of antihemorrhagic activity as the parent vitamin and is superior to it in being more stable to light and heat. The compound is an almost colorless oil that does not respond normally in the Dam-Karrer color test.

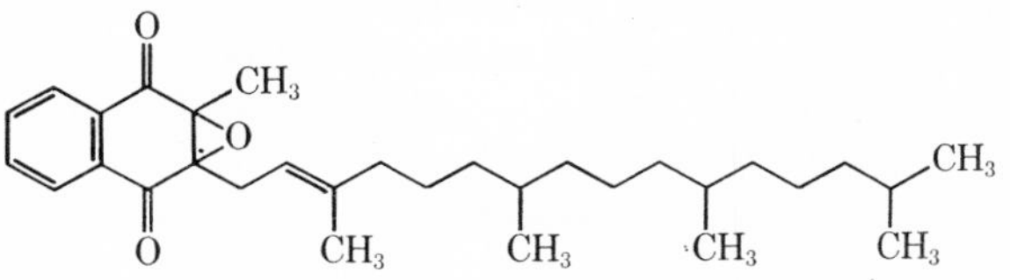

(6)

Warm alcoholic alkali solutions of the oxide, however, take on a red coloration. The uv absorption spectrum of an ethanol solution of the oxide is characterized by maxima at 259 nm (log ϵ 3.79) and 305 nm (log ϵ 3.31) (21).

Dihydrovitamin $K_{1(20)}$ Diphosphate

The bis-phosphate ester (7) of dihydrovitamin $K_{1(20)}$ was synthesized by treating dihydrovitamin $K_{1(20)}$ with phosphorus oxychloride in pyridine solution, followed by hydrolysis. Alkali metal salts of this analog are water soluble.

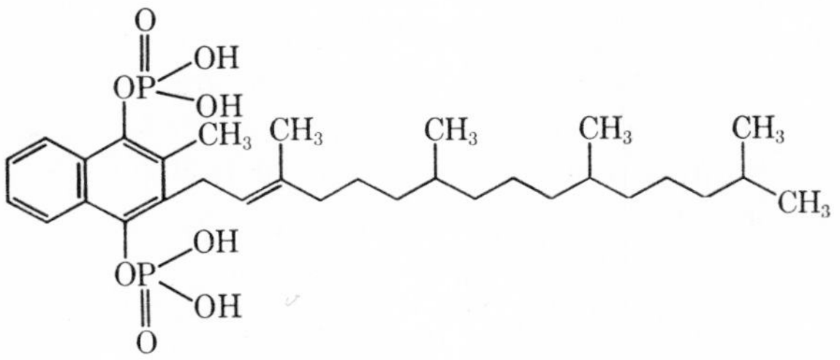

(7)

Vitamin $K_{2(35)}$

Vitamin $K_{2(35)}$ was isolated from putrified sardine meal in 1939. Its structure was erroneously formulated as vitamin $K_{2(30)}$, and the compound was referred to as farnoquinone for almost twenty years. In 1958, however, Isler and his colleagues (29) showed that this original member of the vitamin K_2 series is 2-methyl-3-*all-trans*-farnesylgeranylgeranyl-1,4-naphthoquinone (8).

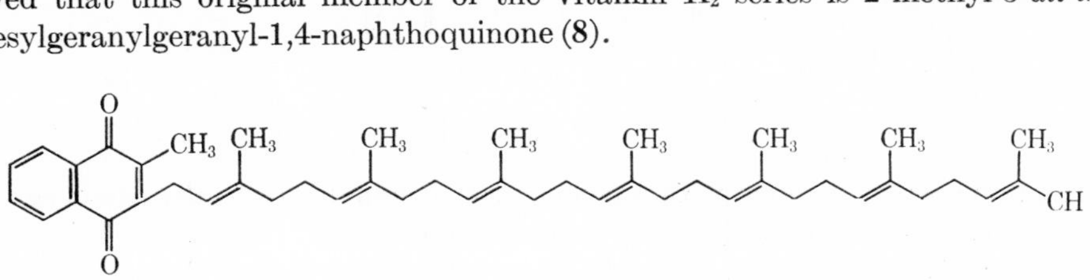

(8)

The compound is a pale-yellow crystalline substance melting at 53.5–54.5°C. In hexane solution, its ultraviolet absorption spectrum is characterized by maxima at 243 nm (log ϵ 4.20), 249 nm (log ϵ 4.24), 260 nm (log ϵ 4.18), 270 nm (log ϵ 4.18), and 325 nm (log ϵ 3.45). On a molar basis, the antihemorrhagic activity of vitamin $K_{2(35)}$ is comparable to that of vitamin $K_{1(20)}$. Except for properties associated with the polyunsaturation in the side chain, vitamin $K_{2(35)}$ also resembles vitamin $K_{1(20)}$ in its chemical and physical properties. Vitamin $K_{2(35)}$ and its homologs with from one to ten isoprenoid units in the side chain have been synthesized (29). More information about this commercially unimportant group of analogs may be obtained by consulting the recommended reading list.

Menadione

PHYSICAL AND CHEMICAL PROPERTIES

Menadione is 2-methyl-1,4-naphthoquinone (**9**) and is the only simple analog that is more active than the natural forms of vitamin K in the chick assay.

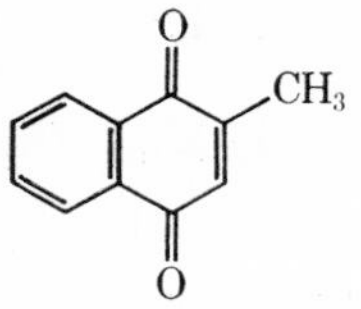

(**9**)

The crystalline compound melts at 105–106°C. It is soluble in benzene and ether, somewhat less soluble in glacial acetic acid, and only sparingly soluble in hydrocarbons. The compound dissolves to the extent of 17 mg/ml in 95% ethanol and 13 mg/ml in sesame oil. It is only slightly soluble in water but dissolves readily in an aqueous solution of sodium dehydrocholate. The ultraviolet absorption spectrum of menadione in hexane or isooctane is characterized by maxima at 244 nm (log ϵ 4.30), 253 nm (log ϵ 4.29–4.31), 263–264 nm (log ϵ 4.22), and 334 nm (log ϵ 3.4–3.5) (30). The oxidation potential of menadione at 25°C in 70% ethanol that is $0.2N$ with respect to hydrochloric acid and $0.2N$ in lithium chloride is 0.408 V (31). In 50% ethanol, $0.1N$ in hydrochloric acid and $0.2N$ in lithium chloride, at 25°C, the redox potential is 0.422 V (32,33). Polarographic reduction of menadione at pH 6.24 gives a step at -0.17 V (34,35). Menadione is sensitive to heat and light; the action of light produces the dimer (**10**).

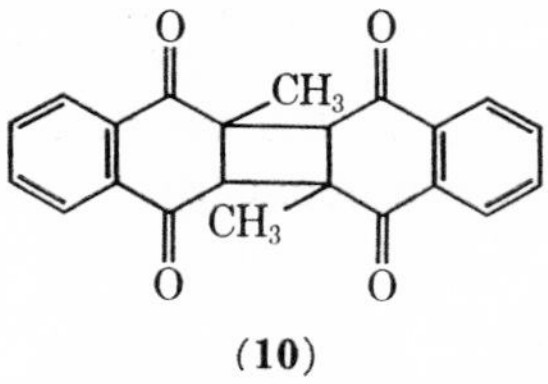

(**10**)

Treatment of menadione with aqueous or ethanolic alkali in the presence of air yields phthiocol (3-hydroxy-2-methyl-1,4-naphthoquinone). Oxidation with dilute nitric acid yields phthalic acid. Menadione is reduced to 2-methyl-1,4-naphthohydroquinone by those methods that apply for vitamin $K_{1(20)}$. Procedures that reduce the phytyl side chain or result in cyclization of vitamin $K_{1(20)}$ are equally applicable for the simple reduction of menadione to dihydromenadione. Mild reductive acetylation of menadione gives 2-methyl-1,4-naphthohydroquinone diacetate, whereas hydrogenation of menadione over platinum in acetic acid yields 2-methyl-5,6,7,8-tetrahydro-1,4-naphthohydroquinone.

Menadione undergoes 1,4-addition reactions with reagents of the type HX; its reactivity toward these reagents is intermediate between that of naphthoquinone and that of vitamin $K_{1(20)}$. For example, it reacts with methylamine to give 2-methyl-3-methylamino-1,4-naphthoquinone. A 1,4-addition does not occur between aniline and menadione (**36**); under these conditions, a red color is generated that is the basis of a colorimetric assay for menadione (**37**). Mercapto analogs undergo 1,4-addition to menadione readily, whereas azides add in a 1,4-manner in aqueous alcohol but not in dilute acetic acid. Two types of compounds are formed in the reaction of menadione with bisulfite salts. One group is comprised of salts such as sodium 2-methyl-2,3-dihydro-1,4-naphthoquinone-2-sulfonate (**11**) which have high antihemorrhagic activity; the other is typified by salts such as sodium 2-methyl-1,4-naphthohydroquinone-3-sulfonate (**12**) which are only weakly active.

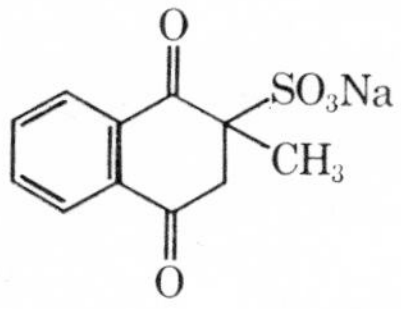

(**11**) 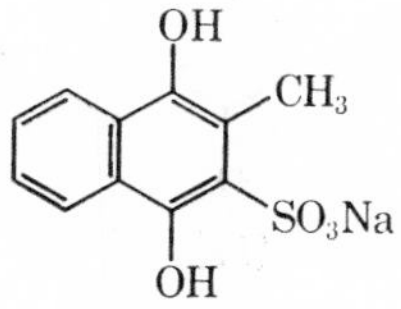   (**12**)

Grignard reagents add to menadione in a 1,4- and a 1,2-manner. It is also possible to add across the 2,3-position of menadione. Among reactions of this type are Diels-Alder condensations with conjugated dienes and addition reactions with unsaturated nitrogen analogs such as diazomethane. The latter reaction yields the analog (**13**), as well as 2,3-dimethyl-1,4-naphthoquinone and bis-(3-methyl-1,4-naphthoquinone-2-yl)-methane. Reactions of the carbonyl groups of menadione are discussed in the section on assay.

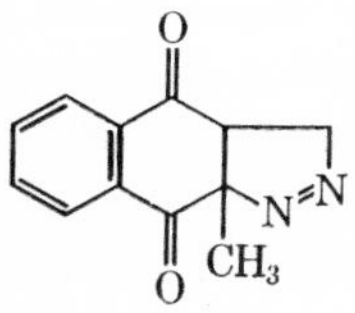

(**13**)

SYNTHESIS

A simple but important synthesis of menadione is accomplished by the chromic acid oxidation of 2-methylnaphthalene (see Vol. 13, p. 678) in acetic acid solution (**38**). The yield is of the order of 50–60% (**39**). Other oxidizing agents that convert 2-methylnaphthalene to menadione include hydrogen peroxide in acetic acid and sodium dichromate in sulfuric acid. Yields with these reagents are of the order of 30%. Analogs of 2-methylnaphthalene that contain a hydroxyl, alkoxyl, or amino group at the 1- and/or 4-position are converted to menadione on treatment with chromic anhydride, ferric chloride, hydrogen peroxide, or nitric acid in acetic acid, or by heating with p-nitrosodimethylaniline in ethanol. For example, treatment of 2-methyl-1-aminonaphthalene with an excess of hydrogen peroxide in boiling acetic acid gives menadione in 81% yield (**36**). Other substrates that have been used in such oxidations include 2-methyl-4-amino-1-naphthol, 2-methyl-4-chloro-1-naphthol, and 2-methyl-1,4-dimethoxynaphthalene.

Other approaches to the synthesis of menadione proceed by a Diels-Alder condensation of an alicyclic or cyclic diene with 2-methyl-1,4-benzoquinone. Scheme 1 shows the condensation of butadiene with 2-methyl-1,4-benzoquinone (14) yielding 5,8,9,10-tetrahydro-2-methyl-1,4-naphthoquinone (15). Acid-catalyzed isomerization of the latter compound yields 5,8-dihydro-2-methyl-1,4-naphthohydroquinone (16), which is oxidized to menadione (9) in about 50% yield by chromic acid. Alternatively, treatment of the naphthohydroquinone intermediate (16) with silver oxide yields 5,8-dihydro-2-methyl-1,4-naphthoquinone (17), which on oxidation with lead tetraacetate in benzene gives menadione (9).

Scheme 1

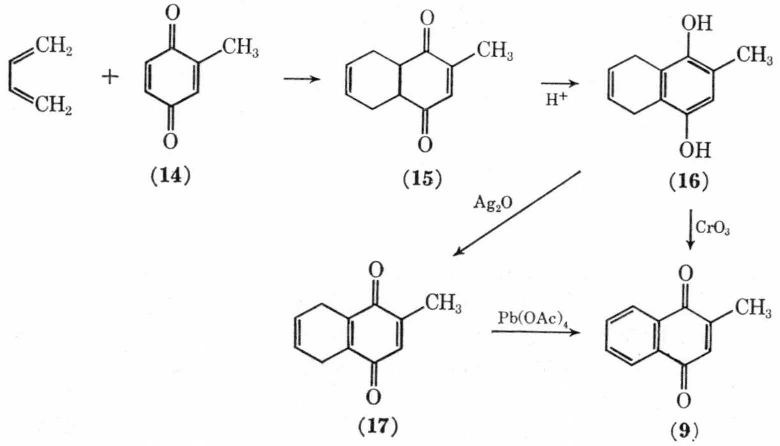

Scheme 2 outlines a synthesis of menadione involving substituted bicyclo [2.2.2.]-octene intermediates (40). Condensation of 1,3-cyclohexadiene (18) with 2-methyl-1,4-benzoquinone (14) yields 5,8-ethylene-5,8,9,10-tetrahydro-2-methyl-1,4-naphthoquinone (19), which in turn is isomerized to 5,8-ethylene-5,8-dihydro-2-methyl-1,4-naphthohydroquinone (20) by aqueous hydrobromic acid. The hydroquinone intermediate (20) is oxidized to 5,8-ethylene-5,8-dihydro-2-methyl-1,4-naphthoquinone (21) by ferric chloride; deethanation of this product by pyrolysis gives menadione in a 75% overall yield.

Scheme 2

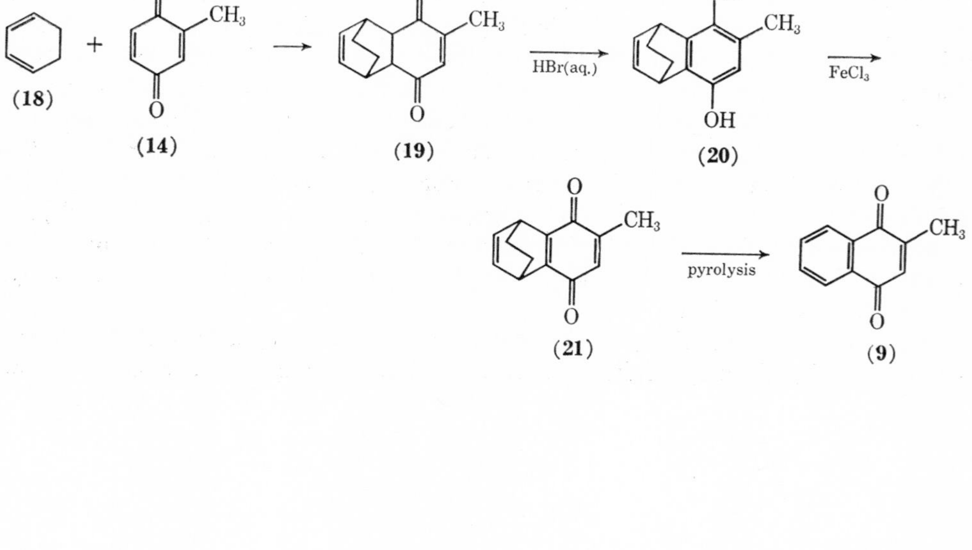

ASSAY

Volumetric Analysis. Menadione can be reductively titrated with titanous chloride using potassium indigosulfonate as indicator (35), or it may be reduced to the hydroquinone form which is then oxidatively titrated with ceric sulfate in the presence of an appropriate indicator (41,42). Other methods include reduction of menadione by potassium iodide under standard conditions and titration of the iodine generated (43). Alternatively, menadione may be treated with an excess of bromine, and the unreacted bromine can be determined by appropriate titration (42).

Colorimetry. Treatment of menadione with 2,4-dinitrophenylhydrazine and then with ammonia produces a green color that is the basis of a useful assay (44); an intense bluish-violet color that develops after the addition of ethyl cyanoacetate and alcoholic ammonia water to menadione forms the basis for another (45,46). When a methanolic solution of menadione is treated with sodium hydroxide and dichlorophenol-indophenol, a green color develops that can be extracted into amyl alcohol and measured (46). A rapid determination involves treatment of solutions of menadione in water, alcohol, plasma, or urine with cysteine and sodium hydroxide to produce a yellow color (47).

Ultraviolet Spectroscopy. This ranks among the more accurate assays of purified samples. The significant maxima are outlined in the section on chemical and physical properties.

Fluorometry. Menadione condenses with *o*-phenylenediamine yielding a fluorescent compound (48). The blue fluorescence of this analog is so intense that it may be detected in very dilute solution.

Gravimetric Analysis. Menadione can be precipitated by 2,4-dinitrophenylhydrazine.

WATER-SOLUBLE MENADIONE DERIVATIVES

Water-soluble derivatives of menadione are commercially available for parenteral administration. In general, solubility is achieved by the introduction of functional groups with salt-forming properties. The activity of the salt, however, is attributed to its conversion to menadione in vivo. A few examples of such analogs are listed below to illustrate the various means used to achieve solubility.

The sodium salt of 2-methyl-1,4-naphthohydroquinone is very sensitive to air oxidation and consequently unsuitable as a water-soluble analog. More stable sodium salts are obtained, however, by reductive acylation of menadione in the presence of anhydrides of dibasic acids. The addition of the theoretical amount of alkali to the resulting half-acid-half-ester analogs yields water-soluble salts.

Salts of the bis-phosphoric acid ester (**22**) constitute a commercially important class of water-soluble menadione derivatives (49,50). Treatment of dihydromenadione with phosphorus oxychloride in pyridine solution yields the bis(dichlorophosphoryl) analog which is readily hydrolyzed to the ester (**22**).

The latter is converted into soluble salts on treatment with base. Disulfate esters are prepared in a similar manner from chlorosulfonic acid and dihydromenadione and converted to soluble salts (49). When 1-acetyl-2-methylnaphthohydroquinone is substituted for dihydromenadione in the above reactions, the 4-disodium phosphate ester and the 4-sodium sulfate ester of 1-acetyl dihydromenadione are obtained (51).

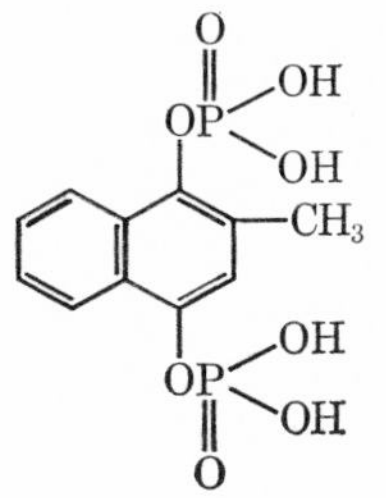

(22)

Menadione sodium bisulfite (11) is a commercially significant compound. This water-soluble sodium bisulfite addition product is equivalent to menadione in physiologic activity and resistant to oxidation.

A number of nitrogen analogs of menadione have been synthesized and owe their water solubility to the propensity of the basic nitrogen function to form salts. For example (52), acylation of dihydromenadione with nicotinoyl chloride yields the dinicotinate ester (23), the dihydrochloride of which is water soluble.

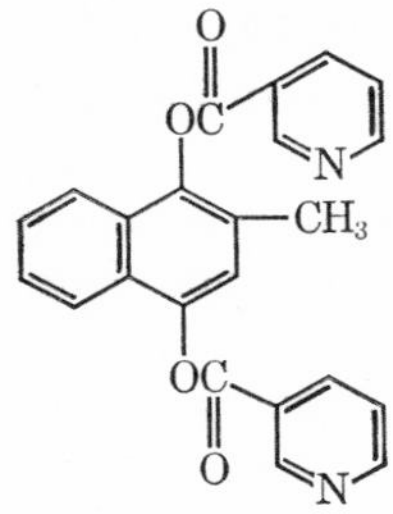

(23)

A water-soluble bis(quaternary ammonium) analog (25) of menadione was prepared by treating the bis-chloroacetate (24) of dihydromenadione with trimethylamine (53).

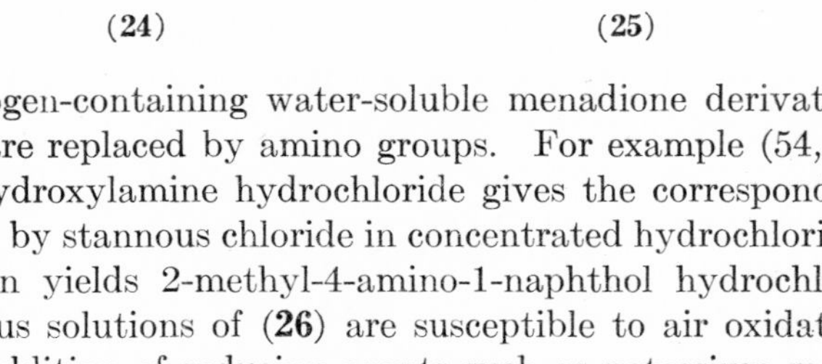

(24)   (25)

In other nitrogen-containing water-soluble menadione derivatives one or both oxygen functions are replaced by amino groups. For example (54,55), treatment of menadione with hydroxylamine hydrochloride gives the corresponding 4-monoxime, which on reduction by stannous chloride in concentrated hydrochloric acid or by catalytic hydrogenation yields 2-methyl-4-amino-1-naphthol hydrochloride (26). Unfortunately, aqueous solutions of (26) are susceptible to air oxidation and must be stabilized by the addition of reducing agents such as potassium metabisulfite. The dihydrochloride of 1,4-diamino-2-methylnaphthalene (**27**) has been synthesized (55) from 2-methylnaphthalene. This analog is more water soluble and its aqueous solutions are reported to be more stable than those of the corresponding 4-amino-1-naphthol (**26**).

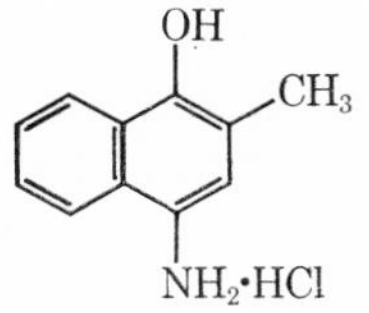

Finally, water solubility has also been achieved by preparing bis glycoside analogs of menadione (56,57). Treatment of dihydromenadione with β-D-glucose penta-acetate in the presence of p-toluenesulfonic acid at 130°C yields the bis(glucoside tetraacetate). Purification of the product, followed by deacetylation yields the water soluble bis glucoside of dihydromenadione.

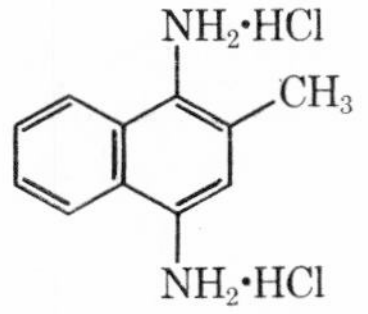

Economic Aspects

The annual production and sales of K-vitamins as reported by the United States Tariff Commission are listed in Table 1. Several facts should be borne in mind when

Table 1. Production and Sales of K-Vitamins

		Sales	
Year	Production, lb	quantity, lb	value, $
1944	400	200	23,500
1947	1,100	1,200	87,400
1949	800	60	17,000
1951	1,900	650	25,000
1953	1,000	1,000	21,000
1958	25,000	7,000	81,000
1962	14,000	5,000	47,000
1965	136,000	73,000	578,000
1966	156,000	72,000	981,000
1967	184,000	56,000	517,000

studying these figures. First, the reporting basis varies from year to year; for one year it may be vitamin $K_{1(20)}$, whereas for others it may be menadione, menadione sodium bisulfite, or compilations thereof. Secondly, sales quantities are usually only a small portion of total production, and the sales value refers only to that portion of total production that is sold. The amount produced and not sold is generally consumed internally. Finally, there are some years for which this information is not available or for which only some categories are available.

Metabolism of the K-Vitamins

Vitamin $K_{2(20)}$ appears to be the significant form of vitamin K activity in mammalian metabolism, and it is suggested that the other K-vitamins of dietary origin are converted to vitamin $K_{2(20)}$ in vivo. When vitamin $K_{1(20)}$ labeled with tritium in the nucleus and C^{14} in the 3-substituent was fed to chicks, doubly labeled vitamin $K_{1(20)}$ and tritium labeled vitamin $K_{2(20)}$ were found in the liver; in organs such as the heart, kidney and skeletal muscle, only tritium labeled vitamin $K_{2(20)}$ was isolated (58). Comparable results were obtained using tritium labeled vitamin $K_{2(30)}$ in feeding experiments with the chick, pigeon, or rat (59). The conversion of menadione to vitamin $K_{2(20)}$ exclusively in vivo was also demonstrated in chicks and rats (60,61). From such studies, it may be concluded that the K-vitamins of dietary origin are stored in the liver and metabolized in this organ by cleavage of the 3-substituent to yield menadione. Part of this menadione is converted to vitamin $K_{2(20)}$ in the liver, and part is transported to other organs before alkylation by geranylgeranyl pyrophosphate to yield vitamin $K_{2(20)}$.

It has been suggested that certain K-vitamins function in respiratory enzyme systems by participation in the coupled electron transport and phosphorylation step. A role for these vitamins in this physiologic process, however, is far from being established.

Therapeutic Role of the K-Vitamins

Participation by the K-vitamins in the process of blood coagulation is the only established physiological role for these compounds (62). Originally, it was suggested that a deficiency of vitamin K in the diet or interference with its absorption resulted solely in a lower blood level of the protein prothrombin, which in turn gave rise to the hemorrhagic syndrome. The evidence still supports the view that the prothrombin level of blood is reduced during vitamin K deficiency, but it has been found that the deficiency also influences the blood levels of other proteins that participate in the coagulation process. The dependence of these activities (Factor VII, Factor IX, and Factor X) on vitamin K is established, but there is disagreement on their importance to prothrombin in the coagulation process. All that should be said, therefore, is that vitamin K plays a role in the blood clotting process by regulating the synthesis of blood clotting proteins in the liver.

The K-vitamins may be administered therapeutically for the prevention of hemorrhage in the newborn since the blood levels of plasma-clotting factors are low during the first few days of life. The deficiency, however, is more likely to occur in premature or anoxic infants or in those born of mothers on anticoagulant therapy. A single dose of 1 mg of vitamin $K_{1(20)}$ is considered adequate to prevent hemorrhage in the newborn (63).

The K-vitamins are also effective in preventing hemorrhage owing to a lack of bile secretion into the intestine. They are especially useful for the prevention of postoperative hemorrhage in patients with obstructive jaundice, since the hemorrhagic syndrome develops because of poor absorption of the vitamin from the intestine. Compounds with vitamin K activity may be administered parenterally in aqueous medium, or if necessary, orally in combination with bile salts.

Certain of the K-vitamins are used therapeutically to prevent hemorrhage that may result from anticoagulant therapy. Vitamin $K_{1(20)}$ in a solubilized form or in the form of an aqueous colloidal suspension is effective, whereas menadione and its water-soluble analogs are not. Vitamin $K_{1(20)}$ oxide is also effective to counteract excessive anticoagulant therapy.

For various reasons, an absolute daily requirement for vitamin K has not been established. The average diet, however, appears to contain adequate quantities of the vitamin.

Bibliography

"Vitamin K" in ECT 1st ed., Vol. 14, pp. 858–874, by Ralph Hirschmann, Merck and Co., Inc.

1. H. J. Almquist, *J. Biol. Chem.* **114**, 241 (1936); **115**, 589 (1936).
2. H. Dam and F. Schønheyder, *Biochem. J.* **30**, 897 (1936).
3. S. A. Thayer, D. W. MacCorquodale, S. B. Binkley, and E. A. Doisy, *Science* **88**, 243 (1938).
4. R. W. McKee, S. B. Binkley, D. W. MacCorquodale, S. A. Thayer, and E. A. Doisy, *J. Amer. Chem. Soc.* **61**, 1295 (1939).
5. H. Dam, A. Geiger, J. Glavind, P. Karrer, W. Karrer, E. E. Rothschild, and H. Salomon, *Helv. Chim. Acta* **22**, 310 (1939).
6. S. B. Binkley, D. W. MacCorquodale, S. A. Thayer, and E. A. Doisy, *J. Biol. Chem.* **130**, 219 (1939).
7. P. Karrer, A. Geiger, A. Ruegger, and H. Salomon, *Helv. Chim. Acta* **22**, 1464 (1939).
8. L. F. Fieser, *J. Amer. Chem. Soc.* **61**, 2559, 3467 (1939).
9. L. M. Jackman, R. Rüegg, G. Ryser, C. von Planta, U. Gloor, H. Mayer, P. Schudel, M. Kofler, and O. Isler, *Helv. Chim. Acta* **48**, 1332 (1965).
10. H. Mayer, U. Gloor, O. Isler, R. Rüegg, and O. Wiss, *Helv. Chim. Acta* **47**, 221 (1964).
11. D. T. Ewing, F. S. Thomas, and O. Kamm, *J. Biol. Chem.* **147**, 233 (1943).
12. P. Karrer and A. Geiger, *Helv. Chim. Acta* **22**, 945 (1939).
13. B. Riegel, P. G. Smith, and C. A. Schweitzer, *J. Amer. Chem. Soc.* **62**, 992 (1940).
14. E. B. Hershberg, J. K. Wolfe, and L. F. Fieser, *J. Amer. Chem. Soc.* **62**, 3516 (1940).
15. R. Hirschmann, R. Miller, and N. L. Wendler, *J. Amer. Chem. Soc.* **76**, 4592 (1954).
16. O. Isler and K. Doebel, *Helv. Chim. Acta* **22**, 945 (1939).
17. J. F. Pennock in R. A. Morton, ed., *Biochemistry of Quinones*, Academic Press Inc., N.Y., 1965, p. 67.
18. B. H. Arison, Personal communication, 1969.
19. H. Lindlar, *Helv. Chim. Acta* **35**, 446 (1952).
20. D. W. MacCorquodale, S. B. Binkley, S. A. Thayer, and E. A. Doisy, *J. Amer. Chem. Soc.* **61**, 1928 (1939).
21. M. Tishler, L. F. Fieser, and N. L. Wendler, *J. Amer. Chem. Soc.* **62**, 1982, 2866 (1940).
22. S. B. Binkley, D. W. MacCorquodale, L. C. Cheney, S. A. Thayer, R. W. McKee, and E. A. Doisy, *J. Amer. Chem. Soc.* **61**, 1612 (1939).
23. D. W. MacCorquodale, L. C. Cheney, S. B. Binkley, W. F. Holcomb, R. W. McKee, S. A. Thayer, and E. A. Doisy, *J. Biol. Chem.* **131**, 357 (1939).
24. U.S. Pat. 2,325,681 (Aug. 3, 1943), O. Isler (to Hoffmann-LaRoche, Inc.).
25. H. Dam and E. Søndergaard, in P. Gyorgy and W. N. Pearson, eds., *The Vitamins*, 2nd Ed., Vol. 6, Academic Press Inc., N.Y., 1967, pp. 245–260.
26. N. R. Trenner and F. A. Bacher, *J. Biol. Chem.* **137**, 745 (1941).
27. F. Irreverre and M. X. Sullivan, *Science* **94**, 497 (1941).
28. L. F. Fieser, M. Tishler, and W. L. Sampson, *J. Amer. Chem. Soc.* **62**, 1628 (1940).
29. O. Isler, R. Rüegg, L. H. Chopard-dit-Jean, A. Winterstein, and O. Wiss, *Helv. Chim. Acta* **41**, 786 (1958).
30. D. T. Ewing, J. M. Vandenbelt, and O. Kamm, *J. Biol. Chem.* **131**, 345 (1939).
31. L. F. Fieser and M. Fieser, *J. Amer. Chem. Soc.* **57**, 491 (1935).
32. K. Wallenfels and W. Mohle, *Chem. Ber.* **B76**, 924 (1943).
33. J. W. H. Lugg, A. K. Macbeth, and F. L. Winzor, *J. Chem. Soc.*, 1457 (1936).
34. J. E. Page and F. A. Robinson, *J. Chem. Soc.*, 133 (1943).
35. J. L. Pinder and J. H. Singer, *Analyst* (London) **65**, 7 (1940).
36. K. Fries and W. Lohmann, *Chem. Ber.* **54**, 2912 (1921).
37. E. E. Martinson and G. I. Meerovich, *Biokhimiya* **10**, 258 (1945); (through) *Chem. Abstr.* **39**, 5274.
38. V. Vesely and J. Kapp, *Chem. Listy* **18**, 201 (1924), *Rec. Trav. Chim.* **44**, 360 (1925).

39. Japan. Pat. 162,647 (March 8, 1944), (to Sakuragaoka Research Inst., Inc.); (through) *Chem. Abstr.* **42**, 4201.
40. M. R. Grdinic and V. K. Jogovic, *Arkiv. Kem.* **23**, 73 (1951) (in German); (through) *Chem. Abstr.* **46**, 11165 (1952).
41. J. Rosin, H. Rosenblum, and R. Mack, *Amer. J. Pharm.* **113**, 434 (1941).
42. E. Schulek and P. Rozsa, *Mikrochemie ver. Mikrochim. Acta* **29**, 178 (1941).
43. M. P. Zakharova and V. A. Devyatnin, *Biokhimiya* **9**, 256 (1944); (through) *Chem. Abstr.* **39**, 3224.
44. A. Novelli, *Science* **93**, 358 (1941).
45. R. Craven, *J. Chem. Soc.*, 1605 (1931).
46. W. Bosecke and W. Laves, *Biochem. Z.* **314**, 285 (1943).
47. J. V. Scudi and R. P. Buhs, *J. Biol. Chem.* **141**, 451 (1941).
48. M. Kofler, *Helv. Chim. Acta* **28**, 702 (1945).
49. L. F. Fieser and E. M. Fry, *J. Amer. Chem. Soc.* **62**, 228 (1940).
50. U.S. Pat. 2,407,823 (Sept. 17, 1946), L. F. Fieser (to Research Corp.).
51. U.S. Pat. 2,334,669 (Nov. 16, 1943), G. H. Carlson and B. R. Baker (to Lederle Labs., Inc.).
52. U.S. Pat. 2,428,253 (Sept. 30, 1947), F. von Werder (vested in the Attorney General of the U.S.).
53. U.S. Pat. 2,372,655 (April 3, 1945), M. Bockmuhl, O. Schaumann, E. Bartholomaus, and H. Leditschke (to Winthrop Chem. Co., Inc.).
54. P. P. T. Sah, *Rec. Trav. Chim.* **59**, 454, 1029 (1940); **60**, 373 (1941).
55. H. Veldstra and P. W. Wiardi, *Rec. Trav. Chim.* **62**, 75 (1943).
56. U.S. Pat. 2,336,890 (Dec. 14, 1943), B. Riegel and P. G. Smith (to Abbott Labs.).
57. U.S. Pat. 2,357,172 (Aug. 29, 1944), G. H. Carlson and B. R. Baker (to Lederle Labs., Inc.).
58. M. Billeter and C. Martius, *Biochem. Z.* **333**, 430 (1960).
59. M. Billeter and C. Martius, *Biochem. Z.* **334**, 304 (1961).
60. C. Martius, *Biochem. Z.* **327**, 407 (1956).
61. C. Martius and H. O. Esser, *Biochem. Z.* **331**, 1 (1958).
62. E. A. Doisy, Jr., and J. T. Matshiner in R. A. Mortin, ed., *Biochemistry of Quinones*, Academic Press, Inc., New York, 1965, Chap. 10, p. 320.
63. *Recommended Dietary Allowances*, 7th Edition, Nat'l. Acad. Sci., Wash., D.C., 1968, p. 29.

RECOMMENDED READING

1. R. A. Morton, ed., *Biochemistry of Quinones*, Academic Press, Inc., N.Y., 1965.
2. W. H. Sebrell, Jr., and R. S. Harris, eds., *The Vitamins-Chemistry, Physiology and Pathology*, Vol. 2, Academic Press, Inc., New York, 1954, pp. 388–447.
3. W. H. Sebrell, Jr., and R. S. Harris, eds., *The Vitamins*, 2nd ed., Vol. 3, Academic Press, Inc., New York, to be published.
4. A. F. Wagner and K. Folkers, *Vitamins and Coenzymes*, Interscience Publishers, a division of John Wiley & Sons, Inc., New York, 1964, pp. 407–434.

Arthur F. Wagner
Merck Sharp & Dohme Research Laboratories
Merck & Co., Inc.

VITRINITE. See Coal.

VM&P NAPHTHA (varnish makers' and painters' naphtha). See Petroleum (products), Vol. 15, p. 81; also such articles as Coatings, industrial; Paint.

VULCACEL, VULCALURE. See Rubber chemicals.

WALLBOARD

Classification

The word "wallboard" is a general term that is applied to large boardlike materials used on or in a wall. Those in the building business, and others, have come to recognize the term as a general classification. Usually excluded from the accepted meaning of the word are built-up combinations of materials. These are referred to more often as "panels," although plywood and gypsum board, well-known wallboards, are in a way laminated products. This article also deals with related materials, which might be described as "building boards," and could be used inside of walls or for other related building purposes such as acoustical treatment, core materials or structural roofing products. See also Plywood; Laminated and reinforced plastics.

Wallboards, in the broader definition, are classified in major types below. Eliminated are boards in minor use and those that would be called panels or some other name. Among those excluded are plastic panels, metal panels, laminated wood fiber products, laminated corrugated panels, and the like. Finishes are not considered in this classification. Plywood, gypsum board, hardboard, and particleboard are all finished with prints on the filled surfaces, plastic overlays, and special factory applied coatings. Resin polymers are involved in all of these finishes.

Building code requirements are changing in such a way as to require walls, floors, and ceilings to have substantially more resistance to flame spread, fire penetration, and passage of sound. Especially in multifamily dwellings and offices, people are demanding more acoustical privacy. In these same structures the suppression of flash fires and fires which might spread from one occupied area to another are causing changes in choice of building materials and their application. They are also encouraging research efforts which have resulted in better products for these uses and better systems of application. Table 1 gives a general classification of wallboards.

Table 1. General Classification of Wallboards

Organic	Inorganic-organic	Inorganic
hardboard	gypsum board	asbestos-cement
particleboard	mineral board	ceramic board
insulating board	wood-inorganic binder	
plywood		

The division between organic and inorganic–organic wallboards is not entirely sharp, since organic boards use a small proportion of inorganic chemicals such as sulfuric acid, alum, or sodium aluminate in the sizing operation. At the same time, most of the inorganic board contains some wax as a sizing agent or paint as a coating. The only true inorganic board listed is the ceramic board whose surface coating is also a ceramic glaze. Table 2 shows characteristic ranges of density and thickness of organic wallboards.

Table 2.   Characteristics of Organic Wallboards

Type of board	Density, lb/ft³	Thickness, in.
hardboard	31–75	⅛–⅜
particleboard	25–70	¼–1½
insulating board	10–30	¼–1
plywood	25–45	⅛–1¼

The common denominator of the boards in Table 2 is that they are all made of wood or wood fiber. (A small amount of insulating board is made from bagasse fiber.) The processes of manufacture vary widely and are covered in the following pages. Table 3 shows characteristic ranges of density and thickness of inorganic-organic wallboards.

Table 3.   Characteristics of Inorganic-Organic Wallboards

Type of board	Density, lb/ft³	Thickness, in.
gypsum board	40–65	¼–1
mineral board	6–35	⅜–1⅛
wood-inorganic binder	20–40	½–3

The above wallboards are made from a very broad range of materials. Gypsum board uses basically paper and gypsum. However, special fire requirements of walls have brought about the use of additives in the board core, such as glass fiber, boric acid, vermiculite, and perlite. Mineral board uses more mineral fiber than any one product. However, other inorganic minerals used in them are clay, expanded clay pellets, vermiculite, perlite, asbestos, mica, and diatomaceous earth. The binders consist of phenolic resins, urea resins, starch, silicate, and gelatinized kraft pulp. Usually one of these binders is used in each process. The bulk of the wood–inorganic binder boards are made from wood excelsior fiber and either portland cement or magnesium oxysulfate as binders. Table 4 shows characteristic ranges of density and thickness of inorganic wallboards.

Asbestos–cement boards are made from asbestos fibers and portland cement. The ceramic boards are made from mineral fiber, perlite, or expanded clay pellets as the aggregate, with clay as the principal binder. The clay fuses and becomes a binder during the firing process.

Table 4.   Characteristics of Inorganic Wallboards

Type of board	Density, lb/ft³	Thickness, in.
Asbestos-cement	40–125	⅛–⅝
ceramic board	8–30	⅜–1

Hardboard

Hardboard is defined as "a panel manufactured primarily from interfelted lignocellulosic fibers, consolidated under heat and pressure in a hot press to a density of 31 lb/ft^3 or greater." William H. Mason, in 1926, made the first American hardboard. He was a pioneer in efforts to utilize waste wood left over from sawmill operations. He was familiar with the softening effect of steam on wood and began experimental work to combine this action with a method of rupturing the fiber bond by explosion with high pressure steam. He worked with steam pressures as high as 1000 psi in the early stages in order to get the type and quality of fiber that he wanted. The high pressure was preceded by a lower pressure steaming time to soften the lignin which held the fibers together.

This exploded fiber was a brown, fluffy mass which is known as *gun fiber*. It was found to be very well suited for the manufacture of structural insulation board, and his first attention was turned toward this market. He also wanted to make a denser, stronger product, but his first attempts were not successful. The story told is that he went to lunch one day leaving a fibrous mat in a press. When he returned, he found that a leaking steam valve had heated the press during this time. When the press was opened, he found he had made his first piece of hardboard. Subsequent further work resulted in his being able to make boards of controlled thickness, density, strength and water resistance. This new product was called "Presdwood." Mr. Mason and some associates formed a company to make hardboard. They adopted the word "Masonite" as a trademark for their product and called their company The Masonite Corporation. In 1968, there were 3.7 billion ft^2 of hardboard made in the United States, and 1.725 billion ft^2 of that was made by the Masonite Corporation.

The Masonite process spread abroad and was used in Sweden's first hardboard plant in 1929. At one time, 80% of the world's hardboard was made by the Masonite process. More recently, the Defibrator process has come into use very extensively, especially in countries other than the U.S.A. The A. B. Defibrator Co. located in Stockholm, Sweden, is basically a manufacturer of machinery for the production of fiber and pulp used in the manufacture of hardboard, insulation board, and paper. They will plan and build a "turnkey" hardboard manufacturing plant for a customer anywhere in the world. They combine their very broad process and mechanical know-how with that of the other companies who supply them with major pieces of equipment for these plants. The result is a very highly automated hardboard plant which produces a broad line of products at a minimum usage of manpower and other costs. The Dry Process of making hardboard was developed by the Plywood Research Foundation in Tacoma, Washington, and the Semi-Dry Process was developed by the Weyerhaeuser Company.

Almost all of the fibrous hardboards have similar properties, regardless of the type of process used to produce them. There is a Product Standard for Hardboard which was developed by the American Hardboard Association in cooperation with the National Bureau of Standards, U.S. Department of Commerce. (CS251-63) Table 5, gives a list of physical properties which a customer can expect to obtain from hardboard made by a company which is a member of the American Hardboard Association regardless of the type of manufacturing process that they use. Many of the imported hardboards do not meet all of these standard properties.

The definitions of hardboard terminology are: *standard hardboard*. Hardboard in

Table 5. Physical Properties of Hardboard

Type	Surface	Nominal thickness range, in.	Modulus of rupture, psi	Tensile Strength, psi		Water absorption 24 hr soak, %
				parallel to face	perpendicular	
tempered	S-1-S	$\frac{1}{12}$–$\frac{3}{8}$	7000	3500	150	30–8
tempered	S-2-S	$\frac{1}{10}$–$\frac{3}{8}$	7000	3500	150	25–10
standard	S-1-S	$\frac{1}{12}$–$\frac{3}{8}$	5000	2500	100	35–12
standard	S-2-S	$\frac{1}{12}$–$\frac{3}{8}$	5000	2500	100	35–12
service	S-1-S	$\frac{1}{8}$–$\frac{7}{16}$	3000	1500	75	30–25
service	S-2-S	$\frac{1}{8}$–1	3000	1500	75	30–12
service tempered	S-1-S	$\frac{1}{8}$–$\frac{3}{8}$	4500	2000	100	20–14
service tempered	S-2-S	$\frac{1}{8}$–$\frac{3}{8}$	4500	2000	100	25–18
special	S-1-S or S-2-S	$\frac{3}{8}$–1	2000	1000	35	25–20

substantially the same form as when it comes from the manufacturing press; *tempered hardboard*. Hardboard which has been impregnated with a siccative material such as drying oil blends of oxidizing resins which is then stabilized by heat to give improved physical properties; *service hardboard*. Hardboard of good strength, substantially as it comes from the press, but is somewhat less strong than standard hardboard; *service tempered hardboard*. Same as Tempered Hardboard except that it starts as a Service Hardboard before being tempered; *special hardboard*. Hardboard in substantially the same form as when it comes from the press except it is of moderate strength and has a low unit weight; *S-1-S*. Means smooth one side, other side has a screen impression; *S-2-S*. Means smooth two sides; *thickness range*. Boards are made in a range from the thinnest to the thickest dimension shown; *water absorption*. The high percentages are for the thinnest board and the low percentages are for the thickest board.

Hardboard differs from insulating wallboard by its greater density, rigidity, resistance to water absorption, harder surface, and greater resistance to impact. Its harder, smoother surface also lends itself much better to its being prefinished. Hardboards normally are in the $\frac{1}{12}$–$\frac{3}{8}$ in. thickness, a width up to 5 ft, and length up to 24 ft. The full density range is 31–75 lb/ft^3.

Uses. Hardboards are too thin and dense to be good heat insulators. They can be used alone or with other boards, such as gypsum or mineral, in walls and floors, to give excellent resistance to sound transmission. A very large market has developed for hardboard in $\frac{1}{4}$–$\frac{1}{2}$ in. thickness as siding for houses, both as vertical panels and as clapboard type. Other residential construction uses for unfinished hardboard are for floor underlayment, closet liners, soffit board, basement walls and when perforated as a support for tools in garages and workshops. In commercial construction it has large usage as a concrete formboard.

Prefinished hardboard with either woodgrain prints or thin plastic film overlays has found a very large market in fine wall paneling in homes, offices, schools, mobile homes, and factory manufactured homes and school units. Hardboard as a raw material has established uses in the automotive industry in door panels and roof liners. A great deal is used in furniture manufacture as core material, drawer bottoms, and in case goods. It is used in the manufacture of toys and signs. In prefinished form it is used in the cabinets for tv sets, hi-fi's and radios.

Consumption of hardboard in the U.S.A. since 1952 has increased from 1000 million ft^2 to 4410 million ft^2 in 1968. The 1968 figure includes U.S. production of 3725 million ft^2 and imports of 685 million ft^2. In 1968 there were 27 plants in the U.S.A., and there are three plants planned or are being built. (All figures are based on $\frac{1}{8}$ in. board thickness equivalent. Thus, 1000 ft^2 of $\frac{3}{8}$ in. material would be quoted as 3000 ft^2.) Manufacturing costs of hardboard vary somewhat by process but are also influenced a great deal by the cost of the wood which goes into them. In 1969, $\frac{1}{8}$ in. standard hardboard packaged and ready to ship at manufacturing plants had a manufacturing cost between \$24 and \$30. $\frac{1}{8}$ in. tempered hardboard on the same basis had a manufacturing cost of between \$28 and \$34 per MSF. $\frac{1}{4}$ in. hardboard used mostly for the manufacture of prefinished grain-printed paneling cost between \$38 and \$42.

The processes for the manufacture of hardboard can be classified as follows: *The wet process*, which includes the Masonite process, the Asplund process, and the miscellaneous wet processes; *the semi-dry process*, and *the dry process*.

The Wet Process. *Masonite Process.* The Masonite process lends itself well to making hardboard from a very wide range of wood species. In all cases, their plants can use nearly any type of wood which is available in the plant's vicinity. They can use either hard or soft woods but only in very rare instances would the two types be mixed. Most of the wood comes in chip form. Its source usually is trim and slab from a sawmill from within a hundred mile radius. A choice source of wood is chips made from unusable veneer from plywood plants. In all cases the chips must be bark-free. A small amount of bark in the wood supply will show up as semi-concealed lumps on the board surface which are especially objectionable if the board is to be woodgrain printed and prefinished.

The Masonite Co. still uses their famous "gun" process in about 85% of their production in the U.S.A. The chips are steamed at about 500 lb of pressure for 90 sec. The steam pressure is raised to 600 lb 5 sec before the quick opening valve located at the bottom of the digestor permits the chips to "explode" into the atmosphere. The last second increase of pressure helps to clear the guns. The very rapid pressure release tears the fibers apart into a fluffy brown mass. This fibrous mass and the steam are separated in a cyclone. (See Vol. 10, p. 340). The fiber bundles go through a disc refining operation which rubs them down to very close to the size of the primary fibers. From there, the fibers go to a series of stock chests where the various chemicals are added. These include about 1% of liquid phenolic resin which is precipitated onto the fibers, a small amount of wax emulsion, and enough alum to give the desired pH.

The fibers and additives end up in a machine chest, a head box from which they go as a thin water slurry to a Fourdrinier forming machine, similar to a standard paper making machine (see Paper). The water is removed from the slurry first by gravity drainage through the continuous wire, then by vacuum boxes below the wire and finally by the press rolls which squeeze out all the water which can be removed in this manner. As the water leaves the fiber, a thick soft felted fibrous mass is formed which becomes increasingly firmer as it progresses through the final squeeze rolls of the press section.

The compressed wet lap has the edges trimmed to a slightly oversize width mat and is then cut to a slightly oversize length by a traveling cross-cut saw. The wet mat is then deposited on a wire-covered steel plate referred to as a caul, and sent to the press. The mats are compressed at from 500 to 1000 psi, depending on the board density required. The press temperature normally is at about 450°F. The time the mats are in the press varies from 7 min for a mat which presses to a ⅛ in. thick board to 15 min for a ⁵⁄₁₆ in. thick board. Press widths are normally either 4 ft or 5 ft. A press will handle up to 30 mats at a time, each being up to 24 ft long.

The raw pressed hardboards are nearly bone dry when they come from the press. In the next step they are heat treated in an oven to be sure the phenolic resin binder is fully polymerized and that the board has reached maximum strength. The resultant board is brown in color, very smooth on one side, referred to as the face and has a screen mark on the back. It next goes to a humidifier where the moisture content of the board is brought up from bone dry to between 4 and 6% moisture content. If the board is to be tempered, it goes from the hot press through a dip tank of a drying oil, thence to a bake oven, and then to the humidifier. Masonite also makes S-2-S board by forming a wet mat, drying the mat in a kiln, and pressing that dried mat into hardboard. Abitibi and U.S. Gypsum use this same process.

Asplund Process. The major difference between the Masonite and Asplund processes is the way in which the fiber is prepared. In the Asplund process, chips, coarse sawdust, or mixtures of these with planer shavings are fed by a screw conveyor into a preheating chamber. The steam pressure in the preheater is normally about 150 psi. The time in the preheater can vary from 30 sec to several min. The softened wood particles still under steam pressure are fed into the Asplund Defibrator which converts them to a very uniformly fine type of fiber. For extra capacity or extremely fine free-draining fibers, the Defibrator can be followed by an Asplund Raffinator. The fiber is discharged through a continuous valve to a cyclone where the fiber and the steam separate.

After the steam has been separated from the fiber, the steam goes through a chamber into which a fine water spray is injected. The steam is condensed by the spray and the heated water is used anywhere in the plant where hot water is required. Most of it goes back into the fiber preparation system. Heating the water in the stock makes it easier to drain away the surplus water as the next mat is being formed. The forming process and additives used from here are the same as the Masonite process. It is possible to make a reasonably good hardboard with either process without the use of the resin binder. However, the resin gives added board strength and dimensional stability.

In both the Masonite and the Asplund process, the solubles formed by the steam preheating are washed from the fiber before it is made into hardboard. The percent of dissolved wood sugars is somewhat higher in the Masonite process due to the much higher steam temperatures used. The higher steam temperatures increase the amount of dissolved sugars which must be removed from the plant-waste water before it can be discharged into a stream. The most common system is the "activated sludge process" which involves aerobic fermentation, thickening of sludge, sludge removal, sludge burning, and aeration of the main liquid stream before it is discharged into a river. This system also usually involves large storage ponds. The storage ponds keep the cleaned-up liquid in such a way that it can be mechanically aerated to add oxygen back into it. They also permit the retention of all cleaned effluent so that it is only discharged into the streams during periods of high stream flow.

Some plants take the total effluent without an attempt at removal of dissolved or undissolved solids and spray it over large field areas to irrigate the fields. Masonite converts their effluent into molasses by a process they have developed. They then sell the molasses as cattle feed and thereby recover a substantial part of their treating costs. One plant also sprays the water on plowed fields and depends on the soil bacteria to tie up the dissolved sugar. The plants that still dump their entire effluent into streams and lakes are living strictly on borrowed time. (See also Water.) The Asplund process gives a somewhat higher yield of usable wood fiber. The Asplund process is used very widely all over the world. About 50% of the world's hardboard is made by this process.

Miscellaneous Wet Processes. Combinations of the two above processes are used, especially in smaller plants. Asplund Defibrators are used in many plants other than regular Asplund process plants as primary refiners. The Bauer Corp. also supplies a large number of primary refiners to the hardboard industry.

The Semi-Dry Process. The Weyerhaeuser Co. developed the first semidry process. They use Asplund Defibrators as primary refiners. They use Bauer units for secondary refining of coarse fiber removed from the original fibers by an air separation system. Phenolic resin at 1% solids is added to the fiber as it comes from the Defibrators. The fiber is flash dried in a matter of seconds in a hot air suspension to about 6% moisture content. The mat is formed in three layers on a continuous wire belt with suction below. This makes possible a board with fine fiber on both surfaces and coarser fiber in the center. A scalping mechanism follows each of the three felting heads to make a level, uniform mat. The felt is steamed slightly just before entering a heavy duty rubber-belt pre-press. The steam destroys the static charge on the fibers so that the mat stays at its pre-pressed level and does not spring back in thickness before entering the press. Each mat is sprayed with water prior to going into the press loader. When the hot platens (about 385°F) close on this wet surface, the steam generated plasticizes the wood fibers, giving a board with a very smooth surface. This process takes a press cycle of about $3\frac{1}{2}$ minutes for $\frac{1}{8}$ in. thick board. Baking or tempering and humidifying are done the same as in the Masonite process. The process generates very little effluent.

The Dry Process. This process was developed originally by the Plywood Research Foundation in Tacoma, Washington. The primary refiners are Bauers. The fiber has a small percentage of phenolic resin along with wax size added to it. The forming is accomplished with dry fiber falling onto a continuous wire in a multihead process with scalping rolls after each head for mat leveling purposes. A typical system of this type has three heads, each of which lays down a part of the total mat. Normally, very fine fiber is used on the top and bottom parts of the mat, with much coarser fiber going into the center layer. This gives a board with a very smooth surface on both faces but also permits the usage of coarse fibers in the center of the board, and coarse fibers are less expensive to make.

The largest plant using this process in the U.S.A. was built by Bowater Co. and later sold to U.S. Plywood. It is at Catawba, South Carolina. It uses Bauer refiners which have Defibrator preheaters and feeders. This process produces S-2-S hardboard. Its press cycle is 3 min for $\frac{1}{8}$ in. thick board. The press platens are at 450°F, and the pressure used is 900 psi maximum.

Asbestos–Cement Board

Asbestos–cement wallboards are unique among the wallboards generally available. They are essentially inorganic and are highly weather resistant without the necessity for any surface treatment. They are entirely incombustible and are not damaged by water. Wallboards are produced in thicknesses from $\frac{1}{8}$ in. to $\frac{5}{8}$ in. and in sheet sizes up to 4 by 12 ft. Their natural color is a light gray. They are also made to simulate wood shakes and used as exterior siding for homes. These siding products are usually finished with baked-on coatings which have many years of color-fast life. The permanent colors and deep wood-like textures make them very attractive.

The physical properties of asbestos–cement boards which meet the use requirements for wallboards are hardness, toughness, flexibility, fire resistance and weatherability. Their inherent strength and hardness make it possible to use them in $\frac{1}{8}$ in. thickness as surface skins over various types of insulation for curtain walls on nonresidential buildings. They are used in large sheet sizes as the external surface of fireproof walls and as door skins. They have also gained quite wide acceptance as formboard under poured gypsum roof decks where resistance to fire and high humidity are a requirement. The thicker sheets are prefabricated for office partitions. They are also used in marine walls for fire safety.

The main chemical reaction in the natural air cured setting of portland cement is the formation of tricalcium silicate.

$$2(3Ca.SiO_2) + 6H_2O = 3CaO.2SiO_2.3H_2O = 3Ca(OH)_2$$

Other end compounds formed are dicalcium silicate, tricalcium aluminate, dicalcium ferrite and tetracalcium aluminoferrite. In all cases, 10–11% of free lime is formed in the setting of the cement. Some asbestos–cement wallboards are cured under the high temperature and pressure of an autoclave. For these curing conditions, additional SiO_2 is added which reacts with the lime to form additional calcium silicate binders. (See also Cement.) The portland cement, asbestos fibers, and other minor ingredients are mixed as a wet slurry. The slurry is made into thin layers on a multicylinder forming machine. This multilayer web is built up on an accumulator roll until the desired mat thickness is achieved. It is cut off at this point and becomes a flat sheet. The master sheet size is a function of the width and diameter of the wet machine cylinders and accumulator roll. After being cut off the accumulator roll, the sheets pass through a surface drying oven which facilitates subsequent pressing of the sheets by smooth or textured rolls. Depending on the end use, sheets are cured under atmospheric conditions or autoclaved after which decorative coatings can be applied.

The following figures are the sales for 1969 for the Asbestos-Cement Industry:

Product	Thousand ft³
formboard (poured gypsum roof deck)	1,140
flat sheets (building construction)	36,180
corrugated sheets (building construction)	15,750
roofing shingles (houses)	950
siding shingles (houses)	150,250

Particleboard

Wood particleboard may be defined as a composition board made up largely of individual, essentially dry, wood particles or fibers which have been coated with a synthetic resin binder and wax, and formed into flat sheets by pressure. Heat is applied with the pressure to cure the resin binder which is normally urea formaldehyde if the board is for interior use, or phenol formaldehyde if the board is to be used for exterior exposure or under severe interior surroundings. Combinations of melamine–urea–formaldehyde resin are often used if the heat supplied to cure the resins comes from a radio frequency source.

The board may be of a homogeneous structure with all particles, all flakes or fibers. Or it may be multilayered with fine surface flakes with a core of coarse flakes, or it may have a coarse-flake core with an overlay of fibers on each of its surfaces. Other combinations are also used. Flakes are made by a machine which shaves off the wood so that its length is parallel to the wood grain. The flakes vary in size. A flake 1 in. $\times$ 1 in. would be exceptionally large. A normal size would be $\frac{1}{4}$ $\times$ 1. Flake thickness varies from 0.005 in. to 0.075 in., depending on their use. The thinner flakes make the smoothest board surfaces.

Board thickness normally starts at $\frac{1}{8}$ in. and goes up to 2 in. This is usually done by pressing a single mat, however, boards of the higher thicknesses can also be made by stacking more than one mat ahead of pressing or by laminating boards together after the press. Widths vary from 3 to 8 ft, and lengths go up to 24 ft. It is interesting to note that in 1955, a particleboard press was considered adequate with 10 openings of 4 ft width and 10 ft length. Press sizes and number of openings have increased rapidly since then. In 1970, presses are being installed with 22 openings which will produce boards 8 ft $\times$ 25 ft. The larger board size helps a great deal in supplying customers economically with cut-to-size pieces.

Wood Raw Materials. Many of the early particleboard plants were built to develop an economic use for very large quantities of green and dry planer shavings. Chips from waste veneer at plywood plants were another source. Bark was carefully explored but has been pretty well discarded. Sawdust was first ignored as being too small in particle size. A lot of it is being used now and some of it is being mechanically refined into fibers first. The flake board type plants rely quite heavily on pulp-size logs (5–12 in. diameter) as the source of the raw material to be converted to flakes. As the pulp mills and hardboard plants bid up the cost of almost all types of small logs and waste wood, even dry furniture waste is being used. Particleboard customers have become more sophisticated and are demanding better grades of board. Smoother surfaces and tighter edges are needed for furniture parts.

Resins. Urea formaldehyde has been the workhorse of the synthetic resin binders used by the industry. Its cost has been cut in half between 1955 and 1969. Amounts used vary from 6 to 10% of dry solids based on the dry weight of the board. In multilayer board, quite often 6% may be used in the core while 10% is used in the two surface layers to give extra strength and hardness. The resin is normally catalyzed to shorten the press cycle. A good press cycle for $\frac{3}{4}$ in. thick board would be 7 min. with 325° press platen temperature. Particleboard press pressures vary from 200 to 1000 psi, depending on the density of the product being made. In nearly all cases, the press platen comes down against a metal stop which sets the thickness of the board.

In the early plants, formaldehyde fumes were extremely bad, causing complaints from the employees, the plant's neighbors, and the customers. Dr. David L. Brink, Weyerhaeuser's chief chemist, came up with the idea of adding raw urea to the resin to tie up the free formaldehyde. It was very successful and the procedure was quickly picked up by the resin manufacturers. Urea resin works very well for board which is not to have exterior exposure or to be exposed to a high moisture interior environment.

Phenol formaldehyde resin is the binder used in particleboard manufacture when the product is to have exterior exposure. The cost is higher but a phenolic bound board will take nearly any type of exposure. Melamine urea formaldehyde blends of resins are finding some use, expecially in those board plants that use high frequency electricity for curing.

Melamine formaldehyde resin when used as a binder in particleboard yields a product comparable to phenol formaldehyde resin with respect to resistance to exterior exposure. It has been found that a mixture of 10–20% of melamine resin and the remainder of urea resin yields a particleboard with relatively more resistance to high humidity or partial exposure than with one made with urea resin alone. This mixture is somewhat more expensive than straight urea but less expensive than straight phenolic resin.

Sizing Materials.   Particle board, like hardboard, has to have a built-in resistance to water absorption. Water entering the boards weakens them, aggravates surface fiber or particle raising, and greatly increases the board's dimensional instability. A "size" is therefore used, which may vary from crude scale wax emulsions to microcrystalline waxes. Hercules' Paracols are of a microcrystalline wax base and find wide usage as a sizing material in particleboard.

Manufacturing Processes.   There are two general methods of making particle board. These are the extrusion process and the mat formed process. In the *extrusion process*, mixtures of wood particles, resin and size are forced by extruders through dies to make flat boards. The extrusion process is used almost entirely by companies who consume all of their own output for furniture cores. The board so made has excellent face screw holding properties but leaves something to be desired in many other physical properties.

The mat process is the common method of laying down particleboard prior to the pressing operation. It involves the placement of flakes, fibers or particles which have had resin and wax sprayed onto them, on a moving wire. If the board has only one type and size of particle, the mat can be formed with a single forming head. If the mat is made to give a multiply board, then three forming heads are normally the minimum that can be used. The first one puts down a fine surface material; the second one puts down a coarser material for the center of the board; and the third one puts down another layer of finer materials. In some of the modern high speed mat formers, there are as many as five forming heads. These are needed because of the excessive bulk of the material which is being used to form the mat.

The mat formed boards are more versatile, and lend themselves to many types of multiply boards. U.S. Plywood's Novaply is a good example of a versatile flake board. Their process is based on the Fahrni patents. Georgia Pacific and Weyerhaeuser make several grades of particleboards with homogeneous and multilayer structures. There are other very large manufacturers of the same types of boards. Celotex Co. makes an excellent board which uses fine fibers throughout. This process was developed by Miller-Hofft and Allied Chemical, with Miller-Hofft holding the patents. The press-curing is done by a combination of hot platens and radio frequency (RF) electricity. This method of board curing can turn out 25 lb density boards $1\frac{1}{2}$ in. thick in 3 min press time. A conventional press time might run to 25 or 30 min. Also, because the high frequency cured board is cured from the center out and at the surfaces, almost no stresses are built into the board. The product as made is in short supply because of its quality and it commands a premium price. (Some people prefer to consider this product as a thick, medium density hardboard.) The RF curing method is particularly applicable to the manufacture of thick boards in a density range from 25 to 50 lb per ft^3.

There are also caul type systems where the prepressed mat is carried in and out of the hot press on metal caul plates. The caulless system does not use the metal plates, but requires a precompressed mat with enough integrity before pressing so that it can be moved into the press by a belt conveyor and left behind intact as the conveyor is withdrawn from the press.

Presses with simultaneous closing, balanced weight platens have come into very large usage since 1960. When all the press openings close at once, it permits the air to be squeezed out of the mats slowly enough so that the particles are not blown out of the press. It also avoids precuring of the resins in the bottom mats while the press is being closed. The balancing of the platens means that the same pressure is applied to all of the mats after the press is closed, even though there may be 30 boards being formed in the press at the same time. It aids in the control of board thickness and simplifies the use of stops between the press plates. Washington Iron Works is the leader in developing and making this type of press.

Table 6. Particleboard Type and Uses

Type	Description	Uses
corestock	products of flakes or particles, bonded with urea-formaldehyde or phenolic resins with various densities and related properties	for furniture, casework, architectural paneling, doors and laminated components
wood veneered particleboard	corestock overlaid at the mill with various wood veneers	for furniture, panels, wainscots, dividers, cabinets, etc
overlaid particleboard	particleboard faced with impregnated fiber sheets, hardboard or decorative plastic sheets	for applications such as furniture, doors, wall paneling, sink tops, cabinetry and store fixtures
embossed particleboard	surfaces are heavily textured in various decorative patterns by branding with heated roller	for doors, architectural paneling, wainscots, display units and cabinet panels
filled particleboard	particleboard surface-filled and sanded ready for painting	for painted end-products requiring firm, flat, true surfaces
exterior particleboard	made with phenolic resins for resistance to weathering	for use as an exterior covering material. See FHA UM-32 or consult manufacturer
toxic-treated particleboard	particleboard treated with fungicides and broad spectrum biocides to resist insects, mold & decay producing fungi	for tropical or other applications where wood products require protection against insect attack or decay
primed or under-coated	factory painted base coat on either filled or regular board, exterior or interior	for any painted products
floor underlayment	panels specifically engineerd for floor underlayment	underlay for carpets or resilient floor coverings. See FHA UM-28
fire-retardant particleboard	particles are treated with fire retardants	for use where building codes require low flame spread material, as in some schools, office buildings, etc

Particleboard Industry Manufacturing Capacity. Particleboard in the U.S. is made in forty-eight plants in eighteen states. In Canada, it is made in six plants in six provinces. The U.S. plant capacity in 1968 was 1,771.5 million ft² based on ¾ in. board thickness, and output was 1,494.6 million ft². Announced plant expansions and increased capacities will bring the Industry capacity in the U.S. to 3,145.5 million ft². It will be by far the biggest increase of plant capacity in the history of particleboard. Canada, which had board output of 115.4 million ft² in 1968, will have plant capacity of 172 million ft² based on ¾ in. board. Very little particleboard is imported into the U.S. except from Canada.

Physical Properties. Physical Testing is covered under ASTM D-1037-65. Much of the data presented here came from items listed in the bibliography. Tables 6 and 7 come (from "Particle Board Design and Use Manual" IAI Manual 23-L), and list various particleboard types and physical properties, respectively.

Table 7. Physical Properties of Particleboard

Type	Grade	Class	Modulus of rupture (min avg), psi	Modulus of elasticity (min avg), psi	Internal bond[a] (min avg), psi	Linear expansion[b] (max avg), %	Screw holdings (min avg) face, lb	edge, lb
	A							
	(high density, 50	1	2,400	350,000	200	0.55	450	
	lb/ft³ & over)	2	3,400	350,000	140	0.55		
1[c]	B							
	(medium density,	1	1,600	250,000	70	0.35	225	160
	between 37 & 50	2	2,400	400,000	60	0.30	225	200
	lb/ft³)							
	C							
	(low density, 37	1	800	150,000	20	0.30	125	
	lb/ft³ & under)	2	1,400	250,000	30	0.30	175	
	A							
	(high density, 50	1	2,400	350,000	125	0.55	450	
	lb/ft³ & over)	2	3,400	500,000	400	0.55	500	350
2[d]	B							
	(medium density,	1	1,800	250,000	65	0.35	225	160
	less than 50 lb/	2	2,500	450,000	60	0.25	250	200
	ft³)							

[a] Internal bond is the inherent resistance of a board to any splitting action which may work upon it.

[b] Linear expansion for particleboard is a measure in percentage of its total change in length when the board is taken from equilibrium moisture content at 75°F and 50% relative humidity, to 90°F and 90% relative humidity.

[c] Type 1. Mat-formed particleboard (generally made with urea–formaldehyde resin binders) suitable interior applications.

[d] Type 2. Mat-formed particleboard made with durable and highly moisture and heat resistant binders (generally phenolic resins) suitable for interior and certain exterior applications.

Wood–Inorganic Binder Board

Wallboards composed of wood particles or strands cemented together with inorganic binders, have a limited but growing production in the U.S. In Europe the production of this type of wallboard is far more extensive due primarily to competitive factors. Germany has many manufacturers in the business, most of them small producers for local consumption. The U.S. had nine manufacturers in 1954 with a comparatively small total production, and in 1969, there were eleven manufacturers.

Wood is used as an aggregate in cement, replacing the normal aggregates of stone or sand, to obtain a material of lighter weight, greater toughness and strength, and much better workability. The wood aggregate is largely of the wood strand or "excelsior" type.

Excelsior is made from wood slabs. The wood slabs are selected to have as few knots as possible. The machine which makes the excelsior holds the wood slabs firmly while the slab is "combed" with a large number of very sharp, pointed knives which strip the wood out of the slab in individual strands. These strands are the length of the slab. This length varies from one to two feet. Sometimes the wood slab is notched crossways of this length at 6 in. intervals so that each strand is only 6 in. long. Sawdust and other granular wood particles do not produce sufficient strength to compete with the strand aggregate for the manufacture of wallboard. Owing to the physical properties resulting from the strand wood aggregate, the material can be formed into relatively thin sheets and is suitable for many wallboard purposes. Each strand of wood is coated with the inorganic binder and thus protected to some extent from fire and from microorganisms that destroy wood. The rough surface customarily produced provides large surface voids. These voids give an excellent key for stucco or plaster application, or, if uncovered, a good degree of sound absorption. The wood aggregate is of sufficient size to hold nails nearly as well as solid wood. Sawing and cutting are not as easily done as in solid wood, and carpenters need carbide-tipped tools to cut it. The board is, however, classed as "workable."

Inorganic–wood strand wallboard is made by processes that lend themselves to large or small-scale manufacture by hand methods. This unusual fact has resulted in the widespread, small-scale manufacture in Europe. The inorganic–wood-strand wallboard is made in 20–48 in. widths and 48–96 in. lengths. The popular thickness is 1 in., but $\frac{1}{2}$–3 in. thicknesses are obtainable. The density varies from 24 to 42 lb/ft^3, the thermal conductivity from 0.45 to 0.55 Btu/(hr) (ft^2) ($^\circ$F/in.), and the noise reduction coefficient may be as high as 0.80.

Raw Materials. The inorganic binders are many and varied. The principal binders are portland cement, calcined gypsum, magnesite, magnesium oxychloride, and magnesium oxysulfate. A binder that hardens rapidly is preferred because of the simplified process that results. Magnesium oxysulfate and calcined gypsum harden quickly but lose strength when they are rewet. Portland cement takes several days to harden, but holds its strength better under adverse conditions. The ratio of binder weight to wood weight ranges from 2:1 to 3:1.

Wood excelsior of widths up to $\frac{1}{4}$ in. and thicknesses up to $\frac{1}{10}$ in. is the principal aggregate. The fiber excelsiors are preferred for acoustical boards and some wallboards because they produce a smoother appearance. The wood itself is of many species depending upon availability and compatibility with the inorganic binder. Certain woods containing acids and soluble sugars, for example, are unsuitable for use with a portland cement binder. Very small amounts of wood sugars will retard or completely inhibit the hardening of the cement. Thus a slight hydrolysis of gums or hemicelluloses may be responsible for unsatisfactory results. Oak, southern pine, and larch are not suitable for a portland cement binder, while cottonwood, aspen, ash, and gum give good results. Treatment of unsuitable woods chemically or by cooking or washing does not often improve their characteristics, and increases costs.

Excelsior has a tendency to tangle and lump, and also has the undesirable quality of springiness. The springiness presents a problem in manufacture due to the difficulty in maintaining the compressed thickness necessary for smoothness and density. The net result is the cement-covered, wood-strand mat must be held under pressure until the cement has gained enough strength to maintain the compressed state. To avoid maintaining compression, some makers have tried to reduce the length of the excelsior strands. This has resulted in some decrease in the difficulty of mixing and spreading the strands; but these benefits are accompanied by some loss in strength, and maintained compression is still necessary although shortened.

A further change in the wood particle shape has been made in Europe. Wood in the form of kraft pulp (see Vol. 16, p. 702), is combined with inorganic but surely the far shorter binders. Often asbestos fiber is included as well. In this case formation is done by a wet-felting machine and repress. The problem of springback is eliminated and maintained pressure is not necessary during curing. A wallboard of the kraft fiber type, known as Emsen board is made by Eternit, S. A., Belgium. An example of excelsior board made with a magnesium oxysulfate binder is the Herkalith board (Herkalith, Germany). In the United States, an excelsior board made with magnesium oxysulphate binder Tectum, is manufactured by the National Gypsum Company; also a portland cement-excelsior board, Insulroc, is manufactured by Flintkote Corp.

Manufacturing Processes. The manufacture of wood–inorganic binder boards is quite varied, and no two processes are alike in more than a general way. This is due to the variation in the requirements of the binder, type of wood aggregate, availability and cost of labor, and the size of the market. The largest group having similar manufacturing methods is that of the makers using excelsior aggregate. Variations are present in this group also, but the processes are basically similar. All begin with excelsior, untangle it to some extent, then apply the inorganic binder. There are many ways in which the binder is applied to the excelsior. One method wets the excelsior with water and causes it to fall through a suspension of dry cement in air; another tumbles the wet excelsior with dry cement; and still another tumbles dry excelsior with a thin slurry of the cement in water. In one way or another the excelsior and cement are combined and passed to the sheet-forming equipment.

The small-scale manufacturers employ molding pans in which is spread, by hand and as evenly as possible, the tangle of excelsior strands which have been previously coated with the inorganic binder. A plate of wood or steel is laid on the pan, the pans are piled one upon another, and pressure is applied until the plates rest on the pan edges or against stops. The pile is then clamped and the stack set to one side until the cement has hardened enough to maintain the compressed size of the mat. If the uncured boards are left in a steam environment, it speeds the cure of the cement by as much as two days. When the cement has hardened sufficiently, the boards are removed from the pans, dried, and trimmed. Drying may be postponed until after a hardening period is allowed for the cement. The pans are cleaned and returned to the process. A variation of this process is one in which a tall stack of pans is formed, the weight of the pile being more than ample to supply the necessary compression. Newly filled pans are introduced to the bottom of the pile and hardened sheets taken from the top of the pile. The pile height is often 24–40 ft. depending upon the pressure and time required for the cement to harden. This is a small-scale process.

In large-scale operations the cement-coated strands of wood are mechanically deposited more or less uniformly on a moving steel or rubber belt. A top-moving belt supplies the compression, and the whole moves forward until the cement has hardened enough to hold the compressed thickness. Fast-hardening cements such as calcined gypsum or magnesium oxysulfate are especially useful in this process as better speeds can be maintained with shorter belts. However, as noted above, strengths are not permanent under adverse conditions, and magnesium oxysulfate is better than calcined gypsum. Heat is sometimes applied to the compressed mat to hurry the hardening of the binder. Herkalith belt-forms the mat using magnesium oxysulfate as the binder. The belts holding the mat in compression pass through a heated tunnel where the binder hardens. The hardened sheet is cut to length and is passed to a dryer. Emerging from the dryer the sheet is cooled, trimmed, and placed in stock. Thus the process is truly continuous. In a variation of the high-stack method, the stack of molds is enclosed, and heated air or steam passed upward through the chimney surrounding the stack of molds to accelerate cement hardening. Carbon dioxide is also used sometimes.

The shorter wood fiber strands do not change the excelsior methods materially. Compression forces are reduced and hold-down times shortened, but otherwise the processes are similar. The fine wood fiber boards such as the Belgium Emsen board require an entirely different process, the only similarity being the ingredients of the board, wood, and inorganic binder. This process consists of drymixing the cement, wood, and asbestos fiber and then adding water in excess to form a thin slurry. The slurry is run to one or more vats, in which wire-faced wheels turn. The hydrostatic head acting on the wheels forces stock against the wire. Most of the water passes through the wire into the interior leaving the stock on the wheel. Pumps remove the water from the wheels maintaining the hydraulic differential inside and out. A felt picks up the stock deposited on the wire of the wheels and transports it to a large accumulator roll where the stock is again transferred, this time from the felt to the accumulator roll. By depositing the stock through a number of revolutions of the accumulator roll, a variety of thicknesses can be made. The sheet is then stripped from the accumulator roll and piled with plates and flat wires for repressing.

Uses. The excelsior type wood–inorganic binder board is the only one produced extensively in the U.S. The major present use is for roof decks under built-up roofing. For this purpose thicknesses of $1\frac{1}{2}$ to 3 in. are common. Thinner boards will serve as formboards in casting concrete or other lightweight inorganic roof slabs. Exterior sheathing under stucco is another application for the thinner boards and as a plaster base for interior plastering. Stucco or plaster base boards require a metal mesh over all joints to prevent cracking. The thinner sheets are also suitable for rough flooring because of good nail holding and spanning strength. A rough floor having a leveling coat of $\frac{1}{2}$ in. of cement makes a good base for floor tile application. In waterproof binder types the board is suitable for roof boarding under shingles.

The unusual properties of the excelsior boards are their resistance to fire and rot, and their acoustical absorption. These properties combined with comparatively good strength and workability indicate the many uses to which they can be put. They also have a very attractive appearance when the exposed surface is made of comparatively fine fibers.

Insulating Board

Origin. Insulating board was invented and developed by Carl G. Muench. His work was really divided into two phases; board from wood fiber and board from bagasse. The first work and the initial invention was achieved while he was employed by the Insulite Div. of Minnesota and Ontario Paper Co. at International Falls, Minn. (later purchased by Boise Cascade Corp.). Mr. Muench's original objective was to develop a profitable use for pulp screenings. This is a reject material from the paper mills consisting of fiber bundles and undissolved pieces of knots screened from high grade pulps. Within a matter of months after he attacked the problem, he designed and built the first insulating board machine. It had a capacity to produce and dry 3,000 ft² of ½ in. insulating board in 24 hr. This was in May, 1914. In 1916, a new building was erected to house a forming machine and dryer with a daily capacity of 60,000 ft² of insulating board. The second phase of Mr. Muench's work started in 1919 when he left Insulite with several associates and established the Celotex Corp. in New Orleans. Here the raw material used was the residue from sugar cane, bagasse. In 1920 they built a plant at Marrero, La., which had a daily capacity of 200,000 ft² of ½ in. insulation board per day. This eventually grew to a factory with a daily capacity of 3 million ft².

Uses. The use of insulating board to give structural strength and conserve heat in homes, offices, stores, and churches caught on very fast. Since that time it has obtained markets as roof insulation, formboard for poured gypsum roof decks, acoustical tile, insulating roof deck, insulation board shingle backer, nail base sheathing for wood and asbestos cement shingles, corner brace sheathing, and rated sound deadening board for walls and floors. Sound deadening board is the fastest growing single use followed closely by finished board for the ceilings of mobile homes.

Plants and Locations. In 1954, there were 15 manufacturers of insulating board in the U.S.A. and 2 in Canada. The total production was about 2.6 billion ft² on a ½ in. thickness basis. In 1968, there were 22 manufacturers in the two countries. Three companies were making board in both countries. These 22 manufacturers operated 34 plants, with Celotex and U.S. Gypsum operating four each in the U.S.A, and Building Products of Canada operating three. The total production in 1968 was 3.6 billion ft² in the U.S.A. and 0.521 billion ft² in Canada. It was produced in 16 states in the U.S.A. and 5 provinces in Canada. Foreign imports were negligible.

Board Characteristics. Insulating board varies in density from 10 to 30 lb/ft³. It is the lightest of the wall boards and can easily be distinguished from hardboard by Its lower density and the lower rigidity of its softer surfaces. It is made in greater thickness than other boards for general production, ½ in. as compared to ⅜ in. in plywood and ⅛ in. in hardboard. The structure is formed by felting the fibers from a water suspension. The bonds for the simpler boards are the result of either hydration of the fibers during the refining step, which will produce acceptable fiber to fiber bonding, or the use of prehydrated fibers such as ground wood or pulp mixed with free draining fibers from steam pressurized refiners. For the more sophisticated boards such as that used for acoustical tile, a small percentage of starch may be used. For the strong boards such as nail base sheathing, asphalt, up to 30% by weight of the finished board, is used to give hardness, nail holding properties, permanence to weathering and dimensional stability. All companies use fire resistant coatings on their acoustical tile. The Insulation Board Institute certifies that all of these products made by its members will give a flame spread below 200 when tested in the ASTM E-84 tunnel test.

Simpson Timber Co. and National Gypsum Co. produce wood fiber board which will give a Class A fire test rating, or a flame spread below 25 when tested in the ASTM E-84 tunnel. The product is classified as incombustible. This result is obtained by adding substantial amounts of fire resistant salts to the board in the manufacturing process. These salts may be mixtures of borax and boric acid, or monoammonium or diammonium phosphate. The problem in the use of the salts is to retain them in the board as the water is drained away from the fibrous stock. This can be done by using a closed water system where all of the water is recycled (except that which is in the board entering the dryer) or by sucking the concentrated chemical into the wet mat after the mat has come through the press rolls.

Raw Materials. Nearly any wood pulp can be made into insulating board. In addition, other materials in use to make board are bagasse, extracted pine-stump waste (see Rosin, also Turpentine), and flax straw. Several manufacturers make their own ground wood which acts somewhat as a binder, while others add substantial amounts of waste newspaper because it is a cheap furnish and performs as groundwood. A very large market has developed for what was previously waste wood. Most saw-mills now bark their logs so that residues are bark-free and can be used for various forms of board. Between pulp mills, hardboard and particleboard plants, and insulating board mills, there is a very large market for chips from sawmill waste, saw-dust, and planer shavings. The cost of the wood used in the board is an increasingly large amount of the overall cost. This is more true with the extensive automation which is reducing manhours per 1000 ft² of board produced. Wax and alum are used for board sizing. The use of starch and asphalt was covered previously. No synthetic resins are used.

Manufacturing Processes. *Fiber Preparation.* Several processes are used in the manufacture of insulating boards. These divide basically into two processes, one mechanical and the other a combination of chemical and mechanical. In the *mechanical process*, logs are ground to a pulp on large, coarse pulp stones using pressure to hold the logs sideways against the surface of the stone and water sprays for cooling. The log bark is often removed before grinding. The pulp thus formed is diluted with a large quantity of water to a consistency of less than 1% and passed over screens. Rejects at the screens are reprocessed, and the accepted pulp is dewatered and re-diluted to a specific consistency, often 3–5%, for sizing and storage in large chests preparatory to mat formation. Another process in chipping of the logs and reduction of the chips to pulp in a disk mill where rotating plates do the work. Not all woods are suitable for grinding. Those not suitable include most hardwoods, the exception being the soft hardwoods such as aspen, cottonwood, and willow. The use of saw-mill waste such as slabs is hampered in the grinding operation by the difficulty and cost of bark removal and of handling such sizes and shapes.

The *chemical–mechanical process* is fitted for agricultural residues such as straw, bagasse, and flax straw, and for hardwoods that cannot be satisfactorily ground. The process consists of pressure-steam treating the material in the presence of lime, sodium hydroxide, or neutral sodium sulfite. Five to ten percent chemical on a dry weight basis is sufficient and one to two parts by weight of water. The cook serves to soften the material so that the fibers are more easily separated with less destruction. The next process is a mechanical rubbing and working of the material in a disk mill or other type of equipment that will produce a mechanical separation of the fiber. The fiber thus prepared is then treated in the same way as the ground wood fiber.

A variation sometimes used in fiber preparation is the Asplund process where the material is fed into a continuous cooker with or without chemicals and subjected to steam pressure for several minutes. The continuous cooker feeds the material still under pressure into a disk mill for reduction of the softened material to a coarse fiber. In the next step the coarse fiber no longer under steam pressure is passed to a disk refiner for further reduction toward fiber size. Bauer now has a somewhat similar unit. A further variation is termed "chemigroundwood," which is adapted to woods that cannot be ground in the usual way. This process is essentially that of ground wood, the only difference lying in the preparation of the pulp logs for grinding. The logs are loaded on cars which are run into a steam tank, much as in the creosoting process, and vacuum is applied. Next the chemical solution is injected, and heat and pressure applied after which the pressure is removed and the available chemical solution drained off. Chemicals are the same as those used for treating chips. The logs are next removed, fed to the regular grinders, and reduced to pulp. The grinding may be done hot or cold any time after log treatment.

Pulp yields in the semichemical processes are about 84%, whereas the groundwood process gives about 95%. The semichemical process is accompanied by a more serious stream pollution problem than the ground wood. The tightening of laws covering the dumping of pulp wastes into streams is making a pollution control system mandatory in all insulating board plants. These cost from $\frac{1}{2}$ million to several million dollars per plant. In all fiber preparation methods, control and adjustment of pH must be made to apply size properly and to obtain a suitable pH at the forming machine. The addition of fungicides and broad spectrum biocides, such as sodium pentachlorophenate, may require further pH manipulation.

Forming and Drying. There are essentially two types of insulating-board forming machines; a cylinder type such as the Oliver board forming machine, and the flat-wire or Fourdrinier machine similar to a paper machine. The cylinder machine is a hollow, wire-faced wheel turning partly submerged in a water suspension of the wood or other fiber. The wheel has a hub vacuum-valve system and is segmented so that a vacuum may be applied to assist sheet formation and mat dry-out, but can be cut off at the point where the mat is removed from the wheel. Fifteen to twenty inches of vacuum are frequently carried on these machines. The cylinder-type forming machine was generally preferred for insulating-board formation because of its ability to form a mat faster than the flat-wire types. However, a number of flat-wire machines are in use and have certain advantages over the cylinder machine such as simpler thickness control and uniformity and access of mat for wet end coatings.

The flat-wire machines are in effect moving horizontal wire belts upon which the waterborne fiber is spread. Dewatering is accomplished by gravity and by suction below the wire. Pressure may be used at the head box to accelerate stock deposition on the wire or by gravity head or pump. The use of suction belts and boxes below the wire on the Fourdrinier type of machines has permitted much faster running speeds and has given them by far the preference in newest plants over the Oliver former. After formation the wet mat passes between a top and bottom wire of a roller press for further dewatering and across a roller table to be cut into wet mats 16–24 ft long. The mats are then placed in a roller dryer, heated by direct-fired gas, direct-indirect oil, or pressure steam. Sheets are dried to 1 or 2% before leaving the dryer, after which trim saws cut the large mat into smaller pieces, the 4 by 8 ft size being the most common. After the trim saws, the sheets are rewetted to normalize their moisture content and placed in stock.

Finishing and Packaging. Treatment of the board from the trim saws onward depends upon the type of product being made and the degree of mechanization the manufacturer has applied to the operation. Coating of interior surfacing wallboard, for instance is usually done by taking board from stock and running it through coating or painting equipment.

In some instances the board passes directly from the trim saws to the painting operation, emerging from the coater ready for packaging or working to other sizes and shapes. In the case of exterior sheathing, some manufacturers pass the sheathing directly from the trim saws to an edge-beveling machine and brand-printing press, while others stockpile the blanks and later draw upon stock for the process from beveling on.

Sheathing, roofing insulation, and like items are not painted. Painted items are passed through a roll, spray or flow coater, coated one side and dried. Speeds above 100 linear feet per minute are sometimes used in the coating operation. The paints are water-vehicle types of a greater or lesser degree of washability and are all flame-retardant. Most insulating board products are packaged by carton or wrapping. Loose shipments are occasionally made in carload lots of items, such as sheathing or roofing insulation.

Bibliography

Hardboard and Semihardboard treated in *ECT* 1st ed. under "Wallboard," Vol. 14, pp. 875–884.

Hardboard

Methods of Test, American Society for Testing and Materials, Philadelphia, Pa., D-1037-65.

Federal Specification LLL-B-8109, Building Board (Hardboard), Washington, D.C., July 7, 1965.

W. J. Baker, A. J. Panshin, E. S. Harrar, and J. S. Bethel, eds., "Wood Composition Board," in *Forest Products*, pp. 247–263, 1962.

Wood Composition Boards: General Characteristics, Uses and Manufacturing Processes. Report by Pacific Power & Light Co.(1955).

H. Lambert, "State of the Board Industry," in *Forest Industries Magazine*, July 1969.

W. C. Lewis, *Insulation Board, Hardboard and Particle Board*, Forest Products Lab. Rept. (1967).

Fiber Board Mills Report, Defibrator AB Stockholm, Sweden, April 1969.

Fiberboard Mills by Regions and Countries, Defibrator AB Stockholm, Sweden, June 1969.

Fiberboard Industry and Trade—Some Statistical Data, Defibrator AB Stockholm, Sweden, May 1969.

U.S. Pat. 1,399,976 (Dec. 31, 1921), G. J. Mason.

U.S. Pat. 2,198,269 (April 25, 1940), H. K. Linzell and J. W. Gill (to United States Gypsum Co.).

U.S. Pat. 2,537,101 (Jan. 1951), 2,552,597 (May 15, 1951), 2,669,552 (Feb. 16, 1954), 2,687,556 (Aug. 31, 1954), 2,692,206 (Oct. 19, 1954), D. F. Othmer, L. G. Ricciardi, and W. R. Smith.

Commercial Standard—C.S. 251-63, National Bureau of Standards, American Hardboard Association, Hardboard Users.

U.S. Pat. 3,429,055 (Feb. 25, 1969), Lars Bergh-Gunnar (to Aktiebolaget Svenska Flaktfabriken, Stockholm, Sweden).

U.S. Pat. 3,372,217 (March 5, 1968), Franciscus Paerels and W. E. Hostettler (to F. Fahrni).

U.S. Pat. 3,367,828, D. C. Carter and D. E. Noyes (to Johns Manville Corp.).

U.S. Pat. 3,414,461, G. E. Brown and R. R. Huff (to Monsanto Co.).

U.S. Pat. 3,383,274, D. W. Craig (to U.S. Plywood, Champion Papers, Inc.).

Asbestos-Cement Board

U.S. Pat. 769,078 (Aug. 30, 1904), Ludwig Hatschek.

U.S. Pat. 984,870 (Feb. 1911), J. A. Wheeler.

U.S. Pat. 1,353,512 (Nov. 21, 1920), L. Baumgartl.

U.S. Pat. 1,790,822 (Feb. 3, 1931), J. W. Ledeboer (to Ambler Asbestos Shingle & Sheathing Co.).

U.S. Pat. 2,015,084 (Nov. 24, 1935), Walter McQuade (to Johns-Manville Corp.).

U.S. Pat. 2,156,383 (May 2, 1939), John Ferla.

U.S. Pat. 2,182,353 (Dec. 5, 1939), E. W. Rembert (to Johns-Manville Corp.).

U.S. Pat. 2,327,706 (Aug. 24, 1943), R. T. Hastead (to Johns-Manville Corp.).

U.S. Pat. 2,347,684 (May 2, 1944), L. A. Hatch (to Minnesota Mining & Mfg. Co.).
U.S. Pat. 2,354,350 (July 25, 1944), C. C. Schuetz (to United States Gypsum Co.).
U.S. Pat. 2,354,351 (July 25, 1944), C. C. Schuetz (to United States Gypsum Co.).
U.S. Pat. 2,422,344 (June 17, 1947), G. L. Easterberg (to Philip Carey Mfg. Co.).
U.S. Pat. 2,446,782 (Aug. 10, 1948), H. J. Otis (to Johns-Manville Corp.).
U.S. Pat. 2,500,923 (March 21, 1950), Wm. Bernard (to American Asbestos Industries, Inc.).
U.S. Pat. 2,738,713 (March 20, 1956), E. J. Buczkowskii (to Keasbey & Mattison).
U.S. Pat. 2,784,650 (March 12, 1957), A. Magnani (to F. C. Smith & Co.).
U.S. Pat. 2,818,824 (Jan. 7, 1958), C. I. Read (to Tilo Roofing Co., Inc.).
U.S. Pat. 2,866,484 (May 12, 1959), C. V. French (to Johns-Manville Corp.).
U.S. Pat. 2,929,735 (March 22, 1960), B. H. Field (to Patent & Licensing Corp.).
U.S. Pat. 3,027,293 (March 27, 1962), P. S. Bettoli (to Ruberoid Co.).
U.S. Pat. 3,197,529 (July 27, 1965), N. S. Greiner (to Johns-Manville Corp.).
U.S. Pat. 3,219,467 (Nov. 23, 1965), F. W. Redican.
U.S. Pat. 3,237,361 (March 1, 1966), C. R. Norman (to United States Gypsum Co.).
U.S. Pat. 3,269,888 (Aug. 30, 1966), Julie Yang (to Johns-Manville Corp.).

Particleboard

The Bartrev Press for Continuous Production of Board, Bartrev, Ltd., London.
Commercial Standard CS 236-66, Mat. Formed Wood Particle Board.
Federal Specification LLL-B-800a, Washington, D.C., May 15, 1965.
"Annual Board Review," *Forest Industries*, **96** (8) (July 1969).
W. C. Lewis, *Insulation Board, Hardboard and Particle Board*, Forest Products Laboratory, March 1967.
H. C. L. Miller, *Hot Plate Press Particle Board*, F.P.R.S. Meeting, N.E. Section, April 1956.
Particle Board Design and Use Manual, AIA File No. 23-L, National Particle Board Assoc. 1967.
Proceedings of First Symposium on Particle Board, Engineering Research Div., Wood Technology Section, Washington State Univ., 1967.
Proceedings of Second Symposium on Particle Board, Engineering Research Div., Wood Technology Section, Washington State Univ., 1968.
B. G. Heebink and F. B. Hefty, "Steam Post Treatments to Reduce Thickness Swelling of Particle Board, J. Forest Prod., Nov. 1969.
Tentative Methods of Evaluating the Properties of Wood-Base Fiber and Particle Panel Materials, A.S.T.M. Designation D1037-65.
U.S. Patent No. 2,993,239, C. C. Heritage, July 25, 1961. Production of Integral Layered Felts.
U.S. Pat. 3,071,822 (Jan. 8, 1963), J. G. Meiler.
U.S. Pat. 2,912,723 (Nov. 17, 1959), J. R. Roberts.
J. Weiner and J. Byrne, Institute of Paper Chemistry, Wood Particle Board, Bibliographic Series Number 10-1964.
Wood Particle Handbook, School of Engr., North Carolina State College, North Carolina (1956).

Wood–Inorganic Binder Board

U.S. Pat. 2,594,280 (April 29, 1952), Julian F. Beaudet.
R. C. Weatherwax and H. Tarkow, *Effect of Wood on Setting of Portland Cement*.
J. Weiner and J. Byrne, *Inorganic and Miscellaneous Boards, Bibliographic Series No. 211–1964, Cement Fiber Board No. 34, 116, 123, 136, 155, 156, 159, 162, 175, 265*, Institute of Paper Chemistry.
Wood Wool Building Slabs, British Standard Institution, 1951.
J. D. Dale, *Durisol Lightweight Pre-cast Concrete, Wallboard Production and Uses, Bulletin No. 31, 79, Northeastern Wood Utilization Council* (1950).
A. E. Elmendorf, Forest Products Research Society, **4** (2), 87 (1954).
U.S. Pat. 3,438,853 (Jan. 1969), Charles Haines, Jr. and R. R. Rothrock (to Armstrong Cork).

Insulation Board

The Story of Insulation Board, Insulation Board Institute.
Fiberboard Mills by Regions and Countries, Defibrator AB Stockholm, Sweden, June 1969.

Fiberboard Industry and Trade—Some Statistical Data, Defibrator AB, Stockholm, Sweden, May 1969.

W. C. Louis, *Insulation Board, Hardboard and Particle Board*, Forest Products Laboratory Report, 1967.

List of Mills Designed by Defibrator AB, Defibrator AB, Stockholm, Sweden, April 1969.

Wood Composition Boards: General Characteristics, Uses and Mfg. Processes, Pacific Power and Light, 1955.

Report 1740—Thermal Insulation of Wood Base Materials, Forest Products Laboratory.

Insulation Board, Thermal and Insulation Block—Federal Spec. LLL-1-535a, Washington, D.C.

Sound Controlling Blocks and Boards—SS-S-118a, Federal Spec. on Fire Rating.

Structural Fiber Insulating Board, CS42-49, Bureau of Standards, Dept. of Commerce, Washington, D.C.

Spec. C209, Standard Method for Testing Structural Insulating Board made from Vegetable Fibers, American Society for Testing and Materials, Philadelphia, Pa.

Spec. D-1037, Standard Methods of Evaluating the Properties of Wood Base Fiber and Particle Panel Material, American Society for Testing and Materials, Philadelphia, Pa.

Specifications 1–7, Insulation Board Institute, Chicago, Ill.

L. O. Anderson, *Agriculture Handbook 364*, Forest Products Laboratory, U.S. Dept. Agriculture Forest Service.

International Fiberboard Process, Pulp and Paper Mag. Can. **50** (5) 78 (1949).

"Insulation Board, Wallboard and Hardboard," *U.S. Forest Products Lab. Rept.*, 1677 (1950).

U.S. Pat. 2,943,965 (July 5, 1960), A. E. Stagne.

"The Story of Insulating Board," *Paper Mill News* **71** (32:12–14) 16 (Aug. 7, 1948).

Stramit Boards, Ltd. Straw and Paper Building Board. *Paper Making and Paper Selling* **67** (3:36, 35) (Autumn, 1948). B.I.P.C. 19:302.

J. R. ROBERTS
United States Gypsum Co.

GYPSUM BOARD

Gypsum board is best described by the essentials of its manufacture. Calcined gypsum (called "stucco" in the industry) is mixed with water to a stiff slurry which is introduced between continuous wide ribbons of chip paper cover sheets and formed to dimension by passage between precision steel rolls or other shaping devices. The sandwich is then supported on a conveyor belt during the few minutes required for the stucco core to hydrate and harden; it is then sheared to length and dried in a multideck kiln.

The term wallboard in its original sense is particularly applicable to gypsum board, since it is used to a far greater extent than the other materials treated in this article for the lining or cladding of walls, partitions, and ceilings. Gypsum board has gained wide acceptance for such use because of its ease of workability, dimensional stability, inherent resistance to fire and relatively low cost. Regular gypsum board is normally employed in applications that do not entail direct exposure to water; special types are available for use where moisture-resistance is required.

The manufacture of gypsum wallboard originated in the U.S. in the early 1900's and has since spread to all highly industrialized regions of the world. The earliest gypsum board was known to the trade as plasterboard because it was intended for use as a base (lath) for plastering. In the 1920's gypsum board in large sheets and with an improved surface finish was brought to market, permitting direct application of decorative finishes without plastering. By far the larger proportion, about 85%, of all gypsum board now goes into such "drywall" construction; most of the remainder is accounted for by various plaster-base (lath) products.

In 1969, 94 gypsum board plants were in operation throughout North America producing about 9,000,000,000 ft² (total area irrespective of thickness) during the year with a market value of about $360,000,000. The United States Gypsum Company and the National Gypsum Company are the largest producers of gypsum board in the United States.

Gypsum wallboard is most commonly made in a width of 4 ft (1250 or 1200 mm where the metric system is used) and in lengths of 8–16 ft. Depending on the intended application, it is made in thicknesses of ¼ in., ⅜ in., ½ in., and ⅝ in. The ½ in. thickness is the most common.

Growth of the gypsum board industry has been accompanied by the proliferation of special-purpose products in the following major classes: (*1*) predecorated boards manufactured by the application of special coatings, and by the lamination of overlay sheet materials, such as vinyl film; (*2*) insulating boards made by lamination of bright aluminum foil to provide heat reflecting and vapor-barrier properties; (*3*) boards with cover sheets or core, or both, treated for enhanced water resistance. Such boards are intended for use as sheathing, on exterior ceilings, or as backer for clay tile or other finish materials; applications where moderate or intermittent exposure to moisture is expected. (*4*) Boards with enhanced fire resistance obtained by use of mineral fibers or other special ingredients in the core; (*5*) boards with kerfed (slotted) or other special edge designs for use with special mechanical erection systems; (*6*) boards for use as permanent forms for poured-in-place roof decks of gypsum concrete.

A special related field of technology has emerged owing to the fact that in most drywall construction it is desired to conceal the joints between adjacent boards completely and permanently. After the wallboard is erected, reinforcing tape of special semibleached kraft paper is applied over the joints, using special adhesives known as joint compounds. The compounds are formulated not only to develop the necessary adhesive bond, but also to provide a plaster-like workability which permits the joint area to be brought to a smooth, level surface. Joint compounds are marketed either as powders which require to be mixed with water, or as ready-to-use pastes. Casein is the conventional adhesive binder for the dry powder types; polyvinyl acetate emulsion is generally employed in the ready-mixed forms. Some of the newer powder types employ a combination of cementitious and adhesive binders to permit use over the widest latitude of drying conditions.

Effort has also been directed to the design of wallboard edges so as to aid concealment of joints with tape and compound. Edges are generally made with a shallow .035 in. recess or taper. A recent significant innovation is disclosed in U.S. patent 3,453,582, assigned to the United States Gypsum Co. covering an edge formation which by virtue of a specific contour provides a joint structure having high resistance to stresses that tend to create a visible ridge along abutting edges.

Composition. Although various additives or modifiers may be used in the core of gypsum board, they ordinarily amount to not more than 2% of the total by weight. The quality of the finished board and efficiency of its manufacture are overwhelmingly dependent on the qualities of the stucco supplied to the board machine. Total impurities in the stucco including naturally occurring inert mineral substances should not exceed 12%. Soluble salts and hygroscopic clays are especially deleterious and their combined total should not exceed 0.2%. The stucco should be ground so that not more than 5% is retained on a No. 100 U.S. Standard sieve (149 μ). The fineness of the stucco in terms of Blaine surface area should be at least 7,000 cm²/g.

By-product or waste gypsum from phosphoric acid production has often been considered as a starting material for board stucco. However, special process steps including rigorous control of residual phosphates are required in order to make a usable stucco from by-product gypsum, and the added process costs can be justified only in the rare situations where good mineral gypsum is not available.

Modified starches derived from maize or from grain sorghum are commonly added in amounts of about 0.6% to reinforce the interfaces between cover sheets and the core. Fine freshly-ground gypsum is used in amounts of about 0.6% to accelerate the set of the stucco. Hammermilled paper fiber is employed in amounts ranging from 0.5% to 1.0%, as required to reduce core brittleness. Lignosulfonates are sometimes used in amounts up to 0.1% to increase the fluidity of the slurry, ie, to reduce the amount of mixing water required.

Even when fluidizing agents are used, board stucco as compared to other cementitious materials requires a relatively large amount of water to make a formable slurry. A typical ratio is 75 parts water per 100 parts stucco. Of these 75 parts water, about 18 are taken up as the stucco ($CaSO_4 \cdot \frac{1}{2} H_2O$) hydrates to gypsum ($CaSO_4 \cdot 2H_2O$), leaving about 57 parts to be evaporated. The void volume left by evaporation reduces the apparent core density but still lighter weights are desired for ease and economy in shipping and handling. Weight reduction is sometimes accomplished by addition of light aggregates such as expanded perlite, vermiculite, or polystyrene, but more often by incorporation of a separately generated aqueous fluid foam. The most widely used foaming agent is a special potassium rosin soap. Synthetics, especially alkali metal salts of alkyl arenesulfonic acids, are frequently employed. Usages range from 0.5 lb to 1.0 lb per 1000 ft^2 of board, to obtain core densities of the order of 43 lb/ft^3, or board weights (including cover sheets) of about 1800 lb/100 ft^2 on a $\frac{1}{2}$ in. basis.

The special chipboard paper used for the cover sheets is made at a caliper of about 0.020 in. and a weight of about 70 lb per 1000 ft^2. The paper must be sized so that it is readily wettable by the stucco slurry, yet resist through-saturation which would cause surface staining. High porosity is also required in order to permit rapid passage of the large amounts of water vapor driven off in the drying operation; if the sheet is too tight, "paper blows" will occur in the kiln.

Manufacturing Process. The board-making process is usually continuous throughout. Minor solid ingredients are metered volumetrically into the main stucco stream which in turn is usually metered by a variable-speed auger conveyor. The stucco, water, and separately-generated foam are introduced through separate ports into a horizontal, high-energy pin mixer. Thorough blending is accomplished even though the residence time in the mixer is of the order of only 2 sec. The slurry is discharged onto the advancing face sheet of paper, a few feet ahead of the "master" roll. Just before the master roll the edges of the face sheet are folded up and inward, forming a flange to receive the back cover sheet. The edges of the back cover sheet are glued to this flange. The core is completely enveloped in the cover sheets, which form what amounts to a permanent mold; once the slurry is discharged from the mixer it does not make contact with any parts of the machine.

The core requires about 4–5 min to harden, and during this time the board is supported on belt and roller conveyors. Hence, at a typical machine speed of 150 linear ft/min, the conveyor must be at least 650 ft long. At the end of the conveyor the board is firm enough to be sheared to length by a rotary knife. The sheets are then inverted and fed to a multideck kiln by means of a tipple. The kiln is usually direct-heated, with two or more heat zones. Even though relatively large amounts of water must be evaporated (about 1 lb./ft², or about 36,000 lb/hr at a speed of 150 ft/min) drying is accomplished in about 1 hr. The air temperature in the first zone is often as high as 600°F, but owing to evaporative cooling the board core temperature does not exceed 212°F, so recalcination of the stucco does not occur. Finally, the boards are end-trimmed and automatically bundled and stacked for shipment.

The principal manufacturer of gypsum wallboard machinery is The Ehrsam Company of Abilene, Kansas. Although practically all gypsum wallboard is made in the sandwich form as outlined above, minor quantities are made with cover sheets omitted and with the core heavily loaded with fibrous material. Boards of this type, called "fibrous gypsum," are produced mainly in Australia. The boards are formed on fixed tables and large quantities of sisal or other fiber are worked into the slurry using a semimechanized process. A fully mechanized process has been developed in West Germany, wherein a wet pulp consisting of stucco and paper fiber is continuously roller-pressed into sheet form. A scaled-down version of the conventional board machine has been developed to make slender shapes simulating decorative wooden moldings, but only a few such machines have been built.

The following ASTM Specifications relate to gypsum board:

C 36-67	Gypsum Wallboard
C 37-67	Gypsum Lath
C 79-67	Gypsum Sheathing Board
C 318-67	Gypsum Formboard
C 442-67	Gypsum Backing Board
C 473-66	Physical Testing of Gypsum Board Products etc.
C 474-67	Joint Treatment Materials for Gypsum Wallboard Construction, Testing
C 475-64	Gypsum Wallboard Construction, Joint Treatment Materials for
C 588-66T	Gypsum Base for Veneer Plasters

As will be noted from the bibliography, the published literature relating to the technology of gypsum board is limited almost entirely to patents.

Bibliography

Gypsum board treated in *ECT* 1st ed. under "Wallboard," Vol. 14, pp. 889–894.

W. B. Lenhart, "Developments in Gypsum Manufacture," *Rock Prod.* **55** (1), 152 (1952).
"U.S. Gypsum's 50 Years of Progress," *Rock Prod.* **55** (2), 94 (1952).
"Gypsum Plant-by-the-Numbers," *Miner. Process.* **8** (1), 16 (1967).
M. K. Lane, "Disintegration of Plaster Particles in Water, Part 1," *Rock Prod.* **71** (3), 60 (1968); "..., Part 2," Rock Products **71** (4) 73 (1968).
U.S. Pat. 1,029,328 (June 11, 1912), C. W. Utzman (to U.S. Gypsum Co.).
U.S. Pat. 1,330,413 (Feb. 10, 1920), C. W. Utzman (to U.S. Gypsum Co.).
U.S. Pat. 1,383,249 (June 28, 1921), C. W. Utzman (to U.S. Gypsum Co.).
U.S. Pat. 1,500,452 (July 8, 1924), J. F. Haggerty (to Universal Gypsum Co.).
U.S. Pat. 1,701,291 (Feb. 5, 1929), M. K. Armstrong (to U.S. Gypsum Co.).
U.S. Pat. 1,758,200 (May 13, 1930), F. D. Pfeffer and F. Trotter (to U.S. Gypsum Co.).
U.S. Pat. 1,840,443 (Jan. 12, 1932), F. J. Gough (to U.S. Gypsum Co.).
U.S. Pat. 2,017,022 (Oct. 8, 1935), C. K. Roos (to U.S. Gypsum Co.).

U.S. Pat. 2,080,009 (May 11, 1937), C. K. Roos (to U.S. Gypsum Co.).
U.S. Pat. 2,180,433 (Nov. 21, 1939), J. Page and C. R. Southwick (to U.S. Gypsum Co.).
U.S. Pat. 2,253,059 (Aug. 19, 1941), T. P. Camp (to U.S. Gypsum Co.).
U.S. Pat. 2,322,194 (June 15, 1943), G. D. King (to U.S. Gypsum Co.).
U.S. Pat. 2,432,963 (Dec. 16, 1947), T. P. Camp (to U.S. Gypsum Co.).
U.S. Pat. 2,464,759 (March 15, 1949), T. P. Camp (to U.S. Gypsum Co.).
U.S. Pat. 2,597,901 (May 27, 1952), W. C. Riddell and G. B. Kirk (to Henry J. Kaiser Co.).
U.S. Pat. 2,604,411 (July 22, 1952), W. C. Riddell and G. B. Kirk (to Henry J. Kaiser Co.).
U.S. Pat. 2,744,022 (May 1, 1956), M. Croce and C. G. Shuttleworth (to Certain-Teed Products Corp.).
U.S. Pat. 2,749,267 (June 5, 1956), J. W. Gill and C. R. Southwick (to U.S. Gypsum Co.).
U.S. Pat. 2,762,738 (Sept. 11, 1956), R. R. Teale (to National Gypsum Co.).
U.S. Pat. 2,806,811 (Sept. 17, 1957), R. S. vonHazmburg (to U.S. Gypsum Co.).
U.S. Pat. 2,965,528 (Dec. 20, 1960), C. J. Loechl (to The Celotex Corp.).
U.S. Pat. 2,985,219 (May 23, 1961), J. M. Summerfield (to U.S. Gypsum Co.).
U.S. Pat. 2,996,811 (Aug. 22, 1961), C. J. Loechl (to The Celotex Corp.).
U.S. Pat. 3,050,104 (Aug. 21, 1962), G. H. Burt (to The Celotex Corp.).
U.S. Pat. 3,190,787 (June 22, 1965), P. L. Henkels (to U.S. Gypsum Co.).
U.S. Pat. 3,297,651 (June 10, 1967), E. Maynard and M. P. Ptasienski (to U.S. Gypsum Co.).
U.S. Pat. 3,359,146 (Dec. 19, 1967), M. K. Lane (to U.S. Gypsum Co.).
U.S. Pat. 3,435,582 (April 1, 1969), V. Disney and M. P. Ptasienski (to U.S. Gypsum Co.).

B. W. NIES
United States Gypsum Co.

WARCOCIDE. See Surfactants.

WARFARIN. See Poisons, economic, Vol. 15, p. 919.

WAR GASES. See Chemical warfare.

WASTES —INDUSTRIAL

A new term has come into common usage in the last few years of the 1960s, "waste management." It is so new that it is not yet completely defined, but it pertains to total waste-handling systems which sooner or later must be employed by government and private industry to solve the problem of man contaminating his own environment.

Because of legislative pressure, public relations and even economic considerations, industry has been forced to assume technological leadership in the solution of its own and the nation's waste management program. As production increases it is obvious that the size of the facilities to produce these products must also increase. This means a greater use of manpower and raw materials. It also means an increase in the number and types of wastes produced by these processes.

The expression "waste" implies that a disposal system must follow, but the disposal methods which have been used in the past are no longer realistic because our environment can no longer sustain continued pollution. Levels of pollution that previously were acceptable are no longer acceptable, and legislation and public opinion are forcing a rapid reevaluation of waste disposal methods.

Industrial wastes are by no means a majority of our environmental pollutants but they are significant, and because industry has the money, the manpower and the technological resources to solve the problem it is perhaps the area of best "problem definition" and the most immediate action.

Classification and Segregation of Wastes

There are many possibilities for the classification of industrial wastes but perhaps the simplest is to classify them into their three natural states: gases, liquids, and solids. All wastes fit into one of these three categories or combinations thereof.

Gaseous Wastes. Industrial gaseous wastes can be subdivided into several major categories: (1) pure gases or vapors, (2) combinations of gases and solids, (3) combinations of gases and liquids, and (4) combinations of gases, liquids and solids. The last three are usually considered gaseous wastes because the gas is the carrier for the solid or liquid phase.

Typical pure gases which might exist from an industrial operation would be hydrogen, hydrogen sulfide, carbon monoxide, carbon dioxide, etc. A pure gas may be considered an air pollutant if it is either toxic or obnoxious in odor. Table 1 shows the limits of various types of gases in these categories, as recommended by the American Conference of Industrial Hygienists.

Table 1. Recommended Threshold Limits for Airborne Contaminants[a]

Substance	Limit, ppm[b]	Substance	Limit, ppm[b]
Ammonia	50	Hydrogen bromide	3
Benzene	25	Hydrogen fluoride	3
Biphenyl	0.2	Hydrogen sulfide	10
Bromine	0.1	Iodine	0.1
Carbon dioxide	5000	Isopropyl alcohol	400
Carbon disulfide	20	Maleic anhydride	0.25
Carbon monoxide	50	Methyl chloride	100
Carbon tetrachloride	10	Naphthalene	10
Chlorine	1	Nitric acid	2
Chloroform	50	Nitric oxide	25
Ethanolamine	400	Nitrogen dioxide	5
Ethyl acetate	1000	Ozone	0.1
Ethyl alcohol	400	Phenol	5
Ethyl ether	50	Phosgene	0.1
Ethylene oxide	0.1	Phthalic anhydride	2
Fluorine	5	Propane	1000
Formaldehyde	5	Sulfur dioxide	5
Formic acid	5	Toluene	200
Hexane	500	Turpentine	100
Hydrogen chloride	5	Vinyl chloride	500
Hydrogen cyanide	10	Xylene	100

[a] American Conference of Industrial Hygienists.   [b] Exposure during 8-hr period.

The second category of industrial gaseous waste is a gas containing solid particles. Generally, solid particles are neither toxic nor offensive in odor and usually the carrier gas is air or combustion products. The solid or particulate matter in the gas must be of such a size that the effluent gas can carry these particles into the atmosphere. An example of such gaseous waste is the effluent from grinding or milling applications such as lime, cement dust, abrasive dust, etc. While many of these solid particles may be greater than one micron in size others are sub-micron in size. They are easily airborne in the atmosphere and when ingested by humans can cause respiratory ailments or, if toxic, have more serious results.

The third classification is a combination of gases and liquids. The liquids are in small droplets and are carried by the effluent gas stream to adjacent areas. Such liquids may or may not be toxic but can cause difficulties similar to particulate matter. Large quantities of water vapor or organic liquids entrained in air are examples of this type of gaseous effluent.

Liquid Wastes. The variety of liquid wastes which can be discharged from industrial processes almost defies classification. Liquid wastes may contain gases or they may contain solid particles. Gases usually are separated and may be treated as a gaseous waste. Solids may be separated by filtration or centrifugation.

Pure liquid waste should first be classified as aqueous or nonaqueous waste. Aqueous wastes are those containing a high percentage of water with small amounts of dissolved inorganic or organic materials. If the dissolved material is primarily organic and biodegradable then biological methods can be used for separation. If the waste is primarily water with dissolved inorganic solids, then chemical treatment, concentration, ion exchange or distillation may be more practical. Liquid waste should be further classified as combustible and noncombustible. Pure organic materials can be burned in air thereby reducing them to little, if any, problem from the point of view of waste.

Solid Wastes. Of the various types of industrial waste, solid waste is the most easily classified. It can be divided into combustible and noncombustible waste. It should also be classified into waste which will be satisfactory for a sanitary land fill and waste which will not. Waste suitable for land fill is stable organic or inorganic material which either will remain inert or degrade biologically in the soil. Finally solid waste should be classified into waste which may be compacted and/or reduced to uniform size by shredding, and waste which can not, since both compaction and size reduction will measurably help in a land fill operation.

The segregation of wastes into their various classifications is an important step in waste management. In many cases segregation is the first step. In general a wide variety of different wastes, whether they be gaseous, liquid, or solid, should not be combined into a single disposal area, when if segregated at their source they could be handled more easily. This general statement applies primarily to liquid and solid waste, and should be the first consideration in all cases. Many plants channel all of their liquid waste into a single lagoon, and then begin to consider the treatment of these liquids, whereas segregation at their source or point of creation as a waste would have allowed different but simpler methods to be employed by treating each waste by methods based on its own properties. For example, it would be foolish to combine a combustible organic waste with a highly aqueous waste before treatment, since the combustible waste may be burned easily in a simple incinerator, but if combined with an aqueous waste it would require a more expensive disposal method. On the same note, an organic waste which would create a toxic vapor when burned (such as a chlorinated hydrocarbon) should not be combined with a simpler organic waste which could be burned to carbon dioxide, nitrogen, and water vapor. Similarly a combustible gaseous waste should not be combined with an air stream containing particulate matter before treatment. The latter could be treated with a separation device such as a cyclone or bag filter, but the addition of a combustible fume would create an explosion hazard. The first should be treated by incineration and the second by a separation method. In general, segregation of any waste in accordance with the proposed final treatment method is the reasonable engineering approach to any waste management problem.

Reuse of any usable material in a waste stream should always be considered before disposal. Many waste streams contain products which if removed from the waste stream can be recycled into the process and saved as a valuable commodity. In the past many industries have not employed such considerations but today, with pollution regulations requiring each industry to take a more careful look at its waste products, recovery of waste will become more practical. For example, recovery by activated carbon of valuable solvents from printing or coating operations can stop air pollution and return a portion of the cost to the user.

Treatment Methods

GASES

Waste-gas treatment methods usually are not complicated from an engineering standpoint, and a number of methods are available. However, due to the large volume of gas emitted from various industrial processes, the equipment is large and the cost is high.

Mechanical Separation of Solids. The first step in the treatment of waste gases is to remove particulate or solid matter which is entrained in the gas due to its velocity. Methods for doing this are discussed under Gas cleaning; Electrostatic precipitation.

Condensation of Vapors. A gas emitted from a process may be a pure vapor which is easily condensed if its temperature is reduced below its normal boiling point. A condenser then serves to recover the solvent completely, or very nearly so. On the other hand, most waste gases are a mixture of air or inert gas and solvent vapors. The solvent content is quite small so that condensation may not be a practical solution. It means that the temperature of the gas must be reduced to the dew point of the solvent concentration and as liquid is removed from the gas the dew point continues to decrease, making 100% recovery possible. Therefore in some cases it is practical to remove a portion of some vapors which are present in air by supercooling if the dew point of the vapor is high enough. This generally requires refrigeration and heat exchangers.

Solvent Extraction. The use of activated carbon for the removal of solvent vapors from air is a practical means of solving an air pollution problem while at the same time recovering a valuable solvent for reuse. Recovery varies from 95% to 99%. See Adsorption; Solvent recovery.

Control of Odor. Although the presence of highly toxic gases in minute quantities in the air which we breathe is hard to detect, the presence of many odorous nontoxic organic compounds is not. Extremely small percentages of certain materials can cause severe complaints. Any one who has lived near or downwind from a fat-rendering plant, for example, is familiar with the sickening odor which permeates the atmosphere. A number of treatment methods can be used for such odors. See Odor control.

LIQUIDS

The treatment of a liquid industrial waste is considerably more complex than the treatment of a gaseous waste, because there are so many more possibilities. Sanitary engineering or treatment of sewage is of course a vast subject with an extensive literature (see Water).

Separation of Solids. The separation of solids from a waste liquid can be achieved by several methods: filtration (qv) utilizes a screen, cloth, or other porous material, for mechanical separation; centrifugal separation (qv) separates the solids from the liquid by centrifugal force; and sedimentation (qv) utilizes the force of gravity causing the heavier solids to sink to the bottom of the liquid container.

All three methods are widely used in the separation of solids from industrial wastes. It is generally preferable to have a solids-removal step early in the treatment of any waste liquid since it simplifies subsequent treatment methods.

Concentration. A second step which can be used in the treatment of waste liquid effluent is for the removal of dissolved solids, particularly those of an inorganic nature. Concentration of the waste is desirable since it does not cause air pollution, emitting only water vapor to the atmosphere. Concentration of the waste liquid reduces its volume and if carried beyond the solubility of the dissolved salt it can also be used to remove at least a portion of the dissolved salt from the solution. See Evaporation; Crystallization.

Distillation (qv). Distillation is a possibility for the recovery of valuable solvents from a waste-liquid stream, either aqueous or nonaqueous. Fractional distillation will permit almost 100% removal of pure solvents from a mixture. Azeotropic distillation (qv) can be utilized for removal of solvents from difficult mixtures. A distillation system normally would be a part of the process equipment rather than a waste control method.

Solvent extraction. Liquid solvent extraction is seldom used in industrial practice for waste disposal. Nevertheless, it is a possibility since a solvent can be removed from a waste stream by contacting it with a third agent in which it is more soluble. The mixer-settler type of equipment is most generally used. However, success has been achieved with equipment such as the Podbielniak centrifugal extractor.

Adsorption (qv). Recovery of solutes can also be accomplished by using liquid-phase activated carbon. Small amounts of organic material such as phenol have been removed from water by passing the water through a bed of activated carbon.

Equalization and Neutralization. Large industrial wastewater treatment systems employ an *equalization basin* as a first step. Once the solids have been removed the wastewater from various sources is collected in a single tank or lagoon and provision is made to mix these wastes to assure homogeneity. The purpose of this equalization step is to obtain a more uniform wastewater for subsequent treatment. This makes chemical treatment a much easier job. For example, if a plant had an acid wastewater and a basic wastewater it would be foolish to treat these separately when if mixed the treatment would be at least partially accomplished. Neutralization involves changing the pH level of an acid or alkaline waste, by the addition of alkali or acid, so that it can be safely discharged into a nearby water source. Neutralization usually is accomplished by measuring the influent and effluent pH and adjusting the reagent feed-rate accordingly. Neutralization usually is carried out on acid wastes by the addition of liquid caustic, or lime slurries, or, in certain cases, the waste may pass through a bed of limestone for partial neutralization.

Biological Degradation. This is a means of removing organic matter from a wastewater stream by the action of bacteria. Such processes are used for the treatment of municipal sewage and the design and construction of the industrial sewage treatment plant follows essentially the same principles (see Water). There are two types of bacteria, aerobic and anaerobic. Aerobic action, which is the most generally used, requires that oxygen be available for oxidation of the organics, that the pH level of the wastewater is nearly neutral, that suitable nitrogen and phosphorous compounds are available, that there are no bactericidal materials present and that the system is adequately mixed. In the industrial plant the organic matter usually is not of a natural origin, however, in some cases a mutation will arise in the bacterial culture which will be capable of dealing with what is actually present.

As oxidation occurs the organic material is converted into carbon dioxide and water and new cells. The waste is the food for the microorganisms and the rate at which the food is converted into carbon dioxide and water usually increases with increasing temperature until a maximum level is attained. Some of the types of biological systems are activated sludge units, aerated stabilization basins, oxidation ponds, trickling filters, anaerobic lagoons and anaerobic contact units.

Aeration is necessary for aerobic oxidation. Aeration methods in industry are similar or identical to those used in the sanitary or municipal disposal industry. Aeration can be obtained either naturally or mechanically, depending upon the requirements of the system. Natural aeration may be achieved by large oxidation ponds for surface aeration of the waste. Trickling filters involve the use of biological growths attached to a contact medium over which the wastewater flows, with sufficient space in the filter to maintain contact with atmospheric oxygen. Spray irrigation onto the surface of land areas has also been used as a means of biological treatment and aeration. Pressure aeration uses mechanical means such as blowers or compressors distributing the compressed air to the points in the treatment mechanism to entrain air in the wastewater through the use of diffusers or pipes with numerous small holes. By dividing air into small bubbles, the volume-to-surface ratio is great and aerobic conditions persist. Sparged-turbine aerators, which pull air through a fine spray of water particles, have proven to be more efficient than the compressed-air type of system. In another design wastewater is pumped through a tube while at the same time drawing air from the surface for entrainment with the liquid. Some aerators are designed with floats so that they can float on top of the aeration ponds.

Ion exchange (qv). Purification of water, especially the removal of dissolved inorganic salts, has been accomplished for some years by ion exchange. The zeolite water softner which has been used on commercial and domestic water systems for many years utilizes sodium zeolite as the ion exchange resin, and exchanges the sodium ion for the calcium or magnesium ion present in the water.

Ion exchange has been most useful in the removal of various metal ions from plating wastes. Chromium, silver, and other compounds which are often used in the plating of metals can be removed by ion exchange with resins developed for this purpose. Ion exchange can be used for the purification of waste acids from pickling operations and similar industrial uses. It has been used in the sugar-beet industry for the removal of color. Ion exchange offers a distinct possibility for the separation of small quantities of inorganic contaminants from a wastewater system.

SOLIDS

There is very little that can be done to solid waste before ultimate disposal by incineration or other methods discussed below, but mention may be made of compaction and shredding.

Compaction. Compaction of waste automobiles has been used for some time as a process that reduces the size and provides a more easily handled product. Similar methods are being used today for the reduction of the volume of all types of waste.

Metal wastes are especially desirable to handle in this matter, since they can be used for land fill, but paper and other refuse is also being compacted to improve the handling characteristics for land fill operations.

Shredding. Shredding of wastes has recently become a popular treatment method because it produces a uniform waste size. This applies not only to combustible waste but also to metals and other noncombustibles. If the next step is incineration, reduction to a uniform size makes it possible to feed the waste at a standard rate into an incinerator rather than using batch methods. The same is also true in the application of shredded waste to land fill. It is easier to distribute and cover up a uniform waste than a nonuniform one.

Both shredding and compaction have the shortcoming of requiring high horsepower and specialized equipment.

Disposal of Wastes

GASES

When it comes to the final disposal of any gaseous waste there are only two possibilities which are available. Once the waste has been treated it is either acceptable for disposal in the atmosphere or it must go to some additional step which we shall call disposal. In many cases the waste is not acceptable in its present concentration for distribution into the atmosphere but by proper dispersion, that is mixing with sufficient air, the concentration of the waste gas can be reduced so as to be acceptable.

Incineration. Although incineration means to reduce to cinders by burning, it is often applied to the burning of gases, or liquids, even though no appreciable amount of cinders is produced. Incineration of a waste gas can be practiced if it contains organic compounds which will rapidly oxidize at high temperatures. There are three basic types of waste gas incineration systems: direct flame, thermal, and catalytic.

Direct-flame incineration is applicable if a combustible substance is present above its lower combustible limit when mixed with air; in other words the waste gas is much the same as a fuel and will burn with a visible flame when mixed with air. This usually requires a waste gas which has a heating value of more than 100 Btu/ft^3 although sometimes gases with a lower heating value can be burned in a commercial combustor or burner with an auxiliary fuel.

Gases having a lower heating value may become combustible when preheated to 600 or 700°F or more, usually with preheated air. Blast-furnace gas is a typical example of a low-heating-value fuel which sustains combustion. Typical waste gases which may be burned in this manner are hydrogen cyanide, hydrogen sulfide, carbon monoxide, hydrogen, etc. Of these hydrogen sulfide is an example of a gas that produces a toxic product, sulfur dioxide, which requires further scrubbing for removal.

The contaminant in the waste gas may serve as part or all of the fuel in the system. Direct-flame combustion should be employed only where the contaminant supplies at least 50% of the fuel value of the mixture. The equipment for direct-flame incineration can be either a conventional industrial burner or combustor utilizing either forced or induced draft, or it may be a flare-type burner as found in many petroleum refineries and chemical plants. Flares are of two basic types: the ground flare and the elevated tower flare. The ground flare as its name implies is used at or near grade level and sufficient space must be provided around the flare for safety purposes.

It is normally used in oil fields, gas fields, or other areas where there are no adjacent buildings. The tower flare is utilized in refineries and chemical plants to keep the flame well above the process equipment, protecting against fire hazards. Tower flares have a pilot burner for continuous automatic ignition. They consist of a pipe with a special flame-holding device at one end discharging the gas directly into the atmosphere. The waste gas is ignited and burned at this point with the flame-holding or retaining device ensuring stable combustion.

Flares often require steam injection to prevent smoking when burning waste hydrocarbon gases. Steam injection is used for hydrocarbons in which the carbon-to-hydrogen weight ratio is greater than three, for example, ethane or propane (C_2H_6 or C_3H_8).

In general flares are a practical method of direct-flame combustion but they have the shortcoming of being affected by atmospheric conditions, especially high winds, and can not be considered an infallible method of waste disposal.

Thermal incineration, the second type of incineration which applies to waste gas, usually is utilized for waste-gas streams containing small amounts of combustible organic material. Here, instead of injecting waste gas directly through a burner along with auxiliary fuel, the burner is fed exclusively with the auxiliary fuel and is used to heat the waste gas to a temperature 200–300°F above the autoignition temperature of the waste organic present in the gas. For most waste organic materials this will be on the order of 1000–1500°F. Most industrial effluents of this type come from ovens, paint spraying operations, etc, and involve organic materials carried in air so that sufficient oxygen is present in the waste gas to supply the oxygen for combustion of the organic material.

Thermal incineration is used to incinerate waste gases containing organic materials at or below 25% of the lower explosive limit for the particular solvent present. Care should be taken in the incineration of wastes containing higher percentages of organic materials, against the possibility of flame propagation from the incinerator back to the source of the waste gas.

The design of the incinerator will greatly depend upon the three T's of combustion, Time, Temperature and Turbulence. It is necessary that the operating temperature be high enough to oxidize the organic contaminants, that the residence time within the incinerator be long enough to complete the reaction and the turbulence sufficient to mix the air and the waste material completely.

Thermal incineration systems may be inline systems as shown in Figure 1 where the burner is actually installed in the waste duct or they may use a tunnel-type burner firing from the outside into the duct. Refractories usually are required at the temperatures involved; however, stainless steel may be satisfactory in certain cases.

Catalytic incineration is another possibility for the incineration of gaseous waste. As with thermal incineration it is used for waste containing low concentrations of combustibles in air. The catalyst is generally a noble metal such as platinum or paladium dispersed on the surface of a catalyst support. One catalyst support material commercially available today is in the form of a silica honeycomb; another consists of a screen of nichrome wire. The advantage of a catalyst is the low-operation temperature which reduces the auxiliary fuel cost. The disadvantage of a catalyst is its cost, and the possibility of poisoning or blanketing agents which will reduce its activity.

In any waste gas incineration system the possibility of heat recovery should be considered. Generally, the waste gas incinerator, whether it be catalytic or thermal in

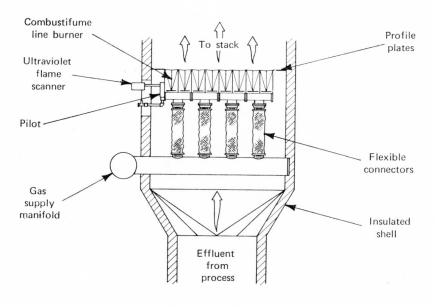

FRONT VIEW

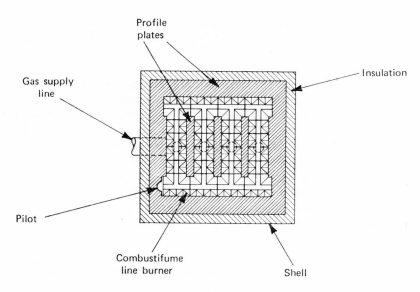

TOP VIEW

Fig. 1. Duct-type fume burner. Courtesy Maxon Premix Burner Co.

nature will have an exit temperature of 800–1600°F. These gases may be discharged directly to the atmosphere, but if they can be used to reduce the auxiliary fuel input then it is advantageous to add heat-recovery equipment. The waste heat from the combustion reaction can be used to preheat the incoming fumes to a temperature at

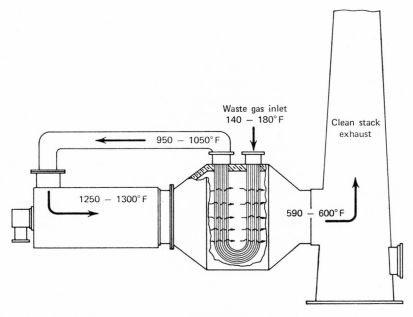

Fig. 2. Waste-gas heat-recovery and incineration unit. Courtesy Thermal Research & Engineering Corp.

least halfway between the outlet temperature of the incinerator and the incoming temperature from the process equipment (Fig. 2).

Heat exchangers for waste gas incineration systems can be of the shell-and-tube type or may be a plate-type system. Pressure drops are usually quite low, on the order of 10 inches of water, and maximum operating temperatures are usually limited to 1400°F.

Another type of heat exchanger which may be used is the rotary-plate type. This utilizes revolving metal plates which first pass through the hot stream absorbing heat and then into the cold stream where heat is desorbed to preheat the incoming waste gases. These units are commonly used for recovery of heat in power boilers, refinery stills and steel-mill furnaces. There is some leakage of gas from one side of the heat exchanger to the other because seals can not be built to withstand the temperatures and pressures normally involved and still prohibit leakage, but by having a higher pressure on the exit gases than on the incoming gases pollution can be avoided.

The refractory regenerative heat exchanger is a cyclic system. A large refractory mass or similar device absorbs heat from the hot gases exiting from the incinerator and once this mass is up to temperature the high temperature exit stream is reversed by means of a switching valve to another refractory mass and the incoming cold waste gas is passed over the hot refractory into the incinerator.

Dispersion Methods. For many years industrial plants with a waste disposal problem involving a gaseous waste would merely vent them into a stack or chimney, regardless of the contaminants or pollutants. With the present emphasis on air pollution abatement it is no longer possible to utilize the same approach; stacks are not *always* the solution. But they can be used in some cases, and there may be situations in which they are the only means available.

A short stack or vent pipe is still considered acceptable for certain waste gases that do not contain pollutants but for some waste gases containing small amounts of pollutants, the tall stack or chimney may be utilized safely and to economic advantage. The purpose of a high stack is to convey waste gases containing either toxic or particulate matter to a point high enough above the ground level so that after normal dispersion into the atmosphere the ground-level concentration of these materials will be well below permissible levels. This not only applies to concentrations directly at the base of the stack, but at any distance from the base of the stack. Today tall stacks or chimneys are utilized to obtain low concentrations of pollutants above the inversion layer so that these pollutants will not be trapped in or under the inversion layer causing high ground-level concentrations.

Stacks three or four hundred feet high are normally required in most areas, however, due to topography in certain areas, it is often necessary to go as high as 1000 or 1100 feet. Many factors influence the acceptability of using a stack as a means of dispersing a pollutant—the relative location of the stack to residential areas and buildings; the prevailing wind direction and velocity; humidity and rainfall; the topography; the amount of pollutant exiting from the top of the stack and its exit velocity.

The permissible ground-level concentration of particular gaseous wastes is generally available from literature, but it may vary from locality to locality depending upon the pollution codes.

As a plume of smoke or waste gas exits from the end of a stack it usually has a conical shape. As it expands it will touch the ground at some distance from the base of the stack. The ground-level concentration of contaminant reaches a maximum at a distance from the stack which is usually about ten times the effective stack height, that is, the height of the stack itself plus the plume resulting from the discharge velocity and buoyancy of the gas.

A stack should be about two and one-half times the height of any nearby buildings. Gas-ejection velocities for the stack should be greater than 60 ft/sec so that the stack gases will escape a turbulent wake of the stack. In many cases it is desirable to have the exit velocity on the order of 100 ft/sec, if possible. There is a critical wind velocity for every stack exit velocity above which there is no corresponding rise of the waste gas due to that exit velocity. Above this critical wind velocity, gas temperatures and flow rate no longer affect the ground-level concentration.

With stacks of diameters less than five feet and height under 200 feet there will be times when the plume will hit the ground leading to excessive ground concentrations. For such small stacks the design basis is unreliable and results are unpredictable. When stack gases have been subjected to atmospheric diffusion and turbulence induced from surrounding buildings is not a factor, the gound-level concentration should be on the order of 0.001 to 1% of the stack concentration, assuming that the stack is properly designed. This is quite a wide range, and it certainly shows that design is not an exact science.

In many industrial complexes, multiple stacks within a short distance of one another can cause problems in ground concentration of various contaminants, because their plumes combine at some point causing an excessive ground concentration in this area.

Absorption methods. Sulfur dioxide, from burning fossil fuels and from certain industrial operations, such as smelting, is poured into the atmosphere as a pollutant at a very large rate. (See Vol. 19, p. 408.) Most of this is dealt with simply by diffusion from stacks, but with the present emphasis on pollution control several methods have been developed for removing it and recovering the sulfur. Most of these methods are by no means economical, but they are a necessity for pollution abatement, and the recovered sulfur (as sulfur, hydrogen sulfide or sulfuric acid) reduces the cost of the treatment. (See Vol. 19, p. 415.)

LIQUID WASTE

There is a wide variety of industrial liquid wastes, with correspondingly complicated disposal methods. The best approach is to segregate each waste at its source so that it can be treated individually before it is necessary to handle a complex mixture. As with gaseous waste it is necessary to define a disposal system as compared with a treatment system. This often is unclear. Assuming that the liquid has been treated in some separation apparatus to remove the solid particles, and other treatment methods previously mentioned have been considered to recover any valuable waste material, there are only a few final disposal methods which are available—oxidation (incineration), deep-well disposal, surface disposal and dispersion in lakes or ponds, rivers or oceans.

Incineration. Obviously, this is only applicable if some combustible material is present in the waste. Many aqueous solutions containing only a small amount of combustible organic waste can be disposed of by incineration. However, the more combustible material there is present, the more economical the incineration method will be.

The same three T's of good combustion, Time, Temperature and Turbulence are important here as they are with a gaseous waste. But, since we are starting with a liquid rather than a gas we must supply the necessary heat for vaporization of the liquid in addition to raising it to its ignition temperature. Liquids vaporize and react more rapidly when they are finally divided into tiny droplets in the form of a spray. Atomizing nozzles are usually employed to inject waste liquids into incineration equipment whenever the viscosity of the waste permits. Slurries, sludges and other materials of high viscosity can be handled in other types of incineration systems. A general rule of thumb which can be used to determine the combustibility or acceptability of a liquid waste for incineration is that the waste should be pumpable at ambient temperatures or capable of being pumped after being heated to some reasonable temperature level. The liquid must be capable of being atomized under these conditions. This usually requires a viscosity under 1000 SUs at the operating temperature.

In order to sustain combustion in air without the assistance of auxiliary fuel, the waste should have a calorific value of at least 8000–10,000 Btu/lb. This is not an absolute rule, but a guide in determining combustibility vs noncombustibility.

Materials which fall in the combustible category include light solvents such as toluene, benzene, alcohol and acetone, and also heavy organic tars and still bottoms which would be similar to residual fuel oil. The waste may also be combinations of both, which would have an intermediate viscosity and heating value. Such wastes come from cleaning operations in chemical plants and refineries or are residues from various distillation processes. They may also be waste oils from a variety of machining or metal-treating applications.

Direct incineration of this type of waste is usually accomplished in a commercial combustor or burner utilizing the waste material as a fuel. Auxiliary fuel is used only to heat up the system or to act as a pilot for ignition. The waste is fed through the atomizing nozzle of the burner at pressures up to several hundred psig, atomized and ignited. Sufficient combustion volume is provided downstream from the burner to effect complete oxidation of the waste within the incinerator. Assuming that the waste is completely organic in nature and contains no halogens, sulfur, or inorganic salts, the resulting effluent should be carbon dioxide, nitrogen and water vapor. Atomization of the waste may be effected mechanically or by using steam or air as the atomizing fluid.

Forced-draft systems are preferable to induced- or natural-draft burners because they are more easily controlled. A closer control over the excess air for combustion will provide for better combustion. Heat release rates in a forced-draft system can be as high as one million $Btu/(hr)(ft^3)$, whereas with natural draft they are limited to 25,000–50,000 $Btu/(hr)(ft^3)$. Residence times in the incinerator are from $\frac{1}{4}$ to 1 second.

The incinerator may be almost any shape, but generally the cylindrical arrangement is the most suitable, and it may be vertically or horizontally disposed. The vertical arrangement has the advantage that the incinerator acts as its own stack. However, if a tall stack is required a horizontal incinerator is more acceptable since it can fire into the base of a tall refractory-lined chimney.

Direct incineration of waste materials in this manner does not provide for the recovery of any heat, and many combustible wastes have excellent heating values. With proper system design waste burners can be fired into a waste-heat boiler or other heat-recovery device such as an air preheater, with subsequent improvements in economy. If, however, the waste liquid is to be burned in this manner, it is important to determine that the waste does not contain any materials which would damage the heat-recovery device. Liquids containing noncombustibles such as inorganic salts or chlorides could create corrosive compounds in the combustion reaction which might damage expensive heat-recovery equipment.

Partially combustible wastes are liquids having heating values below 8000 Btu/lb and generally are aqueous. Auxiliary fuel must be used in sufficient quantities to vaporize the entire mixture and to reach the ignition temperature of the organic present. Operating temperatures for this type of incinerator are several hundred degrees above the autoignition temperature of the organic in the mixture.

The waste material must be atomized as finely as possible, and adequate combustion air should be provided. Generally, several hundred per cent excess air is necessary based on the combustibles in the waste. The heat from the auxiliary fuel must be sufficient to raise the temperature of the waste and the combustion air to a point several hundred degrees above the ignition temperature of the organic material and the waste.

Two approaches are used for the destruction of such waste. The first is to feed the waste together with the fuel into a combination burner which is operating on natural gas or other auxiliary fuel. This will usually require temperatures over 2000°F to maintain combustion in the burner. The other possibility is to inject the waste downstream into the incinerator by separate atomizing nozzles. This arrangement permits the auxiliary fuel burner to operate at its full temperature, mixing its products of combustion with the injected waste. Single or multiple nozzles may be used for the injection of waste into incinerators of this type. Configurations for this type of waste incinerator are generally the same as for combustible waste material. Baffles or a checker wall may be used to induce turbulence and mixing within the incinerator. As long as the organic material in the wastes contains no halogen, inorganic salts, or sulfur, the products of combustion should be carbon dioxide, nitrogen and water vapor. Heat recovery in this type of system is usually not considered because of lower stack temperatures.

Certain special wastes must be handled carefully but can be partially disposed of by incineration. These involve wastes containing inorganic salts, those which contain halogen compounds, and those which contain sulfur compounds.

Wastes containing inorganic salts often give the oxide of the metal in the ash. This oxide is usually in a finely divided submicron form as it exits from the incinerator and it will cause an obvious air pollution problem. The normal method of removal of such oxides is by wet scrubbing and a high-energy scrubber must be used.

Organic compounds containing chlorine (or other halogen) lead to the elemental halogen which must be removed by caustic scrubbing. However, it is much easier to operate the incinerator in such a way that the resulting effluent is the halogen acid, which is highly soluble in water and which can be removed easily. Conversion into the halogen acid can be accomplished by using excess auxiliary fuel such as natural gas or propane to supply enough hydrogen. It can also be accomplished at higher temperatures by the injection of steam. After the conversion, subsequent scrubbing with water will remove virtually all of the halogen acid from the stack gas. An example of such a system would be the incineration of trichloroethylene, which is frequently encountered as a pollutant arising from metal degreasing or drycleaning. The normal reaction without the use of natural gas would produce the following: $CHCl:CCl_2 + 2O_2 = 2CO_2 + HCl + Cl_2$. If, however, we add natural gas to the system in the proper amount the reaction can be as follows: $CHCl:CCl_2 + CH_4 + 3\frac{1}{2}O_2 = 3CO_2 + 3HCl + H_2O$.

Sulfur compounds found in liquid waste are either part of the sulfonated organic molecule or in the form of sulfates or sulfides. Complete combustion of these wastes will result in sulfur dioxide if excess air is kept to a minimum. If high excess air rates are used sulfur trioxide will be formed. See Vol. 19, p. 415.

Other Oxidation Methods. Sludges, slurries and materials that do not lend themselves to atomization in a conventional liquid incinerator may be handled in a variety of other kinds of equipment, some of which have been used in sewage disposal for many years. Rotary kilns fired by a burner, which tumble sludges or a combination of solids and sludges, can be used to advantage for the incineration of many such wastes. Although the combustion within the rotary kiln does not usually burn the waste completely afterburners can be applied to prevent air pollution. Large rotary kilns are presently used in several major chemical facilities for the disposal of all types of waste. They are flexible but expensive and they require large amounts of auxiliary fuel.

The multiple-hearth furnace is another possibility (Fig. 3). This is the outgrowth of sewage treatment and was first used in 1934. It is a vertical, cylindrical furnace with a number of horizontal hearths, each having an opening to the hearth below. The waste sludges move from the top hearth to lower hearths by the plowing action of air-cooled metal rabble arms. The sludge is dried on the first two hearths, burned on the next three and ashes are cooled on the lower hearth. Operating temperatures range from 600°F on the top and bottom hearths to 1800°F on the combustion hearths. Auxiliary fuel is generally required in this type of system, if only for ignition, and the exhaust gases generally require afterburning to prevent air pollution. This is a reliable, but generally expensive system.

The fluidized bed (see Fluidization) has just begun to come into more general use

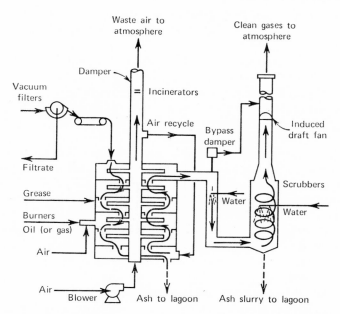

Fig. 3. Multiple-hearth furnace. Courtesy *Chemical Engineering* magazine.

in the last few years. Waste sludge is fed onto a bed of sand that is fluidized with air at a pressure of 3–5 psig. The air is preheated to approximately 1000°F and combustion of the waste takes place on the surface of the bed. Auxiliary burners are often located above to provide supplemental heat and to destroy gases which are evolved from incomplete combustion. The unit is operated at about 25% excess air and combustion temperatures can go as high as 2000°F. The air is preheated by exchange with the stack gases and ash passes out with the effluent gas and is removed in a cyclone separator. This type of incinerator has the advantage of a large heat sink in the sandbed which permits intermittent operation; furthermore it has few moving parts, so that maintenance costs are low. See Figure 4.

Flash-drying systems have been used to incinerate a variety of wet sludges. This system is almost identical to the standard flash dryer used for drying many chemicals. As the solids in the system are dried they are separated in the cyclone and are sent to a secondary incinerator for final destruction. The gases from the dryer may require afterburning. Supplementary fuel is required in every case.

Wet-air oxidation is a process which oxidizes the sludge in the liquid phase without mechanical dewatering. High-pressure, high-temperature air is brought into contact with the waste material in a pressurized reactor. Oxidation occurs at temperatures between 300 and 500°F. Once oxidation has started the need for supplemental heat is minimal. The product is usually a free-flowing liquid. Wet-air oxidation has been used for pulp-mill liquors, plating waste, sewage, cyanide and the treatment of sour waters containing sulfides and phenol and for the recovery of valuable products such as silver from x-ray film.

The atomized suspension technique (AST), a development of the Pulp and Paper Institute of Canada, was conceived to dispose of paper-mill waste but it has never been successful for this purpose because of high-temperature corrosion problems. It is, how-

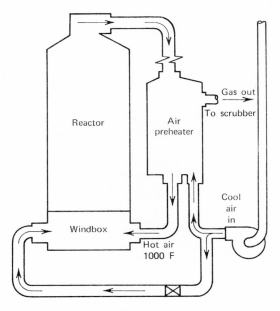

Fig. 4. Fluid-bed incinerator. Courtesy *Chemical Engineering* magazine.

ever, applicable to the disposal of certain liquid wastes. It is a pyrolysis operation rather than a direct oxidation method. Liquid is sprayed into a vertical metallic tube which is externally heated by burners or electric elements to 1500°F or more. The solids are dried and resulting volatile materials are fed into an afterburner. These units have been used in only a few commercial applications.

Deep-Well Disposal. Materials which are not adaptable to incineration or biological processes, for example, large volumes of weak acid or plating wastes where recovery is not economically attractive, represent a distinct disposal problem. While neutralization, ion exchange, or a variety of treatment methods could be used on such waste, they are definitely expensive and involve large, complicated systems. Injection of such waste into deep wells is one possible solution.

This technique is an outgrowth of the oil production industry, which has for many years injected large volumes of saline water into the subsurface strata through old dry deep wells. The salt water which is a by-product of many oil-recovery operations, if allowed to accumulate on the surface, would kill the plant life in the area but injection back into the earth provides a reasonable disposal method. There are some 40,000 brine disposal wells in the United States but only 100 of these wells are presently being used for the disposal of industrial waste. This can not be considered a final solution to the disposal problem, since waste material which is stored under the ground can still constitute a troublesome source of contamination, but it is being used in certain areas where the rock strata is proper for injection. Some states prohibit subsurface injection of waste materials, but in many areas it is still a legal possibility.

A suitable deep well (Fig. 5) should have geological formations at the injection point which have sufficient porosity to act as a reasonable storage reservoir. The injection horizon, which is the level at which the waste will be stored, should be well below the level of fresh-water circulation and the area should be confined by an imperme-

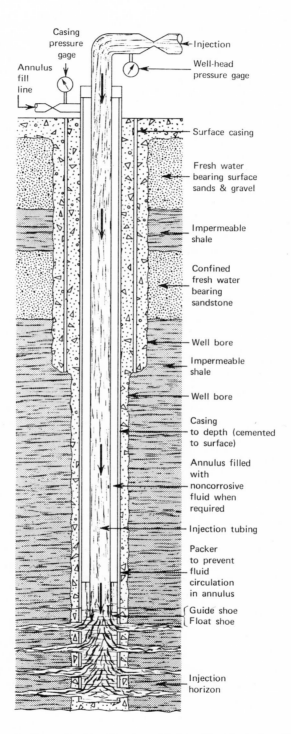

Fig. 5. Deep-well waste-disposal unit. Courtesy *Modern Manufacturing*, © McGraw-Hill, Inc.

able rock enclosure. Sandstone, limestone and dolomitic rock usually are suitable because they contain nothing more than saline water at deep-well levels. Sedimentary rocks may contain oil, gas, coal and sulfur resources and protection of these resources should be observed. A confining layer of shale, clay or some other material should overlie and underlie the injection horizon to prevent vertical escape of the injected waste. Tests should be made to determine the extent of lateral movement of the injected waste from the well since the waste could migrate to other wells in the surrounding area.

The depth of a deep well can vary from several hundred feet to over 12,000 feet. The disposal formation into which the waste is injected has been measured in some cases to be as much as 10,000 square miles and the thickness of the formation up to 5000 feet. Usually only a fraction of the total depth of the well is used for disposal because only a portion is permeable enough to accept the waste. Suitable injection wells have been found in the Gulf Coast and Atlantic Coastal plain because they contain relatively thick sequences of salt-water-bearing sedimentary rock and also because the subsurface geology has been determined through previous oil- and gas-well drillings. Once the well has been drilled the material may be pumped into the well at a pressure high enough to cause it to be injected into the interstitial area of the formation. Injection rates may be either high or low depending on the geology but generally range from 100 to 300 gal/min at pressure levels of 100–500 psig. Obviously the lower the pressure that can be used for injection, the more desirable the operation, because of the lower pumping cost.

Materials containing solids or suspended matter are usually not recommended for deep-well disposal because they can plug the pores of the formation. Similar plugging can also result from entrained gas, bacteria, and mold. Any material which will form an insoluble solid should be removed before injection. The problem of reaction between the injected waste and the interstitial water should also be considered. Precipitates may form which will plug the pores of the injection horizon. This can only be determined by sampling the water and mixing it with the waste prior to injection.

Deep wells generally cost $10–15 per linear foot. The lower figure applies to wells from 1000 to 5000 feet deep and the higher figure applying to wells between 10,000 and 15,000 feet in depth.

Injection of the waste is accomplished by different types of pumps, either centrifugal, turbine or piston type. The centrifugal pump is most suitable for low pressures and the duplex or multiplex plunger pump most suitable for high pressures.

Deep wells are drilled with a cable tool type of drill rather than rotary drills to avoid the plugging of the injection horizon with drilling mud. Once the well is drilled, a well casing must be inserted and cemented in place. Then the injection tubing and well-head facilities can be installed. Many times several casings should be considered. The first may extend several hundred feet or more to protect fresh water at one level from salt-water contamination at the deeper level. The next casing, which is smaller in diameter is run through the surface casing to the top of the disposal horizon or to the bottom of the hole. Corrosion in deep-well injection is often a serious problem. Casings can be selected from a variety of materials and can be coated with plastics, corrosion-resistant paints, etc.

Deep wells can be treated to increase the rate of injection. Chemical treatment using hydrochloric or hydrofluoric acid tends to dissolve calcium carbonate and other soluble materials. Acidizing is effective in limestone or dolomite rock. Hydraulic fracturing, used in the oil well industry to increase the rate of brine injection, can also be used but it must be used carefully in waste injection wells because it can open vertical fissures which extend through the cap rock. After wells have been installed they should be constantly monitored to see that waste material is not escaping into surrounding areas.

Deep-well disposal for industrial waste is being used for such applications as pickle liquor from steel-mill pickling applications, brine disposal in oil fields and potash mines, and for acid waste generated in petroleum refineries and chemical plants. It is being considered for the disposal of radioactive waste. Radioactive waste is mixed with a special cement to form a grout. The technique of hydraulic fracturing is used to inject the waste into the ground at high pressures, which causes the fracture to occur in a shale formation. The grout will then set up at this layer into a hard cement, trapping radioactive material below the earth's surface.

Lagoons and Ponds. One of the earliest disposal methods used for liquid waste was to dig a large earthen basin and to collect the waste within its confines. Any solid materials that may be present gradually sink to the bottom of the lagoon and the clean, or at least less polluted water can be drawn from the surface into a storm sewer or stream. Acid mine-water which is pumped to the surface is neutralized with lime and then pumped into a lagoon where the calcium sulfate and iron hydroxide precipitate.

Lagoons are also used for the separation of oils or floating materials from water. The oil will rise to the surface where it may be removed and burned and the clean water can be drawn from the lagoon near the bottom. Lagoons are used for aeration to accomplish oxidation of organic materials and for solar evaporation to reduce the volume of the effluent. Solid materials which settle must be removed periodically by mechanical methods and reprocessed, if valuable, or disposed of in other ways.

Lagoons also act as equalization and stabilization basins for a variety of liquid wastes. The BOD (biochemical oxygen demand, see Water) is generally reduced, and this action can be accelerated by mechanical aeration. Lagoons should be baffled to prevent cross-flow of the waste from inlet to exit, and also to collect scums and oils at the outlet end. Most lagoon designs are based on a given retention time which may be as long as a week or as short as a day. The major function of the lagoon is the storage of the sludge. The amount of the storage volume required is dependent upon the solid concentration of the feed and the rate of settling and biological degradation. Therefore, lagooning is mostly applicable to wastes containing large quantities of solids.

The wall of a lagoon or storage pond should be made of compacted soil which will withstand the hydrostatic pressure of the impounded material, and not allow leakage to the outside. If the lagoon is to store materials which can seep into the adjacent watershed then plastic lining or similar impervious material must be used to cover the bottom and sides. Sludge removal is usually accomplished by pumping or by mechanical removal such as clam shells and drag lines.

Lagoons have been used for the separation of rolling-mill oil from water, for the separation of calcium sulfate from neutralized mine water and many other solid–liquid separation applications.

Dispersion of Liquid Waste. Dilute waste waters with low levels of pollutants may be discharged into rivers, lakes or streams after appropriate treatment. The proper dispersion of this waste is important. The location of the discharge point and the type of dispersion equipment are significant in their protection of other water resources and in maintaining an overall desirable situation. Plants located near the ocean, large lakes or rivers may discharge waste through a pipe or ditch leading to the shore, but if the discharge of the pipe occurs above the main body of water, the formation of foam due to air entrainment, or incomplete dilution because of a low-water condition, may result in high concentrations near the point of discharge. A properly designed subsurface dispersion system will, however, permit the receiving body to assimilate the waste properly. This reduces treatment requirements considerably.

Liquid-waste dispersion is accomplished with numerous types of devices. A submerged open-end pipe with special nozzles, or diffuser systems consisting of a series of smaller pipes with holes or slots can be used. Waste should be discharged at the best angle to the flow of water in the main body of water to effect rapid dispersion. These pipes should be so located that the discharge point is far enough from the shore line that it will protect plant or other intake systems, and municipal supplies. It should be directed so that existing currents or tides tend to disperse the waste into the main body rather than to bring it back to the shore. The liquid-waste dispersion problem is similar to gas-waste dispersion at the top of a tall stack, and the same considerations are important. The chemical characteristics of the waste, the discharge velocity, the turbulence of the receiving stream, tidal action, temperature variations, density differences and flow patterns of the receiving body are all of utmost importance.

Many wastes are being disposed of today many miles at sea. The wastes are carried by a barge or ship to a point beyond the 3-mile limit, and dumped overboard in containers to sink to the bottom of the ocean. While no serious results have been noticed to date from the piping or barging of wastes to sea, there is evidence of ecological changes, which should act as a warning. Some of these ecological changes are advantageous. Certain types of shell fish thrive on waste which have been dumped into the sea. However, other species have been destroyed by indiscriminate dumping. A careful study of all the ramifications should be made before this type of disposal is considered viable.

Land Disposal. Many industrial waste waters contain constituents which have BOD or COD (chemical oxygen demand) requirements, but which are not toxic. Such waters may be considered for land disposal, which is the spreading of the waste on the land allowing for normal oxidation in the soil. Sewage wastes have been treated this way for many years in other countries, but for aesthetic reasons this is not a practice in the United States. If land can be used as a means of dewatering and oxidizing industrial wastes that contain biodegradable materials, then it is a reasonable disposal method.

In actual practice, waste from food products and paper industries have been disposed of in this manner. Such waste waters are spread over farm land by tank trucks or by means of spray irrigation, or by a network of pipes discharging the waste between ridges or furrows on the ground. Generally, such wastes have a low solids-content.

Paper-mill wastes have been spread over wooded hillsides by spraying from the top of elevated spray nozzles into the top of the trees. This provides for some absorption and evaporation of the waste before it reaches the ground, preventing erosion and excess water runoff. Land disposal can be used beneficially on many crops if it adds soil nutrients.

The economics of such a system are dependent upon the quantities of the waste water and the distance that it must be pumped or transported. It is often more attractive than many alternate methods. There are some shortcomings, however, especially in colder climates because freezing of the wastes may occur and cold weather limits the reduction of BOD significantly.   In warmer climates, a reduction of 60% of the BOD can be accomplished before the water reaches ground-level water.   Loading rates for such irrigation vary quite widely, but they are generally between 10,000 and 100,000 gallons/(day)(acre).   The BOD loading should be kept below 200 lb/(day)(acre).

SOLID WASTES

The solid waste problem facing the world today is perhaps the most difficult because at the present time there are only a few possible systems for disposal.   We are limited to three methods: the first is incineration, the second is composting and the third is sanitary land fill.   Incineration usually leaves products which must be handled by land fill or by reclamation and composting certainly leaves a high percentage of waste for another means of disposal.   On the other hand, land fill is perhaps the most shortsighted of the possibilities, since it requires great acreage to accomplish the task.

For many solid wastes, reclamation of any portion of the waste should be considered first.   For many years it has been cheaper to make new steel rather than to reclaim old.   It has been cheaper to make new bottles than to clean those used.   It has been cheaper to make new paper, rather than reprocess waste paper.   However, the day is coming when the reclamation of such materials must be considered not because of its economics, but because of the necessity of overcoming the pollution problem.

Solids Incineration.   Municipal and industrial incinerators account for 30–50% of the total trash disposal within the United States at the present time.   The object of any incinerator is to provide complete combustion of the material fed into it.   The complication is the wide variety of materials which must be burned.   Waste materials range from wet garbage having heating values up to 2000 Btu/lb to materials such as polystyrene and other plastics which will have heating values up to 19,000 Btu/lb.   It is a difficult problem to ensure the correct amount of air for combustion of such wastes when they are mixed together in a large incinerator.

The design of a solids incinerator, to give minimum pollution problems, should take into account the following considerations.   Incinerators should be designed to operate with excess air, usually on the order to 100 to 150% above stoichiometric requirements.   They should use a minimum of underfire air to keep particulate matter out of the waste-gas stream.   They should use overfire air to provide ample oxygen and turbulence in the combustion space above the fuel.   The temperature in the incinerator should be between 1400 and 1800°F to reduce the rate of smoke formation and to decrease odor.   Sufficient combustion volume should be provided in the incinerator to provide residence time for the burnout of all flying particulate matter.   The average heat release is 5000 Btu/(hr)(ft³).   A secondary chamber or zone should be provided in every incinerator and in fact is required by most municipal and state codes.   The gas residence time in the incinerator should be between one and two seconds.   If the incinerator has a grate, low loading rates per square foot of grate surface should be observed, even in forced-draft incinerators.   This should be no more than 60 pounds of refuse/(ft²)(hr).

Single-chamber incinerators have been outlawed by most modern air pollution codes.   The multiple-chamber incinerator as shown in Figure 6 is employed in most cases.   The incinerator consists of: (*1*) an ignition chamber to preheat the feed and where some combustion takes place, (*2*) a mixing chamber for the addition of secon-

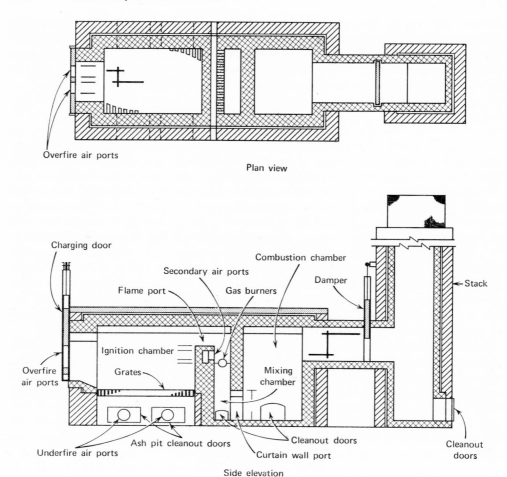

Overfire air ports

Plan view

Charging door

Combustion chamber

Secondary air ports

Flame port

Gas burners

Damper

Stack

Ignition chamber

Overfire air ports

Grates

Mixing chamber

Underfire air ports

Ash pit cleanout doors

Cleanout doors

Curtain wall port

Cleanout doors

Side elevation

Fig. 6. Multiple-chamber incinerator. Courtesy Los Angeles County Air Pollution Control District.

dary air, (3) a secondary combustion chamber for further combustion and expansion of gases, (4) a chamber for settling the flyash and (5) a stack which discharges the gases into the atmosphere (a scrubbing system may also be added at this point).

The solid waste is first charged into the ignition chamber, either manually or automatically, through a charging door onto the grates which are at the bottom of the chamber. In this chamber drying, ignition and combustion of the solid refuse occurs. As the burning proceeds, the volatile constituents and the moisture in the fuel are vaporized and partially oxidized. These gases then pass from the ignition chamber through the flame port which connects the ignition chamber with the mixing chamber. The flame port is an opening over the baffle wall which separates the two chambers. From the flame port the products of combustion from the refuse flow into the mixing chamber, where secondary air is introduced. A combination of temperature and excess air which can be supplied by secondary fuel burners, if necessary, assist in initiating the second stage of the combustion process. Turbulent mixing resulting from changes in direction causes the combustion reaction to go to completion. In the third chamber there is also a refractory wall or curtain wall between the mixing chamber and the final combustion chamber, where flyash and other solid particulate matter are collected by gravity-settling or by impingement. The gas is then finally discharged through a stack or a gas cooler or scrubber, as the case may require.

Solid-waste incineration equipment in small sizes is usually factory fabricated and shipped as a complete package. Such incinerators will handle up to several hundred pounds per hour of waste. In large sizes, the unit is generally installed at jobsite by the contractor. It may have a steel shell lined with refractory or it may be constructed of several courses of brick, the inner course being high-temperature-resistant fire brick. The small, packaged incinerators generally have a combustion chamber and an afterburner section to provide for secondary combustion of the volatile matter in the waste.

Often a pollution-free incinerator is more a function of the operation than of the design. It must be charged properly by the operator in order to reduce the formation of flyash and maintain adequate flame conditions within the unit. Overcharging can produce smoke and undercharging can produce a problem because of insufficient heat. The waste charge should be spread evenly over the grates so that the flames can propagate over the surface of the newly charged material, and variations in the air requirements must be compensated for by operation. Proper charging and excess-air control by the operator are essential for clean operation.

An inexpensive simple incinerator devised in the early 1960s involved the burning of trash or plastics in an open pit blanketed by air from an overfire air-manifold supplied by a centrifugal blower (Fig. 7). While this unit will not necessarily meet all air pollution regulations for normal trash, it will do an excellent job on plastics such as polyethylene and polypropylene. Its cost is $\frac{1}{5}$ to $\frac{1}{3}$ that of a two-chamber incinerator. The pit incinerator is certainly a consideration for the small community and for plant waste in remote areas.

Heat recovery has not been used to any great extent in solid-waste incinerators. The heat-recovery equipment would have to be specially designed and therefore, quite expensive. It can hardly be justified unless there is immediate use for steam or other

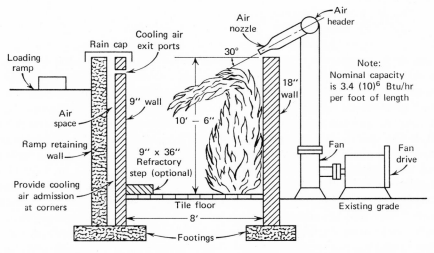

Fig. 7. Cross-section of an open-pit incinerator. Courtesy *Journal of the Air Pollution Control Association.*

source of heat in the area. A typical 5000 lb/hr incineration system for general plant trash releases approximately 32½ million Btu/hr. If this energy could be recovered, then comparing it to natural gas at five cents per therm, annual recoveries would represent heat savings of between $21,000 and $50,000 per year.

One system currently (1970) receiving a great deal of attention is the water-wall incinerator or boiler. There are some problems with this design at the present time since it is designed without refractory and complete destruction of the waste material is difficult. Air pollution requirements are met by flooding the final chambers with water, removing the burned and unburned waste from the flue gases. As more and larger waste incineration systems are being designed and installed, undoubtedly heat recovery will become a more important development and a more exact science.

Composting. Composting is not entirely a waste disposal method. It is a means for reducing the volume of solid waste and returning at least a partly usable product to the industrial plant or community.

Composting is merely accelerated biological degradation of organic waste resulting in a mature compost containing nitrogen, phosphorus, and potassium. It has limited food value to plants but it is a conditioner for layers of soil and does replace certain nutrients that soil may require.

Composting plants have been built throughout the world but very few have been installed in the United States. The cost of land and the problem of odor associated with the degradation of waste and the available market for the product have dampened the enthusiasm of potential users. In Europe, however, where the population density is much higher and where waste recovery, whether heat, water or air, is a fact of life, there are a number of installations. A Danish corporation has been building composting plants with capacities from 5 to 300 tons per day for over 30 years and has installed these throughout the world except the Western Hemisphere. In their system a rotary cylinder similar to a cement kiln is fed waste material at one end and pathogenically pure compost is delivered at the other.

The problem of removing glass and metal particles in municipal or industrial trash has been satisfactorily overcome by special techniques and pretreatment of the dry waste. Shredding or granulation is almost a necessity to ensure homogeneity and definite control of moisture content. Such pretreatment accelerates the rate of decomposition and creates decompostion temperatures which will guarantee destruction of the pathogenic organisms. Once the waste material has been added to the kiln, air is added at low pressure and in controlled amounts over the length of the cylinder. In this manner an environment is created where the action of aerobic microorganisms ensures rapid decomposition of the waste. Normal operating temperatures are about 140°F and are spontaneously developed without additional heat. The weight reduction from the feed to the final compost is 30–40%. The resulting material can be used as fertilizer or soil-conditioner and is usually applied during the autumn and winter months to the ground at rates between 5 and 20 tons an acre.

Sanitary Land Fill. Perhaps the most widely used solid waste disposal method today is the sanitary land fill. Most land fill operations do not fill lands nor are they sanitary. However, properly conceived, land fill operations can be beneficial. Most swampy areas near industrial complexes and not those used for wild life preserves can be filled in and leveled using trash from domestic operations thus recovering valuable land for industry. The earth is removed or scooped out by use of dredges or bulldozers and trash trucks dump the refuse into the void. The dozer then distributes the waste over the area, compacts it, and covers it with six inches of soil and the process is once again repeated.

Land fills cannot be justified only because of a need to dispose of waste, they should also serve a purpose by reclaiming low land in a valuable area. Compacted materials usually provide a better land fill than noncompacted materials, by preventing later problems of erosion. A uniform waste is much more desirable than a nonuniform waste for land fill but unfortunately most land fill material is quite nonuniform in nature. Land fill materials which are not biodegradable, such as glass, plastics, and metals, will not decompose and will always remain in their original form.

Wastes used for land fill, if they contain water-soluble materials which can be leached by rain or surface water, ultimately will contaminate nearby springs and streams. If toxic contaminating materials are present this can be a serious problem. Another serious problem, unless properly handled, is the possibility of harboring rodents and vermin. This can only be overcome by proper and immediate coverage of all material upon which rodents may feed.

Land fill is perhaps the most controversial of all solid-waste disposal methods, because it involves taking one person's trash or garbage and dumping it on another person's property, and whether it is done properly or improperly, the general idea is unacceptable. It can be used however, where proper engineering techniques are applied and it can be an economical means of disposal. The cost of sanitary land fill ranges from $2 to $5 per ton for small operations and from $.75 to $2.50 per ton for large operations.

Economics of Waste Disposal

The disposal of industrial or municipal wastes can seldom be justified from a standpoint of economics. Occasionally, a disposal operation involving a recovery operation, especially heat recovery, can bring a total return on the capital invested, but this is the exception, rather than the rule. The addition of heat-recovery equipment, or any recovery equipment for that matter, usually adds substantially to the capital investment, requiring careful investigation of the economics. In general, a waste disposal system should not be considered as a paying proposition and the incentives for the installation of such systems must come from other sources. These are usually: (*1*) laws and ordinances which forbid improper disposal, (*2*) a desire on the part of the company or municipality to improve or guard environmental conditions in its area; and (*3*) a realization that we can no longer continue to pollute our environment.

There is no rule of thumb which can automatically give the simplest and most economical means of waste disposal for either a gas, liquid or solid. Each situation must be carefully engineered and evaluated on its own merits and the proper method may not always be the most economical method. The following steps should be taken as a normal course in such evaluation:

(*1*) Reuse of any material should be examined and a reclamation of any portion of the waste considered not only as return in dollars but as a reduction of the total waste problem.

(*2*) If reuse or reclamation is impossible then all of the possible disposal methods for the waste should be individually evaluated with respect to economics.

(*3*) Careful design and evaluation of equipment should be made by the engineering department of the company or a qualified consultant.

(*4*) The system should be installed and then maintained with the same degree of interest as any piece of profit-making process-equipment within the plant.

This will *not* be a profitable operation but it will keep the expense commensurate with the results.

Regulations and Standards

The federal government, states, and local municipalities are just beginning to develop codes and ordinances governing air and water pollution. They are doing this because public apathy concerning the problems of pollution is on the wane and indignation concerning pollution, especially industrial pollution, is on the rise. As legislators and lawmakers get more pressure from their constituents, these codes and ordinances are bound to become more complete and more stringent.

Water pollution legislation in most areas has been ahead of air pollution legislation since it was first prompted by the Federal Rivers and Harbors Act in 1899, but air pollution today is getting more attention and soon we will have a more stringent set of regulations than for water pollution. The federal government has begun to set up a series of air quality-control areas throughout the United States, and federal legislation will prohibit certain types of air contamination in much the same way as it has prohibited water contamination. Before any disposal technique is decided on, federal, state, and local codes must all be carefully examined. Thirty-four states now have some type of air pollution regulations and all fifty states have some type of water pollution regulations. On top of this, many cities, towns, and counties have their own codes which limit various contaminants.

Water pollution codes usually specify the amounts and types of contaminants which may be discharged into streams, rivers, lakes, and other water bodies within the legislative area. Air pollution regulations limit the emission of particulate and gaseous effluents as well as placing a regulation on the opacity of gaseous effluent from stacks. The variety of requirements and the types of governing and enforcing bodies vary widely and an entire book could be written on the various regulations throughout the United States. Specific regulations for any area are usually available from the state or local agencies and can be obtained by writing the governing body. Similar information can be obtained on air pollution by contacting the Air Pollution Control Association in Pittsburgh, Pa., or the Water Pollution Control Federation in Washington, D.C.

Bibliography

"Wastes, Industrial," in *ECT* 1st ed., Vol. 14, pp. 896–914, by H. E. Orford, W. B. Snow, and W. A. Parsons, Rutgers University.

1. J. E. Williamson, R. J. MacKnight, and R. L. Chass, *Multiple-Chamber Incinerator Design Standards for Los Angeles County*, Los Angeles County Air Pollution Control District, 1960.
2. *Water Pollution Control Facts, Bull. No. 3M-7-69-GP*, Water Pollution Control Federation, Washington, D.C., 1969.
3. *Controlling Water Pollution*, a collection of pertinent articles from *Plant Engineering* (1968).
4. "Water: The Basic Chemical of the Chemical Process Industries," *Chemical Research Report*, McGraw-Hill Book Co., New York, 1967.
5. "Pollution: Causes, Costs, Controls," *Chem. Eng. News* (June, 1969).
6. William Lund, "Industrial Pollution Control Manual," *Modern Manufacturing*, McGraw-Hill Book Co., New York, 1968.
7. *Air Pollution Control*, a collection of pertinent articles from *Plant Engineering* (1968).

8. "Design and Operation for Air Pollution Control," *Proc. Metropolitan Engineers Council on Air Resources*, 1968.

9. M. M. Feldman, "Handling Cities' Solid Wastes," *Consulting Engineer*, **32** (3), 158 (1969).

10. *Handbook of Air Pollution, Publ. No. 999-AP-44*, U.S. Dept. of Health, Education and Welfare, Public Health Service, Natl. Center for Air Pollution Control, Durham, N.C., 1968.

11. *Air Pollution Engineering Manual, Publ. No. 999-AP-40*, U.S. Dept. of Health, Education and Welfare, Public Health Service, Cincinnati, Ohio, 1967.

12. J. R. Marshall, ed., "Deskbook Issue on Environmental Engineering," *Chem. Eng.* **75** (22) (1968).

13. *Consulting Engineer*, **28** (3), a series of articles on pollution, 1967.

14. R. B. Dean, "Ultimate Disposal of Waste Water Concentrates to the Environment," *Environmental Sci. Technol.* **2** (12), 1079 (1968).

15. *Proc. 15th Ontario Industrial Waste Conf.*, 1968.

16. "Air and Water Pollution Control," *Heating, Piping, Air Conditioning Conf. Proc.*, 1968.

17. *J. Water Pollution Control Federation*, **46** (6), annual literature review issue (1969).

18. R. D. Ross, *Industrial Waste Disposal*, Reinhold Book Corp., New York, 1968.

19. Edmund B. Besselievre, *The Treatment of Industrial Wastes*, McGraw-Hill Book Co., New York, 1969.

20. R. C. Corey, *Principles and Practice of Incineration*, John Wiley & Sons, Inc., New York, 1969

R. D. Ross
Thermal Research & Engineering Corp.

WATER

The question of water supply is, of course, of world wide interest, and the literature on it is world wide. In much of it, however, use has been made of American (or British) engineering units, the foot, the pound, the U.S. gallon, the British gallon (Imperial gallon) and the "acre-foot," the amount of water that would cover an acre to the depth of one foot (1 acre = 43,560 ft^2). Given below are some conversion factors to metric units.

Conversion Factors

to convert	to	multiply by
ft	m	0.3048
ft^2	m^2	0.0929
ft^3	liter	28.316
lb	kg	0.45359
U.S. gallon	liter	3.7853
British gallon	liter	4.546
acre-feet	m^3	1233.48

SOURCES AND UTILIZATION

Water is the most abundant, and the most widely distributed and used chemical compound. The earth is covered over 75% of its surface with water, to an average depth of over two miles. In the summer the air above each square mile of land or sea may contain as much as 100,000 tons of water vapor.

Blood contains about 78% water. About 70% of the weight of the human body is water.

It has been calculated, based on the water required to produce a minimum food requirement of 2.5 lb dry weight (wheat), that something in the neighborhood of 300 gallons of water per day are required to sustain human life. Alternatively, on a diet of beef protein and fat, with some vegetable matter supplement (substantial American diet), the total water requirement per person is 2500 gallons per day. Thus, water supports animal and plant life.

Water furnishes steam for power, electricity and water areas for transportation, and is a solvent for cleaning and washing in the industries. The industries are the prime users of water, to an extent of 160 billion gallons per day (1966) for production processes in the U.S. In 1966 irrigation was the second largest user of water, using 141 billion gallons per day in the U.S., while municipalities used 22 billion gallons per day (1).

It is to be pointed out that much of the water used by industries for washing or cooling, or by irrigation interests, is simply "borrowed," and may be returned in uncontaminated condition for reuse at the site or at a lower level. Of interest to ranches is the high consumption of farm and range animals, which are: horses, 10–12 gal; dairy cows, 12.5 gal; 2-year-old steers, 10 gal; sheep, 0.25–1.5 gal. Cattle on range average (New Mexico) 3 gal/day (2).

Most places of the world with dense population have sufficient water for at least limited domestic use. In the exceptional places supplemental measures may be taken such as saline water recovery, sewage reclamation, reservoir evaporation control, or simply, restrictive usage.

Water makes up the sea, containing about 3.5% NaCl, and other salts. A cubic mile contains 166 million tons of dissolved salts. From the sea are recovered bromine, magnesium, etc (see Ocean raw materials).

The major problem is the poor distribution of water as rainfall, running streams, or as reservoirs. Abundances are not necessarily in the places where water is most needed, such as near population or industrial centers.

Climate and Rainfall

Climates and rainfall vary greatly in places around the world. This is true even for places in the U.S. because of the large expanses in latitude and longitude. For example, within the state of Texas, on the eastern edge of the state rainfall may be 50 in./yr and on the western edge 5 in./yr. Prevailing wind currents, proximity to large bodies of water, mountain barriers, and latitude affect climates and rainfall.

We are dependent on the sea for rainfall. The winds, laden with moisture cross the lands, become cooled, and drop their moisture as rain. The streams and rivers return the water to the sea, along with dissolved salts. This cyclic process has been in operation for millions of years. The saline content of the sea has changed little in a finite short span of time; the first general analysis of sea water was made by Dittmar and his collaborators on the H.M.S. *Challenger* in 1884 (3). No change in sea water composition has been noted from that date to the present.

A light rain may be defined as an accumulation of 0.10 in./hr, a moderate rain as 0.11–0.30 in./hr, and a heavy rain as greater than 0.30 in./hr.

The degree of annual rainfall may be classified into six categories: under 10 in., 10–20 in., 20–40 in., 40–60 in., 60–80 in., and over 80 in./yr (4). In some places in Africa, and in Chile, rainfall is of the order of 0.02 in./yr. The highest rainfall for one year in any area occurred in Cherrapunji, India and was 1041 in. The southern slope of the Himalayas has the heaviest rainfall in the world; it occurs mainly in the summer, and is of a magnitude of 200–600 in./yr. Figure 1 presents a map of U.S. rainfall, and Figure 2 a map of arid regions of the U.S. (5).

Regions with rainfall above 200 in./yr are rare, and comprise just a few areas of the world in New Guinea, Guinea in western Africa, and in Nicaragua. Rainfall of 60–200 in. occurs in the eastern U.S. and Canada, Mexico, central and southern Africa, Central and South America, India, China, and Russia. Rainfall of normal proportions, say, 15–60 in. occurs in the west central U.S., Russia, the coastal perimeter of Australia, and the eastern and northern parts of Argentina. Sparse rainfall (0–15 in.) occurs in the western U.S., north Africa, Mongolia and China, Chile, Peru, interior Australia, and at the tip of Africa. Figure 3 is a world map showing mean annual precipitation.

The annual rainfall is not even approximately constant from year to year at a given location. From data taken from rainfall figures for the eastern U.S. (Philadelphia) maximum rainfall for any year will be 153% and the minimum rainfall 62% of the average annual figure (6). For other localities, similar factors of the same magnitude are applicable. Percentage factors such as these have been found to be accurate to 2% for any locality when based on annual rainfall records of 35 years or more. Hodge and Duisberg point out the variability of rainfall, stating that "average" years of rainfall mean little (7).

Water that falls on the earth (as rain or snow) may take any one or more of several courses (8). It may (1) re-evaporate almost immediately, (2) it may soak into the earth or collect into natural or artificial surface reservoirs and be eventually evaporated, (3) it may fall as snow and remain for some time in this form, (4) it may percolate through surface soil and enter underground porous beds (aquifers), (5) it may run off in streams and rivers or (6) it may be held as polar ice caps. The first two categories are grouped under the heading of *evapotranspiration;* such water is unavailable for man's use.

During the last 40 years, annual precipitation over the U.S. has averaged 30 inches, or a total of 1,564,000 billion gal/yr. The use of this water for irrigation, industrial, public, or other supplies was at a rate of 87,000 billion gal/yr. This is a gross utilization of only 5.6% of annual precipitation (9). The difficulty is that there is a nonideal distribution of rainfall in regard to the requirements of the various areas.

Evapotranspiration contributes its part to this low efficiency. For the U.S. as a whole, about 72% of the precipitation is returned to the atmosphere by evapotranspiration. Such a figure will vary from 40 to 90% depending on the aridness of the region. In the 17 arid western U.S. states, some 20–25 million acre-feet are thus short-circuited and unavailable to meet man's needs.

Rainfall is measured in a collection gage, which is designed with a funnel receiver having an area 10 times that of the collection cylinder. Thus, the reading taken is

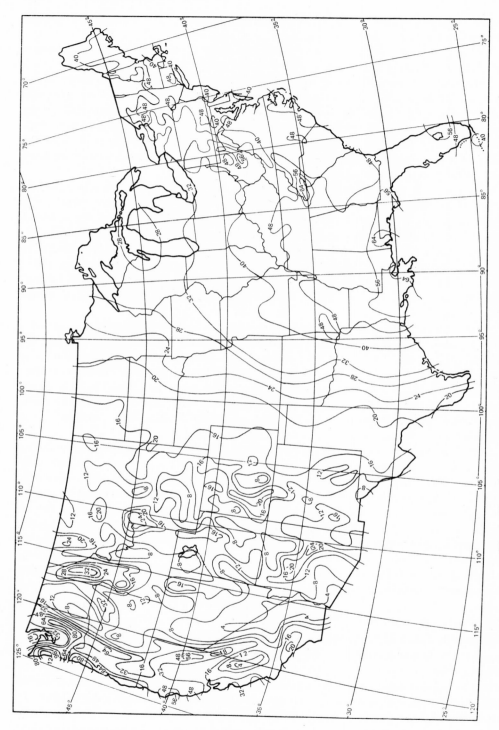

Fig. 1. Normal annual amount of precipitation in the United States (in in.).

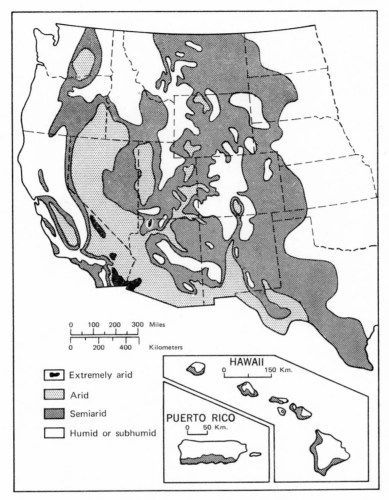

Fig. 2. Arid regions of the United States.

magnified by this factor to improve accuracy.   By dividing the actual depth collected by 10, the true inches of rainfall is found.   Snowfall is converted to the equivalent amount of rainfall before reporting.   Usually 10 inches of snow (depending on degree of packing) is considered to be the equivalent to one inch of rain.

Water Supplies and Quality

Earth's estimated available water resources, including polar ice and glaciers, are presented in Table 1 (10).   An alternative breakdown of the world's water supply is given in Table 2 (11).

Water may be utilized from stagnant or flowing reservoirs on the surface or underground.   The reserve life of fresh water reserves (12) for the U.S. is given in Table 3.

Many waters may be reused with or without treatment.   Water that has been used in industry in condensers or heat exchangers, which are not contaminated, may be reused directly.   Other waters, which are used where contamination exists but to a

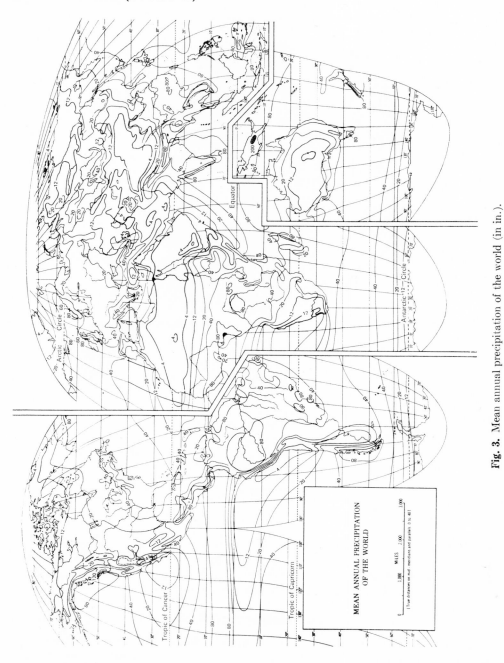

Fig. 3. Mean annual precipitation of the world (in in.).

minor extent, may be reused with minor treatment or returned to the normal water supply source.

The Federal Water Pollution Control Act as ammended by the Water Quality Act of 1965, authorizes the state and federal governments to establish quality standards for interstate waters. As an example, figures for the state of New York give an evaluation of quality required. These standards set minimum requirements not for effluents

Table 1. Estimated Relative Quantities of Water Available in the earth's Hydrosphere

Source	Million acre-feet	Index of amt relative to soil moisture	Total estimated fresh water present, %
oceans	1,060,000,000	51,960	
atmosphere, earth's crust[a]			
fresh-water bodies	33,016,084	1618	100
polar ice and glaciers	24,668,000	1209	74.72
hydrated earth minerals	336	0.16	0.001
lakes	101,000	5	0.31
rivers	933	0.046	0.003
soil moisture	20,400	1	0.01
ground water:			
fissures to 2500 ft	3,648,000	179	11.05
fissures 2500 to 12,500 ft	4,565,000	224	13.83
plants and animals	915	0.045	0.003
atmosphere	11,500	0.56	0.035
hydrologic cycle, annual:			
precipitation on land	89,000	4.4	
stream runoff	28,460	1.4	

[a] To 12,500 ft depth only. Based on data in C. S. Fox, *Water*, Philosophical Library, New York, 1952; R. L. Nace, *Water Management, Agriculture and Ground Water Supplies*, U. S. Geological Survey, Washington, 1958 and R. L. Nace, personal communication, 1959.

Table 2. Distribution of World's Estimated Supply of Water (11)[a]

Location	Surface area, 1000 sq mi	Volume of water, 1000 cu mi	Total water, %
world area, total	197,000		
land area	57,500		
surface water, continental			
polar ice and glaciers	6900	7300	2.24
fresh-water lakes	330	30	0.009
saline lakes, inland seas	270	25	0.008
stream channels, average		0.280	0.0001
total surface water	7500	7350+	2.26
subsurface water, continental			
soil root zone	50,000	6	0.0018
ground water above 2640-ft depth		1000	0.306
ground water, 2640 to 13,200-ft depth		1000	0.306
total subsurface water	50,000	2000	0.61
total water on land		9350+	2.87
oceans	139,000	317,000	97.13
atmospheric moisture		3.1	0.0001
total world supply of water		326,000	100

[a] All quantities rounded-off.

to be discharged into a stream, but rather for the resultant quality of the stream itself. Streams are classified as Class AA (for drinking, culinary or food processing and other such uses); Class A (same as Class AA, except in some details standards are lower); Class B (bathing); Class C (fishing); Class D (agriculture, industrial cooling and processing).

Table 3. U.S. Fresh-Water Reserves (9,10)

Source	Amount, 10^6 acre-ft	Reserve life, yr[a]
atmosphere	176	0.6
soil moisture	635	2.2
rivers	45	0.1
reservoirs	365	1.3
lakes	13,000	46
ground water	47,500	168
total	61,721	215

[a] To support entire national water withdrawal

Table 4. Quality Standards for Class AA Waters (13)

Items	Specifications
floating solids; settleable solids; oil; sludge deposits; taste- or odor-producing substances	none attributable to sewage, industrial wastes or other wastes
sewage or wastes effluents	none that are not effectively disinfected
pH	to range between 6.5 and 8.5
dissolved oxygen	for trout waters, not less than 5.0 ppm; for other waters, not less than 4.0 ppm
toxic wastes, deleterious substances, colored or other wastes or heated liquids	none alone or in combination with other substances or wastes in sufficient amounts or at such temperatures as to be injurious to fish life; make the water unsafe or unsuitable as a source of water supply for drinking, food processing purposes; or impair the waters for any other best usage as determined for the specific waters that are assigned to this class
ammonia or ammonium compounds	not greater than 2.0 ppm at pH 8.0 or higher
cyanide	not greater than 0.1 ppm
ferro- or ferricyanide	not greater than 0.4 ppm $(Fe(CN)_6)$
copper	not greater than 0.2 ppm
zinc	not greater than 0.3 ppm
cadmium	not greater than 0.3 ppm

Quality standards for Class AA waters are given in Table 4 (13), and may be taken as representative for waters intended for potable use.

U.S. Geological Survey Water-Supply Papers give chemical analyses of water used by 73 irrigation network stations in operation west of the Mississippi River (14). Similar data are given for eastern U.S. locations reporting on 141 stations on 88 streams (15).

The U.S. Public Health Service has established standards for drinking water (16). Specifications as to sampling and tolerances of injurious substances are given. In summary, the principal contaminants are given in Table 5 with their limits of permissible concentrations.

When fluoride is naturally present in drinking water, the concentration should not average more than the appropriate upper limit in Table 6. Presence of fluoride in average concentrations greater than two times the optimum values in Table 6 constitutes grounds for rejection of the supply. Where fluoridation (supplementation of

Table 5. Concentration Limits for the Principal Contaminants of Drinking Water

Substance	Concentration, mg/l
alkylbenzenesulfonate (ABS)	0.5
arsenic	0.01
chloride	250.0
copper	1.0
carbon content of chloroform extract (CCE)	0.2
cyanide	0.01
iron	0.3
manganese	0.05
nitrate	45.0
phenols	0.001
sulfate	250.0
total dissolved solids	500.0
zinc	5.0
Grounds for Rejection	
arsenic	0.05
barium	1.0
cadmium	0.01
chromium, hexavalent	0.05
cyanide	0.2
lead	0.05
selenium	0.01
silver	0.05

Table 6. Recommended Control Limits of Fluoride Concentrations

Annual average of max daily air temperatures[a]	Lower, mg/l	Optimum, mg/l	Upper, mg/l
50.0–53.7	0.9	1.2	1.7
53.8–58.3	0.8	1.1	1.5
58.4–63.8	0.8	1.0	1.3
63.9–70.6	0.7	0.9	1.2
70.7–79.2	0.7	0.8	1.0
79.3–90.5	0.6	0.7	0.8

[a] The recommended fluoride concentration depends on the temperature as the temperature affects the amount of water ingested.

fluoride in drinking water) is practiced, the average fluoride concentration must be kept within the upper and lower control limits in Table 6.

Utilization of Water

Figure 4 shows the overall U.S. water demand projected to 1980.

In many of the process industries water is used for suspending, washing, condensing, or cooling operations. Water requirements may be from 20,000 gal to 2,000,000 gal/ton of finished product. The aluminum companies report that 320,000 gal are required for each ton of metal produced (17). It takes 770 gal to refine a barrel of petroleum, 200,000 gal to make a ton of viscose rayon, and 600,000 gal to make a ton of synthetic rubber. Table 7 shows the water intake, as of 1960 and projected into the future, for some water-using industries in the U.S. (18).

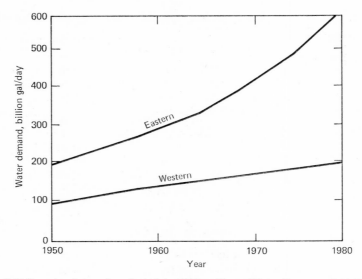

Fig. 4. Overall U.S. water demand projected to 1980. (From *Water Resource Activities in the U.S.*, Res. 48, Committee Print No. 1, 86 Congress, Aug. 1959.)

A detailed tabulation of water uses in manufacturing is given in a U.S. Department of Commerce publication (19). Koenig gives a tabulation of water usages by categories (20).

One of the big potential uses for water is for cooling water in the chemical industry and for nuclear reactors. As of 1969, the chemical process industries are expected to spend $1.8 billion for water cooling between 1969 and 1974 (21). Some 50 trillion (1 trillion = 10^{12}) gallons were used for cooling in 1964, and probably 60 trillion in 1969. However this is liable to cause thermal pollution, which is defined as the discharge of waste heat into natural waters with deleterious effects on aquatic life. Cooling ponds and evaporative cooling towers are the most widely used methods of dissipating heat.

For a future nuclear plant of 1000-megawatt size, a pond of 2000 surface acres may be required. For the 514-MW Vermont Yankee plant at Vernon, Vt., cooling

Table 7. Water Requirements of the Principal Water-Using Industries, 1954–2000[a]

Industry	Year							
	1954		1959		1980		2000	
	Amount	Percent	Amount	Percent	Amount	Percent	Amount	Percent
iron and steel	2812.0	36.6	3500.0	35.7	7123.7	26.8	12,826.5	20.8
chemical	2380.0	30.9	3210.0	32.7	10,800.0	40.7	29,700.0	48.2
pulp and paper	1607.0	20.9	2140.0	21.8	6893.0	26.0	16,173.0	26.3
food	590.0	7.7	630.0	6.4	1000.0	3.8	1500.0	2.4
aluminum	263.0	3.4	278.0	2.9	605.2	2.3	1138.8	1.9
copper	42.0	0.5	49.0	0.5	114.0	0.4	247.0	0.4
total	7694.0	100	9807.3	100	26,535.8	100	61,585.3	100
index	100		128.0		345.0		800.0	
daily average	21.1		26.9		72.7		168.7	

[a] All figures in billions of gallons

tower operation and amortization were estimated at \$800,000/yr for one design, and \$2,100,000 for another.

One aspect of water recovery that is interesting (but apparently at this time of limited application) is that of treatment and utilization of sewage water (22). Sewage effluent (treated), with or without further industrial treatment is used for industrial coolant or for long-distance transporting of solids, such as powdered coal. Bethlehem Steel has used effluent since 1942. Potable water, also, may be produced from sewage effluent by coagulating, settling, filtering, passing through activated carbon, and disinfecting. At Lake Tahoe and at Windhoek, South West Africa, potable waters are made from sewage effluent.

Water is also used for hydroelectric power generation. The potential hydroelectric power of the world is roughly estimated at $2^3/_4$ billion kilowatts. Of this, about 181 million kilowatts are developed. The U.S. has 1/5 of the world's developed hydroelectric power (35% of the country's potential). Canada, Australia, and Europe make up most of the balance (23).

Irrigation is the second largest user (after industry) of water. Various waters may be used for this purpose, including treated sewage effluents, providing that the waters meet standards regarding, principally, mineral content. The following gives the properties of waters suitable for irrigating most plants under most conditions (24):

(*1*) boron—generally <0.5 mg/l
(*2*) chlorides— <2 to 5.5 milliequivalents/l
(*3*) sulfates— <4 to 10 milliequivalents/l
(*4*) specific conductivity $\times 10^6$ at 25°C—$\leq$500 to 1000 ohm^{-1}cm^{-1} $\times 10^{-6}$
(*5*) total salts— $\leq$700 mg/l
(*6*) sodium—ratio [Na] $\times$ 100/([Na] + [Ca] + [Mg] + [K]) (conc in milliequivalents/l) 30% or less, to 60%.

Weather Modification

The term weather modification has broad meaning and covers the increasing of precipitation (cloud seeding), the suppression of hail, the dissipation of fog, the modification of storms and hurricanes, the suppression of lightning, and other effects.

For example, during 1967 approximately 111,000 sq mi in the U.S. were treated experimentally or commercially in 23 states in various weather modification activities by 44 operators (25). Tabulation of these activities is shown in Table 8. Of these treatments, the major effort over some years has been directed toward influencing the control of the amounts and the distribution of rainfall through cloud seeding.

Cloud Seeding. The earliest scientific trials of cloud seeding date back to 1839 (26). J. P. Espy at that time investigated the production of rain by fire, and as a result made certain conclusions concerning the role of heat, dewpoint, and other factors. Another concept, in 1871 (27), was based on the possible production of rain by use of explosives. In 1891 Congress appropriated \$10,000 to carry on rain-production trials of this type in Texas (unsuccessful). Beginning in 1921, many proposals were made in an effort to make rain, including the use of dust, electrical methods, and explosives. The first successful attempt was achieved by A. W. Veraart in 1930 using dry ice (solid carbon dioxide) and applied from an airplane. Later, both dry ice and water ice were used (28).

Table 8. Weather Modification Field Operators' Reports for Fiscal 1967

Modification	Area treated, sq mi	Projects, number	States, number	Operators, number
rain augmentation	40,771	31	16	19
snowpack augmentation, to provide increased runoff	21,250	10	4	6
hail suppression	20,556	4	3	4
fog dissipation	118	15	13	15
experimental	25,874	14	9	11
lightning suppression	314	1	1	1
atmospheric research	1000	3	2	2
severe storm modification	1500	1	1	1
total	111,383	79	23	44

Later work by Schaefer (29), Langmuir, and Vonnegut began in 1946. The date is significant because it marks the verification of the process of atmospheric ice-crystal formation (in a cold box experiment using dry ice), as well as the first scientific field experiments on a large scale for modification of supercooled clouds. Since 1946, cloud modification by artificial seeding with chemicals and other agents has been tested extensively in the U.S. and in other countries. More than $10 million and much time has been devoted to these studies, including thousands of field tests in more than a score of countries. These tests have served to place weather modification and rain-making on a scientific basis. The background of cloud seeding is given in a 1957 report (30), which points up the relative activity of silver iodide as compared with numerous other substances.

A 1960 weather modification summary report (31) lists several techniques for cloud seeding; they are (1) seeding from aircraft, (2) seeding from ground generators, and (3) seeding by ballons and rockets. Most of the attempts have been made from ground generators, using silver iodide.

In general, silver iodide is the favorite seeding material of commercial operators, and it appears to have produced significant increases in precipitation when used intelligently by qualified personnel. However, no one has yet devised a scheme for producing rainfall where normal moisture is lacking in the atmosphere.

The silver iodide is applied by ground operated nuclei generators, and particles of about micron size (10^{-6} cm) have been found to be more effective than most other sizes. An evaluation of 14 operational silver iodide seeding projects in the eastern U.S. indicated rainfall increases averaging about 10–20% in the nominal target areas (32).

In 1965, the support of U.S. government agencies to weather modification amounted to about $5 million, of which about 30% was for academic research.

The theory of rain formation is presented in a 1966 report (32). It has been shown that small droplets of uniform size on colliding in an electrical field will coalesce in the majority of cases. It is in this manner that small droplets coalesce to form rain.

The alleviation of droughts and the supplying of water systems at a higher-than-natural level have been the objectives in essentially all of the many attempted experiments and field operations in rain-making over the past 20 or so years.

Latest reports (1960–69) in general indicate favorable but not conclusive results, with possible rainfall increases of 15–20%. Similar tests to those described have been conducted in Switzerland, African Congo Basin, East Africa, U.S.S.R., United Kingdom, Canada, Argentina, France, India, Tunisia, etc.

There are legal aspects to weather modification. In reviewing the laws there has been an attempt to define the rights and duties of those engaged in changing the weather, as well as the rights of those affected by such changes. It can be understood, for example, that inducing rain at one locality may preclude rain at other localities. Another possibility is that misoperation could do flood or hail damage in adjacent areas.

Hail Suppression. Hail suppression is another phase of weather modification, with an objective of saving some $200 million of agricultural crop damage each year in the U.S. The approach is similar to that used in rain-making, ie, to treat the entire storm area with massive doses of silver iodide, from ground to air. The intent is to augment the freezing nuclei by heavy seeding so that only small ice particles will form. The Russians have used artillery shells fired from the ground, each containing 100 grams of silver iodide. The Russians are the most enthusiastic, and have large programs in the Caucasus and Transcaucasus. In Argentina results are somewhat inconclusive showing decreases of hail damage of 10–70%, but in some cases, an increase of 100%. The Swiss project (1960, end of 7 years' operation) indicated a 68% greater frequency of hail-days (33). Thus, results to date have not been completely beneficial.

Fog Dissipation. Fog dissipation is another type of weather modification. Fog may be thought of as a cloud on the ground, where the water droplets are small (25–50 μ in diameter) and are sufficiently dispersed so that they cannot touch and coalesce. Fog scatters light and reduces visibility. The application of heat from oil or gasoline burners has been effective in its dispersal, but is far from ideal because of cost, and the turbulence of air created. Other means include the application of high-frequency sound, electrical charge injection, chemical sprays, or large fans to blow hot air into the fog.

Although little practical progress has been made, a method of seeding warm fog with common salt in concentrations of 2–5 mg/m³ has increased visibility by a factor of 3 to 20 in some cases. The technique may be practical for clearing the lower 300 ft of fog at airports.

Seepage and Evaporation Control

A surface reservoir may lose water from the top and from the bottom. These are termed *evaporation* losses and *seepage* losses, respectively.

In order to determine evaporation losses and seepage losses from reservoirs, and to distinguish one from another, a mathematical approach has been used. The procedure is based on U.S. Geological Survey correlations (36). The method involves the plotting of the drop in level of a reservoir per unit time vs. a factor derived from climatological observations, which is $U^{3/4} (E_w - E_a)$. U represents the wind velocity in miles per hour; E_w is the vapor pressure of water in inches of mercury corresponding to the surface temperature of the water; and E_a is the vapor pressure of air corresponding to the dewpoint. Examples of the use of this formula are given in calculating the seepage loss and the theoretical evaporation loss from a 50-acre reservoir (37).

Seepage Control. Seepage is important, especially in reservoirs or farm ponds, etc, which have natural soil bottoms. Attempts to halt or to alleviate seepage losses have used several approaches (34,35a,35b). Polyethylene sheets have been used to line the bottom of small reservoirs (up to several acres). Alternatively, filler material, such as bentonite, has been used to expand in the leakage fissures and thereby seal the reservoir. One product, an oil-soluble resinous polymer, in a diesel-oil carrier, has been recommended for this purpose, and reductions of seepage of up to 80–95% are claimed. Another product on the market forms a wax membrane in the soil.

The actual loss from a pond or lake due to seepage (alone) may be as little as 0.05 in./24 hr, or as much as 6.0 in./day in some reservoirs when sand or gravel bottoms or sides are involved. Ponds and lakes when first built have higher seepage, but the amount declines after the first year, due to natural filling of cracks and porous openings.

Reservoir Evaporation Control. The conserving of evaporation losses from reservoirs (farm ponds to large-capacity storage reservoirs) by use of floating monomolecular films is of large potential. Although great concentration, almost worldwide, has been placed on this phase of water conservation, other means of solving this problem exist. Freese (38) has summarized other methods as follows:

> (1) construction of reservoirs with maximum average depth
> (2) concentration of water into single reservoirs
> (3) elimination of shallow-water areas
> (4) elimination of water growths
> (5) storing of water in ground-water reservoirs
> (6) roofs and floating covers
> (7) windbreaks

As an example of water losses through evaporation, it is stated (39) that Mead Lake (146,000 surface acres), on the Colorado River, loses 830,000 acre-ft/yr. From all reservoirs combined, evaporation losses in years ahead will average 1.9 million acre-ft/yr in the U.S. An acre-ft of water will support 5 people for a year in normal conditions in the U.S. Thus, the above amount of water would supply the requirements of 9.5 million people.

The actual vertical loss from ponds and lakes will vary according to the climatic conditions, and losses may be as high as 0.7 in./24 hr in hot weather, with low humidity and strong winds. Figure 5 shows the annual evaporation losses from locations in the western U.S. (40). Most ocean areas evaporate 1500–3000 mm (60–120 in.) per year. Most of the rainfall originates from the evaporation over the ocean, thus, artificial retardation of evaporation from inland water bodies has little influence on rainfall.

It is to be pointed out that water saved by any means of evaporation control is of the purest quality, being, in essence, distilled water. The residual water has less dissolved salts, and some marginal brackish waters become potable or usable. Another factor is that compounded water already in a reservoir and retained through evaporation control represents no pumping costs. The cost of pumping, if needed, might amount to 5–10¢/1000 gal.

The principle of using monomolecular films was first examined from the theoretical viewpoint by Hedestrand (41), Rideal (42), and Langmuir (43). Dressler and Johanson (40) point out the early history of monolayer film research. A complete review is presented by LaMer (44).

Briefly, the technique consists of putting a film of organic chemical material on the surface of water and thereby retarding the evaporation losses. The rate of evaporation from a water surface is governed by four factors: wind velocity, its humidity, its temperature, and the water surface temperature.

Early laboratory studies indicated that the normal saturated alcohols (C_{16} and C_{18}) were efficient for this purpose. The practical problem of applying the alcohols (solid) to the water surface in the minute amounts required, was frustrating. Attempts to apply the fatty alcohol from moving boats, from caged floats, in dispensers, as a solution, etc, were made.

The application of the basic principle of conserving water of evaporation on outdoor reservoirs was initiated by Australian workers (45) in 1955, using floating cages containing hexadecanol. Although this method of application eventually proved unsuitable, the activity brought worldwide attention to the possibilities.

Dressler points out eight conditions that should be met in an ideal method for applying films to a reservoir surface (37). The method most highly developed is that of making a dispersion of powdered alcohol in water and pumping the slurry through orificed lines along the water's edge. The method, called the suspension process, is

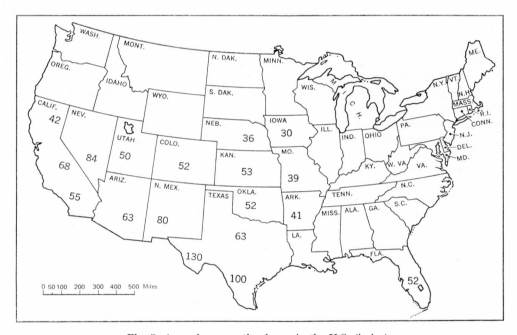

Fig. 5. Annual evaporation losses in the U.S. (in in.).

described in a 1964 publication (46,47). This approach has been tried on several U.S. Western reservoirs, ie, Lake Hefner, Okla. (2500 acres), Elephant Butte Reservoir, New Mexico, (36,000 acres), and on a portion of Lake Mead, Nevada, with some measure of success. The government of Israel applied films to two 200-acre reservoirs in 1967, using pumps and lines as proposed by the suspension process, with good results.

With some of the commercially available alcohols, such as hexadecanol ($C_{16}H_{33}OH$) or octadecanol ($C_{18}H_{37}OH$), about 50–55% savings of water can be expected under closely-controlled laboratory conditions. In outdoor trials, with winds and waves, 40% is typical. However, recent laboratory research, using mixtures of odd and even chain alcohols, has achieved 85% reduction in evaporation (48,49,50). Likewise, the use of ethylene oxide-fatty alcohol adducts and other similar compounds has been proposed to achieve 85–90% efficiencies (51).

In the United States, there are some 17 western states where evaporation control would be of special benefit. A Geological Survey paper lists 1300 major reservoirs and lakes in the U.S. totaling 11 million surface acres (52).

The importance of evaporation control is summarized in a Department of Interior release (53) which states that evaporation losses from large lakes and reservoirs in the U.S. (500 or more surface acres) in the 17 western-most states alone, exceed 14 million acre-ft each year, enough to serve the domestic needs of 84 million people.

The principle of using monomolecular films has been applied to the growing of flooded crops. In paddy rice culture, it is found that due to retarded evaporation, when films cover the water surface, the flood water temperature is 5–7°C higher on a clear day and 2–3°C higher on a cloudy day, as compared with free water. This procedure results in an accelerated growth of rice (54).

Monomolecular films have been suggested for use in preventing shoreward drift of oil slicks on the ocean. Also application has been studied as a means of controlling the magnitude of hurricanes by coating ocean surfaces at the time of storms thus preventing evaporative heat from accumulating in the rising vapors, and preventing a rush of tangential incoming air.

The question frequently has been asked: How do the films work on saline or brackish waters? This question is answered in a report on research by Cheves (55). Solutions of salts, $NaCl$, $CaCl_2$, $AlCl_3$, and $Al_2(SO_4)_3$, in concentrations of 0–20 wt % were tested. The effectiveness is less than with pure water, but is sufficient to justify applications of films on saline or brackish water bodies.

The economics of evaporation control are presented in a 1958 publication (40), and also in a 1969 article (49). The cost of saving water is of the magnitude of 1.5–2¢/ 1000 gal. This compares extremely favorably with saline-water recovery methods with costs ranging, as of 1970, from at least 25–50¢/1000 gal.

Reiser (56) presents a basic design for equipment for applying suspensions by the suspension (Dressler) process. He also estimates the cost of equipment and costs of saving water. Assuming a lake of circular shape with savings of 35 in./yr (ie, 70 in./yr evaporation with 50% savings) he obtains the figures presented in Table 9.

Table 9. Estimated Cost of Water Saved by the Suspension Process

Size of lake, acres	Cost of Water saved, $/acre-ft[a]
1000	10.00
2500	6.00
10,000	2.65
35,000	1.30

[a] $3.25/acre-ft = 1.0¢/1000 gal

Reiser states that optimistic estimates for large desalting plants show a cost of $33/acre-ft (10¢/1000 gal) at the ocean shore. (See also under Desalination, below.)

There are many areas of the world where evaporation control might economically be used. Theoretical studies or outdoor trials have been performed in Australia, British East Africa, South Africa, India, Argentina, United States, Israel, etc. Other regions where the process might be used to advantage would include northern Africa, Egypt, Saudi Arabia, Spain, Peru, Mexico, etc. The economic criterion might be arid countries having reservoirs and with annual evaporation rates in excess of 40 in./yr.

Bibliography

1. G. A. Nikolaieff, *The Water Crisis*, H. W. Wilson Co., New York, 1967.
2. M. W. Talbot, *Range Watering Places in the Southwest, Dept. of Agriculture Bull. No. 1358*, U.S. Govt. Printing Office, Washington, D. C., 1926.
3. W. Dittmar, "Challenger Reports," *Phys. and Chem.*, **1**, 1–251 (1884).
4. B. Espenshade, Jr., Goode's World Atlas, Rand McNally, 1957, pp. 14–15.
5. The Times Atlas of the World, Comprehensive Edition, Houghton Mifflin Co., 1967, plate 4.
6. W. A. Hardenbergh and E. B. Rodie, *Water Supply and Waste Disposal*, International Textbook Co., Scranton, Pa., 1966.
7. C. Hodge and P. C. Duisberg, *Aridity and Man, Publ. No. 74*, Am. Assoc. Adv. Sci., Washington, D.C., 1963.
8. J. Hirschleifer, J. C. Dehaven and J. W. Millimin, *Water Supply*, University of Chicago Press, 1966.
9. K. A. Mackichan, *Estimated Use of Water in the U.S., 1955, Geol. Survey Circ. No. 398*, Washington, D.C., 1957.
10. E. A. Ackerman and G. O. G. Lof, *Technology in American Water Development*, Johns Hopkins Press, Baltimore, 1958.
11. R. D. Hockinsmith, *Water and Agriculture*, Horn-Schafer Co., Baltimore, 1960.
12. L. Koenig, "Economics of Ground Water Utilization," *J. Am. Water Works Assoc.* **55** (1) (1963).
13. Lawrence K. Cecil, "Water Reuse and Disposal," *Chem. Eng.*, 92–104 (1969).
14. S. K. Love et al., *Quality of Surface Waters for Irrigation, Western States, U. S. Geol. Water-Supply Paper*, 1962.
15. L. B. Leopold, K. S. Love, *Quality of Surface Waters of the U.S., U.S. Geol. Water-Supply Paper*, 1968.
16. U.S. Public Health Service, *Drinking Water Standards*, **1962**, *Dept. of Health, Education and Welfare*, Washington, D.C., 1962.
17. Anonymous, *Chem. Week*, June 29, 1957.
18. *Water Resources Activities in the U.S.*, S. Res. 48, 86 Congress, *Committee Print No. 8*, U.S. Govt. Printing Office, Washington, D.C., 1960, p. 2.
19. *Census of Manufacturers–Water Use in Manufacturing*, Dept. of Commerce, Bur. of Census, Washington, D.C., 1963.
20. L. Koenig, *Development and Exploitation of Water Resources*, presentation reprint, Am. Petrol. Inst., San Antonio, 1960.
21. Anonymous, "Cooling Will Be Costly," *Chem. Week*, March 15, 1969.
22. T. H. Y. Tebatt, "Reclaimed Water for Industry," *Effluent Water Treatment J.* **7** (11), 587–589, 591–593 (1967).
23. *The World Book Encyclopedia*, Field Enterprises Educational Corp., 1969 Vol. 20, p. 110.
24. J. E. McKee and H. W. Wolf, *Water Quality Criteria*, 2nd ed., State Water Quality Control Board (Cal.) *Publ. 3A*, 1963.
25. *Weather Modification*, 9th Annual Report, Natl. Sci. Foundation, U.S. Govt. Printing Office, Washington, D.C., 1967, p. 75.
26. *Weather Modification and Control., U.S. Senate Rept. No. 1139*, U.S. Govt. Printing Office, Washington, D.C., 1966, pp. 11–15.
27. E. Powers, *War and the Weather, or the Production of Rain*, S. C. Griggs and Co., Chicago, 1871.
28. A. M. Borovikov and A. Khrgian et al., *Cloud Physics*, Leningrad, p. 281 (Eng. trans 1963).
29. V. J. Schaefer, *Sci.* **104**, 457 (1946).
30. I. P. Krick, *The Theory and Practice of Weather Modification*, Water for Texas 3rd Annual Conf., Texas A. and M., 1957.
31. *Water Resources Activities in the U.S., Weather Modification Committee Rept. No. 22*, Select Committee and Natural Resources, U.S. Senate, U.S. Govt. Printing Office, Washington, D.C., 1960.
32. *Weather and Climate Modification, Problems and Prospects, Natural Resource Council, Publ. No. 1350*, Washington, D.C., 1966, Vol. 1.
33. *Weather and Climate Modification, Problems and Prospects, Natural Resource Council, Publ. No. 1350*, Washington, D.C., 1966, Vol. 2.
34. Anonymous, *Chem. Week*, Nov. 4, 1967.

35a. Seepage Control, Inc., P.O. Box 4425, Phoenix, Ariz., 85030, booklet.

35b. U.S. Pat. 3,108,441 (California Research Corp.).

36. W. B. Langbein, C. H. Haines, and R. C. Culler, *Hydrology of Stockwater Reservoirs in Arizona*, *U.S. Geol. Survey Circ. No. 110*, 1951.

37. R. G. Dressler, "*An Engineering Approach to Reservoir Evaporation Control*," in V. K. LaMer, *Retardation of Evaporation by Monolayers*, Academic Press, Inc., New York, 1962.

38. S. W. Freese, 1st International Conf. on Reservoir Evaporation Control, S.W. Research Inst., 1956, p. 45.

39. Anonymous, *U.S. News and World Report*, March 31, 1969, p. 86.

40. R. G. Dressler and A. G. Johanson, *Chem. Eng. Prog.* **54** (1) (1958).

41. G. Hedestrand, *J. Phys. Chem.* **29**, 1244 (1924).

42. E. K. Rideal, *J. Phys. Chem.* **29**, 1585 (1925).

43. I. Langmuir and D. B. Langmuir, *J. Phys. Chem.* **31**, 1719 (1927).

44. V. K. LaMer, *Retardation of Evaporation by Monolayers*, Academic Press, Inc., New York, 1962.

45. W. W. Mansfield, *Nature* **175**, 247 (1955).

46. R. G. Dressler, *Ind. Eng. Chem.* **56**, 36–39 (1964).

47. U.S. Pat. 2,903,330 (Sept, 3, 1959) to R. G. Dressler.

48. E. L. Foulds and R. G. Dressler, *Ind. Eng. Chem., Prod. Res. Dev.* **7**, 75 (1968).

49. A. J. Simko and R. G. Dressler, *Ind. Eng. Chem., Prod. Res. Develop.* **8** (4), 000 (1969).

50. U.S. Pat. 3,450,488 (June, 1969) to R. G. Dressler.

51. Y. Mihara, *Practical Uses of Newly Created Evaporation Retardants for Evaporation Control and Transpiration Control*, Natl. Inst. Agric. Sci., Tokyo, presented at Symp. Water Evap. Control, Poona, India, 1962.

52. N. O. Thomas and G. E. Harbeck, Jr., *Reservoirs in the U.S.*, U.S. Geol. Survey Water-Supply Paper 1360-A, U.S. Govt. Printing Office, Washington, D.C., 1956.

53. News release, May 27, 1969, U.S. Dept. of Interior, Bureau of Reclamation.

54. Y. Mihara, *The Microclimate of Paddy Rice Culture and the Artificial Improvement of the Temperature Factor*, Natl. Inst. Agric. Sci., Div. Meterol., Tokyo, 1961.

55. F. A. Cheves, R. G. Dressler, and W. C. McGavock, *Ind. Eng. Chem., Prod. Res. Develop.* **4**, 206 (1965).

56. G. O. Reiser, *Ind. Eng. Chem. Prod. Res. Develop.* **8** (1) (1969).

R. G. Dressler
Trinity University

WATER PROPERTIES

Liquid water is easily the most extraordinary substance known. It is anomalous in all of its physical-chemical properties. To take a well known example, the melting and boiling points of the group VI A hydrides decrease in an orderly progression with decreasing molecular weight (Fig. 1)—all, except water. Water, with the lowest molecular weight has by far the highest melting point and boiling point.

Our example is a good one for it illustrates two extremely important points. Water is a necessary condition for life. Life began in the oceans of Earth and evolved there. As a consequence the peculiarities of water have been utilized by and incorporated into the fabric and workings of the life processes in the most delicate and intricate way. Water, to a considerable extent, determines not only the structural configuration, but the biological function of biomacromolecules. On the basis of the trends shown in Figure 1, we would expect water to be a gas at room temperature. If such were the case, the Earth would have no hydrosphere and life would be impossible. Can anything be more important?

The second value of our example is that it points rather directly to the cause of water's peculiar behavior. The anomalously high transition temperatures clearly indicate that the water molecules are loath to separate from one another. Some force is holding them together, preventing their escape. That force, the hydrogen bond, has now been well characterized and it is by no means unique to water. Other liquids such as hydrogen fluoride, ammonia, and the lower alcohols also exhibit hydrogen-bonding, but in none of these do the forces of aggregation give rise to such an astounding complex structural polymerization as in the case of water.

Pure water is a polymer. Properly its formula is $(H_2O)_n$ where the value of n is presently unknown. It is the subject of intense controversy and, to make matters

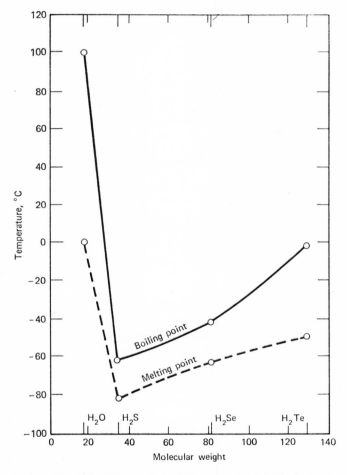

Fig. 1. Transition temperature of the group VI A hydrides (15).

worse, the value of n depends on environmental conditions such as temperature and pressure. The value of n even differs depending on where the water sample is averaged, being different near interfaces than at some great distance from them, and, in the case of aqueous solutions, different near dissolved species, both ionic and nonpolar solutes.

Liquid water is not one—it is several, possibly many—species. In an attempt to try to give an exceedingly complex situation the appearance of simplicity, the following, somewhat arbitrary, system of nomenclature is presented:

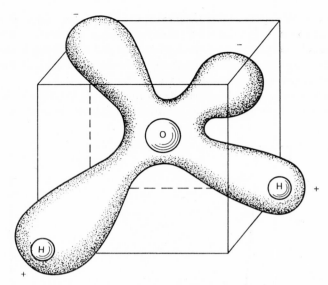

Fig. 2. Electronic cloud of the water molecule (17).

In bulk, pure liquid

    water-i   the monomer

    water-ii   small polymers $(H_2O)_n$ where $n = 2$ to 4

    water-iii   larger polymers where $n > 4$; (a) randomly hydrogen-bonded; (b) hydrogen-bonded with non-ice-I-like, at least near-neighbor, order

    water-iv   ice-I-like

Near solutes

    water-v   electrostricted water of hydration

    water-vi   enforced water structures near ions (except water-v)

    water-vii   broken water structure near ions

    water-viii   "icebergs" or clathrate structures near nonpolar solutes or nonpolar segments of macromolecules

Near interfaces

    water-ix   near neutral and nonpolar interfaces

    water-x   near silica

    water-xi   absorbed or chemically-bound water

Subsequent research may show that some of these species (i, ii, and iv) are non-existent in the liquid, that some are synonymous (iii and iv; ii and vi; viii, ix, and x) and some can be further subdivided (xi).

In order to avoid the implication that these forms represent phases in the thermodynamic sense, in contrast to the case of the ices, lower rather than upper case Roman numerals have been used.

Although not repeated in the above scheme, a given water species may occur in more than one of the three location categories: water-iv, for example, may be found in bulk solution, near solutes, and near interfaces; waters-v, -vi, and -vii will surround charge sites on a surface as well as ions in solution; and water-x may exist in bulk solution as well as near silica surfaces.

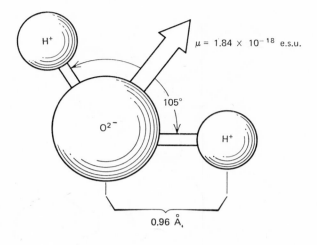

$\mu = 1.84 \times 10^{-18}$ e.s.u.

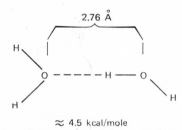

Fig. 3. Schematic summary of the structural features of the water molecule and hydrogen bond (17).

The polymeric nature of water has been recognized since the 19th century, and in the intervening span of time a great many theories of the nature of the structure of liquid water have been advanced. Early theories, reviewed by Chadwell (1), are largely in terms of the mixture of the monomer (water-i) and small polymers (water-ii), a concept once discarded which may now be in process of partial revival (2). The proliferation of modern theories began some years ago with the now-classic paper by Bernal and Fowler (3). But before getting lost in this labyrinthine topic attention may be turned for a moment to the subject of the structure of the individual water molecule and of ice.

Structure of the Water Molecule. The aspect presented by the electronic cloud of the water molecule is that of a sort of truncated, distorted jack (Fig. 2). The oxygen atom occupies the center of the not-quite cube; the hydrogens, opposite corners of one face. Directed towards the corners of the opposite face, most removed from the hydrogens, are two arms of negative electrification—possibly the most important chemical feature in all of creation—for these arms are responsible for the hydrogen bonding which in turn makes the phenomenon of life possible. Other structural features of the molecule and of the hydrogen bond are shown schematically in Figure 3.

The Structure of Ice. A number of theories of the structures of liquid water take as their point of departure the structure of ice. In the lattice of ordinary ice, ice-I, a given water molecule is H-bonded to four tetrahedrally arranged neighbors. The hydrogen in the H-bond between the oxygen centers is distributed between two loci

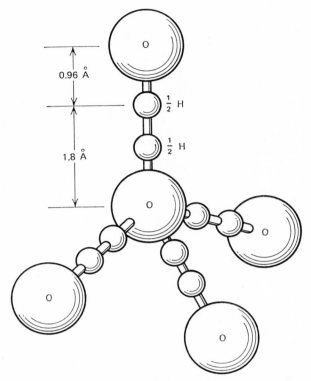

Fig. 4. Positions of the hydrogens in the ice-I lattice (17).

(Fig. 4) (4).   The oxygen centers in the ice lattice are arranged in the same way as the silicon atoms in tridymite, more or less in a layered network of puckered hexagonal rings (Fig. 5).   The interoxygen distance in ice is 2.76Å.

Although simple by comparison with the liquid state, ice-I is a fascinating substance in its own right.   Table 1 summarizes some of its properties.   Despite careful study for some years now, it has not been completely stripped of its secrets, in particular, the details of the mechanism of transport processes within it (5), and of the mysterious fluid film which appears to cover its surface (6).

The phase diagram for water (Fig. 6) is characteristically complex, with no less than eight different solid forms being reported.   The structure of ice-VI is particularly magnificent (Fig. 7) (7).   It is a self-clathrate consisting of two elaborately interpenetrating but nowhere interconnecting lattices.   The structure of the high-pressure ices may prove to be of a wider interest than hitherto suspected for the suggestions have been made that they may also be encountered at 1 atm in the bulk liquid (8) and in water-x (40).

Structure of Pure, Bulk, Liquid Water. So many theories of the structure of liquid water have been advanced (Table 2) and so often have they been reviewed in detail (10–17) that no more than a few brief remarks will be presented here.   Customarily the theories have been divided into "mixture" models and "continuum" models although the exact distinction between these two classifications has not always been entirely clear.   Walrafen's careful experimental demonstration (18) that the Raman spectrum exhibits an isobestic point, characteristic of mixtures, and indicating a *step-*

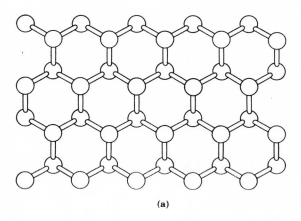

(a)

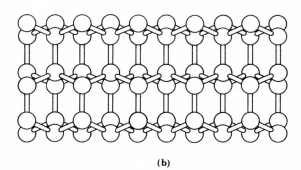

(b)

Fig. 5. Schematic representation of the structure of ice-I (44).

wise rupture of hydrogen bonds, has dealt a body blow to the latter model. Yet **it is** probable that some time will pass before its proponents admit defeat, particularly inasmuch as they shortly before tasted apparent victory (19–21).

The difficulty with the proliferation of liquid water models is not that the theories fail, but that they all work well. There would seem to be many points of truth in each of them. There is much to be said for an "everything" model (Table 2) looking upon liquid water as an exceedingly complex porridge containing "monomers" (water-i) existing in a broad spectrum of degrees of freedom; small polymers (water-ii) (2), including chains and perhaps even rings (22); and larger H-bonded regions, perhaps with a modified ice-I structure, or the structure of one of the high-pressure ice-forms, or randomly arranged (15). Such an inexact, all-embracing picture (Fig. 8), however, is not very satisfying to the scientific penchant for neat categories, and for this reason until a clearly superior model is forthcoming, the "flickering cluster" theory of liquid water structure advanced by Frank and Wen (23) and developed in a quantitative manner by Nemethy and Scheraga (10) will continue to enjoy popularity among scientists. This model visualizes liquid water as consisting of a mixture of hydrogen-bonded polymers or clusters (see Fig. 8) and more or less free or monomeric water. At room temperature an average cluster contains about 40 water molecules. Small polymers tend to be rare because of the cooperative nature of the hydrogen-bonding process. The clusters are in a highly dynamic state, "flickering" or constantly forming

Table 1. Selected Properties of Water Vapor and Ice-I (9)

Property[a]	Value
water vapor	
molecular weight	18.0154
heat of formation at 100°C, kJ/mole	242.49
viscosity at 20°C, mP	96.0
sound velocity at 100°C, m/sec	405
diffusion coefficient in air at 100°C, cm²/sec	0.380
specific volume at 100°C, cm³/g	1729.6
specific heat at 100°C, J/(g) (°C)	2.078
thermal conductivity at 110°C, watt/(cm²) (°C/cm)	2.44×10^{-4}
ice-I	
heat of formation at 0°C, kJ/mole	292.72
lattice parameters, hexagonal, Å	
a	4.535
c	7.14
Young's modulus of elasticity at −10°C, kg/mm²	967
density at 0°C, g/ml	0.9168
coefficient of linear thermal expansion at 0°C, °C⁻¹	5.89×10^{-5}
isothermal compressibility at 0°C and 300 bars, bar⁻¹	12×10^{-6}
specific heat at 0°C, J/(g) (°C)	2.06
thermal conductivity at 0°C, watt/(cm²) (°C/cm)	21
dielectric constant at −1°C and 3000 cps	79

[a] At 1 atm pressure unless otherwise specified.

and fragmenting with an estimated half-life of about 10^{-10}–10^{-12} sec. Very brief to be sure, but sufficiently long compared to molecular vibrations to enable us to speak meaningfully of the clusters' existence. Liquid water continues to exhibit many of the structural features of ice-I. Pauling (24) has estimated that only 15% of the hydrogen bonds are broken upon melting, while the x-ray radial distribution curve indicates that, not only is the tetrahedral spatial arrangement of ice largely retained, but that the "radius" of the water molecules is little altered by the phase change. Nevertheless, Frank and Wen (23) and especially Nemethy and Scheraga (10) are very careful to qualify their description of the clusters as "ice-like" by pointing out that other lattice arrangements and even a random structure are quite commensurate with the model. These reservations are not without basis, for the remarkable tendency for liquid water to supercool to temperatures far below its freezing point, indicates either that the structure of the clusters is not that of ice-I or, if it is, the concentration of clusters large enough to serve as nucleation centers is below the critical limit. Similarly, they exercise equal caution in describing the "free" water, again being careful to point out that even the free waters are exerting forces on one another. Again, failure to recognize this qualification has led to considerable controversy in the literature (20,21,25).

The structure of heavy water, D_2O, is very similar to H_2O although with a somewhat greater fraction of hydrogen-bonding, at least at room temperatures and below, the O—D—O bond being stronger than the O—H—O bond (11,26).

The Properties of Pure, Bulk, Liquid Water. Liquid water is indeed anomalous in every single one of its physical and chemical properties. It is in fact, difficult even to find substances to compare its behavior with, since materials of comparable molecular weight, such as NH_3, CH_4, and Ne are gaseous. Perhaps the only comparable compound is HF, which is also strongly hydrogen-bonded. The properties of water, heavy water, HF, and NH₃, are compared in Table 3 along with n-pentane, which, since liquid ammonia also is hydrogen-bonded, can be taken as our sole example of a "normal" liquid, albeit of a much higher molecular weight.

Since the value of n in $(H_2O)_n$ depends upon temperature and pressure, the dependence of liquid water properties on these parameters can be very useful in giving clues as to what is happening to the structure. Estimates of the fraction of hydrogen-bonding in liquid water vary widely (2,21), but there is agreement that as temperature increases the clusters fragment and dissolve. Nemethy and Scheraga (10), for example, have estimated that the average number of water molecules (n_{av}), in a Frank-Wen

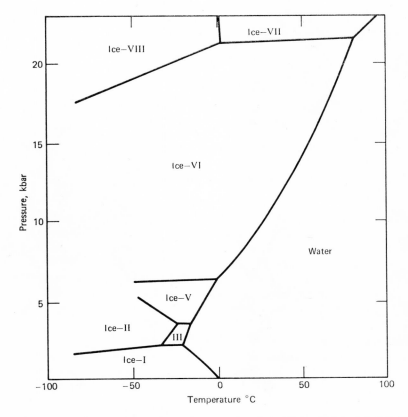

Fig. 6. Phase diagram of water. (An ice-IV was reported but was subsequently found to be identical to ice-V.)

cluster, decreases from about 65 to 12 in going from 0 to 100°C. This thermally induced disruption of the structured regions gives rise to a decrease in viscosity (Table 4).

Similarly the destruction of the less-dense clusters superimposed upon the expected thermal expansion gives rise to the well-known anomalous maximum in the density of water (Table 4) near 4°C. The lower density of the clustered regions also makes them subject to destruction by the application of hydrostatic pressure. These structural changes, again, are nicely reflected in the viscosity of water which, unlike any other liquid, first decreases as the clusters break up, goes through a minimum and then increases with increasing pressure like a "normal" liquid should (Fig. 9), suggesting that at normal temperatures but at pressures in excess of about 1000 kg/cm² water is a "normal," unassociated liquid (28). Raman spectra and infrared data however, indicate that the step-wise destruction of hydrogen bonds with increasing pressure, however, is less pronounced than with temperature, the water going from only 4-bonded to 3-bonded (18).

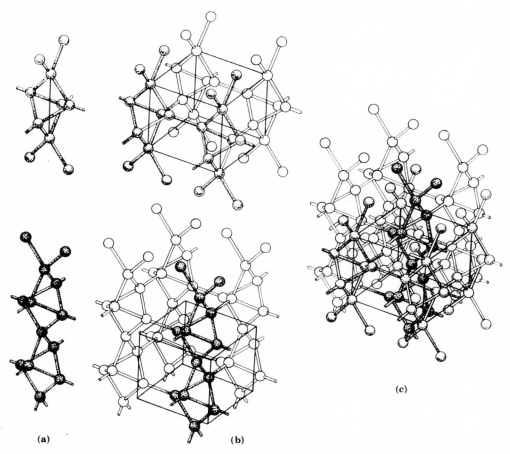

(a)   (b)   (c)

Fig. 7. Structure of ice-VI, shown by assembly of its component parts. In column (**a**) are shown hydrogen-bonded chains of water molecules running parallel to the c-axis of the crystal. The chain below is identical with the one above but is rotated through 90°. Oxygen atoms of water molecules are represented by balls, and hydrogen bonds by sticks; the hydrogen atoms are not shown. In column (**b**) such chains are linked sideways to form two framework structures and are placed in their proper location relative to the unit cell (outlined). In column (**c**) the two framework structures are combined within a single cell to make the complete structure of ice-VI (7).

Under ordinary conditions water is ever so slightly dissociated, $H_2O \rightleftharpoons H^+ + OH^-$, the product of the hydrogen and hydroxide ion concentrations being about 10^{-14}. A few years back it was fashionable to represent the aquated proton or hydronium ion as H_3O^+ although in the light of more recent work $H_9O_4^+$ would seem to be more

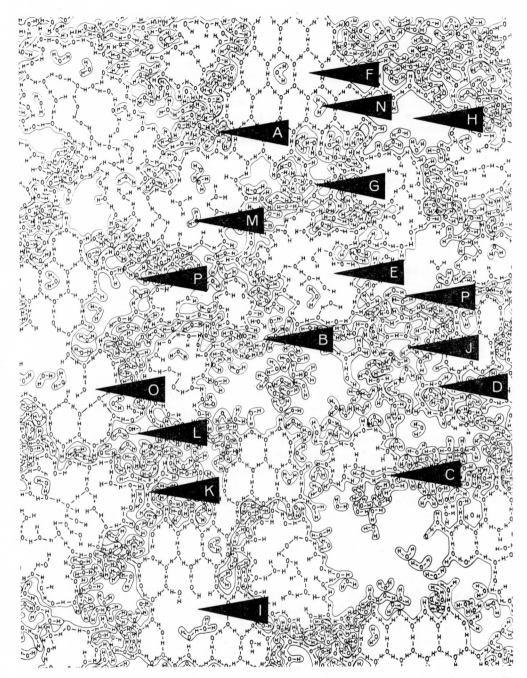

Fig. 8. Possible structural representation of liquid water: A, free or monomeric water; B, dimer (water-ii); C, trimer (water-ii); D, small ring; E, Frank-Wen cluster (water-iii$_a$); F, ice-like region (water-iv); G, chain; H, hole; I, vacancy in an ice-like lattice; J, branched chain; K, hydronium ion, $H_9O_4^+$, from self-ionization of water; L, hydroxide ion, OH^-, from self-ionization of water; M, trapped monomer in a Frank-Wen cluster; N, interstitial monomer in an ice-like lattice; O, degenerate ice-like lattice region; P, notice that even the "free" monomers interact with one another and tend to mutually orient in accord with their dipolar nature.

Table 2. Some Theories of Liquid Water Structure

Date	Field encompassed	Reference
before 1927	water-i, water-ii	H. M. Chadwell, *Chem. Rev.* **4**, 375 (1927)
1933	water-iii$_b$, water-iv	J. D. Bernal and R. H. Fowler, *J. Chem. Phys.* **1**, 515 (1933)
1946	water-i, water-iv	O. Y. Samoilov, *Zh. Fiz. Khim.*, **20**, 12 (1946)
1951	water-iv	J. A. Pople, *Proc. Roy. Soc.* (*London*) **205A**, 163 (1951)
1952	water-i, water-iv	E. Forslind, *Acta Polytech.* **155**, 9 (1952)
1957	water-i, water-iii ($n \approx 40$)	H. S. Frank and W. Y. Wen, *Disc. Faraday Soc.* **24**, 133 (1957)
1961	water-i, water-viii	L. Pauling, *Science* **134**, 15 (1961)
1961	water-i, water-viii	H. S. Frank and A. S. Quist, *J. Chem. Phys.* **34**, 604 (1961)
1961	water-iv, ice-I	G. Wada, *Bull. Chem. Soc. Japan* **34**, 955 (1961)
1962	water-i, water-iv	M. D. Danford and H. A. Levy, *J. Am. Chem. Soc.* **84**, 3965 (1962)
1963	water-i, water-iii ($n \approx 100$)	I. Pelah and J. Imry, *Israel At. Energy Comm.* **IA**, 875 (1963)
1964	ice-I, ice-III, water-i	R. P. Marchi and H. Eyring, *J. Phys. Chem.* **68**, 221 (1964)
1965	water-iv, ice-I	C. M. Davis, Jr., and T. A. Litovitz, *J. Chem. Phys.* **42**, 2563 (1965)
1968	water-iii, ice-II	S. R. Erlander, *J. Macromol. Sci.* **A2**, 595 (1968)
1968	Everything	R. A. Horne, *Sur. Prog. Chem.* **4**, 1 (1968)
1968	water-i, water-ii, water-iii	G. E. Walrafen, *J. Chem. Phys.* **48**, 244 (1968)

appropriate (Fig. 8) (29). At very high pressures (>100,000 kg/cm²) the dissociation of water increases by a factor of 10^{12} and water becomes a good electrical conductor (30,31).

Local Water Structure near Solutes. Water in nature is rarely pure. Sea-water for example is a moderately concentrated, mixed electrolyte, aqueous solution roughly 0.5M in NaCl, 0.05M in MgSO₄ and containing a trace of everything else imaginable. The addition of a solute to water, in addition to altering the gross properties of the fluid, such as raising the boiling point and lowering the freezing point, profoundly alters the structure of the water in the immediate neighborhood of the solute species. In addition, the solution process is accompanied by temperature changes—evidence that chemical bonds are being broken and new ones formed. The remarkable solvent properties of water are familiar to everyone; it is the closest approximation to the alchemist's dream of a universal solvent.

It is convenient to classify solutes into two types: electrolytes which dissolve to form ionic species and nonpolar solutes. Some solutes when added to water make it more viscous, others slightly less (Fig. 10), that is, some seem to reinforce the water structure and others to break it. In order to account for these and other varied phenomena we must theorize that an ion in solution has changed the local water structure around it and has formed a complex hydration envelope (Fig. 11) which we will call *coulombic hydration*. This consists of an inner structure-enforced zone, where, in the terminology of Samoilov (12), a given water molecule is resident longer than at a given point in pure water (*A* and *B* in Fig. 11), surrounded by a region of broken water structure (water-vii, and *C* in Fig. 11) where the coulombic field of the ion is sufficient to disrupt the structure of bulk water but too weak to reorient the water dipoles into some

Table 3. A Comparison of Some Selected Properties of Liquid Water and Other Substances[a]

Property	Value				
	H_2O	D_2O	HF	NH_3	$n-C_5H_{12}$
molecular weight	18.015	20.031	20.01	17.03	72.15
melting point, °C	0	3.82	−92.3	−77.7	−131.5
boiling point, °C	100	101.42	19.4	−33.35	36.2
density, g/ml	0.998	1.105	0.987 (l)	0.817(−80°C)	0.626(20°C)
temperature of max density, °C	3.98				
surface tension, dyne/cm	71.97			23.4(11°C)	
index of refraction, n_D	1.3325	1.3388	1.90(g)	1.325(17°C)	1.357(20°C)
viscosity, mP	8.949	12		2.55(−33°C)	2.89(0°C)
self-diffusion coefficient, cm^2/sec	2.57×10^{-5}		3.5×10^{-4}		
ionic dissociation constant	10^{-14}				
heat of ionization, kJ/mole	55.71				
heat of formation, kJ/mole	285.89(18°C)	294.59(18°C)	268.6(18°C)	46.2(18°C)	0.116(−130°C)
heat of fusion, kJ/mole	6.010(0°C)	6.345(0°C)	4.579(−83°C)	7.7(−75°C)	
heat of vaporization, kJ/mole	40.651(100°C)		0.3(17°C)	23.3(−33°C)	1.84(20°C)
dielectric constant	77.94	77.94	84(0°C)	16.9	
apparent dipole moment, cgseu/mole	5.59×10^{-19}		5.74×10^{-19}	3.89×10^{-19}	
sound velocity, m/sec	1496.3				
isothermal compressibility, atm^{-1}	45.6×10^{-6}				
specific heat, J/(g)(°C)	4.179				
thermal conductivity, watt/(cm^2)(°C/cm)	$5.98 \times 10^{-3}(20°C)$			0.50×10^{-3}	1.35×10^{-3}
electrical conductivity, $ohm^{-1} cm^{-1}$	$<10^{-8}$				

[a] At 25°C and 1 atm unless otherwise specified.

Table 4. Selected Physical Properties of Pure Water (27)

Temp. °C	Density, g/ml	Specific volume, ml/g	Vapor pressure, mmHg	Dielectric constant	Viscosity, cP
0	0.99987	1.00113	4.580	87.740	1.787
5	0.99999	1.00001	6.538	85.763	1.516
10	0.99973	1.00027	9.203	83.832	1.306
15	0.99913	1.00087	12.782	81.945	1.138
20	0.99823	1.00177	17.529	80.103	1.002
25	0.99707	1.00293	23.753	78.303	0.8903
30	0.99568	1.00434	31.824	76.546	0.7975
40	0.99224	1.00782	55.338	73.151	0.6531
50	0.98807	1.01207	92.56	69.910	0.5467
60	0.98324	1.01705	149.57	66.813	0.4666
70	0.97781	1.02270	233.81	63.855	0.4049
80	0.97183	1.02899	355.31	61.027	0.3554
90	0.96534	1.03590	525.92	58.317	0.3156
100	0.95838	1.04343	760.00	55.720	0.2829

new pattern. Thus, a given water molecule is resident at a given position for a shorter time than in pure water at the same temperature. The structure-enhanced region may be divided still further into the innermost, strongly bound electrostricted water molecules (water-v), surrounded by what seems to be in effect a Frank-Wen cluster (water-iii$_a$). The water-v is sometimes called the "primary hydration" and this water commonly remains attached to the ions when they are crystallized out of solution in solid form, the so-called "water of hydration."

An ion's character as a water structure-maker or -breaker is determined by the relative importance of regions A, B, and C. Generally speaking, we know much less about the hydration of anions than cations, but it can be said that anionic hydration is less extensive and anions tend to be structure-breakers. Inasmuch as it is more difficult to crowd water dipoles about an anion with their regions of positive electrification (the protons) pointing in, than with the oxygens pointing in as in the case of cations, this is not surprising.

The thermodynamics of the dissolution of nonpolar solutes such as the noble gases, hydrocarbons, etc, is markedly different from that of electrolytes (32). Therefore, while they clearly structure the local water about them, this structure must be quite different from that surrounding ions. This structured zone (water-viii) has been called an "iceberg," but inasmuch as there is no evidence that its form is indeed ice-I-like, the description *hydrophobic hydration* may be preferable to distinguish it from the coulombic hydration of ions. The nature of this structure is not known, but since many of these solutes can be crystallized out of water as well-characterized clathrates, or guest molecules trapped in a water network cage (Fig. 12), a natural guess would seem to be that the icebergs in solution are clathrate-like.

A complex molecule such as a biopolymer in solution may have a complex hydration sheath with its polar regions surrounded by coulombic hydration and its hydrocarbon or nonpolar segments encased in hydrophobic hydration.

Entirely too much information exists for the properties of aqueous solutions to be briefly and usefully summarized here. However, as an isolated example, it is instructive to compare the properties of a typical sea-water sample (Table 5) with pure water (Table 3) (18,33).

Local Water Structure near Interfaces. The water structure and the resulting disinclination of water molecules to be separated becomes even more marked as a neutral interface is approached. Not only is the surface tension of liquid water enormous, but the temperature-dependence of this near-interfacial or vicinal-water-structure-dependent property is very much less than that of the viscosity, a bulk-water-structure-dependent property (Fig. 13). The most strange behavior of liquid water near interfaces, notably in capillaries and in porous media, has been recognized

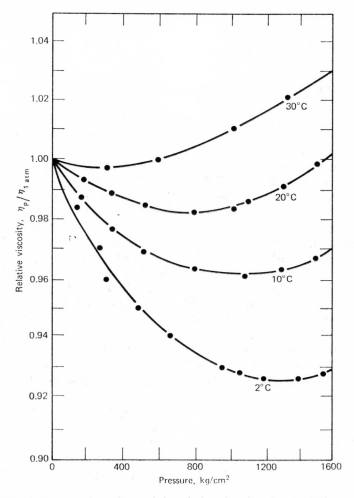

Fig. 9. Pressure dependence of the relative viscosity of water (45).

for some time (34). Among the physical properties of vicinal water are a viscosity greater than "normal" bulk water and a freezing point below −40°C (35). Hitherto the prevailing opinion seems to have been that the surface layer of vicinal water was thin, that the enhanced structure of water near an interface decayed exponentially and penetrated into the bulk water for only about a dozen or so water molecules deep (36). Recently, however, the suspicion has grown that the surface structure may extend hundreds and even thousands of water molecules deep. The implications of this

Table 5. Selected Properties of Sea Water of Salinity 35 o/oo (g/kg) (18)

Property[a]	Value
freezing point, °C	-1.91
boiling point, °C	100.56
density, g/ml	1.02412
index of refraction, n_D	1.3389
viscosity, mP	9.02
sound velocity at 0°C, m/sec	1450
isothermal compressibility at 0°C, atm^{-1}	46.4×10^{-6}
specific heat at 18°C, J/(g)(°C)	3.898
temperature of max density, °C	-3.52
electrical conductivity, ohm^{-1} cm^{-1}	5.3×10^{-2}
surface tension, dyne/cm	72.8

[a] At 25°C and 1 atm pressure unless otherwise specified.

possibility are staggering. Most suspended liquid water in the atmosphere, most ground water and water in marine sediments, and most water in living organisms would then be vicinal water, exhibiting the peculiar properties and chemistry of vicinal water.

What is the structure of vicinal water, of water-ix? At the present time only one structural feature seems to be nailed down with any degree of finality: the water dipoles at the interface are oriented with their negative (oxygen) vertex aimed outward (Fig. 14) (36). Again, the drastic supercooling of water in capillaries would seem to preclude an ice-I structure (as drawn in Figure 14, purely for the sake of two-dimensional convenience). On the other hand we have attributed a clathrate structure to the hydrophobic hydration of nonpolar solutes such as hydrocarbons, and we can avoid the awkward necessity of postulating a structural change when concentrations of these species become sufficiently large to form a second phase. Perhaps, then, the structure of water-ix is the same as that of water-viii. Similarly, there is no reason to assume that the local water structure near a charged site on an interface is any different from the coulombic hydration (water-v, -vi, and -vii) surrounding an ion in solution, even including the structure-broken (water-vii) region (37). A complex macromolecule, surface, or membrane will have its polar regions sheathed by coulombic hydration and its nonpolar regions by hydrophobic hydration (Fig. 15). Some of the situations are incredibly complex: for example, so complex is the spatial distribution of ions at the electrical multilayer at an aqueous solution-electrode interface, that few have had the courage to start to worry about the configuration of the water molecules. Even more complex, and of far greater current interest, is the hydration crust of biopolymers and the state of water in living cells and tissues (38). It is indeed true that water structure plays a crucial role in determining both the configuration and the physiological function of many biopolymers.

In its more general sense, the term vicinal water designates water near any kind of an interface, not merely a neutral one. Mention may be made of two further forms of vicinal water—ortho-water (water-x) and absorbed or chemically-bound water (water-xi).

The announcement of the discovery of a new form of liquid water formed in quartz capillaries in the laboratory of the Soviet academician B. V. Derjaguin (39) created a mixed response in western scientific circles ranging from skepticism to intense interest. Variously named "ortho-water," "anomalous water," "thick water," "super water," and "polywater," the reported properties of water-x are indeed remarkable: it is

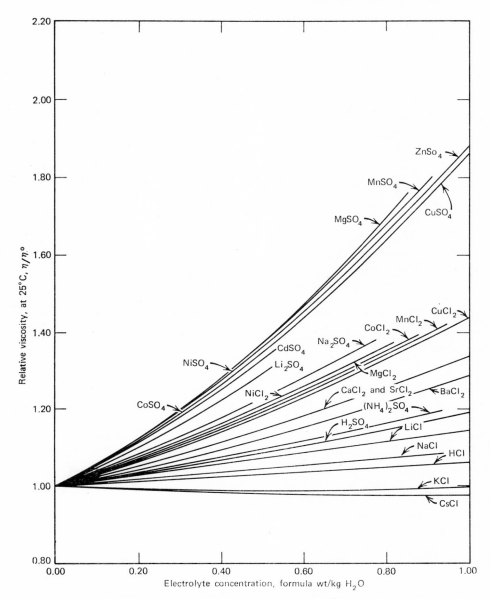

Fig. 10. The effects of various electrolytes on the viscosity of water (46).

10–15 times more viscous than "normal" liquid water, its density is 1.4 times greater, its refractive index is 1.44 [(compared to 1.33 (Table 3)], it remains liquid to −40°C, it is stable in the liquid phase above 300°C, it is miscible with ordinary water, and its lower vapor pressure suggests greater stability. Again the details of the structure of water-x are not known. Erlander (40) has identified it with that of ice-II, while it has been suggested, on the basis of its high density and viscosity, that it might be identical with water-viii (41). However, Bolander, Kassner, and Zung (42) have rejected this suggestion and propose instead a tetrahedrally-bonded, four-molecule cluster, which

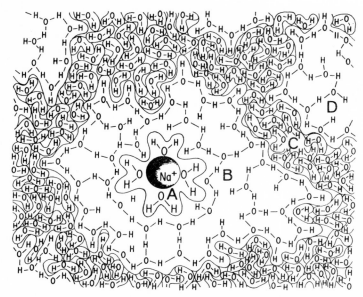

Fig. 11. The local water structure near a cation in aqueous solution: A, innermost, dense, tightly bound, electrostricted region (primary hydration); B, rarified, bound, Frank-Wen cluster-like region; C, broken water structure region; D, bulk water—Frank-Wen clusters and monomeric water. $D_A > D_D > D_B$ (47).

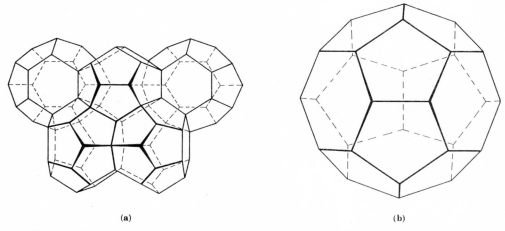

(a)   (b)

Fig. 12. Clathrate hydrate polyhedra: (a) part of the unit cell of structure I hydrate. The lines correspond to hydrogen bonds, and the corners correspond to oxygen atoms. The pentagonal dodecahedron (upper center) is formed by 20 water molecules. The tetrakaidecahedra (2 hexagonal and 12 pentagonal faces) are formed by 24 water molecules. (b) The large cavity of the structure II hydrate. This hexakaidecahedron is formed by 28 water molecules. There are 4 hexagonal and 12 pentagonal faces (48).

would be computable with the reported molecular weight of 72. Water-x, unlike water-viii, appears to be peculiar to quartz surfaces which apparently catalyze its formation (43). Lippincott et al. (43) have proposed a network of hexagonal units and such a model has the advantage of suggesting analogous structures for the polymerized forms of OH-containing substances other than water that have been reported.

All the forms of water mentioned above have been local perturbations of the structure of bulk water by the presence of a second substance. For the sake of completeness mention should be made of one more form of water, especially important in mineral substances and possible tissue, and that is chemically-bound or absorbed water (water-xi). This water is strongly bound, even more strongly than water-v, and in some instances it is not readily removed even at elevated temperatures. Water-xi forms a layer on the substrate substance, one, or at most, a few water molecules thick, and sometimes it is possible to further classify this water on the basis of the strength of the bonding and the immobility of the molecules.

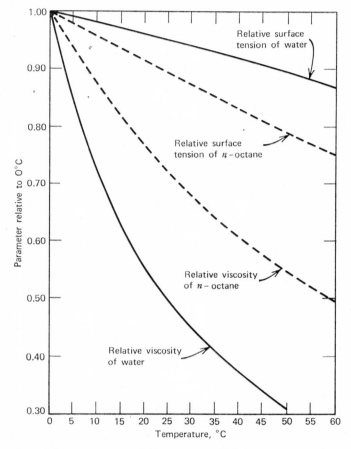

Fig. 13. Comparison of the temperature-dependence of a bulk (viscosity) with a vicinal (surface-tension) water structure parameter and with a "normal" liquid (n-octane) (41).

Historical Footnote. With the growing interest in oceanography and the recognition of water's central role in a host of ecological problem areas, and even more especially, with the advent of molecular biology and analytical physiology, water in the past few years has stepped front and center in the most stimulating vanguard areas of science. This is not a new role for water. Water has always been an eminence when science is on the move. The first scientist, Thales of Miletus (*ca* 624–545 B.C.) in-

SURFACE

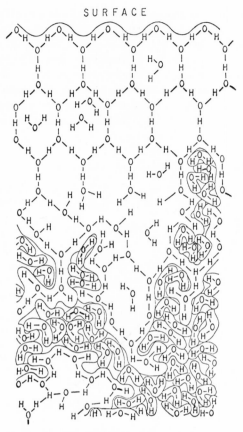

Fig. 14. Water structure near a neutral interface. (Note that contrary to the figure, the surface structure is probably not ice-I-like and it may very well have a density greater, rather than smaller, than bulk water.)

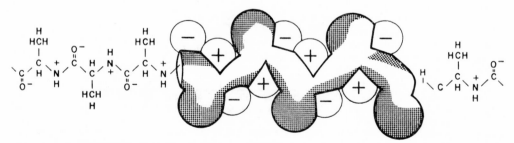

Fig. 15. Hybrid hydration sheath of a biomacropolymeric molecule. Coulombic hydration (open circles) surrounds polar sites and hydrophobic hydration (shaded circles) encases nonpolar regions.

sisted that "water is best" and that all substances are forms of water. For many centuries water had to share honors with three other Aristotelian elements, but with the alchemical speculations of Sir Isaac Newton, water resumed its position of primacy. The ancient all-is-water hypothesis was replaced by the all-is-hydrogen hypothesis of

Prout, the final resolution of which was accomplished in that great triumph of 19th century chemistry—the exceedingly accurate determination of atomic weights—a labor of enormous proportions which, ironically enough, led to the discovery of isotopes and thus, by the sometimes strange and devious dialectic of history, helped to usher in the Atomic Age in which we live.

Bibliography

1. H. M. Chadwell, *Chem. Rev.*, **4**, 375 (1927).
2. E. Wicke, *Angew. Chem.*, **5**, 106,122 (1966).
3. J. D. Bernal and R. H. Fowler, *J. Chem. Phys.*, **1**, 515 (1933).
4. E. O. Wollan, W. L. Davidson, and C. G. Shull, *Phys. Rev.*, **75**, 1348 (1949).
5. J. Jaccard, "Transport Properties of Ice," in R. A. Horne, ed., *Structure and Transport Processes in Water and Aqueous Solution*, Interscience Publishers, a division of John Wiley & Sons, Inc., New York, N. Y., in press.
6. H. H. G. Jellinek, *J. Colloid Interface Sci.*, **25**, 192 (1967).
7. B. Kamb, *Science*, **150**, 205 (1965).
8. B. Kamb, "Ice Polymorphism and the Structure of Water" in A. Rich and N. Davidson, eds., *Structural Chemistry and Molecular Biology*, W. H. Freeman & Co., San Francisco, 1968.
9. N. E. Dorsey, *Properties of Ordinary Water-Substance*, Reinhold Publishing Corp., New York, 1940.
10. G. Nemethy and H. A. Scheraga, *J. Chem. Phys.*, **36**, 3382 (1962).
11. J. L. Kavanau, *Water and Solute-Water Interactions*, Holden-Day, San Francisco, Calif., 1964.
12. O. Y. Samoilov, *The Structure of Aqueous Electrolyte Solutions and the Hydration of Ions*, Consultants Bureau, New York, N. Y. 1965.
13. W. Drost-Hansen, in W. Stumm, ed., *Equilibrium Concepts in Natural Water Systems*, Adv. Chem. Sev. No. 67, Amer. Chem. Soc., Washington, D. C., 1967.
14. F. Franks, *Chem. & Ind.*, 560 (May 1968).
15. R. A. Horne, *Sur. Prog. Chem.* **4**, 1 (1968).
16. D. Eisenberg and W. Kauzmann, *The Structure and Properties of Water*, Oxford Univ. Press, 1969.
17. R. A. Horne, *Marine Chemistry*, Interscience Publishers, a division of John Wiley & Sons, Inc., New York, N. Y., 1969.
18. G. E. Walrafen, *J. Chem. Phys.*, **48**, 244 (1968); see also *Hydrogen-Bonded Solvent Systems*, A. K. Covington and P. Jones, eds., Taylor & Francis, London, 1969, pp. 9–29.
19. K. J. Mysels, *J. Am. Chem. Soc.*, **86**, 3503 (1964).
20. T. T. Wall and D. F. Hornig, *J. Chem. Phys.*, **43**, 2079 (1965).
21. M. Falk and T. A. Ford, *Canad. J. Chem.*, **44**, 1699 (1966).
22. T. T. Wall, *J. Chem. Phys.*, **43**, 4187 (1965).
23. H. S. Frank and W. Y. Wen, *Disc. Faraday Soc.*, **24**, 133 (1957).
24. L. Pauling, *The Nature of the Chemical Bond*, 2nd ed. Cornell Univ. Press, Ithaca, N. Y., 1948.
25. D. P. Stevenson, *J. Phys. Chem.*, **69**, 2145 (1965).
26. R. C. Bhandari and M. L. Sisodia, *Indian J. Pure Appl. Phys.*, **2**, 266 (1964).
27. R. A. Robinson and R. H. Stokes, *Electrolyte Solutions*, 2nd ed., Butterworths Publications Ltd., London, 1959.
28. R. A. Horne and D. S. Johnson, *J. Phys. Chem.*, **70**, 2182 (1966).
29. M. Eigen and L. DeMaeyer in W. J. Hamer, ed., *The Structure of Electrolyte Solution*, John Wiley & Sons, Inc., New York, N. Y., 1959, Ch. 5.
30. H. G. David and S. D. Hamann, *Trans. Faraday Soc.*, **55**, 72 (1959).
31. S. D. Hamann and M. Linton, *Trans. Faraday Soc.*, **62**, 2234 (1966).
32. H. S. Frank and M. W. Evans, *J. Chem. Phys.*, **13**, 507 (1945).
33. R. A. Horne, *Adv. Hydroscience*, **6**, in press (1970).
34. J. C. Henniker, *Rev. Mod. Phys.*, **21**, 322 (1949).
35. J. A. Schufle and M. Venugupalan, *J. Geophys. Res.*, **72**, 3271 (1967).
36. N. H. Fletcher, *Phil. Mag.* **7**, 255 (1962).
37. W. Drost-Hansen, *Ind. Eng. Chem.*, in press, 1970.

38. C. Berendsen, *Biol. Bull.*, **119**, 287 (1960); J. Cerbson, *Biochim. Biophys. Acta*, **88**, 444 (1964); I. M. Klotz, *Proc. Fed. Amer. Exper. Biol.*, **24**, No. 2, Part III, Suppl. 15, 24(1965); I. D. Kuntz, Jr., T. S. Brassfield, G. D. Law, and G. V. Purcell, *Science*, **163**, 1329 (1969); G. N. Ling, *Amn. N. Y. Acad. Sci.*, **125**, 405 (1965).
39. B. V. Derjaguin, *Disc. Faraday Soc.*, **42**, 109 (1966).
40. S. R. Erlander, *Phys. Rev. Lett.*, **22**, 177 (1969).
41. R. A. Horne, A. F. Day, R. P. Young, and N. T. Yu, *Electrochim. Acta*, **13**, 397, (1968).
42. R. W. Bolander, J. K. Kassner, and J. R. Zung, *Nature*, **221**, 1233 (1969).
43. E. R. Lippincott, R. R. Stromberg, W. H. Grant, and G. L. Cessac, *Science*, **164**, 1482 (1969).
44. C. M. Davis, Jr. and T. A. Litovitz, *J. Chem. Phys.* **42**, 2563 (1965).
45. E. M. Stanley and R. I. Batten, *J. Phys. Chem.* **73**, 1187 (1969).
46. R. A. Horne and D. S. Johnson, *J. Chem. Phys.* **44**, 2946 (1966).
47. R. A. Horne in *Chemical and Microbiological Analysis in Water and Water Pollution*, L. L. Ciaccio, ed., Marcel Dekker, Inc., New York, in press 1970.
48. S. L. Miller, *Proc. 2nd Symp. Underwater Physiology, Publ. No. 1181*, Natl. Acad. Sci. U.S.A., Washington, D.C., 1963.

Ralph A. Horne
Woods Hole Oceanographic Institution

WATER ANALYSIS

The analytical operations performed on samples of environmental water are determined by the purpose for which the data are needed. Analytical data are essential to evaluate the physical and chemical characteristics of a source for some particular purpose or purposes, such as a public or domestic supply, industrial uses, manufacturing processes, steam generation, or simply for cooling. Other uses of water which may require special analytical tests include agricultural uses, such as irrigation and stock watering, recreation, including bathing and fishing, and even commercial fishing. Water is an important geologic weathering agent and its role in altering the nature of the earth's crust can best be studied by means of chemical and physical analyses of environmental water, both surface and ground water.

A great deal of information concerning the nature of a water supply or source can be obtained by performing a few rather simple chemical and physical tests. For example, most fresh water is composed of a solution of the bicarbonates, chlorides, and sulfates of just three metallic ions: calcium, sodium, and magnesium. If the bicarbonate, chloride, and sulfate ion concentrations are determined, and calcium and magnesium concentrations (together these constitute the major components of water hardness, a fairly simple test), one can estimate the concentration of sodium that must be present on the basis of equivalent amounts of cations and anions, assuming that the concentrations of potassium and other cations are so small as to be negligible. These data, together with such additional information as pH, specific conductivity, and perhaps turbidity, reveal much important information about the quality of the water and its suitability for many purposes.

On the other hand, because water is an active solvent attacking nearly every substance to at least some degree, given proper conditions and enough time, the list of analytically determinable solutes in environmental water can be very long. To be sure, many solutes are present in almost vanishingly small concentrations. Nevertheless, with continuing improvement in analytical techniques, exceedingly small quantities of a great many different solutes are being measured almost routinely in samples of environmental water. Very small amounts of certain radionuclides, for example, can be readily measured, and extremely sensitive tests have been developed for detecting minute quantities of pesticides and other organic solutes.

While a few simple tests suffice to identify the major constituents, and provide a considerable amount of valuable information about a given water source, the suitability of water, or rather more often its nonsuitability, for a special purpose often can be determined only by means of some precise tests—frequently depending on very sensitive measurements of trace amounts of certain constituents.

For convenience of discussion and for management of analytical facilities, water-analysis operations can be divided into seven categories based on similarity of the nature of the solutes or properties measured, as well as certain similarities in the types of laboratory equipment and facilities needed for the analyses. These analytical categories are:

> Physical properties
> Major mineral constituents and gases
> Minor mineral constituents and gases
> Organic solutes
> Radioactive solutes
> Biologic and microbiologic substances
> Suspended solids

Several manuals of recommended procedures for analyzing water for a variety of constituents and purposes have been published (1,2,3).

Physical Properties

Temperature. See also Temperature measurement. Water temperature must be measured in situ, or immediately upon withdrawal of a sample, for obvious reasons. A standard mercury thermometer, or a thermistor or resistance thermometer, is commonly used to determine water temperature, which is reported in either Fahrenheit or Centigrade units. It is commonly reported to the nearest tenth of a degree, although data to the nearest one-half degree are also common.

Specific Conductance. See also Electrical testing; Electroanalytical methods. The specific conductance of a water sample is an indication of the concentration of dissolved, ionized solutes. The test is easily and quickly performed using simple, reliable, and relatively inexpensive equipment, and provides a convenient indication of the relative salinity of a water source. The measurement is particularly useful in identifying brines and highly mineralized waters, as well as in following the encroachment of sea water into ground-water aquifers along the coastline, and the movement of salt-water fronts in estuaries.

Conductivity measurements are made by means of an ac Wheatstone bridge circuit. The equipment consists of a low-voltage, ac generator, a measuring cell, and either a null indicator and balancing system or a direct-reading microammeter to indicate the extent of imbalance in the arms of the bridge. The measuring cell may be quite simple, consisting merely of two small, platinized-platinum electrodes mounted in a protective and isolating sheath of glass or plastic, which serves both to isolate a portion of the sample for measurement and to insulate the electrodes from the container sides or other nearby objects. Each cell has a constant which, multiplied by the reciprocal of the resistance measured, gives the specific conductance. The constant depends upon the size and the shape of the cell; for a cell which is exactly a 1-cm cube, the constant would be 1.00. Cells having constants ranging from 0.0100 to 100 are available for special purposes. The cell should be selected so as to give, with the solution being measured, a resistance of from 500 to 10,000 ohms, since this is the range of greatest accuracy of the electrical measurement.

The specific conductance of water is conveniently reported in $ohm^{-1} cm^{-1} \times 10^{-6}$, a unit which is referred to as "micromho/cm." Most fresh water samples have specific conductances ranging from 10 or 20 up to several thousand micromho/cm.

A rough approximation of the dissolved-solute load of a fresh-water source in mg/l can be obtained by multiplying the specific conductance in micromho/cm by 0.65, although the factor may range from 0.55 to as high as 0.80 for water from different sources.

Specific conductance is temperature dependent. It is, therefore, necessary to either measure all samples at exactly 25°C, or measure the temperature of the sample at the time of test and apply an appropriate correction factor, usually a direct comparison with a solution of potassium chloride of known concentration. Because the latter is the more convenient, a thermistor or resistance, or combination temperature probe may be built into the conductivity cell both to automatically correct for temperature deviations from 25°C, and also to indicate simultaneously the temperature of the sample.

Density (qv). The density of a water sample is determined by carefully weighing a known volume at a known temperature—usually 20°C. A 25–50 ml sample is a convenient volume to weigh. If a thermostat bath is not available, the temperature of weighing can be noted and corrected to 20°C by reference to relative density tables. The density is commonly reported to three decimal places.

Pure water has a maximum density of 1.0000 at 4°C, and a density of 0.9982 at 20°C. The presence of 1,000 mg/l of dissolved solutes increases the density of pure water by about 10^{-3} g/ml or about 0.1 percent, so that waters containing less than 5,000 mg/l of dissolved solutes have nearly the same density as pure water. For this reason, density measurements are ordinarily of significance only when dealing with highly mineralized waters and brines.

Color. See also Color measurement. The color of environmental water samples is a difficult property to measure objectively by any simple method. Many samples, however, have a brownish-yellow color as a result of contact with natural organic materials and their decomposition products. When such is the case, the sample color is determined by direct comparison with standard colored solutions containing known amounts of a colored substance. A prepared solution whose color closely matches many environmental-water samples is the platinum–cobalt standard, prepared by dissolving appropriate amounts of potassium chloroplatinate (K_2PtCl_6) and cobalt chloride ($CoCl_2 \cdot 6H_2O$) so that a solution containing exactly 500 mg/l of Pt and 250 mg/l of Co is obtained. A series of dilutions of this stock solution down to 1.00 and 0.50 mg/l of Pt and Co is prepared as required for direct comparison with colored-water samples. The latter solution exhibits 1 unit of standard color (1.00 mg of Pt/l).

The occurrence of turbidity in a sample makes the measurement of color somewhat uncertain. Unfortunately, most simple methods of removing turbidity, such as filtration or flocculation, also tend to remove the color. Turbidity removal by centrifuging is best, when possible.

Light colors are more reliably compared than deep colors. For that reason, samples whose color exceeds about 70 color units are diluted before measuring their color.

Glass color disks which match the Pt–Co standards are available and have an advantage over standard solutions in that they are less subject to change with time and are more convenient to handle.

Although rarely used, a more exact color description, especially of heavily polluted water, can be obtained by evaluating the tri-stimulus color values to obtain the dominant wavelength (of transmission), the hue, the percent luminence, and the percent purity. The method, while more exact than the direct color-comparison method, involves considerably more work, and the added precision is seldom worth the greater effort required.

Turbidity. See also Turbidimetry and nephelometry. A turbid appearance is due to the scattering of light by particles suspended in the sample. Large, heavy particles settle out quickly in a quiet sample, and the turbidity may change rapidly on some samples following vigorous agitation. On the other hand, finely divided and lightweight particles may remain in suspension indefinitely and cause more or less permanent turbidity.

Turbidity may be measured by observing the depth of solution which just causes the disappearance of a target object under conditions of normal daylight. The target object may be a bright platinum wire, a disk divided into black and white alternate quadrants, or a standard candle flame. Such depths are calibrated in terms of the number of milligrams per liter of a standard clay material which just causes disappearance. The candle unit (Jackson) is a common standard for turbidity; most other turbidity units relate to the Jackson Candle Unit. Such measurements are simple and are widely used, although their application lacks a certain objectivity.

A more exact and objective measurement of turbidity is achieved by a nephelometric measurement, that is, by measuring the amount of light scattered at 90° to the beam direction by the suspended particles, the intensity of the scattered light being measured photometrically. Even though this measurement is strictly nephelometry rather than turbidimetry, it is commonly calibrated in the same Jackson Units.

Optically clear water has zero turbidity. Good drinking water should have no discernible turbidity or, at the most, no more than about 3 units of measurable turbidity.

Taste and Odor. See also Organoleptic testing. Taste and odor cannot be measured completely objectively because both depend on personal judgment as to the sensations of taste and smell, and individuals vary widely in their sensitivities to these sensations. A panel of from 5 to 10 persons is much better than a single evaluator of taste and odor. The ratio by which an odorous sample has to be diluted with odorfree water for the odor to be just detectable is called the "threshold odor number" and is designated by the abbreviation T.O. Obviously, the mean T.O. of a panel of several observers or testers is a more reliable measure than the T.O. based on the observation of just one person.

Odors are sometimes characterized as being aromatic, flowery, medicinal, earthy, or musty, or by some similar categorizing characteristic.

A similar dilution technique is used to detect a distinct taste in water.

Dissolved Solids. Evaporation of a known volume of a filtered sample followed by heating for an hour at 180°C provides a dry residue, which, when weighed, is a measure of the dissolved-solids content of the sample. A platinum dish is commonly used for the evaporation. A final drying temperature of 180°C is a compromise between a temperature high enough to drive off most of the water of hydration from hydrated salts, and yet low enough to avoid volatilization or thermal decomposition of salts and organic and labile materials.

Suspended Solids. Suspended solids are determined by weighing the residue retained on a filter disk after filtering a carefully measured volume of sample. Fairly dense, glass-fiber filters are usually used. The residue must be thoroughly dried at 105–110°C before weighing.

pH. See also Hydrogen-ion concentration. The pH of water is determined electrometrically with a pH-sensitive glass electrode. The pH of well-buffered samples can be measured with high accuracy, to 0.01 pH unit or better. The pH of environmental waters is subject to change once a portion has been isolated for measurement. Because of the convenience and comparative simplicity of the glass-electrode measurement, an in situ determination is preferred.

Alternatively, the pH may be estimated by means of color-indicator papers, strips of paper impregnated with pH-sensitive dyes. Wide-range papers permit estimation of pH to within 0.2 to 0.5 pH unit, and narrow-range papers will permit a measurement to within 0.1 unit. Solutions of weak organic acids or bases of intense color are also used as pH indicators. The addition of a few drops of a suitable indicator solution to the test sample and comparison with standards enables a reasonable estimate of pH.

Major Mineral Constituents and Gases

Alkalinity. The pH of most unpolluted, environmental waters lies between about 6.0 and 8.0, and is controlled primarily by the chemical equilibria involving bicarbonate and carbonate ions. An alkalinity determination involves titration with a standard acid, either sulfuric or hydrochloric, and reflects the overall concentration of substances capable of reacting with hydrogen ions under the conditions of the titration. Because of the dominance of the carbonate–bicarbonate equilibria, the titration shows two end-points if the sample pH is greater than 8.3, and two alkalinity values may be reported: one based on titration to pH 8.3 or the phenolphthalein end-point, the other based on continued titration to pH 4.5 or the methyl-orange end-point. These alkalinity values are reported as phenolphthalein and methyl-orange alkalinities, respectively. When the phenolphthalein alkalinity exceeds the methyl-orange alkalinity, the presence of free hydroxyl ions is indicated, the amount of which is equivalent to the excess of phenolphthalein alkalinity over the methyl-orange alkalinity. A sample whose pH is less than 8.3 cannot contain free OH^- in the presence of bicarbonate and will have only methyl-orange alkalinity. Because water alkalinity is commonly a function of its carbonate- and bicarbonate-ion content, the alkalinity titration is frequently used to estimate the concentration of these constituents.

While water alkalinity is most commonly due to bicarbonate and/or carbonate ions, the presence of other titratable substances should not be precluded. Several minor acid radicals such as borate, phosphate, and silica, if present, will contribute to the measured alkalinity.

Acidity. Acidity is a measure of the amount of solute which reacts with a strong base and is determined by titration to a phenolphthalein end-point, or electrometrically to a pH of 8.3. Thus, a given sample may simultaneously have both acidity and alkalinity. Ordinarily, the water acidity is not determined unless the sample pH is considerably below 4.5, indicating the presence of a significant amount of free hydrogen ions. The titration of strongly acid samples frequently involves reactions of a variety of other solutes in addition to the strong acids, and the titration may be neither simple nor straightforward. For example, the neutralization of dissolved gases such as CO_2 and H_2S, and of hydrolyzable metals such as Fe and Al, frequently complicates the titration. Moreover, many of these complicating reactions are temperature-sensitive, and either involve oxidation and reduction or result in the formation of insoluble oxides, hydroxides, or hydroxy salts. As a result, the best determination of acidity requires construction of a titration curve and determination of the true end-point by inspection of the plotted curve.

Because of the nature of substances contributing to water acidity, other special precautions and treatments are sometimes advisable. For example, loss of dissolved gases, including dissolved oxygen, from warmed or hot solutions will result in a smaller acidity value than if heating were avoided. Conversely, heating the sample, particularly at or near the end point, favors completion of those reactions, especially hydrolysis and oxidation reactions, which take place only slowly at room temperature. In order to assure oxidation of hydrolyzable metals, especially iron, a strong oxidizing agent, such as H_2O_2, may be added prior to completing the acidity titration.

Hardness. Water hardness is due to the presence of dissolved alkaline earths. Calcium and magnesium are the principal alkaline earths in environmental water; strontium is sometimes a significant contributor to hardness but barium only rarely. Hardness is expressed as an equivalent amount of calcium carbonate in mg per liter. The U.S. Geological Survey classifies water hardness as shown in Table 1 (4).

Table 1. Classification of Water Hardness (4)

Hardness (mg/l as $CaCO_3$)	Water classification
0 to 60	soft
61 to 120	moderately hard
121 to 180	hard
more than 180	very hard

The alkaline earths can be titrated with disodium ethylenediaminetetraacetate (EDTA) solution with Eriochrome Black T indicator. The optimum pH for the titration is 10.4 or above. Heavy metals, if present, are also titrated. Their interference can be eliminated or minimized by adding hydroxylamine, ferrocyanide, and cyanide to reduce and/or complex these materials. Aluminum reacts slowly with the titrant and causes a fading end-point.

Alternatively, hardness can be calculated by totalling the concentrations of individually determined alkaline earths, as by atomic absorption spectrophotometry. The sum of calcium and magnesium concentrations is a sufficiently accurate measure of water hardness except when significant amounts of strontium and barium are present.

Hardness, Carbonate and Noncarbonate. That fraction of the hardness which is exactly equivalent to the alkalinity of a sample is called *carbonate hardness* and can be thought of as caused by the presence of alkaline-earth bicarbonates. It is sometimes called *temporary hardness*, in that boiling decomposes the bicarbonates resulting in formation of insoluble alkaline-earth carbonates, thus effectively removing hardness.

Any hardness in excess of an equivalent amount of alkalinity is termed *noncarbonate hardness* and can be thought of as *permanent hardness*. It is due to alkaline-earth salts other than bicarbonates, for example sulfate.   Carbonate- and noncarbonate-hardness values can be computed from determined values for hardness and alkalinity and are expressed as equivalent concentrations of calcium carbonate.

Calcium and Magnesium.   Almost all environmental waters contain appreciable amounts of calcium and magnesium, usually more of the former.   Calcium may be determined directly by titration with disodium ethylenediaminetetraacetate (EDTA) solution at a pH of 10.4 or above and with murexide as indicator.   In this titration, magnesium and barium do not react with the indicator, although strontium, if present, will be titrated along with the calcium.   Magnesium concentrations are commonly determined by difference—calculating the difference between the hardness titration (Ca + Mg + Sr) and the calcium titration (Ca + Sr).

Both calcium and magnesium can readily be determined directly by atomic absorption spectrophotometry.   Alkaline samples (pH > 7.0) must first be acidified and lanthanum chloride solution added to minimize interferences due to phosphate, sulfate, and aluminum.   An air-acetylene flame is used.

Sodium and Potassium.   The alkali metals, sodium and potassium, are conveniently determined by either flame photometry or atomic absorption spectrophotometry.   Both techniques give reliable data if proper precautions are taken.   Sodium interferes with the determination of potassium by both methods, and an excess of sodium is therefore commonly added to all potassium standards and samples to minimize minor variations of sodium in the samples being analyzed.   Sodium interference in the atomic absorption spectrophotometric determination of potassium varies with the position in the flame and careful positioning of the flame will reduce the effect of sodium.   Use of a reducing flame is also helpful.

There are few, if any, interferences with the determination of sodium.   When high concentrations are being measured, dilution of the sample lowers the concentration of other substances to below levels at which interference might occur.

Chloride.   Chloride is a common constituent of environmental water and may be determined by the well-known Mohr titration with silver nitrate using potassium chromate indicator.   A mercurimetric titration is generally preferable, however, because of its sharper and more easily recognized end-point.   The mercurimetric titration is considered superior to the Mohr titration when determining chloride at concentrations of less than 10 mg per liter.   Diphenylcarbazone is used as an end-point indicator.   Bromide and iodide titrate along with chloride, but their concentrations are usually insignificant by comparison.

Sulfate.   High concentrations of sulfate may be determined by precipitation of barium sulfate followed by the separation, drying, and weighing of the precipitate. Lower concentrations are determined either turbidimetrically, observing the turbidity of the suspended, freshly precipitated barium sulfate, or volumetrically by titration with standard barium chloride solution using thorin as an end-point indicator.   Thorin and barium react to form a deep-red complex.   The color intensity is greater in organic solvents, so the titration is usually carried out in a mixture of about 66% dioxane by volume and the mixture adjusted to a pH of between 2.2 and 5.0.   Several multivalent cations besides barium form intensely-colored complexes with thorin.   To eliminate this interference, the sample aliquot is treated with Amberlite IR-120 (Rohm and Haas) or similar cation-exchange resin, prior to titration.

Nitrate and Nitrite.   Nitrite is determined directly by diazotization with sulfanilamide and subsequent coupling of the resulting diazo compound with 1-naphthylethylenediamine to form an intensely-colored red dye. Nitrate can be determined similarly if it is first reduced to nitrite by a suitable reducing agent. Hydrazine or a cadmium amalgam work satisfactorily.

Low concentrations of nitrate (less than 5 mg/l) may also be determined spectrophotometrically, based on the intensely yellow-colored compound formed when nitrate ion reacts with the alkaloid brucine in a strongly acid medium. Organic color, nitrite ion, and all strong oxidizing and reducing agents interfere to a greater or lesser degree. Sulfanilic acid may be added to eliminate interference from small amounts of nitrite. The interference of chloride is effectively masked by adding an excess of this ion to the test solution.

High concentrations of nitrate may be determined by reduction to ammonia with deVarda's alloy (Al, Zn, Cu), distillation of the resulting ammonia into a boric acid solution, and titration of the resulting ammonium borate with standard acid, usually sulfuric.

Fluoride. The determination of fluoride concentrations has assumed importance because of the widespread use of fluorides to reduce dental caries. Two methods are in general use for determining fluoride concentrations. One is based on the bleaching effect of fluoride ion on the red complex formed by zirconium and Eriochrome Cyanine R dye in a strongly acid solution. If the sample has natural color, or if interferences are present, it is advisable to distill the fluoride from a concentrated sulfuric acid solution before carrying out the determination. Distillation effectively separates the fluoride from all substances which might otherwise interfere with the determination.

A second and similar method is based on the bleaching of a dye formed by reaction of zirconium and SPADNS reagent, which is sodium 2-(p-sulfophenylazo)-1,8-dihydroxy-3,6-naphthalenedisulfonate. The method is rapid and results in analytical data which are satisfactory even without resorting to the lengthy distillation step to remove interferences.

Fluoride can also be determined electrometrically in a buffered sample by means of an ion-selective (fluoride) electrode in conjunction with a calomel reference electrode and an expanded-scale pH meter. The fluoride electrode is a laser-type, doped, lanthanum fluoride crystal across which a potential is developed by fluoride ions in a manner similar to that of the glass electrode for measuring hydrogen-ion activities. The ion-selective electrode, however, measures fluoride-ion activities, rather than concentrations, so high concentrations of dissolved solutes and substances which form stable fluoride complexes cause an apparent error in the determination. It is therefore necessary to add a highly concentrated salt-buffer solution which effectively masks minor variations in the salt content of the samples, and greatly minimizes this error. The optimum pH for measurement is between 5.0 and 8.5 Citrate ion is also added with the salt-buffer mixture to minimize preferential complexing of fluoride by ferric ion and by other polyvalent cations which show a strong tendency to complex with fluoride.

Phosphate. Orthophosphate is the most common ionized form of phosphorus in water. The method of analysis is highly specific for orthophosphate, but does not distinguish between the three ionization products of phosphoric acid, $H_2PO_4^{-1}$, HPO_4^{-2}, and PO_4^{-3}. The relative concentrations of these ionic forms in any given sample are a function of pH. Orthophosphoric acid reacts with molybdate ion in acid solution to form the yellow phosphomolybdate ion which can be reduced to form an intensely-colored (blue) complex. Ascorbic acid, in the presence of antimony ion as a catalyst, is commonly used as the reducing agent.

In addition to orthophosphate, water samples may contain other forms of phosphorus which may be converted to orthophosphate under special conditions. For example, the polyphosphates, such as hexametaphosphate and certain simple organic phosphates, hydrolyze readily in hot sulfuric acid solution. After hydrolysis, these acid-hydrolyzable phosphates may be determined by the same phosphomolybdate method. The result, of course, represents the sum of ortho- plus acid-hydrolyzable phosphorus.

More drastic conditions, such as heating or digesting the sample with sulfuric acid–ammonium persulfate solution, decompose most of the more refractory organic phosphates. An analysis preceded by such vigorous digestion will include the refractory organic phosphates, as well as the more readily hydrolyzed and orthophosphate forms.

Boron, Borates. Many environmental waters contain traces of borates. Excessive amounts of boron in irrigation water are harmful to many plants. At least three different methods are in common use for determining the trace amounts ordinarily encountered. One is the dianthrimide method, wherein the sample is heated with 1,1'-dianthrimide in the presence of concentrated sulfuric acid. The intensity of the resulting deep-blue color is proportional to the original boron concentration. In a second method, heating of the sample with carmine, a biological stain, in sulfuric acid solution yields a complex, again of deep-blue color. In the third method, the sample is reacted with curcumin in an acid solution. Upon evaporation, a red-colored product is formed called rosocyanine. The rosocyanine is dissolved in ethyl alcohol and the red color of the resulting solution compared with similarly prepared standards, either visually or spectrophotometrically. Close attention must be given to operational details to achieve reliable results.

There is little to recommend any one method in preference over the others. The carmine method is less sensitive, but on the other hand it is the simplest, and for a quick and easy estimation of boron it is frequently satisfactory.

Silica. Silica may occur in environmental water as both dissolved silicate or silicic acid, and as colloidal or so-called unreactive silica, SiO_2. Dissolved silicate or silicic acid reacts with the molybdate ion in a strongly acid medium to form a yellow silicomolybdate compound. This substance reacts further with the sulfite ion to form a reduced compound of intense blue color, whose intensity is proportional to the silica content of the sample. The silicomolybdate complex is known to exist in two different polymorphic forms, α and β, of which the latter has the deeper color. Formation of the β-polymorph is favored if the solution is kept strongly acid. In some cases, all or most of any unreactive silica present can be converted to the determinable, reactive form by digesting with a rather concentrated solution of NaOH. Silica may also be determined directly by atomic absorption spectrophotometry provided a high-temperature burner, such as the nitrous oxide-acetylene burner, is used.

Oxygen. The solubility of oxygen in water is dependent upon the partial pressure of oxygen in air, the temperature of the water, and the mineral content of the water. The presence of at least 5 mg/l in surface water is desirable from the standpoint of supporting fish life. Organic and certain other pollutants deplete the oxygen in water.

The Winkler titration, or a modification thereof, has long been synonymous with the determination of dissolved oxygen. In this method, the oxygen in the sample is immediately fixed by absorption on freshly precipitated manganese(II) hydroxide formed by adding manganese(II) sulfate and potassium hydroxide to the sample.

Upon later acidification in the presence of an excess of iodide, iodine is released in a quantity equivalent to the dissolved oxygen present. The liberated iodine is then titrated with standard thiosulfate solution. One- or two-tenths of a milligram per liter can be detected.

Dissolved oxygen (DO) can also be determined simply and quickly by means of one of the membrane-covered electrode systems commonly available. Their use is to be recommended in polluted and highly colored waters. Membrane electrodes of the polarographic as well as the galvanic type are used. Being rugged in construction and completely portable, the electrode systems are particularly suitable for DO measurements in situ, and for continuous monitoring.

Oxygen-sensitive electrodes consist basically of two solid metal electrodes in contact with a volume of supporting electrolyte separated from the test solution by a polytetrafluoroethylene (or polyethylene) membrane which is permeable to oxygen. In the polarographic type, a voltage of about 0.6 V is applied to the indicator electrode. When oxygen is present in the test sample, a small current flows which is limited by the rate of diffusion of oxygen through the membrane material. The diffusion rate, in turn, is a function of the oxygen concentration, so that the observed current is proportional to concentration.

The galvanic-type electrode, on the other hand, requires no externally applied voltage; the electrode system operates spontaneously. The current, however, is again limited by the rate of diffusion of gaseous oxygen through the membrane. There is a significant temperature coefficient of diffusion and a correction must be applied if the temperature of the test sample differs significantly from that at which the instrument was standardized. The instrument meter may be calibrated directly in units of mg/l of dissolved oxygen.

Membrane electrodes offer the distinct advantage of permitting analysis in situ and therefore eliminate all errors which might be introduced through sampling and sample-handling.

Minor Mineral Constituents and Gases

METALS

Manganese, Iron, Aluminum. Traces of aluminum, iron, and manganese are present in nearly all environmental waters. More than a few tenths of a milligram per liter of either iron or manganese make the water unsuitable or at least undesirable for many purposes. Most people become aware of the presence of excessive iron and manganese when white porcelain plumbing fixtures become stained or when white laundered items turn yellow. The undesirable properties of waters containing aluminum are less serious, although some industrial and manufacturing uses require water of quite low limits of aluminum concentration.

Both iron and manganese can be determined by atomic absorption spectrophotometry, the former usually directly, and the latter either directly or after separation and concentration by chelation with ammonium pyrrolidine dithiocarbamate (APDC) and extraction of the chelated complex into methyl isobutyl ketone (MIBK). Atomic absorption spectrophotometric methods lack sensitivity for the very small concentrations of aluminum usually encountered and a spectrophotometric method is used, based on the formation of a soluble, intensely-colored compound when ferron (8-hydroxy-7-iodo-5-quinolinesulfonic acid) reacts with aluminum. The interference from iron is significant and a complexing agent, *o*-phenanthroline, must be added to minimize its effect. Incorporating *o*-phenanthroline in the procedure has the added advantage that iron may be determined simultaneously, inasmuch as this reagent forms an intensely-colored complex with iron. The interference of manganese and fluoride may be significant when appreciable amounts of these substances are present and aluminim determinations of highest accuracy require corrections to allow for their effect.

Rare Alkali and Alkaline-Earth Metals. Trace amounts of the elements Li, Rb, Cs, Sr, and Ba are best determined by direct-flame photometry or by atomic absorption spectrophotometry, the latter being especially suitable for strontium, and even for barium using a nitrous oxide-acetylene burner.

Transition Metals. To determine most of the important transition metals, one usually has a choice between (*1*) a sensitive spectrophotometric method, (*2*) an atomic absorption spectrophotometric method (usually by chelation and extraction prior to the flame-absorption measurement), or (*3*) an emission spectrochemical method.

When the number of metals to be determined is small, either atomic absorption or spectrophotometry may be the choice, mainly because the required equipment is less elaborate and expensive, and the methods are correspondingly simpler. Although atomic absorption spectrophotometric instrumentation is more costly than that required for ordinary spectrophotometry, it is generally preferred because of the greater simplification of preparation of samples for analysis. Standard methods are available by both techniques for all the transition metals of principal interest, including Cr, Mn, Fe, Co, Ni, Cu, Zn, Cd, and Pb.

When it is necessary to determine many transition metals in a single sample, an emission spectrochemical method is preferred. This technique has the unique advantage of providing quantitative or qualitative data on up to thirty-five minor metallic elements in a single sample (5). Moreover, the sensitivity of emission spectroscopy for most transition metals is very good. Only for zinc and cadmium, of the most important metals, is the method seriously lacking in sensitivity. Fortunately, the sensitivity of atomic absorption spectrophotometry is usually adequate for these metals, so the two techniques complement each other in this regard. The sensitivities of emission spectrochemical techniques may be further enhanced by preconcentration of the elements of interest, such as by precipitation (6,7).

NON-METALS

Arsenic. Arsenic rarely occurs at harmful levels of concentration in environmental waters. It is determined by reduction to arsine (AsH_3) by zinc metal in acid solution, distillation of the arsine through a lead acetate scrubber to remove sulfides, and collection of the arsine in a solution of silver diethyldithiocarbamate, which reacts to form a compound of deep-red color. The intensity of color formed is proportional to the arsenic content of the sample. The method is exceptionally free of interferences if carefully performed. Only antimony interferes seriously, and its occurrence in water is even more rare than arsenic.

Selenium. Selenium is a cumulative poison to man and animals, and potable water should not contain more than 10 μg/l of selenium. The diaminobenzidine method of Rossum and Villaruz (8) is commonly used to determine this element. Selenium is quantitatively separated from most other elements by distillation of the volatile tetrabromide from an acid solution containing bromine. The bromine is generated in situ by a hydrogen peroxide–bromide reaction. Selenium tetrabromide is volatilized and absorbed in water, the excess bromine precipitated as tribromophenol, and the selenium determined by reacting the resulting selenous acid with 3,3'-diaminobenzidine to form a piazselenol of intense yellow color. The yellow color of the resulting solution is compared spectrophotometrically with standards prepared in a similar manner.

Bromide and Iodide. Environmental fresh waters normally contain only trace concentrations of bromide and iodide ions although many natural brines contain rather high concentrations. When concentrations in excess of 1 mg/l are to be determined, it is common to determine both bromide and iodide collectively and then determine iodide alone; bromide is then calculated by difference. In this method, bromide and iodide are oxidized to bromate and iodate, respectively, by hypochlorite and the excess hypochlorite subsequently destroyed with sodium formate. Iodine equivalent to the combined bromate and iodate is subsequently liberated by the addition of iodide to the acidified solution, and the liberated iodine then titrated with standard thiosulfate solution. To determine iodide alone, bromine water is added to oxidize only the iodide to iodate; no oxidation of bromide takes place. The iodine equivalent to the iodate formed is then determined by thiosulfate titration as in the combined determination. Preliminary treatment of the sample with calcium oxide minimizes interferences which would otherwise occur due to the presence of iron and manganese in the sample.

Trace concentrations of bromide are determined by a method based on the catalytic effect of bromide on the oxidation of iodine to iodate by permanganate in acid solution, as described by Fishman and Skougstad (9). Two reactions occur consecutively: (1) iodide oxidizes rapidly to iodine, and (2) the resulting iodine oxidizes slowly to iodate. It is the latter reaction that is catalyzed by traces of bromide, and at any given temperature the oxidation rate is directly proportional to the amount of bromide present. In the determination, the reaction is stopped after a suitable time by extracting the unreacted iodine with carbon tetrachloride and measuring the absorbance of the extract. The ratio of the absorbance of the extract of an unknown sample to that of a reference sample containing no bromide is inversely proportional to the bromide concentration.

Similarly, trace concentrations of iodide are determined by a method based on the catalytic effect of iodide on the oxidation of arsenic(III) to arsenic (V) by cerium(IV) in acid solution. Details of the procedure are given by Rossum and Villarruz (10) and by Mitchell (11). At a given temperature, the reaction rate is directly proportional to the amount of iodide present. The reaction may be stopped at any suitable time by adding silver nitrate solution. The absorbance of the solution at 450 nm (nanometers), when compared with the absorbance of a sample containing no iodide, provides a ratio whose value varies inversely with the iodide concentration of the sample.

Both bromide and iodide may also be determined by appropriate ion-selective electrodes of the solid-state membrane type. However, certain ionic species, such as sulfide and cyanide which form either highly insoluble silver salts or stable silver complexes, interfere and render the electrode unsatisfactory for many environmental water samples.

Cyanide. Environmental waters may be expected to contain no more than trace amounts of cyanides from natural sources. Polluted waters, however, frequently contain significant concentrations from industrial wastes. Hydrocyanic acid and its salts, the cyanides, are important industrial chemicals and are widely used in industrial processes. Cyanide is determined spectrophotometrically based on the chlorination of cyanide and subsequent reaction of the product with a mixed solution of pyridine-pyrazolone to form a stable dye complex of intense color (12).

Because only simple cyanides react, it is necessary to decompose complex cyanides by an acid reflux and distillation prior to the colorimetric procedure. Distillation also has the added advantage of removing certain potential interferences. If it is known that interferences are absent and that no complex cyanides are present, the refluxing and distillation may be omitted. The common interferences include sulfide, heavy metals, fatty acids, steam-distillable organic compounds, cyanate, thiocyanate, glycine, urea, and oxidizing agents, as well as appreciable sample color or turbidity.

GASES

Hydrogen Sulfide. Waters containing hydrogen sulfide are generally undesirable because of their disagreeable odor. There is no simple chemical method that is entirely satisfactory for the determination of very low concentrations of sulfides. Higher concentrations, of the order of several tenths of a milligram per liter or more, may be determined by adding a known excess of standard iodine solution, a portion of which is reduced by the sulfide present in the sample, and then titrating the unreduced iodine with standard thiosulfate solution.

Concentrations of sulfides down to 0.1 mg per liter may also be determined by potentiometric titration with standard silver nitrate in strongly alkaline solution using an ion-selective sulfide electrode in conjunction with a double-junction reference electrode and an expanded-scale pH meter as the end-point detector.

Ammonia. In the absence of gross pollution, only small amounts of ammonia and/or ammonium ions are found in environmental waters. To determine ammonia (or ammonium ion) the sample is buffered to a pH of 9.5 to prevent hydrolysis of organic nitrogen compounds and yet ensure complete recovery of ammonia. Ammonia is then distilled from the buffered solution, and an aliquot of the distillate is "Nesslerized," or reacted with a reagent consisting of mercuric iodide in a potassium iodide solution. The reaction yields the intensely red-colored compound mercuric amidoiodide, $Hg(NH_2)I$, whose color intensity is directly proportional to the ammonia concentration of the sample.

Accurate determinations can be made only on fresh samples. When delay is unavoidable, acidification of the sample and storage at 4°C will minimize risk of loss of ammonia.

Organic Materials

Oxygen Demand—Biochemical. The biochemical oxygen demand (BOD) test is an empirical laboratory test used to indicate the oxygen requirements of polluted waters. In conducting the test, the amount of oxygen lost from a sample stored at 20°C for five days is measured. Chemical reducing agents may account for some loss of oxygen, but the chief loss is due to bacterial action. Hence, the designation of the test as *biochemical oxygen demand* (BOD).

Water saturated with air at 20°C contains slightly less than 10 mg/l of dissolved oxygen, and, as a polluted river-water sample will absorb more than this in five days, the sample must be diluted with well-oxygenated water before conducting the test. A properly diluted sample will result in about 50% of the oxygen being consumed in about five days. The actual dilution is determined by experience and by trial and error, although as a general guide a river-water sample will require about a 1-to-3 dilution. A synthetic, buffered dilution water is recommended, and is prepared by adding phosphate buffer, magnesium sulfate, calcium chloride, and ferric chloride solutions to good-quality distilled water.

To carry out the test, an appropriate dilution of the sample is made from which the two bottles required for the determination are filled completely and stoppered without air bubbles. One bottle is incubated for five days at $20 \pm 0.5°C$. The oxygen in the other bottle is determined immediately by either a Winkler titration or, preferably, electrometrically by means of one of the commercially available membrane-electrode instruments. Following incubation, the oxygen remaining in solution in the other bottle is similarly determined. The difference between the two determinations multiplied by the dilution factor gives the BOD value of the sample.

Oxygen Demand—Chemical. An important means of assessing organic pollution is provided by the chemical oxygen demand (COD) test (13). A sample is refluxed with known amounts of potassium dichromate and sulfuric acid, usually for a period of 2 hours. At the end of this time the amount of unreacted dichromate is determined by titration with standard ferrous sulfate solution. The oxygen equivalent to the dichromate consumed is taken as a measure of the chemical oxygen demand of the sample. It is desirable to include a catalyst such as silver sulfate in the digestion mixture to facilitate oxidation of certain resistant substances, for example straight-chain aliphatic compounds. The interference of moderate amounts of chloride, which is also oxidized by dichromate in acid solution, can be greatly minimized by adding mercuric ion to the digestion mixture. When excessive chloride is present, however, it is necessary to make a correction based on determinations made on similar samples containing known added amounts of chloride. Both $0.250N$ and $0.025N$ dichromate reagents are used to cover the wide range of COD values encountered.

Another useful technique for measuring chemical oxygen demand is based on the injection of a microsample into a heated combustion tube through which dry carbon dioxide is flowing (14). Reducing materials in the sample react with the latter to form carbon monoxide which is measured specifically while passing through an infrared stream analyzer. The increase in carbon monoxide content of the gas stream is thus directly proportional to the chemical oxygen demand of the sample.

The two methods do not always provide identical oxygen demand values because of the differing oxidation conditions employed.

Organic Carbon. The amount of organic carbon in a sample can be determined by injecting a microsample into a combustion tube swept by a stream of oxygen (15). All carbon-containing organic matter is oxidized, the carbon going to CO_2. The stream containing the oxidation products is then passed through a nondispersive infrared analyzer to detect carbon dioxide. The increase in CO_2 content of the stream is directly proportional to the organic carbon in the sample. To distinguish between nonvolatile organic materials and carbonates or other volatiles, nitrogen is first bubbled through an acidified portion of the sample.

Pesticides. Fractional $\mu g/l$ concentrations of various chlorinated hydrocarbon pesticides, including some pesticidal degradation products and related products, are determined by gas chromatographic techniques. The specific compounds included are: BHC, lindane, heptachlor, aldrin, heptachlor dioxide, dieldrin, endrin, Perthane, DDE, TDE, DDT, methoxychlor, endosulphan, γ-chlordane, and Sulfenone. The method consists of four basic steps: (1) extraction of the pesticide material into an ethyl ether–hexane solvent mixture, (2) a cleanup and separation procedure to remove the major portion of interferences, (3) gas–liquid chromatographic separation of the pesticide components, and (4) detection, identification, and measurement of the individual components.

As much as 3 liters of sample may be extracted with successive portions of solvent, either diethyl ether–hexane or hexane alone, and the volume of extract subsequently evaporated to a small volume. If the concentrated extract is known or suspected to contain oily or waxlike materials which will later interfere with the measurements, special cleanup and separation procedures must be followed. These involve Florisil column chromatographic separations followed by thin-layer chromatography, techniques which effect not only a separation of the pesticides from interferences, but also, by choice of eluent, certain group separations of pesticides, should mixtures of two or more be present.

Following cleanup and preliminary separations, the concentrated extract is placed on each of two different gas–liquid chromatographic columns for final separation of the component pesticide materials. Both a nonpolar and a relatively polar packing are used, the former usually the designated OV-17, and the latter a mixture of Qf-1(FS-1265) and DC-200. Identification and measurement of the resulting chromatograms is either by microcoulometric titration, which is highly specific for chlorinated compounds, or by electron capture, which is exceedingly sensitive for minute concentrations. The pesticide concentrations are determined by direct comparison of the chromatograms with those of known amounts of the pure compound obtained under identical conditions. The extreme sensitivity of the method and the high degree of preconcentration of the sample permit reporting concentrations on the order of nanograms per liter.

Methods for the determination of pesticides appear in several publications (16,-17,18,19). See also Soil chemistry of pesticides.

Detergents. Synthetic detergents that are made up of linear alkyl sulfonates (LAS) (see Surfactants) are readily biodegradable and rarely pose any problem in fresh water supplies. They are tested for, however, by the methylene-blue test, based on their reaction with this reagent to form a blue salt which is soluble in chloroform and whose color intensity is proportional to the surfactant concentration. Because not a single compound, or even a single type of compound, but several surface-active materials react with methylene blue in a similar manner, the test is designated as one for methylene-blue active substances (MBAS). If a significant amount of surfactant is shown to be present, additional confirming tests should be made to distinguish and confirm the presence of LAS as opposed to other similarly behaving substances. Chromatographic separation of the components in such mixtures and infrared spectrophotometric identification may be necessary to adequately characterize such samples.

Oil and Grease. Most heavy oils and greases that are present in water result from industrial pollution. In most cases these materials are insoluble although detergents and other chemicals may emulsify (or in some cases saponify) the oils and greases. The amount of oil and grease in a sample is determined gravimetrically by weighing the residue obtained by evaporating the organic solvent used to extract the materials from an acidified portion of the sample. Commonly used solvents are petroleum ether, hexane, benzene, chloroform or carbon tetrachloride. A 1-liter sample is commonly taken for the extraction. Particular care must be observed to collect a representative sample inasmuch as oils tend to float on the water surface. Many oils are volatile at the temperature needed to distill off the last traces of extraction solvent, and special care must be taken to avoid prolonged heating of the solvent at the end of the evaporation.

Radioactive Materials

A number of naturally occurring radioactive elements, including uranium, thorium, radium, potassium-40, rubidium-87, and others, emit alpha or beta and gamma radiations and are potential contributors to the radioactivity of environmental waters. Additionally, artificially produced radioactive isotopes may find their way into environmental water systems. Thus it becomes of significant importance in some instances to be able to determine the overall radioactivity of a water sample as well as the concentrations of certain specific isotopes. Frequently, the determination of the gross alpha and gross beta activity of a sample serves to indicate radioactive contamination and the need for additional data on the presence of radioactive nuclides.

To determine gross alpha or beta activity, an appropriate volume of sample is acidified with nitric acid and evaporated to 1–2 ml. The evaporated sample is transferred quantitatively to a flat sample-mounting dish and carefully evaporated to dryness. The volume chosen for evaporation must be such as to provide a minimum amount of residue having measurable alpha or beta activity. Some residues are very hygroscopic and must be retained in an efficient desiccator until the actual counting is begun. Alpha-counting is preferably done in an internal proportional detector or a scintillation detector. Beta-counting may be either by an internal or external proportional gas-flow chamber or by means of an end-window Geiger-Müller tube. Beta-radiation detectors are also sensitive to alpha-, gamma-, and x-rays and their efficiency for beta radiation depends on the type of detector. Most alpha counters are not sensitive to beta-, gamma-, and x-radiations. The radioactivity is reported as gross beta or alpha activity in units of picocuries per liter.

Bacteriological Examination

A bacteriological examination is essential to determine the suitability of water for drinking and bathing, for cooking, and for confirmation of evidence of pollution of the water by pathogenic organisms. A chemical analysis cannot always be counted on to reveal evidence of pollution. Bacterial testing, however, reliably indicates the possibility of pollution of a water source with harmful organisms.

Routine bacteriological or sanitary examination of water is concerned with the detection of a group of more or less harmless bacteria which are known inhabitants of the intestinal tracts of both man and animals. The organisms which are identified and measured are the colon bacilli, commonly called the coliform group. These bacteria are so extensively abundant in the excreta of man and animals that the pollution of water by extremely small amounts of sewage or waste materials can be readily demonstrated bacteriologically. Their presence in any significant amount in a water sample constitutes evidence of pollution and the presence of other harmful, pathogenic bacteria must be assumed until proved otherwise. Conversely, the absence of the coliform group above a specified tolerable maximum, constitutes adequate evidence of the absence of pollution, and may be taken as an indication of the suitability of the water for drinking, bathing, etc. Methods of testing for the coliform groups are specified by the U.S. Public Health Service; detailed procedures are given in Standard Methods (20).

The coliform group is defined as "the aerobic and facultative anaerobic, Gram-negative, nonspore-forming, rod-shaped bacteria which ferment lactose with gas formation within 48 hours at 35°C" (21). These criteria form the basis for one of the two standard tests used to confirm their presence, the *multiple-tube fermentation test.* In this test a series of five fermentation tubes containing either lactose broth or lauryl tryptose broth are inoculated with an appropriately graduated volume of the sample to be tested. The volume of sample used for inoculation (a multiple or submultiple of 1 ml) is based on a prior knowledge of the probable bacteria count in the sample, and is chosen so as to provide a test of at least 50% positive results whenever possible. The inoculated tubes are incubated at 35 ± 0.5°C for 24 hours and then examined for the presence of gas formation in the tubes. If no gas is observed, incubation is continued an additional 24 hours and the tubes again examined. The presence of gas in any amount after 48 hours constitutes a positive presumptive test and requires that those samples be submitted for confirmation of the presence of the coliform group. The *"Confirmed Test"* is performed by transferring a portion of the medium from each primary tube showing gas formation to a second fermentation tube containing confirmatory brilliant green lactose bile broth, or to plates containing either Endo medium or eosin methylene-blue agar. The bile broth fermentation tubes must be incubated at 35 ± 0.5°C for 48 hours, at the end of which time any observed gas formation can be construed as indicating a positive Confirmed Test. The agar plates must be streaked with a portion of the material from the primary tubes which showed gas formation and incubated for 24 hours at 35 ± 0.5°C. The colonies, if any, developing within this time are examined with a microscope and classified as to being typical, atypical, or negative. The presence of typical colonies constitutes a Confirmed Test for the coliform group. If only negative colonies develop, or if no colonies at all develop, the Confirmed Test is considered negative. If atypical colonies are present, the Confirmed Test cannot be considered negative and a further test is necessary.

"The Completed Test is used as the next step following the Confirmed Test. It may be applied to the brilliant green lactose bile broth fermentation tubes showing gas in the Confirmed Test, or to typical or atypical colonies found on the plates of solid differential medium used for the Confirmed Test. If the brilliant green lactose bile broth tubes used for the Confirmed Test are to be employed for the Completed Tests, streak one or more Endo or eosin methylene blue plates from each tube showing gas, as soon as possible after showing gas. Incubate the plates at 35 ± 0.5°C for 24 hours. From each of these plates or from each of the plates used for the Confirmed Test, fish one or more typical coliform colonies or, if no typical colonies are present, fish two or more colonies considered most likely to consist of organisms of the coliform group, transferring each fishing to a lactose broth fermentation tube or a lauryl tryptose broth fermentation tube and to a nutrient agar slant.

"The agar slants and secondary broth tubes incubated at 35 ± 0.5 °C for 24 ± 2 or 48 ± 3 hours, and Gram-stained preparations from those corresponding to the secondary lactose broth tubes that show gas, are examined microscopically.

"The formation of gas in the secondary lactose broth tube and the demonstration of Gram-negative, nonspore-forming, rod-shaped bacteria in the agar culture may be considered a satisfactory Completed Test, demonstrating the presence of a member of the coliform group in the volume of sample examined.

"If, after 48 ± 3 hours, gas is produced in the lactose and no spores on the slant, the test may be considered "completed" and the presence of coliform organisms demonstrated" (22).

The number of positive findings of coliform group organisms—resulting from multiple-portion decimal dilution plantings—should be computed and recorded in terms of the "most probable number" (MPN) index. "It should be realized that this is merely an index of the number of coliform bacteria which, more probably than any other number, would give the results shown by the laboratory examination. It is not an actual enumeration of the coliform bacteria in any given volume of sample. It is, however, a valuable tool for appraising the sanitary quality of water and the effectiveness of the water treatment process" (23).

A second standard test for coliform bacteria is called the *membrane filter technique*. In this test, a suitable volume of the sample is filtered through a sterile, micropore membrane filter of 0.45-μ average pore diameter. The filter disk containing the microorganisms collected on its surface is placed in a petri dish containing M-Endo nutrient broth, M-Endo medium, and incubated for 24 hours at 35°C. At the end of the incubation period any coliform colonies present show a characteristic green sheen which can be taken as positive evidence of the presence of members of the coliform group. Because of the simplicity of the operations, this test can be conveniently carried out in the field using a portable filter kit and incubator. This is a distinct advantage in view of the importance of beginning a bacteriological analysis promptly. The membrane filter technique also has the advantage of generally shortening the time required to confirm the presence of coliform bacteria in the samples under consideration. In the test, any colonies that appear doubtful as coliforms can be confirmed by transfer to lactose broth in a fermentation tube and observed for the formation of gas bubbles on incubation at 35°C and, if necessary, further confirmed in brilliant green bile broth as in the multiple-tube fermentation test.

Results of the standard membrane filter test are reported as the number of colonies formed per 100 ml of sample filtered. When very low bacterial counts are encountered, a large volume of sample may be filtered. The membrane filter method is limited in application when very turbid samples are to be analyzed because of the difficulty of filtering such samples through a micropore membrane filter.

Fecal Coliforms. The foregoing tests detect "total" coliforms. Additional valuable information about the source of pollution can be obtained by conducting a test for fecal coliforms, those exclusively from a fecal source. This test is applied to a positive presumptive test or a coliform colony from solid medium. A loopful from the lactose broth tube is transferred to a fermentation tube containing EC medium (24), and incubated at 44.5 ± 0.5°C for 24 to 48 hours. An exact incubation temperature is important and is so selected as to favor the growth of fecal coliforms while discouraging growth of other bacteria. The formation of gas during incubation is taken as positive evidence of fecal coliform pollution.

A fecal coliform test may also be performed by the membrane filter technique. In this procedure, the membrane filter containing the microorganisms filtered from the sample is placed in a petri dish containing MC broth. The entire dish is then wrapped in a plastic bag and immersed in a water bath at a temperature of 44.5 ± 0.5°C for 24 hours. Fecal coliform colonies will be blue in color when the filter is examined following incubation.

Suspended Sediment

Load. Surface waters, especially, may contain significant amounts of suspended, particulate matter which render the water turbid. The amount of suspended matter is determined by filtering a measured volume of sample through a weighed filter disk and weighing the disk plus suspended matter removed after drying at 103–105°C. The difference in the two weights represents the amount of particulate matter removed. An asbestos mat in a Gooch crucible, a glass-fiber filter, or a micropore membrane filter of 0.45μ pore size, is used to filter the sample.

Sometimes it is important to distinguish between volatile and nonvolatile particulate matter. In this case, the weighed, dry residue is ignited in a muffle furnace regulated to 600°C, and reweighed. The loss in weight represents volatile materials, largely organics.

Mineral Characterization.   In addition to information about the total amount of sediment, it is sometimes of interest to obtain information about the various mineral constituents which comprise the sediment load.   x-Ray diffraction studies of the suspended sediment material provide a convenient means of both identifying the specific minerals present in the various fractions of sediment material and, providing satisfactory standards can be prepared, of supplying at least semiquantitative estimates of the relative amounts of each.   Not only can the various minerals be identified, but the several clay substances present can also be verified.

Sea Water, Brines

There is no universally-accepted classification of waters with respect to their load of solutes.   The U.S. Geological Survey provides a classification as shown in Table 2 (25).

Table 2.   Classification of Saline Waters (25)

Solute load (mg/l)	Classification
Less than 1,000	Nonsaline
1,000 to 3,000	Slightly saline
3,000 to 10,000	Moderately saline
10,000 to 35,000	Very saline
More than 35,000	Brine

The analytical methods discussed above are suitable for the analysis of nonsaline and slightly saline water samples.   They may not be suitable for analyzing samples containing more than a few thousand milligrams per liter of solute material.   Sometimes the method can be used if a dilution is first made or a much smaller sample aliquot is taken for analysis.   It must be recognized, however, that the use of small aliquots or large dilution factors increases the error of the analysis proportionally.   Interferences become a more serious problem in analyzing saline waters and brines. Often a simple dilution prior to analysis reduces the concentration of an interferring substance to a harmless value.   In general, however, the application of methods for the analysis of fresh waters to the analysis of brines must be done with a great amount of care.

Bibliography

In *ECT* 1st ed. under "Water (municipal)," Vol. 13, pp. 946–962, by H. O. Halvorson, University of Illinois.

1. *Standard Methods for the Examination of Water and Wastewater*, 12th ed., American Public Health Association, New York, 1965.
2. *Manual on Water, STP 442*, American Society for Testing and Materials, Philadelphia, 1969.

3. *1969 Book of ASTM Standards, Part 23, Water: Atmospheric Analysis,* American Society for Testing and Materials, Philadelphia, 1969.

4. H. A. Swenson and H. L. Baldwin, *A Primer on Water Quality,* U.S. Govt. Printing Office, Washington, D.C., 1965, p. 17.

5. J. Haffty, *Spectrographic Analysis of Natural Water: Residue Method for Common Minor Elements, U.S. Geol. Survey Water-Supply Paper 1540-A,* U.S. Govt. Printing Office, Washington D.C., 1960.

6. W. D. Silvey and R. Brennan, *Anal. Chem.* **34,** 784–786 (1962).

7. E. C. Mallory, Jr., in *Advances in Chemistry Series: Trace Inorganics in Water,* Robert F. Gould, ed., American Chemical Society, Washington, D.C., 1968, pp. 281–295.

8. J. R. Rossum and P. A. Villarruz, *J. Amer. Water Works Assoc.* **54,** 746–750 (1962).

9. M. J. Fishman and M. W. Skougstad, *Anal. Chem.* **35,** 146–149 (1963).

10. J. R. Rossum and P. A. Villarruz, *J. Amer. Water Works Assoc.* **52,** 919–922 (1960).

11. C. G. Mitchell in *Selected Techniques in Water Resources Investigations, 1965, U.S. Geol. Survey Water-Supply Paper 1822,* compiled by Glennon N. Mesnier and Edith Becker Chase, U.S. Govt. Printing Office, Washington, D.C., 1966, pp. 77–83.

12. F. J. Ludzack, et al., *Anal. Chem.* **26,** 1784–1792 (1954).

13. W. A. Moore, R. C. Kroner, and C. C. Ruchhoft, *Anal. Chem.* **21,** 953–957 (1949).

14. V. A. Stenger and C. E. Van Hall, *Anal. Chem.* **39,** 206–211 (1967).

15. C. E. Van Hall and V. A. Stenger, *Anal. Chem.* **39,** 503–507 (1967).

16. J. I. Teasley and W. S. Cox, *J. Amer. Water Works Assoc.* **55,** 1093–1096 (1963).

17. W. S. Lamar, D. F. Goertlitz, and L. M. Law, *Identification and Measurement of Chlorinated Organic Pesticides in Water by Electron-Capture Gas Chromatography, U.S. Geol. Survey Water-Supply Paper 1817-B,* U.S. Govt. Printing Office, Washington, D.C., 1965.

18. A. W. Breidenback, et al., *The Identification and Measurement of Chlorinated Hydrocarbon Pesticides in Surface Waters, Publication WP-22,* U.S. Dept. of the Interior, Federal Water Pollution Control Administration, Washington, D.C., 1966.

19. D. F. Goerlitz and W. S. Lamar, *Determination of Phenoxy Acid Herbicides in Water by Electron-Capture and Microcoulometric Gas Chromatography, U.S. Geol. Survey Water-Supply Paper 1817-C,* U.S. Govt. Printing Office, Washington, D.C., 1967.

20. Reference 1, pp. 567–628.

21. Reference 1, p. 594.

22. Reference 1, pp. 598–599.

23. Reference 1, pp. 567–568.

24. Reference 1, p. 584.

25. Reference 4, p. 20.

MARVIN W. SKOUGSTAD
U. S. Geological Survey